FLORICULTURE IN INDIA

FLORICULTURE IN INDIA

DR. G. S. RANDHAWA
DR. AMITABHA MUKHOPADHYAY

ALLIED PUBLISHERS PRIVATE LIMITED
NEW DELHI MUMBAI KOLKATA LUCKNOW CHENNAI
NAGPUR BANGALORE HYDERABAD AHMEDABAD

ALLIED PUBLISHERS PRIVATE LIMITED

Regd. Off.: 1/13-14 Asaf Ali Road, **New Delhi**–110002
Ph.: 011-23239001 • E-mail: delhi.books@alliedpublishers.com

17 Chittaranjan Avenue, **Kolkata**–700072
Ph.: 033-22129618 • E-mail: cal.books@alliedpublishers.com

15 J.N. Heredia Marg, Ballard Estate, **Mumbai**–400001
Ph.: 022-42126969 • E-mail: mumbai.books@alliedpublishers.com

No. 25/10 Commander-in-Chief Road, Ethiraj Lane
(Next to Post Office), Egmore, **Chennai**–600008
Ph.: 044-28223938 • E-mail: chennai.books@alliedpublishers.com

P.B. No. 9932, 1st Floor, Sunkada Chickkana Complex,
Loop Lane Off Race Course Road,
Next to Hotel Maurya, Gandhi Nagar, **Bangalore**–560 009
Ph.: 080-22262081 • E-mail: bngl.books@alliedpublishers.com

Sri Jayalakshmi Nilayam, No. 3-4-510, 3rd Floor
(Opp. State Bank of India) Barkatpura, **Hyderabad**–500027
Ph.: 040-27551811, 040-27551812
• E-mail: hyd.books@alliedpublishers.com

Website: www.alliedpublishers.com

First Published, 1986

Reprinted 1994, 1996, 1998, 2001, 2002, 2004,
2007, 2010, 2012, 2015, 2016, 2017, 2018, 2019 2021 2022, 2023, 2025

ISBN : 81-7023-057-8

Cover design: V. Padmanavan

Published by Sunil Sachdev and printed by Ravi Sachdev at Allied Publishers Pvt. Ltd., 1/13-14 Asaf Ali Road, New Delhi–110002

Foreword

Private citizens, institutions and governmental units on the large Indian subcontinent – with its considerable diversity of climate, soils, and vegetation – have a need for a reference on landscape horticulture and floriculture that is at once broadly encompassing and detailed. In writing *Floriculture in India*, Dr. G.S. Randhawa and Dr. A. Mukhopadhyay appear to have achieved the production of such a book. May this volume serve as the beginning of a series of several in this field; it is to be hoped that these and other enterprising authors will follow with additional, intensive coverage of landscape architecture and landscape design applied to specific needs in the widely varying parts of India.

Globally, it is to be noted that, beyond the strictly aesthetic aspects of landscape horticultural plants, serious consideration is now increasingly being given to their environmental functions. Plants are being evaluated for their potential part in noise abatement, in traffic control, as bioindicators of air pollution, as well as for other climatic, architectural and engineering uses, including temperature modification and glare control. With this in mind, the reader of *Floriculture in India* will appreciate that numerous plants used as "ornamentals" may have much utilitarian value for these other purposes as well. Indian horticultural researchers of the stature of the authors will undoubtedly give attention to the roles of plants in environmental enhancement in upcoming years. The importance of such studies, with implications for quality of life for urban and rural residents, should be apparent.

In addition, use of some of the trees, shrubs, ground covers and vines listed in this book, for agroforestry, social forestry, and horti-pastoral purposes, may prove to be quite viable for marginal lands in rural areas. This is another landscape horticultural research frontier in which much research remains to be done under Indian conditions.

At the same time as he or she notes the emerging areas concerning environmental uses of plants, the landscape horticulturist retains an appreciation of the beauty of form, texture, and colour in plants, often also fragrance and smoothness to touch. These are given due coverage in *Floriculture in India*. The reader is exposed to a wealth of information about plant characteristics, site preferences, and cultural techniques, among a myriad of other details. As a result, even with casual perusal of the book, one is exposed to highly useful material on which to base decisions and maintenance practices.

All who are interested in the advancement of landscape horticulture (including floriculture) and environmental horticulture in India must express gratitude to the authors for their contribution to the literature in this field. The book will be very useful to amateur gardeners, students, and teachers.

Prof. Arthur S. Lieberman
Landscape Architecture Programme
Cornell University
Ithaca, New York, U.S.A.

Bangalore, India
July, 1985

Introduction

The importance of Ornamental Horticulture, from both aesthetic, environmental and economic points of view, is not properly understood in India and this subject remained more or less neglected. However, in the past two decades Ornamental Horticulture has attained importance and made considerable progress. Until recently, there was hardly any University or Institute in the country offering postgraduate teaching on the subject. Even today, with the exception of the Indian Institute of Horticultural Research, Hessaraghatta, Karnataka, where a full-fledged Division of Floriculture and Landscape Gardening has been created, there is possibly no Institute or Agricultural University which has a separate division of Ornamental Horticulture. In most of the Institutes and Agricultural Universities, Floriculture is clubbed with either fruits or vegetable crops in one department. Nevertheless, it is encouraging to note that during the past decade or so, almost all the Agricultural Universities have started to teach Ornamental Horticulture as one of the major subjects for postgraduate studies leading to M.Sc. and Ph.D. degrees. Recently, a Centre of Advanced Studies in Tropical Horticulture has been established at the Indian Institute of Horticultural Research, Hessaraghatta, Karnataka, in collaboration with the University of Agricultural Sciences, Bangalore, under an UNDP/ICAR project. In this programme Floriculture has been included as one of the subjects for advanced postgraduate teaching leading to M.Sc. and Ph.D. degrees. With this advancement a number of scientists will be trained in Ornamental Horticulture abroad which will go a long way in improving research and education, and also beautifying cities, towns, and even rural areas through well-conceived bio-aesthetic planning using ornamental trees, shrubs, and other plants. In fact, the second author has received such training in the U.S.A., and the field of his training was nursery and greenhouse management, landscape horticulture and landscape ecology.

Another lacuna in this field is the non-availability of a comprehensive and standard book covering all aspects of floriculture and landscape horticulture, thus depriving the students as well as the amateurs of guidance and knowledge on the various aspects of Ornamental Horticulture. Keeping in view the needs of postgraduate and undergraduate students, it was decided to write the present book. The book covers the syllabi of several Agricultural Colleges, Universities, and other Institutes which are offering degree and diploma courses in Ornamental Horticulture including Landscape Horticulture. This book, therefore, will serve as a textbook covering all aspects of Ornamental Horticulture at one place. It would also be of immense use for amateur ornamental horticulturists and general public interested in beautifying their homes through gardening.

We have taken particular care regarding the spellings of botanical names. Wherever possible, the more common synonyms have also been used. The following books were consulted for the spellings of botanical names: *Manual of Cultivated Plants* by L.H. Bailey (1951), *The Standard Cyclopaedia of Horticulture* by L.H. Bailey (1963), *The Royal Horticultural Society Dictionary* edited by Fred J. Chittenden (1956), *Exotica* by Alfred Byrd Graf (1973), and *Plant Groups* by H. Mukherjee and A.K. Ganguly (1964). For the spellings of Cacti, the book entitled *The Cactaceae* by N.L. Britton and J.N. Rose (1963); for Succulents the book entitled *Succulent Plants* by H. Jacobsen and translated by Vera Higgins (1959), and for Orchids the book *Encyclopaedia of Cultivated Orchids* by Alex D. Hawkes (1970) were consulted.

We add with pleasure that recently the Indian Agricultural Research Institute, New Delhi, also created a separate division of Floriculture and Landscape Gardening.

During the preparation of the book several scientists and other staff of this Institute have helped us in many ways. Particularly, the assistance of Dr. M.P. Alexander, Dr. Foja Singh, Dr. R. Doreswamy, Dr. K.S. Amin, Dr. S.R. Sharma, Dr. S.V. Sarode, Dr. S.P.S. Raghava, Messrs. B.G. Bagle, G.J. Bankar, R.C. Alandkar, and V. Padmanabhan is gratefully acknowledged. Thanks are also due to Mrs. Sukla Mukhopadhyay, Mr. S.R. Nagabhusan, Mr. A.R. Venkatesh, and Miss Madhura for their kind help.

Lastly, we hope that this book will be useful to everybody concerned.

G.S. RANDHAWA* AMITABHA MUKHOPADHYAY

Indian Institute of
Horticultural Research, Hessaraghatta
Bangalore, INDIA
1985

***Present address: 34/1, Meanee Avenue Road, Bangalore-560042.**

Contents

Chapter I

History of Gardening

The history of gardening in India is as old as its civilization. The first recorded history of civilization in India dates back to 2500 B.C. before the Aryans came. The civilization prospered in Punjab, Sind, Rajasthan, and Gujarat. They used to cultivate food crops such as wheat and barley, and fruits such as date. The first evidence of an ornamental plant, the *Pipal* (*Ficus religiosa*), comes from a seal from Mohen-jo-daro, of the third millennium B.C. Another seal from Harappa of the same period depicts a tree similar to that of a weeping willow (*Salix babylonica*).

The Aryan civilization dates back to about 1600 B.C. The Aryans of the *Vedic* period were great lovers of trees and flowers. The lotus has been mentioned frequently in the Sanskrit scriptures of *Vedic* times. The *Atharva Veda* mentions the *Asvatha* (*Pipal*) or *Ficus religiosa*. Possibly, the *Rig Veda* also refers to this tree when it describes a tree with fair foliage. The epics of the Aryans, the *Ramayana* and the *Mahabharata*, believed to have been compiled during 500 B.C., mention about gardens, trees, and flowers. In the *Ramayana* it has been narrated that the palaces had nice gardens with numerous flowers and shady groves. It is frequently mentioned that lakes were full of lotuses. Every Indian must be knowing about the *Asokavana* where Sita was held in captivity by Ravana, the demon king. This grove was composed mostly of *Asoka* (*Saraca indica*) trees. The *Mahabharata*, written by sage Vyasa, deals with the conflict between the Kauravas and the Pandavas. The Pandavas had their capital in Indraprastha. In the *Sabhaparva* of this epic, detailed descriptions have been given of the pleasure gardens, parks, and lakes around the palaces in Indraprastha. Both the epics mention the following trees and creepers – *Ficus benghalensis*, *F. religiosa*, *Saraca indica*, *Michelia champaca*, *Terminalia arjuna*, *Butea monosperma*, *Mesua ferrea* (*Nagkesara*), *Cassia fistula*, *Shorea robusta* (*Sal*), Palmyra palm (*Borassus flabellifer*), Screwpine (*Pandanus*), *Bignonia*, and Oleander. Special mention can be made of the tree *Kadamba* (*Anthocephalus cadamba*; Syn. *A. indicus*) as it was closely associated with the life of Lord Krishna. This tree appears to have been a very popular one during the period of the *Mahabharata*. The lotus was a popular flower and was regarded by both the Hindus and the Buddhists as a symbol of purity. The flower grows in mud but is free from any 'mud-sticking', as no dirt can stick to it. It is like a sacred and virtuous man who remains unaffected even in the filthiest surroundings. In the *Mahabharata* it is mentioned that Bhima brought a lotus for his consort which had one thousand petals, glowing like sun, and with sweet heavenly fragrance which could prolong youth and revive beauty.

During the later Aryan period the Aryans developed some wishful thinking in the form of *Kalpavriksha*, the wish-granting tree, and *Kalpalata*, a similar creeper. The faith goes that anybody, standing under these creepers or trees, wishing anything from jewellery, beautiful clothing, good food, or even beautiful maidens, would get the same. One sculpture depicting such a tree dating back to the third century B.C. and now preserved in the Calcutta Museum was identified as the banyan tree. Another legend of that time also makes interesting reading. According to the legendary story, once Lord Vishnu was bathing in a lake on earth when a lotus bloomed and from within came out *Pitamaha* Brahma, who claimed that the lotus was the prettiest flower one had ever seen. Although Vishnu agreed that it was a beautiful flower, he guided Brahma to his own paradise *Vaikuntha* and showed him a rose bloom as pale as moonbeam and full of sweet fragrance. Brahma had to change his view and accept the rose as the prettiest bloom in the universe.

The association of different trees with the life of Lord Buddha is well known. The Buddha was born in 563 B.C. His birth took place under the *Asoka* (*Saraca indica*) tree. Other trees such as *Sal* (*Shorea robusta*) and *Plaksha* (*Butea monosperma*) were also there. It is said that at the moment of his birth the Queen seized the branch of a *Plaksha* and not of a *Sal*. The great Chinese traveller, Hiuen Tsang, who came to India (630-645 A.D.) confirmed that the Buddha was born under the *Asoka* tree. The Buddha attained his enlightenment under a *Pipal* tree, spread his new teachings under shady banyans and mango trees, and breathed his last in a *Sal* grove.

The great Emperor Asoka (264-227 B.C.) adopted arboriculture as one of his state policies. He encouraged the planting of avenue trees. His son Prince Mahendra took a sapling of the great *Bodhi* tree (*Ficus religiosa*) and planted it at Anuradhapur in Sri Lanka (250 B.C.). The unearthing of the *stupa* at Bharhut (184-72 B.C.) gives us an insight into the ancient history of the Buddhist India. Besides the *Jatakas* which deal with the previous births of the Buddha, there are a number of sculptures which depict different *Bodhi* trees of the other *Bodhisattas* (the Buddhas of previous births). These trees are identified as *Ficus religiosa*, *Ficus benghalensis*, *Ficus glomerata*, *Albizzia lebbek* (*Sirisa*), *Shorea robusta*, and *Saraca indica*. The pipal was regarded as a sacred tree even before the birth of the Buddha. The *Atharva Veda* says that the Gods of the Third Heaven used to sit under the *Asvatha* tree.

The sculptures of the Kushan period (78-200 A.D.) found at Mathura had many trees inscribed on them. The trees identified from the sculptures are *Saraca indica*, *Kadamba*, and *Ixora parviflora*. The other trees on the sculpture are possibly *Michelia champaca* and *Mesua ferrea*.

The golden age of Hindu culture is considered to be between 100 and 600 A.D. during the rules of Kanishka and the Gupta kings. Just prior to this period we come to know about gardens and flowers from *Mrichchakatikam* written by Sudraka (100 B.C.). Poet Asvaghosha mentioned a number of trees in his *Sundara Nanda* including *Mesua ferrea*, *Kadamba*, and lotus. He also described *Nandanavana* in which the Buddha saw flowering trees and lotuses. In the works of Kalidasa during the rule of Chandragupta II we find the mention of several flowering trees. In most of his plays we find the mention of the *Asoka* tree. The name of other trees found in his works are – *Kadamba*, *Arjuna* (*Terminalia arjuna*), *Butea monosperma*, *Parijata* (*Nyctanthes arbor-tristis*), *Bauhinia variegata*, *Mimusops elengi*, *Albizzia lebbek*, and screwpine. The creeper *Madhavi* (*Hiptage madablota*) occupied a prominent place in his play *Sakuntala*. In the same play we find the mention of pleasure gardens. About women of Alakapuri he writes in *Meghduta*, "In the monsoon *Kadamba* flowers decorated the hairs of these charming women and they carried pink lotuses in their arms."

The Gupta era brought long-lasting peace and the people enjoyed their days merrily. Vatsayana (about 300-400 A.D.) in his book *Kamasutra* gives a glimpse of the joyful civic life of that period. He narrates four kinds of gardens: *Pramododyan* meant for the enjoyment of the royal couples; *Udyan* where the kings played chess, enjoyed the dances of the maids and jokes of the court jesters; *Brikshavatika*, the garden where high-placed persons in the king's court enjoyed life with courtesans; and *Nandanavana* which was dedicated to Lord Krishna. Vatsayana also described the duties of a virtuous wife, a few of which regarding gardening can be mentioned here. It is said that a loyal and affectionate wife should lay out a garden around the house. Besides other plants the garden should be planted with jasmine, *Tabernaemontana coronaria*, China rose, etc. It was also advised that a good housewife should construct arbours and seats in the gardens.

Hiuen Tsang, the famous Chinese pilgrim, travelled widely in India from the North to the South in Andhra and to Bodh Gaya in Bihar. He had given a detailed account of life in India at the time. About flowers he wrote that the Hindus needed flowers every morning for worshipping and they used flowers to adorn their hair.

The different frescos at Ajanta (100-600 A.D.) also are indicative of the importance of flowers and gardening in ancient India.

Before the Mughals came to India there was a long blank in the gardening history of India, with the exception of Firoz Tughlaq (1351-1388 A.D.) who developed about twelve hundred gardens around Delhi and more elsewhere. Babur laid the foundation of the Mughal empire in India in the year 1526 A.D. He was a great lover of flowers and gardens. Immediately after his victory over Ibrahim Lodi at Panipat (1526 A.D.) he laid out a garden there. Thereafter he laid a garden at Agra, now known as Rambagh. From the *Babur Nama*, the memoirs of Babur, we find the names of the following flowers: oleander, hibiscus, white jasmine, and *keora* (screwpine). Babur is credited with the introduction of Persian rose in India.

Akbar (1556-1605 A.D.) built a new capital at Fatehpur Sikri, complete with gardens, trees, and flowers. He was the first Mughal to enter Kashmir and establish a garden, Nasim Bagh, close to the Dal lake. The tomb garden of Akbar at Sikandra is an example of the gardening art during that period.

Jehangir (1605-1627 A.D.) was also a great admirer of gardens and flowers, and so was his famous wife Nur-Jahan. The gardens at Shalimar, Achhabal, and Verinag in Kashmir were created by him. The tomb garden at Itimad-ud-Daulah in Agra, in memory of his father-in-law, was also his creation.

Shah Jahan (1627-1658 A.D.) also established several beautiful gardens. He built the Red Fort in Delhi which had a beautiful garden. The gardens around Taj Mahal in Agra, in memory of his beloved queen Mumtaz Mahal, was also his creation. The small but pretty garden at Chashma Shahi in Kashmir was laid by one of his governors. The best garden created by Shah Jahan was the Shalimar garden in Lahore (Pakistan).

Sher Shah Suri (1540-1544 A.D.), who ruled for a short period after defeating Babur, constructed the famous Grand Trunk Road and planted avenue trees on both sides.

The Nawabs of Oudh also established a number of gardens in their empire, especially in

Lucknow. One of these was Sikandar Bagh which was established by the last Nawab Wajid Ali Shah (1847-1856 A.D.), which is now the National Botanical Research Institute.

The Rajput Rajas in India also established several gardens. The garden at Amber Fort near Jaipur was started by Man Singh (1590-1615 A.D.) and completed by Jai Singh II (1699-1743 A.D.). This beautiful garden is laid in three terraces, and has a lake. The Mandor garden near Jodhpur is a really beautiful garden still today, and was built by Raja Abhai Singh (1724-1749 A.D.). The city of Jaipur was founded by Jai Singh II (1727 A.D.) who built a palace in the heart of the city with beautiful gardens. The garden palace at Deeg in the erstwhile Bharatpur State was founded by Suraj Mal. With fountains, water courses, and other features, it is one of the most beautiful gardens ever created by any Rajput king. The gardens and the palaces in Chittor, the Capital of Mewar rulers, were also famous. The rulers of Bundi in Rajasthan also were fond of gardening.

The arrival of the British in India changed the whole pattern of gardening. The Mughal gardens in India were laid out in symmetrical patterns, but the English changed the style into informal patterns by laying out parks and gardens. Until then Indian gardens had only scented trees and shrubs, climbers and bulbs, but had no flowering annuals. The English brought with them beautiful annuals and biennials such as phlox, carnation, verbena, dahlia, larkspur, etc. Before the arrival of the English the Indians knew how to make garlands and *Guldasta* (bouquets), but had no idea of flower arrangements for interior decoration, which was taught by the British. But the greatest contribution by the British was the establishment of botanical gardens starting from the one established near Calcutta.

For details of the gardens and botanical gardens established by the British in India, Chapter XXVII on "Some Famous Gardens of India and their Salient Features" may be consulted.

The present-day garden styles in India are more or less on the pattern of the English style.

Chapter II

Importance and Scope of Floriculture and Landscape Gardening

The role of flowers and gardens in the life of the nation and the people is not properly understood in India. Largely, floriculture and gardening are being neglected by the administrators and get last priority in any process of planning or allotment of funds. Even whatever meagre funds are allocated to floriculture, it is only this branch that is made the first target of any economy drive.

It is also sad to note that though we have at least one or more Agricultural Universities in each State, none of them can boast of having a professor of floriculture and landscape gardening. Though the Indian Council of Agricultural Research has done a lot for floriculture, one of its leading institutes, the Indian Agricultural Research Institute, New Delhi, which has bred several beautiful rose cultivars, does not have a full-fledged Division of Floriculture and Landscape Gardening. But such a division has been established at the Indian Institute of Horticultural Research, Bangalore, where a number of cultivars of hibiscus, croton, and bougainvilleas have been bred and released for cultivation.

Though floriculture and gardening have to go a long way to develop to a significant level, fortunately the position is not as bad as it used to be some time back. The credit for giving floriculture and gardening a good start and consequently some status goes largely to the noted administrator-cum-scientist, Dr. M.S. Randhawa. His work in this field was carried further by another noted scientist and great lover of flowers, particularly roses, Dr. B.P. Pal. But for the encouragement given by these two scientists, floriculture would not have reached the present level of achievement and gained the status that it is having today. Of course, the patronization and encouragement given to floriculture and gardening by the late Prime Minister Jawaharlal Nehru and also the late President Dr. Zakir Husain must also be recalled with gratitude.

Floriculture

It is not possible to define 'Floriculture' in the sense as one can define a triangle in the language of geometry. But possibly the closest definition is, "Floriculture is the art and knowledge of growing flowers to perfection." But this definition is also not complete as floriculture includes not only flowers but also many ornamental foliage plants, cacti, and others. The art of growing *bonsai*, the dwarfing of plants, also come under floriculture. Often the term 'Ornamental Horticulture' is used to include the culture of flowers, foliage plants, trees, and possibly gardening also. But if the term 'Floriculture' is preferred, it should mean not only the culture of flowers but also other ornamental plants such as foliage plants, trees, shrubs, palms, ferns, bamboos, cacti, other succulents, landscape gardening, and so on.

Significance and Importance

Though many people are nostalgic about floriculture, there are others who brush away the idea as a luxury or even a wastage of money. But the significance and the importance of floriculture cannot be ignored or bypassed so lightly. Flowers symbolize purity, beauty, peace, love, and passion. To a Japanese flower-arranger each flower expresses one or more meanings. To an Indian, especially to Hindus, flowers have a much greater significance. A devoted Hindu needs flowers every morning for religious offering to the family deity. It is estimated that about 30 per cent of the total cut flowers consumed in the city of Calcutta are used for worship (Anonymous, 1976). The aesthetic value of flowers in our daily life cannot be over-emphasized. In our society no social function is complete without the use of flowers. Floral garlands, *gajras* and *venis*, are needed for marriage ceremonies. The use of *gajra* and *veni* is not limited to such occasions as marriage only; these are also used as adornment for hair by our women of all ages, especially in the South, every morning, evening, or even all through the day. Floral ornaments, bouquets, or flower arrangements also find a pride of place in social gatherings, birthday parties, welcoming a home-coming friend or relative, and honouring dignitaries. The arrival of new-borns is rejoiced with flowers, the sick are wished speedy recovery by offering flowers, while the dead are bidden farewell with flowers along with tears of sorrow. Even those who dismiss floriculture as a luxury cannot afford to do without flowers when the occasion comes.

The potential of floriculture as an industry has not been exploited properly. Floriculture is an intensive type of agriculture and the income per unit area from floriculture is much higher than any other branch of agriculture. In markets such as Delhi and Bombay a single spike of gladiolus may sell for up to three rupees. Similarly, cut blooms of roses, jasmines, tuberoses, marigolds, some other flowering annuals and *Michelia champaca* have a very good market in the metropolitan cities of India. It is estimated (Anonymous, 1976) that 50 per cent of the flowers sold in the cities of Madras, Bangalore, and Bombay are utilized for making *gajra* and *veni* and up to 70 per cent of the flowers sold in Calcutta are for decoration purposes. Tuberose cut stems are sold in large numbers in the Calcutta market for decoration purposes and this flower is grown commercially in the neighbouring district of Midnapur. The business is said to be quite profitable and the nationalized banks advance loans to the cultivators to promote this cut flower industry. Rose is another important commercial flower the cut blooms of which are highly prized in the large cities. A dozen of good rose stems may sell for up to Rs.36 in a city like Bombay. *Rosa damascena* and Edouard rose are grown in Kanauj, Jaunpur, and Ghazipur in U.P. for the manufacture of *attar* and rose water. Roses are also grown near Amritsar for preparing conserve or *gulkand*, a product mainly used in sweets, and for chewing in combination with *pan* (betel leaf). The flowers of *R. damascena* and *R. borboniana* are used for these purposes. Long back, the business of growing these roses for the above purposes was a flourishing one. The largest rose-growing area (1,000 acres) is situated in the Aligarh district with an annual flower production of 728 tons. The price of rose flowers in these areas fluctuates widely. In 1969, *R. damascena* flowers used to be sold at Rs.105 to Rs.150 per 40 kg, and that of Rose Edouard at Rs.120 to Rs.180 per quintal. In recent years, this industry is on the decline because of unremunerative prices. The main cause behind this decline is the unscientific method of cultivation adopted by the rose-growers.

The sale of loose flowers of jasmines is a roaring business, especially in Southern India, and also in Bombay and Calcutta. The flowers are used mostly for making *gajra veni* and garland, as well as for worship. Not much flower is utilized for the extraction of jasmine oil though this is a highly remunerative product. One ton of jasmine is said to yield about 3 kg of concrete which sells at Rs.8,000 to Rs.10,000 per kg.

There is a great demand for good quality flower seeds which are at present in short supply. Some of the seed firms are selling one packet of seed containing about 100 seeds for Rs.5. F_1 hybrid seeds sell at a much higher price. Nursery stocks such as bulbs, budded roses, and potted plants have a good internal as well as export market. A named gladiolus corm may be as costly as Re.1 to Rs.2 and a hybrid Dutch amaryllis bulb sells at an astronomical price of Rs.10, even in India. Any good quality and named budded rose may fetch from Rs.5 to as much as Rs.15, depending on the rarity of the cultivar. Some nurseries are exporting potted plants, especially foliage plants, to Arab countries at remunerative prices.

Role of Floriculture in a Developing Country

In India, floriculture is not getting the priority it deserves, though it has a great role to play, as shall be clear from the foregoing discussion. It is an intensive type of agriculture and the income per acre is much higher than any other agricultural product if it is done in a scientific way.

First, we should not neglect the great aesthetic value of flowers in our daily life. Growing colourful flowers in the house compound and in the parks will greatly enhance the beauty of the surroundings in the countryside, the towns, and the cities. They will also meet the daily necessities of our life such as adorning the hair or for indoor decoration or for offering to gods and goddesses. Flowers also bring happiness to life and boost the sagging spirits. Our old cities and towns are shabby-looking. If the citizens and the municipalities encourage the growing of flowers, at least partly, the cities will look much more gay. This is more true for the new industrial townships that are coming up so frequently and which offer a much greater opportunity for growing flowers for beautification and usage.

Commercially, floriculture can open up great opportunities to our poor farmers. Our country has diverse climatic conditions which offer the scope for growing several kinds of commercial flowers. The cultivators can deploy a part of their land for growing commercial and common flowers such as marigold, China aster, etc., which do not require much care and generally earn more profit than many other crops. The government can organize the flower growers into societies and offer them help in selling their produce through a network of retail stores in the big city hotels and markets, thus eliminating the menace of middlemen. In the recent times, India has made good progress in industry and agriculture. As a consequence, there are now more affluent people who can afford to buy flowers and floricultural products. This has also enlarged the markets for cosmetics, a wide variety of which use floricultural by-products such as perfumes and aromatics as their ingredients. Some of the essential oils from such flowers as jasmine, rose, and tuberose used in the manufacture of these cosmetics are priced as high as Rs.8,000 to Rs.12,000 per kg. Now the manufacturers get their supply mostly from Bulgaria and France. The government can organize this trade on a cooperative basis and establish modern extraction plants for essential oil, thus bringing a good income to the poor flower-growers.

Similarly, the business of nursery plants should be encouraged and organized on sound business lines so as to earn for the country a good amount of foreign exchange through the sale of plants and seeds. Recently, quite a good number of plants are being exported to countries in the Middle East.

One vital step in this direction will be to set up an organization for certifying the seeds and standardizing the nursery stocks. At present there is no such organization and the customers are cheated with sub-standard products.

Moreover, our growers lack the scientific knowledge about growing flowers. The consequence of this is that our yields are much lower compared to other countries. For example, while the yield of *J. grandiflorum* in our country is only about 1,600 kg per hectare, the yield in France is about 3,000 kg, and in Morocco, Sicily, and Italy it is as high as 4,500 kg. Therefore, efforts should be made to disseminate modern methods of cultivation to increase the yield. Efforts should also be made to standardize improved crop production technology in commercial flowers to obtain maximum yield compatible with high quality blooms.

Scope for Floriculture

In India, floriculture is only a developing subject and as such offers much scope for improvement. This problem can be tackled from several angles.

Conservation, Domestication, and Introduction: India is rich in its plant resources, many of which are of ornamental value and some are potentially ornamental. Much of this wealth is wasted as a result of rapid urbanization, industrialization, and unscrupulous collection. For example, some unscrupulous nurserymen are selling one beautiful *Paphiopedilum insigne* orchid plant for the price of a single flower of this species in foreign markets. As a result of such wanton collection many species of orchids are threatened with extinction. To conserve such rare orchids, the I.C.A.R. has now started some orchid sanctuaries, which is a correct step in this direction.

Domestication of wild plants with potential ornamental value is another way of improving garden wealth. In the process of domestication, possibly in an altogether different climate, the wild plants generally pass through many changes, which itself may cause some chance improvement. We have some very good plants in the wild with potential ornamental value, especially ferns, which can be acclimatized and domesticated under quite different climates. In this regard, the National Botanical Research Institute (formerly the National Botanic Gardens), Lucknow, has done some good work in acclimatizing the Himalayan ferns in the sub-tropical climate of Lucknow.

Introduction of plants either from the wild with potential ornamental value or from abroad offers a good scope for enriching ornamental plants. Initially we should introduce from the wild, plants with potential ornamental value, which can be improved upon by systematic breeding. Simultaneously, plants of proven quality can be introduced from abroad and acclimatized in our country. The process of introduction has improved our ornamental horticulture to a great extent. To cite a few examples, we may quote the instance of the introduction of *Mussaenda philippica*, the double-bracted mussaenda with white, pink, and red colours, from Thailand, which have acclimatized very well in the humid and mild climates of Calcutta, Bangalore, and Kerala. The introduction of many beautiful bougainvilleas from abroad, e.g., 'Lady Mary Baring', 'Golden Glow', and the multi-bracted cultivars such as 'Mahara', 'Cherry Blossom', etc., has enriched our ornamental horticulture to a very great extent. A great variety of the so-called Hawaian Hibiscus introduced by Lalbagh, Bangalore, has completely changed the complexion of many gardens in and around Bangalore.

Advanced countries such as the U.S.A. and Canada introduce ornamental plants in large quantities. If we want to enrich our collection of ornamental plants, we should at least introduce on a limited scale some selected ornamentals from all over the world for meeting the requirement of various agro-climatic regions of the country.

Prospects for Improvement

(a) *Hybridization:* Acclimatization and introduction are not enough for improving the plant wealth. The tastes of the people change very frequently, and there is always the craze for new things. One of the best ways to improve is through hybridization between varieties of species, or interspecific hybridization within a genus or even intergeneric hybridization. For example, in orchid hybrids in between the genera *Cattleya*, *Brassavola*, and *Laelia* are available. The hybridization work done in India, especially on roses, has enriched our floriculture to a great extent. Similarly, new hibiscus cultivars released by the Indian Institute of Horticultural Research, Bangalore, and others also enriched our collection of ornamental plants. Our efforts in this regard should be directed towards developing new hybrids suiting the different agro-climatic conditions prevailing in the country.

(b) *Mutation:* Another important way of developing ornamental wealth is by mutation breeding. The natural mutant bougainvillea 'Mary Palmer' evolved spontaneously in a Calcutta garden has been acclaimed all over the world. But since natural mutants develop only by chance, other physical and chemical mutagens could also be employed to get new attractive mutants. Again, mutation can also be combined with hybridization to further enrich the floricultural wealth.

(c) *Polyploidy:* Polyploidy is another method of plant improvement. Polyploidy can be induced by colchicine and other chemicals. Polyploids generally have larger flowers and intense-coloured petals sometimes with ruffling at the edges. Polyploidy combined with hybridization may result in remarkable improvement.

(d) *Propagation:* Easy and rapid propagation of ornamentals will go a long way in spreading the cult of floriculture in India. The use of mist propagation units along with the application of root-promoting hormones has enabled many hitherto difficult-to-root trees and shrubs to root from cuttings. This method has improved the scope of supplying rare plants to garden lovers at comparatively cheap rates. Another field where not much work has been done is the production of disease-free plants. The tissue culture method offers the production of virus-free plant material. This method should be employed increasingly in ornamental plants not only for getting disease-free stocks but also to have rapid multiplication. For example, the process of multiplying orchids and *Anthuriums* through vegetative means is very slow and once a mother plant is divided for propagation it takes some time to flower again. Moreover, only a limited number of plants can be raised by this method. If the technique of tissue culture is employed, rapid multiplication is possible.

(e) *Dwarfing of plants:* Dwarfing of ornamentals by growth-retardant sprays is being exploited commercially in advanced countries. In India also growth regulators have been used to dwarf ornamentals (Bose, 1972). In the present-day society more and more people are living in flats in cities where the scope of gardening is limited or is not there at all. With the help of growth retardants it is possible to dwarf large plants to manageable sizes, so that the flat dwellers in cities can grow them in their limited spaces. Plants sprayed with growth retardants become tolerant to adverse growing conditions, the leaves become darker in shade and shining, and on many occasions early flowering is induced. In chrysanthemum, the growth-retardant spray can obviate the process of pinching. The commonly used growth retardants are B-Nine, Phosphone, and Cycocel. Maleic hydrazide (MH) in high concentrations (10,000-15,000 ppm) also acts as an effective retardant. This method should be increasingly employed by the nurseries to encourage flat dwellers to take to gardening.

(f) *Extending Blooming Period:* It often happens that some of the commercial flowers bloom for only a short period of the year irrespective of the planting date. For example, in the North, chrysanthemum starts blooming by November 15 and completes flowering by the first week of January. This causes a glut in the flower market. Therefore, staggering of the blooming period of such crops will go a long way in solving the marketing problem for the growers. The blooming of chrysanthemum can be prolonged by manipulating the photoperiod. This crop needs longer photoperiods for vegetative growth but shorter photoperiods for flowering. It is possible to extend the blooming period of chrysanthemum from the usual 45 days to five months under Lucknow conditions (Kher, 1969).

Scope and Prospect for Domestic and Export Market

(a) *Domestic market:* A conservative estimate puts the land under floriculture in India at around 3,000 hectares. Considering the size and the population of our country, this area is not at all impressive. There is scope for putting more area under flowers provided modern scientific methods of cultivation and marketing are adopted.

According to a survey conducted by Sampath (1962), cut flowers weighing 10,500 tons and worth Rs.9.26 crores are sold annually in five metropolitan cities of Calcutta, Delhi, Bombay, Bangalore, and Madras. This survey is back-dated and the volume of trade now must have increased manifold. If the volume of trade of nursery stocks such as potted plants. trees, shrubs, bulbs, and seeds are also taken into consideration, the total turnover in the trade of ornamentals will be quite significant.

In the South, the small-flowered chrysanthemum, especially a yellow cultivar, is very popular and the turnover is quite large. Research may be directed towards producing a high-yielding strain acceptable to the producer and the consumer. At the Indian Institute of Horticultural Research, Bangalore, some promising small-flowered hybrids have been developed, which may be released shortly for commercial cultivation. The cultivators should also be helped by such research institutes to prolong blooming periods to avoid market gluts. Similarly, in other popular cut flowers such as tuberose, jasmine, China aster, etc., research should be done in this direction.

A proper method of packaging and prolonging the cut flower life will also help enlarge the domestic and the foreign markets.

There is unlimited scope for increasing the sale of nursery stocks such as budded roses, indoor plants, bulbs, and seeds, provided the customers are assured of good quality materials. The immediate task will be to set up a certification organization to standardize the quality of these products.

Some of the horticultural by-products are of immense commercial value. The essential oils from jasmine, rose, and tuberose sell at prohibitive prices. For example, 1 kg of jasmine concrete may cost as high as Rs.10,000 and one acre of jasmine can yield up to 6 kg of concrete. At present, most of our essential oil requirements for the perfume and soap industry are being met by import. With a little help this industry can develop not only to meet the internal demand but also to feed the export market. It will also bring substantial income to the farmers.

The dried fruit of *Vanilla planifolia* (Syn. *V. fragrans*) yields the vanilla of commerce which is now produced synthetically. In Kenya, this orchid is grown commercially for the extraction of this commercial product which is used to flavour desserts and other products. With our favourable climatic conditions it is possible to grow this orchid commercially in some regions of the country.

Some of the ornamental plants yield products which are eaten as vegetable or otherwise. There is ample scope to popularize these products and consequently increase their cultivation. For example, the underground stem of *Nelumbium* (lotus), commonly called *bhasinda*, is a popular vegetable in North India. The same product is also used to make pickles. The stems of water lilies are also eaten as vegetable by many Bengalees, especially those from Bangladesh. The prickly aquatic plant resembling lotus, *Euryale ferox*, and commonly found to grow in the tanks of Bangladesh and parts of India, yields seed of the size of a pea which when parched or roasted over hot sand gives a puffed-up product, commonly called *makhana*, and is a delicacy for the table. The fresh seed of lotus fruit is popularly eaten in Bengal after removing the green plantlet from inside which is bitter, but the seed itself is tasty. With so many water tanks in the Eastern part of the country and elsewhere it is possible to extend the cultivation of these products commercially.

(b) *Export market:* The possibility of exporting cut flowers, plants, bulbs, etc., has not been explored properly though there is a very good potential. An UNCTAD Survey (Anonymous, 1969) conducted in 1968 showed that seven European countries, namely, West Germany, France, Switzerland, U.K., Norway, The Netherlands, and Italy, imported cut flowers worth over Rs.88 crores of which the share of West Germany alone was over Rs.70 crores. Unfortunately, among the exporting countries the name of India was absent, although we have got favourable and diverse climatic conditions ideal for growing varied types of flowers required by the westerners. In 1972, India exported about 15,000 rose blooms to Europe and the quality of our flowers was much appreciated. But, at present, export of flowers is negligible. The European countries need cut blooms mostly during November to March when these countries remain largely under snow and hence field production of flowers is not possible. In fact, during the winter, flowers are produced under glass which increases the total production cost, adding to the already high cost of labour there. In most parts of India the winter is the best period for producing most of the commercial flowers. Therefore, with our abundance of cheap labour we are in an enviable position to build our cut flower trade with these countries. The best way to organize this trade is to develop the centres of cultivation with prospective export-type flowers, nearer to international or domestic airports, so that the flowers can be air-lifted without much delay to the places of consumption. The cut flowers which can find ready markets, especially during the winter, are roses, chrysanthemums, orchids, gerbera, gladiolus, carnation and tuberose.

The sale of cut blooms of orchids is a multimillion dollar business in the U.S.A. alone. India being so rich in its natural orchid resources and with its favourable climatic conditions should be able to capture a portion of this market. At present though small countries such as Thailand and Singapore are doing good business in this line, unfortunately India is not exporting cut blooms of orchids.

The export of ornamental bulbs, when they are dormant, offers good scope as these can be transported by sea instead of by air and thus can be sold cheaper than fresh flowers which are to be air-lifted. In Europe, the Dutch have a very organized trade in bulbs of tulip, gladiolus, iris, etc., and we cannot compete with them in this field. Our export effort should be concentrated on sub-tropical and tropical bulbous plants such as *Amaryllis*, *Eucharis*, tuberose, gerbera, *Crinum*, lilies, canna, football lily, etc. The price of one Dutch hybrid *Amaryllis* (*Hippeastrum*) bulb in the U.S.A. is between Rs.10 and Rs.35, whereas one Indian bulb of similar quality should not cost more than Rs.5 to Rs.15. One more word of caution is that we should not unload our export products during the summer months when the Dutch are there to flood the market.

The export of F_1 hybrid seeds and potted foliage and other plants is also viable commercially. These have been dealt with in detail in Chapter XX on "Commercial Floriculture".

Prolonging Cut Flower Life

If we are to succeed in the cut flower trade, research must be directed towards prolonging the life of cut flowers and scientific packaging so that the flowers may reach the consuming centres in garden-fresh condition. The specification for exporting cut flowers is that the intervening period between cutting the bloom in the field in India and its arrival in the European market should not be more than 36 to 48 hours. The technique of pre-cooling of the bud before shipment for prolonging its life should be mastered for each flower.

Flowers dried in their natural colours offer a very good scope for export to Europe, the U.S.A., and Japan. Australia has specialized in this trade and is exporting these to Japan. Borax is used in drying flowers and foliage. The flowers are dried by keeping them in a standing position in a container and covering the whole flower carefully with borax. The flowers remain in borax till all the moisture is absorbed.

Landscape Gardening

The importance of gardening is also not understood. in India. The people should be educated to realize the importance of gardening, by providing good parks and gardens in cities for a large population to relax and enjoy the beauty of nature.

Bio-aesthetic Planning

The term bio-aesthetic planning, a concept of Prof. Lancelot Hogben, means the proper utilization of the available flora and fauna in the beautification of the surroundings. In India, the theme of bio-aesthetic planning was propagated by Dr. M.S. Randhawa who gave a practical shape to it in planning Chandigarh along with the famous architects Le Corbusier and Pierre Jeanneret. The aim behind this concept is to plant ornamental flowering trees along roads, in parks, house compounds, public places, and also to develop national parks where non-carnivorous animals and beautiful birds will find sanctuary along with beautiful flora. This term also includes landscape gardening though in a wider sense. Bio-aesthetic planning can be extended to the entire country. The bio-aesthetic planner can be described as a master artist who uses the whole country as his canvas and his paints are the rich colours of red, blue, orange, and white of the different flowers. It is said that the untouched nature is quite monotonous. It is only with a touch of bio-aesthetic planning that the countryside and other places will look not only natural but pleasing too to the eye.

Bio-aesthetic planning should run hand-in-hand with town planning. Our new townships should not be allowed to grow as mushrooms in dung-heaps as our older towns are. The roads in towns and cities should be broad, planted with flowering and shade trees, and there should be spacious parks along with conservatories for harmless animals and birds.

The older congested cities and towns also should be retrieved from their present state by bio-aesthetic planning. One way of doing this is to acquire by compulsion the ugly areas of the towns in some centrally located pockets and to convert them into parks. The displaced persons may be accommodated in multi-storey buildings which occupy less land. But many planners are against vertical growth in our towns; a compromise must be found somewhere.

In some of the most congested cities such as Calcutta, Kanpur, Bombay, etc., it may be really impossible to get land in the city proper for bio-aesthetic planning. In such cases 'garden suburbs' should be developed on the outskirts of the city easily approachable by vehicles or even bicycles. The garden suburb concept has been discussed in detail in Chapter XXIV. The real estate value of any property goes up if it has a compound properly landscaped and planted with trees.

Air Pollution

Air pollution, one of the most-talked-about problems in the present age, has reached disturbing proportions in some of the largest cities of the world and also in some of the metropolitan cities in India. The unplanned growth of the cities has resulted in congestion of houses, factories in residential areas, and proliferation of motor vehicles. The smoke from the coal ovens (*chulas*) of the residential houses, the dust and the smoke from the grinding mills and chimneys of the factories, and the exhaust fumes from the motor vehicles – all add to the pollution of our cities. Pollution is causing an increase in the diseases of the respiratory tract, cancer, and many other ailments. Unless something is urgently done the health of our citizens may deteriorate rapidly.

The role of open spaces such as parks and of living plants in checking air pollution is well known. The parks are considered as the lungs of a city. The barrier of trees checks noise pollution, dust pollution, and air pollution.

Human Welfare

The role of landscape gardening in human welfare cannot be overlooked. Even in an under-developed country as India, people do not live by bread alone. They also need some finer things of life. It is a great tragedy that most of our children in big cities do not have any open space to play and to see colourful flowers, birds, and butterflies. It is the moral duty of our government, through the municipalities, corporations, and such other bodies, to provide the citizens with spacious parks having beautiful trees and flowers where they can relax, find peace of mind, and breathe fresh air after a day's hard work. The children will also be able to play freely in such parks. It is a common sight in congested cities that groups of youngsters play football, cricket, or hockey in the by-lanes in the absence of playgrounds and parks. Such a state of affairs should not be allowed to go on indefinitely. The wealth of any nation is linked with the health of its people. Unless we can ensure the healthy development of our citizens, especially the younger generation, by providing for them open breathing places through bio-aesthetic planning and landscape gardening, we cannot expect to build up a healthy society and a prosperous nation.

Chapter III

Nursery Management

PROPAGATION

The living organisms reproduce by producing off-spring or by multiplying themselves. In nature plants propagate themselves by seeds and vegetative parts. The gardener has to learn the different methods of propagation to replace the old or the diseased stocks and to meet the increasing demand of a growing garden.

Methods of Propagation

There are different methods of plant propagation, which can be classified under two heads: (A) Sexual propagation and (B) Asexual or vegetative propagation.

(A) *Sexual Propagation*

Plants are propagated sexually by seeds and spores. Actually, a spore (as in ferns) is an asexual body. But, when this falls on a moist surface it produces small plant bodies (*prothallia*), which develops the sex organs (*archegonia* and *antheridia*) and these in due course develop the sex elements (*gametes*), which fuse and the resulting body develops into the fern plant.

It is generally observed that plants raised from seeds are more vigorous in growth than those raised by vegetative methods. A seedling may not be an exact replica of the parents as a result of natural cross-fertilization or segregation of characters, whereas plants raised by vegetative means resemble the mother plant in all plant characters unless the plant part used for propagation turns out to be a natural 'sport'. Propagation by seed is necessary to obtain new plants or hybrids. One disadvantage of propagation from seeds is that the seedling may take years to produce flowers. For example, an orchid seedling may take 5-8 years to produce flowers. But a plant raised by vegetative propagation from another mature plant flowers within a season or between 1 and 2 years.

Factors controlling germination of seeds: It is difficult or rather impossible to get 100 per cent germination, even when the seeds are sown in a perfect condition, for a variety of reasons. The factors which control germination are age, stage of maturity and viability of the seeds, water, free supply of oxygen, and heat or temperature. A combination of most favourable conditions of these factors for each kind of seed is thus called the 'optimum' for that species.

If conditions are normal, most seeds will germinate within a short time, but some may fail to do so because of immature embryos or a rest period or after-ripening period. Normally, an embryo has sufficient food material stored for germination, but in such isolated cases as the orchids, nutrients have to be provided externally to ensure germination.

The factors such as age, stage of maturity, and viability of seeds are interrelated. Viability of seeds determines their power to germinate. The poor viability of a seed may be due to the fact that it was harvested before full maturity or it might have been harvested from a weak plant and from plants grown under improper humidity and temperature. Viability also greatly depends upon the age of the seeds. Most of the annuals lose viability within 6-12 months. Seeds of some plants and trees should be sown immediately after harvest. Seeds of some kinds of lotus are said to have germinated even after 150-200 years!

Oxygen is essential for initiation of growth and this is generally present in sufficient quantities unless the soil is poorly prepared or is compact or water-logged, or the seed is too deeply buried.

Water is necessary since plant food should be made available to be embryo in a dissolved state. Moreover, water also softens the hard seed coat and hence encourages early seed germination. A quantity of water enough to completely saturate the seed coat should be used as too much watering is not good for germination. Excess of water may block the supply of oxygen or help the growth of disease-causing organisms. On the other hand, some of the xerophytes and annuals such as portulaca and eschscholtzia need small quantities of moisture for germination, whereas seeds of aquatics such as water lilies and lotus remain submerged.

The temperature should not be too high or too low. Some of the annuals such as nemophila and larkspur germinate better when the nights are really cool.

Light is not essential for germination for most of the species. Seeds of some species are indifferent to light, while others are slightly inhibited (e.g., larkspur); some are completely or partially inhibited (e.g., *Nigella arvensis*); still some others require light and germinate without cover (e.g., members of *Gesneriaceae*, Ranunculus, and Veronica).

Delayed germination and pre-treatment of seeds: Some seeds do not germinate easily for a variety of reasons, the most common causes being dormancy, rest period, and presence of hard seed coat.

Seeds with hard coats require some kind of external treatments for germination (e.g., seeds of palm and canna). The methods commonly employed are cracking of the coats by mechanical means, abrasion, soaking in water or acid, and stratification. Sulphuric acid is often used at full, half, or one-fourth strength for periods varying from 10 to 30 minutes. Some seeds when soaked in water for 12 to 24 hours or more germinate better. Stratification consists of fully exposing the seeds to the action of the weather by placing them in alternate layers of moist sand or sifted soil either in shallow boxes or in flats or in trenches outdoors. The flats and boxes are placed outdoors to be exposed to rain and snow and to freeze and thaw. In places not subjected to snowfall, the freezing treatments are given under artificial cold storage. The after-ripening period and the rest period can be shortened by treatment with low temperatures of 35 to 41°F for periods ranging from one to four months. Rose seeds are thus treated to get better germination. Larkspur also germinates better during the summer if stored in a cool moist place for several days before sowing.

Orchid seeds are so minute in size that they have very little or no stored food to nourish the growing embryo. Hence these are germinated in special media, the details of which have been discussed in Chapter XX on "Breeding of Orchid".

Seed testing: Before sowing on a large scale, it will be worthwhile to test the viability of the seeds. The simplest indoor method is, first, to sterilize by boiling in water or under pressure a pair of petridishes and two pieces of filter paper cut to size. Then one piece of filter paper is laid over the petridish and the seeds spaced over it and the other piece is used to cover the seeds. The filter papers are moistened with sterilized water and then the lid is put over this. The petridishes are kept in a dark, warm place. Some sterilized water may have to be supplied in between, till the seeds germinate. If most of the seeds germinate, then it is assumed that the stock is of good viability.

(B) *Vegetative Propagation*

By vegetative propagation is meant the production of a complete plant from one vegetative bud or several such buds. This definition includes all asexual methods such as cutting, layering, division, separation, budding and grafting. Tissue culture method of propagation is also included under this.

Cutting

This may be defined as a process by which a plant is produced by severing a vegetative portion from the plant and rooting it in a favourable medium under optimum conditions. The plant parts that are used for this purpose are stems, roots, leaves, and modified stems such as tubers, corms, rhizomes, runners, and bulbs. In general, propagation by cutting is the cheapest and the most convenient method and, hence, this is used more popularly to raise new plants. It is probably possible to raise most of the species of plants by one method of cutting or the other, but in the case of annuals, biennials, and some perennials some other methods such as seedage, layering and grafting are easier and more economical.

(1) *Stem cuttings:* Stem cuttings are of three types: (a) soft wood or herbaceous, (b) semi-hard wood, and (c) hard wood. Experimental evidence shows that cuttings, with a few leaves at the top, root better than leafless cuttings.

(a) *Soft wood cuttings:* Soft wood stem cuttings may again be separated into two distinct categories: (i) those taken from herbaceous plants such as carnation, chrysanthemum, coleus, dahlias, delphiniums, petunia, etc. (Fig. III.1), and (ii) the unripened tips of woody plants such as most of the ornamental shrubs and some trees. Soft wood cuttings are taken from below a node and the bottom leaves are removed.

(b) *Semi-hard wood cuttings:* The portions of the stems which have passed the soft wood stage but are not yet mature are referred to as semi-hard wood cuttings. Semi-hard wood cuttings are used for propagation in jasmine and hydrangeas.

(c) *Hard wood cuttings:* These are cuttings of shrubs and trees taken from mature current year's growth. The length of cuttings varies with the type of plants and weather conditions. One metre to one and a half metres long hard wood cuttings of some trees and shrubs such as *Ficus benghalensis*, *Citharexylon*, *Gliricidia maculata*, and species of *Erythrina*, when planted out in the open, root easily. But, generally, hard wood cuttings are 15-30 cm long.

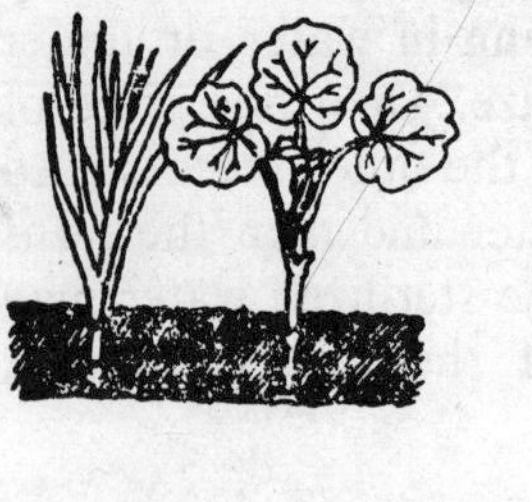

Fig. III.1

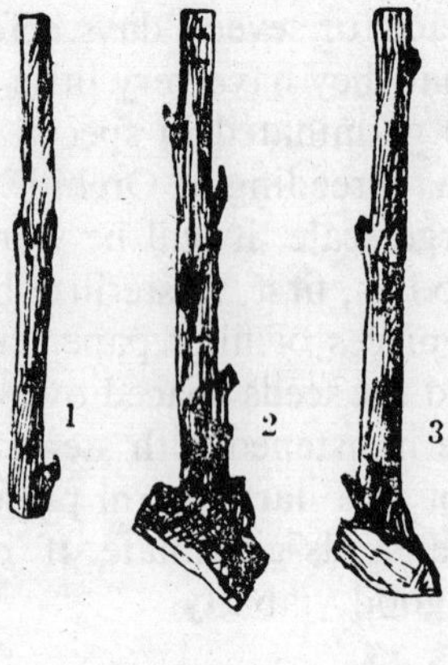

Fig. III.2

Fig. III.3

Fig. III.1. Soft wood or herbaceous tip cuttings of carnation (left) and geranium (right).

Fig. III.2. Stem cuttings (leaves removed for clarity, but in practice one or a few leaves should be retained, wherever feasible): 1. An ordinary stem cutting, 2. Mallet cutting, 3. Heel cutting.

Fig. III.3. A leaf cutting of Begonia rex laid in rooting medium.

The soft wood or herbaceous cuttings are generally 2.5 cm to 10 cm long.

Water-shoots, which are over-vigorous and over-luxuriant growths from the base, do not make good cuttings. Similarly, in carnation the basal cuttings are not good, as these produce more vegetative growth than flowering. The cuttings of middle portions and tips are better in carnation. In some plants, the terminal tip cuttings produce better plants.

Stem cuttings are of *three kinds:*

(i) *Terminal cuttings:* These are obtained from the tip portion of the shoots. The lower leaves are removed by snipping (not tearing off) and only about four terminal leaves are left in such cuttings. If the terminal leaves are very large, they are cut to half of their length to minimize transpiration. Before inserting the succulent cuttings of cactus, cotyledon, etc., these are allowed to dry up for 3-4 days or put in dry sand and not watered for a number of days.

(ii) *Heel cuttings:* The lateral shoots which are pulled off from the stem and contain a portion of the stem are known as heel cuttings; these root more successfully than terminal cuttings. The rough wood and bark at the bottom may be smoothened with a sharp knife. Cuttings having a small portion of the woody branch attached to the base are similar to heel cuttings but are termed as *mallet cuttings* (Fig. III.2).

(iii) *Node cuttings:* Single or multiple node stem cuttings are also obtained from plants such as *Dieffenbachia* and *Dracaena* and these are inserted horizontally in sand.

With the present-day facilities such as mist propagation, propagation frames, bottom heat, etc., it is possible to raise cuttings any time of the year. But, it has been found that the best time for taking cuttings corresponds to the time of the growing season of the parent plant. In most parts of India, the duration of this season is from July to September. In the hill stations, the corresponding time will be the spring. But, soft wood cuttings can be propagated any time during the year.

All cuttings should be inserted fresh without delay except in the case of cactus and some succulents. If there is a considerable time lag between obtaining the cuttings and planting, these should be stored in shade, packed in either moist cloth or sphagnum moss. Instead of forcing the cutting through the medium, a hole of equal thickness should be made with the help of a stick and the cutting inserted in the hole. The medium (soil, sand, or vermiculite) around the cutting should be made firm to avoid failure.

Another important point is to insert in the rooting medium that end of the cutting which was nearest to the root of the parent plant. Root forms in a stem cutting at the end which is nearest to the root of the parent plant and shoot develops at the other end. This phenomenon is called 'polarity'. Some species may be forced to root when planted invertedly, but such growth will be weak and short-lived. Sometimes a slanting cut below the node is given at the rooting end, thus exposing more tissue for callus formation. If the cuttings are raised in earthen pots, a common practice is to insert the cuttings around the edge of the pot as this allows good aeration.

Soft wood cuttings are inserted in the medium to a depth of 2.5 to 3 cm after removing the bottom leaves. The same procedure is followed for semi-hard wood cuttings. About one-third of the length of the cutting is inserted in the medium in the case of hard wood cuttings. Soft wood and semi-hard wood cuttings are generally raised in partial shade, but the response to shading varies according to the season and the species. It has been observed that the shade provided by a muslin or cheese-cloth or roughly woven gunny cloth is good for an average plant. The grower will learn by experience the amount and the kind of shade needed for various plants during various seasons. When cuttings are grown in a mist-chamber the need for shading becomes irrelevant. If planted other than in a mist-chamber the cuttings will need syringing 2-4 times a day depending on the weather. Hormones or growth substances play an important role in rooting of cuttings. The role of hormones has been discussed in Chapter V.

Rooting media: A number of rooting media and their combinations have been suggested by various authors. A good rooting medium should be well aerated, but at the same time should hold enough moisture. While a wide range of plants root well in a light soil, a large number of others need some special media or compost. The general practice is to use sterilized sharp sand, especially for soft wood or semi-hard wood cuttings, or a combination of sand and sphagnum peat. The use of sand alone as rooting medium is questioned by many and they suggest instead a mixture of sand and sphagnum peat especially for hard wood cuttings. Other media used for rooting are fine cinders, sphagnum moss, and coconut fibre. Vermiculite (expanded and granulated mica) is a good medium if used with care. This retains a lot of moisture and, hence, overwatering should be avoided. In temperate climate, the rooting medium is heated by bottom heat provided through fermenting manure, electricity, or steam. In most parts of India, except in some hill stations, no bottom heat is needed.

(2) *Root cuttings:* Quite a large number of plants can be grown by root cuttings. The general criterion is that any plant that readily produces sucker may be propagated by root cuttings. Root cuttings should be taken from young roots of average thickness and the length may vary from 2.5 to 15 cm with one or a few buds. The cuttings are placed horizontally or vertically in light soil or sand. The longer (7 cm or more) cuttings are placed vertically. In the case of root cuttings too the phenomenon of polarity should be taken into account. For easier identification, an oblique cut is given at the lower end of the cutting. Horizontally placed cuttings are covered adequately and firmed. The adventitious buds give rise to new shoots and roots grow from the base of these buds or from the cut sections of the old roots. Examples of ornamental plants produced from root cuttings are Achillea, Anemone, Gypsophila, Gaillardia, *Phlox decusata*, Salvia, etc., which belong to the herbaceous group. The woody plants produced by this method are *Aralia*, *Campsis*,

Clerodendron, *Lagerstroemia*, some rose species, wisteria, yucca, etc.

(3) *Leaf cuttings:* Propagation of plants by leaf cuttings is followed in such plants as *Peperomia*, rex begonia, *Saintpaulia*, gloxinia, gesnera, *Bryophyllum*, *Kalanchoe*, and *Sedum spectabile*. In *Peperomia* and *Saintpaulia*, the leaf-stalk is inserted in moist sand or compost (sand and peat) with the leaf-blade above. The leaf-stalk will produce new roots and plantlets. In Rex Begonia the entire leaf with stalk is taken and the large veins are cut or notched. The leaf is then placed on the rooting medium with small amounts of soil or compost placed over the cut portions. New plants will be produced at the cut places (Fig. III.3). Sometimes the leaves are slit in strips, tapering towards the base, and each strip is inserted vertically. Begonia Rex can also be raised in the same way as *Peperomia*. When *Bryophyllum* and *Kalanchoe* leaves are placed on moist medium with their underside downwards, small plants develop all along the edges. Leaves of Sedum are placed vertically to obtain new plants. A few shrubs such as *Viburnum rhytidophyllum* can also be raised by leaf cuttings.

(4) *Leaf-bud cuttings:* In alternate-leaved plant, a leaf-bud cutting is removed by starting the knife 1.5 cm above the bud and leaving the stem by about 2 cm below the bud, cutting through about one-third of the stem's thickness. In opposite-leaved plants, the stem is cut off about 1 cm above and 2 cm below the pair of leaves. Afterwards the stem is split into half and thus two leaf-bud cuttings are obtained (Figs. III.4 and III.5). The cuttings are

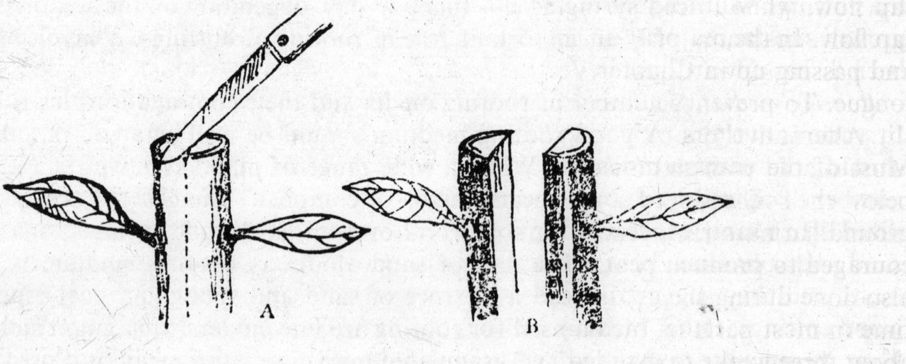

Fig. III.4

Fig. III.4. Leaf-bud cutting (for opposite-leaved plant): A. Before taking cutting, B. Two cuttings obtained by splitting the stem in halves.

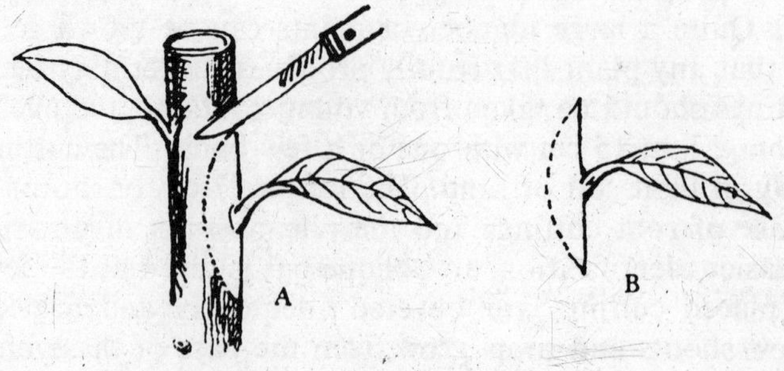

Fig. III.5

Fig. III.5. Leaf-bud cutting (for alternate-leaved plants): A. The twig with leaf, B. Leaf-bud cutting ready for planting.

then inserted with the bud going just below the surface of the rooting medium or buried deep enough to prevent it from toppling down while watering. To accommodate more cuttings leaves may be trimmed but not removed completely. Root-promoting hormones may be used for proper rooting. Once sprouted, the bud quickly develops into a shoot. The plants commonly raised by this method are poinsettia, hydrangea, geranium, camellias, and many others.

Leaf cuttings or leaf-bud cuttings are raised in hot, humid, and partially shaded places.

Layering

The method of inducing roots in a stem which is still attached to the plant, and then detaching it after the root is formed for transplanting is known as layering or layerage. Mostly shrubs, creepers, and trees are raised by this method, but some herbaceous plants such as carnation, verbena, and chrysanthemum can also be raised by layering.

(i) *Simple or ordinary layering:* This method requires just bending down a flexible branch to the ground and covering a portion of it with 6-12 cm deep soil at the back of the tip, leaving 10-30 cm of the tip above ground (Fig. III.6A). The branch is held in position with the help of a stone piece or hooks or pegs. The aim is to hinder sap flow at the bend to accelerate root formation. In jasmine and oleander a little twist will be enough to interrupt sap flow, while in others tying a wire tightly around the bent stem will help in stopping sap flow. In the majority of the plants a slit is made with a knife starting 2 cm under a node and passing upward just below the node, cutting half way through the wood, thus forming a tongue. To prevent healing together again, a match-stick or a tooth-pick is inserted at the slit. Alternatively, a ring can be made by removing the bark, 3 cm in diameter, below a node. Most of the plants root better by the tongue-slit method. The injured portion is placed below the soil and held in position. Leaves are removed from the portion that goes underground. In nurseries, stock or mother plants are kept ready for this purpose and are encouraged to produce basal shoots by cutting back the top shoots. Like cuttings, layering is also done during the active season when the mother plants are growing. The corresponding time in most parts of India is between July and September. Root formation takes place in about three weeks to three months, depending on the species. After 2-3 weeks when callus formation starts a notch is made just below the layer in the parent stem, thus interrupting further sap flow. This notch is deepened a little at fortnightly intervals till the layer is detached from the parent plant.

(ii) *Mound or stool layering:* Short-stemmed and stiff-branched shrubs are raised by this method. Generally, such plants are encouraged to produce many shoots from the ground level by cutting back the main stem the previous season. All the branches are wounded near the ground and earthing, up to a height of about 10-20 cm, is done to form a mound around the branches (Fig. III.6B). One season's growth is enough for this purpose. This method is most suitable for raising rootstocks for fruit trees. But shrubs with stiff upright stems such as *Deutzias* can be raised in this way. It may be worthwhile to employ this method in some species of *Cestrum* also.

(iii) *Compound or serpentine layering:* In this method the stem is covered at several points with soil, alternating with other points which are not covered. A similar slit as in simple layering is made at each node covered. Long running shoots of vines which are used in this method give rise to several plants (Fig. III.6C).

(iv) *Tip layering:* There is another method of layering known as tip layering, but this is applied only in a few fruit plants. This is similar to simple layering, the only difference being that while in simple layering the tip must remain above the ground, in this method the tip is buried in the ground (Fig. III.7A).

(v) *Continuous layering:* In this method a long branch close to the ground or stock plants having single stems only are used. The entire branch is placed horizontally and completely covered with soil except the tip. New shoots develop from the nodes and these are also covered with soil to a depth of 10-15 cm as they grow (Fig. III.7B). Each of these shoots will develop into a new plant. As buds in most of the plants will not sprout into shoots if buried, in such plants the branch is kept in a horizontal position without covering it with soil. Only when the shoots start growing, these are covered with soil as described above. *Hydrangea arborescens*, *H. paniculata*, dianthus, sedum, and some other plants can be raised by this method. This method is also known as trench layering.

(vi) *Air layering or gootie:* The Chinese or the pot layering is also similar to this. The

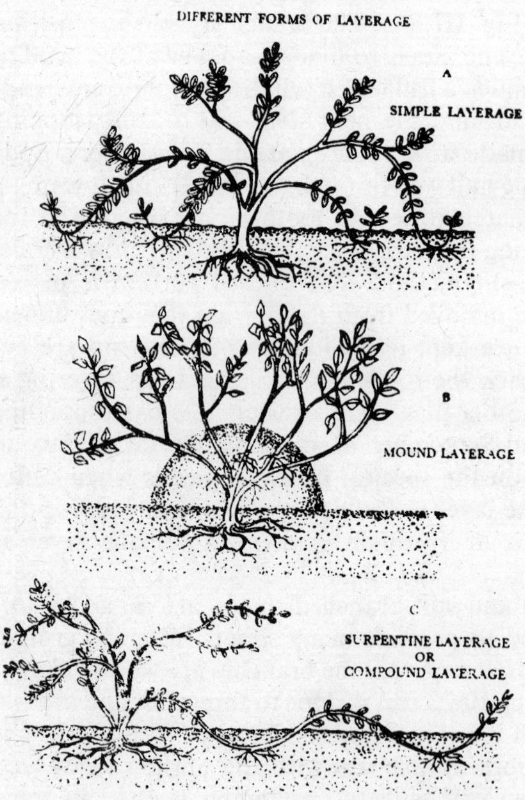

Fig. III.6

Fig. III.6. Different forms of layerage: A. Simple, B. Mound, C. Serpentine or Compound.

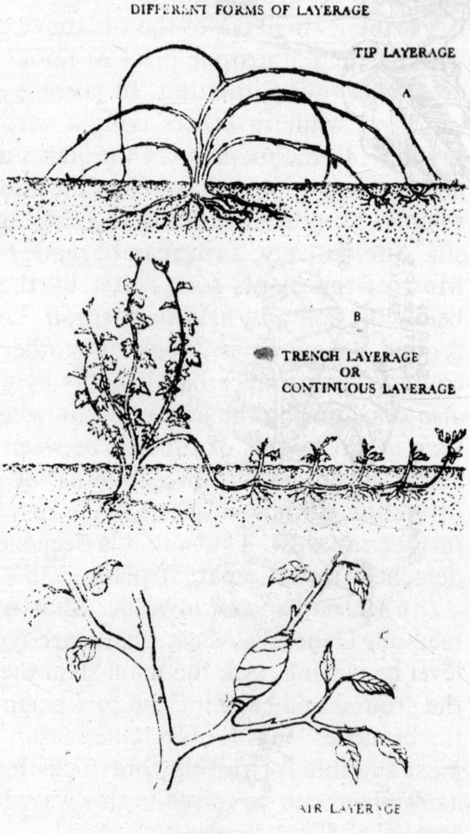

Fig. III.7

Fig. III.7. Different froms of layerage: A. Tip, B. Trench or Continuous, C. Air or *Gootie*.

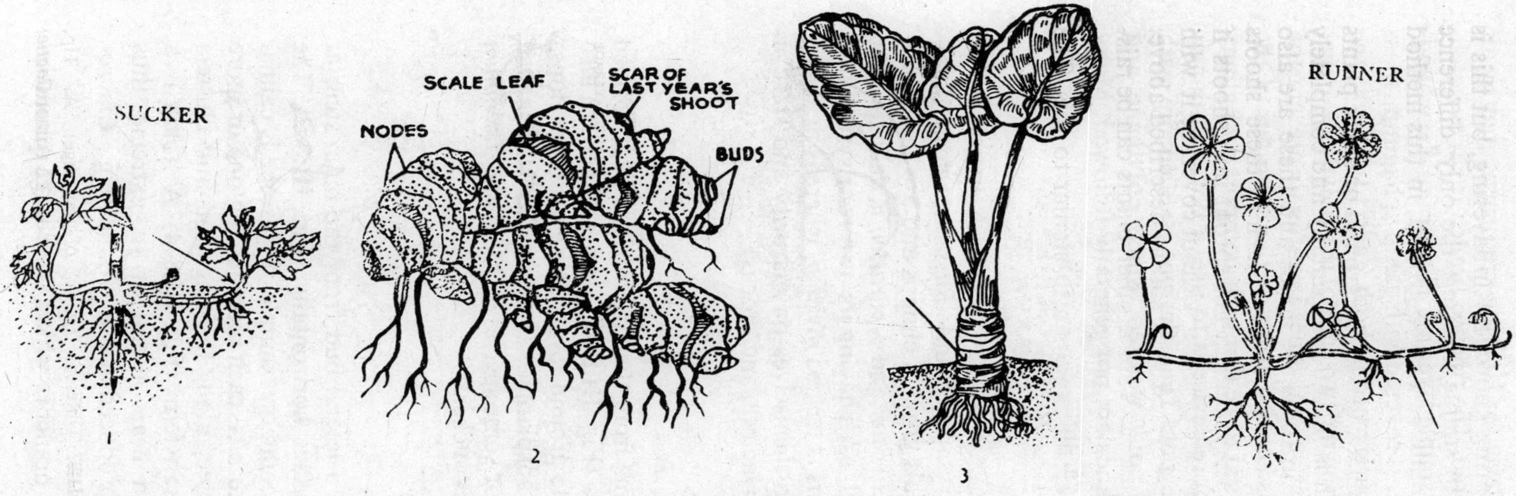

Fig. III.8

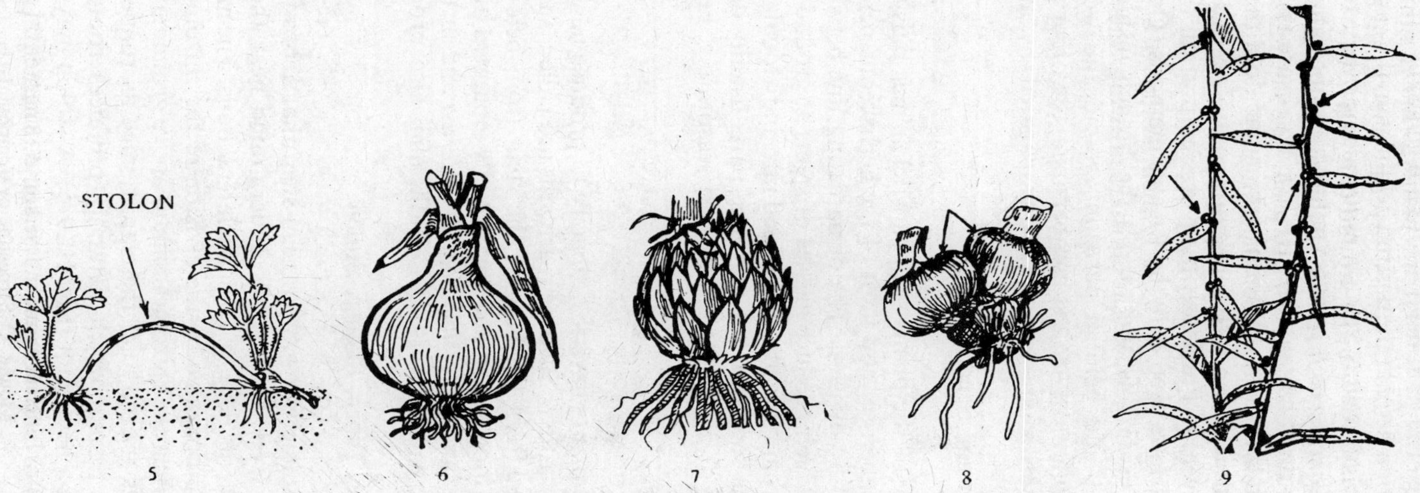

Fig. III.9

Figs. III.8 & III.9. Vegetative propagation: 1. Sucker (arrow), 2 & 3. Rhizomes [in 3 the vertical rhizome of *Alocasia* is known as Rootstock (pointed with arrow)], 4. Runners (arrow), 5. Stolon (arrow), 6. Tunicated bulb, 7. Scaly bulb, 8. Corm of gladiolus (the two daughter corms are marked with arrows, while at base the old shrivelled mother corm is visible), 9. Bulbils (pointed with arrow).

branches of plants which cannot be bent to ground level for layering are propagated by air layering. The terminal portions are generally air-layered. The ripened branch is slit or notched below a node at the desired height as in the case of simple layering. The part of the wounded stem, as well as 5 cm on either side of the wound, is covered with damp sphagnum moss and this is covered over by polythene paper and securely tied at both ends to prevent the moss from drying up (Fig. III.7C). To ensure this, bring the two ends of the polythene paper around the moss to form a bag and fold the ends several times inwards to prevent the moisture in the moss from coming out. The two ends are tied firmly on the two sides of the stem layered. Previously the wounded portion was used to be tied with compost and gunny cloth and watered daily by various methods. One more device is to use a pot with longitudinal side slits and to slip the branch inside the pot. Then the pot is filled up with compost and held in position around the wound and watered. Since polythene and sphagnum moss are less cumbersome, the other methods are not generally followed. Many of the shrubs and creepers are propagated by air-layering.

Division and Separation

Many of the plants such as Michaelmas daisy, chrysanthemum, *Russelia juncea*, plumbago, tuberose, and most of the herbaceous perennials grow in clumps and are easily propagated by division. The demarcating line between *division* and *separation* is delicate. The plants which produce masses of stems at ground level, each having its own root system, are lifted from ground and divided into individuals. This is termed *division*, whereas in *separation* the rooted or the unrooted parts usually detach themselves on maturity and start or develop as a new individual next season (e.g. bulb hyacinth, and crocus).

Suckers

Some plants have the capacity to produce new stems from the adventitious buds formed on their roots. Such new growths are called suckers (Fig. III.8–1). Stem suckers are also produced from the base of the stem from below the ground level. These soon develop a healthy root system and are carefully separated and transplanted. Trees such as *Millingtonia hortensis* and *Holarrhena antidysenterica* and shrubs of some species of *Ixora* and *Jasminum* produce root suckers. Chrysanthemum also produces suckers.

Rhizome, Tuber, Runner, Stolon

A rhizome is a subterranean stem that arises from a lateral bud from the main stem at ground level and extends underground near the surface horizontally (Fig. III.8–2). The aerial shoots develop from the lateral buds. The most common example of rhizome is canna. This is propagated by the division of the rhizome into bits, each piece having one or more buds. When the rhizomes remain above ground and creep along forming roots and clusters of leaves at each joint these are termed as runners as in ferns (Fig. III.8–4). A stolon is a slender branch which naturally produces roots and bears a bud at its extreme, thus forming a new plant (Fig. III.9–5). A stolon may be produced above ground (e.g., *Chlorophytum*) or underground as in some ornamental grasses.

Tubers are short, thickened parts of subterranean branches which store large quantities of plant food, mainly starch. Tubers contain buds in the axils of reduced leaves. Some species of *Begonia* produce tubers, but these do not reproduce naturally like potato tubers. Dahlia roots are not true tubers and cannot reproduce unless a portion of the stem is attached to it with present or adventitious buds.

Bulbs

In bulbs, the stem is reduced extremely to a disc around which fleshy leaf scales remain attached. These scales may be in continuous layers around the axis (tunicate or laminate, Fig. III.9–6) as in amaryllis or loosely arranged (scaly, Fig. III.9–7) as in lily. In bulbs the root development takes place from the basal plate, whereas new shoots arise from the axils of the scales from the central axis. In some plants such as amaryllis and tuberose smaller bulbs are produced round the base of the mother bulb. These are separated from the mother or the older bulbs and started individually.

Narcissus develops buds in the axils of the fleshy leaves and as these grow they form "slabs" which are small flattened bulbs. The slabs take about three years to develop into a mother bulb.

A ring of small bulbs is formed about the basal plate in hyacinth. These bulbs are removed and planted, and they take four years to mature. To hasten the process of bulb formation in hyacinth the bulbs are 'scooped' or 'scored'. Scooping consists of removing the entire basal portion and making a concave cut through the bases of the scales. In the case of scoring three cuts are made across the base of the plate and through the growing point. The injured bulbs are then stored in well-ventilated places at a temperature of 70-90°F. In about 4-6 months new bulbs are formed at the injured portions and these along with the mother bulb are planted together. The bulblets are separated after one year and planted individually.

Scaly bulbs of lilies are propagated by removing a number of the outer scales and planting them singly 5 cm deep in light soils. Small bulbs will develop after one season, but the bulbs will take 2-5 years to mature.

Corms

A corm is an underground stem with a short, fleshy, vertical axis, covered with dried leaf bases. It differs from a bulb in being solid, having no stems (Fig. III.9–8). As the leaves are produced the old corm starts shrivelling and on top of it and at the base of the leaves the new corm starts developing. At lifting time the old corm will almost disappear and in its place the new corm surrounded by numerous cormlets or coremels at the base will develop. If some new cultivars are to be propagated rapidly a corm may be cut into two with one bud and a portion of the stem in each portion and then planted. This is not an ideal procedure and is used only for the above purpose and to rejuvenate flat corms into higher crowned ones. Sometimes, gladiolus develops aerial corms.

Bulbils

Narcissus and some of the lilies produce tiny buds or small bulbs in the axils of the leaves which when planted in the ground develop into bulbs. These structures are called 'bulbils' (Fig. III.9–9). Some other plants such as the fern *Asplenium bulbiferum* and *Begonia evan-*

siana also produce similar structures for propagation. In the flower stalks of Agave a number of young Agaves are formed like seeds.

Grafting

The operation of inserting one part of a plant in another in such a manner that a union takes place between the two parts is called grafting. The plant or the part of the plant on which the grafting is done is known as rootstock or 'stock' and the part which is united with the stock is called 'scion'. If only a single bud is used as scion the operation is called budding, but if the scion consists of a twig with more than one bud the operation is termed as grafting.

Factors for successful grafting: The first important factor on which the success of grafting depends is that the 'stock' and the 'scion' should be related botanically, that is, in general they should belong to the same family and sometimes even the same genus. Sometimes even one species will have specific preference. For example, the deciduous *Daphnes* grow best on the species *D. mezereum*, whereas the evergreen ones should be grafted on *D. laureola*.

In budding and grafting operations it is the cambium surfaces of the stock and the scion that unite with each other. The cambium layer is situated just beneath the rind. While attempting budding or grafting the aim should be to ensure a fair union between the two cambium layers. Only through this fair union does the scion receive the sap needed to keep it alive. During budding time the stock should be in active condition and the bud should preferably remain dormant.

Another important factor is to maintain the moisture content of both the scion and the stock. However efficiently the union between the stock and the scion has been effected, these should be protected from drying up, by proper wrapping with polythene tapes and maintaining humidity, especially in herbaceous grafts.

Methods of grafting: There are several methods of grafting most of which are employed for propagating fruit trees. Grafting, except budding, which is also a form of grafting, is not widely used in ornamental horticulture except in a few cases. The types of grafting which are used in ornamental plants are limited to inarching, side grafting, splice grafting, saddle grafting, flat grafting, and cleft grafting. Many other forms of grafting like bridge, veneer, whip grafting and top working employed for fruit trees are generally not used for ornamental plants under Indian conditions.

(1) *Approach or inarch grafting:* This method is followed in the propagation of roses in West Bengal and Bihar. A kind of briar rose is used as rootstock. The stock is prepared from cuttings in nursery pots and these are placed near the scion plant. The thorns from the portion of the scion that will unite with the stock and those of the stock are removed. The stock and the scion should preferably be of the same thickness. To facilitate union the bark is removed by cutting through one-third of the stem in both stock and scion to a length of about 5-7 cm. The cut surfaces are placed as closely as possible and then are tied securely with the help of raffia or polythene tape. After union, which may take 4-6 weeks, the scion is cut below the union and the stock above the union. Instead of severing the scion by one cut, this may be done at intervals of 10 days as in the case of layers. Some other shrubs which do not grow well on their own root (e.g., *Allamanda violacea* is grafted on vigorous growing species of *A. cathartica* var. *schottii*) are inarched on cuttings of a

vigorous species. Similarly, shrubs difficult to propagate by other methods (e.g., *Petrea arborea* is grafted on *P. volubilis*) are grafted on some easy-to-root species.

(2) *Side grafting:* A single oblique or angular cut is made with the help of a knife on one side of a branch, intended to be the stock. A slanting cut 5-7 cm long is made into the scion wood having not more than 4-5 buds (10-15 cm long). A slanting cut is also made on the opposite side to make the scion double-tapered. The cut in the stock is opened up by pressure and the double-tapered scion is just inserted between the bark and cambium and as soon as the pressure is released the scion tightens up. This method is followed in the case of roses, camellias, etc. In side graft, the stock is generally thicker than the scion. Sometimes, even a 'T'-cut is made as in budding to insert the scion. The scion is tied tightly with polythene tape.

(3) *Splice grafting:* This method consists of putting a long slanting cut at the base of the scion wood and a similar cut at the top of the stock. The two are united and tied with the help of polythene tapes. The stalk is cut close to the ground and the scion is obtained from a flowering wood (Fig. III.10).

(4) *Saddle grafting:* This method is used generally to propagate Rhododendron and lilac (*Syringa* sp.). In this method the stock is cut in the form of an inverted wedge with one side more slanting than the other. A similar cut in the reverse direction is inflicted on the scion and this is fitted over the inverted wedge of the stock securely. The jointed portion is tied with raffia or polythene tape.

(5) *Flat grafting:* This is a method suitable for grafting cactus. The rootstock used are *Cereus*, *Cephalocereus* and other species. In this method the stock and the scion should be of the same diameter. A smooth transverse cut is inflicted on both scion and stock and the flat surfaces are held firmly in position by thread or raffia. The authors have found that one of the best ways to keep these in position is to prepare two U-shaped split bamboo pegs and to fix them to the ground so that the bent portions cross over the scion, the free ends pushing inside the ground on both sides of the stock. The pressure of the two inverted U-shaped stakes will keep the scion tightly fitted against the stock (Fig. III.10). To make the split bamboos bend easily in the middle, scrape the middle of the stake to a length of 5-7 cm, depending upon the diameter of the scion and to a depth of two-thirds of the original thickness. Different species of *Mammillaria*, *Echinocactus*, and *Gymnocalycium* which are usually globular-shaped are grafted in this manner.

(6) *Cleft grafting:* This method is used for grafting fruit trees. But in ornamental horticulture this is mainly used to graft cacti, particularly those having finger-like stems. Any stock having a diameter of about 2.5 cm or more is used as stock. First a clean horizontal cut is made across the stock. Then a cleft of about 3-6 cm is opened across the centre inside which the scion, tapered to a point (to facilitate maximum cambium contact) from both sides, is inserted. The stock may be tied with raffia to keep the scion secured (Fig. III.10). Another easier method is to hold the two together by passing a cactus thorn through both stock and scion at the middle of the union. Plug grafting, quite similar to cleft grafting, is also practised in cactus (Fig. III.10).

Budding

This is nothing but a form of grafting, the only difference being that instead of grafting a scion twig only a single bud is implanted in the stock. Since budding involves the hand-

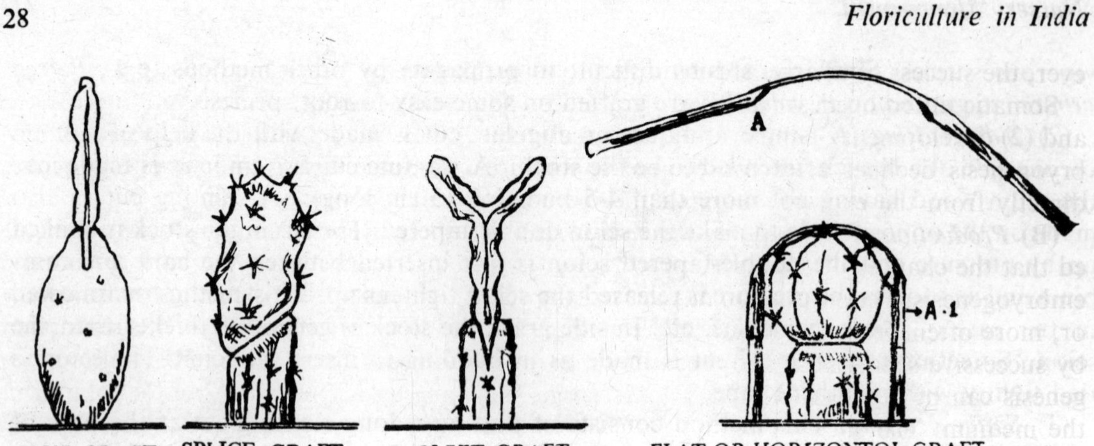

Fig. III.10. Different types of graft: A. Split bamboo support. Notice the central portion thinned out to facilitate bending and pass over the graft joint. A.1 Support fixed over graft and inserted in ground. Two such supports may be put in a crisscross manner.

ling of a single bud, some precautions are to be taken. The first point is to select the right type of bud and to conduct the operation in the right season. In ornamental horticulture, mostly 'T'- or 'Shield'-budding is employed for propagation. For details of 'T'-budding consult Chapter XX on "Rose".

Although *patch budding* is not generally used in India to propagate ornamental plants, there is a possibility to use this method to multiply hibiscus hybrids which fail to grow well on their own roots and to propagate some flowering trees which do not propagate by cutting or layering with a view to perpetuating some attractive seedling variations, as in the case of *gulmohar* (*Delonix regia*). Patch budding is done in plants having thick barks. This method consists of removing a patch of bark containing a bud and relocating it on the stock by removing a bark of corresponding shape. This is then tied with polythene tape leaving the bud open.

Tissue Culture

The propagation of orchid clones through meristem culture ('mericlones') was the first commercially successful venture in tissue culture. With the success obtained in orchids, scientists all over the world started to propagate several genera by tissue culture. In spite of the several drawbacks inherent in the method, the work done indicates that the principles of this method can be successfully employed in the propagation of ornamental plants. The work of Pierik (1975) has shown the effectiveness of this method even for shrubs and trees.

Methods of Plant Multiplication in Vitro

(A) *Somatic cell embryogenesis:* It has been demonstrated that somatic embryos are organized structures similar in nature to those obtained in zygotes (Steward *et al.*, 1969), but, as the name implies, these are asexual cells produced from somatic tissues. Somatic embryos have the capability of germinating *in vitro* and subsequently developing into plantlets. This procedure is the fastest among the clonal multiplication methods. However, the success of the method is limited to a few genera.

Somatic embryogenesis involves two steps: (1) callus induction and multiplication, and (2) development of embryos and plantlets. But it has been noticed that the rate of embryogenesis declines after consecutive subcultures. In some cases, embryogenesis occurs directly from the explant, skipping the intervening callus phase.

(B) *Production of adventitious shoots and related organs:* Research findings have indicated that the chances of development of these organs *in vitro* are more compared to somatic embryogenesis. The production of these organs may result either from the callus intermediary or, more often, from the explant directly. The propagule number may further be multiplied by successive separations and repeated culturing of emerged organs. The process of organogenesis can be accelerated by the application of 2-3 mg/1 each of auxin and cytokinin in the medium. The shoots obtained by this method are rooted and then transferred to the soil. The plants raised by this method may show genetic variability and, further, repeated re-culturing results in decline in the rate of organ multiplication. It is not advisable to propagate chimeral plants by this method as the chimeral components are likely to segregate.

(C) *Enhanced axillary branching:* It is common knowledge that axillary shoots start developing as soon as the apical dominance is removed. It is possible to accelerate the axillary branching process of quiescent or active meristems obtained in the lateral buds and stem terminals by the application of higher concentrations of cytokinins (10-30 mg/1). Within a period of 4-6 weeks miniature branchlets develop, which are excised and recultured individually in fresh media. When a large number of plants are thus obtained, they are rooted and transferred to the soil. Murashige *et al.* (1974) were able to increase clonal plants by a millionfold in gerbera compared to the traditional method. Plants raised by this method are also less variable.

The Procedures in Clonal Multiplication

There are several steps in the process of plant multiplication *in vitro* with each step having specific objectives and often specific requirements. The various steps involved are discussed below.

Step I: In this step the excised tissue or the explant is obtained. The main criterion in this step is to obtain a clean, microbial infestation-free explant. Though generally there is no problem in achieving this, sometimes it may become difficult to get one hundred per cent disinfection especially when the microbial infestation is deeply embedded in the tissue. Occasionally some explants also show discolouration after treatment with surface sterilization agents. Another important point in this step is to select the appropriate media and growth adjuvants.

The procedure of preparing the explant in gerbera is described in brief here. In the first step the gerbera plants should be separated into divisions and then leaves are stripped off completely from each division reducing it to a bud piece measuring only 0.5 to 1.0 cm in length. This piece is composed of the terminal bud and a portion of the subjacent stem. The next step is to disinfect this piece by immersing in dilute laundry bleach (5.25 per cent sodium hypochlorite) diluted 10-20 fold with water and containing 0.1 per cent Tween 20 (0.1 per cent polyoxyethylene sorbitan) emulsifier for about 10 minutes. The pieces are then rinsed thrice with autoclaved water and cut further till a shoot-tip explant measuring 2-3 mm is obtained. To achieve this the help of a dissecting microscope is needed. The explant is again dipped in a 20 times diluted bleach solution and then transferred to the nutrient solution.

Step II: In this step the actual propagule multiplication is obtained. The plant materials, as per the three major methods of multiplication mentioned earlier, are now sub-cultured again and again till the required number of embryos, shoots, or organs are obtained. In this step the different media with the growth adjuvants are also formulated.

Step III: The clonally multiplied plants are then transferred from media to soil. The transferring of *in-vitro* plants to soil sometimes poses ticklish problems, which may result in mortality. To overcome this *in-vitro* plants are first hardened and then gradually exposed to outside environment. The requirements of plants produced *in vitro* will vary between different types of plants, and while transplanting, knowledge of individual requirements of various plants will play a major role in achieving success.

Basic Parameters for Propagation in vitro

Before achieving a successful programme for tissue culture propagation of ornamental plants it is necessary to study carefully and then standardize certain parameters such as initial explant characteristics, nutrient media formulations, and cultural environment.

(1) *The explant:* From the point of view of both quantity and quality, the choice of the right type of explant is of utmost importance in clonal multiplication. Improper selection of the explant will result in limited success even if other procedures are followed scrupulously. The explant chosen should possess the inherent capacity for totipotency. The shoot tip with the resident apical meristem is most frequently used for propagation. Even if explants of all other categories fail to multiply, it may still be feasible to regenerate plants from the apical meristem. The normal propagation technique of any individual ornamental plant will be a useful guide in selecting the explant. For instance, in plants such as begonia which are normally propagated from leaves, excised leaf portions have been found to be useful (Chylah, 1972). In other species sections of root, bulb, stem, flower, ovary, nucellus, cotyledon and other structures may be useful (Murashige, 1974). It has been shown in orchids (Rao, 1977) that small shoot apices containing meristematic dome and a few leaf primordia can be used successfully in clonal propagation. Similar findings are reported with chrysanthemum, petunia and some other ornamentals (Holdgate, 1977). Murashige (1977) observes that debladed nodal explants with axillary bud can also be used. In perennial plants the juvenility of the plant part from where the explant is taken is an important aspect. It has been observed that tissues excised from more recently produced parts of herbaceous plants are more regenerative than older regions (Murashige and Nakano, 1967). The embryogenesis obtained in cultures established from cells of embryo explants was found to be copious (Steward *et al.*, 1969). It is a common phenomenon that cultures established from reproductive tissues regenerate much better than those obtained from vegetative tissues.

(2) *The nutrient medium:* The success of *in-vitro* propagation depends much on the chemical composition and the physical qualities of the nutrient media. The minimal medium

is composed of macro and micro elements, carbohydrate (generally sucrose), vitamins, and growth hormones. In the U.K. and the U.S.A. several of these formulations are available as ready-mix powders and the grower has only to reconstitute it by dissolving in required quantity of distilled water. Out of the various formulae available, the one suggested by Murashige and Skoog is most widely used. There are other media formulated for some specific purposes (Gamborg *et al.*, 1976). The media can be either in liquid form or solidified with purified agar-agar (Bacto or Noble Agar). In liquid media it is necessary to provide supports such as filter paper bridges or glasswool. If supports are not used, aeration will be required. By altering the levels of growth regulators used it is possible to induce callus or manipulate organogenesis. It is generally accepted that by altering the auxin-cytokinin ratio, shoots and/or roots can be obtained (Skoog and Miller, 1957). Besides growth substances, addition of accessory substances such as L-tyrosine, adenine, or adenine sulphate may stimulate organogenesis. Nevertheless, the requirement of growth regulators in organogenesis is by and large empirical and a proper selection has to be made by trial-and-error method.

(3) *The culture environment:* Light plays a major role in the success of tissue culture. A lower light intensity (1,000-3,000 lux) may be quite suitable for propagule multiplication and culture initiation, but for transfer to soil much higher intensities of light may be used. Not only the quantity but also the quality of light is important in initiating organogenesis. Seibert *et al.* (1975) have observed that mainly blue and near-ultraviolet regions are more helpful in the initiation of shoot, whereas Letouze and Beauchesne (1969) have found that to a greater extent red light and to a lesser extent blue light enhance rooting. It is not certain whether photoperiodism in the true sense of the term affects organogenesis, but it has been observed that a 16-hour period of light has been found to be beneficial for shoot and root initiation. In nature the bulbs and the tubers which are generally formed under short days might also require long dark periods *in vitro*.

Besides light, temperature and humidity also play an important role. Tropical plants grow better in warm temperatures (27°C). It is also believed that seasonal fluctuations in temperature may have some effects. It is imperative to study the chilling requirements of some bulbous ornamental plants. But experiments show that the regenerants of bulbous plants formed *in vitro* should be chilled to ensure successful transfer to soil. A proper balance of the relative humidity has to be maintained, as higher relative humidity may cause the growth of fungi and bacteria, and lower humidity may result in the drying of the media.

Conclusion

Propagation of ornamental plants by tissue culture method is gaining popularity. Nurserymen in the U.S.A. are using this method on a commercial basis. While producing plants *in vitro* it must be kept in view that, in spite of the various precautions taken, some plants may be produced which are generally different from the original. Such deviants should be screened thoroughly. One should gain sufficient knowledge regarding the trend in abnormal plant production before going in for large-scale production of plants through tissue culture. One more point is that unless actual pathogenecity tests are made, the claim for disease- or pathogen-free plants produced by the method of tissue culture should not be made.

A list of ornamental plants having demonstrated potential for clonal multiplication is given below:

Family	*Genera*
Agavaceae	*Cordyline, Dracaena*
Amaryllidaceae	*Alstroemeria, Amaryllis, Hippeastrum, Narcissus, Nerine, Clivia*
Asteraceae	*Chrysanthemum* and other composites; *Gerbera*
Begoniaceae	*Begonia*
Bromeliaceae	*Aechmea, Ananas, Billbergia*
Caprifoliaceae	*Weigela* hybrids
Caryophyllaceae	*Dianthus*
Euphorbiaceae	Various *Euphorbia* sp.
Geraniaceae	*Pelargonium*
Gesneriaceae	*Gloxinia, Saintpaulia*
Iridaceae	*Freesia, Gladiolus, Iris, Schizostylis, Sparaxis*
Liliaceae	*Aloe, Haworthia, Hemerocallis, Hyacinthus, Lilium, Muscari, Ornithogalum, Scilla*
Orchidaceae	So far many genera have been propagated *in vitro*
Polemoniaceae	*Cyclamen, Phlox*
Scrophulariaceae	*Antirrhinum*
Solanaceae	*Petunia*

Advantages and Disadvantages of Sexual and Asexual Propagation

Both sexual propagation and asexual or vegetative propagation have points in favour and against them. Propagation by seeds is the most common and the easiest method and the only method for propagating annuals and biennials with few exceptions. However, some of the biennials and annuals such as carnation, petunia, and marigold can also be propagated asexually. In raising new hybrids by cross-fertilization and obtaining new species or cultivars, propagation by seeds plays a prominent role, although now-a-days it is also possible to raise new plants by induced mutation and raising polyploids induced by chemicals. The plants raised from seeds are generally vigorous and of increased vitality. The main disadvantage of sexual propagation is that the plants produced from seeds do not generally resemble the parents. Moreover, the seedlings of perennial plants take longer to flower. For example, an orchid seedling may take 5 to 10 years to come to blooming, whereas a plant produced by the division of an old clump will flower in the same or the next season. Plants not producing viable seeds, 'sports' and mutants, can be propagated true to the type only by vegetative propagation.

The negative points of sexual propagation are the positive points or advantages of vegetative propagation. Vegetative propagation perpetuates a species, a variety, or a cultivar as plants thus raised have the characteristics similar to those of the parent except in some cases of incompatibility between the stock and the scion. The disadvantages are that this is not as rapid, simple, and cheap as propagation by seeds. Further, it is not possible to raise any hybrid or new plants by this method of propagation.

Besides the advantages of vegetative propagation over those of seed propagation, one method of vegetative propagation has certain advantages over the other methods, as for example, cutting over layering or grafting over cutting, or *vice versa*.

Cutting is the cheapest and the most convenient method of vegetative propagation and hence is the most popular. When a plant material is scarce or where a population is to be built up rapidly, leaf-bud cuttings (as in Hydrangea, Poinsettia, Geranium, etc.) and leaf-cuttings (as in African violet) are of great importance. Some of the plants which are difficult to root by cutting often root by layering, the reason being that the plant part propagated still remains attached to the parent plant and the sap flow is not interrupted. Again, different methods of layering have advantages of their own. Plants having drooping branches are generally ground layered (simple layering) and those which cannot be sent to the ground level are propagated by air layering. It is most advantageous to propagate many of the creepers by serpentine layering. The short stemmed and stiff-branched shrubs (e.g., *Deutzia*) are easily raised by mound layering. Moreover, this method of layering produces strong, stocky plants.

Those plants which do not root easily and cannot be propagated by any of the above methods of vegetative propagation are raised by grafting. Moreover, plants which do not survive or thrive on their own roots are propagated by grafting on strong and sturdy rootstocks. For example, modern cultivars of roses are propagated by grafting or budding only. Some stocks are resistant to certain diseases and the scions which are susceptible to such diseases (e.g., root-rot, wilt, powdery mildew, etc.) thrive well if grafted over such stocks. For example, *Rosa indica* is resistant to powdery mildew and when a susceptible cultivar is budded over it, it is said that this character is transmitted by the stock to the scion. Sometimes bridge grafting is used to repair a damaged portion of a plant. Generally, herbaceous plants are not grafted commercially. But sometimes this is done for transmitting diseases, indexing for virus diseases, for novelty and for inducing flowering for research purposes.

Whenever feasible, budding is preferred to other methods of grafting. The principal reason is that in budding only one bud is taken in contrast to grafting, where a scion containing several buds is grafted and hence budding is more economical.

Propagation by bulbs, tubers, rhizomes, and corms has one advantage over other methods of vegetative propagation: these can be stored for a longer period of time and planted in a suitable season.

Rest period: Many of the corms (e.g., gladiolus) and bulbs (e.g., hyacinth, narcissus, tulip, etc.) have a specific rest period during which they will not grow. This explains why many freshly harvested corms and bulbs fail to grow.

Harvesting, Packing, Storage, and Marketing of Nursery Stock

Harvesting

The proper stage of harvesting is important for seeds, bulbs, etc. Seeds are generally harvested when they are completely ripe on the plant. Some seeds of annuals and perennials fall from the plant at ripening either due to bursting of the fruit or for some other reason such as blowing off due to wind of light seeds (e.g., balsam, calendula, etc.). Such seed capsules should be covered with muslin cloth or butter paper bag before ripening, so that in the event of possible shedding these will be collected in the bag.

Bulbs, corms, and tubers should be harvested when properly matured. Corms and bulbs are generally harvested when the leaves start yellowing or when they dry up. These are carefully dug out without injury, along with the cormels or the bulblets. In the case of gladiolus if the leaves have not sufficiently dried up to break off easily from the corm, they are left alone with the corm to be detached at a later date.

Before harvest, nursery stock should be mature in the sense that the tissues are hardened against water loss and shrinkage. It is a common practice to defoliate shrubs and trees some days before they are to be dug out. This can be done by chemical defoliants, by withholding water, and by hand. Some plants such as orchids are packed for marketing when these are in a dormant condition.

Live plants intended for transport are sent with a ball of earth around their roots.

Packing and Handling

(i) *Seeds:* After harvest the seeds are cleaned and stored in closed bottles or tins. Various types of packages are also used for storing seeds. The most common method is to use ordinary paper packets. Paper packets lined with aluminium foil and polythene papers are also in common use. Sometimes, before putting the seeds in packets, each seed is coated with a pellet to keep it fresh for a longer period. In advanced countries seeds are put in containers complete with compost, fertilizers, etc., and the grower has only to open the lid and water the seeds to get the seedlings.

Before packing, the seeds are dried first in shade for 2–3 days and finally in the sun for a couple of days. In husked seeds, the husks are removed before packing.

(ii) *Plants:* The object of packing is to deliver the plants or other nursery stock at their destination without rendering them useless in transit. The method of packing will depend upon the plant and the distance to be covered. One important rule in packing is to see that the container is neither over-packed nor packed in such a manner that the contents will move about inside the container. All spare space should be filled up by some packing material such as straw or dried grass.

Ordinarily bamboo baskets of different sizes and shapes are used. The common variety is round or flat box-type baskets with lids. If the plants are to travel a short distance taking about two to three days in transit, these are first watered and then taken out of the pots or lifted from the ground with a ball of earth around the roots. The ball of earth is soaked in water and made into an oval shape covering the roots and then wrapped in straw or gunny cloth. These are then placed upright in the bamboo-matted basket. Several split-bamboo stakes, a little longer than the packed plants, are firmly fixed on the sides of the basket interlocking it in between the woven surface. These are then bent inwards and tied together over the basket to form a balloon-like structure. The plants are then tied with strong thread (*sutli*, i.e., gunny-twine) along the edges of the basket to hold them in position. The stakes are also tied with each other by passing such threads in two to three parallel lines. All extra space around the ball of earth is to be packed with straw to prevent the plants from moving around. Finally, the whole thing is covered with gunny cloth or mats made of palm leaves.

If the consignment is to travel a long distance taking a week or two in transit, much more care should be taken in packing. The ball of earth is soaked in water and covered with a thick layer of wet moss. Only plants having a well-developed root system should be selected for long-distance travel. The plants are placed upright in wet sand for about

15-20 days, their earth balls remaining immersed in sand. The plants which withstand this ordeal and still remain fresh are then taken out with their water-soaked earth balls surrounded by wet moss. The earth ball is then covered with polythene paper or placed in polythene bags and tied to prevent water loss. These are then packed as above.

When plants are to be sent to still longer distances the earth balls are packed with moss and polythene as above and these are then placed in wooden boxes having a thin layer of moist sawdust. Only two small holes are made – one at the top and the other at the bottom. In advanced countries greenhouse potted plants are transported in Wardian cases, the slanting roofs of which are made of glazed glass to allow light inside. Two small holes are left – one at the top and the other at the bottom – for aeration. The pots are kept in position by putting packing material in the vacant spaces.

Many living plants are transported in bamboo-matted boxes which may be lined on all sides with polythene sheets to check evaporation. Plants such as orchids, bromeliads, cacti, and succulents which live for a long time without moisture are transported in this way. The plants are packed in between layers of dried sphagnum moss, paper or wood clippings. Two to three such layers can also be packed depending on the depth of the box and the size of the plants. Each individual plant may also be wrapped in tissue paper to prevent injury. After packing, the top should be covered with a polythene sheet. The boxes are tied with strong threads or wire, after closing the lid.

For short-distance travel sometimes roses are sent bare-rooted without earth ball packed in bamboo baskets and gunny cloth. When plants are sent by air a minimum amount of earth is used to prepare the ball and only light packing materials are used.

(iii) *Bulbs:* Bulbs, tubers, and corms withstand rigours of handling and these are packed in bamboo-matted boxes in between layers of straw. Rhizomes of water-lily and lotus are wrapped in moist sphagnum moss and polythene to keep them moist during transit and then packed in baskets or cardboard cartons.

Storage

Seeds are stored in a cool, dry place or kept in a desiccator. Seeds needing long stratification are subjected to such treatment immediately after harvest.

Living plants are not stored for a long time for obvious reasons. But during the time lag between lifting and packing, these should be kept in shade. Similarly, plants received from nursery are shifted to a shady place and watered. If a considerable delay is anticipated, say, from a few days to one week, the plants are shifted into a cool, humid glasshouse or plastic-house or a mist-chamber.

Bulbs, corms, and tubers are stored in single layers over dry sand, flat wooden trays, or racks in a well-aerated store room with low temperature and low humidity. Before storing, they should be treated with 0.1 per cent benlate or 0.1-0.2 per cent captan or any other suitable fungicidal solution for 15-30 minutes and then dried. The bulbs and corms are also dusted with 5 per cent DDT or BHC to prevent insect attacks. The roots of dahlia which tend to dry up in storage are sprinkled with water occasionally.

A better but costlier method of storage is to keep the bulbs, the corms, and the tubers in cold storage. The requisites of ideal storage are proper aeration, and optimum temperature and relative humidity.

Marketing of Nursery Stock

By the term marketing we mean buying and selling. The price of a nursery stock depends upon supply and demand. If the particular stock is in short supply and the demand is high, the price will naturally be higher. If the supply is plentiful and there is no demand the price will be low. But there are other factors also which influence the relationship between supply and demand. These are:

(1) *Distance between the centres of production and consumption:* If the distance between these two is long, the transportation cost is added to the selling price of the produce.

(2) *Purchasing power of consumer:* In cities having more buying power a produce will sell at a higher price compared to a countryside market where the buying power is low.

(3) *Advertising and salesmanship:* These also play an important role in pushing up sale and profit.

(4) *Consumer preference:* Some consumers feel that plants are a luxury item and they defer the purchase of such items. But others feel that planting a garden will increase the resale value of their property and hence rush through the purchase of plants whatever the price.

(5) *Quality of produce:* This is possibly the most important factor which influences a prospective buyer, and, hence, a seller should ensure quality produce. Quality of nursery stocks can be maintained by proper grading and careful handling to ensurc that leaves and branches are not damaged during transit. A lot of plants or plant products offered for sale should be of uniform quality because even a single bad item in the lot will bring down the price. The plants should have clean and neat foliage unaffected by insects or diseases. The potted plants should not be pot-bound. The pot should be of the right size with sufficient soil. The evergreen and deciduous plants should be well 'balled-and-burlapped'. Seeds should be of good quality and bulbs and tubers also should be good in size and shape, free from diseases, and well packed.

Consumer appeal and satisfaction also play a great part in marketing. The nursery area, green houses, etc., should be neat and clean and the plants should be orderly arranged, properly labelled, and attractively displayed. The arrangements should be changed periodically to break the monotony. The buyer should be given catalogues and pamphlets containing instructions for handling, planting, and management. One more important factor in marketing is courtesy and politeness of the seller towards the customer. A smiling and polite seller definitely creates a host of goodwill towards the prospective buyers.

Nursery Structures

In a large nursery certain structures are absolutely necessary. The structures are storehouses, potting sheds, nursery beds, mist chambers, propagation frames, greenhouses, etc.

Store-house: A store-house is absolutely necessary for storing implements, fertilizers, sprayers, and nursery stocks such as seeds, bulbs, corms, tubers, etc. For large gardens, a storehouse made preferably of brick and mortar with RCC or CI or asbestos sheet roofing having two compartments is constructed. One of the compartments, which is comparatively small, is meant for storing implements, fertilizers, etc., while the other, a large and well-ventilated one, is meant for storing bulbs, corms, etc. Along the walls wooden or split bamboo shelves in several layers should be constructed one above the other for storing bulbs, corms, and other planting materials. The rest of the space may be reserved for other purposes.

Potting and Packing Shed: Along with the store-house on one side a CI sheet or asbestos sheet shed with a rough brick-and-mortar floor is constructed for the purpose of packing of nursery stocks and for potting during rainy season or hot weather. The length and the breadth of the shed will depend upon the volume of work. Inside the shed, on one side, two to three underground storage tanks, made of brick and mortar, should be constructed for storing compost, manure, and garden soil. The shed will be useful mostly during the rains, as during that period potting or packing operations will be impossible to perform in the open.

In addition, a pergola-like structure should also be constructed to keep the newly planted potted plants or pricked seedlings. Shade is provided by training a creeper over the pergola or by putting on the top bamboo mats or palm leaf mats.

Nursery Beds: For a large garden it is advisable to have permanent nursery beds, mostly for sowing seeds. These are raised boxes made of brick and mortar, provided with drainage holes at the bottom. The dimensions of the boxes are 60 cm high, 120 cm broad, and the length may be as required but preferably not exceeding 10 metres. Roof structures, slanting on both sides and forming ridges at the centre, are constructed on the tops of the nursery beds. These structures may be made parmanent with angle-irons, or may be made of wood. Movable bamboo mats, palm leaf mats, or *sirki* made of *Saccharum spontaneum* leaves or mats made of typha (*hogla*) leaves are placed over these structures to protect the seedlings from hot sun and heavy rains. The height of the roof structure from the seed-bed will be 60-90 cm on the sides and 90-120 cm at the centre. The boxes are first filled with broken bricks and crocks for drainage and the top layers are filled with sterilized soil or compost. Whenever sunlight is required for the seedlings the mats on the tops are rolled or removed, else these are kept in position.

Mist Chamber: This is a novel structure in a nursery where propagation of leafy soft wood cuttings is achieved with great success. Many difficult-to-root plants and shrubs root successfully under mist.

The principle is to spray the cuttings with a minimum quantity of water. This is best achieved by providing the cuttings a series of short bursts of spray, termed as intermittent spraying, rather than a continuous spray. Such intermittent spraying can be done easily by means of a high pressure pump and a time switch. The pump leads to a pipeline system inside an airy and well-lighted glasshouse. The mist nozzles are fitted to these pipelines and suitably spaced over the propagating boxes. During very hot weather the roof should be shaded with green-painted gunny cloth and proper ventilation should be arranged. During very cold weather bottom heating should be provided to aid in rooting.

This mist arrangement can also be provided in specially prepared outdoor propagating beds. The drifting of the mist by air can be checked by providing shelters around the beds. Plastic sheets supported by wire-frames can also be used.

One main defect of the mist chamber or mist propagation with time switch is that it continues to work with the same schedule irrespective of the external conditions. If the weather outside suddenly changes from a sunny to a very cloudy humid weather the mist will continue to operate at the same rate. To overcome this shortcoming, an ingenious mechanical device, called the 'electronic leaf', has been developed. This is placed among the cuttings and as soon as this and the cuttings go dry, the leaf is activated and starts the water mist in a series of short bursts until the mist is adequate. This ensures supply of mist according to external conditions.

The compost used in a mist chamber may consist of one part peat and three parts of sand or vermiculite alone.

In mist chamber cuttings can be raised any time of the year, except during December-January and April-May in northern and eastern India. In southern India, it is possible to raise cuttings throughout the year, although during the cooler months of December and January the time taken for rooting may be longer. The ideal temperature range in a mist chamber should be about 22-35°C.

Mist chamber-raised cuttings have thick and fleshy roots which are not ideal for outdoor growing. Therefore, cuttings raised under mist are acclimatized by gradually exposing them to outside weather.

Propagating or Cold Frames: Large gardens and nurseries have one or more frames earmarked for propagation by cuttings. Such frames may be permanent or movable. The permanent ones are constructed by masonry walls on the sides with slanting (either on one side or on both sides) covering of one or more glass sashes. The frames of sashes should be made of light material so that they may be easily opened or closed by one person. The sashes are fitted on hinges and these can be manoeuvred to allow more air into the frame by allowing them to rest over removable stays. The bed inside the frame should be well drained by providing broken bricks and crocks at the bottom and the top 45-60 cm layer is filled with washed sand. If the frame is not situated in a place where it should receive only morning sun, shade is provided by covering the sashes with green-painted gunny cloth. The difference between a permanent frame and a portable frame is not much, except that in the latter the sides are made of wooden planks. This will look like a box without a bottom.

In temperate countries, such frames can be situated in the corner of a glasshouse and heated by means of steam-pipes or with electricity.

Hotbeds: Heated frames are used for propagating cuttings and sowing seeds in the cold season. Hotbeds heated by steam or electricity are easy to operate and the temperature can be regulated according to need. But the conventional type of hotbed is quite useful. This consists of a bed of stable manure mixed with equal volumes of leaf-mould, compost, or peat. The beds are 45-60 cm larger in size on all sides than the frame covering it. The manure should be well pressed to bring the final depth to 60 cm. Generally, such beds are made on the surface, but, in light, well-drained soils sunken pits can also be prepared. Mixing of stable manure or farmyard manure alternately with leaf manure ensures a steadier and longer heat. No portion of the manure is allowed to dry up. When the temperature becomes steady around 70-75°F, a layer of 10-15 cm well-sifted soil is placed over the manure. The temperature of such beds lasts for two to three months, although the temperature will gradually decline. But, if this decline is rather rapid, more fresh manure should be added outside the frame. The sashes are opened in the morning and in the evening to allow fresh air and to wipe out moisture from the inner side of the glass. The seeds or cuttings are placed directly in the soil. No watering is needed for about a month, but afterwards the soil should be kept moist. The rooted cuttings or seedlings are taken out of the hotbeds and placed in a cold frame for a few days. These are then gradually hardened to open-air conditions. After the hotbed ceases to generate heat this is worked up and used as compost.

Greenhouse: See Chapter XXIII.

Chapter IV

Some Routine Garden Operations

In a garden there are certain operations that are to be followed judiciously for successful cultivation of ornamental plants. Most of these operations, such as transplanting, pruning, pinching, etc., are of vital importance for the growth of the plants. Some other operations, such as topiary or shearing of hedges, are designed to enhance the ornamental or aesthetic value of the plant material.

Soil Sterilization

This is an essential operation to eliminate soil-borne diseases, caused mainly by fungi. Soil sterilization can be done either with steam or with chemicals. The soils meant for sowing are generally sterilized to save the seedlings from fungus attacks or else soil fungi may play havoc with germinating seedlings. Steam sterilization may be done in a sterilization chamber with steam or heat under pressure at about 100°C. The seed pans or boxes or flats containing the soil are inserted into the sterilization chamber and kept for half an hour. The other method consists of passing perforated pipes under the beds to be sterilized. The pipes are connected with a high pressure boiler. The surface of the beds is covered with canvas and then steam is released to raise the soil temperature to about 100°C.

For home gardens, where only a limited quantity of soil is required, the soil can be sterilized by a simple method. A sauce-pan is half filled with water and boiled to about 95°C on a stove. The soil is then taken in a metal container smaller in diameter than the sauce-pan and placed in the boiling water, taking care that water from the sauce-pan does not enter the soil. The soil in the container is heated to about 95°C and kept for 30 to 40 minutes. The soil is then used in the seed-pan for sowing.

The other method is to sterilize the soil with 2 per cent formalin (or formaldehyde). The formalin solution is mixed thoroughly with the soil and this is covered with tarpaulin or gunny bag for 48 hours. Afterwards when the trace of formalin smell has gone the soil is worked up thoroughly, dried up, and used for sowing.

In recent years, some fungicides such as benlate, captan, bavistin, brassicol, and blitox are also used for soil sterilization. A solution of 0.1 to 0.2 per cent is used to drench the soil to be sterilized.

Sterilization by steam is by far the best method as it kills all fungal, bacterial spores, nematodes and destroys all unwanted seeds and tubers in the soil.

Seed Sowing

Sowing of seed is an important operation for annuals, biennials, and some herbaceous perennials, as seeds of most of them are very minute and need special care during sowing. The seeds of trees are quite bold and do not require as much care as is required for annuals, although there is no harm in treating such seeds in the same way. The same procedure should be followed for seeds of shrubs. Orchid seeds are sown in a completely different manner, which has been dealt with in Chapter XX on "Orchid". The details of sowing of annual seeds have been dealt with in Chapter VI which should be followed while sowing seeds.

Pricking

This is another important operation for seedlings of flowering annuals. For details regarding this, consult Chapter VI.

Planting and Transplanting

Planting and transplanting are two important operations. The time of planting depends on the climate of the area. Deciduous plants are transplanted during the dormant season when they are in leafless condition. This is done only on the hills and at places having real winter. The other season of planting is spring when new growth takes place, provided the plants can be given sufficient care and irrigation during the ensuing summer. Under Indian conditions, the best period is to transplant during the rainy season, provided there is no water-logging. The pits should be dug before the rains start. Under a mild climate such as of Bangalore, planting can be done almost all the year round except during the hot months of March to May. The idea is to plant in a safe period when there will be minimum casualties.

The terms potting, repotting, shifting, balling, and barlapping or setting out are all actually methods of transplanting.

Planting of shrubs and trees: The details of planting of shrubs and trees have been discussed in the respective chapters. While digging the pits for planting, the good surface soil is kept separately and not mixed with the rest of the soil. Sometimes, it may be necessary to replace the subsoil if it is not of good quality. After digging the pit the soil is returned to the pit mixed with the requisite quantity of manure, the surface soil going on the top. While planting a small hole is made at the centre of the pit and the plant is placed with the ball of earth. Plants of eastern India have balls of earth which are clayey in nature. These should be broken carefully before planting and the roots spread out evenly and placed in the central hole. The plant should not be placed deeper than when it was in the nursery. After transplanting, the earth around the root should be firmed thoroughly. A basin should be made on the ground around the plant for watering.

Effect of transplanting on plant behaviour: The herbaceous plants, evergreen trees, and shrubs always have leaves and most of the water is lost through the leaves. During transplanting, specially of herbaceous plants, a part, if not whole, of the region of absorption is destroyed. Therefore, during transplantation more water is lost than absorbed, causing deficiency of water in the tissues. This causes two adverse effects: (1) in the region of elongation the cells fail to get the necessary amount of water and as a result they are reduced in size, and (2) the photosynthetic activity is reduced or stopped. This happens because the guard cells loose turgour and as a result, the stomata close partially or completely. Thus, transplanting causes reduction in growth and development of plants. The extent of the injury depends upon the duration of the water shortage in the tissues.

Factors influencing rate of recovery: There are certain factors which determine the rate of success in transplanting. The first factor is the age and the size of the plant at the time of transplanting. It is common knowledge that the older and larger a plant is during transplanting the greater are the chances of its failure. The reason behind this is that larger and older plants have more extensive root system and while transplanting the more tender root tips get damaged, and as a result the region of absorption is reduced considerably. Moreover, larger plants have more leaf area and lose water more rapidly. This causes more transpiration than absorption during transplanting older and larger plants.

Another factor is the rate at which the root system regenerates. The root regeneration depends upon the amount of carbohydrates stored in the tissue of the plant at the time of transplanting.

The last factor is the climate. During winter, spring, or rainy season, the rate of water loss will be lower than in summer and consequently the success will be more during these seasons. But it may be noted that winter is not an ideal time for transplanting most of the plants.

The other external factors that influence success or failure include the time of planting. Planting during late afternoon or evening is better than in the morning hours as the plants get enough opportunity to recoup during the cool night hours. For the same reason transplanting on a cool and cloudy day is beneficial.

Transplanting

There are generally four types of transplanting in a garden:

(1) *Transplanting a potted plant to ground:* This is one of the simplest transplanting operations. In the afternoon preceding the morning when actual transplanting is to be done, the pot is watered copiously. At the time of transplanting the compost in the pot should not be too wet or too dry. The pot is held upside down, placing the left hand over the surface of the soil and the stem of the plant passing between the extended fingers. The outer surface of the pot is held with the right hand and one end of the brim of the pot is gently tapped on the edge of a wooden stool or a raised platform or a log of wood or an inverted earthen pot, if the plant being transplanted is smaller in size. The ball of earth will come out intact without falling apart. Before tossing out the plant in this way one should first observe if a portion of the tap root has come out of the drainage hole. In that case this root is cut away or it will hamper the tossing out operation. After taking out, the crocks are removed and the mesh of roots outside the ball is loosened before putting the plant in the prepared hole in the ground. Any plant with a tap root should not be kept in a pot too long as this becomes useless due to twisting.

(2) *Transferring a potted plant to another pot:* See potting and repotting in Chapter XVIII.

(3) *Potting a ground plant:* This operation is very simple in a clayey soil, but not very effective in red or sandy soils. The plant is lifted with a ball of earth. One day prior to lifting, the plant is watered thoroughly. Next day, a circular trench is dug around the plant cutting through the roots, gradually tapering below. The pot is prepared by putting crocks and a little of soil. Then the plant is lifted carefully with the ball of earth which should be slightly less in diameter than the pot and placed in the pot. The gap between the ball of earth and the side of the pot is filled with more compost and gently rammed and copiously watered. Roses and chrysanthemums (in the *Sen Rin Tsukuri* system) are transplanted like this.

(4) Transplanting of large trees and shrubs from the ground to another suitable location poses problems. For lifting very large trees the help of tree-lifters or cranes and trucks will be needed. The procedure is as follows. The plant to be lifted is watered thoroughly a couple of days prior to the operation. The object will be to lift the plant with minimum root injury and with as many roots as possible. A trench of comfortable width (45-60 cm) is dug in a circle around the plant so that one can work conveniently inside the trench. The diameter of the circle will vary according to the size of the tree and may be between 60 cm and 2.40 m. The trench should taper towards the centre of the circle as it deepens forming a cone-shaped ball of earth around the roots. The trenching operation can be taken up in three stages. First, one-third of the circle is dug up and then this is lightly filled with soil and kept like that for a week or a fortnight to observe the reaction of the tree. If there is not much adverse effect, again one-third portion is dug out and left to rest in a similar manner. After the final digging the plant is ready for transplanting. At this stage the plant should be held in position with the help of ropes tied to strong stakes driven into the ground. To prevent the ball of earth from breaking this is held in position by burlapping, i.e., by wrapping with gunny cloth or polythene sheet and securely tied with ropes. If the ball is very heavy this is to be lifted with the help of a crane or a tree-lifter and placed in a tractor trailer to shift to the place of transplanting. Smaller plants can be lifted and shifted with manual labour. The plant with the ball of earth is then lowered in the prepared pit at the same depth or a few centimetres more than the original depth. If in the process of lifting the earth ball breaks, the roots are immediately pasted with a thin paste made up of two-thirds clay and one-third cowdung. The space between the earth ball and the trench is filled up with compost and firmed by pressing. Shading and frequent watering are needed for some days. The trunk may be wrapped with moist straw to prevent transpiration loss. If the tree has a heavy head of foliage, it may be reduced before transplanting by cutting the branches to two-thirds of their length. For a tree having long branches with smaller lateral branches, many of the lateral branches can be pruned back to 10-12 buds. Some trees and shrubs do better if the top is beheaded at transplanting. The cut ends of the branches are to be painted with white lead or thin solutions of coal-tar to prevent insect or disease attack. The transplanted plant will need strong staking with ropes and pegs driven into the ground as described earlier.

Newly planted trees and shrubs also need shading, which is provided by thatching or by guard plants as described in Chapter XI on "Management and Care of Transplants".

Transplanting of a large tree becomes necessary when a tree is uprooted by storm or killed by disease. It will be necessary to replace the loss with a reasonably equal-sized tree though it will not be possible to replace a large tree of comparable size.

Use of waxes: It has been demonstrated experimentally that use of emulsified waxes at transplanting time increases the chances of survival of leafy plants. These waxes are water-soluble. The wax reduces the rate of transpiration, but does not interfere with the rate of photosynthesis or respiration. The emulsified wax is sprayed once before lifting and again after transplanting.

Now-a-days many nurseries are growing shrubs and trees in large tins or plastic containers. The advantage is that large-sized shrubs and trees can be kept in such pots for a longer period without any damage to the tree. Such plants are easy to transplant as one has only to cut open the container and put the plant in the pit.

Shading

Shading and protection of plants is an important operation in a garden, especially in places having hot summers and severe winters. The newly planted seedlings or plants need shade from the scorching summer sun. Unless properly shaded the leaves will get scorched and sunburnt and the delicate plants may be killed. In places where frost is likely to kill the plants these should be given proper protection. Some delicate plants such as *Magnolia grandiflora* when grown in the hotter parts of north or east India will also need shading.

Different shading materials are used in different parts of the country. The seedlings in the nursery need shading and the procedure has been described in Chapter III under the heading "Nursery Beds". The newly planted seedlings may need shading. This is done by placing a 15 cm pot over the seedling, the northern side of the pot raised slightly from the ground to allow the hot air inside to escape. If the seedlings are pricked in 7 cm pots and carefully transplanted, no shading may be needed.

For shading of trees and protection of shrubs from frost injury consult Chapter XI on "Shade, Ornamental and Flowering Trees" and Chapter VII on "Ornamental and Flowering Shrubs", respectively.

Stopping or Pinching

The operation of pinching or stopping involves the removal of the growing-point of a shoot along with a few leaves. The two main purposes of this operation are to encourage branching to produce a bushy growth, and/or the production of flower-buds on the branch which is pinched. Pinching is done mostly in annuals and herbaceous perennials and is hardly required for any flowering trees. The plants which need pinching include dahlia, chrysanthemum, carnation, brachycome, and marigold. Some plants have the habit of growing lanky, which presents an unpleasant look. Such plants should be pinched when they are still young, thus causing the dormant buds in the leaf axils to grow, forming a bushier and dwarfish growth. Pinching is done at a stage when the plants are young and between 7 and 15 cm in height, depending on their habits of growth. It often happens that one shoot of a plant starts growing vigorously at the expense of the others. Stopping of this strong shoot at the proper stage is likely to encourage the growth of the weaker shoots and finally a shapely plant is obtained. Since the shoots pinched are of herbaceous nature, no special tool is needed for pinching and in most cases this is effected by snapping with fingers. Sometimes the help of a knife or a shaving blade may be needed.

Stopping may involve only the pinching of the main stem, as in chabaud carnation, or both the main stem and the side shoots to obtain a more bushy plant, as is done in *Sen rin tsukuri* method of growing chrysanthemum.

One main disadvantage of stopping is that some plant diseases such as those caused by virus may get transmitted through the pinching blade from a diseased plant to a healthy one. Therefore, as far as possible pinching should be done by snapping with hand.

Pinching procedures for different plants have been dealt with in the respective chapters.

Deshooting

Deshooting involves the removal of shoots that are not wanted. Some flowering annuals and herbaceous perennials produce numerous side shoots and if all of them are allowed to flower, the size and quality of the flowers will be greatly reduced. Only a specific number of side shoots are allowed to flower and the others are removed at an early stage. Deshooting of carnations grown for cut flower trade and chrysanthemum for exhibition purposes are common practice. The operation involves removal of the unwanted shoots by hand or with a blade. The shoots thus removed are often used as cuttings for raising new plants.

Disbudding

In many plants several superfluous buds, flowers, and shoots develop in the axils of the main stems and branches. These superfluous growths share the sap of the plants. To divert this sap to a few stronger branches which are required for the production of branches and flowers, the superfluous growths are removed at an early stage of growth. This operation is known as disbudding. Therefore, in a broader sense, even deshooting is a part of disbudding. But, actually, in ornamental horticulture disbudding refers to removal of the superfluous flower buds only. This operation is done mainly for annuals, herbaceous perennials, and in roses for exhibition. This is followed in large-flowered cultivars of zinnia, marigold and also in carnation, chrysanthemum, dahlia, and many other flowers. The main purpose is to obtain quality blooms. The general practice is to retain only one bud per shoot.

Defoliation

The removal of foliage is known as defoliation. This is done mainly with a view to inducing flowering in certain plants. Sometimes, this can also be done to reduce transpiration loss during periods of stress and strain and also during transportation of certain plants such as roses. Defoliation can be achieved by the removal of leaves by hand or by the use of chemicals and by withholding water. In jasmine, it is a common practice to defoliate the plants after pruning just prior to the flowering season. This can be achieved by withholding water for a few days when the leaves drop or by removing the leaves by hand or a combination of both. Chemical defoliants are not commonly used in India. But chemicals such as pentachlorophenol has been used with success as a defoliant in jasmine. In nasturtium sometimes an excess of foliage is produced and as a result flowering becomes scarce. Therefore, it is a common practice to cut the excess foliage to induce flowering in nasturtium. Moreover, if there is more foliage than necessary, the flowers remain hidden inside the foliage. In roses if any leaves are left in the pruned branches, these are removed by hand. While transporting or transplanting roses and some other shrubs it is necessary to remove the foliage to reduce transpiration loss.

Staking

Plants in the garden, either in pots or on ground, need support at least for a part of, or throughout, its life. Selecting the proper stakes and fixing them aesthetically as well as purposefully is an art. Stakes may be of various kinds. The herbaceous plants do not require very strong stakes. The most common stakes used in India are made of either whole bamboo or split bamboo of various sizes depending upon the type of plants to be staked. Some other types of stake which are commonly available under Indian conditions and can be used as stakes for herbaceous plants are the dried stems of jute, cotton and pigeon pea (*Cajanus cajan* syn. *C. indicus*). For carnations thick gauge G.I. wire stakes, either simple or fitted with a loop or a mobile ring, are used. For annual creepers such as sweet pea, wire-netting or wire trellis is used as stake. In areas where high winds prevail, large trees should be staked with strong bamboo stakes; sometimes 5 to 7.5 cm diameter G.I. pipes have to be used to keep the plants straight. For example, *Araucaria cookii* tends to bend too much as it grows. To keep such a strong tree straight, a strong stake such as a G.I. pipe will be needed. Sometimes, the twiggy growth of a few woody plants is also used as stake. Before fixing, the stakes are painted green, especially those used for herbaceous plants, so as not to look very conspicuous.

For a single stemmed plant a single stake will suffice as in the case of dahlia or standard rose. For bushy plants three, four, or even more stakes may be required. The stakes can be interlinked with strings, forming an interlacing mesh as is done in the case of *Sen rin tsukuri* method of growing chrysanthemum. When a single stake is fixed it should be done at the time of planting. Unless this is done the stake may be driven at a later date too close to the plant, injuring the best roots. In areas subject to high winds often the stake is planted at an angle with the tree staying clear of the roots. When properly tied this will give good support against the wind. If the stake is forked at the apex the support will be further strengthened.

The importance of staking cannot be over-emphasized, as it not only supports a newly planted or a weak-stemmed tree but also prevents the plants from rocking about during high winds. To secure the plants strongly the stakes are to be firmly tied to them with the help of strong threads that can stand long against the weather. Too severe tying will injure the growing and soft tissues. First the tying material such as thread, raffia, etc., is tied with the stake by a knot and this is then passed over the branch in a loose loop allowing enough space for the tissues to grow but at the same time not too loose to allow the plant to rock about during the winds. For tying trees a sheet of rubber cut out from old tubes of pneumatic tyres or a pad of gunny-cloth should be placed at the place of tying to avoid injury to tissues. The tying should be checked at least once a year to see that it does not obstruct the growth of the stem. The knots should be so made that they do not draw over the stem.

Pruning

The planned removal of branches, twigs, limbs, shoots, or roots is termed as pruning. Even the removal of a dried flower can be termed as pruning. Each pruning is done with a view to increasing the usefulness of a plant. Pruning is a major horticultural practice and skill which can be acquired through experience, observation, and practice.

Principles of pruning: One of the main principles of pruning is to reduce the apical dominance. If the apical shoot is removed the lateral branches will start growing.

The ratio between roots and tops and *vice versa* influence to a considerable extent the vegetative growth, flowering, and fruiting of a plant. The principle is to strike a balance between shoot and root growth. If the leaf area is large there will be more transpiration and the roots will be taxed to absorb more water. But if some leaf area is pruned, a part of the reserved food will be diverted towards new root growth. Root pruning is done to reduce the absorbing area of the plant which in turn reduces top growth. The extra stored food is then utilized for new root growth.

If a grown-up plant is pruned heavily each year the phase of growth will be altered as the plant will tend to produce only vegetative growth while flowering or fruiting will be delayed. Therefore, to keep a balance between vegetative and reproductive growth pruning is done judiciously. A certain amount of annual vegetative growth is essential for the production of flowers. When the plants grow older, new vegetative growth slows down and to encourage new growth moderate to heavy pruning is resorted to.

Some environmental factors also determine the principle of pruning. A densely foliaged plant deprives the inner branches or leaves of the necessary amount of light. The principle is to well distribute the available light by judicious pruning and training of the branches. Air movement is also influenced by pruning. An open structured plant will be less attacked by fungi than a compact plant. Temperature also has its influence on pruning. Over-pruning may result in soft succulent growth which may be injured due to summer heat or severe winter temperature.

Objectives or purposes of pruning: Pruning is done with some definite objectives. Therefore, before pruning one should ascertain the reasons for or the objectives of such action. The main purposes of pruning are as follows:

(1) The primary objective should be to give a definite direction and shape to the plant. Each plant has a characteristic natural form. If training and pruning are done for maintaining the natural form, only a small amount of pruning is needed to keep the plant in shape. All growth which departs from the desired shape should be removed.

(2) Another objective will be to develop a strong framework. The plant should be pruned and trained to have a balanced branching on either side so that the plant does not become lop-sided. Narrow-angled branches are easily blown off by strong winds. So at the early stages of growth only wide-angled scaffold branches should be retained by proper training and pruning.

(3) In some cases pruning is also done with the aim of utilizing the available space effectively. For example, rambler roses are pruned, trained, and staked to utilize the available space most efficiently. Pruning of hanging branches also facilitates cultural operations such as hoeing or weeding.

(4) It is possible to control the growth by judicious pruning. The result of pruning is actually both dwarfing the plant and invigorating its growth. The type, the time, and the manner of pruning influence the objective of a reduced or an increased growth. Frequent trimming will result in dwarfness, whereas pruning of older and dormant wood will invigorate growth.

(5) Pruning also influences productiveness and improves the quality of the product. An unpruned rose bush produces fewer flowers and even their quality is poor. A well-pruned

rose plant will produce the required vegetative growth and flowering. A harder pruning in roses increases the size of the flower. The manner of pruning to achieve these objectives will depend upon the flowering habit and the vegetative growth of the species concerned.

(6) All ornamental plants are grown for a definite objective, e.g., shading, flowering, decoration, or fencing. It should be the endeavour of the grower to attain the maximum usefulness of the plants keeping the goal in mind. For example, a tree grown for shade purposes should be encouraged to develop a dense top growth. A hedge can be kept neat and made impenetrable by proper training and clipping.

(7) Removal of all dead or diseased wood at the time of pruning is another objective which should be kept in view.

Time of pruning: Pruning should be done in the right season or the whole purpose will be defeated. A grower will know by experience the time a plant has to be pruned by acquaintance with the flowering and growth behaviour of the plant. A plant may bear flowers on the last season's growth or the current season's growth and hence the pruning time should be adjusted accordingly. The plants bearing flowers on the last season's growth are generally pruned immediately after flowering, whereas those flowering on the current season's growth are pruned sufficiently ahead of the flowering season. Plants belonging to the latter category such as those flowering during the summer are pruned either during the winter or early spring (December to February).

Precautions during pruning: The following points should be kept in mind during pruning.

(1) It is absolutely necessary to give a clean-cut during pruning. Avoid bruising the bark or splitting the wood while pruning. For this purpose, a sharp pruning knife and secateurs should be used. For removing larger branches a pruning saw is used.

(2) When removing large branches from a tree or a shrub, they should be cut close to the trunk, but not through the trunk, so as not to leave any stub. This helps quicker healing.

(3) All cut ends should be treated with some antiseptic solution such as carbolic acid, fungicides, etc., and sealed off with tar solution. For further details see Chapter XI.

Methods of pruning: The method of pruning varies with the plant specimen. The ornamental trees hardly need any pruning except the removal of dead or broken branches. For pruning of shrubs consult Chapter VIII on "Ornamental and Flowering Shrubs". For other methods of pruning consult stopping (pinching), disbudding, and defoliation in this chapter.

Root pruning: Root pruning is done to a limited extent in ornamental horticulture, as in the case of rose and jasmine. The main effects of root pruning are (1) reduced absorption, (2) reduced top growth, and (3) more branching in the root system.

For smaller herbaceous plants root pruning to some extent is an automatic process as during cultural operations such as hoeing many roots are damaged. For large shrubs and plants a trench is dug around the tree, far enough from the trunk, so as to remove only the periphery of the feeder roots. Often root pruning of large flowering trees is done to prevent such roots from invading a rose or a flower bed by digging a trench about 60-75 cm deep on the periphery of the bed facing the trees.

Wintering

This may be considered as an alternative to root pruning. In the hotter parts of India it may not be wise to resort to root pruning. In such places ornamental plants are "wintered". Flowering shrubs such as rose and jasmine can be wintered in northern and eastern India, though some experts have reservations regarding the wintering of roses and feel the operation is useless.

During resting period the water supply to the plant to be wintered is stopped for a few days and the roots are exposed to the sun by removing the surface soil around the trunk. The duration of exposure varies from three to fifteen days depending upon the age, the nature, and the hardiness of the plant. After this the roots are covered with the same soil enriched with farm yard manure and copiously watered.

Clipping or Cutting of Hedges and Edges

To keep the hedges in symmetry, good health, and beauty, constant vigilance and regular shearing or pruning are needed. It has been acknowledged that hedges benefit from frequent clipping. There are some hedge plants such as *Pandanus* and species of *Euphorbia* and *Opuntia* which do not need regular clipping, but only occasionally the side growths or the central heads are removed to keep them in check.

The quick-growing low hedges (50 cm to 1 m) need frequent pruning during the rainy season. When such hedges are overgrown, the old woods are to be cut back below the maximum height to encourage new shoot growth. No hedge should be allowed to grow taller and thicker than the normal size each year. If a hedge has overgrown by neglect it should be cut back in the same manner as described above.

A sharp pair of pruning shears are used for clipping the herbaceous twigs. But to cut woody branches secateurs or pruning saws will be required.

Hedges can be cut to different shapes, but it is easier to maintain a square-cut hedge. Other shapes that can be tried are rounded top, wavy top, columns, etc. Sometimes slightly raised columns are left at regular intervals to an otherwise square-cut hedge. Often the *mali* deshapes the hedge by uneven pruning. To prevent this the following procedure may be followed. Green-painted iron or bamboo stakes of equal size are sunk along either side of the hedge at equal intervals. All the stakes are to be marked at equal height on the top and a string or a wire can be tied along these marks to form the desired frame of the hedge. Then whatever hedge is left outside this frame is clipped.

Coniferous hedges do not look nice nor stand well if they are clipped to a square shape. They are cut to a pyramidal shape, which is the natural form of such plants, though on many occasions the *Casuarina equisetifolia* is cut to a square shape to form a strong hedge.

Since the hedging plants are to be kept low they are trimmed very often with pruning shears. Hence, to maintain the geometric shape the help of a stake and a wire will be needed.

Mulching

The purposes of mulching are varied. The intention is to regulate soil temperature, to conserve moisture, and to control weeds. There are several kinds of mulch such as straw, leaves, pulverized corn cobs, peanut hulls, cottonseed hulls, stable manure, etc. Plastics (polythene sheet) and aluminium foils are also used as mulch. In ornamental horticulture generally none of the above mulches is used. But a dry mulch may be applied when the soil becomes friable after irrigation. This consists of lightly breaking up the top 2-3 cm surface of the soil by hoeing, raking, or cultivation. This stirring-up breaks the capillary tubes in the soil and thus preserves the moisture at the lower layers for a considerable period of time. This operation also helps check weed growth. It is, therefore, always advisable to follow each irrigation with light hoeing.

Topiary

The art of clipping and shearing shrubs and small trees and sometimes even herbaceous perennials into ornamental or abstract shapes is known as topiary. The term is derived from the Roman words *ars topiaria* meaning ornamental gardening. This art started possibly in the first century A.D. or even earlier.

Plants which have numerous dark-green foliage and stand frequent clipping and shearing are suitable for topiary work. It takes many years to train and shape plants to desired shapes and sizes. A lot of attention and a heavy amount of labour are involved in raising such plants. Since topiary plants will last for a long time, thorough preparation of the ground by deep digging is needed before planting.

It is possible to attain shapes of different types. Simple shapes such as a globe, sphere, dome, table, chair, cube are not very difficult to achieve, but when figures of birds, beasts, or human beings are to be shaped, a lot of patience and perseverance is needed. Shapes such as globe and dome can be given without the help of any pre-fabricated moulded wire model. But difficult shapes such as bird, animal, etc., are to be obtained by preparing a rough outline with wires and then training the shrub along the frame and clipping carefully over years to get the desired shape. The branches are tied carefully to the frames without causing any injury. The instructions given for tying under the heading "Staking" should be followed.

It should be borne in mind that whatever form is chosen, it should have a broader base. For example, a peacock standing over a substantial base such as a dome or a cube is more impressive than standing on its own. Once the base is made, the bird is shaped along the outline of a wire frame. It may take about 3 to 4 years of clipping and training to obtain the general shape. The objective should be that any shoot or branch which can be trained into the framework should not be pruned.

Topiary gives the impression of the old world and goes well with old buildings and cottage gardens. Topiary work is not to be scattered all over the garden, but should be fitted into a formal garden.

Topiary, though considered as "tree sculpture", is not free from controversy. The antagonists feel that twisting and trimming of trees are nothing but waste of time and a cruelty. Whatever may be the opposition, topiary has come to stay and there are many gardens all over the world containing fine specimens of topiary. In India, the hanging gardens (or Kamala Nehru Park) of Bombay is famous for its topiary work.

The shrubs most commonly used in India for topiary are *Clerodendron inerme* and *Duranta plumieri*. Other shrubs used are Bougainvillea, *Murraya exotica*, and species of *Hibiscus*. Generally, the difficult shapes of birds, animals, etc., are done with *Clerodendron inerme*, *Duranta plumieri*, and Bougainvillea. Thuja, *Cupressus macrocarpa*, *C. sempervirens*, *Putranjiva roxburghii*, and *Polyalthia longifolia* are also used to shape domes, cones, spheres, umbrellas, etc. In temperate regions, Box (*Buxus sempervirens*) and English Yew (*Taxus baccata*) are widely used for topiary work.

Chrysanthemums, especially the small flowering types, can be trained into various shapes such as table, chair, bird, animal, etc.

Chapter V

Role of Plant Growth Regulators in Floriculture

To remove the confusion regarding the terminology of auxins and growth hormones that prevails with the amateurs or even the physiologists, let us begin this chapter with proper definitions of these terms.

In the first place let us understand the meaning of the term hormone. *Hormone* can be defined as a substance produced in one part of any organism but later on transferred to another part of the same organism where it affects a specific physiological action. It must be noted that a hormone has to be produced in an organism, and it is transported from the place of formation to the actual site of action. The hormones are required only in minute quantities to produce large "specific physiological process". Therefore, strictly speaking, chemical substances produced entirely artificially or compounds such as amino-acids, sugars, and other such substances cannot be called hormones. None of these substances are specific in their action.

The term *phytohormone* is applied to those hormones which are exclusively produced inside a living plant. The growth hormones and the hypothetical flowering hormones come under this group. Some vitamins are also included in this class by some authors. The term *growth hormone* is ascribed to hormones essential for growth by cellular enlargement. These hormones are also necessary for the growth of such organs as roots, stems, buds, and fruits.

Auxin is defined as an organic substance which when applied in dilute concentrations to plant shoots promotes irreversible growth in volume along a longitudinal axis and acts almost independently of the growth-promoting substances present in the plant. Therefore, as per this definition, those substances including growth hormones and many other chemical compounds which cause similar growth to take place can be referred to as auxins.

The term *growth regulator* is applied to organic compounds, other than plant nutrients, which when used in minute quantities can either inhibit, stimulate, or alter growth. In this group are included auxins, auxin synergists, anti-auxins, epinastic agents, and some other substances such as gibberellins, cytokinin, and abscisic acid.

There are certain substances which when used on their own in auxin-free tissues cannot stimulate growth but can do so when used in association with auxins. These are termed as *auxin synergists*. Conversely, these substances are capable of inhibiting growth if present in comparatively large quantities. The compound 2, 3, 5-triiodobenzoic acid (TIBA) is one of the best known auxin synergists. Some other commonly known auxin synergists are coumarin, protoanemonin, indole, some vitamins such as niacin, ascorbic acid, pyridoxine, pantothenic acid, *p*-aminobenzoic acid, salicylic acid, tannic acid, *p*-hydroxybenzoic acid, gallic acid, etc. The action of auxin synergists in promoting rooting of cutting in association with auxins is well known.

There are a variety of substances called *anti-auxins*, which can inhibit growth in the presence of auxins. The first such compound reported was possibly γ-phenylbutyric acid. It is also accepted that *trans*-cinnamic acid inhibits growth. Many of the anti-auxins also show weak auxin activities. Besides inhibiting auxin-inspired growth, anti-auxins can neutralize auxin inhibitions. For example, they remove the apical-dominance properties of auxins.

Epinastic agents are those which cause abnormal swelling of cells thus causing monstrosities in leaves and stems of plants. Examples of such compounds are acetylene, ethylene, carbon monoxide, many auxins and their related compounds. An epinastic agent may cause distortion in one species but may not have any such effect on another species.

Role of Growth Regulators

The major areas where plant growth regulators are used in floriculture are plant propagation, plant height control, regulation of flowering, and prolonging the life of cut blooms, besides some other minor uses.

Propagation: A large number of ornamental plants are propagated by vegetative methods such as stem and leaf cuttings, by bulbs, corms, and tubers. The use of auxins in rooting has brought about a vast quantitative change in plant propagation.

The most commonly and widely used auxin in promoting rooting is Indolebutyric acid (IBA). Next in importance are Naphthaleneacetic acid (NAA) and Indoleacetic acid (IAA). But NAA and IAA are less uniformly effective compared to IBA. It has been observed that often mixtures of two auxins are much more effective than when applied singly.

Many synergists when used in combination with auxins induce much more rooting than when the auxin is used singly. Basu *et al.* (1969) observed that tannic acid and gallic acid promoted rooting in leafy-cuttings of *Eranthemum tricolor* in combination with NAA and IBA. Salicylic acid in combination with IAA, IBA, and NAA greatly promoted rooting in the same species. Synergism was also noted between *p*-hydroxybenzoic acid and IAA or NAA. Similarly, Bose *et al.* (1972) found that *p*-hydroxybenzoic acid and salicylic acid proved to be effective synergists of IAA, IBA, and NAA in forming roots in cuttings of a number of ornamental shrubs.

There are generally three techniques of auxin application to cuttings. The more orthodox method is the prolonged soak treatment for 24 hours at low concentrations (25-100 ppm). An improved technique is to treat the cuttings in high concentrations (1,000-10,000 ppm) of auxin solution by quick-dip method, i.e., dipping the basal portion for 5 seconds to 2 minutes depending upon the nature of the cuttings, whether they are soft or hard wood cuttings. For this treatment, the auxins are dissolved in 40-50 per cent alcohol. The third technique is to dip the wet basal portion of the cuttings in talc mixed with auxin ranging in concentrations of 500 ppm to 12,000 ppm. Bose and Mondal (1973) obtained not only high percentages of rooting but also large numbers of roots in many difficult-to-root trees, shrubs, and climbers by treating these with IBA and NAA in talc. The concentrations of auxins used were 3,000 ppm and 6,000 ppm for semi-hard wood and hard wood cuttings, respectively. This latter method has a limitation as when in commercial propagation a large number of cuttings are dipped in talc, the cuttings in the centre of the bundle may not get the talc uniformly. In difficult-to-root plants the layers can also be treated with auxins either mixed in talc as above or in a paste in lanolin at concentrations of 1,000-3,000 ppm. Recent researches at the Indian Institute of Horticultural Research, Bangalore, showed that even such difficult-to-root plants as *gulmohar* can be induced to form roots by treating them with IBA in lanolin paste. The method consists of first making a ring in the undetached cutting and applying IBA in lanolin paste (8,000-12,000 ppm) in the ring. Afterwards when callusing starts in about a month, the cutting is detached from the mother plant below the ring and rooted in a mist chamber.

The success of rooting depends upon some external and physiological factors. The external factors are season, photoperiods, light, temperature, aeration, and humidity. In our country, the rainy season appears to be the best time for the rooting of cuttings or layers. Photoperiods, either long or short, influence rooting depending upon the species. The optimum photoperiod appears to be that which helps in the accumulation of carbohydrates in the cutting prior to its detachment from the mother plant or at the time of rooting process. Cuttings taken in complete shade root poorly compared to cuttings inserted in cutting beds in a lighted place. But if the portion of the cutting that produces roots is exposed to light, poor rooting may result. It has been observed that though callus formation may be favoured under a cool condition, root growth is promoted by higher temperatures. Proper aeration near the rooting zone is also necessary for root formation. As the cutting is handicapped for not being able to draw sufficient water, a high humid atmosphere is favourable for root formation.

The physiological factors that influence rooting are polarity, presence of leaves and active buds in the cutting, nutrient status of the cutting, and auxin level in the cutting. Cuttings placed upside down will not root as the movements of auxins are from the apex to the base (polarity) of the original position of the cutting in the mother plant. The leafy cuttings root much better than leafless cuttings as the leaf provides some nutritive materials beneficial for root formation. Presence of active buds in the cutting helps rooting and this effect is ascribed mainly to the production of auxins in the buds. But, at least in certain species, the flower buds have antagonistic effect on rooting. Occurrence of comparatively low levels of soluble nitrogen in the cutting and presence of traces of boron in the rooting media are beneficial for root formation. The total absence of boron may result in complete failure in root formation. Relatively high level of nitrogen in the cutting inhibits root formation, especially when the carbohydrate level is also low. Other plant growth regulators recently tested for propagation are ethephon and retardants such as SADH and chlormequat.

Gibberellins, cytokinins, and morphactins can play a great role in the regulation of seed and bulb production. Bose and Kapur (1969) observed that spraying of gibberellic acid at low concentrations (10 ppm) on pansy immediately after pollination resulted in early maturity of fruit. GA_3 (1 ppm), applied 24 hours after pollination, increased the percentage of seed set. Bose and Mukherjee (1968) observed that application of GA_3, IBA, and Naphthalene acetamide (NAd) spray at concentrations of 1, 10, and 100 ppm after 6, 24, and 48 hours of pollination, respectively, improved the yield of seeds in cineraria, depending upon the concentration and time of application. Agnihotri and Srivastava (1964) found that when seeds of *Impatiens balsamina* were soaked in 5 ppm IBA solution before sowing for 20 hours, the resultant plants produced more seeds over control plants and plants produced from seeds treated with higher concentrations. Asahira *et al.* (1968) observed that quantitative variation in number and weight of cormel yield could be caused by changes in daylength and by gibberellic acid treatment. Gibberellic acid application has resulted in increased percentage of germination in seeds of some species. In an experiment with five ornamentals Thompson (1969) found that the seeds which germinated poorly at 25°C responded positively with gibberellic acid treatment in solutions varying in concentration from 1mg/l to 1,000 mg/l. Hogg and Orr (1969) obtained limited success with GA_3 treatments used, instead of cold stratification, to promote germination of seeds of some shrubs and trees. Thompson (1969) obtained further success in germination of seeds of several species with gibberellin A_3 and A_4 treatment under conditions in which untreated controls showed complete or partial failure to germinate. It is reported that cytokinin, SD 8339, promoted bulb formation in lilies. Kinetins are reported to have induced cormel formation in gladiolus *in vitro.* Here the work of Ginzburg and Ziv (1973) may be cited, where they observed the influence of four plant hormones on cormel development in gladiolus stolon tips grown *in vitro.* They found that kinetin induced cormel formation while gibberellin A_3 inhibited it at low kinetin levels. Abscisic acid did not affect cormel formation but inhibited its growth. NAA had no direct effect on tuberization. Anatomical observations revealed no difference between cormels formed *in vitro* and *in vivo.*

Plant height control: This is another important area where good work has been done regarding the role of plant growth regulators. In modern days, city dwellers living in flats can hardly afford to grow tall plants because of lack of space. The use of growth retardants to reduce plant heights brings hope for such people. Growth retardants not only make available to the consumer plants of various sizes and shapes but have also evolved new plant types which were hitherto not available. For example, there was no genetic dwarf plant in poinsettias, but application of growth retardants has produced mini poinsettias.

Sometimes, the retardants are specific in their action. A particular retardant may dwarf only a particular plant species while it may not be effective on others. For example, SADH is very effective in controlling the height of chrysanthemum and other bedding plants, although Cycocel and Alar are also quite effective in chrysanthemum. Phosphone is very effective in controlling heights in lilies, while Cycocel acts effectively against poinsettias. The selective actions of retardants have been further established by the findings of Bhattacharjee *et al.* (1971). They found that while B-Nine (N-dimethylaminosuccinamic acid) at concentrations of 5,000 and 10,000 ppm retarded plant growth effectively in 10 cultivars of dahlia, CCC (2-chloroethyle-trimethyl-ammonium chloride) had no inhibiting effect. Bose *et al.* (1968) observed that Cycocel (CCC) was very effective in inhibiting growth in many malvaceous plants. The leaves of the treated plants were thick in texture, small in size, and dark-green in colour. A recently marketed retardant in the U.S.A., namely Ancymidol, has been found effective on chrysanthemum, other bedding plants, lilies, poinsettias, and a variety of other plants. In chrysanthemum, the retardants are sprayed on the foliage two weeks after pinching to have the maximum effect. A second application after a fortnight from the first may be necessary for the taller cultivars. In bougainvillea various growth retardants have been applied with success. Valmayor *et al.* (1963) produced dwarf bougainvilleas in the Philippines by applying two sprays of maleic hydrazide (MH) at concentrations of 1,000 ppm and 1,500 ppm at the interval of one week. Bose and Hore (1967) produced dwarf bougainvilleas in Calcutta, by the soil application of dusts of B-Nine, Cycocel and Phosphone. These growth retardants produce compact bushy plants of bougainvilleas which are suitable for pot culture. The works of Hackett and Sachs (1967), who used Cycocel and Alar on bougainvillea, further confirmed the dwarfing effects of these retardants.

The ideal growth retardant of the future should be one which will be universally effective, non-phytotoxic, and will prolong the post-harvest life without leaving any harmful residual effect.

Another area where the plant growth regulator can play an effective role is promoting plant growth, though most of the current research is on retarding the growth. Some of the research findings where auxin applications promoted growth are discussed here. Agnihotri and Srivastava (1964) found that *Impatiens balsamina* plants grown from seeds soaked in 5 ppm IBA produced increased shoot length. Mittal (1967) observed that GA_3 treated dahlia plants were taller in height than untreated ones. Similarly, GA_3 increased the extension growth (height) of *Impatiens balsamina* till 56 days under 8- and 24-hour photoperiods (Nanda *et al.*, 1973). Shanmugam *et al.* found that GA_3-treated (100 to 400 ppm) chrysanthemum varieties had increased plant height, internode length, and flower stalk length.

Some research findings point out that chemicals which are normally used for retarding growth can also stimulate rooting in stem cuttings. The works of Read and Hoysler (1969) demonstrate that herbaceous cuttings of geranium, dahlia, and chrysanthemum dipped momentarily in solutions of several concentrations (1,000 and 5,000 ppm were effective, while 2,500 ppm was optimum) of B-Nine produced significantly greater weight and numbers of adventitious roots than untreated cuttings. But Cycocel treatment caused marked depression, as this chemical behaved as an anti-auxin.

Another interesting finding regarding the use of various growth retardants on some flowering annuals has been reported by Powell (1974). He found that growth regulators like ethrel, B-Nine, and Cycocel not only caused dwarfness but also extended the shelf-life (saleable life) of these ornamentals by several days. The chemicals were applied either as foliar spray or soil drenching and the effective concentrations for different chemicals were ethrel 500 to 800 ppm, CCC 3,000 ppm, and B-Nine 5,000 ppm depending upon the annual. The potential of gibberellins and cytokinins in initiating early growth in woody ornamentals should also be studied properly.

If plant growth regulator can be utilized to eliminate the operation of hedge pruning, considerable saving in terms of labour can be achieved. Using a new growth retardant (NIA 10637) on the campus of the University of Arakansas, U.S.A., it was found that 37 per cent of labour required for pruning was saved over a period of two years with the hedges of several plant species. This new product not only delayed pruning time but also caused a deeper green colour and repelled insects. With most of the retardants it was found that foliage colour became deeper.

Recent researches in the U.S.A. with lawn grass have revealed that gibberellin application on grass during cool weather stimulated growth; use of retardants delayed spring growth and antagonized seed-head production.

Regulation of flowering: When growers alter temperature and photoperiod to control blooming time, in the process they thus alter the balance of endogenous growth and flowering regulators. But artificial application of plant growth regulators can also enhance, hasten, or delay the flowering time in some plant species. Ram *et al.* (1970) found that spraying of GA_3 on antirrhinum, in the concentration range of 5-25 ppm resulted in early blooming. Mukhopadhyay *et al.* (1975) observed that foliar application of GA_3 (10 ppm) on carnation 140 days after germination increased the number and size of flowers. According to some other reports, gibberellins induced early flowering and prolonged flower life. It is also known that TIBA can increase the flowering in many species.

Auxins may slightly hasten or strongly inhibit flowering in many species. The inhibitory effects of auxins are more pronounced in plants sensitive to photoperiodism or vernalization and intermediate species. The general finding is that comparatively higher concentrations of auxins inhibit flowering in most of the species tested. It is reported that in many species the vernalization requirement may be replaced by gibberellin application. Application of gibberellin induces flowering in some long-day plants under non-inductive short photoperiods, but gibberellin spray fails to induce short-day plants to flower under long-days.

The application of plant growth regulators to promote flowering may result in some unwelcome side effects. Often it may so happen that the flower stems become weak making them vulnerable to lodging. It has been proved experimentally in certain plants that a second application of the plant growth regulator used for flower induction, soon after the first application, strengthens the flower stalk.

A review of some more works on regulation of flowering in different flowering plants by the application of various growth regulators will be very helpful in understanding this phenomenon. Pre-sowing treatment of seeds of *Impatiens balsamina* in 5 ppm IBA solution caused the production of more flowers in plants produced from such seeds (Agnihotri and Srivastava, 1964). Mittal (1967) observed that application of 100 and 200 ppm GA_3 in dahlia plants induced early flowering. The works of Ravindran and Narsimham (1970) suggest that in rose cultivar "Edward" the application of GA_3 (100 and 150 ppm), 2,4-D (1 ppm), and IAA (150 ppm) at early bud stage, and HQC (200 ppm) and 2,4-D (3 and 5 ppm) at half bud stage, promises to be commercially useful in increasing the flower size and weight. In bougainvillea Alar and Cycocel induced flowering on smaller plants under short-day conditions (Hackett and Sachs, 1967). Application of B-Nine (5,000 and 10,000 ppm) on dahlia proved to be effective in increasing the size and improving the quality of the flowers on shorter stalks (Bhattacharjee *et al.*, 1971).

In *Impatiens balsamina*, which is a qualitative short-day plant, GA_3 increased the extension growth (height) till 56 days under 8- and 24-hour photoperiods. Cycloheximide which decreased height slightly under inductive conditions, at a later stage did not affect the GA_3-promoted extension growth. Both GA_3 and cycloheximide caused enhancement of the rate of differentiation, although this effect was temporary in the case of GA_3. Cycloheximide does not affect photoperiodic induction, whereas it hastens and increases the magnitude of GA_3-induced flowering (Nanda *et al.*, 1973). In another experiment with the same plant, Nanda *et al.* (1969) observed that gibberellins A_3, A_{4+7}, A_{13}, (−) kaurene or TIBA initiated floral buds under strictly non-inductive conditions. Sawhney *et al.* (1970) obtained floral induction in the same plant by 2,3,5-triiodobenzoic acid.

Goh and Seetoh (1973) induced flowering by decapitation in an orchid hybrid of *Aranda*. They further observed that flowering responses of the decapitated plants were inhibited by a continuous supply of 10^{-4M}, IAA solution. However, once auxin supply was interrupted, floral initiation took place. It was also found that gibberellic acid had no significant effect on the flowering response of decapitated plants. CCC at 10^{-2M} concentration caused a slight decrease in the number of floral initials developed under decapitation. They thought that GA_3 probably played only a secondary role in controlling flowering. In an experiment with *Impatiens balsamina* on the effect of decapitation and growth retardants (Phosphon-D and Cycocel), Nanda *et al*.(1969) noticed inhibition of flowering by auxin. The auxin control of flowering is also confirmed by Bonner and Bandurski (1952) and De-Zeeuw and Leopold (1956). The work of Cajlahjan (1970) should also be mentioned here. He demonstrated the role of endogenous gibberellins in relation to photoperiodism in a number of decorative and other plants for both long- and short-day species. In long-day plants the increase in GA content led to the formation and growth of stems and subsequently to the formation of flower, while in short-day plants GA stimulated stem growth only. CCC at various concentrations delayed flowering in long-day plants but had no effect on neutral or short-day plants. Phenolic auxins and inhibitors had an indirect effect on flowering, either stimulating or inhibiting its onset. The existence of "anthesins" or metabolites inducing flower initiation was established by indirect methods, but their chemical composition could not be determined. Fowering in the seedlings of winter and biennial plants was related to preliminary vernalization at low temperature, at which time "vernalins" were formed. Transformation of vernalins into gibberellins took place in long-days only.

Prolonging the cut flower life: This is a field where the successful use of plant growth regulators can help the commercial growers greatly. Recent researches, largely by Marousky (1973), have not only opened up the possibility of prolonging cut flower life but also made it possible to harvest blooms in the tight bud stage and allowing them to open normally in the vase.

Einert (1971) demonstrated experimentally that stem weaknesses in cut tulips can be reduced by pre-harvest sprays with Ancymidol. The same growth retardant, when applied to vase solutions, also produces the same results (Einert, 1975). For further details on extension of cut flower life consult Chapter XX on "Commercial Floriculture".

Dormancy: To counteract the apical dominance and to encourage branching it is a customary practice with many floricultural crops to remove the growing tip mechanically, an operation known as pinching or stopping. But experimental findings have shown that auxins can remove apical dominance in many plants. Beach and Leopold (1953) found that maleic hydrazide in concentrations of 600 or 1,000 mg/1 promoted as much branching in chrysanthemum as was obtained by pinching. Maleic hybrazide had no adverse effect on flower quality. Asen and Hamner (1953) observed that ethylene and TIBA (2,3,5-triiodobenzoic acid) encouraged basal growth in roses. Bhujbal and Zurange (1973) also observed that application of TIBA on marigold at concentration of 25 ppm at 5-leaf stage produced more branches compared to pinching.

Another important field of plant growth regulators is the prevention of sprouting of nursery stock of trees and shrubs during storage period. This method has a great practical utility. If nursery stock can be transported in dormant condition without sprouting, mortality in transit can be reduced to the minimum. Possibly, Marth (1942) was the first scientist to demonstrate that methyl ester of naphthaleneacetic acid applied in a wax emulsion could prevent bud sprout of nursery stock of shrubs and trees during storage. Further researches have shown that naphthaleneacetic acid applied as an emulsion, spray, dip treatment, and ester vapour can reduce bud sprout in storage of several plant species, provided proper concentration is applied. The optimum effective concentration of auxin varies from species to species. The efficacy of maleic hydrazide to prolong dormancy of woody plants in storage has also been demonstrated. The efficiency of the auxins in this direction depends much on the time of application and temperature. It appears that early foliar application, i.e., some time before lifting the plant for storage, is good but at the same time may be demaging to the foliage. The right time of application has to be determined for each species.

The plant growth regulators are also helpful in breaking dormancy. Ethylene chlorohydrin is used to break dormancy in tuberose tuber. Kinetins have been proved to be effective in breaking dormancy of gladiolus corms.

Herbicides: The effectiveness of auxins in controlling weeds is well known. The commonly used auxin herbicides are 2,4-D (2,4-dichlorophenoxyacetic acid), 2,4,5-T (2,4,5-trichlorophenoxyacetic acid), and MCPA (2-methyl, 4, chlorophenoxyacetic acid).

Chapter VI

Flowering Annuals

Introduction

The general landscape design, lawns, trees, shrubs, and other permanent features are of fundamental importance for any garden. Nevertheless, proper introduction of annual flowering plants in the garden adds immensely to the decorative value of the garden, whether it is a large public garden or a small private garden, and they are indispensable. Annuals are a group of plants which grow from seed, flower, and die within one year or one season. In addition, there is a group called plur-annuals, which are perennials but flower in the first year and do not survive usefully another year. Besides, in the temperate region, there are many plants which are perennial in their natural habitat but are grown as annuals to avoid the trouble and the expenses to grow them under glass in the winter when they may be killed by frost. In tropical conditions also, many biennials and perennials can be grown as annuals when raised from seeds. In fact most of the flowering annuals complete their life cycle within 3-6 months.

Annuals provide a beautiful display of colour in the garden. Many of these plants are unequalled for their length of flowering, colour range, ease of cultivation, general adaptability, height and shape of plant. In addition, many of the annuals possess good fragrance.

Annuals can be grown in the ground or as potted plants. The uses of annuals, as well as biennials, are many and varied in a garden. They can be grown individually or with perennial plants as mixed borders. Many of these plants can also be grown in the cool green house almost throughout the year. Again, in a new garden, till the permanent features such as shrubbery or trees get established, the annuals can be used to fill the gaps. Many of the climbing annuals such as sweet pea, nasturtium, and morning glory are suitable for covering trellis, fences, and old tree stumps. If selected with some thought, annuals also suit very well in rock gardens. In a short tenure garden, as, for example, in a rented house, where the length of stay is uncertain, the annuals come in handy and offer a cheap but the most effective mode of gardening.

General Hints for Cultivation

With a few exceptions, annual flowers need full sun for their perfect growth. Shade in the annual beds even for a couple of hours in the morning or mid-day will greatly affect the vegetative growth and flowering; the latter may be delayed by a few days to a few weeks. The blazing afternoon sun, of course, is less important.

In the temperate countries of Europe, annuals are put into the following three groups for the purpose of cultivation as the temperatures there are very low. The same classification may hold good for hill stations in India:

(1) *Hardy Annuals:* They need no artificial aid for their growth and flower freely in the open.
(2) *Half-Hardy Annuals:* These need sowing under glass but are later planted out in beds.
(3) *Tender Annuals:* They are cultivated under glass protected from the vagaries of adverse weather conditions.

In the plains also, the annuals may be divided into groups; for example, those flowers which need long cool winter are placed in one group. The plants of this group, e.g., antirrhinum, sweet pea, pansy, carnation, cineraria, and violet, cannot be grown with success in such areas as Madras, Coimbatore, coastal areas of Kerala and Andhra where there is practically no winter season. But in northern India and in the mild climate of Bangalore almost all the annuals can be grown. But these annuals perform much better in the cool winter of northern India than in the milder climate of Bangalore. The hardy annuals are those which can be easily grown in areas where there is no winter, e.g., zinnia, tithonia, and marigold. Of course, there are annuals in the winter-loving group, which are not so hardy as to withstand the frost of the northern plains. Some of these are salvia, dahlia, and zinnia.

In India, the annuals are generally classified according to the season when they are grown, viz., winter annuals, summer annuals, and rainy season annuals. The list of winter annuals is lengthy, but a few of the important ones are antirrhinum, China aster, ageratum, arctotis, brachycome, carnation, calendula, companula, cineraria, dimorphotheca, dianthus (pink), larkspur, lupin, mignonette, nasturtium, nigella, pansy, petunia, phlox, salvia, stock, sweet pea, verbena, and viola. The summer annuals include cosmos, coreopsis, gaillardia, portulaca, sunflower, tithonia, and zinnia. The rainy season annuals, in addition to the above summer annuals, include balsam, cockscomb, gomphrena, and marigold.

The sowing of summer annuals starts from January and ends in June depending upon the climatic zones of the country. The rainy season annuals are sown with the break of the monsoon or a little earlier, in June or July. The winter season annuals are sown during September-October in the northern plains. But some late flowering species such as carnation, cineraria, and hollyhock and some which are damaged by frost such as salvia should be sown earlier in August-September. Many of the winter annuals such as petunia, salvia, carnation, and dianthus (pink) flower during August-September in Bangalore and in areas having similar climate and these should be sown as rainy season crops. In the hills, the sowing time is February-May for the tender and the half-hardy annuals and August to October for the hardy annuals. Where there is danger of frost in the hills, the annuals will need protection. Many of the annuals can be grown under glass with controlled temperature at other than normal season of growth.

In India, the annuals and the biennials are grown in the open ground, whereas in temperate countries these are also grown in greenhouses. Annuals thrive best in the ground and look natural. But there are annuals such as brachycome, carnation, cineraria, pansy, petunia, and phlox which form beautiful potted plants also. There are also annuals such as candytuft, *Hymenatherum tenuifolium*, nasturtium (creeping), petunia, phlox, portulaca, verbena, and *Zinnia linearis* which are grown in the window boxes and hanging baskets.

For pot culture, the sizes of the pots vary between 15 and 30 cm, depending on the type of the annual. Before using the pot, it is cleaned by removing the old soil mixture. The potting mixture generally consists of two parts of good garden soil, one part well-rotten cowdung manure, one part of coarse (for larger plants) or sieved (for delicate plants) leaf-mould, and 1/4 part of sand, if the soil is heavy. The alluvial soil of northern and eastern India does not need any sand. For red and black cotton soils, a pot mixture of 1 part of soil, 3 parts of well-rotten horse manure, and $1\frac{1}{2}$ parts of sand is recommended. To this one part of leaf-mould may or may not be added. Since the organic matter content of the red and black cotton soils is often low, the potting mixture should be given liberal doses of organic manure.

Soils, Sites and Layouts

In general, the annuals thrive best in deep, well-drained, and well-cultivated soil which is not acidic in nature. If the soil is acidic, addition of lime will be necessary to make it neutral as far as possible. In clay soil, the addition of lime will break up the stiff soil and an application of sufficient organic matter will make the soil porous. In sandy soil also, a lot of organic matter should be added, and wherever possible this should be improved upon with the addition of clay as it is costly to improve the texture of such soil with manure only. The beds should be worked up thoroughly to a depth of 30 cm but the digging work should not be undertaken in a very wet soil. The digging should be repeated 2-3 times and the soil pulverized thoroughly to obtain the best results. Before planting, the surface should be firmed by treading or drawing a light roller.

The seasonal flowers do best in full sun. But many of the annuals such as ageratum, sweet alyssum, calendula, clarkia, forget-me-not, larkspur, and salvia can be grown in semi-shade also, although the quality and the quantity of the flowers may suffer.

It is advisable to locate the beds with a strip of lawn in front and between two beds, having enough running and turning space for the lawn mower to work. While situating beds in the midst of the lawn, care should be taken that the annual flowers do not have to compete with the invading roots of the grasses. This is ensured by running a sharp spade deep into the ground around the edges of the bed every two or three weeks. Another precaution in this regard is to make the beds at least one metre wide. In the shrub border, this precaution is not needed where the intention is only to cover the vacant spaces with the annuals. The beds may be formal or informal. For mixed borders, large informal beds are better. In smaller beds, only one kind of annual with attractive colour and profuse blooming habit is grown. For the mixed borders, the taller growing plants such as hollyhock, tithonia, and sunflower are put in groups in the background. The intermediates such as antirrhinum, clarkia, poppy, salvia, and lupin are put in middle rows. The third clump may be made up of dianthus (pinks), aster, and calendula, and the edge may consist of the dwarf types such as sweet alyssum, pansy, and candytuft.

Sowing

It is the general practice to raise the seedlings in nursery and then transplant them in beds or pots. The only exceptions are those which have either bold seeds such as sweet pea, nasturtium, sunflower, morning glory, etc., or those annuals which are difficult to transplant, such as Californian poppy (*Eschscholtzia* or *Eschscholzia*) and linaria. These are sown directly at places where they are to be grown and later thinned out to proper spacings depending upon the type.

The seeds can be sown in earthen seed pans or pots, wooden seed trays, raised nursery beds, or seed frames, which are generally 1 to 1.2 m broad, 60 cm deep and of a convenient length, which may be anything from a few metres to 20 m or more. For details, see Chapter III on "Nursery Management". For the amateurs, the best thing is to sow seeds in seed pans. The compost for the seed beds or pans consists of one part sifted garden soil and one part of finely sieved leaf mould and if needed 1/2 part of sand may be added, especially for clayey soil. The compost should be sterilized (see Chapter IV for methods of sterilization) before sowing to eliminate soil-borne fungi, especially those causing "damping off". Before filling up the pans or the trays, adequate drainage facility is to be ensured by providing holes at the bottom. The seeds are sown in raised beds for this purpose only. The bottom layer in the pans is filled with broken bricks or crocks to prevent the possibility of the drainage hole getting clogged. The compost is filled and made firm with the help of a rammer made of a flat wood fitted with a handle at the centre.

Before sowing, the seeds are treated with a little of fungicide such as captan, arsan, or cersan to prevent the seedlings from getting fungal diseases. The seeds are sown in the pans in rows placed 3-6 cm apart made with the help of a pencil or a stick of similar thickness at 0.3 to 0.6 cm depth. Even and thin sowing is of particular importance as crowded seedlings often perish by "damping off". Small seeds such as petunia may be mixed with a little quantity of fine silver sand to get uniform sowing. Before sowing the seeds, the compost should be made slightly moist as the water applied after sowing will not be readily absorbed and the seed may float. Immediately after sowing the seeds watering is done carefully with a fine rose-can. A better way of watering is to plunge the seed pans in a large bath of water, taking care that the water level remains below the rim of the pan, till the water reaches the surface through the drainage hole. The pots are then covered with glass pane or newspaper sheets and kept in partial shade. Further watering is done, if needed, keeping in mind that over-watering is harmful. After germination, the seedlings are exposed to sunlight. Otherwise, they may become weak and lanky. The exposure to sunlight is gradually increased, starting from a couple of hours a day, till transplanting time when these should become sufficiently strong to withstand the full sun.

Pricking

This is the term given to the operation of transferring the young seedlings to another pan or tray. As soon as the seedlings are sufficiently large to handle, these are pricked with the help of a pointed stick or a metal label, at 5-7 cm apart in previously prepared holes and the compost around the collar (neck) and the roots of the seedlings is firmed carefully. The compost in this case will be a little richer than that of the seed pans. This operation ensures healthy growth of seedlings. The whole operation should be carried out with extreme care so as not to damage the very fragile stem and roots. Often the seedlings are pricked singly in small nursery pots (7.5 cm) and transplanted in permanent beds at 4- to 6-leaf stage along with the compost which comes off intact when the pot is inverted with a little jerk, if the compost is just sufficiently moist. While inverting the pot the outcoming compost is to be supported with the fingers of the left hand to prevent cracks in the compost. This method eliminates the chances of mortality in transplanting as the roots are not disturbed.

Transplanting

The time of transplanting the seedlings is generally one month after sowing when they have developed four leaves. A day or two before the actual transplanting operation, water is withheld to harden the seedlings. This operation reduces the chances of mortality. Transplanting is done on a cool cloudy day or in the evening which enables the seedlings to get established during the cool night. The bed for transplanting should be prepared thoroughly to a depth of 30-45 cm and the soil pulverized to a fine texture. The soil should be enriched with the addition of 4-8 kg of well-rotten cowdung or horse manure or any other good compost and four tablespoonfuls of superphosphate or bone meal per square metre area. A small hole is made at the place where the seedling is to be inserted and the soil around the roots and the neck of the plant is pressed firmly. If the seedlings are grown in nursery pots, a large enough hole is dug to accommodate the seedlings with the mass of compost. The transplanting is done as per the method described in the last paragraph under pricking. In this case, the compost is pressed against the ground to leave no loose soil around the seedling. To get the seedlings established, light watering every day in the early morning and the late afternoon will be required for about a week. In foreign countries, ready-made containers are available complete with sterilized soil and seed. The grower has just to open the lid and water the containers.

Watering and Feeding

The most common mistake done by the amateurs is daily watering of the established seedlings. This is extremely harmful. Daily light watering is needed only for about a week after transplanting. Thereafter, the correct method is to flood the beds first, and then weeding and hoeing are done after a couple of days when the soil is just moist. The soil is then allowed to dry before the next flooding.

It has already been mentioned that the ground should receive at the time of soil preparation sufficient organic matter, which is vital for plant growth and development. The chemical fertilizers are added in small doses during this time.

During vegetative growth, a mixture of chemical fertilizers and organic manure, in the form of a weak solution, is given once in a fortnight. About 4-5 kg of fresh cowdung manure or oil-cakes such as *neem*, groundnut, and mustard, and a handful each of ammonium sulphate and superphosphate are dissolved in about five litres of water and allowed to ferment for a few days; it is diluted to tea colour and then applied at the rate of 3-4 litres per square metre area followed by flooding with water. Alternatively, a dry mixture is prepared with ammonium sulphate, superphosphate, and potassium sulphate in the proportion of 2 : 1 : 1 respectively and added at the rate of 60 gm per square metre and forked in, followed by flooding of the beds. If the fertilizer mixture comes in direct contact with the foliage, it may cause scorching. When the flower buds appear, a fertilizer mixture consisting of superphosphate and potassium sulphate in the proportion of 2 : 1, respectively, is applied at the same rate as the above mixture. Alternatively, the liquid manure may be continued.

Staking, Pinching and Disbudding

The plants with weak, slender, or straggling stems will need staking when they are 15-20 cm tall. Creepers such as sweet pea, morning glory, nasturtium, etc., will also need necessary supports which has been dealt with separately along with each flower later in this chapter. Stakes prepared with split bamboos are painted green. Painting of the sticks with green helps conceal them among the green plants. Jute, cotton, and pigeon pea sticks are also useful as stakes. Many seedlings such as carnation, marigold, and dianthus (pink) need pinching at this stage. Aster, cosmos, and zinnia develop flower buds as early as when these are only 5-10 cm high, and are pinched off. The number of buds per stem is reduced by disbudding the axillary buds, if large flowers are desired.

Classification of Annuals

(A) *Early-Blooming Annuals*

Ageratum, *Alyssum maritimum*, *Amaranthus*, *Browallia*, *Celosia*, *Coreopsis drummondii*, *Cosmos*, *Gomphrena*, *Impatiens balsamina*, *Salvia*, *Tagetes*, and *Zinnia*.

(B) *Late-Blooming Annuals*

Abronia umbellata, *Althaea rosea*, *Anchusa capensis* cv. 'Blue bird', *Antirrhinum*, *Arctotis grandis*, *Centauria cyanus*, *Cheiranthus cheiri*, *Chrysanthemum carinatum* (annual tricolour chrysanthemum), *Dianthus* sp. (carnation, pink, sweet William), *Gilia capitata*, *Gilia tricolor*, *Mathiola* (also spelt *Matthiola*), *Salvia coccinea*, *Salvia horminum*, and *Viscaria*.

(C) *White-Flowered Annuals*

Alyssum, *Callistephus chinensis* (China aster), *Convolvulus major*, *Dianthus* (pink) cv. 'Double White Margaret', *Mathiola* (various white cvs.), *Nigella*, *Phlox* (various white cvs.), *Papaver*, and *Zinnia*.

(D) *Blue, Purple, and Lavender-Flowered Annuals*

Ageratum mexicanum, *Anchusa*, *Browallia elata*, *Callistephus chinensis* (various blue cvs.), *Centaurea cyanus*, *Clitoria ternatea*, *Convolvulus minor*, *Delphinium ajacis*, *Delphinium hybridum*, *Echium plantagineum* cv. 'Blue Bedder', *Gilia capitata*, *Heliotropium arborescens*, *Iberis umbellata*, *Ipomoea rubro-caerulea*, *Lobelia erinus*, *Myosotis*, *Petunia hybrida* cvs. 'Blue Lace', 'Black Magic', *Salvia farinacea*, *Viola tricolor hortensis*, and *Verbena* × *hybrida*.

(E) *Yellow and Orange-Flowered Annuals*

Cacalia lutea, *Calendula officinalis*, *Dimorphotheca aurantiaca*, *Eschscholzia californica*, *Tagetes*, *Tropaeolum*, and *Zinnia*.

(F) *Annuals for Edging of Beds and Walks*

Ageratum 'Blue' and 'White', *Alyssum* (sweet), *Brachycome*, *Dianthus* (pinks), *Iberis* (candytuft), *Lobelia*, *Nigella*, *Portulaca*, *Saponaria calabrica*, and *Torenia*.

(G) *Annuals for Hanging Baskets*

Ageratum (dwarf), *Hymenatherum tenuifolium*, *Lobelia* (basket type), *Petunia*, *Phlox*, *Portulaca*, *Tropaeolum canariense*, *Torenia*, *Verbena*, and *Zinnia linearis*.

(H) *Annuals for Rockery*

(i) For humid regions – *Ageratum*, *Alyssum* (sweet), *Brachycome*, *Phlox*, *Portulaca* and *Tropaeolum* (nasturtium)

(ii) For drier and cooler parts of India – *Collinsia bicolor*, *Cuphea*, *Dimorphotheca*, *Echium*, *Gilia*, *Godetia*, *Limonium sinuatum*, *Linum*, *Nemesia*, *Phacelia*, *Saponaria*, *Scabiosa*, *Schizanthus*, *Ursinia*, and *Venidium*.

(I) *Fragrant Annuals*

Alyssum, *Cheiranthus cheiri*, *Centauria moschata*, *Dianthus caryophyllus* (carnation), *Heliotropium*, *Lathyrus odoratus*, *Mathiola* (stocks), *Phlox*, and *Reseda odorata* (Mignonette).

(J) *Annuals Suitable for Shade*

There is no annual which can survive in full shade. However, there are annuals which thrive well in depleted light or partial shade, but the quality and the quantity of the flowers get a set-back. These are *Ageratum*, *Alyssum*, *Calceolaria*, *Calendula*, *Clarkia*, *Delphiniums* (including larkspur), *Godetia*, *Iberis*, *Impatiens*, *Lobelia*, *Lupin*, *Myosotis*, *Nemophila*, *Nicotiana*, *Phlox*, *Salvia*, *Senecio cruentus*, *Verbena*, and *Viola tricolor hortensis*.

The above list in the various categories are not exhaustive but only indicative and the grower could add many more in each case.

Colour Scheme and Grouping

The annual flowers consist of all imaginative hues and throw an opportunity to a person to decorate his garden most brilliantly in an artistic manner. The arrangement of different colours varies with the taste of individuals but basically it is an artistic job. Any arrangement that does not spoil the beauty of the garden but has an appeal to the casual visitor is welcome to the extent that the scheme does not look patchy, but brings some harmony. A basic study of the natural colours will be helpful in formulating colour schemes in the garden. The three basic colours are red, yellow, and blue, while the secondary colours are orange, green, and violet. Red, orange, and yellow are considered as hard or warm colours, whereas blue, violet, and green are considered soft or cool. In the same reckoning, white, black, and grey are accepted as neutral colours.

The basic arrangement of colours in annual beds may be classified as follows, but one should remember that no rigid formula can be provided for all the places for all times and may vary according to situation and convenience.

(1) *Monochromatic:* It is arranging different tones of the same colour either of the same species or in combination with different species. For example, a group of *Salvia horminum* in colours of deep-violet blue, with deep blue larkspur or blue petunias with blue lobelia, pink antirrhinum with salmon pink candytuft; a mixture of two cultivars of dimorphotheca of orange and flame colour. A yellow monochromatic colour scheme with variation in height is possible with yellow marigold, yellow antirrhinum, nasturtium, Californian poppy, and viola or *Hymenatherum*.

(2) *Analogous:* This can be achieved by using closely related or harmonious colours such as violet with blue and light blue with green, and yellow with green, or yellow with yellow-orange. This may be done using either soft or warm analogous colours. A soft or cool harmony can be achieved by using the blue colours available in sweet peas, China asters, and petunias; whereas a warm harmony is possible by using yellow antirrhinum with orange dimorphotheca or yellow marigold with orange antirrhinum, and so on.

(3) *Complementary or Contrasting:* Two opposite or contrasting colours – blue and orange, red and green, violet and yellow – are called complementary. This can be sub-divided further to split-complementary with one dominant and two complementary colours (violet with yellow-green and yellow-orange) and triads such as red, yellow, and blue; or violet, pink, and yellow. The red salvias grown against a lush green hedge form a beautiful contrast. With thoughtful imagination, one can arrange a series of such pleasing contrasts. Here also a contrast may be soft or warm.

The proper combination of colours is very important in annuals and the same is true in the case of plant height and time of flowering. The tall plants such as hollyhock, sunflower, tithonia, and marigold are put in the background, antirrhinum and larkspur coming next in order. Phlox, dianthus (pinks), sweet alyssum, and pansy are arranged next in accordance with their height. While arranging the annuals in a herbaceous border, the time of flowering should be kept in mind. The late flowering types such as hollyhock should be sown at least a month earlier than the early flowering sweet alyssum, otherwise a part of the bed will look patchy if all are sown together. The annual beds should also harmonize with the other plants of the garden such as shrubs, trees, hedges, etc.

A basic guideline has been provided above for the arrangement of different annual flowers and one can improve upon this with experience gained.

Some Annuals and Their Culture

Abronia umbellata (Syn. *A. rosea*); Common Name: Sand-verbena; Family: *Nyctaginaceae*; Origin: California (U.S.A.)

This seasonal flower is most suitable for the rockery. It takes about 3-4 months from sowing to flowering. This is a climbing or trailing annual growing up to a height of 45 cm and trailing up to 90-150 cm. The flowers are rosy lilac, sweet-scented during night and somewhat resemble verbena.

For cultivation as a climber, this plant needs staking. In rockeries, they may be allowed to trail along a rocky slope. Seeds are sown in October in the plains and in March in the hills. Seedlings can be transplanted a month after sowing. The husk (calyx) is removed before sowing.

Acroclinium roseum (Syn. *Helipterum roseum*); Common Names: Acroclinium, Immortelles, everlasting flower; Family*: *Compositae*; Origin: Australia.

The plants are 45-60 cm tall bearing double daisy-like flowers of pure white, pink, and shades of rose with bright golden yellow and dark coloured centres. The flowers are borne singly at the end of long stems and are papery in texture. It is useful for cutting and lasts long when cut and dried. The leaves are narrow and glaucous green.

Cultural hints: The plants are easily grown and prefer light sandy soil and a warm sunny situation. The seeds are sown in September-October in the plains and March-April in the hills. The plants do not like root disturbance and it is, therefore, advisable to sow the seeds *in situ* and thin out the seedlings 15-20 cm apart. The plants flower in about 6-8 weeks after sowing. Acrocliniums are suited for medium and high altitudes and places having cool winter. The plants prefer a dry climate and hence are not very suitable for growing in the humid climate of Calcutta and places having similar climate.

Acrocliniums do well in pots. The pot mixture consists of half part of sandy loam and half part of stable or cow manure mixed with sufficient coarse sand

Actually, the genus *Helipterum* includes both *Acroclinium* and *Rhodanthe*. *Acroclinium roseum* bears rose coloured flowers, *A. roseum* var. *album* pure white flowers, and *A. roseum* var. *grandiflorum* larger flowers than commonly seen in the type; *A. roseum* var. *plenum* is with double flower heads. Both the latter varieties have pink flowers.

A species, *Helipterum manglesii*, also known as *Rhodanthe manglesii*, is a bushy, free-flowering plant, 30-40 cm tall, with globular heads of flowers in pink and white. It makes an attractive pot plant. Like the flowers of *Helichrysum*, the flowers of this species can also be preserved for winter decoration, in dried form.

Adonis aestivalis; Common Name: Pheasant's Eye; Family: *Ranunculaceae*; Origin: Central Europe.

Branching annual and perennial rock plants which are 25-40 cm tall and suitable for cultivation in mixed borders, open woodland, and wild gardens. The plant has a deep-green, finely-cut, fern-like foliage and bears small dark crimson buttercup-like flowers of about 2.5 cm across.

Cultural hints: The plants can grow in ordinary soil, but prefer a moist and shady situation. The seeds are sown during October in the northern plains and the plants flower in February-March. In the hills, sowing can be done in spring or autumn. A spacing of 25-30 cm is recommended.

Ageratum (*Ageratum conyzoides* and *A. houstonianum* or *A. mexicanum*); Common Name: Floss Flower; Family: *Compositae*; Origin: Central America.

From the above two species, garden forms, ranging from 15 to 45 cm in height, have been bred. The dwarf types are especially suitable for beds, borders, or rockeries. The branches are covered with fluffy heads of flowers with colour ranging from pale lavender to deep mauve. Some cultivars bear white flowers also. The flowers remain fresh for quite a long time and for this reason it is named *ageratum*, meaning ever-young. The taller forms make

*Now the family *Compositae* is known as *Asteraceae*.

good cut flowers.

Cultural hints: Ageratum prefers light soil and a warm sunny situation. It is transplanted and pinched several times to keep it dwarf. Seeds are sown in September to October or even June to September for early flowering. In the hills the seeds are sown in February-March. In Bangalore, *Ageratum* can be grown almost throughout the year. The fine seeds should be sown thinly and just covered with fine sifted soil or leaf-mould. The seedlings are transplanted four weeks after sowing. They take about 2½ to 3 months to come to flowering. Faded flowers should be removed frequently to prolong flowering. Spacings of 15 cm for dwarf and 25 cm for tall cultivars are recommended.

Ageratums are good as potted plants. When grown in pots, they should be syringed occasionally with clean water to keep down red spiders.

The important cultivars are 'Blue Ball' (dark blue), 'Blue Miniature', and 'Fairy Pink' (light pink and salmon). Also available for cultivation are a tetraploid cultivar named 'Blue Mink' having very large flowers and some F_1 hybrids such as 'Summer Sky' and 'Blue Blazer' with uniform flowers.

Ageratum is very easy to grow and becomes a weed in the garden.

Agrostemma githago 'Milas'; Common Name: Corn Cockle; Family: *Caryophyllaceae*: Origin: Turkey and Europe.

This is a hardy annual which was found in the town of Milas in Turkey and hence the name. This is a tall, slender plant growing to a height of 60-90 cm. The large *Viscaria*-like flowers are 5 to 7.5 cm across and pale lilac in colour with radial lines of dark dots on each petal. Though the plant is slender in habit, it stands well to strong winds and rains and makes a beautiful cut flower along with its long, narrow and graceful foliage. This is useful for mixed border and pots also.

Cultural hints: It is very easy to grow. The seeds are sown in nursery beds in October in the plains and during spring or autumn in the hills, and transplanted 25 to 30 cm apart.

Althaea rosea; Common Name: Hollyhock; Family: *Malvaceae*; Origin: Orient, China.

In temperate countries and in our hill stations, this is treated as a perennial, whereas in the plains it behaves like an annual. Hollyhock is a majestic plant, attaining a height of 1.5 to 2.5 m or more under favourable conditions, and with rough, large, round, and kidney-shaped leaves. The large, single, semi-double, or double flowers are borne in the axils of the leaves all along the length of the stem. The double-flowered cultivars frequently throw out singles and semi-doubles. There are cultivars having fringed-petalled and carnation-like flowers. The shade of the flower varies from white to cream, yellow, scarlet, rose, lilac, mauve, deep violet, near-black, etc. There are cultivars in which the centre of the flower is differently coloured.

It is suitable to form a background in the mixed or annual border and useful for covering open spaces in a shrubbery, more so in a newly planted one. Hollyhock can also be used as a screen.

Cultural hints: A deep, fairly rich but well-drained soil, with a little of lime rubbish, is ideal for its cultivation. It is generally grown from seeds, but young shoot cuttings, 7.5 cm to 10 cm in length, can also be rooted in pure sand or sandy soils, especially in temperate regions. Seeds are sown in September or October or even earlier in the plains. For annual borders, seeds should be sown at least one month earlier than other annuals, so that the flowering time coincides with the others. The plants take 3½ to 5 months to flower. In the hills and temperate countries, the plants take one year to come to flowering. The seeds are sown in June-July in the nursery. Unless the winter is very severe, the plants are put in position in early autumn. In case of severe winter, it will need protection under frame. The plants will flower next June or July. If flowers are required in the first season, the plants should be planted under glass in February and they will flower from July onwards. Plants grown from imported seeds take longer to flower than those grown from acclimatized seeds. The plants are likely to be damaged while transplanting. It is, therefore, advisable to sow the seeds directly and thin out the seedlings 45-60 cm apart. Transplanting should be done at the 4-leaf stage. The plants need a lot of root space and, therefore, the soil should be well dug up. As the plants are very tall they will need staking against strong winds. If so desired, hollyhock may be forced to branch by pinching at one metre height. In poor soil, the plants should be provided with adequate cowdung manure. In low rainfall areas, they can be grown in the rainy season also. In places having mild climate, such as Bangalore, they can be grown throughout the year.

There are many cultivars; some of them are suitable for the plains, while others are good for the hills. The recent powder puff cultivars with extra large flowers are becoming popular. In this type, the flowers look more showy as the leaves are small compared to leaves in the older types.

Hollyhock rust is a serious disease which is checked by dusting sulphur (300 mesh) at fortnightly intervals.

Alyssum maritimum (Syn. *Lobularia maritima*); Common Names: Sweet Alyssum, Madwort, Sweet Alison; Family: *Cruciferae*; Origin: Europe and West Asia.

This forms a low-growing, spreading plant with narrow grey foliage. The plants branch freely and each branch ends in a spike of usually white flowers which are honey-scented. When in flowering, the compact plant looks like a carpet of flowers. There are cultivars having rose-pink, lilac, violet, or yellow flowers.

The uses of sweet alyssum in a garden are varied. They are suited for beds, borders, window boxes, hanging baskets, and as the first row in a herbaceous border. It is good for covering the open spaces under blooming bulbs, as edges in rose beds or under standard roses. It is an attractive plant for the rockeries, paved gardens, and dry walls.

Cultural hints: Seeds are sown during September-October in the plains. In mild climates (Bangalore, Poona, etc.) this can be grown for a larger part of the year. In hill stations, the seeds are sown in spring or early summer. While sowing, the seeds should be mixed with an equal quantity of sand to obtain uniformity. For edging of beds or for growing under standard roses seeds can be broadcast directly and then thinned out. To have a mass effect closer planting, 15 to 30 cm apart, is recommended. It can also be sown in succession from September to January in the plains and March to June in the hills. To obtain a second flush of blooms, the fading flowers should be removed. The plants flower within six weeks after sowing.

Amaranthus; Common Names: Amaranth, Love-lies-bleeding, Joseph's Coat, Prince's Feather; Family: *Amaranthaceae;* Origin: Tropics.

This group comprises some useful garden annuals, with bright foliage or blooms or both.

The 'Love-lies-bleeding' or *Amaranthus caudatus* grows to a height of 1 to 1.5 m. It bears tail-like, often with knots at short intervals, drooping flower spikes, white, pale green, or crimson in colour. It is useful for cutting, especially if the leaves are striped. The 'Prince's Feather' (*Amaranthus hypochondriacus* or *Amaranthus hybridus* var. *hypochondriacus*) bears upright crimson plume-like spikes above the green foliage and grows to a height of 60 to 90 cm. *Amaranthus tricolor* (*A. melancholicus* var. *splendens*, *A. gangeticus*) or 'Joseph's Coat' is a brilliantly coloured foliage plant growing to a height of 50 cm. The leaves are bright scarlet-crimson marked with bronze and yellow. *Amaranthus tricolor* var. *ruber* is an ornamental foliage plant with red foliage. *Amaranthus salicifolius* is a tall plant growing to a height of 1 m. Commonly called 'Chameleon' or 'Fountain Amaranthus', it has long undulating foliage which is marvellously coloured orange, carmine, and bronze. 'The Molten Fire' or 'Summer Poinsettia' (*Amaranthus bicolor*) grows to a height of 60 to 120 cm, the foliage is deep maroon, and each foliage has a crimson top like poinsettia. All the species are good as pot plants. They are suitable for growing in groups or in rows in an annual border.

Cultural hints: All the above species are annuals and very easy to grow. They do well in a sunny situation and deeply worked out soil. In shady or semi-shaded places, the colour of the foliage does not develop properly. In a very rich soil also, the colour tends to become dull.

The rainy season is the best time for their cultivation. In winter, the plants remain dwarf. The seeds are sown in January-February or June-July directly at permanent places and then thinned to 30 to 45 cm. Seedlings can be raised in seed beds or seed pans also and are transplanted when about 2.5 to 4 cm tall. In the hills, sowing should be done in February-March. In Bangalore, sowing can be done for a greater part of the year. Plants take about 2½ to 3 months to grow to their full beauty.

Anchusa capensis; Common Names: Alkanet, Bugloss, Cape Forget-me-not; Family: *Boraginaceae*; Origin: South Africa.

In its place of origin, the plant grows as a biennial or a perennial but can also be grown as an annual in the garden. It grows well in the hills and also in the plains having a long and cool winter.

The plant is compact and grows to a height of 30 to 45 cm. The leaves are spear-shaped and rough in texture. It is free flowering. The flowers are deep blue with a white centre and 0.6 to 0.7 cm across.

When planted in combination with dwarf yellow-flowered annuals such as yellow French marigold, it looks very attractive.

Cultural hints: The seeds are sown directly in the beds or seedlings are raised in seed beds for transplanting later. The best time for sowing seeds is September-October in the northern plains and spring or autumn in the hills. This annual may not do well in such places as Bangalore. A well-drained soil and a sunny situation are good for its cultivation. A spacing of 30-45 cm is sufficient. The plant may need staking. The cultivar 'Blue Bird' has blue flowers with a white centre.

Anchusa italica (*A. azurea*), a native of Caucasus, is a perennial tall plant—growing to a height of 150 cm. This can be grown as a biennial also. The sowing is done in April in the nursery beds and transplanting is done in autumn. The flowers are available in the next season. 'Morning Glory', 'Pride of Dover', and 'Dropmore' are some of the well-known cultivars.

Antirrhinum majus; Common Names: Antirrhinum, Snapdragon; Family: *Scrophulariaceae;* Origin: Europe.

The ancestor of the modern garden antirrhinum came from the Mediterranean region. This is a branching perennial plant, 30 to 90 cm tall, with numerous spikes, mostly of crimson and white colour. The species *Antirrhinum majus* has been developed from this wild plant by natural selection. The modern garden forms may be classified as Tall (90 to 125 cm), Intermediate (60 to 75 cm), and Dwarf or Bedding (30 to 45 cm). Subsequently, the dwarf and the spreading types with a wide range of flower colours came into being, known as 'Rock Hybrids' or 'Hybrid Gem' as a result of crossing with *Antirrhinum molle* and *A. glutinosum.*

Use: It is one of the most popular and widely grown garden annuals. The tall cultivars can be used among the shrubs, in mixed borders, and also with herbaceous plants. The intermediate cultivars are very useful for bedding and pot culture. Both these types are excellent for cutting. The dwarf cultivars are good for edging and bordering formal beds. The dwarf rock hybrids are specially suited for covering the patches in rock gardens.

Colour: The colour of the flower varies widely in all the types ranging from white and cream to yellow, shades of pink, scarlet, red, orange, brick red, mauve, rose, maroon, etc. Now-a-days, the variegated and "eyed" types with marked throats are very popular. The shape of the flower resembles a dragon's jaw and the lip can be opened widely by pressing gently with the fingers.

Cultural hints: Strictly speaking, this is a perennial plant, but may be best treated as an annual. It grows well in a rich light soil enriched with lime. The plant does not thrive well during the rains as a certain amount of dryness around the root region is preferred by the plant. A sunny situation is best for its growth.

The seeds are sown thinly in nursery beds or seed pans mixed with 3 to 4 times their quantity of fine sand during September-October in the plains. Pricking of seedlings when large enough to handle in seed pans at 5 cm apart is helpful. Final planting in beds, 30 to 40 cm apart, is done when the seedlings are 7-8 cm tall. For pot culture, they are grown in 25 cm pots. Some people prefer to grow them first in 15 cm pots and then transfer them to 25 cm pots. In low rainfall areas, sowing can be done as early as in June-July. In places having a mild climate, such as Bangalore, flowers can be obtained at least twice a year, first during December-January and then during August-September. In the hills, sowing can be done during spring and autumn. The latter sowing is meant for pot culture. Plants take about 2½ to 4 months to come to flowering.

Over-manuring may result in poor quality flowers. In poor soil, the application of one tea-spoonful of ammonium sulphate and sulphate of potash to each plant as top dressing, about a month after transplanting, is found to be beneficial for vegetative growth and flowering. Over-watering should be avoided as this may cause rotting. Some people prefer pinching of the main shoot to obtain a bushy growth. Subsequent pinching of side shoots, when they are about 1.5 cm long, will give further bushy growth. But, if tall and quality spikes are preferred to quantity, pinching should not be resorted to.

Cultivars: There are numerous cultivars of antirrhinum: Tall, Giant or Grandiflorum, Intermediate or Semi-dwarf, Dwarf or Tom Thumb, and Miniatures. Recently, some tetraploids, with double flowers, have been introduced. The F_1 and F_2 hybrids are also very popular. Tetraploid cultivars such as 'Glacier', 'Rosabel', 'High Noon', etc., with their super-sized ruffled flowers, are also rust-resistant. In recent times, two cultivars, namely, 'Tinkerbell' and 'Juliana', with flowers of different shapes, have been introduced.

Rust is a serious disease in antirrhinum. This disease seems to develop in the hotter and drier regions. With the development of rust-resistant cultivars, it is now possible to grow antirrhinum with greater success.

Arctotis grandis (Syn. *A. stoechadifolia grandis*); Common Names: African Daisy, Blue-eyed or Transvaal Daisy; Family: *Compositae*; Origin: South Africa.

Arctotis grandis is common in the Indian gardens. The species is vigorous growing and bushy in habit; the flower stem attains a height of 45 to 60 cm and is thus suitable for cutting. The colour of the flower is white with a lilac centre and the back side of the petals is steel-grey. The *A. grandis* hybrids have flowers of pale pastel shades of white, ivory, buff, orange, bronze-red, and salmon.

Cultural hints: Arctotis is suitable for daytime decoration only as the flowers close the petals during the night. The plant needs a fairly rich soil and a sunny situation. The seeds are sown in September-October in the northern plains in seed beds. Direct sowing in permanent sites is also possible. In the hills, seeds are sown during March-April. *Arctotis* is quite easy to grow and does not require any special attention.

Bartonia aurea (*Mentzelia lindleyi*); Common Names: Blazing Star or Golden Star; Family: *Loasaceae*; Origin: California, U.S.A.

An easy-to-grow, hardy, and showy annual, 45 to 60 cm in height. The flowers are large buttercup-like and glistening orange in colour, with a central mass of feathery stamens. It is useful as a bedding plant.

Cultural hints: A well-drained soil and a sunny situation are good for its cultivation. Seeds are sown in October in the plains and in April in the hills. Spacing followed is 25 cm.

Bellis perennis; Common Name: Daisy or English Daisy; Family: *Compositae*; Origin: Europe including the British Isles.

Strictly, it is a perennial plant but can be best grown as biennial in the hills and as annual in the plains. It is a dwarf plant, 20-30 cm in height. The leaves are strap-shaped, tapering towards the tip and form into tuft at the base. The flowers are 2.5 to 5 cm across and the flower stalk is about 15 cm long. The flowers are single or double and the petals may be flat or quilled. The flat-type is available in white, crimson, and pink; the quilled in a beautiful salmon-pink shade and in crimson. Now a miniature up to 15 cm in height and a double-flowering cultivar, 'Lilliput', is available in such varied colours as white, light to dark rose, and dark red.

The daisy is useful for bedding, edging, for rock gardens, and for covering open spaces in beds of bulbs. All types are suitable as pot plants and cut flowers look delightful in low bowls.

Cultural hints: Plants raised from seeds are said to be better as they bear more flowers than those raised from suckers. Seeds are sown in September-October for flowering in the following February-March. In the hills, this is grown as a biennial and the seeds are sown in February-March. Propagation by division is also practised. The seedlings are transplanted at the 6-leaf stage, 15 to 20 cm apart, when the plants are about 5 cm tall. It can be grown in 15 to 25 cm pots also. Daisies like a light but rich soil. They may be grown in a situation sheltered from the afternoon sun. Daisies are not very suitable for places not having long cool winter.

Brachycome iberidifolia; Common Name: Swan River Daisy; Family: *Compositae*; Origin: Australia.

It is a beautiful dwarf annual (30-50 cm) with a slender stem and finely divided leaves. The flowers are 2.5 cm across, borne in profusion and are white, blue, pale blue, mauve, or purple.

The Swan River Daisy is a wonderful pot plant and is useful for edging. It is also an appropriate plant for rock gardens and herbaceous borders. The cut flowers are useful for vase decoration.

Cultural hints: Plants are raised from seeds which are sown in September-October. Seedlings do not stand transplanting well and seeds should preferably be sown directly *in situ*. The seedlings may later be thinned out 15 cm apart. The sowing in the hills is done in April. For pot growing, 25 cm pots are quite suitable. Because of their thin and slender stems, the plants have a tendency to lodge and hence they should be planted closely at 15 cm from each other for self-support. The plants need pinching to make them bushy. They need a medium soil and a sunny situation.

Browallia demissa (Syn. *B. elata*, *B. americana*); Common Names: Browallia, Amethyst Flower; Family: *Solanaceae*; Origin: South America.

It is a plant of medium size, rather moderately tall, attaining a height of 30-45 cm. Flowers resemble Lobelia, 1 cm long, and are borne in the axils of narrow, oval, green leaves. There are two colours – blue and white.

Browallia speciosa major is another species which is strictly a perennial and a useful pot plant. It grows to a height of 45-60 cm and has large bright blue flowers. They are effective for mass planting in beds or borders.

Cultural hints: Both the species are easy to cultivate. The plants do well in partial shade. A warm climate and an adequate soil moisture are beneficial.

The seed is sown in June for flowering in the rains in low rainfall areas. For winter flowering, September is the best time for sowing seeds. In the hills, sowing may be done in March-April. It takes about 2½ to 3 months to come to flowering. Spacing is 20-25 cm.

Browallia speciosa major can be sown in spring for late summer and autumn flowering in the hills. For flowering in greenhouse during winter and spring, sowing is done in July-August. This species can be raised by cuttings also.

Cacalia coccinea (Syn. *Emilia flammea*); Common Names: Tassel Flower, Flora's Paintbrush; Family: *Compositae*; Origin: Tropical America.

A hardy annual, 30 to 45 cm tall. The leaves are thin, long, and oval and remain attached to the wiry and hairy stem. The small flowers, which are 1.5 cm across, are borne in clusters on the wiry stems and look like a tassel. Flowers are commonly orange-scarlet or dull red in colour, but there is one variety (*C. lutea*) with golden-yellow flowers. A strain with white flowers is also reported. It is very graceful as cut flower, bedding, and border.

Cultural hints: Seeds can be sown round the year, in January-February, June-July, or September-October. The plants do well in all climates. They grow well without much care, like a weed, in a warm sunny place and a well-drained light soil. In the hills sowing is done in spring (April) and autumn (early September). Spacing allowed is 15 to 25 cm, depending on the season.

Calceolaria; Common Names: Slipperwort, Pouch flower; Family: *Scrophulariaceae*; Origin: Central and South America.

The plants are remarkable for their brightly-coloured flowers and a peculiar shape as that of circular pocket purses. *Calceolaria* includes many species of annuals and perennials. They may be herbaceous or shrubby in habit. The species *C.* × *herbeohybrida*, *C.* × *profusa* (*C. clibranii*) and *C. multiflora nana* are suitable for growing in higher elevations only. The species *C. mexicana* and *C. pinnata* (30 cm high) can be grown only in the northern plains and at medium elevations. They are excellent as pot plants.

Cultural hints: These species need cool and moist conditions and grow better in partial shade. In the northern plains, seeds are sown in September-October. Greenhouse *Calceolaria* is sown in June-July for flowering in the following spring or early summer in the hills. Sowing can also be done in February-March for pot culture. Seeds are very fine; therefore, extra care is to be taken for such seeds. The shrubby types are propagated from stem cuttings also. For pot culture, 25 cm pots may be used. The plants should be pinched for bushy growth and they take about 8-10 months to come to flowering.

Calendula officinalis; Common Names: Calendula, Pot Marigold; Family: *Compositae*; Origin: Southern Europe.

One of the most commonly cultivated seasonal flowers, its name is derived from the word *calendae* meaning the first day of the month, as it flowers almost throughout the year, especially in protected places. The name Pot Marigold is given probably because its petals were used in England in the earlier days for flavouring soups.

The flowers measure 4-5 cm across and their colour varies from straw colour to deep orange. The flowers may be single or double. Calendula is useful for bedding, cutting, potting, and for window boxes.

Cultural hints: A moist well-drained soil and full sunshine are the requirements for its growth. The seeds are sown in the plains in September-October or earlier in July, especially in low rainfall areas. In the hills, sowing is done during autumn and spring. In places having a mild climate, as Bangalore, sowing can be done almost round the year. Normally, the seedlings are transplanted, but direct sowing on the beds is also possible. Pinching the plants for a bushy growth and removing the faded flowers are recommended for calendula. Imported seeds do not germinate well except in cool weather. It is, therefore, advisable to collect own seeds.

Callistephus chinensis: See Chapter XX.

Calliopsis: See *Coreopsis.*

Campanula medium; Common Name: Canterbury Bells; Family: *Campanulaceae*; Origin: South Europe.

It is a biennial, bearing bell-shaped flowers of white, blue, pink, mauve, or violet colour. The flowers are single or double, and the variety *calycanthema* has cup-and-saucer shaped flowers. The species *C. ramosissima* (*C. loreyi, C. drabifolia*) is an annual, 22-45 cm tall. *C. pyramidalis*, commonly called 'Chimney bell flower', is another biennial species. *C. persicifolia* and *C. carpatica* (commonly called 'Tussock Bellflower') are two perennials. 'Venus's Looking-Glass' (*C. speculum*) is a dwarf annual, *Campanula*, and is now known as *Specularia speculum*.

The 'Canterbury Bell' is a useful plant for annual and mixed borders and also for pots. The flowers are good for cutting.

Cultural hints: Canterbury Bell is suitable for the hill stations, but the biennial and annual species do well in many parts of northern India. In low rainfall areas, sowing may be done as early as in July, but otherwise the sowing season is late September or October. The plant has a long growing season and flowers in six months after sowing. In the hills, sowing is done in autumn or spring. Partial shade is ideal for its proper growth.

Celosia argentea (*C. plumosa*); Common Name: Cockscomb; Family: *Amaranthaceae*; Origin: Tropical Asia.

A very popular and easy-to-grow annual, 25-50 cm in height. In some parts of the country, especially eastern India, the plant may grow up to 150-180 cm during the rainy season. It has a branching growth habit and produces feathery flower spikes in a pyramidal fashion resembling Ostrich-plumes, in various colours of silver, yellow, golden-yellow, red, orange, etc.

The cockscomb is useful as a bedding plant or for borders and pots. It can be used as a cover plant in the shrubbery, especially a newly planted one. The flowers are good for cutting and can be dried for table decoration as the dried flower heads retain the colours for a long time.

A variety, *cristata* (*Celosia argentea* var. *cristata*), produces large flower heads which are crested and resemble cock's comb. It grows to a height of 20 cm to 1 m, depending upon the season. The large, compact, velvety head may be of yellow, crimson, purple or other colours. *Celosia childsii* (Chinese wood flower) bears a rounded head of flower, of the size of a tennis ball.

According to Bailey (1964), different Latin-form names have been ascribed to the cultivated races, e.g., *C. childsii, floribunda, plumosa, pyramidalis, spicata*, and *thompsonii*; and a group of hybrids are referred to as 'Gilbertia'.

Cultural hints: They are generally grown in the hot and the rainy seasons. Sometimes, they are grown in the winter season also. In Bangalore, these can be grown almost throughout the year and they perform superbly. Seeds can be sown as early as in January or February for flowering in the summer months or during May-June for flowering in the rainy season. Sometimes cockscomb fails to grow properly as the seedlings cannot stand transplanting. These are to be transplanted when very small (2.5-3 cm) to avoid transplantation shock. In *C. cristata*, the spacing should be 20-30 cm for dwarfs and 30-45 cm for tall ones. In the case of *C. plumosa*, the spacing is 45 to 60 cm, depending upon the weather. Cockscomb responds well to the application of liquid manure. In *C. cristata*, flowering takes about three months, but the super giant cultivars take 4½ to 5 months to develop their heads fully. *C. plumosa* flowers within 2½ months. In the hills, sowing is done between February and April.

Centauria cyanus; Common Names: Cornflower, Blue-bottle, Bluet, Bachelor's Button, Rogged Robin; Family: *Compositae*; Origin: Europe including the British Isles.

A hardy upright annual. A very old plant of the British gardens and a well-known weed of the corn fields. The plant is slender and 60-90 cm tall. The original blue flower colour sported to a multiple range of colours such as white, pink, maroon, and purple. Many of these colours come true from seed. There is also a dwarf (30 cm) cultivar, named 'Jubilee Gem'.

It is useful as cut flower and can also be grown in annual and mixed borders.

Cultural hints: Seeds can be broadcast or raised in seed beds and transplanted but it is advisable to avoid transplanting as far as possible. Sowing is done in June-July (low rainfall regions) and in October in the plains. In the hills, seeds are sown during March-April and August-September. Spacing allowed is 20 to 25 cm. The plant flowers in 3 to 3½ months. It is advisable to remove the faded flowers to prolong flowering.

Centaurea moschata (*C. suaveolens*, *C. odorata*, and *Amberboa moschata*); Common Name: Sweet Sultan; Family: *Compositae*; Origin: Eastern Mediterranean.

Bears fragrant flower-heads of white, delicate mauve, deep mauve, purple, yellow, or deep red on long stems. Plants are about 90 cm to 1 m tall. The flowers are excellent for cutting and look like powder puffs.

Cultural hints: It thrives in ordinary garden soil. Seeds are sown in nursery beds or *in situ*. Sowing is done during September-October in the plains and March-April or August to October in the hills. The plant flowers in about 3½ months. A spacing of 30 cm is recommended.

Centaurea americana is another species which is 90 cm tall and produces large thistle-shaped flowers. It is also good for cutting. The cultural requirements are identical with those of *Centaurea moschata*.

Cheiranthus cheiri; Common Name: Wallflower; Family: *Cruciferae*; Origin: Europe, but naturalized in Britain.

Wallflower is a medium tall (30-45 cm) shrubby biennial or perennial. There are some early flowering annuals also. Most of the giant strains are biennials bearing flower stems as long as 45 cm, and suitable for hill stations. Among the Giant flame red strains, 'Giant Fire King' is outstanding. In the early flowering annual sections 'Persian Wallflowers', 'Early Wonders' are important. They flower several weeks earlier than the biennial.

Wallflowers are popular for spring bedding and cutting because of their wide colour range and sweet scent. A place near the bedroom window should be preferable for them. All the types, especially the early flowering ones, are also suitable for pot culture. The dwarf cultivar 'Tom Thumb' may be grown in window-boxes.

Cultural hints: The biennial varieties can be cultivated with success only in the hills or places such as Delhi having a long cool winter. In other places, annuals will do better. The biennials will take 6-8 months to flower, whereas the annuals flower within 3 to 3½ months. Sowing is done in September in nursery beds in the plains and February-March or July or September-October in the hills. They need protection from frost. In fact, they like a sunny situation and soil having sufficient lime.

Another biennial species, *Cheiranthus* × *allionii* (probably *Erysimum asperum*), is known as Siberian Wallflower and bears trusses of bright orange flowers. It can be grown in Delhi and other areas having long cool winter and in the hills. Sowing is done in September in the plains and in July-August in the hills. For a second crop, pinching back is done as soon as the first flowering is over.

Chrysanthemum (Annual); Common Names: Annual Chrysanthemum, Corn Marigold, Crown Daisy, Tricoloured Chrysanthemum; Family: *Compositae*; Origin: *Chrysanthemum segetum* – Britain; *Chrysanthemum coronarium* – South Europe; *Chrysanthemum carinatum* – Morocco.

Out of the three species, the Corn Marigold (*Chrysanthemum segetum*) is a weed in the cornfields in England. It is a well-branched plant, 30-60 cm tall, with flowers having shades of yellow and ivory-white. The important cultivars are 'Morning Star', 'Evening Star', and a beautiful cultivar 'Eastern Star'. It is not suitable for cultivation in humid eastern India.

The 'Crown Daisy' or 'Garland Chrysanthemum' (*Chrysanthemum coronarium*) is a taller (60-90 cm) and more vigorous plant. The flower shade is yellow and white. 'Golden Crown' is a double-flowered cultivar and 'Golden Glory' is the commonly grown single cultivar.

Chrysanthemum carinatum or Tricoloured Chrysanthemum is undoubtedly the best of the annual types. The flower colour is rich crimson, but a mixture of colours is available under the name 'Eclipse'. Some flowers are double. The plant grows to a height of 60 cm. The large flowers, about 7 cm across, show a bright coloured ring at the base of the petals and a dark coloured disc at the centre, thus producing a triple ring effect.

A hybrid, *Chrysanthemum* × *spectabilis*, has been developed from a cross between *Chrysanthemum carinatum* and *Chrysanthemum coronarium*. It is very vigorous growing, 90-125 cm tall, and produces double and single yellow, primrose, and white flowers.

Annual Chrysanthemum produces excellent cut flowers for table decoration. They are suitable for growing in beds, pots, and mixed or annual borders.

Cultural hints: It is easy to grow and all that is needed is an open sunny situation. In areas having a cool climate and scanty rainfall, sowing is done during June-July for rainy season flowering. For winter flowering, seed is sown during September. In the hills, seeds are sown in the spring. Seedlings can be transplanted or sown *in situ*. The spacing varies from 50 to 75 cm; the maximum is given for *Chrysanthemum carinatum* and the minimum for *Chrysanthemum segetum*. The plants flower in about 3½ months.

Cineraria: See *Senecio cruentus*.

Clarkia elegans; Common Names: Clarkia, Mountain Garland; Family: *Onagraceae* (*Oenotheraceae ?*); Origin: California (U.S.A.).

Clarkia is a hardy annual with slender but graceful branches and grows to a height of 60 to 90 cm. The single or double flowers are borne in the axils of the leaves and the petals are beautifully ruffled. The colour range has been greatly widened and the size increased by selection. Now, flowers of various colours such as white, salmon pink, light pink, scarlet, rose, and deep purple are available.

The long spikes of flowers are most attractive and the buds open in water, which makes them very useful for cutting. Clarkia is also useful for grouping in clumps, in borders and beddings. This can be grown as a pot plant.

Cultural hints: The plant does not like too much moisture and will do well in semi-shade. The seedlings are difficult to transplant and, hence, the seeds are sown *in situ* and then thinned out 20-30 cm apart. If done with proper care, the seedlings also can be transplanted. Sowing is done in September-October in the plains and in spring or in August-September in the hills. The spring-sown plants flower early during June-July while the autumn-sown plants flower in late May. Flowering commences 7-8 weeks after sowing but lasts long. It is also possible to grow them in 22 cm pots.

Clarkia pulchella is dwarfer (30-45 cm) and flowers for a longer time. Flowers are single or semi-double.

Cleome spinosa (Syn. *C. pungens*); Common Names: Spider Plant, Giant Spider-Flower; Family: *Capparidaceae*; Origin: Tropical America.

It is a tall plant (90-120 cm), with sharp spines all over, including the leaves. When bruised, the leaves emit a foul smell. The large clusters of four-petalled flowers have long prominent stamens and these give the plant its name. In spite of its drawbacks, the plant is worth growing. The flowers are scented and may be pink, light purple, rose, or white in colour. The prominent cultivars are 'Pink Queen' and 'Helen Campbell'.

The plant is grown in beds, borders, and also used as a background or a temporary hedge.

Cultural hints: For summer flowering, seeds are sown in January-February and for blooms in the rainy season sowing is done in June-July in the northern plains. In the hills, seeds are also sown in March-April for flowering in May-June or until late fall. The seeds should be soaked in water for three hours before sowing. The plants flower for several months during the summer season in the hills. The plant flowers in about three months after sowing of seed. Spacing allowed is 45-60 cm. The plants do well in a sunny situation and well-cutivated soil.

Clianthus dampieri (Syn. *C. formosus*); Common Names: Lobster's Claw, Parrot's Bill, Glory Pea; Family: *Leguminosae*; Origin: Australia.

Clianthus is one of the most showy winter annuals. The plants are branching and grow up to a height of 80 cm. The flowers are quite large, 6 to 8 cm across, and are borne in clusters of 4-6 flowers. The flowers are pendulous and dark crimson in colour with a raised black base which resembles the eyes. The flower itself has the shape of a parrot's bill or a lobster's claw and is very pretty in appearance.

Cultural hints: It thrives in a light, well-manured and well-drained soil and in a sunny situation. In excessive rainfall areas such as Calcutta, it fails to perform well. It grows exceedingly well under climatic conditions prevailing in Delhi. Before sowing, the hard seed coat may be punctured at one end. Sowing is done in September-October, preferably *in situ.* Seeds may also be sown in 7 cm nursery pots and the small seedlings are transplanted along with soil without disturbance. Plants need staking. It flowers in about 3½ months.

Collinsia bicolor (Syn. *C. heterophylla*); Common Name: Collin's flower; Family: *Scrophulariaceae*; Origin: California, U.S.A.

An easily-grown beautiful hardy annual of branching habit. Plants are slender and grow up to 40 cm. The flowers are borne in long spikes and are lilac, white, and salmon in colour. The flowers resemble Snapdragon. Suitable as a pot plant and woodland garden.

Cultural hints: Easily grown in any type of garden soil and does not need any special care. The plants stand light shade. Spring sowing is good in the hills, but autumn sowing is also possible in favourable situations. They are good for the drier parts of India. Sowing is done in the plains in September. Seeds are sown directly and thinned out 20 cm apart.

Coreopsis; Common Names: Tick-seed, Calliopsis; Family: *Compositae*; Origin: North America.

There are two commonly grown species, namely, *Coreopsis tinctoria* (Syn. *C. elegans*) and *Coreopsis drummondii* (Syn. *C. diversifolia*). In *Coreopsis tinctoria*, the flower colour is mainly yellow and crimson brown. In *C. tinctoria* var. *marmorata* the flowers are curiously marbled and striped. The plants may be tall (60-80 cm) or dwarf (20-30 cm). *C. drummondii* has yellow flowers with dark crimson centre. The plants are medium tall (45-60 cm). A tetraploid cultivar, 'Golden Crown', with large flowers is available. Two other species, namely, *C. stillmanni* and *C. maritima*, commonly included in the genus *Leptosyne*, are also grown in the garden and their culture is the same as for Calliopsis.

Coreopsis is good for beds, background borders, and cutting. Dwarf forms are good for edging.

Cultural hints: In the plains seeds may be sown thrice during January-February, June-July, and September-October. In the hills, sowing is done during spring and autumn. It flowers throughout the year except during severe cold. Seeds are sown in seed pans or nursery beds and transplanted after about a month. A spacing of 30-45 cm is allowed depending upon the variety. Flowering takes about four months.

Cosmos bipinnatus; Common Names: Cosmea, Cosmos, Mexican aster; Family: *Compositae*; Origin: Mexico.

A delightful annual, mainly grown in the rainy season. The plant is 50-120 cm tall and bears rose, pink, crimson, purple, lavender, and white flowers, with yellow centre, on long stalks. Flowers are large, 10-15 cm across, and may be single or double, including some tetraploids.

A yellow or orange flowering species, with a protruding yellow and black centre, is also grown in the garden and is known as *Cosmos sulphureus* or Klondyke. The plants may be dwarf or tall. It is good for cut flowers, borders, bedding, and also for growing in a mass in the midst of tall shrubs.

Cultural hints: Seeds are sown during January-February, May-June and September-October. In the hills, sowing is done during spring or autumn. Seedlings are transplanted or seeds are sown in permanent beds and thinned out 25-45 cm apart. It thrives in sunny or semi-shaded situations and takes about 2½ to 3 months to come to flowering.

Cuphea ignea (*C. platycentra*, *Parsonsia ignea*); Common Name: Cigar Plant; Family: *Lythraceae*; Origin: Mexico.

A dwarf (30 to 35 cm), bushy annual, having unusual-shaped scarlet flowers with a dark ring at the base resembling a cigar. *Cuphea miniata* (Syn. *C. llavea* var. *miniata*) is another species with open scarlet flowers, having purple bases and white mouth.

It is grown in beds or pots during the winter season. In the hills, it flowers in late summer or early autumn. A spacing of 20-30 cm is given. It flowers in about three months.

Cynoglossum amabile; Common Names: Hound's Tongue, Chinese Forget-me-not; Family: *Boraginaceae*; Origin: China.

A hardy perennial, about 30 cm tall, but is grown as an annual. The tiny blue flowers are borne in clusters and are sweet-scented. A white cultivar is also available.

Cultural hints: It is grown in beds, borders, edges, and in the rock-garden. It can also be grown in partial shade under shrubs or trees. The seeds are sown *in situ* and thinned out 30 cm apart. In the northern plains, it is grown as a winter annual and in the south it can be grown in the rainy season. In the hills, seeds are sown in spring or autumn. It takes about 3½ months to flower.

Dahlia: See Chapter XX.

Delphinium ajacis and *Delphinium consolida*; Common Name: Larkspur; Family: *Ranunculaceae*; Origin: South Europe.

The first species mentioned above is known as 'Rocket' or 'Hyacinth-flowered Larkspur', and the latter is known as 'Stock-flowered' or 'Imperial Larkspur' or 'Forking Larkspur'. The rocket type includes white, blue, mauve-purple, pink, and red shades. The plants grow up to 90 cm in height. A dwarf type (30 cm) is also available in this group. The stock-flowered types are usually tall, 90-150 cm, and branching.

Larkspur is very popular as cut flower and for growing as a background or in front of a tall hedge or a border and in pots.

Cultural hints: It is generally sown *in situ* in September-October in the hills and thinned out 20-30 cm apart. When grown for cut flowers, it is sown in rows 60 cm apart and 5-8 cm apart within the rows. This will force the plants to produce single long stems for cutting and more production can be obtained from a unit area. In Bangalore, sowing can be done during May-June for flowering in August-September. The plants may need staking, especially when grown in pots. It flowers in about four months.

The other two not so common species are *Delphinium paniculatum* (native of Balkan), commonly called 'Macedonian' or 'Siberian Larkspur' and *Delphinium grandiflorum* (*Delphinium sinense*), the 'Chinese Delphinium'. Both are dwarf.

Delphinium hybridum; Common Name: Perennial Delphinium; Family: *Ranunculaceae*; Origin: Garden Origin.

The perennial, sometimes biennial, *Delphiniums* are of great beauty in the garden. The modern Delphinium arose from crosses between *Delphinium elatum* and other species such as *Delphinium cardinale*, *Delphinium formosum*, *Delphinium belladonna*, etc. Perennial Delphinium thrives best in hill stations and can be grown as annual only in the northern plains in and around Delhi, with long cool winter. *Delphinium hybridum* is a robust, 150-180 cm tall plant, with flowers in several colours including pure white, deep bright blue, and sky-blue), which come true from seeds. The other flower colours are purple, lavender, lilac, and mauve. The most recent colours added are from the University Hybrids developed by Dr. Legro in Holland and these are red, scarlet, yellow and orange shades. *Delphinium belladonna* is a dwarf plant (90 cm) and bears blue flowers.

The best use of Delphinium is as cut flower. It is also grown in mixed or annual borders. The dwarf Chinese Delphinium (*Delphinium grandiflorum* or *Delphinium chinense*) can be used in pots or window boxes and rock gardens.

Cultural hints: The plant needs rich and well-drained soil and a sunny situation. It is propagated from seeds. In the hills, propagation is possible from division or cutting as it is a perennial there. Seeds take a long time to germinate. Sowing in the plains is done in late September or early October and flowering is in late March or early April. Due to hot weather at flowering time, the flowers suffer badly in the plains. If divisions or cuttings are procured from the hills and planted in July-September good flowering can be obtained. If seeds are sown in gentle heat in February or March in the hills, early flowering can be obtained in the following August-September. For autumn sowing, the plants should be given protective covering during winter. In spring, these are planted out in permanent position for flowering. A spacing of 25-60 cm is recommended, depending on the species.

Dianthus barbatus; Common Name: Sweet William; Family: *Caryophyllaceae*; Origin: Northern France.

This is a biennial plant but an annual strain is also available. Some of the early flowering biennials can also be grown as annuals. Sweet William is closely related to pink. The plant is about 30-45 cm tall with double or single fragrant flowers borne at the apex of the branches in rounded clusters. The flowers which are white, scarlet, dark red, and in different shades of pink, are self-coloured, edged, mottled, or with prominent white eyes.

Sweet William is suitable for beds, borders, and for cutting. The dwarf varieties can be used in the rock garden or window boxes.

Cultural hints: The biennials do best in the hills and in areas with long cool winter such as Delhi. The annuals can be grown anywhere in India.

The seeds are sown in September-October in the plains and transplanted after about a month. The annual types flower in about 3½ months after sowing of seed and the biennial types take 6-8 months. If sown under glass in February-March in the hills, the plants flower in late summer or September. A spacing of 15-25 cm is sufficient. A richly-manured soil and an open situation are preferable for its cultivation.

Dianthus caryophyllus (Carnation): See Chapter XX.

Dianthus chinensis (Syn. *D. sinensis*); Common Name: Indian, Chinese, or Japanese Pink; Family: *Caryophyllaceae*; Origin: East Asia.

The plants are 30-45 cm tall and bear brightly-coloured scented flowers. The flowers are self-coloured, exquisitely marked, edged or blotched. Some of them have petals beautifully fringed. There are cultivars in which blooms are borne in upright stems. The flower colour ranges from white to red, scarlet, pink, lilac, violet, maroon black, and mauve, The two common varieties in pink are *heddewigii*, 15-30 cm tall, known as 'Japanese Pink' or 'Rainbow Pink' (*D. chinensis heddewigii*) and the *laciniatus*, or 'Fringed Pink' (*D. chinensis laciniatus*). Single or double flowered cultivars are available in both the groups. In the hills, perennial species such as *D. plumarius* ('Cottage Pink') and *D. allwoodii*, which are hybrids of Carnations and Pinks with Sweet William, etc., can be grown.

The pinks are useful for bedding, borders, rock garden, edging, and pots, as well as for cut flowers.

Cultural hints: They need cool weather and a sunny position for their growth. The seeds are sown in September-October or even June-September in areas such as Bangalore. In the hills, sowing is done during spring or autumn for flowering in early spring. The seedlings are transplanted after about a month. The perennials can be propagated by pipings (which are actually cuttings, but instead of cutting with knife these are pulled from the parent plants), by division of old plants, or layering. Pinks take about 3½ months to flower. Spacing varies from 15 to 25 cm, depending on the variety. For pot growing, 25 cm pots may be used.

Digitalis purpurea; Common Name: Common Foxglove; Family: *Scrophulariaceae*; Origin: Great Britain.

Foxglove is a biennial, rarely perennial, but sometimes treated also as annual. It is grown in the hills in India. It is also possible to obtain blooms in places such as Delhi, but with difficulty. The plant, 60-150 cm tall, is also cultivated for the production of digitalin, a medicine. The flowers are bell-shaped, slightly drooping, and open at the mouth, which is sometimes spotted. Through natural variation (sport), selection, and crosses with some European species (*D. lutea*), several cultivars with various shades such as white, apricot, pink, crimson, etc., are available. The most important cultivar is 'Excelsor Mixed'.

Foxglove is excellent for woodland and semi-wild gardens. It can also be used as cut flower.

Cultural hints: A well-drained moist soil with abundant humus is good for its cultivation. Seeds are sown in May-June for flowering in the following year. The seedlings are transplanted in permanent beds afterwards. It is possible to transplant them even at the flowering stage. Sowing can be done in February-March also and a spacing of 25-40 cm is adequate. In the plains (Delhi, etc.) sowing is done in September.

Dimorphotheca sinuata (Syn. *D. aurantiaca C. calendulacea*); Common Names: Star of the Veldt, African Daisy, Namaqualand Daisy, Cape Daisy, and Cape Marigold; Family: *Compositae*; Origin: South Africa.

It will be interesting to trace the meaning of the name *Dimorphotheca* which is as follows: *Di*=two, *morphos*=form, *theca*=box or seed vessel; i.e. producing two different types of fruits or *achenes*. It is surprising that the disc florets produce the common flattened fruit, whereas the ray florets produce rod-like undulated fruits or seeds quite different from the former but producing identical plants.

The plants are about 30 to 60 cm tall, branching, and bearing daisy-like large flowers of various shades of colour such as white, bluish white, shades of yellow, salmon, rose, reddish-purple, etc. The plants are free-flowering and the flowers remain open during day time only. Even on cloudy days, the flowers remain closed.

Several hybrids have been evolved by crosses between different species such as *Dimorphotheca sinuata*, *D. ecklonis*, *D. chrysanthemifolia*, *D. calendulacea*, etc. There is a tetraploid variety named 'Goliath', with large orange flowers.

The plants provide beautiful cut flowers for daytime decoration, since flowers close at night. The flowers last long. The plants are also good for low edging along pathways, in beds, borders, rockeries, and pots.

Cultural hints: This plant requires a sunny situation and a well-drained rich soil. Seeds are sown in September-October and even earlier in July where rainfall is low. In Bangalore, *Dimorphotheca* can be grown for a greater part of the year. In cool greenhouses, flowers bloom for a long period. In the hills, sowing is done during March-April. The plants start blooming in about 6-8 weeks after sowing. A spacing of 25-40 cm is allowed. In Calcutta, the plants tend to spread more and hence need a little more spacing.

Echium plantagineum (Syn. *E. violaceum*); Common Names: Viper's Bugloss, Tower of Jewels; Family: *Boraginaceae*; Origin: Chiefly Mediterranean region.

In recent years, a lot of improvement has been made in *Echium*. It has both tall (60-90 cm) and dwarf (25-30 cm) cultivars. The small bell-shaped flowers have a wide range of soft pastel shades of blue, mauve, purple, rose-pink, and white.

It is useful for mixed borders, rock-gardens, and for edging. The taller varieties are also useful for semi-wild gardens. This can be grown in pots also.

Cultural hints: This plant has the added advantage of succeeding under diverse conditions such as cold, wet, windy, as well as in dry and sunny weather. Seeds are sown in spring, late summer (June-July), and autumn for obtaining flowers during summer, autumn, and winter respectively. Seeds can be sown directly or transplanted. In the hills, sowing is done during spring. A spacing of 25-30 cm is recommended.

Emilia flammea: See *Cacalia coccinea*.

Eschscholzia californica; Common Name: Californian Poppy; Family: *Papaveraceae*; Origin: California (U.S.A.).

In its natural home, the plant is a perennial but is grown in the garden as an annual. The plants are 30-45 cm tall, with blue-green, finely cut foliage and slender stem, somewhat resembling a carrot. Some varieties are dwarf, 20-30 cm tall. The flowers are large, about 10 cm in diameter, saucer-shaped, single, semi-double or double, and have shades of bright yellow, lemon, orange, crimson, cream, creamy-white. Some are bicoloured, - outer side bronze, inner side orange. The petals have a satiny *sheen* and the flowers open in bright sun only. The petals fall off on withering.

The plant can be grown as edging along walks, in borders, beds, pots, rock gardens, and old walls. For cutting, the flowers are cut at the bud stage and are immediately immersed in ice-cold water. It is the most popular and hardy annual.

Cultural hints: Light and sandy soil and a sunny situation are good for its growth. The seedlings do not stand transplanting well; therefore, seeds are sown directly in permanent beds and then thinned out 20-25 cm apart. Seeds are sown in September-October in the plains and even earlier (June-July) in areas such as Bangalore. In the hills, seeds are sown in March-April. Only acclimatized seeds give good results. The plants take about four months to flower.

Eschscholzia caespitosa (*E. tenuifolia*) and *E. maritima* are two other species which could be grown. (Sometimes spelt as *Eschscholtzia.*)

Euphorbia heterophylla (Syn. *Poinsettia heterophylla*); Common Names: Spurge, Annual Poinsettia or Mexican Fire Plant; Family: *Euphorbiaceae*; Origin: Southern States of the U.S.A.

It is a hardy annual growing to a height of 60 to 90 cm. The flowers, orange-red in colour, are borne in clusters at the end of branches. The leaves surrounding the flowers also get attractively, coloured during the hot season and resemble Poinsettia.

The bushy plants are excellent for growing in beds and pots during the hot months and the coloured leaves along with flowers are cut for decoration.

Cultural hints: The plants are hardy and require very little care, but should be grown in a sunny situation. Seeds are sown in March-April and the plants come into flowering in two months. A spacing of 30 cm is adequate.

Euphorbia marginata (*E. variegata, Lepadenia marginata*) is another species in which light green leaves are veined and margined with white. This is commonly known as 'Snow-on-the-mountain'.

Felicia bergeriana; Common Name: Kingfisher Daisy; Family: *Compositae*; Origin: South Africa.

A dwarf (15-20 cm), useful annual, bearing numerous small, bright metallic-blue flowers, resembling daisy. Flowers are apt to close in cloudy weather and at night. Plants are useful for rock gardens, for front position in borders and pots.

Cultural hints: A semi-shaded situation and a light soil are suitable. Seeds are broadcast in permanent beds in September-October in the plains and later thinned out 15-20 cm apart. March-April is the best time for sowing in the hills. The plant flowers in about three months. The other synonyms are *Cineraria bergeriana, Aster bergerianus.*

Gaillardia pulchella; Common Name: Blanket Flower; Family: *Compositae*; Origin: America.

A free-flowering annual which blooms in succession for quite some time. This is very easy to grow and most popular. The plants are medium, 40-45 cm tall. The flowers are large, 5-7 cm across, single or double and are available in many pretty colours and their combinations such as yellow, lemon, yellow-purple, cream, orange, maroon, etc.

The two most common varieties are *Gaillardia pulchella* var. *picta* and *G. pulchella* var. *lorenziana*. In the first group are included the single flowering types and in the latter the double flowers with quilled petals. It is very useful for bedding, borders and cuttings.

Cultural hints: It can be grown for a greater part of the year and the flowering season can further be extended by removing the faded flowers. Seeds are sown in the plains during February-March, May-June, and September-October for flowering in summer, the rainy season, and winter, respectively. In the hills, seeds are sown in March under glass and planted out in May for flowering throughout summer till autumn. The seedlings are transplanted about a month after sowing. The plants can tolerate heat and drought and come into flowering 3½ to 4½ months after sowing. Spacing allowed is 30 cm. The plants can also be grown in 25 cm pots. In Bangalore they can be grown all the year round.

Gazania splendens; Common Name: Treasure Flower; Family: *Compositae*; Origin: South Africa.

A dwarf (15-22 cm), trailing plant with narrow, long, pointed, silvery and hairy leaves. The flowers are in shades of red, pink, orange, yellow, and white and are daisy-like in shape. The ray petals have contrasting zones of colouring, sometimes marked conspicuously. The colours come true from seeds. The modern *Gazanias* have been developed by selection and hybridization. *Gazania* is grown in rock gardens, in front rows of borders, and for edging.

Cultural hints: The plants need a sunny situation. The flowers open in sunshine only and close by afternoon. Seeds are sown directly in permanent sites and are thinned out 30 cm apart. Sowing time is September-October in the plains and March-April in the hills. The plants flower in about 3 to 3½ months. They can also be propagated by cutting, layering, or division of sucker. They can be grown in pots in a mixture of loam, leaf-mould, and sand.

Gilia capitata; Common Names: Gilia, Ipomopsis; Family: *Polemoniaceae*; Origin: California (U.S.A.).

The plants are 45-60 cm tall with very finely-cut foliage and small (2.5 cm across), lavender-blue flowers.

The other two annual species are *Gilia* × *hybrida* (*Leptosiphon hybridus*) which is dwarf (7.5-15 cm), with a cluster of minute flowers having yellow, red, orange, or purple shades and *G. tricolor* (*G. nivalis*), commonly known as 'Bird's Eye', 30-45 cm tall, with dense flower-heads having white, pink, lavender, and white with golden centre flowers. *G. rubra* (*G. coronopifolia*) is a tall (90-120 cm) biennial plant with scarlet flowers. Gilia is suitable for growing in front rows in borders and rock gardens.

Cultural hints: Seeds are broadcast directly and thinned out 15-20 cm apart. Seeds are sown in September in the plains and March-April in the hills. These are grown in a sunny situation.

Godetia grandiflora and *G. amoena*; Common Names: Satin Flower, Farewell-to-Spring; Family: *Onagraceae* (some authors say *Oenotheraceae*); Origin: Western States of North America.

The garden *Godetias* have been developed from two American species, namely, *Godetia grandiflora*, a dwarf (30 cm) plant with large flowers produced in clusters, and *G. amoena*, a comparatively tall (45-70 cm) and slender plant producing flowers in large spikes. Tall bushy hybrids were produced by crossing the above species. The present *Godetias* may be categorized in the following groups:

(1) Dwarf or Grandiflora single (20-35 cm)
(2) Dwarf or Grandiflora double or Azalea flowered (37-45 cm)
(3) Tall single (60-75 cm)
(4) Tall double (60-75 cm)
(5) Lavender or Mauve coloured varieties.

The flower colour in the first four groups may vary from pure white to rose, deep red, crimson, orange-red, cattleya-blue, and blotched shades.

Godetia, owing to its long season of blooming and for the bright colour of its flowers, is ideal for bedding, rock gardens, window boxes, pots, and cut flowers, especially the tall double types. The flowers of grandiflora section are sweet-scented.

Cultural hints: The plants need a sunny but cool situation, and a well-drained soil for proper growth. The land should be cultivated thoroughly before planting. The seeds are sown in September-October in the plains and August to October and March-April in the hills. The seedlings are transplanted at the 4-leaf stage. For growing in pots in the hills, sowing is done early in October and potted in 20-25 cm pots. The pots are kept in sufficiently cool conditions and protected from frost. The plant flowers in about 3½ months after sowing. A spacing of 20-30 cm is recommended. The plants are treated as Clarkia.

Gomphrena globosa; Common Names: Globe Amaranth, Bachelor's Button, Everlasting;

Family: *Amaranthaceae*; Origin: India.

Gomphrena is a rugged-looking annual, and does not require any special attention. The plants are bushy in nature and may be tall (30-50 cm) or dwarf (up to 25 cm). The flowers are round and clover-like or button-like (1.5 to 2 cm) in shades of white, orange, magenta, purple, violet, rose, and pink, and are "everlasting", i.e., even the dried flowers look nice.

The plants are suitable for beds and borders and very useful as cut flowers. The dwarf varieties are used for edging or rock gardens. The dried flowers retain their colour and are useful for table decoration.

Cultural hints: A very hardy annual which can be grown in any type of soil and in hot weather. A sunny situation is required for best growth. Seeds are sown in February-March for flowering in the summer season and in May-June for flowering in the rainy season. The seedlings are transplanted 30-40 cm apart when 5 cm high. For indoor decoration, the plants are grown in 15 cm pots. The plants flower in about 2 to 2½ months after sowing.

Gypsophila elegans; Common Names: Baby's Breath, Chalk plant; Family: *Caryophyllaceae*; Origin: Caucasus.

A dwarf (30-45 cm), slender, lance-shaped, grey-leaved annual, bearing clusters of small (0.5 to 0.6 cm across), white or pink flowers.

The flowers look like sprays of mist and are extensively used for flower arrangements and bouquets. It is grown in beds, borders, rock gardens and for covering open spaces in beds of bulbous plants.

Cultural hints: A light soil having sufficient lime and a sunny situation are best for its growth. Seeds are sown *in situ* during September-October in the plains and August-October or March-April in the hills and the seedlings are thinned out 20-25 cm apart. The plants are short-lived. To obtain a succession of bloom, sowing should be repeated at 2-3 weeks' intervals. The plants flower in about 2½ months after sowing. As rabbits are fond of this plant, necessary protection should be given.

The perennial species, *Gypsophila paniculata*, with single or double flowers, and *G. repens*, a plant with trailing habit and small pink flowers, grow only in the hills.

Helianthus annuus; Common Name: Sunflower; Family: *Compositae*; Origin: Western U.S.A.

Helianthus is derived from the Greek words *helios* meaning the sun and *anthos* a flower, suggesting that the flowers follow the course of the sun.

The giant (1.2 to 3 m), common yellow sunflower is a coarse-looking plant with rough-textured leaves. The flowers are 20-30 cm across and are single or double. The double chrysanthemum-flowered variety is a tall plant, 1.8 to 2 m, with large flowers, 15-20 cm across. There are dwarf double cultivars like 'Sun-Gold' (90-120 cm) and 'Yellow Pygmy' (60 cm). Later, from a natural variation, a red-coloured cultivar, 'Sutton's Red', was developed in England. When this was crossed with other cultivars, several new forms of flowers with shades of chocolate, red, bronze, and wine-red were developed.

The Japanese Sunflower, or Miniature, with cucumber-like leaf, belongs to the species *Helianthus debilis* (*H. cucumerifolius*). The plants are 90-120 cm tall, branching out to a bushy form, and flower freely, bearing numerous small yellow, golden-yellow, or yellow with a dark eye flowers on long stalks. A cultivar with bronze-coloured flowers is also available.

Recently, a cross between the above and 'Red Sunflower' (*Helianthus annuus*) resulted in a new hybrid known as 'Sutton's Autumn Beauty' which is medium tall (120-150 cm). The colour range in this hybrid is wide such as pale primrose yellow, bronze, orange-yellow, and maroon. The flowers have long ray petals and many are brightly zoned. A white flowering cultivar, 'Italian White', is also available.

The common giant sunflower is not a very artistic plant, especially for the small gardens. But, sometimes, this may be useful as a screen or against a hedge or a wall. The miniature sunflowers are small-growing and are useful for cultivation in mixed borders and for cut flowers. The Russian types are economically important as the seeds yield edible oil.

Cultural hints: Sunflowers are easy to culture and not very fastidious regarding soil. They, however, need abundant light and air. Sunflower is a deep-rooted plant and hence the soil should be well-cultivated and manured. The plants also need staking as they grow tall. Seeds are either sown in permanent sites or transplanted. Seeds can be sown successfully from January to June for flowering during summer and the rainy season. This can also be grown in the cold season. In the hills, sowing is done during April. For the dwarf types, a spacing of 50-60 cm is allowed, whereas, the tall types should be spaced at 75 cm. The plants flower in 2-3 months.

Helichrysum bracteatum; Common Names: Strawflower, Everlasting Flower; Family: *Compositae*; Origin: Australia.

It is a commonly grown plant, 50-80 cm tall, bearing everlasting flowers of 7-10 cm diameter. The flowers last long and give a papery feeling when touched with hand. The plants flower without inhibition. The flowers exhibit a long range of colours from silvery-white to rich yellow, rich red, salmon-pink, and deep red. Many of the colours come true from seeds. A dwarf bushy cultivar is also available.

The dried flowers are used for winter decoration. Flowers are cut at half-open stage before the yellow centres are showing and are kept in hanging positions in a cool, airy place for drying. Both dried and fresh flowers are used for table decoration.

Cultural hints: It is sown in a warm sunny position and does not need any special attention or heavy manuring. Seeds are sown in October in the plains and transplanted after about a month. In the hills, sowing is done in March, if necessary under glass. A spacing of 30-40 cm is sufficient. The plant flowers 3½ months after sowing.

Heliotropium arborescens (Syn. *H. peruvianum*); Common Names: Cherry Pie, Heliotrope; Family: *Boraginaceae*; Origin: Peru.

Heliotropium corymbosum and *H. peruvianum* are the two species from which the present garden Heliotropes have arisen. Strictly speaking, Heliotrope is a perennial plant but can be grown as an annual, especially some modern strains. The plants are branching and compact, 30-45 cm tall. The small tubular flowers are borne in long trusses well above the foliage. The plants flower freely and the flower colour ranges from white to blue, purple and various shades of purple. The flowers are vanilla-scented.

Heliotrope is a greenhouse perennial and thus a favourable pot plant. It is sold in large numbers as pot plants in the U.S.A. and other countries and the sale ranks next only to geranium. It is used as a cut flower during the winter but the lasting quality is uncertain. This is suitable for annual borders and beds also. It is good for window gardens.

Cultural hints: Any light soil, which need not be very rich initially, is suitable for its cultivation. The plants need a sunny situation. Seeds are sown in September-October and after about six weeks these are transplanted in permanent beds. In the hills, seeds should be sown in March-April. Plants can also be raised from stem cuttings and layering. For growing in pots, final repotting is done in 25 cm pots. The cuttings root easily at 60°F. Heliotropes may be grown in the same pot for several years by using top-dressing and weak liquid manure. At around freezing point, heliotrope is likely to be damaged and so should be shifted to a glass greenhouse. Heliotrope can be grown as standard by removing the side branches up to a height of 60 cm. The plants are stopped (pinched) for bushy growth. Spacing allowed for the annual type is 45 cm. The plants flower in about 3½ to 4 months.

Helipterum roseum: See *Acroclinium roseum.*

Iberis umbellata **and** ***Iberis amara*****; Common Name: Candytuft; Family:** ***Cruciferae*****; Origin: Europe, including South Britain.**

The common name Candytuft originated from the fact that one of the species was brought from Candia and the flowers appear in tufts. The popular Candytuft (*Iberis umbellata*) is about 15-30 cm tall and small flowers are borne in broad clusters, known as Umbels, with a wide range of colours including white, pink, carmine-rose, purple, and lilac. A dwarf cultivar, 15-20 cm tall, is available in which the plants remain completely covered with flowers.

Iberis amara is taller (35-45 cm) than the above and bears 15 cm long spikes of white flowers which resemble hyacinth.

The plants are useful for cut flowers, for massing in beds, in annual borders, and edging along paths. The hyacinth types are suitable for growing in pots.

Cultural hints: The plants do well in a light, well-drained soil containing lime and a sunny situation. The coloured cultivars may not do well in places having mild winter. The seeds are sown in September-October in the plains and transplanted after about a month. The seeds may be sown directly also in permanent sites, thinned out 15-22 cm apart. In places with moderate climate such as Bangalore, sowing can be done in June also. In the hills, seeds are sown in spring. For obtaining continuous blooms, sowing may be done at fortnightly intervals. The plants flower in about three months after sowing.

Impatiens balsamina**; Common Names: Balsam, Touch-me-not,** ***Gulmendi*** **(Hindi),** ***Dopati*** **(Bengali); Family:** ***Balsaminaceae*****; Origin: India, Malaysia, China.**

The name is derived from the Latin word *impatiens* (impatient), an allusion to the behaviour of the pods which, when ripe, burst open on a slight pressure, scattering the seeds. The plants are 22-60 cm tall, bearing Camellia-like double or single flowers, with varying colours such as rose, pink, scarlet, red, violet, salmon-pink, purplish-red, lilac, blue, mauve, white, etc. In some cultivars the flowers are spotted, blotched, or variegated.

Balsam is best suited for growing in beds and pots, especially the dwarf ones. This is also grown along paths and borders.

Cultural hints: It makes a showy, summer and rainy season flowering plant. It requires a warm sunny situation and rich, moist soil. A heavy dose of manure and plenty of watering during growth are good for its cultivation. The seeds are sown in January-February for summer flowering and May-July for flowering in the rainy season. In the hills, sowing is done in March-April. In places such as Bangalore, flowering can be obtained throughout the year. Seeds are shaded till germination and transplanting is done at the 6-leaf stage. Seeds may be sown *in situ* also and then thinned out 22-30 cm apart. But when the plants are allowed to grow bushy, a wider spacing of 40-50 cm may be allowed. Overcrowded plants do not produce good blooms. About three shoots are allowed to grow on each plant. For obtaining large blooms all side shoots are removed, keeping only a single stem. The lower leaves tend to cover the blooms and, hence, these should be removed. In dry localities, the leaves wither and drop off. Balsam is a gross feeder and should be given liquid manure once a week. The plants grow well under partial shade. For pot culture, single plants are grown in 25 cm pots. Flowers start appearing 60 days after sowing and the blooming period lasts for about 15 days.

The two perennial species, namely, *Impatiens holstii* (Patience Plant) and *I. sultanii*, which are the natives of Zanzibar, and some of their hybrids can also be cultivated as annuals. They do well in semi-shade and thrive well inside the rooms. The plants are pinched for bushy growth. The culture is the same as that of Balsam.

Kochia scoparia var. *trichophila* (Syn. *K. trichophylla*); Common Names: Summer Cypress, Mock Cypress, Burning Bush, Belvedere; Family: *Chenopodiaceae*; Origin: South France, Eastward to Japan.

The name *Kochia* is after W.D.J. Koch, a Professor of Botany at Erlangen. It is a dense bushy symmetrical, oval or round annual foliage plant, 60-90 cm tall and 30 cm across. The whole plant is covered with finely-cut, narrow pea-green ornamental foliage. On maturity, the plant turns into rich coppery-red and at this time the red flowers also appear giving the plant the look of a burning bush.

Kochia is ideal for growing in pots, for planting along paths, and can be used as temporary hedge, edge, or screen.

Cultural hints: A rich well-manured and well-drained soil and a sunny position are best for its growth. Seeds are sown during February to June. In the hills, sowing is done in March-April. A spacing of 30 cm is allowed between the plants. The maximum size is attained in 3-4 months after sowing. The plants can be clipped to various forms. For pot culture, only one plant is grown in a 25 cm pot.

Lathyrus odoratus; Common Name: Sweet Pea; Family: *Leguminosae*; Origin: Sicily and South Italy.

Sweet pea is valued and widely cultivated for its range of flower colours, elegant form, fragrance, and utility as a cut-flower. Recently, sweet pea has ranked among the first leading commercial cut flowers in the U.S.A. From its ancient home in Sicily, the seeds were first brought to England for cultivation in 1699.

Sweet pea is very much like the common garden pea in growth habit and may creep to a height of 2-3 m in good soil.

Types: Numerous cultivars of sweet pea have been developed in recent years with flower colours ranging from white to almost any colour except yellow and deep blue. These cultivars have developed as a result of natural mutation, chance crossing, and hybridization effected by man. Henry Eckford of England and Luther Burbank of America were pioneer scientists in this field. There are now six distinct strains of sweet peas, which are listed below:

(1) The Giant Frilled or Spencer
(2) The Early-flowering (winter) frilled
(3) The Cuthbertson (Spring-flowering) frilled
(4) The Multiflora
(5) The Dwarf Cupid
(6) The Dwarf Cupid Giant Frilled (Little Sweetheart).

In the plains of India, only the early flowering strains do well while the others adapt and perform well on the hills.

Uses: The best use of sweet peas is undoubtedly as cut flowers. Their wide range of colouring, which come true from seeds, sweet scent and long duration of flowering make them suit every taste. Sweet peas are commonly used as temporary screens to hide unsightly objects or to separate the kitchen garden from the flower garden or alongside a path. They are good for growing on trellises. Sweet peas can also be grown in clumps of 5 or 6 plants in mixed borders. The dwarf cupid cultivars, especially the new Little Sweetheart, are useful for edgings. Sweet peas can also be grown thickly in window boxes and are allowed to droop.

Sowing of seeds: In northern India, seeds are sown in September and this should not be delayed beyond the third week of October if good results are desired. In areas having less rainfall, sowing can be undertaken in August. In the plains, near the coastline, sweet peas do not perform well. To obtain early flowering in spring in the hills, seeds are sown during August to October. In times of severe frost, plants need protection. For late summer or early winter flowering, in the hills, the seeds are sown during March-April. In areas where winter is severe, seedlings are raised in pots in cold frames and planted out in congenial weather during March for flowering. For growing as a hedge or a screen, seeds are sown *in situ*, 5-7 cm apart and 2-5 cm deep. It is advisable to sow seeds in double rows spaced 30-40 cm apart and thinned out 15-22 cm apart in rows after germination. When grown in clumps, the seeds should be sown in circles at a distance of 15-22 cm and not more than 5 or 6 plants should be grown at each clump. Seedlings can also be raised in 7.5 cm pots under glass or cold frame and these may be transplanted in permanent sites when large enough and when the weather is favourable. The wrinkled seeds generally give rise to bluish, lavender, and mauve coloured cultivars and the scarlet, pink, and rose coloured types are obtained from smooth and plump seeds. Acclimatized seeds give better performance than imported seeds.

Cultural hints: The first essential requirement of sweet pea culture is an open sunny location. Plants grown in excessive shade become weak and spindly and produce few flowers. Any garden soil which is well drained will be all right for sweet pea cultivation provided it receives manure in ample quantities. Sweet pea is a deep-rooted plant and the success of its culture will depend largely on the preparation of the soil. The soil is dug to a depth of 60 to 75 cm and some sand is added if the soil is heavy. Sweet pea is partial to lime and a dusting of lime (30 g per metre row), followed by watering, should be given at the time of land preparation and once a month afterwards. This is necessary because in acid soil a legume such as sweet pea will not thrive well. Moreover, lime makes clayey soil open and porous. As the plants flower best in full sunshine, the rows should be arranged from north to south, so that both sides of the rows get full sun. It is better to make slightly raised ridges, when the plants grow, to prevent rotting by over-watering.

It should be remembered, while watering sweet peas, that mere driblets that damp the surface to a depth of only 2-3 cm is rather harmful than good. A thorough soaking once a week or once in 10 days, depending on the weather, should be given. At the same time, over-watering which often results in shedding of flowers should be avoided. The soil should be stirred with a fork in between two irrigations and in hot weather a mulch with straw or lawn clipping will help conserve soil moisture.

As stated earlier, sweet peas need heavy manuring. At the time of soil preparation, add one part of well-rotten cowdung manure to two parts of soil. A little quantity of bone-meal should also be incorporated in the soil. There is a controversy whether fertilizer should be applied as a basic dose or in the form of liquid manure. Some experts suggest that chemical fertilizers are generally not required. S. Percy Lancaster has advised the application of 2 kg of superphosphate, 2 kg of bone-meal, and ¼ kg of sulphate of potash for each 20 running metres of trench. The inclusion of potash makes the plants sturdy and nullifies the adverse effects of an overdose of nitrogen, thus preventing the plants from being prone to diseases. The same author also suggests the application of liquid manure by dissolving 30 g of ammonium sulphate, 30 g of superphosphate, and 15 g of sulphate of potash in 20 litres of water and about 250 ml of this solution is applied once a week when the buds start to appear. Alternatively, an application of well-diluted liquid manure prepared by dissolving cowdung (the dissolved liquid to be fermented for 3 to 4 days before application) in water, once a week or a fortnight after the buds appear, is also beneficial.

As the plant attains a height of 15 cm the main shoot is pinched to make the plant bushy. However, pinching is not necessary all the time. The training on Cordon System is necessary to obtain large exhibition blooms. In the Cordon System, a single main stem, sometimes two, is retained on each plant and all the side shoots that appear from leaf axils and tendrils are removed, keeping only the leaves. This will force the sap into the flower. In this system, the plants are not allowed to flower before they attain a height of about 90 cm. This is a tedious and time-consuming method but the result will be seen in the production of large-sized blooms, though fewer in number. But when the aim is to have a mass effect for garden display, this method need not be followed.

When the plants are 30 cm tall, they need staking. This is done by fixing green-painted thin bamboo poles (about 120 cm long) in the ground along the rows, 150-180 cm apart, and strong threads or thin wires are stretched in parallel lines 45 cm apart and fixed with the poles. If one can afford, instead of bamboo poles, galvanized pipes may be used and chicken wire netting or thin gauze wire may be fixed to these pipes. When grown in clumps, the plants are supported by split-bamboo sticks or dried pigeon-pea, cotton or jute sticks.

The plants start flowering in February, from seeds sown in September-October, and continues till March or the middle of April. The plants should not be allowed to seed as this shortens flowering time. The faded flowers and seed pods should be removed regularly to encourage more flowering.

Pests and Diseases: See Chapter XXVI.

Limonium sinuatum; Common Names: Statice, Sea Lavender; Family: *Plumbaginaceae*;

Origin: Mediterranean region.

Previously, the genus *Limonium* was also known as *Statice*, but due to some ambiguous use by the botanists the latter generic name is not commonly used now.

The species is actually a biennial but is generally grown as an annual. The plants grow to a height of 45 to 60 cm and the stems are hardy and angular. The foliage is grey-green and deeply cut. The flower stalk is winged and the calyx is toothed and straw-like. The flower colours available are white, rose, mauve, lavender, and dark blue. The flowers retain their shades for a long time when dried and are used for dry decoration as 'Everlasting' flowers.

Another species, with large yellow flowers, which is a native of North Africa, resembles the above, and is known as *Limonium bonduellii*.

The other popular species, a native of Turkestan, is known as *Limonium suworowii*. The plants are 45-60 cm tall, with a rosette of long narrow leaves. The beautiful, showy, bright rosy-pink, branching flower scape (or spike) rises 20-30 cm above the foliage.

The statice is good for edging and cutting. The dried flowers can be used for decoration for a long time ('Everlasting'). The species *Limonium suworowii* is a very good pot plant and when closely grouped in beds or borders provides a beautiful effect. The spikes are very popular as cut flower.

Cultural hints: A cool climate, plentiful sunlight, and a well-drained soil are good for its cultivation. The seedlings are difficult to transplant because of their taproots. However, if transplanted at an early stage with care, they will get well-established. Sowing *in situ* is done during September-October in the plains and March-April in the hills. A spacing of 45 cm for *Limonium suworowii* and 30 cm for *L. sinuatum* is recommended. The plants flower in 3-4 months after sowing. Pots of 25-30 cm are used for pot cultivation.

Linaria maroccana; Common Names: Toad Flax, Linaria; Family: *Scrophulariaceae*; Origin: Spain, Portugal and Morocco.

A commonly grown annual, with small pretty flowers. The plants are 20-35 cm tall, bushy, having erect branches. The leaves resemble Flax. The mini antirrhinum-like flowers are produced in abundance. The flowers are white, yellow, pink, purple, lavender, rose, blue, mauve, bicoloured, having blotches of different colours on the lip.

The plants are suitable for growing in pots, edgings, borders, beds, and rock gardens. They are useful for covering open spaces, especially in beds of bulbous plants. The flowers are used for cutting also.

Cultural hints: The seeds are sown directly in permanent beds. The fine seeds should be mixed with fine-grained sand before sowing and then covered with a thin layer of sieved soil or leaf-mould and watered carefully after sowing. After germination, seedlings are thinned out 15-25 cm apart. The sowing is done during September-October and March-April in the hills. In places such as Bangalore, sowing may be done earlier during June-July. The plants like a sunny situation. To obtain a bushy growth, the main shoot should be pinched. The plants flower in about three months after sowing of seeds.

Another species, *Linaria alpina*, bearing violet-blue flowers with an orange patch on the lower lips, is also grown as an annual.

Linum grandiflorum var. *rubrum*; Common Name: Scarlet Flax; Family: *Linaceae*; Origin: North Africa and Europe.

The plants are 30-60 cm tall with narrow pointed foliage. The scarlet flowers with dark centres are 2.5 to 3.5 cm across and are borne in loose panicle. The variety *rubrum* is the best among the varieties. *Linum usitatissimum* or the 'Common Flax', 40-60 cm tall, with light blue flowers, is also cultivated as an ornamental. *L. perenne*, a perennial species with azure-blue flowers (var. *album* is white flowered), is suited for the hills.

This showy annual can be used in annual borders or beds for a fine display.

Cultural hints: A light, well-drained soil and a sunny situation are the factors required for its best performance. Seeds are sown directly in permanent beds in September-October and later thinned out 15-20 cm apart. In places such as Bangalore, sowing can be advanced to June. In the hills, seeds are sown during either August to October or March to April. The plants flower in about 3½ months. The plants benefit by feeding with liquid manure once a fortnight when buds start to appear.

Lobelia erinus; Common Name: Lobelia; Family: *Campanulaceae*; Origin: South Africa.

A low-growing, 15-30 cm tall, exceedingly beautiful and profusely flowering plant of great garden value. The small flowers (1.9 cm) are white, different shades of blue, crimson, and purple in colour, with a yellow or white eye. The plant may be either compact or trailing in habit. It is a perennial but can be cultivated as an annual.

If sufficient moisture is available, the plant flowers for a long time and is thus invaluable as a bedding plant. It is also good for pots, borders, rock gardens, and edgings. The trailing types are useful for hanging baskets and window boxes. Lobelia forms a good combination with blue ageratum, pink sweet alyssum, blue petunia, and others.

Cultural hints: Lobelia does not perform well and flowers very late in the plains except in some parts of northern India. It thrives best in the hills. A well-cultivated, rich, and moist soil is best for successful cultivation of lobelia. The fine seeds are mixed with sand before sowing and lightly covered with sieved leaf-mould or soil. The best time for sowing is September-October in the plains and March-April in the hills. The seedlings are transplanted when 2.5 cm tall. The plant can be grown in partial shade also. Pots of 25 cm are good for pot culture. Potted plants need staking. The plants come to flowering in 3½ months.

Lupinus; Common Name: Lupin; Family: *Leguminosae*; Origin: America and Southern Europe.

A genus comprising several annual and biennial ornamentals, commonly grown in the garden. The most important annual species is the *Lupinus hartwegii*, a native of Mexico. The plants are 45-75 cm tall, with foliage deeply-lobed, and the long flower spikes, growing well above the foliage, bear numerous small pea-like flowers of white, pink, and pale blue colours. The other annual species are: (1) *L. luteus* (40-60 cm) which bears sweet-scented yellow flowers, (2) *L. subcarnosus*, commonly called 'Texas Blue Bonnet', a dwarf plant (20-30 cm) with blue flowers, (3) *L. mutabilis*, a tall plant (120-150 cm), which bears dark blue (marked with yellow and rose), blue, white, and pink fragrant flowers, and (4) *L. tricolor* (*L. hybridus*) which is a bushy plant, 45-60 cm tall, bearing white flowers which change to pink with age.

The most common perennial species is *Lupinus polyphyllus*, the flowers of which have a broad range of colours.

Lupins are best suited for cutting, beds, borders, and pots. The dwarf varieties (*Lupinus subcarnosus*) are suitable for rock gardens.

Cultural hints: A well-drained, well-prepared moist soil and partial shade are good for the cultivation of lupins. Seeds should be soaked in water for a few hours before sowing. The best sowing time is October in the plains or even earlier in June in places such as Bangalore. The perennials are sown from April to June in the hills. The annuals are sown in March-April and August-October. The perennials are also propagated by division or cuttings.

Matricaria eximia (Syn. *Chrysanthemum parthenium*); Common Name: Feverfew; Family: *Compositae*; Origin: North Temperate Regions.

Though commonly called *Matricaria,* it is known botanically as *Chrysanthemum parthenium.* The plant is a perennial, 30 cm tall, but, when raised from seed can be grown as an annual. The flat heads of small double flowers may be ivory white or yellow in colour. The other common species is *Matricaria maritima* (*M. inodora, Chrysanthemum inodorum*) which bears double white flowers.

Matricaria is good for bedding, edging and pot culture. *M. maritima* is also useful for cutting.

Cultural hints: The plants thrive well in a richly-manured soil and in a sunny situation. They thrive best on the hills or places having long cool winter such as Delhi, Punjab, etc. Seeds are sown in September-October in the plains and in March-April in the hills. The plants are spaced out 15-20 cm apart.

Matthiola incana; Common Names: Stock, Gilliflower; Family: *Cruciferae*; Origin: Europe, including Britain.

Originally spelled as *Mathiola*, it is a genus of annual, biennial, and perennial flowers. The small, rosetted and rounded single or fully double, scented flowers are borne on long spikes. There is a wide range of flower colours, from white to crimson, dark red, pink, rose, mauve, lavender, blue, etc.

The stock throws single as well as double flowers. Though single flowers are also used as cut flowers, only the doubles are preferred. The character doubleness is a genetic factor. Only single plants produce seeds and only 50 per cent of the seedlings can be expected to produce double flowers. The recent race, Hansen's, developed in Denmark, can be made to produce 100 per cent doubles. The character doubleness has been linked to the colour of the cotyledon. The light green cotyledon seedlings are the doubles and the dark green ones are the singles. To obtain only double flowers, the dark green cotyledon seedlings should be uprooted. The difference in cotyledon colours is noticed only when the seedlings are raised under cool conditions.

The stocks may be non-branching or column and branching. The other types are: Wall-flower-leaved stock known as 'All the Year Round', which has smooth green leaves with white flowers, and East Lothian or Intermediate (dwarf, 30-40 cm), also known as 'Autumn Flowering Stock' has a wide range of flower colours. The biennial Bromptom Stock (30-45 cm) is a branching plant and bears double flowers with various shades of colours.

Stock is an excellent flower for cutting, especially the column type. The life of the cut flower can be prolonged by bruising the cut ends. They are also grown in beds, borders, and pots.

Cultural hints: All stocks prefer a rich, well-cultivated soil, with proper drainage and a sunny location. Seeds are sown in September-October in the plains and August to October (biennials) or in March and April (early flowering types) in the hills. Early sowing in June-

July is possible under such climatic conditions as of Bangalore. The seedlings are transplanted when 5 cm tall and spaced out 20-30 cm apart. Seedlings suffer from damping off; therefore, great care is to be taken in watering. The annuals take 3-4 months and the biennial 7-8 months to come to flowering. For potting, 20-25 cm pots are used.

Matthiola bicornis is a night flowering, single, lilac-coloured and fragrant stock.

Mesembryanthemum criniflorum (Syn. *Dorotheanthus bellidiformis*); Common Names: Livingstone Daisy, Fig Marigold, Mesembryanthemum; Family: *Aizoaceae*; Origin: South Africa.

A handy, spreading, 10 cm tall annual, bearing daisy-like flowers, 2.5 to 3.5 cm across, with a wide range of colours including white marked with crimson, pink or buff at the edges of petals or self-coloured types in shades of yellow, crimson, apricot, orange, red, white, etc.

The plants are useful for edging or growing in the front rows of borders, under standard roses, and in rock gardens, dry walls, hanging baskets, and pots.

Cultural hints: The plants need a light, well-drained soil and a sunny situation as the flowers close in cloudy weather and also at night. They thrive better in a cool dry place. The seedlings are sown directly in permanent beds during September-October in the plains and are later thinned out 15 cm apart. In the hills, sowing is done during March-April. It flowers in about three months after sowing.

Mimulus tigrinus; Common Names: Monkey Flower; Family: *Scrophulariaceae*; Origin: Western parts of South and North America.

The term *mimulus* is derived from the word *mimo*, meaning ape, an allusion to the form of the corolla. Most of the garden forms have originated from *Mimulus luteus* and its varieties and *M. guttatus* or *M. cupreus*. Actually, the large-flowered, bright-coloured crosses of *M. luteus* and its varieties and *M. guttatus* are offered as *M. tigrinus*. Although they are perennial, they could be conveniently grown as annuals. The plants are 30-45 cm tall and moderately branched. The flowers with two lips are gloxinia-like, open-mouthed, hence known as monkey flower. Flowers are 2.5 to 5 cm across with many colours including crimson, pink, yellow and other intermediate shades, and spotted and blotched, generally with brown and maroon.

Cultural hints: Mimulus is good for bedding and pots. The plants need a moist semi-shaded position for proper growth. The seeds are sown in September-October in the plains and in March-April in the hills. The minute seeds are mixed with sand before sowing. The plants thrive well in places having a cool climate. They are planted 15-25 cm apart. For pot culture, 20-25 cm pots are used. The plants flower in 3 to 3½ months after sowing. *Mimulus cupreus* is also cultivated in the garden.

Mirabilis jalapa; Common Names: Marvel of Peru, Four O'Clock; Family: *Nyctaginaceae*; Origin: Tropical America.

A tall (60-90 cm), bushy perennial, though generally grown as an annual. The flowers are fragrant, five-lobed, and 2.5 to 5 cm across, with colours of red, yellow, or white, and often striped. The plants flower in the afternoon or in cloudy weather.

The plants are grown in the shrubbery or formed into a hedge (30 cm apart) at the rear of the flower garden. This plant can also be grown in the semi-shaded corners of the garden.

Cultural hints: Any ordinary well-drained garden soil, preferably loam, and a semi-shaded or sunny location are needed for its cultivation. In the plains, sowing can be done almost in any part of the year, except early winter. The seeds are sown *in situ*, and later thinned out 45 cm apart. March-April is the best time for sowing in the hills. Sometimes, the tuberous roots can be lifted and stored like dahlia for future planting.

Molucella (Syn. *Moluccella*) *laevis*; Common Names: Bells of Ireland, Molucca Balm, Shell-Flower; Family: *Labiatae*; Origin: Mediterranean region.

It is not commonly grown in our gardens but is quite attractive and grows to 60-90 cm in height. The small white flowers, which are inconspicuous, are produced in successive whorls in the axils of the leaves. It is the enlarged calyces which are decorative and persist even when the plants dry up and the flowers drop. This is useful for table decoration.

Cultural hints: Plants are grown in beds or pots. The best performance is obtained in a moist, well-drained, rich soil and a sunny situation. The plants benefit by an occasional addition of cowdung or leaf-mould. Seeds are sown in September-October in the plains and in March-April in the hills. The seedlings are transplanted after about a month, 25-30 cm apart. The plants flower in about four months.

Myosotis alpestris; Common Name: Forget-me-not; Family: *Boraginaceae*; Origin: Europe, including Britain.

A perennial plant but is generally grown as an annual. Both dwarf and tall types are available. The tiny flowers are borne in clusters and are white, sky-blue, dark blue, or pink in colour, with a small yellowish eye, and fragrant in the afternoon.

They can be grouped with pansies and tulips or grown into attractive masses by themselves. It is grown in drifts of wild gardens or as cover in bulb beds, rock gardens and borders.

Cultural hints: The plants need a long cool winter and can be grown only in the northern plains and hills. A well-manured soil and a semi-shaded situation are best for its growth. The seeds are sown in September-October in the northern plains and in March-April in the hills and later transplanted 15-20 cm apart. The plant can be grown in seed pans. The plant flowers after about six months. The perennials are propagated by cutting or division.

Myosotis sylvatica, with dark blue flowers is also cultivated. But in the garden this is often known as *M. alpestris* and *M. dissitiflora*. But the true Forget-me-not is *M. scorpioides*.

Nemesia strumosa; Common Name: Nemesia; Family: *Scrophulariaceae*; Origin: South Africa.

The garden Nemesia has originated from *Nemesia strumosa* and *N. versicolor*. The flowers are white, rich blue, primrose, yellow, orange, scarlet, or crimson in colour and are marked with contrasting colours.

The large-flowered varieties are good for cutting and last as long as 2-3 weeks when kept in low bowls and cut at bud stage. They are also good for bedding and borders. The dwarf types can be used with advantage in rock gardens, edgings, and window boxes.

Cultural hints: An open sunny situation and a moist light soil are good for the cultivation of Nemesia. Nemesias need long cool winter to thrive. The seeds are either sown in permanent places and later thinned out 20-25 cm apart or are transplanted. The time of sowing is September-October in the plains and March-April in the hills. The plants flower in 3½ months after sowing. The plants are pinched back for a bushy growth.

Nemophila menziesii insignis; Common Names: Nemophila, Baby Blue-Eyes, Californian Blue-Bell; Family: *Hydrophyllaceae*; Origin: California (U.S.A.).

A slender growing and spreading plant, 15-20 cm tall. The plants get covered with numerous small bell-shaped and sky-blue flowers, with a white eye. There are some cultivars having deep purple, pure white, and spotted flowers.

The plants are useful for edgings, uncovered patches in annual borders, rock gardens, and bedding. They are also suitable for pot culture in cool glass greenhouse.

Cultural hints: A light well-manured and moist soil and light shade are good for its successful cultivation. Seeds are sown in September-October in the plains and later transplanted. In the hills, seeds are sown during March-April and under favourable conditions during August-October also. The plants are spaced out 20 cm apart. Their growth is quite fast and the plants flower in three months.

Nicotiana alata grandiflora; Common Names: Tobacco Plant, Flowering Tobacco; Family: *Solanaceae*; Origin: Tropical America.

Strictly speaking, it is a perennial species, but is best grown as an annual. The plants are 60-90 cm tall, branching, with large, coarse, and hairy foliage. The flowers are large tubular or trumpet-shaped and white, blotched with various shades such as violet, lavender, yellow, crimson, pale green, maroon, etc., and are sweet-scented. The blooms ordinarily open in the evening but a modern cultivar, a hybrid between *Nicotiana alata* and *N. langsdorffii*, with smaller flowers, opens during the day. The variety *N. alata grandiflora* is also known as *N. affinis*.

Nicotiana suaveolens is an annual strain from Australia, with slender stems and bearing charming, small, greenish-purple and exceedingly sweet-scented flowers which open at night.

Nicotiana is good for bedding, borders, and also for growing in groups in the shrubbery. *N. suaveolens* is particularly good for pots.

Cultural hints: A richly-manured, well-drained light soil, containing sufficient lime, is good for the cultivation of this plant. This can be grown in full sun or in partial shade under trees. In the latter case, the flowers remain open during day time also. The seeds are sown in September-October in the plains and in March-April in the hills. It can be grown in hot weather also. In places such as Bangalore, flowers can be obtained at any time of the year. The seedlings are transplanted at the 6-leaf stage. A spacing of 30 cm is allowed between the plants. Flowering commences within 3 to 3½ months after sowing. For pot culture 25 cm pots may be used.

Nierembergia caerulea (Syn. *N. hippomanica*); Common Name: Cup flower; Family: *Solanaceae*; Origin: South America.

A dwarf (15-20 cm) perennial plant, but grown as an annual. The small (2-3 cm across) petunia-like or cup-shaped, flattish flowers are violet-blue in colour with a yellow eye and are produced in abundance.

The plants are particularly useful for sunny patches in rock gardens, in the front row of annual borders and in pots.

Cultural hints: The plants need a sunny position and a light, well-drained soil. Seeds are sown during August-October in the plains and in March-April in the hills and are transplanted after about 30 days, 20 cm apart. The flowering starts in about 3½ months.

Nigella damascena; Common Names: Love-in-a-Mist, Devil-in-the-bush, Fennel Flower; Family: *Ranunculaceae*; Origin: Europe and North America.

A pretty, upright, 30-45 cm tall plant, with finely-cut foliage and attractive semi-double white, deep blue or rose coloured flowers which remain surrounded by the leaves. The best cultivar is 'Miss Jekyll' with sky-blue flowers.

The plants are excellent for cut flowers and for growing in pots. They can be grown alone or among other tall-growing annuals. The persistent stigmas in the seed pods resemble horns and are used for decoration.

Cultural hints: Nigella is grown in any moderately good garden soil containing little lime and an open situation. Seeds are sown directly in permanent beds, as they do not transplant well, and thinned out 20-30 cm apart. The time of sowing is August to October in the plains and March to April in the hills. The flowering starts in about 3-3½ months. For pot culture, 20 cm pots may be used.

Nigella hispanica, a species from Spain, is also grown in the gardens.

Oenothera; Common Name: Evening Primrose; Family: *Onagraceae*: Origin: Temperate regions of North and South America.

The *Oenothera biennis*, the common evening Primrose, is 60-120 cm tall. It is a biennial but can be grown as an annual also. The flowers are yellow, tubular in shape, up to 15 cm across, and open in the evening. The other biennials which are treated as annuals are *O. drummondii* (30 cm), *O. odorata* (60-90 cm), which is fragrant, *O. trichocalyx* (30-45 cm), and *O. historta*. The last one is possibly annual.

Oenothera is suited for beds, borders, edgings, and rockeries.

Cultural hints: A well-drained soil and an open situation are good for its growth. The flowers remain closed during the day and open during evening. If grown in semi-shade, flowers open during the day also. They do well in places having a cool winter or at elevations above 500 metres. Seeds are sown *in situ* in September-October or during June-July in places such as Bangalore. Sowing is done during March-April in the hills. The seedlings are transplanted 25-30 cm apart. The biennials can also be propagated by division or cuttings. The plants take 3-3½ months to flower.

Papaver rhoeas; Common Names: Shirley Poppy, Corn Poppy; Family: *Papaveraceae*; Origin: Europe, including Britain.

The poppies are among the most popular flowers in cultivation. *Papaver rhoeas* is about 60 cm tall and bears cup-shaped single or double flowers with frilled silken petals. The flower colour may be white, all shades of pink including salmon, pale mauve, blue and dark maroon, all with a white base, and many have attractive picotee edging. The other commonly grown species are: *P. glaucum* or Tulip poppy, which is a spreading annual from Asia Minor, bearing red flowers; *P. somniferum* or the Opium poppy which is a tall (60-90 cm) plant, having blue-green shiny foliage and attractive large double white, pink, scarlet with white, carnation- (or paeony-) like flowers. The Iceland poppy (*P. nudicaule*) grows to a height of 35-45 cm, bearing flowers with a wide range of colours, thrives well in the hills and is a short-lived perennial. The other perennial species, *P. orientale* (or Oriental poppy), is 60-90 cm tall and bears scarlet, crimson-red, pink, and orange-scarlet flowers.

The poppy is useful as a cut flower or for growing in beds, borders, or drifts. The flowers last long if the cut ends of stems are dipped in boiling water (avoid the steam reaching the flowers) or charred in flame. Only the half-open flowers are cut for decoration. Dried seed heads of *Papaver somniferum* are good for winter decoration.

Cultural hints: Poppies are very easy to grow and are very hardy, and can withstand severe frost. An open situation and a well-drained soil are best suited for their growth. Seeds are sown directly in permanent beds, as seedlings do not transplant well, and later thinned out 25-30 cm apart in September-October in the plains. In the hills, sowing is done either in August-October or in March-April. Transplanting also is possible with a little care. The seed pods should be removed as soon as they are formed to have a continuity in bloom. The flowers do not last long in the field as the petals are easily blown off. Poppies take about 2½-3 months to come to flowering after sowing.

Penstemon; Common Name: Beard Tongue; Family: *Scrophulariaceae*; Origin: North America.

Penstemon (spelled as *Pentstemon* by Bailey, 1963) is a perennial but very often grown as an annual. The best form suitable for borders has been raised by a cross between *Penstemon hartwegii* and *P. cobaea*. The two perennial species are *P. barbatus* (Syn. *Chelone barbata*) and *P. diffusus*. The hybrids are branching and free-flowering and grow up to 45-60 cm in height. The plants come true from seeds and are covered with open-mouthed, tubular Gloxinia-like blooms borne on long spikes and are available in white, shades of pink, purple, and crimson, and many are edged with contrasting shades.

Penstemon is suitable for beds and borders. In Bangalore, these are also grown as potted plants.

Cultural hints: A well-drained and well-manured soil and an open position are best for the cultivation of *Penstemon*. The plants thrive best in the hills where they can be treated as perennial also. The seeds are sown in September-October in the plains with a long cool winter and in March-April in the hills. In Bangalore, sowing can be advanced to June-July. The seedlings are first transferred singly to 7.5 cm pots and later shifted to 25 cm pots, with a good potting mixture. The dried flower spikes are removed to allow the secondary growths to come up, which in their turn will come into bloom. On completion of flowering, the flowering shoot is cut back to 5-6 nodes from the ground level when new growth will be produced; out of these, only 6-8 shoots are retained and treated the same way until another flush of flowering is obtained. A top dressing of two parts of cowdung manure or horse manure and one part of soil is given every three months. Repotting may be done once in six months. The seedlings flower in about four months, but may take much longer to flower in a mild climate as that of Bangalore. The plants can be propagated by division and cuttings, and these flower earlier than seedlings in Bangalore. When grown in the ground, a spacing of 30 cm is allowed.

Petunia hybrida; Common Name: Petunia; Family: *Solanaceae*; Origin: South America.

The modern petunias are the result of crosses between two South American species, namely, *Petunia integrifolia* and *P. nyctaginiflora*. From these hybrids many cultivars having diverse growth habits, size, and colour of flowers have been isolated for cultivation. As a result of all these efforts, the petunias of today are the most popular, loveliest, and a valuable race of garden plants. The height of the plant may vary between 20-30 cm and 30-45 cm, depending upon the type. The flowers are trumpet-shaped and may be small or large, up to 15-17 cm across at the widest point. The flowers are single or double and the petals may be frilled and ruffled in various ways and may be veined or marked beautifully. The flower colours available are pink and red, pale blue, violet, yellow, cream, purple, and mauve, and some cultivars are bicoloured. Petunia is a free-flowering plant.

The petunia is classified into the following types:

(1) *Large-flowered-grandifloras:*
   (a) *Smooth-edged:* Petals are smooth-edged and both self-coloured and bicoloured cultivars are available.
   (b) *Frilled or Fringed:* Petals are ruffled, slightly or more pronouncedly, and frilled. The plants are either tall or dwarf. The 'Giants of California' and 'Leviathan' cultivars have petals fully ruffled and in the latter the petals are veined and marked. Some extremely frilled cultivars in this group give the appearance of double flowers.

(2) *Compact Single or Bedding:* As the name suggests, they are good for bedding and borders and dwarf in habit. A wide range of self-coloured cultivars are available.
(3) *F_1 hybrids:* These include Multiflora and Grandiflora F_1 hybrids, and are available in various self-coloured and bicoloured (starred) cultivars.
(4) *F_2 hybrids:* Grandiflora F_1 or Multiflora F_2 are included here.
(5) *Balcony and Pendula:* These are slightly taller-growing, spreading, and with small single flowers.
(6) *All Double:* These were first developed in Japan but are now available in many other countries, including India. All are F_1 hybrids with smooth-edged and ruffled fully double flowers, either self-coloured or bicoloured.

The double flowering types do not produce seeds and the F_1 singles also degenerate after one or two sowings and hence to obtain quality flowers, seeds are to be purchased each time from reputed dealers or nurserymen.

The single bedding petunia is one of the best and most valuable bedding ornamental flowers, as this has a long flowering season under favourable conditions. It flowers best as the hot weather approaches and can be continued till the rains in the hills. Petunias are also good for borders, mass plantings, edgings, rock gardens (especially the dwarf types), window boxes, and hanging baskets (balcony type), and pots. Petunia is also good for vase decoration as cut flowers and when grown in mass under the shrubs they display a striking effect.

Cultural hints: Petunias grow well in very rich soil (too rich for other flowers). Some of the bedding types grow well even in the poorest soil with plenty of moisture. The seeds are sown in August-October in the plains or even during March-July in places having a mild summer and less rainfall. In the hills, sowing is done during March-April or even early in February under glass, if necessary. The seeds are very small and need extra care for sowing. These are bulked with fine sand during sowing. Watering of the seedlings should be done very judiciously with a fine rose can. The very small seedlings are the likely ones that will produce the choicest blooms and double flowers. Generally, the weak ones produce full doubles. The strongest seedlings tend to make weedy plants. Petunias, especially the doubles, are also propagated from cuttings or slips. The seedlings make slow growth in short-days. For obtaining a bushy growth, seedlings are pinched when they are about 15 cm tall. A spacing of 25-30 cm is allowed each way and for very fertile soil it may be increased to 40-45 cm. For pot culture, 15-25 cm pots are suitable. While watering, too much of splashing of water on the leaves is to be avoided. An occasional fortnightly feeding with liquid manure, after the buds appear, is helpful. The faded flowers should be removed from time to time to prolong the blooming period. Sometimes, the bedding types make straggling growth when these should be cut back. A similar practice may be followed when the flowering is almost complete as the cut plants will throw new growth which in turn will flower again. If the potted plants are pinched once or twice during early stages, no staking is needed. Otherwise, 3-4 slender bamboo stakes, painted green, are inserted along the periphery of the pots and jute threads (*sutli*) are tied around them to help the plants to get support. Petunias flower in 3-3½ months after sowing.

Disease: See Chapter XXVI.

Phacelia campanularia; Common Name: Phacelia; Family: *Hydrophyllaceae*; Origin: California (U.S.A.).

A hardy annual, 20-25 cm tall, having attractive grey toothed foliage. it bears clusters of small (2.5 cm across), bell-shaped, deep gentian-blue flowers with conspicuous anthers which are primrose coloured. The bees are very fond of these flowers. Another species, 40-75 cm tall, with lavender-blue flowers and curled flower-head, is *Phacelia tanacetifolia*, which is often grown for attracting bees. *P. viscida* (Syn. *Eutoca viscida*) is 30-60 cm tall, with green foliage and blue flowers.

It is grown in rock gardens, pots, and used for edging. It flowers during winter in the plains. Seeds are sown in spring in the hills. It prefers a light soil and a sunny position.

Phlox drummondii; Common Name: Annual Phlox; Family: *Polemoniaceae*; Origin: Texas, New Mexico (U.S.A.).

One of the most popular and favourite annuals grown for its brilliant display and long lasting blooming period. It is easy to cultivate, free-flowering and ranges in height from 15 to 45 cm. The flowers are delicately scented, with a wide range of colours such as pure white. several shades of pink, vermilion, crimson, violet and purple, and many are with contrasting "eyes".

There is a variety, *cuspidata* (Syn. *P. cuspidata*) or var. *stellaris*, the Star Phlox, with fringed flowers and pointed petals, giving the flowers a star-like shape. The other well-known varieties are *grandiflora*, with large flowers, and *nana compacta*, which is a dwarf and compact plant. In recent years, some tetraploid cultivars, with large flowers and strong stems, have been introduced. Two such types are Tetra Red and Giant Tetra Mixed.

The large-flowered phloxes are excellent for beds, borders, and as cut flowers. The compact ones are grown in the window boxes, pots, formal beds, and also in edgings. Phlox is also suitable for rock gardens and dry walls (especially the dwarf types).

Cultural hints: Phlox grows well in a calcareous loamy soil, rich in humus, and a sunny position. But the plant can be grown successfully under partial shade also. A cool weather during the seedling stage and full sun, when coming to flowering, are the ideal conditions needed for its success. Seeds are sown in seed pans or nursery beds during August-October in the plains and transplanted after about a month. In places having a mild climate, such as Bangalore, sowing can be started from June-July. In the hills, sowing is done in March-April. In areas having severe cold, seeds are sown in August-September for pot culture. The seedlings are kept in 12 cm pots singly in the glasshouse shelf during winter for flowering during April-May. Seeds can be sown directly in permanent beds also and this results in early flowering. Watering sufficient enough to keep the soil moist is given. The seedlings are transplanted 15-30 cm apart. In the northern plains and in the hills where winter is long and cool, phlox grows luxuriantly, whereas in the southern plains plants are less luxuriant. For pot culture, 25 cm pots are used and 1-2 seedlings are sufficient per pot in the hills or the northern plains. In the southern plains, about five seedlings may be grown per pot. The seedlings are pinched to form a bushy growth. It also checks untimely flowering. The plants flower within 3-3½ months after sowing.

Disease and Pest: See Chapter XXVI.

Pimpinella monoica; Common Name: Lady's lace; Family: *Umbelliferae*; Origin: India.

A tall (100-150 cm) annual with deeply-cut foliage bearing numerous white flowers in large spray on a long stem. The umbels resemble the carrot or the fennel.

The plant is grown in the borders and sometimes in the shrubbery. The cut flowers are used extensively for preparing bouquets and for table decoration.

Cultural hints: A sunny situation or partial shade and a moist soil are good for its growth. It is a winter annual for the plains and a summer annual for the hills. The plants are spaced 30 cm apart. Flowering starts within 3½ months after sowing.

Portulaca grandiflora; Common Names: Portulaca, Rose Moss, Sun Plant; Family: *Portulaceae* (genus named as *Portulacaceae* by Bailey, 1963); Origin: South America.

This is a low-growing (10-15 cm) annual with a trailing habit. The leaves are cylindrical and fleshy. The flower colours are brilliant and available in shades of white, purple, sulphur-yellow, lavender-pink, salmon, scarlet, crimson, and orange. There are cultivars having striped petals of contrasting colours. The flowers may be single or double. The flowers open late in the morning, at about 9 a.m., and close in the afternoon, around 3 p.m., and also on cloudy days. There is a perennial portulaca also with deep-rose coloured double flowers.

Portulaca is widely used as a ground cover. This is a good plant for the rock garden, for edging along paths, and for growing in shallow pots or containers, such as seed pans and bowls. This is also suitable for carpet bedding.

Cultural hints: A light well-drained soil and a sunny position are necessary for the proper growth of this plant. The soil should not be allowed to go dry completely or become too damp. The plants are hardy and need very little attention. The seeds are sown directly or transplanted. The best sowing time is March to August in the plains and March-April in the hills. Portulaca is a summer or a rainy season plant and does not do well in areas having severe cold. This can be grown throughout the year in places having a mild climate such as Bangalore or Poona. When grown in shallow pots, the surface should be lightly covered with crocks or broken bricks or stone-chips to prevent the leaves and stems from rotting by coming in contact with the damp soil. The same can be followed in beds. Portulaca is planted closely for a mass effect. The plants flower in 3-3½ months after sowing of seeds.

Reseda odorata; Common Name: Mignonette; Family: *Resedaceae*; Origin: North Africa, Egypt.

Mignonette, a great favourite in the western countries mainly because of the unique fragrance of its flowers, is valued for cutting. In India it is not grown widely. The plants are 30-40 cm tall and bear small flowers of bronze, white, red, crimson, and yellow shades.

The cut flowers last long and emit an agreeable perfume. It is also an excellent pot plant and is suitable for beds, shrubbery, or near the bedroom windows.

Cultural hints: Mignonette prefers a rich heavy soil and a cool moist situation. Hot sunshine is detrimental to its growth. The plants are delicate and resent transplanting. The seeds are sown *in situ* in August-October in the plains and in March-April or July-October in the hills and thinned out 15-25 cm apart. It is possible to raise plants from cuttings. As soon as the flower spikes start to appear, all side shoots around the top of the stem are removed. About 2-3 healthy side shoots from the bottom are allowed to grow for a second flush of blooms. As the plants grow, wire-ring supports are provided around each plant. The plants flower in 3-4 months. One or more plants may be grown in a 25 cm pot.

Rudbeckia bicolor; Common Name: Cone Flower; Family: *Compositae*; Origin: North America.

It is a very attractive annual, 30-60 cm tall, bearing large (about 5 cm across) flowers resembling daisies. The flowers are borne on long stalks and are of yellow shade having a mahogany-crimson central zone and a dark brown central disc.

The plants are ideal for cut flowers, beds and borders.

Cultural hints: It is easy to cultivate in any type of soil and situation, from semi-shade to full sun. Though inhabitants of moist locations, they thrive well under ordinary garden conditions. Seeds are sown during August-September in the plains and in March-April in the hills. Plants can be raised by division of clumps. After blooming is over, plants should be cut back severely to have a second crop. Plants are grown 50 cm apart and flowering starts 4-5 months after sowing.

Rudbeckia hirta, Black-eyed Susan, is a biennial species but can be grown as an annual. The plants are vigorous and bear flower stalks as long as 60-90 cm and flowers as large as 7-12 cm across. The flower colours include orange, bronze, nearer rich crimson, lemon-yellow and some others, with beautiful contrasting zones of colours. In this, tetraploid cultivars having giant single (12-17.5 cm across) and double flowers are also available. *R. hirta* thrives well even in the driest and hottest places. The other cultural requirements are the same as above.

Salpiglossis sinuata; Common Names: Painted Tongue, Velvet Flower; Family: *Solanaceae*; Origin: Chile.

A tall-growing (60-90 cm) beautiful annual, bearing petunia-like tubular flowers, white, crimson, shades of pink, gold, golden-yellow, chamois, scarlet and gold, blue and gold, rich violet, etc., in colour. The plant flowers profusely. The garden cultivar *S. superbissima* is columnar in the habit of growth with thick unbranched stem.

The flowers are good for cutting, bedding, and pots. They form a contrasting background against some annuals such as petunia, nemesia, and linaria.

Cultural hints: The plants of this species grow well in a rich, well-drained soil and in an open sunny situation. They cannot stand strong winds. Seeds are sown directly in permanent beds in September-October in the plains and in August-September or March-April in the hills. Transplanting of seedlings is possible under proper care. The seedlings are thinned out 30 cm apart. *Salpiglossis* does not do well unless the winter season is long and cool. A

25 cm pot can accommodate three plants. The plants come into flowering in about 4½ months after sowing.

Salvia; Common Name: Sage; Family: *Labiatae*; Origin: Mexico, California, Brazil.

There are five species of *Salvia* which are grown as annuals:

(a) *Salvia splendens* (Scarlet Sage): The plants may be tall (60-90 cm) or dwarf (20-30 cm) and bushy. The long terminal spikes with their tubular bright scarlet flowers are borne above the foliage. There are cultivars with salmon, pink, and violet flowers. The Scarlet Sage can be grown in beds or in a patch against green grass or a background of trees. In var. *splendens*, the corolla, as well as the calyx, are equally bright scarlet.

(b) *Salvia horminum:* Often known as Blue Beard, it is a native of South Europe and grows 30-45 cm tall, with small purple blue, pink, or white flowers. A popular cultivar is 'Blue Beard'.

(c) *Salvia farinacea:* It is a perennial species but is grown as an annual. The plants grow to a height of 60-90 cm with shining green leaves, bearing long spikes of lavender-blue flowers. There are two varieties, namely, var. *alba*, the white flowering one, and var. Blue Beddar, with deeper blue flowers.

(d) *Salvia coccinea:* It is an annual (sometimes perennial) Salvia which is very popular in the south. The plants are 30-60 cm tall, and sometimes up to 70 cm. The flower spike is long, erect, and bears small crimson, scarlet flowers. White and pink-flowered types are also available.

(e) *Salvia patens:* This is another perennial species but is best grown as an annual. The plant is 30-75 cm tall, bearing long spikes (45 cm) of bright blue flowers. The var. Cambridge Blue has pale blue flowers, whereas var. *alba* has white flowers.

Salvia is ideal for growing in mass, in beds, borders, pots, as background and in rock gardens (dwarf types). This can be grown in the shrubbery, under trees and semi-shaded places. Sometimes they are used as house plants also but the flowers drop off quickly inside the room.

Cultural hints: Salvias like a rich soil and in pot they may be grown in a rich mixture of loam and manure in the proportion of 1 : 1. Too much of nitrogen fertilization will produce leafy growth and delay flowering. *Salvia splendens* and *S. farinacea* thrive best in the plains and the other species grow well in the hills or at medium elevations. Salvia cannot stand frost at all and in northern India they should be grown in sheltered places (for example, under a large tree) protected from north and west. Seeds are sown in August-September in the plains and in March-April and September-October in the hills. In the hills, seeds raised in October should be protected against frost. In Bangalore, and other places with a similar climate, sowing may be started from June onwards. The fine seeds need good care during sowing. The germination is slow and takes 10-15 days. The seedlings are transplanted when the first two true leaves develop. For pot growing, seedlings are first planted in 7.5 cm pots and then shifted to 15 cm and finally to 30 cm pots. Application of liquid manure, once a week, is recommended. A spacing of 30-60 cm is recommended in the ground. After the first flush of bloom is through, the plants are cut back and a top dressing of manure is given. After some time, new growths will appear for a second flush of flowering. This practice is useful in places with a moderate climate, such as Bangalore and Poona. The plants like liberal watering. Salvia flowers in 3½-4 months after sowing. Salvias can also be raised by cuttings or division of clumps, but it is best to raise them from seeds, when grown as annuals.

Saponaria calabrica; Common Name: Soapwort; Family: *Caryophyllaceae*; Origin: Europe.

It is a dwarf annual (15-30 cm), bearing in abundance sprays of starry pink or white (var. *alba*) flowers. The flowers somewhat resemble *Gypsophila*. Another species much used for cutting is *Saponaria vaccaria* (Syn. *Vaccaria vulgaris*) which bears sprays of pink (var. *rosea*) or white (var. *alba*) flowers and grows to a height of 60-90 cm.

Saponaria is grown in beds, as background, in rock gardens and as edging. The flowers are good for cutting.

Cultural hints: The plants need a well-drained rich soil and a sunny location. The seeds are sown directly in permanent beds or transplanted. The best time of sowing is September-October in the plains and either September or March-April in the hills. Seedlings are thinned out 30-45 cm apart. *Saponaria* flowers in about three months.

Scabiosa atropurpurea; Common Names: Pincushion Flower, Sweet Scabious, Mournful Widow, Mourning Bride, Egyptian Rose; Family: *Dipsaceae*; Origin: Southern Europe.

A colourful branching annual, growing to a height of 60-90 cm. There are dwarf growing (30 cm) varieties (var. *nana*) also. The plants bear handsome globose or rounded heads of flowers borne on long stiff stalks (45 cm) and are available in beautiful shades of pink, pale mauve, blue, scarlet, dark maroon, and white. The light-coloured stamens protrude out of the flower and create a pincushion appearance.

Scabious should be grown in clumps in mixed borders and in beds and pots. The flowers are excellent for cutting. When cut at almost open stage and the cut ends burnt, the flowers will last for several days in water. Scabious looks well in combination with dianthus (pinks), phlox and gypsophila.

Cultural hints: The plants need a light, well-drained soil and a sunny location for their proper growth. It grows well only in places where winter is long and cool. The seeds are sown in September in the plains and in autumn or spring in the hills. The seedlings are transplanted 30 cm apart. To prevent the flower-heads from bending or falling down, the plants may be planted closely. The seedlings should be pinched when about 15-20 cm tall. Scabious flowers late and takes about 4-4½ months for flowering.

Schizanthus × wisetonensis (Syn. *S. hybridus*); Common Names: Butterfly Flower, Poor Man's Orchid; Family: *Solanaceae*; Origin: Chile, Peru.

The modern garden varieties of *Schizanthus* have been derived from crosses between two species, namely, *Schizanthus grahamii × S. pinnatus*. The name comes from two Greek words, *schizo* (to cut) and *anthos* (flower) referring to the deeply cut petals of the flower. The plants are vigorous, branching, and grow up to 90 cm or more. The colour range of the flower is wide and varying and mostly blotched, veined, and spotted exquisitely in contrasting colours, thus earning the name of Butterfly flower. The small flowers are borne in abundance, measuring 2.5 cm across, resembling orchids. The flower colours include various shades of pink, yellow, apricot, carmine, salmon, mauve, and purple, all in numerous combinations and markings. There are cultivars known as 'Cattleya Orchid' and 'Pansy-Flowered' resembling the respective flowers.

It is excellent as cut flower and for pots. The cut flowers last long and open up to the last bud in the spike. The dwarf cultivars are good for rockeries.

Cultural hints: The plants thrive well in light, well-drained, moist soil and sunny or semi-shaded situations. The plant is at its best in hill stations and areas having a long cool winter. The seeds are sown in September-October in the plains and in March-April in the hills. Acclimatized seeds give better results. The seedlings are transplanted at 4- to 6-leaf stage. The young plants are pinched once or twice for bushy growth. As the stems are delicate, staking is needed. The plants benefit from feeding with liquid manure. A spacing of 30-45 cm is recommended. Plants take 4-5 months to come to flowering. For pot culture, the plants are finally transplanted in 20 cm pots after pricking.

Senecio cruentus; Common Names: Cineraria, Florists' Cineraria; Family: *Compositae*; Origin: Canary Island.

The modern garden Cineraria is developed from *Senecio cruentus* (*Cineraria cruenta*) and is one of the best shade-loving annuals. The large dark-green heart-shaped leaves are most attractive. Plants grow to a height of 30-60 cm and the daisy-like brightly coloured flowers are borne in clusters and may be self-coloured or with contrasting centres of white or prominently marked or circled with various shades. Flower colours are white, scarlet, red, pink, purple, lavender, and blue. The flowers are grouped in various categories. The Large Flowered Single, Stellata or Star Singles (flowers star-like), Intermediate (similar to Large Flowered but with smaller flowers), Feltham Beauty Strains having white centres and Multiflora Nana or Dwarf (compact plants).

Cineraria is exceedingly suitable for growing in pots and beds, either in shade or in semi-shaded position.

Cultural hints: A light, well-drained and well-manured soil and a semi-shaded situation or even full shade are recommended for the successful cultivation of this plant. Cineraria cannot be grown with good success in places where winter is not cool and long. It is damaged by frost, especially at a young stage. Cineraria is raised as an annual from seeds, though it is possible to multiply by division of old plants and root or stem cuttings. Seeds are sown in early September or even in August to obtain early flowering. In the hills, sowing is done in March-April. Seeds are minute and special care should be taken while sowing. The seedlings are pricked in 7.5 cm pots when they develop the first rough leaf. Later, these are transplanted in 25 cm pots, with a soil mixture consisting of two parts of good loam soil, one part leaf-mould or well-rotten cowdung/horse manure, half part fine sand, and a handful of wood ash. The plants should be shaded from strong sunlight and sheltered from wind. Careful frequent watering and feeding with liquid manure (weak cowdung solution) are recommended. After the first flowering is over Cineraria can be cut back to about 2.5 cm from the ground level and top dressed with soil and manure to obtain a second flush of flowering. This can be practised only in the hill stations.

Senecio elegans (*Jacobaea elegans* or *J. erecta*) is an annual species. It is tall (45-60 cm) and often known as Jacobaea. Some cultivars in this species are double flowered. The cultivation is the same as above.

Disease and Pests: See Chapter XXVI.

Silene pendula; Common Name: Catchfly; Family: *Caryophyllaceae*; Origin: South Africa and Southern Europe.

A bushy dwarf annual (15-22 cm), bearing single or double flowers in sparse clusters and in colours ranging from white to rose, pink, purple, crimson, and salmon.

Silene can be utilized as a ground work for wallflower and bulbs as well as for bedding.

Silene armeria is another annual having blue-green foliage and bright rose-pink flowers, growing to a height of 30-35 cm.

Cultural hints: The seedlings are sown in September-October and March-April in the plains and hills, respectively, and later transplanted in permanent beds. *Silene* needs a long cool winter for proper growth and flowering. The plants flower in 3-3½ months after sowing of seeds. *Silene oculata*, commonly known as 'Viscaria' (*Viscaria aspera*, *V. oculata*), has a wide range of flower colours including pure white, pale blue, rich delphinium blue, pink carmine, and red. The plants may be tall (30-35 cm) or dwarf (15 cm). Viscarias make good potted plants.

Tagetes (Marigold): See Chapter XX.

Tithonia rotundifolia (Syn. *T. speciosa*); Common Name: Mexican Sunflower; Family: *Compositae*; Origin: Mexico.

A free-flowering tall-growing (120-180 cm) plant with rough, large, heart-shaped foliage and large single flowers (7-10 cm across), resembling single dahlia. The flower colour is orange with a tufted yellow centre. There is another dwarf variety (120 cm) having large orange-red flowers.

Tithonia flowers are borne on long stalks and are useful for cutting. Unless the flowers are cut at the right stage, when almost open, the flowers drop. The stalks are brittle and should be handled carefully. It is suitable for growing as a hedge for screening and as background.

Cultural hints: The plants thrive well in a richly-manured light soil and a sunny situation. The plants are hardy and easy to grow. The seeds are sown from March-June to get flowers during the monsoon. It is also possible to sow seeds during August-September. In the hills, seeds are sown in March-April. Later, the seedlings are transplanted 45-75 cm apart. The plants flower in about three months. Seeds can be sown *in situ* also.

Torenia fournieri; Common Name: Wish-bone Flower; Family: *Scrophulariaceae*; Origin: Tropics, including India and Africa.

Some species of *Torenia* are of Indian origin. But the above species is the most commonly grown annual in the garden. This is an erect, much-branched, glabrous annual growing to a height of 30 cm. The flowers, looking somewhat like antirrhinum in form, are violet and pale blue in colour with a blotch of golden-yellow in the throat. There is a cultivar, 'White Wings', having ivory-coloured flowers tinted with rose.

Torenia is very decorative and useful for borders, beds, pots, rock gardens, hanging baskets, and window boxes. It is also suitable for edging and can be grown in drifts or below large trees to create an informal effect.

Cultural hints: The plants need a lot of water and humidity for their growth. The plants should be grown in a well-drained and richly-manured soil and in a shaded position. The seeds are sown either directly in permanent places or in nursery beds or seed pans and later transplanted 20 cm apart. The best time for sowing is March-April in the hills. In places such as Bangalore, seeds can be sown any time of the year. For pot culture, it is advisable to use 25 cm pots. The seedlings should be pinched 2-3 times to obtain a compact and bushy growth. The plants flower in about three months after sowing, and continue to flower for 6-8 weeks.

Trachymene caerulea (Syn. *Didiscus caeruleus*); Common Names: Lace Flower, Blue Lace Flower; Family: *Umbelliferae*; Origin: Western Australia.

The plants are 45-60 cm tall, with finely divided leaves and small (5-7 cm across) lavender-blue flowers which are borne in umbrella-like clusters on long stalks. There are other types with white and pink flowers, which are not so common.

It is a beautiful plant for cut flowers, pots, and mixed borders.

Cultural hints: The plants can be grown in any well-drained soil and a sunny situation. The seedlings do not stand transplanting well and should be sown directly in permanent beds and thinned out. Seeds are sown in Spetember-October in the plains and in March-April in the hills. The plants bloom in about three months after sowing.

Tropaeolum majus; Common Names: Garden Nasturtium, Indian Cress; Family: *Tropaeolaceae*; Origin: Mexico, South America.

The garden Nasturtium (*Tropaeolum majus*) was developed by selection of natural variations and as a result of crosses involving the species *T. minus* and *T. peltophorum* (*T. lobbianum*). *T. majus* has two variations, namely, var. *flore-pleno* having double flowers and var. *nanum*, known as Tom Thumb nasturtium which is dwarf and non-climbing. The four main garden types of nasturtium are: (a) Tall Climbing Singles which trail or climb to a height of 180-250 cm, (b) Dwarf Single (25-30 cm), (c) Semi-Double Gleam, which are semi-climbing in nature and the flowers are scented delicately, and (d) the Dwarf Double. The nasturtiums have wiry stems, large attractive rounded leaves, and spurred flowers, 6-7 cm across. The flower colours in each section include orange, primrose, salmon, orangered, scarlet, and deep mahogany. Many flowers are marked and spotted superbly with contrasting colours. There are many cultivars in each group, but mention may be made of two, viz., 'Cherry Rose' and 'Jewel Mixed' where the flowers show up above the leaves unlike the other cultivars in which the flowers remain covered by the foliage.

The garden nasturtium is most valuable as it can be used in various ways. The tall nasturtium, with the help of the long stalks of the leaves twisting round the support, can be made to climb or trail stumps of trees through hedges, fences, or some other supports. These can be grown in pots or containers and then trained over triangular supports (frames). The creeping types are also allowed to sprawl over rough sunny banks or slopes. The dwarf Tom Thumb and other compact cultivars are grown as edgings, in window-boxes, hanging baskets, beds, borders, and pots. The semi-tall nasturtium is useful as a ground cover in shrubbery. The cut flowers, along with the decorative leaves, are much used for decorations.

Cultural hints: Nasturtiums need a dry soil and a sunny position for their growth. The soil should not be manured much as this results in more foliage growth and less flowering. The seeds are sown in September-October in the plains, in permanent beds, 15 cm apart, and then thinned out 30 cm apart, but less for dwarf cultivars. The extra seedlings may be used for other beds. In places such as Bangalore, sowing is possible during June-August also. The seeds are sown in March-April in the hills. The leaves, covering the flowers, are clipped off. If plants show weak growth an earthing-up is given and may also be fed with liquid manure once or twice. In frost-prone areas, nasturtium should preferably be grown in the shrubbery for protection, as these are damaged by frost. The plants flower in about $2\frac{1}{2}$ months after sowing.

Ursinia anethoides; Common Name: Jewel of the Veldt; Family: *Compositae*; Origin: South Africa.

A medium tall (30-45 cm) shrubby plant with deeply-cut foliage and beautiful daisy-like flowers of glowing orange colour with crimson centre and borne on long stalks. The flowers are produced for a long duration and are excellent for cutting, bedding, and pots. The plants need a well-drained soil, a sunny situation, and long cool winter for growth. This cannot be grown successfully in south or east India. The seeds are sown in permanent beds and thinned out 22 cm apart. The plants flower in 3-$3\frac{1}{2}$ months after sowing.

The other two species grown in the gardens are *Ursinia pulchra* (Syn. *U. versicolor*), 22 cm tall, with bright orange flowers, and *U. pygmaea*, 15 cm tall, with rich orange flowers.

Venidium fastuosum; Common Name: Namaqualand Daisy; Family: *Compositae*; Origin: South Africa.

A bushy showy annual, 60-90 cm tall, bearing large, about 10 cm across, deep orange flowers with a black zone and a shining black centre, borne on long stalks. The whole plant is covered with white hair. The *Venidium fastuosum* hybrids have cream, ivory, yellow, and straw coloured flowers with shiny black centre and maroon markings. The 'Sutton's Dwarf Hybrid', a cross between *V. calendulaceum* and *V. fastuosum*, is 35-45 cm tall, and bears deep orange flowers having a crimson zone.

Venidium is very useful for cutting, beds, and borders. The plants are not suitable for growing in the plains except in some cooler parts of northern India. In the hills, *Venidium* flowers in summer, about $3\frac{1}{2}$ months after sowing.

Verbena × *hybrida* (Syn. *V.* × *hortensis*); Common Names: Verbena, Florists' Verbena; Family: *Verbenaceae*; Origin: South America.

The garden Verbenas are compact bushy plants, 20-30 cm tall, with a little tendency to trail. The plants are perennial but treated as annual. The flowers are borne in clusters above the grey-green foliage having toothed margins. The flowers have a wide range of colours including white, different shades of pink, purple, mauve, brilliant blue, red, etc., with a contrasting centre, and some are delicately scented.

Owing to the prolonged flowering period and a wide range of colours, this annual is most suitable for bedding and as ground cover under tall plants and in shrubberies. They are also suitable for edgings, rock gardens, window boxes, pots or seed pans, hanging baskets. and for cutting.

Cultural hints: A well-manured soil and an open situation are best for the cultivation of Verbena. Verbena comes true from seeds which are sown in September-October and also March to June in areas such as Bangalore having low rainfall and moderate temperatures. In the hills, sowing is undertaken in March-April. The seedlings are transplanted, after about a month, 30 cm apart. Plants can be raised from cuttings or layering also. The plants root at nodes when they come in contact with the soil. The flowering starts in about three months after sowing.

Viola tricolor hortensis; Common Names: Garden Pansy or Pansy, Heart's Ease; Family: *Violaceae*; Origin: Southern France.

Pansy (Syn. *Viola × wittrockiana, V. tricolor maxima*) is an all-time favourite perennial bushy trailing herb, 15-22 cm tall, but treated as an annual in the plains and as a biennial in the hills. The beautiful butterfly-like flowers are available in almost all shades of colours and their combinations and blotched, variegated, marked, striped or "pencilled" in contrasting colours and sometimes the edges of the petals are wavy or crinkled. The flowers are delicately scented, especially during the early hours, and the size may be up to 10-12 cm across in the giant cultivars.

The giant-flowered types include cultivars such as 'Swiss Giants', 'Masterpiece', and 'Super French Giants'. The Butterfly Hybrids have ruffled petals marked and blotched beautifully. In the cultivars 'Felix' and 'Queen Alexandrine', the flower petals are "pencilled" like cat's whiskers. In 'Clear Crystal', the flowers are of one colour without any markings.

The pansies are very effective when grown under standard roses or in beds of bulbs. Pansies can also be grown under trees or shrubs as they thrive well under partial shade. Besides, these are also suitable for edging, beds, borders, rock gardens, window boxes, hanging baskets, and pots. The cut flowers are good for decoration as button-holes or for arranging in low bowls and last for 3-4 days.

Cultural hints: Pansies do well in a light, well-drained, and richly-manured soil and a sunny or semi-shaded position. At the time of land preparation, liberal quantities of cowdung manure or leaf-mould, 6-8 kg per square metre, should be incorporated in the soil. Pansies respond well to applications of pigeon droppings. The seeds are sown in September-October and the seedlings are transplanted at the 4-leaf stage. In Bangalore, sowing is possible during June-August also. Pansy seeds germinate irregularly; therefore, while transplanting seedlings from seed pans, the soil should be disturbed the least as more seeds are likely to germinate later. For pot growing, the seedlings are first pricked in 7 cm pots and finally transferred to 20 cm pots. The plants are watered frequently and carefully in the afternoon. The plants are given liquid manure prepared from cowdung or horsedung for red soils, once every 15 days, especially during flowering. Hoeing or stirring the soil once a fortnight is beneficial. The plants are pinched once or twice to obtain a bushy growth. The faded flowers are removed to prolong the flowering period. To obtain exhibition-type flowers, only one bud is allowed to flower on each shoot. Pansy seeds degenerate quickly; therefore, seeds should be obtained every time from reputed nurserymen. It is possible to multiply pansy by cuttings also, especially in the hills.

Viola cornuta; Common Names: Viola, Tufted Pansy, Horned Violet, Bedding Pansy; Origin: Spain, Pyrenees.

The garden Viola has originated from the above species as a result of crosses involving a variety (var. *alba*) of this species and *Viola × wittrockiana*. The plants are tufted and the flowers are smaller in size (2.5-5 cm across). The flowers borne on short stalks are self-coloured and include white, blue, violet-blue, yellow, apricot, and ruby.

Viola is good for edging, window-boxes, pots, rock gardens, hanging baskets, beds, and borders. Viola in combination with Sweet Alyssum and dwarf Candytuft forms a beautiful setting for edging.

The cultivation is similar to that of pansy.

Zinnia elegans; Common Names: Zinnia, Youth and Old Age; Family: *Compositae*; Origin: Mexico.

Zinnia is a popular garden annual and varies in height from very dwarf (15 cm) to tall (90 cm) types. As a result of intensive breeding various flower forms – single, semi-double, and double – varying in size from 2.5 to 15 cm across, have been obtained in various colours such as white, cream, primrose, golden-yellow, orange, scarlet, crimson, pink, lilac, purple, and chocolate. The cultivars of Zinnia are numerous. Among the tall (75-90 cm) cultivars, 'Giant Dahlia-Flowered', 12.5 cm large and 5.0 cm deep double flowers borne on long stalks; 'Giant of California' with flowers 12-15 cm across, and 'Super Giants', with flowers 15-18 cm across are important. Among the medium tall (45-60 cm) cultivars, the following are important: 'Burpeeana Giant' or 'Giant Cactus-Flowered' or 'Chrysanthemum-Flowered', 'Cut and Come Again' or 'Dwarf Pumila', 'Gaillardia-Flowered' or 'Navajo', and 'Scabiosa-Flowered' with Scabiosa-like flowers having pin-cushion centres. In dwarf types (15-45 cm) the following cultivars are popular: 'Lilliput' or 'Pompon' or 'Baby' with flowers 2.5-4 cm across and dobule 'Cupid' (small button-like flowers) and 'Tom Thumb', with small flowers. F_1 and Tetraploid hybrids are also available with large double flowers.

Zinnia haageana (Syn. *Z. angustifolia*), also called Persian Carpet, is a dwarf (25-30 cm) annual with single and semi-double flowers in a wide range of colours and the ray petals tipped with yellow and lemon with a broad zone of maroon at the base.

The modern Zinnias are one of the most brilliant coloured annuals grown in the garden and are most suitable as cut flowers which last long. They are also suitable for growing under glass, in borders and beds. The dwarf cultivars are espcially suited for edgings and window boxes. The dwarf cultivars stand adverse weather conditions better than their taller counterparts. The dwarf cultivars also yield good flowers for cutting.

Cultural hints: Zinnias are very easy to grow. They require a well-drained light medium loamy soil rich in organic matter and plenty of sunshine. There is a controversy regarding the manuring of Zinnias. Some experts prefer to grow them in rich soil with a top dressing of manures, while others recommend their cultivation in poorly-manured soil. The personal experience of the authors shows that Zinnias do better in a comparatively poor soil with some mortar from old buildings mixed with it, and an occasional application of liquid manure made out of cowdung. This makes them less susceptible to the deadly virus disease. However, 30-40 days after transplanting, the plants may be given a top dressing with 50 gm of ammonium sulphate and 30 gm of potassium sulphate per square metre to obtain good quality blooms.

The seeds are sown during February-March and August-September in the plains. In places such as Bangalore, it is possible to grow Zinnia at any time of the year with varying degrees of success. To avoid the attack of the virus disease, the seeds in the northern plains should be sown either early or late. Zinnias are likely to be damaged by frost. The seedlings are pricked and then transplanted at the 4-leaf stage. Direct sowing in permanent beds should be done wherever possible and then thinned out 25-40 cm apart for the dwarf cultivars and up to 60 cm for the taller ones. The plants have a tendency to flower at a young stage. The seedlings should be pinched to delay flowering and make them bushy. In the large double flowering type, usually the first bud is of inferior quality, which should be removed. Seeds should be collected from the best flowers. But, Zinnias grown from such collected seeds degenerate gradually. Hence, it is advisable to purchase seeds from reputed nurseries each time. For pot culture 20-30 cm pots are suitable, depending on the cultivars.

Annual Climbers

Annual climbers are also annual flowers but these are being dealt with separately to facilitate their easy identification. The annual climbers serve the same purpose as other annuals, such as providing colour to the garden, for growing in pots, for cut flowers, etc. But the climbers have the added advantage in that they can be trained over trellis, walls, pergolas, arches, poles, and pillars. They can also be grown in pots and, therefore, are useful for terrace gardens. These climbers also make good plants for window boxes. All the annual climbers can be grown from seeds. Most of the climbers prefer a sunny situation for their growth. The cultivation is similar to that of other annuals.

Some of the important annual climbers are discussed below.

Calonyction aculeatum; Common Name: Moonflower; Family: *Convolvulaceae*; Origin: Tropics.

This is a heavy climber, growing to a height of 3 m or more. The leaves are heart-shaped and the plant has a milky-juice. The flowers are small or large, white or purple, sweet-scented and open during the evening or at night and closes in the morning. This can be trained over a low roof or against a pillar. The flowers are suitable for cutting in the evening. The var. *grandiflorum* has large flowers, in var. *heterophyllum* the leaves are lobed, and in var. *variegatum* the foliage is variegated. The plants do well in any ordinary soil. Seeds are sown in July-August for flowering in winter. If the plants are left in the ground these flower again next May or June. This will flower in the hills in July-August.

Clitoria ternatea (Syn. *Ternatea vulgaris*); Common Names: Mussel-shell Creeper, Butterfly Pea; Family: *Leguminosae*; Origin: India.

A twiner, growing to a height of 5 m with showy blue flowers. In var. *alba* the flowers are white. It is very easy to cultivate and can be trained on wire-trellis, pergolas, poles, etc. A compost of humus, garden loam and soil suits the plant well. Plants are raised from seeds which are sown in July-August in the plains for flowering in winter. It flowers in July-August in the hills. The side shoots can be rooted under a bell-glass or mist chamber. A more or less double flower is known to exist for a long time.

Cobaea scandens; Common Name: Cobaea; Family: *Polemoniaceae*; Origin: Central and South America.

A glabrous climber with angled stem, glossy foliage, large bell-shaped flowers, which are at first greenish, then becoming violet. In var. *flore albo* the flowers are white and in var. *variegata* the leaves are variegated. An interesting morphological feature is that the terminal leaflet is replaced by a tendril and occasionally there are indications of tendrils on other leaflets also. The plant grows up to a height of 7-8 m. A fairly rich soil suits this well. An addition of lime and brick rubbish improves the flower colour. Fresh seeds, as old seeds are not reliable, are sown in September-October for flowering in March-April. In the hills, this annual flowers between July and October. The seeds are flat and are sown edgeways. The variegated form may be raised from cuttings. It can be treated as a perennial in a cool glass greenhouse. Sometimes it flowers after one year. This is not very suitable for the plains except in the north where winter is long and cool.

Lathyrus odoratus (Sweet Pea): See pages 88-90.

Maurandia (Syn. *Maurandya*) *barclaiana*; Family: *Scrophulariaceae*; Origin: Mexico.

This is also known as *Asarina barclaiana*. A light climber, excellent for growing on trellis or dwarf wall or in pots. The flower colours vary from deep purple to rose or white and the flowers resemble antirrhinum. The climber flowers freely almost throughout the year. The plants are raised from seeds in September-October in the plains and in March-April in the hills. The young shoots can be rooted in a mist chamber or a bell-glass.

Pharbitis × imperialis; Common Name: Japanese Morning Glory; Family: *Convolvulaceae*; Origin: Hybrids developed in Japan (parents from Australia and Tropical America).

Japan has developed the Morning Glory to perfection having various shades of flower colour and a large size. The Japanese Morning Glory is probably a cross between *Pharbitis tricolor* and *P. hederacea*. It is a tall twiner with larger flowers, up to 20 cm across, resembling Morning Glory, in a variety of shades ranging from white to deep crimson. The plants are raised from seeds any time of the year but do best during the rains. The plants are pinched to make them bushy. The flowers open in the morning and fade by afternoon.

(i) *Pharbitis purpurea*, (ii) *P. tricolor*; Common Name: Morning Glory; Family: *Convolvulaceae*; Origin: Tropical America.

It is an easy-to-grow climber which flowers quickly within 2-2½ months after sowing seeds. Botanically, Morning Glory belongs to two species, namely, *Pharbitis purpurea* (Syn. *Convolvulus major*, *Ipomoea purpurea*) and *P. tricolor* (Syn. *P. rubro-caerulea*, *Ipomoea hookeri*, *I. rubro-caerulea*, *I. tricolor*). The flowers are funnel-shaped, 7-15 cm across and in various shades of white, blue, purple, and scarlet. The flowers open in the morning and last till afternoon in semi-shade after which they fade. These can be grown in pots also. The seeds are sown in August if a large area is to be covered or in October to grow as a dwarf climber. In the hills, *Pharbitis purpurea* flowers in summer and *P. tricolor* in November-December. The seeds may be chipped before sowing. Under favourable climatic conditions, *P. purpurea* and *P. tricolor* grow to a height of 3 m. The plants are liable to be damaged by frost. Some important cultivars are 'Scarlet 'O Hara' (wine-red), 'Flying Saucers' (Light blue with blue streaks), 'Darling' (dark purple), and 'Double White'.

Quamoclit; Common Names: Star Ipomoea, Cypress Vine, Cardinal Creeper, Indian Pink; Family: *Convolvulaceae*; Origin: Central America.

Quamoclit (Greek, meaning dwarf kidney bean) consists of three important species which are cultivated in the garden.

(i) *Quamoclit pennata* (Syn. *Q. quamoclit*, *Ipomoea quamoclit*). Commonly called Cypress Vine or Indian Pink. It is a light smooth twiner growing up to 2-2½ m with finely divided foliage and tubular or trumpet-shaped scarlet flowers. In var. *alba* the flowers are white. It flowers well during the rains as well as in winter. In the hills, flowering is between July and October.

(ii) *Q. coccinea* (Syn. *Ipomoea coccinea*). Commonly called 'Star Ipomoea'. It is a twiner growing to a height of 2 m or more, having slender pointed leaves and fragrant scarlet flowers with yellow throat. Seeds are sown in July-August in the plains. In areas such as Bangalore, it can be grown at other times also. In the hills, it flowers during August-September.

(iii) *Q. lobata* (Syn. *Mina lobata, Ipomoea versicolor*). A vigorous and free-flowering climber, 5-6 m tall. The flowers are bright crimson during opening but change immediately to orange-yellow and then yellow, and are borne in sprays. The leaves are tri-lobed. The seeds are sown in July-August in the plains. In Bangalore, it flowers for a greater part of the year, whereas in the hills the flowering time is between June and September. The plants start flowering in 2-2½ months after sowing seeds.

Thunbergia alata; Common Name: Black-eyed Susan; Family: *Acanthaceae*; Origin: South Africa.

A softly hairy annual twiner. The flowers vary greatly in colour and may be orange, yellow, buff, or white and sometimes with a black eye. This is easy to cultivate and need a well-drained loamy soil. This is ideal for growing in window boxes, pots fitted with a balloon structure, and hanging baskets. This creeper flowers during winter.

T. fragrans; Family: *Acanthaceae*; Origin: India.

An annual to perennial climber. The flowers are white and fragrant. The cultural requirements are the same as above.

Tropaeolum majus.

For details see under Nasturtium (*Tropaeolum majus*). The tall single-flowered nasturtiums are used as climbers and the double-flowered 'Gleam' has a semi-trailing habit.

Tropaeolum peregrinum (Syn. *T. canariense*); Common Names: Canary Creeper, Canary-Bird-Flower; Family: *Tropaeolaceae*; Origin: Peru.

It is a tall, elegant climber, annual or perennial, with deeply-lobed leaves and prettily-fringed golden yellow flowers. The plants are grown in semi-shade or in the open and require a rich soil. The seeds are sown in September-October in the plains and in February-March in the hills. The plants flower in four months after sowing. The plants are susceptible to damage by frost.

Chapter VII

Herbaceous Perennials

There are some plants with soft stems that cannot be included in the shrub group. They are termed as herbaceous perennials. They live year after year, whereas annuals complete their life cycle within one season. Among the many herbaceous perennials available for cultivation, only a few do well in tropical climate. The majority of the others thrive best in temperate climate. The one distinct advantage of this group is that, unlike the annuals, no pains are to be taken to raise them again and again every year or every season. In the hill stations and under temperate climate, some of them need only heading back to the ground level for rejuvenation. The herbaceous perennials are propagated through seeds, offsets, division, or cuttings. The plants, being comparatively hardier than the annuals, are easy to grow and some of them bloom for a considerable part of the year, if not throughout. The herbaceous perennials have been classified below into two groups, according to their climatic requirements: (A) Perennials for the plains and (B) Perennials for the hills.

(A) Herbaceous Perennials for the Plains

Angelonia grandiflora; Family: *Scrophulariaceae*; Origin: South America.

A very popular, hardy, dwarf (60 cm) perennial which bears terminal racemes of sweet-scented purple or white (var. *alba*) flowers.

Cultural hints: Plants grow well in rich sandy soils. The compost for pots (25-30 cm) consists of a mixture of garden soil, peat, and leafmould. The plants thrive best under partial shade or greenhouse, but can be grown in the open under mild climatic conditions of Bangalore and Poona. It is likely to be injured by frost. Easily propagated from seeds and stem or tip cuttings. There is a variegated form also with purple and white flowers.

Aster amellus and others; Common Names: Michaelmas Daisy, Perennial Aster; Family: *Compositae*; Origin: See below.

The Michaelmas daisy consists of three species, namely, *A. amellus*, *A. novae-angliae* (New England Aster), and *A. novi-belgii* (New York Aster).

The first species is a native of Italy and the others are from North America and the eastern

United States of America, respectively. The first one grows to a height of 60 cm, whereas the other two species are much taller and may reach up to a height of 150 cm and 90 cm, respectively. The single flowers are small, 2.5-5 cm across, daisy-like, borne in large heads (30-45 cm) of corymbose panicle and are white, lavender, mauve, deep blue, lilac, and pink.

Cultural hints: The plants throw numerous suckers and the main blooming seasons are the summer and the rains. But in Bangalore, these flower all the year round. The plants do better at medium and high elevations. In hotter climate, aster should be grown in partial shade. After each flowering, the faded blooms are removed, the soil is raked up, and F.Y.M. is added at the rate of 4-6 kg per square metre. The same process should be repeated after each flush of bloom to maintain continuity of flowering. But it is desirable to dig up the clumps every third year and replant them afresh for vigour. Propagation is easy through division of suckers. Application of liquid manure, once a fortnight, is beneficial. The plants like plenty of moisture. These are grown in 25 cm pots also. In ground, these are planted 25-30 cm apart. Michaelmas daisies are good as border, bedding, or pot plants and ideally suitable for covering open spaces in the shrub garden.

Coreopsis; Family: *Compositae*; Origin: U.S.A. (East and South).

The commonly grown perennial species are:

(a) *C. grandiflora* (Syn. *C. longipes*)*:* It is 30-60 cm tall, bearing bright yellow flowers.

(b) *C. lanceolata:* It is also of the same height but sometimes branches at the base and the flower colour is bright yellow.

(c) *C. verticillata* (Syn. *C. tenuifolia*)*:* The plants are 30-90 cm tall, sparingly branched, with furrowed stems and golden yellow flowers.

All the *Coreopsis* species flower during the summer and are propagated from seeds or by division of roots. For further details, readers may see page 78.

Gaillardia or Blanket Flower.

A herbaceous perennial but is usually grown as an annual in the hills. The plants are propagated by seed, cuttings, and division of roots. For further details consult the chapter on Flowering Annuals, page 83.

Gerbera: See Chapter XX.

Geranium: See *Pelargonium.*

Impatiens; Common Names: Perennial Balsam, Touch-me-Not; Family: *Balsaminaceae*; Origin: As given in text below.

There are several perennial species of *Impatiens* which can be grown in India.

(a) *I. hawkeri:* A bushy plant, about 60 cm tall, succeeding well in the plains and bearing deep carmine flowers, about 7 cm across, with a white eye. It flowers during summer. Origin of this species is South Sea Isles.

(b) *I. holstii:* For details see page 88.

(c) *I. hookeriana* (Syn. *I. biglandulosa*, *I. sultani alba*): A much-branched, fleshy stemmed perennial, about 90 cm tall, and leaves having long petioles. The flowers are large and white with purple markings on the larger lower petals. It thrives best at 500 m above the sea level.

Origin is Sri Lanka.

(d) *I. oliveri:* A large, 120-240 cm tall, bushy perennial which bears large flowers, 5-6 cm across, which are pale rose in colour. Plants are propagated from seeds or cuttings. Origin is Tropical Africa. For further details, consult the chapter on Flowering Annuals.

Mirabilis jalapa: See Chapter VI, page 94.

Nierembergia hippomanica (Syn. *N. caerulea*); Common Name: Cup flower; Family: *Solanaceae*; Origin: Argentina.

A handsome branched perennial, with thin stems which are stiffly erect, and grows to a height of 15-30 cm. Flowers are about 2 cm across, numerous, bright blue-violet in colour with darker lines and yellow throat. Flowering is from June to September. They grow best in moist but well-drained soil and are easily multiplied by cuttings.

Orthosiphon stamineus; Family: *Labiatae*; Origin: Tropical Asia (including India).

A herbaceous perennial, 30-60 cm tall, bearing pale lilac-blue flowers, 2.5 cm across, and the stamens are much exerted (projecting) giving the appearance of a cat's whiskers. A compost made of fibrous loam and well-decomposed manure in equal proportions suits the plant well. It is generally grown as a house or greenhouse plant. The plants are grown in semi-shade, guarded from afternoon sun, and are propagated from seeds or cuttings of half-ripe shoots.

Pelargonium; Common Names: Geranium, Pelargonium, Stork's Bill; Family: *Geraniacea*; Origin: Mostly South Africa.

Geraniums are herbaceous perennials, mostly grown as potted plants for their handsome flowers borne in clusters. The leaves are also showy. Botanically, these are known as *Pelargonium.* Most of the modern garden varieties are crosses between various species. The *Pelargoniums* are classified as follows:

(a) *P. zonale* (Zonal or Horseshoe Geranium): The plants grow up to 45 cm in height, sparsely branched with succulent young branches, becoming woody with age. This is the most popular garden type which is characterised by a horseshoe marked darker zone on the upper face of the leaves. The leaves are roundish and wavy at the edges and hairy. The flowers may be single or double and in shades of white, pale rose, pink, scarlet, crimson, vermilion, purple, salmon, and sometimes veined distinctly. There are variegated leaved varieties also. Modern hybrid cultivars of zonal class are grouped as *P.* × *hortorum* recently.

(b) *P. peltatum* (Syn. *Geranium peltatum*): It is known as Ivy-leaved Geranium and is shrubby in habit with slender straggling glabrous (or almost so) branches up to 90 cm long. The flowers are carmine-rose or whitish with deep purple veins and the upper two petals have red lines. There are varieties with variegated foliage. These are suitable for hanging baskets and window boxes.

(c) *P. domesticum:* Commonly called Show, Fancy, Common, and Lady Washington Geraniums. This race is believed to be derived from *Pelargonium cucullatum*, *P. angulosum*, and *P. grandiflorum*, but Bailey (1963) finds no blood of the latter in this class. Unlike the other geraniums, this class has funnel-shaped flowers and palmate and toothed leaves.

(d) The scented-leaved *Pelargoniums* represent several species: *Pelargonium* × *fragrans* is 30 cm tall and somewhat shrubby, having Nutmeg scent. *P. graveolens* (*Geranium graveolens*) is 90 cm tall and grey-green in colour. It is called rose-geranium because of the rose scent. *P. tomentosum* is also 90 cm tall and has peppermint scent. Another scented geranium is *P. quercifolium* though the scent is not quite agreeable. Many of these produce ornamental flowers also.

Pelargoniums are greenhouse plants grown in pots. In a mild climate, it is possible to grow some varieties in beds also.

Cultural hints: Pelargoniums thrive best in the hill stations and in milder climates of Bangalore and Poona. In eastern and northern India, the plants die due to either heavy rains or summer heat. If protected from heat and rains, these survive for some time. A few hardy zonal pelargoniums do well in the plains. In the plains, these are grown in pots, whereas in temperate climate growing in beds is possible. They thrive best in semi-shade, but can be kept in the open also if the sun is not strong. Geraniums are propagated from stem cuttings (10-20 cm long) and which are not so woody or very tender, in a mixture of equal proportions of sand and sifted leafmould. The pots containing the cuttings are kept in semi-shade or in the open. If the sun is mild and the compost is maintained on the drier side, cuttings root in 4-5 weeks. The final potting is done in 20-25 cm pots containing a compost consisting of 2 parts of soil, 1 part sand, and 1 part cowdung manure (horse-dung for red soils). Geraniums are easily raised from seeds also. Watering of geraniums should be done very carefully as over-watering may kill the plants, especially the seedlings which have very few fibrous roots besides the tap-root. All the flower buds should be pinched till the plants make a healthy growth. The excess foliage may be clipped to divert energy towards flower production. After each flush of flowering, plants are allowed one month's rest and then the branches are pruned to one-third length. After the plants have again made sufficient growth, these are repotted or top-dressed with fresh compost.

Phlox decussata (Syn. *P. paniculata*); Family: *Polemoniaceae*; Origin: Eastern North America.

This is commonly known as Border Phlox or summer perennial phlox, which is mostly unbranched and 45-120 cm tall. The leaves are oblong lanceolate and flowers are like those of common phlox with white, pink, mauve, violet, purple, crimson, and red colours.

Cultural hints: A light well-drained soil is best suited for border phlox. They thrive best in the hills. The plants are grown in pots and borders. The plants are propagated from seeds, suckers, and tip and root cuttings. The cultural practices are similar to those of Michaelmas daisy. The plants need protection from the afternoon sun. Plants grown in 20 cm pots flower within 3½-4 months from planting.

Platycodon grandiflorum (Syn. *P. grandiflorus, Campanula grandiflora*); Common Names: Chinese or Japanese Bellflower, Balloon flower; Family: *Campanulaceae*; Origin: China, Manchuria ,and Japan.

It is 30-60 cm tall, bearing bell-shaped flowers of blue or white (var. *album*) shade, single or semi-double and about 7 cm across. The shoots die after each flowering, but sprout again from the thick roots which survive a number of years underground. It can be propagated from seeds also, which germinate slowly.

Portulaca grandiflora; Family: *Portulaceae* or *Portulacaceae*; Origin: Brazil.

A perennial portulaca, with light deep rose double flowers, is often cultivated in the garden. Plants are propagated from seeds, division, and stem cuttings. The cultural requirement is the same as in the case of annual species discussed in the chapter on Flowering Annuals.

Saintpaulia ionantha; Common Names: African Violet, Usambara Violet; Family: *Gesneriaceae*; Origin: Eastern Tropical Africa.

African violets are stemless, hairy perennial, with ovate or subcordate leaves (3.5 cm long) having many small and a few long hairs on both surfaces. It is a plant of great beauty bearing similarity in some respects to Gloxinia to whose family the plant belongs. The flowers are of similar shape as violets but larger in size. In var. *albescens* flowers are white tinged with bluish-pink; in var. *grandiflora* the flowers are much larger than the usual and deep violet in colour; in var. *purpurea* the flowers are dark violet and in var. *variegata* the leaves are variegated light yellow turning to white. There are several hybrids with brilliant colours.

Cultural hints: African violets are grown in 12 cm pots, in the conservatory or greenhouse. In most parts of our country lath house is best for growing African violets. In the conservatory, these can be planted in raised beds. As the plants grow larger they are transferred to 20 cm pots or shallow pans. The compost for growing *Saintpaulia* (the name is from the discoverer Baron Walter von Saint Paul) consists of 1 part of good garden soil (not clay), 2 parts of sifted leafmould, 2 parts of fine sand, and 1 part of well-rotten cowdung manure (or horse manure for red soils). A layer of small brick-pieces or boiler ash should be spread over the surface of the pot so that the fleshy leaves do not come in contact with the soil and rot. Watering is to be done carefully as too much moisture may cause damping-off. The plants are propagated as in Gloxinia from seeds, which are very minute, or from single-leaf cuttings. The mature leaves, but not over-mature, are cut with 2-3 cm stalk and inserted in sand beds or pots containing sand with only a portion of the blade inside the sand. Too close planting is to be avoided. The sand should not be kept too wet during the process of rooting. The plants perform extremely well in Bangalore under cover. In hotter parts of the country, extra protections in the shape of shade during the summer and shelter from frost in the winter should be provided. In the heavy rainfall areas of eastern India, they would need protection from the rains also.

Salvia

Generally it is grown as an annual but can be treated as a perennial also. For details see Chapter VI, page 103.

Solidago canadensis; Common Name: Golden Rod; Family: *Compositae*; Origin: Eastern North America.

Solidago comprises a number of perennial species of which *S. canadensis* is most commonly grown in India. The plants are hairy, 90-180 cm tall, stout, and with much branched stems. The pretty golden-yellow flowers are feather-like and crowded in a long terminal panicle.

Cultural hints: The plants do well in the tropical climate. Golden rod is recommended for growing in rough places, borders of shrub gardens and in large borders. Their bright yellow flowers contrast well with those of Michaelmas daisy both of which flower at the same time, during September-October, in the northern plains and in the summer in the hills. In eastern India and Bangalore, flowering may start earlier during the rains. Plants perform well in semi-shade. They are multiplied from suckers. Other cultural requirements are similar to that of Michaelmas daisy.

Verbena erinoides; Common Name: Moss Verbena; Family: *Verbenaceae*; Origin: South America.

There is some controversy regarding the name *Verbena erinoides*. Some authors claim that it is a misnomer and, strictly speaking, it is nothing but a synonym of *Verbena laciniata*. The other synonyms are *V. multifida*, *V. pulchella*, and *V. tenuisecta*.

The plants cover the ground thickly as moss and hence the name. It is a low-growing, 15-25 cm tall, spreading perennial with graceful finely-cut foliage. The plants are free-flowering and the flowers are borne in clusters with light mauve as the common colour, though pink and white flowering types are also available. The flowering is during the summer, but under a mild climate the blooming period is stretched.

The moss verbena is excellent for growing under standard roses, in small circular or oval beds, as a ground cover in shrubbery, in rock garden and hanging baskets.

Cultural hints: The plants are easy to propagate. The branches, wherever they come in contact with the soil, form roots, which are separated and planted. After each flush of flowering, the plants are cut back to ground level, spaced out 20-25 cm apart, and well-rotten manure is worked into the soil with fork at the rate of 4-6 kg per square metre. The plants flower again in about 3½ months' time if treated this way. Uprooting and replanting the plants once a year is recommended.

Vinca rosea (Syn. *Catharanthus roseus*) and *Vinca alba* (Syn. *Vinca elegantissima* var. *alba*); Common Names: Periwinkle, Madagascar Periwinkle; Vernacular names: *Nayan Tara, Sudukadu Mallikai*; Family: *Apocynaceae*; Origin: Tropics.

The plants are 30-60 cm tall and covered with bright green smooth oblong leaves. The plants grow like weeds in many parts of the country. The flowers are rosy purple (*V. rosea*) or white (*V. alba*), the latter with or without a very small reddish eye. The plants are free-flowering and bloom round the year. The species. *V. major*, with larger rose-coloured flowers, can be grown in the hills only.

Cultural hints: The plants are hardy and grow in any soil. They are propagated from seeds or stem, including tip cuttings. The plants are headed back every six months and replaced every second year. The spacing recommended is 30-45 cm. It has anti-cancer medicinal properties.

Viola odorata; Common Name: Sweet Violet; Family: *Violaceae*; Origin: Europe (including British Isles), Asia, North Africa..

Violets are small herbaceous perennials prized for their sweet-scented flowers. The plants are trailing in habit, with round leaves, and bear single or double flowers in shades of white, blue, and purple.

The flowers have long stalks and are cut for button-hole because of the sweet-scent. The plants can be grown in beds also.

Cultural hints: The plants can be grown only as seasonal plants in the plains, where they do not flower freely and the flower size also becomes small. They do well in the mild climate of Bangalore and in the hills. They need a good garden soil, preferably sandy loam and like plenty of moisture. A shady, cool situation receiving only the morning sun, under the trees, suits the plants very well. The soil should be well prepared by deep digging. Clean cultivation, proper watering (beds should never go dry), and ample aeration are the points to be kept in view for its successful cultivation.

The plant is propagated by division of roots, which can be done in September-October. Young and vigorous offshoots, selected from free-flowering plants, can be rooted in sand. The compost for pot consists of 1 part loam, 1 part sand ($\frac{1}{2}$ for non-clay soils), and 2 parts of well-rotten cowdung (horse manure for red soils). The runners that come up are removed occasionally. The plants need protection from excessive rains. The flowering season is from February to April.

Zinnia linearis; Family: *Compositae*; Origin: Mexico.

This is a perennial zinnia, with neat, bushy erect habit, up to 30 cm tall, producing bright golden-yellow flowers, 3-5 cm across, with a dark centre and stripes of lighter shade. There is a white flowering type also and both the types are free-flowering. The plants flower very early in about six weeks after sowing. The plants flower during the summer, the rainy season, and the winter. *Z. linearis* is free from the attack of virus.

(B) Herbaceous Perennials for the Hills

There are certain other perennials, besides those discussed above, which thrive well only in the hills. Some selected plants in this category are described below.

Achillea; Common Name: See below; Family: *Compositae*; Origin: See below.

The two most commonly cultivated species are *Achillea filipendulina* (Syn. *A. eupatorium*) and *A. millefolium*.

(a) *A. filipendulina:* A native of Caucasus. It is a robust plant, growing to a height of 150 cm, with pinnately lobed, toothed, and pungent foliage. The bright yellow flowers are densely borne in somewhat convex heads which sometimes measure 12 cm across. The plants need staking.

(b) *A. millefolium:* It is commonly known as Yarrow or Milfoil and is a native of Europe, Asia, and America. The plant is 30-90 cm tall and bears white flowers in flattish heads. The vars. *rubrum* and *roseum* produce red or purple flowers.

The plants can be grown in ordinary garden soil in open sunny situation. The flowers are good for cutting. Achillea can be grown in the wild garden (tall types) and for bedding (dwarf types). The flowers last long. The plants are propagated by division or can be raised from seeds which germinate within a fortnight after sowing.

Anchusa italica: See page 69.

Aquilegia vulgaris; Common Name: Columbine; Family: *Ranunculaceae*; Origin: Europe.

There are numerous species which are cultivated in the garden, but the most common is the 'Columbine' (*A. vulgaris*). There are many hybrids also involving different species. These are pretty plants, about 60 cm tall, with handsome fern-like foliage consisting of three leaflets in a leaf. The flowers are double or single and large, 10-12 cm across, with a long spur. The flower colour ranges from white to red, yellow, pink, crimson, blue, and lavender.

The plants are hardy and can be cultivated in partial shade also. Besides hill stations of the north, *Aquilegia* performs well at medium to high elevations in south India. The plants are suitable for growing in the borders. The seeds are sown in March-April. To perpetuate a strain, plants are propagated by division. A spacing of 25-30 cm is recommended.

Bellis perennis: See page 71.

Bergenia cordifolia; Common Name: Elephant Ear; Family: *Saxifragaceae*; Origin: Siberia.

The plants are 30 cm tall, with large heart-shaped foliage and rose-pink flowers measuring 2 cm across, which are borne on medium stalks above the foliage. The other species, *B. ligulata*, a native of Nepal, is also grown in the garden. The flowers are pink or white. The plants are grown in rock-gardens and pots. Propagated by division of roots or from seeds. The flowering is between March and May.

Campanula; Common Name: Bell-flower; Family: *Campanulaceae*; Origin: Europe, Asia (including India), Africa.

The plants bear cup-and-saucer shaped flowers. The common perennial species of the garden are: *C. burghaltii* (Syn. *C. pallida*), flowers grey-blue; *C. carpatica* (blue and white flowers); *C. glomerata* (violet-blue or white flowers); *C. lactiflora* (milky-blue flowers); *C. latifolia* (blue, white, pale mauve flowers); *C. latiloba* (*C. grandis*, flowers blue or white); and *C. persicifolia* (*C. crystalocalyx*) with flowers, double or single, white or lavender. The plants are propagated in the spring by division of roots or from seeds. For cultivation see *Campanula* on pages 73-74.

Chrysanthemum maximum; Common Name: Shasta Daisy or Moon Daisy; Family: *Compositae* (*Asteraceae*); Origin: Pyrenees.

The plants are 30-90 cm tall and the flowers are white with yellow central discs resembling daisy. It is a good border plant and the flowers are cut for decoration. A well-drained soil and a sunny situation suit the plant well. Propagated by division of roots and from seeds. The flowering is during the summer. For cultural hints see Annual Chrysanthemum, page 76.

Delphinium: See *Delphinium hybridum*, page 79.

Dianthus caryophyllus (Carnation): See Chapter XX.

Digitalis purpurea: See Chapter VI, page 81.

Gazania: See details on page 83.

Most of the *G. splendens* hybrids can be grown as perennial in the hills. *G. pavonia*, the Peacock gazania, is also a perennial.

Geum; Family: *Rosaceae*; Origin: Europe, Asia Minor.

A herbaceous perennial, suitable for medium and high elevations. The commonly grown species are *G. chiloense* var. *atrosanguineum* with coppery-red, semi-double, rose-like flowers, and *G. coccineum* having brick red flowers. Both the species flower freely and are best treated as annuals. The plants should be raised from seeds every year. The flowering is from summer to autumn. A cool, semi-shaded location suits them best.

Gypsophila paniculata: See *Gypsophila elegans*, Chapter VI, page 85.

Heuchera sanguinea; Common Names: Coral Bells, Crimson Bells; Family: *Saxifragaceae*; Origin: Mexico, Arizona (U.S.A.).

A herb, 30-45 cm high, with heart-shaped or roundish, much lobed leaves. The flowers are small, bell-shaped and numerous, borne on long stalks. These are suitable for table decoration. In var. *alba* the flowers are white and in var. *rosea* the flowers are rose red. The plants need a light soil and semi-shade for proper growth. Propagated from suckers or seeds. The flowering is from June to September.

Hunnemannia fumariifolia (also spelt *H. fumariaefolia*); Common Name: Mexican Tulip Poppy; Family: *Papaveraceae*; Origin: Mexico.

A perennial, much allied to Californian poppy (*Eschscholzia*). It grows to a height of 40-90 cm and bears yellow flowers which are 5-7 cm across. The flowers are good for cuttings and for planting in herbaceous border. They need a rich, well-drained soil for proper growth. The plants are raised from seeds or suckers and can be treated as annual. The flowering is in July.

Linum perenne: See *Linum grandiflorum*.

Lupinus: See page 92.

Oenothera: See page 97.

Paeonia: Common Name: Peony or Paeony; Family: *Ranunculaceae*; Origin: Europe, Temperate Asia, North-West America, and China.

There are both herbaceous and shrubby paeonies. The commonly cultivated species are *P. officinalis*, *P. lactiflora*, *P. mlokosewitschii* (also spelt with single 'i' at the end), and *P. wittamanniana*. *P. suffruticosa* (Syn. *P. moutan*) is the tree peony. Peonies are grown in beds, borders, and shrubberies. A well-drained soil and a sunny situation suit them best. The plants are propagated by division during the autumn. Plants raised from seeds flower after 6 to 7 years. Flowering season is during the summer. The flowers are prized for cutting.

Penstemon barbatus: See page 98.

Primula: Common Name: See below; Family: *Primulaceae*; Origin: See below.

Primula includes a large number of species which are cultivated in the garden. These are herbaceous perennials, of which one or two may be cultivated as annuals.

The Baby Primrose or Fairy Primrose (*Primula malacoides*), a native of China, is a 30 cm tall plant with a rosette of large broad leaves at the base. The flowers, single or double, are borne on a long slender stalk in trusses and the flower colour in modern horticultural forms may be white, lavender-mauve, pale pink, deep pink, salmon rose, mauve, rose-red, crimson. This plant is good for growing in pots, beds, shrubbery, and for cut flowers. A well-drained rich soil containing sufficient humus is suitable for its cultivation. The plants can be grown in the sun or in a semi-shaded situation. They do best in the hills but can also be grown in areas having a long cool winter. The seeds are small and special care should be taken as recommended for sowing small seeds. Watering of seed pans should preferably be done from below, i.e., pan is placed in a trough of water and water is absorbed through the drainage hole. The germination, which is irregular, is completed in two to three weeks. When the seedlings are large enough to be handled with ease, they are first transplanted to 7.5 cm pots, then to 12.5 cm pots, and finally in beds or 20 cm pots. The compost during each repotting should be a little richer than the preceding one. The spacing should be 20-25 cm. The seeds are sown in September-October in the plains and in March-April or August-September in the hills. The plants flower in March in the plains and in July-October or during the spring in the hills. The healthy old plants are repotted with new compost to obtain flowering again. Young plants of Primula are more productive of blooms and hence it is desirable to raise these from seeds every year, although propagation from cuttings is also feasible. If the winter is very severe, plants scheduled to flower in the spring have to be kept in a glasshouse.

The other species of *Primula* which are commonly cultivated are *Primula* × *kewensis* (*P. floribunda* × *P. verticillata*), *P. obconica* (Syn. *P. poculiformis*), a plant of Chinese origin, *P. sinensis* of China, *P. polyantha*, possibly a hybrid, and *P. vulgaris* (Syn. *P. acaulis*), a native of West and South Europe, including Britain. The last three species do well in the hills only.

Two primulas, namely, *P. denticulata* and *P. rosea*, are natives of the Himalayas and are quite popular in England, other parts of Europe, and some other countries. They grow only at higher elevations. The cultural requirements for all the primulas are similar to those of *P. malacoides*.

Pyrethrum roseum (Syn. *Chrysanthemum coccineum*); Common Name: Pyrethrum of Gardens; Family: *Compositae*; Origin: Caucasus, Persia.

The plants are 30-60 cm tall with finely cut leaves and big flowers resembling daisies in colours of white, lilac, rose, and red. There are innumerable horticultural varieties including hybrids with shades of pink, salmon, red, scarlet, crimson, etc. Sometimes flowers are tipped yellow and the disc also is yellow.

The plant is suitable for growing in borders and produces flowers for cutting. The plant requires a rich, well-drained soil and a moist climate. Propagated from seeds or by division. The spacing recommended is 25-30 cm. The plants flower in the summer.

Rudbeckia: See page 102.

Streptocarpus; Common Name: Cape Primrose; Family: *Gesneriaceae*; Origin: Mostly African.

There are several species, including hybrids, that are in cultivation. These are perennial stemless herbs, suitable for medium and high elevations, with one leaf or a rosette of several leaves, more generally prostrate and a single leaf may measure from 7.5 cm to 15.0 cm. The showy blooms are borne in erect scapes and the shades include blue, lilac, mauve, pink, and white. There are large flowering hybrids showing a diversity of colours.

The plants are suitable for pots and for growing in rockeries in a semi-shaded position. The plants are shallow rooted and need a rich compost. The seeds are minute and should be sown with care as recommended for such seeds. These can be propagated by division or from leaf cuttings made by nicking the main vein and laid on the surface of a compost and sand mixture. Shoots taken from the base of the cutback cotyledon can also be rooted. The seedlings take 8-15 months to flower. The old plants are discarded after some time.

Torenia: See page 106.

Chapter VIII

Ornamental and Flowering Shrubs

A shrub may be defined as a perennial plant having many woody branches arising from the base of the plant. The plant whose basal portion is woody and the upper shoots are soft is called a sub-shrub. A shrub is generally erect and bushy, attaining a height of 0.5 to 4 metres. There are some shrubs such as *Lantana sellowiana* and *Juniperus horizontalis* which grow prostrate.

Utility

The utility and importance of a shrub in a garden or a landscape is universally acknowledged. A garden or a landscape, without shrubs, will lose much of its charm, attraction, and beauty. Even in small gardens where planting of trees is not possible some selected shrubs must find a place.

Shrubs are important because they are available in various sizes, shapes, and forms and the flowers are of different colours and shapes. Possibly, the most important point is that, unlike the seasonal flowers, they are perennial in habit, and need little care. If a proper selection is made the shrubs will provide flowers in the garden round the year. Unlike the trees, most shrubs are quick-growing and flower within a year after planting.

The shrubs are available in various heights from dwarf to tall and thus can be put to various uses in the garden. They can be planted in the shrubbery, border, or as specimens in lawns and in a landscape. Some shrubs can be grown along the paths in a landscape or near the gateway or the staircase.

Prostrate or trailing shrubs such as *Lantana sellowiana* and *Juniperus horizontalis* are suitable for planting in rock gardens. Many shrubs such as *Duranta plumieri*, some species of *Cotoneaster*, *Carissa carandas*, *Ochna squarrosa*, and *Rauwolfia canescens* bear most ornamental-looking berries. The shrub *Fortunella japonica* gets covered with small orange-like fruits and looks beautiful. There are shrubs such as Crotons, *Eranthemums*, and *Acalyphas* with beautiful coloured foliage. Some shrubs such as *Duranta plumieri variegata*, *Citrus limonia* var. *variegata*, *Nerium oleander* var. *variegata*, *Manihot*, etc., have attractive variegated leaves. Many shrubs such as *Cestrum nocturnum*, Jasmines, *Portlandia grandiflora*,

Cananga kirkii, *Citharexylon fruticosum*, *Nyctanthes arbor-tristis*, and *Oncoba spinosa* bear fragrant flowers.

Again, shrubs form a good backdrop for growing seasonal flowers. Shrubs can effectively break the monotony of a large open space or lawn or can separate one part of a garden from another. The low-growing shrubs such as *Lantana sellowiana*, *Sanchezia nobilis*, and some species of *Eranthemums* can be used for edging the lawns. Other shrubs such as *Bougainvillea*, *Duranta plumieri*, *Malvaviscus conzattii*, *Malpighia glabra*, *Murraya exotica*, etc., are used as hedges. *Callistemon lanceolatus*, *Brya ebenus*, and *Catesbaea spinosa* look beautiful when planted along a water front.

Classification

From the horticultural point of view shrubs may be classified as (1) shrubs that are grown for their ornamental foliage and beautiful form, (2) flowering shrubs, and (3) shrubs bearing attractive berries.

But from a practical point of view, shrubs are classified into three groups according to their requirement of sunlight under tropical conditions. Most of the tropical flowering shrubs grow best in full sun and their flowering is affected if they fail to get the morning sun. The classification is as follows: (i) shrubs requiring full sun for normal growth and flowering, (ii) shrubs requiring semi-shade for proper growth and flowering, and (iii) shrubs of the intermediate group which grow well in both semi-shade and sun.

The following shrubs should always be grown under partial shade in a typical tropical climate, otherwise they may show signs of scorching and may not survive during hot summer: *Beloperone amherstiae*, *B. guttata*, *Calliandra brevipes*, *C. houstonii*, *C. inaequilatera*, *Excoecaria bicolor*, *Magnolia fuscata*, *M. grandiflora*, *M. mutabilis*, *M. pumila*, *Mussaenda erythrophylla*, *M. philippica*, *Nandina domestica*, *Olea fragrans*, various species of *Pentas* and *Sanchezia nobilis*. The list is not exhaustive, as the capacity of a shrub to grow in a certain place varies greatly according to the climate of a particular place. For example, in a mild climate as of Bangalore many of the above shrubs will grow in the open, whereas in a dry climate as of Delhi, many more shrubs such as *Ixora singaporensis* and *Ardisia umbellata* will have to be added to the list.

The following shrubs are not very particular about sunlight and can be grown well under partial shade or full sun: *Acalyphas*, *Brunfelsia americana*, *B. grandiflora*, *Cerbera fruticosa*, *Cestrums*, *Crossandras*, *Daedalacanthus nervosus*, *Eranthemum laxiflorum*, *E. reticulatum*, *Hamelia patens*, *Jatropha pandurifolia*, *J. rosea*, *Lemonia spectabilis*, *Mussaenda frondosa*, *Plumbago capensis*, *Thunbergia erecta*, and *Turnera ulmifolia* var. *elegans*. Shrubs such as croton and *Eranthemums* with their beautifully coloured foliage thrive very well in shade where other shrubs will not grow properly.

Selection of Shrubs

General Criterion: The selection of a shrub for a particular locality will depend on several factors such as availability of space, sunlight, humidity, temperature, spread and height of the shrub, the type of the shrub, i.e., whether flowering, foliage, evergreen or deciduous, colour of flower, fragrance, etc.

In a small compound with limited space large and spreading shrubs such as *Callistemon lanceolatus*, *Vitex agnus-castus*, *Tecoma stans*, and *Memecylon edule* cannot be grown. But one can very well grow *Allamanda nerifolia*, *Beloperone amherstiae*, *Crossandras*, *Iochroma tubulosa*, *Lantana sellowiana*, *Plumbago capensis* and many others in such limited space. For places not receiving full sun one can select shrubs which are shade-loving, a list of which has been provided in the preceding pages.

The following shrubs are mainly grown for their foliage: *Acalyphas*, Crotons, *Duranta plumieri variegata*, *Eranthemums*, *Excoecaria bicolor*, *Graptophyllum hortense*, *Manihot*, *Nandina domestica*, *Phyllanthus nivosus* var. *roseo-pictus*, and *Sanchezia nobilis*. The great majority of the other shrubs are grown for their ornamental forms and beautiful flowers.

Climatic factors: Some shrubs do better in certain regions, and perform poorly or do not survive at all in certain other regions. The foregoing passages will be helpful for the garden lovers in selecting the right shrub for their locality. Areas having humid and warm climates as around Calcutta, Bombay, Orissa, Assam, Kerala, and other coastal areas and places having a mild climate such as Bangalore and Poona will have no difficulty in growing most of the shrubs, provided the soil is not saline and is well-drained. If the soil is water-logged shrubs such as *Beloperone*, *Hibiscus*, and *Plumbago* will not survive. Many shrubs such as Bougainvillea may produce only vegetative growth in such soils. Bougainvillea flowers best in the dry climate of north India and other places and in the milder climate of Bangalore. The period and intensity of flowering is much less in more humid areas such as Calcutta.

Many of the shrubs like humidity and grow best in Calcutta, Assam, Orissa, Bangalore, and other places having a similar climate or a mild climate, provided drainage is good. A list of such shrubs is given below:

Ardisia umbellata, *Asystasia coromandeliana*, different *Barleria* species, *Beloperones*, *Cananga kirkii*, *Daedalacanthus nervosus*, *Franciscea bicolor* and others, different species of *Gardenia*, *Gustavia insignis*, *Ixora singaporensis*, *Ixora superba*, *Justicia gendarussa*, *Malvaviscus conzatii*, *Magnolia fuscata*. *M. grandiflora*, *M. mutabilis*, different species of *Mussaeanda* and *Pentas*, *Poinesettia pulcherrima*, and *Sanchezia nobilis*. It is possible to grow some of these shrubs in the dry regions under partial shade and by providing artificial humidity by growing them near a pool. When grown in half shade they fail to produce flowers. For example, the authors have the experience of growing some species of *Gardenia* in a humid greenhouse meant for orchids in Delhi. But the plants failed to flower. There are a few *Gardenia* plants in Talkatora Gardens of Delhi, which rarely flower because of shade. *Ixora singaporensis*, a nice *Ixora*, performing beautifully in West Bengal, Bangalore, and other humid areas fail to grow in the hot and dry climate of Delhi and surrounding areas.

Again many of the shrubs are affected by frost in northern India. But the severity of damage varies from shrub to shrub. Some of these are severely affected and get killed, while others suffer injuries from burning of leaves to defoliation and drying-up of young shoots. In spite of special care the following shrubs do not survive the severe winter frosts of north Indian plains in the open or in partially shaded places: *Beloperone amherstiae*, *B. guttata*, *Brya ebenus* *Cananga kirkii*, *Excoecaria bicolor*, different species of *Gardenia*, *Ixoras*, especially the species *I. singaporensis* and *I. superba*, *Manihot utilissima* var. *variegata*, different species of *Mussaenda*, *Pentas*, and *Portlandia*. All the young shrubs of the above species will perish in the winter unless protected from all sides. The younger branches of well-grown-up plants of *Brya*, *Cananga kirkii*, and *Manihot* get killed in partially protected places but new sprouts come up from the survived matured canes on the onset of hot weather.

The following shrubs are less affected and are not generally killed by frost if grown under trees, only their leaves get scorched and in many cases the shrubs get defoliated: *Ardisia umbellata*, *Barleria* different species: *Citrus limonia* var. *variegata*, *Duranta plumieri*, *D. plumieri* var. *variegata*, *Hamelia patens*, *Iochroma tubulosa*, *Justicia carnea*, *J. gendarussa*, *Phyllanthus nivosus roseo-pictus*, *Poinsettia pulcherrima*, and *Sophora tomentosa*.

One way of protecting the frost-susceptible plants is to grow them under other tolerant tall shrubs or trees. When planted in such a manner the susceptible plants should be situated on the eastern side and guard plants should shelter the western and the northern sides. The dwarf shrubs can be grown under the protection of frost-hardy tall shrubs such as *Callistemon lanceolatus*, *Dombeya*, etc., whereas the susceptible taller shrubs may be grown under such flowering trees as *Cassia fistula*. Another way of growing the less susceptible shrubs is to plant them in front of a wall or in a corner of a building facing east, protected from west and north. A cultural method of frost protection is to flood the soil and spray the plant thoroughly with water late in the evening and if possible once at night. Whenever possible, young plants should be covered with *Sirki*, mat made of *Saccharum munja* leaves (known as *Kash* or *Sar* in Bengali, *Munja* or *Sarkanda* in Hindi, *Kharkara* in Punjabi, and *Gundra* or *Ponika* in Telugu), or thatched around with straw and bamboo during nights. Even the grown-up plants of the most delicate shrubs will need this protection.

Planting, Cultivation, and Propagation: As in the case of large trees or roses, shrubs will also need initial preparation of the soil. When planted in a shrubbery, the entire bed should be dug up thoroughly. For the larger shrubs, which grow 1.5 m and above, pits measuring 75 cm × 75 cm should be dug at the places where the shrubs are to be planted. For dwarf shrubs thorough digging up to a depth of 45 cm will normally be sufficient. The dug-up soil should be mixed thoroughly with well-rotten cowdung manure at the rate of 4.5 kg per pit for larger shrubs, and 4-5 kg per square metre for other shrubs. If cowdung is not available, well-decomposed leafmould should be added. In areas such as Bangalore, having red soil, well-rotten horse manure may be used in place of cowdung. It will be beneficial if a handful of bonemeal is added to each shrub at the time of planting. The same dose of manure and bonemeal should be repeated once a year, preferably before the rains.

The digging up of pits or beds should start just at the beginning of rains, so that the dug-up soil settles down with the first two or three showers. After planting the shrub, which should be done during the rains, the soil around it should be raised to a height of 15-20 cm to prevent water-logging. The spacing in the shrubbery depends upon the size and nature of growth of the plants. When planting singly in rows, the distance will also depend on the growth habit of the shrub, but to some extent this is conditioned by the availability of space and the personal preference of the grower. Large, medium, and bushy shrubs are planted at a distance of 1.5 to 2.5 metres depending on the nature of growth. Many people do not like the vacant interspaces in the shrubbery. This can be filled up with different sub-shrubs or herbaceous plants such as *Angelonia*, *Asystasia coromandeliana*, *Impatiens*, and *Vinca rosea* or with trailing or low-growing herbs such as the different species of *Pileas*, perennial verbena and *Zebrina pendula* or by covering the spaces with bulbs such as *Haemanthus multiflorus*, *Zephyranthes* species, and *Oxalis* species and sometimes by growing shade-loving annual flowers.

Most of the shrubs are propagated from cuttings, by air layering or ground layering and from seeds. Budding or grafting is done in some delicate shrubs such as camellia and flowering cherries.

Irrigation

Shrubs need watering regularly. During the summer months, watering should be frequent, once in 7-15 days depending upon the severity of the summer and the age and nature (dwarf or tall) of the shrub. In the winter also, watering once in 10-15 days will be necessary. In the milder climate of Bangalore, watering once in a fortnight will be sufficient. During the rains it may not be necessary at all to provide irrigation. It is to be remembered that one deep soaking at an interval is far better than frequent sprinkling. Sprinkling of water will result it the growth of a shallow root system. Shortage of water in the soil will result in wilting and defoliation and ultimately the plant may die.

Generally speaking, shrubs are quite hardy and are not attacked by many pests or diseases. Sometimes, leaf-eating caterpillars may infest the shrubs. In humid areas, the shrub *Nycatanthes arbor-tristis* gets infested with swarms of hairy caterpillars. There is hardly any serious threat from diseases. For the control of insect pests, the chapter on Pests and Diseases may be consulted.

Pruning

There is a general practice of pruning the plants at the planting time. It may be well assumed that plants received with bare-roots must have lost about one-third of its root system while digging it out of the soil. So it will be appropriate to remove about the same length in the branch system to draw up a balance between root and shoot growth. If the shrub is a bushy one the nature of pruning may be to remove about one-third of the branches, preferably the weaker ones. In the case of a shrub having a thin branch system, cutting back of one-third of the branches will be necessary. But generally for the bushy shrubs also, thinning out of weak branches and cutting back of other branches are both done simultaneously.

The second pruning is done to remove all the damaged roots by cutting back to healthy tissues. The long straggling roots devoid of proportionate fibrous roots should be reduced to encourage the growth of fibrous roots.

There is no hard and fast rule for pruning of established shrubs and it is difficult to give any general pruning advice. Familiarity with the shrubs will help one to standardize the pruning methods for each individual shrub during the planting time or in subsequent years. The only general suggestion that is possible is against shearing, unless, of course, the shrubs have been grown as hedge. There are many shrubs such as *Hibiscus syriacus* and *Viburnum carlcephalum* which may need little or no pruning.

In cold countries, most shrubs are pruned after spring flowering. Some dormant shrubs are pruned during the winter also. But if there is a danger of frost damage, the pruning may be delayed till February and March. In tropical regions, the best time for pruning is before the rains or just after flowering is over. Large shrubs such as *Dombeya*, *Vitex*, etc., will need pruning to keep their growth in check. Weak growing shrubs may be pruned to encourage growth and flowering. The principle of pruning is to remove dead wood and twiggy and overgrown branches. All the suckers and branches are removed from the base to open up the shrub. *Russelia juncea* often gets crowded and may be thinned by separation of root suckers. In many shrubs, pruning will remain limited to cutting off the dried flowers to conserve plant energy, as in the case of azaleas. Some berry-bearing plants such as *Cotoneasters* are also not pruned heavily. During the rains, cuttings and air-layers are raised in the garden and in the process shrubs undergo a sort of pruning. If done judiciously, this type of pruning will be sufficient for most of the shrubs.

Some Important Shrubs

Abelia × grandiflora (Syn. *A. chinensis*); Family: *Caprifoliaceae*.

This is a hybrid between *A. chinensis* and *A. uniflora*. The shrub grows to a height of 1.5-2 m. The white scented flowers are borne in clusters during the summer. The persistent brown sepals are attractive. The plant prefers a semi-shaded and well-drained soil. The plant is propagated from cuttings. The shrub is suited only for medium and high elevations.

Abutilon; Family: *Malvaceae*; Origin: Mostly Brazil; Common Names: Flowering Maple, Chinese Lanterns, or Chinese Bell Flower.

Abutilons grow better on medium to high elevations and attain a height of 1-1.5 m. It is propagated from seeds or cuttings. Seeds should be soaked for a couple of hours before sowing. They prefer partial shade and shelter from wind and excessive rains.

The flowers are hibiscus-like. In *A. striatum* var. *thompsonii* (Syn. *A. thompsonii*), the leaves are richly mottled yellow and green. It is grown for its foliage. *A. indicum* has yellow flowers and can be grown in the plains. *A. pictum* is a short shrub with orange-yellow veined crimson flowers. *A. megapotamicum* has rich red calyx and yellow corolla. *A. venosum* is a short growing shrub, having golden-orange flowers, veined blood-red and brown.

Acacia farnesiana (Syn. *A. leptophylla*); Family: *Leguminosae*; Origin: U.S.A., Mexico, Asia, Australia; Common Names: Fragrant Acacia, Opopanax, Popinac; Vernacular Names: Hindi – *Vilayati kikar* or *babul*, Kannada – *Kasturi jali*.

A spiny, much-branched, spreading shrub, bearing tassel-like scented yellow flowers in the cold season. Dried flowers are used for scenting clothes. The shrub forms a good hedge when trimmed. It is easily propagated from seeds and cuttings.

Acalypha; Family: *Euphorbiaceae*.

Acalyphas are a group of shrubs bearing colourful leaves, and are used as hedge and also for providing colour in the shrubbery. These are quick-growing and can be grown in any soil. When used as hedge *Acalyphas* are pruned regularly and pruning is done on the onset of monsoon. Propagated from cuttings. The following species are important:

(a) *A.* 'Batik': A beautiful shrub of garden origin, having dark streaks on the leaves.

(b) *A. godseffiana:* A bushy shrub, the leaves of which are green with a prominent yellow coloured margin. Origin: New Guinea. In var. *heterophylla* the narrow lacy leaves have wavy edges and are coloured green with pale yellow.

(c) *A. hamiltoniana:* The leaves are green, the rounded teeth yellow. Plants are taller than the above. A shrub of garden origin.

(d) *A. hispida* (Syn. *A. sanderi*)*:* Commonly called red hot cat's-tail or chenille plant. In Bengali this is known as *Hatisunr*. A bushy shrub, but may grow tall (3 m), and is cultivated mainly for its long, red, drooping amaranthus-like flower spikes. In var. *alba* the spike is creamy-white, and in var. *ramosa*, the upper spikes are branched.

(e) *A. mortfontanensis* (*A. hamiltoniana × A. wilkesiana marginata*)*:* The leaves are bigger than *A. wilkesiana marginata*.

(f) *A. torta*. The dark olive-green leaves are cut into blunt segments and curiously contorted. A shrub of West Indies origin.

(g) *A. wilkesiana* (Syn. *A. tricolor*): Several varieties of shrubs, 1.8-3 m tall, and with leaves blotched and mottled with various shades of crimson, red, and brown. Var. *aliporensis*, a sport of *musaica* with creamy margins; var. *macafeeana*, leaves cordateovate, russet brown, blotched with paler spots. It is a very handsome plant. Var. *marginata* has olive-brown edged rosy-carmine leaves. Var. *musaica*, leaves bronzy-green with orange and red markings; var. *obovata*, leaves are obovate, green edged with white when young but changing to bronzy-green with rose pink margins.

Acanthus montanus; Family: *Acanthaceae*; Origin: Tropical Africa.

A shrub about 1 m tall. The leaves are up to 30 cm long and 15 cm broad, spiny margined and lobed. The rose pink flowers are borne on a terminal spike of 25 cm length. Propagated from seeds and cuttings.

Achania: See *Malvaviscus*.

Adenium obesum (Syn. *A. coetanum*); Family: *Apocynaceae*; Origin: Aden.

An erect shrub, growing to a height of 2-3 m. The stem and branches are succulent and the stem swollen at base. Flowers are borne during June to September in terminal many-flowered corymbs. Flowers are funnel-shaped, pinkish-crimson, yellow inside with reddish longitudinal stripes. A slow-growing shrub, suitable as a pot plant and ideal for the rockery. The plants are propagated from seeds and by air layering.

Aerva tomentosa; Family: *Amaranthaceae*; Origin: Tropical Asia and Africa.

A handsome dwarf shrub, 50 cm to 1 m tall, with woolly, silvery leaves. The flowers are white and borne on long spikes. The shrub is suitable for pots, shrubbery and herbaceous border. Propagated from seeds or cuttings.

Aesculus parviflora (Syn. *A. macrostachya*); Family: *Hippocastanaceae*; Origin: South U.S.A.

A bushy shrub, 2-3 m tall. Leaflets 5-7, oval to obovate, 10-25 cm long. White flowers are borne in panicles during August. Propagated by layering and from seeds.

Aglaia odorata; Family: *Meliaceae*; Origin: China.

A shrub about 2 m tall, bearing small, yellow, fragrant flowers during the summer and the rains. The Chinese use the flowers for scenting tea. Suitable for shrubbery and as a hedge. Propagated from cuttings.

Allamanda cathartica (Syn. *A. linnaei*, *Orelia grandiflora*); Family: *Apocynaceae*; Origin: Brazil.

A common quick-growing shrub for the tropics. It can also be grown as a climber. The large golden-yellow flowers are marked white in the throat. The plant flowers for a prolonged period during the summer and the rains. This should be pruned after flowering. Propagation is by cutting or layers. In var. *schottii* (Syn. *A. schottii*) the leaves are broad and the corolla tube is rich yellow and long. It is a vigorous climber. Var. *hendersonii* (Syn. *A. hendersonii*, *A. wardleyana*) has orange-yellow flowers.

Allamanda neriifolia; Family: *Apocynaceae*; Origin: South America.

A shrub or a half climber, about 1 m tall. The flowers are smaller than the above, rich yellow in colour. Propagation is the same as in the above species.

Allamanda violacea (Syn. *A. blanchetii* or *A. blanchettii*); Family: *Apocynaceae*; Origin: South America.

A shrub about 1.5 m tall. The branches are hairy when young. Flowers are showy, light violet in colour. The shrub when grown on own roots is a poor grower and hence it is grafted on *A. cathartica* var. *schottii* or var. *hendersonii*. The shrub needs a sunny position and a well-drained soil. The flowering is in rainy or hot season.

Angelonia: See Chapter VII, page 114.

Aphelandra tetragona (Syn. *A. cristata*); Family: *Acanthaceae*; Origin: West Indies.

A straggling shrub about 1 m tall. The leaves are lanceolate, tapering at the apex, and are up to 20 cm long. Brilliant spikes of scarlet flowers are borne during March. It grows in partial shade where many other shrubs may fail to grow. Cuttings of half-ripened wood strike roots best. Young shoots with a heel are also useful for cuttings. It prefers a cool and humid atmosphere.

Aralia: See Chapter XIII.

Ardisia crispa (Syn. *A. crenulata*); Family: *Myrsinaceae*; Origin: East Indies.

A shrub 90-120 cm tall. The shrub is chiefly grown for clusters of small, coral red, long lasting fruits, which are preceded by fragrant flowers. This can be grown in rock gardens, pots, or as border.

Ardisia humilis; Family: *Myrsinaceae*; Origin: East Indies, India; Common Names: Hindi – *Kadna, Banjam*, Kannada – *Kantena bodina gida.*

A tall (up to 3 m), erect shrub, with leathery leaves. The umbels of rose-pink flowers are followed by shining black berries.

Ardisia paniculata; Family: *Myrsinaceae*; Origin: India.

A large shrub. The flowers are rose coloured and the berries bright red.

Ardisia umbellata; Family: *Myrsinaceae*; Origin: India.

The flowers are purplish and the berries purplish red. All the *Ardisias* are propagated either from seeds (to be sown soon after harvest), suckers, or cuttings of half ripened shoots. They thrive better in a humid tropical climate but are susceptible to frost. The flowers in all the species are fragrant and the flowering time is during the summer and the rains, but sporadic flushes are continued throughout the year.

Argyreia speciosa; Family: *Convolvulaceae*; Origin: India.

A strong grower which needs frequent pruning to keep growth in check. Mostly used as a hedge. The bell-shaped large deep rose flowers are produced all the year round, but more profusely during the rains. Propagated from cuttings.

Artabotrys: See Chapter X, page 182.

Asclepias currassavica (also spelt *curassavica*); Family: *Asclepiadaceae*; Origin: Tropical America; Common Names: Blood flower, Milkweed or Silkweed.

The herbaceous shrub grows to a height of 90 cm, having downy, lanceolate leaves. The erect umbels of orange-yellow flowers bloom during the summer and the rains. This is good for borders or as specimen bush in the lawn. Once planted, it starts to spread like a weed. Propagated from seeds or cuttings. It is better to replace old plants by new cuttings.

Asystasia coromandeliana (*A. violaceae*); Family: *Acanthaceae*; Origin: India.

A straggling herbaceous plant rather than a shrub, attaining a height of 45 cm. It flowers all the year round. The white or cream flowers have a yellow tube. Thrives best in partial shade. A good herb for ground cover and rock garden. The other species are *A. travancoriana* which bears purple flowers and *A. bella* which is a sub-shrub, 90-120 cm tall, with bell-shaped lilac flowers. Propagated from tip cuttings or seeds.

Atalantia monophylla (Syn. *Limonia monophylla*); Family: *Rutaceae*; Origin: India.

A tall, usually spiny, shrub which looks more like a tree because of the stout woody stem. The white, small, scented flowers are borne in small clusters, more profusely in the summer and the rains. Difficult to propagate where seed is not produced. Air layering with the help of hormone is successful.

Aucuba japonica; Family: *Cornaceae*; Origin: Himalayas to Japan.

A bushy, tall (2-4 m) shrub, best suited for the most shady situation. The long (7-20 cm) leaves are leathery, narrowly oval and shining. The bright scarlet berry-laden potted plants are popular as house plant. Most of the garden forms have variegated leaves (e.g., var. *latimaculata* and var. *variegata*). Propagated from half-ripened green-wood cuttings or freshly harvested seeds.

Azaleas; Family: *Ericacea*.

Botanically the azaleas are included in the genus *Rhododendron*. Some of the Indian and Chinese species are *R. indicum*, *R. kaempferi*, *R. ledifolium*, and *R. poukhanense*. There are innumerable varieties and hybrids which are in cultivation. Some other hardy deciduous azaleas are *R. arborescens*, *R. canadense*, *R. japonicum*, *R. nudiflorum*, *R. occidentale*, and *R. rhombicum*.

Azaleas can be grown only at high elevations in India. In other places they do not do well. Azaleas do not like lime soils. They need a lot of care in cultivation and should be grown in a sheltered position. A light soil with plenty of leafmould is good for azaleas. Use of any strong manure is to be avoided. The Chinese azaleas are free-flowering and the flowers are either single or double and come in white, pink, rose, scarlet red and purple colours. Azaleas are multiplied by division, layering, or seeds. Some delicate varieties are raised by veneer- or tongue-grafting on hardy azalea stocks such as *R. luteum*.

Barleria; Family: *Acanthaceae*; Vernacular Names: Hindi & Bengali – *Jhati* or *Jhate*, Tamil & Kannada – *Spatica*.

A group of evergreen bushy shrubs, attaining heights of 60-150 cm. The flowering is mostly during the rains but continues if pruned after each flowering. These are suitable in mixed borders or as ornamental hedges. The flowers are used for *puja* and hair decoration. The shrub is raised from cuttings and seeds. The common species are:

(a) *B. cristata:* Origin India. Flowers are violet-blue. Hybrids with white, rose, pink coloured flowers. Flowers with white and mauve variegation are also available.

(b) *B. flava* (Syn. *B. gentianoides, Dianthera flava*): Origin Arabia. Shrubs are without spine, 90 cm tall, and bear yellow flowers.

(c) *B. gibsonii:* Origin India. A glabrous, branching shrub with pale bluish-purple flowers.

(d) *B. strigosa:* Origin India. Bears large blue flowers.

(e) *B. tomentosa:* Origin South India. Flowers are rose coloured.

Bauhinia acuminata; Family; *Leguminosae*; Origin: India, Burma, China, Malaysia; Common Name: Orchid Tree.

A quick-growing shrub, 2.5-3 m high. A hardy shrub, suitable for home gardens, parks, and public places. Flowers (white) pofusely in the summer and continues till September. Propagated from seeds which are produced in abundance.

Bauhinia galpinii; Family; *Leguminosae*; Origin: Tropics and South Africa; Common Name: Pride of the Cape.

A semi-climbing shrub, 2-3 m tall. Bears brick-red flowers in the rains. Flowering continues till October. The petals are clawed. It is slow-growing and is raised from seeds. Seedlings flower in the third year.

Bauhinia tomentosa; Family: *Leguminosae*; Origin: Sri Lanka; Common Names: St. Thomas Tree, 'Bell-bauhinia'.

The shrub grows to a height of 2.5-3.5 m. The drooping large sulphur-yellow flowers are produced during August-October.

Beloperone amherstiae; Family: *Acanthaceae*; Origin: South America.

A small shrub (90 cm tall), very ornamental and always in bloom. The flowers are insignificant but the large showy bracts are most attractive. Likes semi-shaded situation. Propagated from cuttings and by air layering.

Beloperone guttata; Family: *Acanthaceae*; Origin: Mexico.

A small shrub which rarely grows beyond 90 cm. The terracotta-red orderly arranged bracts are very showy. Propagation and habit are the same as above.

Berberis; Family: *Berberidaceae*.

A genus of decorative spiny shrubs, varying in height from 30 cm to 360 cm. Flowers are attractive, though not conspicuous. Most of the shrubs bear berries which are ornamental. Flowering is during the spring and early summer. Propagated from cuttings or by air layering. Some rare kinds are grafted on *B. thunbergii* or *B. vulgaris*. Suitable for growing at higher elevations. Some common species are:

(a) *Berberis asiatica:* A yellow flowering shrub, 2-3 m tall, bearing purple berries. A shrub of Indian origin (Himalayas).

(b) *B. bhutanensis:* The shrub grows up to 2 m, and bears yellow flowers and black fruits. Centre of origin is Bhutan and Burma.

(c) *B. vulgaris:* Common Berberry. Flowers are bright yellow and fruits are scarlet at first but finally turn purple.

Bougainvillea: See Chapter IX, page 171.

Brugmansia: See *Datura*.

Brunfelsia; Family: *Solanaceae*.

A genus comprising a group of handsome, free-flowering, evergreen shrubs, blooming scented, funnel-or salver-shaped flowers. The larger shrubs bear more flowers and hence the annual pruning should be minimal. They grow in moist tropical conditions and can be grown in a semi-shaded situation also. A weak manure solution, applied during the growing season, is beneficial. Plants are propagated by air layering, and from cuttings, or seeds. This genus is allied to *Franciscea*. The important species are:

(a) *B. americana:* A handsome, perpetually flowering, medium tall (1-1.8 m) shrub, bearing white flowers which turn yellow. Centre of origin is West Indies.

(b) *B. calycina* var. *macrantha* (Syn. *B. grandiflora, B. lindeniana*)*:* The leaves are as long as 20 cm and 5-6 cm broad, and rich dark green. Flowers are purple with a prominent lavender blue ring around the white eye in the mouth of the corolla tube. Most floriferous. Origin is Peru. The var. *floribunda* has large violet flowers with a white eye, but quickly fading to white and called 'Yesterday-today-tomorrow'.

(c) *B. hopeana* (Syn. *Franciscea hopeana, F. uniflora*)*:* Common Name: Yesterday-today-tomorrow. A bushy shrub, 2 m tall. Flowers profusely during the summer and the rains. It grows well both in the sun and in the shade. The scented flowers open as violet with a yellow centre, then change to purple, finally fade into white. Thus, in the same bush three types of flowers are seen and hence the common name. Propagated by air or ground layering. Origin is Brazil.

(d) *B. latifolia* (*Franciscea latifolia*)*:* A small (60-90 cm) neat shrub. Flowers open as lavender with a white eye and later change to white and are very fragrant. Sometimes this shrub is also known by the common name Yesterday-today-tomorrow, but more popularly as 'Kiss-me-quick'.

Brya ebenus; Family: *Leguminosae*; Origin: Jamaica.

A most beautiful shrub with its pendulous and drooping branches. It grows to a height of 2 to 3.5 m. The small leaves are ivy-green in colour. There are small spines at the axils of the leaves. The drooping branches get covered with small yellow flowers during the summer and the rains. To make it dwarf and bushy the top of the main shoot may be cut. It is good as a backdrop in a shrubbery, well-suited for rock gardens, and is planted as specimen plant. This is also suitable for planting near a water surface. The plant is injured or even killed by frost. Propagated by air layering or from seeds. Seedlings produce flowers in the second year.

Buddleia (Syn. *Buddlea, Buddleja*); Family: *Loganiaceae*.

A group of large attractive shrubs with long perfect lanceolate leaves, the undersurface of some of which is silvery. The branches are quadrangular. The flowering is profuse in long dense panicles, composed of small flowers. The plants thrive well in a rich, well-drained soil and a sunny position. The shrubs need a lot of space. Propagated from seeds, cuttings, and layers. *Buddleias* do not bloom in the plains as profusely as they do at high altitudes. The following species are important:

(a) *B. asiatica* (*B. neemda*): Origin – China, India, Java. A shrub 1.5-2 m tall. The leaves are perfectly lanceolate, 15-20 cm long, whitish underneath. The branches are covered with white or buff coloured tomentum when young. The sweet-scented white flowers are produced in profusion in drooping terminal or axillary panicles, during February-April. Propagated from cuttings. The plant should be pruned after each flowering.

(b) *B. davidii* var. *veitchiana* (*B. variabilis* var. *veitchiana*): Origin China. Flower panicles are dense and large with bright mauve-coloured flowers with an orange eye. Sometimes also spelt as *B. davidi.*

(c) *B. lindleyana* (Syn. *B. salicifolia*): Origin China. A small shrub, 1 to 1.5 m high. The purplish violet flowers are borne in erect, dense racemes, which are scented. The flowering is in great profusion during the summer months. After each flowering the shrub is pruned to 60 cm from ground. In the plains, the shrub may be grown as a biennial and replaced by fresh rooted cuttings. It cannot stand water-logging.

(d) *B. madagascariensis* (Syn. *B. heterophylla*): Origin Madagascar. A straggling shrub, 2-6 m tall, with densely tomentose branches. Leaves are dark green above and silvery or yellow underneath. Orange-yellow flowers are borne in large terminal panicles during January to February. The shrub is suitable for large shrubberies and need severe pruning after each flowering.

(e) *B. officinalis:* Origin China. The shrub grows up to 2.4 m. The nearly-terete branches are greyish tomentose. The lilac flowers with orange eye are borne in 10-15 cm long panicles during the autumn.

Caesalpinia gilliesii; Family: *Leguminosae*; Origin Argentina.

An erect bushy shrub with handsome bipinnate feathery foliage. The yellow flowers with brilliant red stamens are borne in terminal racemes during the hot and rainy seasons. Needs pruning. Propagated from seeds.

Caesalpinia pulcherrima (Syn. *Poinciana pulcherrima*); Family: *Leguminosae*; Origin: Tropics. Vernacular Name: Bengali – *Krishnachura.*

A beautiful bushy shrub, with a few scattered prickles, growing up to 2.5 to 3 m. The mimosa-like leaves are handsome. The orange-scarlet flowers are borne in profusion throughout the year. A very hardy shrub, it grows in any soil receiving full sun. Pruning is done to make it bushy. It is a good shrub for hedge and adds colour to shrubbery. Propagated from seeds.

Caesalpinia pulcherrima var. *flava* (Bengali – *Radhachura*).

A variety with yellow flowers. If pruned every year the plant can be given a round shape. Each branch bears flowers during the summer and the rains. Propagated from seeds.

Calliandra; Family: *Leguminosae*; Origin: Mostly of Tropical America but also India.

A group of beautiful shrubs with handsome bipinnate feathery leaves and powder-puff like flowers. The following species are important:

(a) *C. brevipes:* A Brazilian bushy shrub, 70 cm to 1.2 m tall, which thrives best in semi-shade and well-drained soil. Flowers come in clusters with fine pink stamens. Plants are propagated from seeds or layers. Flowering is almost all the year round.

(b) *C. haematocephala* (Syn. *Inga pulcherrima*): A bushy shrub attaining a height of 2.5 m but extensively spreading in habit. The double pinnate leaves are most handsome. The bright crimson-coloured stamens give the flowers a powder-puff like look. Flowering time is the autumn and the winter. Seedlings grow slowly, hence propagated by air layering. A native of Brazil.

(c) *C. houstonii:* A free-flowering, tall (2.5 m) shrub, which grows better in semi-shade. The flowers are scarlet and tassel-like and is in bloom throughout the year. Propagation is by layers.

(d) *C. inaequilatera:* Commonly known as 'Red Powder-puff', it is a beautiful tropical *Calliandra*. The inflorescence is large, globose, and dense, with silky bright red stamens. A native of South America. This is a large shrub.

(e) *C. speciosa:* A beautiful shrub which deserves a place in every garden. The scarlet flowers are borne all over the plant. The flowers appear in flushes at regular intervals almost throughout the year. It prefers full sunshine and grows best in a rich soil. The plants are raised from seeds as vegetative propagation is difficult.

(f) *C. tweedii* (also spelt as *C. tweedyi*): A very free-flowering, hardy shrub, blooming almost round the year. The flowers are red to red-purple. Seedlings are slow-growing and hence should be propagated by air or ground layering. A native of Brazil.

A white flowering species is also in cultivation lately. In *Calliandra* the petals are usually small and it is the bristle-like stamens which are most handsome, and when the colour of the flowers is referred to it is actually the colour of the stamens. *Calliandras* are good for shrubbery and also for planting as individual specimens in lawns or along paths.

Callicarpa cana; Family: *Verbenaceae*; Origin: China, East India, the Philippines.

A quick-growing bushy shrub, attaining a height of 2.5 m. The small lilac flowers are borne in the axils of the leaves during the spring, the summer, and the rains. This also bears deep purple fruits. It is advisable to prune after flowering to keep it in shape. Raised from cuttings. *C. americana* with bluish flowers and *C. japonica* with pale pink flowers are the other important species.

Callistemon: See Chapter XI, page 211.

Calluna vulgaris (Syn. *Erica vulgaris*); Family: *Ericaceae*; Origin: North Europe, Asia Minor, Eastern North America.

A dwarf (15-90 cm tall), spreading, from almost prostrate to erect shrub, suitable for acid soil. The pink flowers are borne along the branches in long erect racemes from the late summer to the fall. A white-flowered variety (var. *alba*), suitable for high altitudes, is also available.

Camellia: See Chapter XI, page 212.

Cananga (also written as *Canangium*) *kirkii*; Family: *Annonaceae*; Origin: India, Malaysia.

A bushy shrub, between 2.5 and 3 m tall. The solitary, greenish-yellow, long-stalked, fragrant flowers start blooming from March and continue to bloom till August-September. Propagated by air layering.

Carissa carandas; Family: *Apocynaceae*; Origin: India, Java; Common Names: Carunda, Christ's Thorn; Vernacular names: Bengali – *Karamcha*; Hindi – *Koronda, Karaunda*; Tamil – *Perungala*; Telugu – *Oka, Vaka*; Kannada – *Heggarichige*; Marathi – *Karavanda, Boronda*.

The plant bears scented jasmine-like flowers. A thorny, bushy plant (2.5-3 m tall), grown mainly for the berries (black when ripened), which matures during June. A good hedge plant and can be planted in a large shrubbery. Propagated from seeds. The fruits are made into pickles.

Caryopteris incana (Syn. *C. mastacanthus*, *C. sinensis*); Family: *Verbenaceae*; Origin: China, Japan.

The shrub grows to a height of 1.8 to 2.5 m and produces lavender-blue flowers profusely during the winter and the spring. The shoots are pruned to 50-70 cm during the rains. Propagated from cuttings.

Cassia alata; Family: *Leguminosae*; Origin: Tropical America.

A quick-growing erect shrub, 2-2.5 m tall. Bears yellow flowers during the winter. Propagated from seeds.

Cassia artemisioides; Family: *Leguminosae*; Origin: Australia.

A much-branched, bushy and showy shrub, attaining a height of 2.5 m or more. The silvery, grey-green foliage is very decorative and the leaflets turn into needle-like structures. The numerous yellow flowers appear during the winter and emit fragrance. This can be grown in any part of India, especially in the dry tracts. The shrub cannot stand water-logging and frost at young stage. Propagated from seeds which are produced in plenty.

Cassia biflora; Family: *Leguminosae*; Origin: West Indies, South America.

A beautiful shrub with handsome feathery foliage. The leaflets are small (2-4 cm long), oval-oblong. The numerous large (3.5-5 cm across) yellow flowers are borne freely during July-August, but sporadic flowering in other seasons is also noticed. Propagated from seeds.

Cassia glauca; Family: *Leguminosae*; Origin: India, Sri Lanka.

A handsome, tall (3 m) shrub or small tree. Clusters of yellow flowers appear in large numbers throughout the year. Propagated from seeds.

Cassia laevigata; Family: *Leguminosae*; Origin: Tropical America.

An erect (3 m tall), glabrous shrub. Can be grown in full sun as well as in partial shade. The shrub blooms a mass of yellow flowers 8-10 months in a year. Propagated from seeds.

Catesbaea spinosa; Family: *Rubiaceae*; Origin: West Indies; Common Name: Lily Thorn.

A slow-growing, spiny shrub, 1.5-2 m tall. The small shiny leaves are attractive. The terminal branches droop downwards and the funnel-shaped pale yellow or creamy-white flowers hang down in a conspicuous way as if these have been just suspended from there. The shrub is planted as specimens in the lawn, rock gardens, and near water surfaces.

Cerbera fruticosa (Syn. *Kopsia fruticosa*); Family: *Apocynaceae*; Origin: India, Malaysia.

A shrub of compact and bushy growth, attaining a height of 1.75 to 2.5 m. The funnel-shaped pinkish-white flowers, with red tips and centre, are borne throughout the year. It grows well in both sun and semi-shade. Suitable for growing in shrubbery or can be planted for screening purposes. Propagation is by air layering.

Cestrum aurantiacum; Family: *Solanaceae*; Origin: Guatemala.

A shrub of rambling habit, attaining a height of 2.5 m. It can be grown in half-shade. The branches droop on all sides and the bright orange-yellow flowers are borne in great profusion during the cold months. After flowering the plants are pruned. In climates as of Bangalore it blooms in the summer. It is propagated from cuttings.

Cestrum diurnum; Family: *Solanaceae*; Origin: West Indies; Common Names: Day Jasmine or Day Queen.

An evergreen, quick-growing, bushy shrub, which blooms white scented flowers during day time in the summer months and continues to bloom till the rains. The shrub needs pruning after flowering. It is propagated from seeds or cuttings.

Cestrum elegans (Syn. *C. purpureum*); Family: *Solanaceae*; Origin: Mexico.

A plant suitable for temperate climate and high altitudes. The purple-red flowers appear in September and December and are followed by purple berries.

Cestrum nocturnum; Family: *Solanaceae*; Origin: West Indies; Common Names: Queen of Night, Night Blooming Jasmine; Vernacular Names: Hindi – *Raat-ki-Rani*; Bengali – *Hasnahena*.

A quick-growing bushy shrub, it is grown most widely in Indian gardens for its scented flowers. Several main branches arise from the ground level and these again divide into sub-branches on the top. The shrub may be given a bushy appearance by proper pruning. The long, lanceolate leaves, pointed at the apex, are attractive. The creamy-yellow or greenish-yellow small tubular flowers are produced in clusters in great profusion and are highly fragrant during night. It is a hardy plant and is propagated from cuttings.

All the *Cestrums* can be grown up to an altitude of 900 m.

Chaenomeles lagenaria: See *Cydonia japonica*.

Citharexylum (sometimes spelt *Citharexylon*) *fruticosum*; Family: *Verbenaceae*; Origin: West Indies.

A large shrub or small tree with long lanceolate leaves. The shrub bears highly scented white flowers. A hardy shrub, suitable for large shrubbery or planted singly. Propagated from long cuttings. This is also referred to as *C. cinereum*.

Citrus limonia var. *variegata*; Family: *Rutaceae*; Origin: Of garden origin.
A lemon with variegated foliage, and about 1.5 m tall. A bushy, quick-growing shrub for sunny location and needs pruning to keep in shape. The plant is injured by severe frost. Propagated by air layering.

Clerodendron (Syn. *Clerodendrum*) *fallax* (Syn. *C. speciosissimum*); Family: *Verbenaceae*; Origin: Java.
A handsome shrub with large, green, cordate-ovate leaves. The bright scarlet flowers are produced in clusters of 45 cm length. The plant grows best in semi-shade and is propagated from seeds.

Clerodendron fragrans; Origin: Japan, China.
A hardy shrub of 1-2 m height, which grows best in moist and semi-shaded places. The shrub bears usually double, white, highly fragrant flowers in large terminal clusters during the summer and rainy seasons. Propagated from root suckers or cuttings. In var. *leniflorum*, the white flowers are suffused with pink and double.

Clerodendron inerme; Origin: India; Common Name: Forest Jasmine.
A straggling shrub, which forms a compact and clean hedge. The shrub can be pruned very low and it does not produce any leafless woody stem, but remains evergreen. The white fragrant flowers appear during the rains. Easily propagated from cuttings. It is widely used for topiary.

Clerodendron kaempferi (Syn. *C. squamatum*); Origin: China.
A showy shrub with large rounded leaves, 1.5 to 2 m tall. The scarlet flowers are borne in large clusters and are long lasting. Prefers partial shade. A good pot plant which is propagated by tip cuttings.

Clerodendron macrosiphon (Syn. *C. incisum*, *C. incisum* var. *macrosiphon*); Origin: East Africa.
An elegant erect shrub which never grows beyond 1 m. The white flowers are borne in terminal clusters during the summer and the rains. The stamens are red purple. The flowering is in several flushes, but the flowers are short-lived. Suitable for growing in the front row of a shrubbery and rock garden. Propagated from cuttings or by layering. Susceptible to frost.

Clerodendron siphonanthus (Syn. *Siphonanthus indica*); Origin: East India; Common Names: Turk's Turban, Tube-flower.
An erect-growing shrub, about 3 m tall. The leaves are narrowly lanceolate. The long, white, tubular flowers hang from the upright branches. Bears red and purple berries. Requires partial shade. Propagated from cuttings.

Clerodendron thomsonae (Syn. *C. balfouri*); Origin: West Africa.
A tall twining evergreen shrub or a climber. The showy flowers, the petals being scarlet and calyx white, are borne in large numbers during the rains. It is propagated by layering.
Some other species are *C. nutans*, with pale white flowers; *C. paniculatatum* – showy

scarlet panicles; and *C. ugandense* – a blue flowering species.

Coffea bengalensis; Family: *Rubiaceae*; Origin: India, Malaysia; Common Name: Bengal Coffee.

A shrub of deciduous habit, growing to a height of 1.8 m. A hardy shrub which grows in full sun or partial shade. It bears black fruits which persist for a short period. The white flowers appear in the winter in a leafless condition. Propagated from seeds or by air layering.

Cordia sebestena (*C. speciosa*); Family: *Boraginaceae*; Origin: West Indies; Common Name: Geiger Tree.

A quick-growing shrub which produces flowers in the second year when raised from seeds. Clusters of orange or scarlet flowers appear during the summer and the rains. Propagated from cuttings.

Cotoneaster; Family: *Rosaceae*; Origin: See below.

A group of hardy shrubs, mostly grown for their ornamental berries. They thrive only in the hill stations and temperate regions. Propagated from seeds or cuttings. There are numerous ornamental species but only a few of Indian and Chinese origin are discussed below.

(a) *C. frigida:* A large, deciduous, spreading shrub of Himalayan origin. The numerous white flowers are borne in clusters, followed by abundant scarlet fruits in June. It is one of the most beautiful species when in flowering and fruiting.

(b) *C. horizontalis* (Syn. *C. davidiana*)*:* A shrub which grows flat and grows hardly more than 1 m in height. The branching is herring-bone-like and can be used as wall shrub even without nailing. The flowers are pinkish and the fruits bright red. It is of Chinese origin. In var. *variegata* the leaves are prettily edged with creamy-white.

(c) *C. microphylla:* A low, prostrate and densely-branched Himalayan shrub. The white flowers are followed by globose scarlet-red fruits. The shrubs are suitable for rock gardens, especially the prostrate ones.

Crossandra; Family: *Acanthaceae*; Origin: See below.

A group of small evergreen shrubs, between 45 and 60 cm in height. The shrubs are quite hardy and can be grown in shrubbery or cultivated for the flowers, which are used by ladies for adorning hair or for *puja*. Sometimes, the old shrubs are attacked by bugs and scales. It is better to replace the shrubs every two or three years. These are raised from cuttings or seeds. The important species are:

(a) *C. flava:* Unbranched, short-stemmed shrub, with bright yellow flowers and yellow-green bracts which are spine-pointed. Origin: Tropical West Africa.

(b) *C. guineensis:* A free-flowering crossandra, 30-60 cm tall. It grows in full sun and partial shade. Flowers are pale lilac, bracts not spiny. Origin: East Africa, Mozambique.

(c) *C. nilotica:* A 30-60 cm tall shrub which bears brick-red or orange flowers inside bracts which are not spiny. A native of East Africa, Mozambique.

(d) *C. undulaefolia* (Syn. *C. infundibuliformis*)*:* A shrub, 30-90 cm tall. The scarlet-orange flowers overlap one another in dense close-bracted spikes. Origin: India, East Indies. Also spelt *C. undulifolia*.

Cuphea jorullensis (Syn. *C. micropetala, C. eminens*); Family: *Lythraceae*; Origin:

Mexico.

A dwarf (30 cm), much-branched shrub, with yellow-tipped scarlet, minute flowers. Useful as a pot plant and in herbaceous borders. Propagated from cuttings and by division.

Cuphea miniata; Family: *Lythraceae*; Origin: Mexico.

A shrub about 60 m tall, bearing pale vermilion flowers during June to September. Propagated from cuttings and by division.

Cydonia japonica (Syn. *Chaenomeles lagenaria*, *C. japonica*, *Pyrus japonica*); Family: *Rosaceae*; Origin: China, Japan; Common Name: Japanese Quince.

A spreading shrub, 90-180 cm tall, with spiny branches. Scarlet red flowers are borne, in 2-6 flowered clusters, during March-April. It is propagated from seeds or by layering.

Cytisus scoparius (Syn. *Sarothamnus scoparius*, *Spartium scoparium*); Family: *Leguminosae*; Origin: Central and South Europe; Common Name: Common or Scotch Broom.

A deciduous, erect shrub, growing up to 2.5 m. The bright shining yellow flowers are solitary and appear in the summer. Generally raised from seeds but also from cuttings. Suitable for high altitudes.

Daedalacanthus nervosus (Syn. *Eranthemum nervosum*, *E. pulchellum*, *Daedalacanthus pulchellum*, *D. purpurascens*); Family: Acanthaceae; Origin: India.

A shrub about 1.5 m high, bearing indigo-blue flowers during the winter. This can be used for ornamental hedging. The shrub grows well under partial shade. Pruning to control size is done during the rains. Easily propagated from cuttings. The shrub is attacked by scale in some parts of the country.

Daphne cannabina (Syn. *D. bholua*, *D. papyracea*, *D. papyrifera*); Family: *Thymelaeaceae*; Origin: Nepal, Bhutan, Sikkim, Assam.

An evergreen erect shrub, 2-2.5 m tall. The shrub bears fragrant purplish-pink flowers from January onwards. It is propagated from cuttings and by layering.

Daphne cneorum (Syn. *D. odorata*); Family: As above; Origin: Central Europe and Central Russia; Common Name: Garland Flower.

A low, procumbant, evergreen shrub which bears dense scented clusters of small pink flowers during early to middle of the spring and sparingly in the summer. Good soil-drainage is essential for successful cultivation. Propagated from cuttings and by layering.

Daphne sureil; Family: As above; Origin: Sikkim, Himalayas.

An erect, deciduous shrub, 2-3.5 m tall. The young branches are densely white-downy. The green, paling to white, slightly fragrant flowers appear from January onwards.

All *Daphnes* like a lime-free, well-drained soil. They grow at medium to high altitudes in India.

Datura (Syn. *Brugmansia*) *suaveolens* (Syn. *D. gardneri*); Family: *Solanaceae*; Origin: Mexico; Common Name: Angel's Trumpet; Vernacular Names: Hindi – *Dhatura*;

Bengali – *Dhutura.*

A large tree-like shrub which bears long and wide, trumpet-shaped, white, fragrant flowers very freely. The flowers hang downwards and the shrub looks beautiful when in full bloom. Pruning after each flowering is essential. The shrub prefers partial shade. Easily propagated from cuttings. Old plants should be replaced once every 2 to 3 years. *D. sanguinea* is smaller than the above and bears orange-red flowers. Seeds of *dhatura* are poisonous.

Deutzia corymbosa; Family: *Saxifragaceae*; Origin: Himalayas.

A vigorous, tall (3 m) shrub, suitable for growing at medium to high elevations. The pure white, small, fragrant flowers appear in dense panicle or corymb. Propagated from cuttings. Another species, *D. scabra* (Syn. *D. crenata*), bears double, white flowers, sometimes tinted with pink or lavender, in profusion during the summer. *Deutzias* can be grown easily in most soils and under most conditions, but are susceptible to frost.

Dombeya; Family: *Sterculiaceae.*

A group of large shrubs, which, with their coarse, cordate-shaped, many-lobed leaves, are not very attractive or ornamental as such, but once in bloom the clusters of flowers cover the whole shrub and make them quite attractive. They are suitable for large shrubberies and as specimen shrubs. Propagated by air layering. The shrub needs drastic pruning to keep it in good shape. The more important species are:

(a) *D. acutangula* (Syn. *Astrapaea tiliifolia* or *tiliaefolia*): A native of Mauritius, it grows to a height of 3 m. The flowers are crowded, red in colour, and appear during the winter.

(b) *D. gagiana:* A large shrub (up to 3.5 m tall) which does well in partial shade. The bright pink flowers appear in huge bunches hanging at the end of branches.

(c) *D. mastersii:* A native of Tropical Africa, it grows to a height of 1.5 m and bears white, slightly rose-tinged fragrant flowers.

(d) *D. natalensis:* A native of Natal, 1.5 m tall. Pure white, large and sweet-scented flowers appear in umbels.

(e) *D. rotundata:* A large (3.5 m), bushy shrub or a small tree, bearing numerous tufts of white flowers. This native of Africa is also known as *D. rotundifolia.*

(f) *D. spectabilis:* A native of Tropical Africa, it grows to a height of 2.5 m and bears white flowers in axillary and terminal cymes.

The *Dombeyas* are in flower during the winter months. One disadvantage with *Dombeyas*, especially with *D. gagiana* and *D. rotundata*, is that the faded flowers hang in ugly bunches till they are blown off. A hybrid, *D. lancasterii*, bears deep rose pink flowers.

Duranta plumieri (Syn. *D. repens*, *D. spinosa*, *D. inermis*, *D. ellisia*, *Ellisia acuta*); Family: *Verbenaceae*; Origin: Mexico, Brazil, West Indies; Common Name: Golden Dew Drop.

A large (2.5 to 3 m tall), woody shrub with axillary spines. Small, bright green leaves are attractive. The blue flowers appear in abundance, in drooping racemes, succeeded by yellow berries in hanging clusters. A most suitable shrub for hedge and also for topiary as it stands trimming well. Propagated from cuttings and seeds. There are two varieties – var. *alba*, with white flowers and var. *variegata*, with blue flowers and yellow variegated leaves. Duranta is suitable for planting in shrubberies also. The variegated form can be planted as specimens in lawns, rock gardens, and Japanese gardens. Susceptible to severe frost.

Eranthemum; Family: *Acanthaceae*.

A group of shrubs which grow in shady and semi-shady places. The shrubs are grown mainly for their colourful foliage except in *E. laxiflorum* which bears conspicuous flowers. But all other *Eranthemums* also bear flowers in shade during the summer when many other plants even fail to grow. *Eranthemums* need a well-drained soil. The plants can be made bushy by pruning. Propagated from cuttings. Tip cuttings produce the best plants. In humid and mild climates they can be used as ornamental hedge. The nomenclature in *Eranthemum* is quite confusing. The most important species are:

(a) *Eranthemum albo-marginatum* (Syn. *Pseuderanthemum albomarginatum*): A native of Polynesia. It is a bushy shrub, or a foliage plant, growing to a height of 1.5 m. The upper surface of the leaves is broadly margined with white and irregularly suffused with gray, the lower surface being green.

(b) *E. bicolor* (Syn. *Pseuderanthemum aspersum*): A native of Solomon Islands. The plants attain a height of 1.5-1.75 m. A part of the leaf lamina and margins are yellow, whereas the central portion is green, suffused with greenish-yellow patches. The flowers are white with pink lobes.

(c) *E. laxiflorum* (*Pseuderanthemum laxiflorum*): A straggling shrub from Fiji which thrives best in partial shade and bears purplish flowers constantly but more profusely during the rains. If grown in complete shade during the monsoon, it may not survive.

(d) *E. nigrum* (Syn. *E. nigrescens*, *Pseuderanthemum atropurpureum*): A native of Pacific Islands. The shrub attains a height of 1.5-1.8 m. The upper surface of the leaves is blackish purple and the lower surface purplish with dark veins. The flowers are in terminal erect spikes, white and spotted rose at the base.

(e) *E. pulchellum:* See *Daedalacanthus nervosus*.

(f) *E. reticulatum* (*E. schomburgkii; Pseuderanthemum reticulatam*): A native of Australia. It attains a height of 120-150 cm. The leaves are green, netted or reticulated with golden-yellow, which is conspicuous in young leaves only. In older leaves, though not netted, the veins are prominent yellow. White flowers are speckled with blood-red around the mouth of the tube and appear in panicles.

(g) *E. tricolor* (*Pseuderanthemum tricolor*): A straggling, dwarf (1.5 m tall) shrub from Polynesia. The upper surface of the leaves is greyish-purple on a dark green base, the lower surface is reddish with violet spots. The flowers are red, spotted with white.

Euonymus (Syn. *Evonymus*); Family: *Celastraceae*.

A group of shrubs which are not grown for any beauty of their flowers, but many of them are extremely fine in fruit and develop good autumn colouring. These are mainly grown in the temperate region. Some important species are *E. bungeanus* (*bungeana*), *E. chinensis*, and *E. radicans* (Syn. *E. japonica* var. *radicans*). The deciduous species are propagated from seeds and cuttings or by air layering. The evergreens easily root from cuttings.

Euphorbia bojeri; Family: *Euphorbiaceae*; Origin: Madagascar.

A succulent shrub similar to *E. splendens*, with thinner stems and spines. Floral bracts are vermilion coloured and half the size of *E. splendens*. Easily propagated from cuttings.

Euphorbia fulgens (Syn. *E. jacquinaeflora*, also spelt *E. jacquiniaeflora*); Family: *Euphorbiaceae*; Origin: Mexico; Common Name: Scarlet Plume.

A leafy shrub with slender branches. The flowers are borne along the branches and are conspicuous by their scarlet bracts. Propagated from cuttings.

Euphorbia splendens; Family: *Euphorbiaceae*; Origin: Madagascar; Common Name: Crown of Thorns.

A spiny succulent shrub, attaining a height up to 90 cm. The irregularly branched stems are brown in colour. The plant flowers profusely. The flowering is in branching cymes and the bracts are scarlet. A rare yellow flowering type is also in cultivation. Propagated from cuttings. The shrub is suitable for bordering along pathways and in rock gardens.

Excoecaria bicolor (Syn. *Croton bicolor*); Family: *Euphorbiaceae*; Origin: Asia.

A small (up to 150 cm tall), bushy shrub with handsome foliage. The upper leaf surface is glabrous and olive-green and the lower red. The flowers are inconspicuous. Propagated from cuttings or by air layering. Grows well in partial shade or sun. It is usually planted in shrubbery.

Forsythia; Family: *Oleaceae*.

A group of shrubs which grow in any soil, provided it is well-drained. These are suitable for medium and high elevations. The shrubs are cut back after flowering is over in March-April. Propagated from cuttings. Some important species are:

(a) *F.* × *intermedia* var. *densiflora:* The flowers are pale yellow and rather crowded.

(b) *F.* × *intermedia* var. *spectabilis* (Syn. *F. spectabilis*)*:* The flowers are bright yellow and crowded. A beautiful shrub, which flowers in a leafless condition.

(c) *F. viridissima:* A shrub from Eastern China which grows to a height of 150 to 240 cm. The bright yellow flowers are borne solitary or in clusters of three.

Fortunella japonica (Syn. *Citrus japonica*); Family: *Rutaceae*; Origin: Japan; Common Name: Kumquat; Vernacular Name: *Narangi*.

A handsome much-branched ornamental shrub which grows up to 2.5 m. The small scented flowers appear during the rains and the bright-orange coloured fruits cover the whole plant during the winter, turning the plant most beautiful. Propagated by air layering. It can be planted singly in the lawn or in a row along a broad pathway.

Galphimia gracilis; Family: *Malpighiaceae*; Origin: Mexico.

A quick growing, medium-sized (1.5 m tall), bushy shrub mainly grown for hedges, but also very suitable for planting in a group. The numerous yellow flowers are borne throughout the year but more profusely during July-November, and look beautiful against the green ovate leaves. Easily raised from seeds. The seedlings flower within a year.

Gardenia jasminoides (Syn. *G. florida*, *G. radicans*); Family: *Rubiaceae*; Origin: China, Japan; Common Name: Cape Jasmine; Vernacular Name: Bengali – *Gandharaj*.

It is a very popular shrub, especially in West Bengal. It is grown for its highly fragrant flowers. The above species grows to a height of 3 m. The leaves are lanceolate to elliptic, narrowed at both ends. The large, white, double, highly fragrant flowers cover the plants during the summer. The plant is hardy and grows in any soil with good drainage. It loves moisture and does not thrive in the cold climate of northern India. Propagated by air layering but cuttings may also root during the rainy season. The shrubs can be kept in good shape by pruning. Var. *fortuniana* (Syn. *G. fortunei*) is more dwarf than the above and the branches start from ground level. Sweet-scented white flowers are larger than the above and are produced in early summer. Var. *veitchiana* (Syn. *veitchii*) is a dwarf and produces white, scented, double flowers in large numbers. In var. *variegata* the leaves are variegated.

Gardenia lucida; Family: *Rubiaceae*; Origin: India, Burma.

A large shrub, which bears large white fragrant flowers ultimately turning yellow. It is propagated from cuttings or by air layering. Gardenias can be planted singly or in a large shrubbery.

Gmelina asiatica; Family: *Verbenaceae*; Origin: India, Sri Lanka.

A tall, erect-growing shrub, with irregularly bell-shaped yellow flowers. It is propagated by air layering.

Gmelina hystrix; Family: *Verbenaceae*; Origin: East Indies.

A large, climbing, spiny, quick-growing shrub with the habits of Bougainvillea. The yellow flowers with large, reddish-purple, showy bracts appear during the summer and the rains. Severe pruning is done after flowering to keep the shrub in shape. Occasionally some odd branches should also be removed. The shrub is suitable for growing in a large shrubbery or singly in a spacious lawn. It is propagated by air layering or from cuttings.

Goldfussia anisophylla: See *Strobilanthes*.

Graptophyllum hortense (Syn. *G. pictum*, *G. picturatum*, *Justicia picta*); Family: *Acanthaceze*; Origin: Uncertain, but widely cultivated in India; Common Name: Caricature Plant.

A small (120 to 150 cm), bushy, beautiful shrub, valued for its ornamental foliage. The leaves are large, elliptic, curiously marbled white, often resembling faces, and hence the common name. This plant grows better under partial shade and is easily propagated from cuttings. The shrub can be planted in the shrubbery or used as a pot plant for decoration.

Grislea tomentosa: See *Woodfordia floribunda*.

Guettarda odorata; Family: *Rubiaceae*; Origin: Tropical America.

A shrub, 2-3 m tall, bearing reddish flowers which are very fragrant at night. It is propagated from cuttings.

Gustavia insignis (Syn. *G. superba*, *Japarandiba superba*); Family: *Lecythidaceae*; Origin: Tropical America.

A low-growing, 1.2-2 m tall, rare shrub of very slow growth, but bears 12-15 cm diametre large showy flowers throughout the year. The flowers are cream-white and tinged rose outside. It is difficult to propagate even by ground layering. The plant loves a moist climate.

Gynura: See Chapter XIII.

Hamamelis mollis; Family: *Hamamelidaceae*; Origin: China; Common Name: Chinese Witch-hazel.

A woody shrub about 3 m tall, bearing yellow flowers in late winter during January-March. Flowers are set close to the bare branches. The leaves turn bright yellow in late fall. Grows well in light shade. The seeds take two years to germinate. It is propagated by layering and by grafting on *H. virginiana*. It is suitable for growing in hill stations.

Hamelia patens (Syn. *H. coccinea*); Family: *Rubiaceae*; Origin: South America.

A perpetually flowering, handsome shrub, up to 2.5 m high, with branching habit. It has ornamental shining foliage. The small tubular (pipe-like) orange-red flowers are borne in profusion especially during the rains and the summer. It forms a good ornamental hedge. In fact, a well-trimmed hedge along pathways is an object of beauty. It goes very well in a shrubbery and grows well either in light shade or in full sun. Easily propagated from cuttings or by layering. The branches at the centre require pruning every year.

Hamiltonia suaveolens; Family: *Rubiaceae*; Origin: India to China.

A large, up to 3 m tall, stout shrub which starts branching at ground level and the branches with side branches hang down towards the ground. The lavender-white highly scented flowers appear in terminal inflorescences in profusion from September to February. The plant is pruned after flowering and the pruned branches are used as cuttings for propagation. It prefers light shade.

Hibiscus; Family: *Malvaceae*.

A group of flowering shrubs consisting of several species and numerous hybrids. It is possibly the most popular and widely-grown shrub in Indian gardens, especially in eastern India and in Bangalore. The most popular species that are in cultivation in India are discussed below.

(a) *H. mutabilis*; Origin: China; Common Name: Changeable Rose; Vernacular Names: Hindi – *Thalkamal*; Bengali – *Sthalpadma*; Oriya – *Thalopodmo*; Kannada – *Bettada tavare*.

It is a popular shrub in West Bengal, grows to a height of 3 m but may attain the size of a tree, and has a few erect branches. The large double flowers, borne throughout the year, resemble lotus; hence the vernacular name of 'land lotus'. The flowers open white in early morning but change to pale pink and then to deep rose as the day advances. The plant prefers high humidity and a well-drained soil. It should be pruned drastically every year around April to keep it in shape and for regular flowering. Propagated from seeds and cuttings, but a large number of plants will bear single flowers when grown from seeds.

(b) *H. rosa-sinensis* (Syn. *H. sinensis*); Origin: China; Common Names: China Rose; Chinese Hibiscus; Shoeflower; Vernacular Names: Hindi – *Gurhal, Jasut*; Bengali – *Jaba*; Marathi – *Jassvandi*; Oriya – *Mondaro*; Assamese – *Joba*; Kannada – *Dasavala*; Tamil – *Sembaruthi, Mandaram*.

A large, 1-2 m tall, bushy shrub having numerous branches. The broadly-ovate, sharp-toothed green leaves are ornamental. The plant likes a sunny situation, porous soil, moderate temperature, and relatively high humidity. In poorly-drained soils, the shrub is seriously affected. In heavy rainfall areas with sticky soils the plant should preferably be planted at the beginning of the winter. At the time of planting sufficient well-rotten cowdung manure should be applied to each pit. A handful of bonemeal, sterameal or hoof and horn meal should be applied twice during the rainy season to improve flowering. The five-petalled, large, single or double, rose-scarlet coloured flowers are borne terminally on the branches. The petals are slightly arched back, and the attractive columns of pistils and stamens protrude out of the centre of the flower. The shrub is very attractive when in bloom. It does not respond well to pruning. However, removal or shortening of branches is desirable to keep it in shape. Though the shrub is sun-loving, a few cultivars such as 'Netaji' perform well under partial shade. Propagated by air layering, ground layering, budding, and from cuttings. New hybrids are raised from seeds. A host of insects attack the shrub. These are mites, red spiders, coccids, and weevils. There are numerous hybrid cultivars of *H. rosa-sinensis* type with single and double flowers. Some of the important cultivars are described below:

(A) *Single flowered*

(1) 'Agnes': Large cyclamen-pink flowers with a deep-pink centre.
(2) 'Australian Rose' (or 'Celestial Rose'): Flowers are deep rose with a maroon centre.
(3) 'Glowing Sunset': Deep shining salmon-orange.
(4) 'Lip Stick': Bright red flowers with a dark centre.
(5) 'My Beauty': Very large flowers of pink colour with a prominent maroon centre.
(6) 'Netaji': White flowers with a crimson centre.

(B) *Double flowered*

(1) 'Alipore Beauty': A very free-flowering cultivar of prolific growth, bearing medium-sized reddish-pink flowers.
(2) 'Aurora': Large flowers of flesh pink.
(3) 'Bhaskar': Large golden-yellow flowers which turn golden-orange with a flush of pink.
(4) 'Chitra': Handsome flowers of marigold orange.
(5) 'Daffodil': Large flowers of true daffodil yellow.
(6) 'Mahatma': Large crested paeony-type double flowers of cadmium orange with a mandarin red centre.
(7) 'Syriacus double': There are two types, one white double and the other mauve double.

(C) *Some of the hybrids released by the Indian Institute of Horticultural Research*

(1) 'Arunodaya': A vigorous and floriferous plant. The basal part of the single saucer-shaped flower is "Rose Bengal" in colour, radiating into mauve-pink stripes along veins, while the upper part is "Nasturtium Orange".
(2) 'Basant': A moderately floriferous cultivar with dark green foliage. The saucer-shaped single flowers are large, 17-20 cm across, "Sulphur yellow" in colour, without a centre.
(3) 'Be-nazeer': A highly floriferous plant with shining foliage. The single flowers are

large, 17-20 cm across. The basal part of the petal is "Neyron Rose" and the upper part "Capsicum Red".

(4) 'Geetanjali': A moderately floriferous plant with single flowers, measuring 15-18 cm across. The basal part of the corolla is "Dark Red" and the upper part "Turkey Red".

(5) 'Nartaki': Bears multi-whorled flowers, 14-16 cm across. The petals are "Cardinal Red" at base and "Marigold Orange" near the tips.

(6) 'Pakeezah': A moderately floriferous plant with single flowers which are 15-18 cm across. The basal part of the corolla is "Claret Rose" and the upper part "Carmine Red".

In recent years a lot of hybrid cultivars have been introduced in India with super-sized and brightly-coloured flowers. The flowers are of many contrasting and varied colours and the shape may be flat or saucer-shaped. In some cultivars the petals are recurved, ruffled, or wavy. There are about two thousand hybrids in Hawaii and these are becoming popular in many parts of our country. The thick, roundish, or ovate leaves are very ornamental. The hybrids are free-blooming and are grown either in pots or on ground. These hybrids do not fare well in the north Indian climate, but are at their best under such climatic conditions as of Bangalore. Propagation is by budding or side grafting. But in certain humid regions of eastern India, such as Calcutta, the budded plants do not grow so well as in Bangalore. Plants raised by air or ground layering do better.

(c) *H. schizopetalus:* Origin: Eastern Tropical Africa; Vernacular Name: Bengali – *Jhumko jaba.*

A tall, much-branched, glabrous shrub, with ovate-elliptic, toothed foliage. The orange-red flowers are pendulous and look like ear-drops because of their much arched back, deeply cut petals. The long staminal tube with red anthers hangs down from the centre of the petals.

(d) *H. syriacus* (Syn. *Althaea frutex*, *A. fruticosa*): Origin: Syria, etc.; Common Names: Shrubby Althea, Rose of Sharon.

A much-branched, erect shrub, 1.5-2 m high, with smaller rounded-teethed leaves. The flowers may be single or double and white, blue, or mauve in colour. This shrub does best in the hill stations, but can be grown in the plains also. Easily propagated from cuttings.

Holmskioldia sanguinea; Family: *Verbenaceae*; Origin: India; Common Names: Parasol Flower, Chinese-hat-plant.

A large, 1-2 m tall, shrub with a tendency to straggling. The sepals fuse and form a saucer-shaped structure and the petals, a tubular structure, placed at the centre of the saucer. Sometimes the petals fall off but the sepals persist, giving the shrub its beauty. The orange-red flowers bloom from February to May and are borne along almost the entire length of the branches. Grows best in full sunshine and is propagated from seeds or by air layering, while cuttings root with difficulty. *H. sanguinea* var. *acuminata* bears pale apricot flowers. *H. coccinea* has crimson flowers. Recently a yellow coloured species has also been introduced.

Hydrangea; Family: *Saxifragaceae*.

A group of shrubs which can be cultivated successfully at medium and high altitudes. The leaf is handsome, large, oval to roundish in shape, and toothed at margins except in one species. At high elevations, the shrubs grow very large but at medium elevations they are mainly grown as potted plants. After each flowering the shoots are pruned low, as only new branches bear flowers. The shrubs grow best in a well-drained loamy soil, in a sheltered position, preferably receiving only the morning sun. The flowers are borne in large (20-30 cm in diametre) round heads of white, pink, or blue. The pink variety may be induced to bear blue flowers by application of a weak solution of iron sulphate to the pots once a week. The flowering time is during the rains. The shrubs are propagated from seeds or cuttings. The common species *H. hortensia* (Syn. *H. macrophylla*, *H. opuloides*, *H. hortensis*) is the common hydrangea, which grows well at medium elevations, bearing pink or blue flowers. *H. paniculata* is another beautiful species which grows best in the hill stations. The flowers are white, turning to purplish.

Hypericum; Family: *Guttiferae*.

This genus comprises a large group of shrubs suitable for growing at medium elevations. *H. hookerianum* (Origin: South India, Sikkim, Himalayas) bears yellow flowers. The species is also known as *H. oblongifolium*.

Iochroma tubulosa (Syn. *I. tubulosum*); Family: *Solanaceae*; Origin: Tropical America. Common Name: Purple Cestrum.

A bushy shrub, 1.5-2 m high. The shrub prefers a well-drained soil and a sunny situation. The violet, drooping flowers are borne in branches throughout the year but more profusely during the summer and the rains. Propagated from cuttings. Suitable for shrubbery.

Ixora; Family: *Rubiaceae*; Vernacular Names: Hindi – *Raktak*, *Kotagandhal*, *Rangan*; Bengali and Assamese – *Rangan*.

A group of popular shrubs suited for the tropical regions of India. But most of the *Ixoras* fail to grow in the dry and frost-prone areas of north India. The plants grow in any soil, but respond to application of manure. The large terminal trusses of flowers mostly appear during the summer and the rains, but at places such as Bangalore the flowering is continuous throughout the year. The colour may be white, yellow, pink, orange, scarlet, and orange-scarlet. The shrubs can be pruned after flowering.

The shrubs can be grown as single specimen plants (e.g., *I. singaporensis*) or in the shrubbery and as hedge (e.g., *I. chinensis*). *Ixoras* are propagated by layering or from cuttings. Some of them produce seeds. There are several hybrids. The more important species are described below according to the colour of their flowers.

(A) *White flowers*

(1) *I. acuminata:* Origin North India. 1-2 m tall. Bears large fragrant flowers.
(2) *I. barbata:* Origin Andamans. A large (2-3 m tall) shrub, bearing greenish-white scented flowers during the summer and the rains.
(3) *I. parviflora:* See Chapter XI, page 217.
(4) *I. superba:* A rare shrub with large-sized, 30 cm long and 10 cm broad leaves and huge heads of white flowers.
(5) *I. undulata:* Origin Bengal. Flowers are white and sweet-scented.

(B) *Yellow and orange flowers*

(1) *I.* × *aliporensis:* A hybrid between *I. coccinea* and *I. venusta*, it is about 1.5 m tall

and bears deep orange flowers.

(2) *I. armeniaca:* A shrub of garden origin. The dwarf shrub is of compact habit, bearing pale yellow, flushed salmon flowers.

(3) *I. chinensis* (Syn. *I. stricta, I. blanda, I. crocata, I. rosea*)*:* A dwarf shrub, 60-90 cm tall, of Chinese origin. The flower colour is very variable from orange to pink, yellow, or even white.

(4) *I. lutea* (Syn. *I. coccinea* var. *lutea*)*:* A shrub of garden origin, bearing pale-yellow flowers.

(5) *I. venusta:* A plant of garden origin with straggling habit. Bears flowers which at first open orange but turn buff.

(C) *Orange-scarlet shades*

(1) *I. chinensis* 'Prince of Orange': A compact shrub of garden origin. The flowers are deep buff while opening but change to orange-scarlet.

(2) *I. coccinea* var. *dixiana:* A compact and bushy (1.5 m tall) shrub of garden origin. Flowers are orange-scarlet, and are produced in large corymbs.

(3) *I. coccinea* var. *pilgrimii:* A shrub of garden origin, bearing orange-scarlet shaded crimson flowers, in 15-18 cm diameter round corymbs.

(4) *I. fulgens* (Syn. *I. salicifolia*)*:* Origin Java. A shrub, 90-120 cm tall. Bears flowers which are clear orange-scarlet or orange while opening but turn to scarlet.

(5) *I. singaporensis:* It is one of the best ixoras in cultivation. The plant is bushy, 2 to 2.5 m tall, having long shining green leaves. The brilliant orange-scarlet flower-heads are 15-20 cm across, and are borne in profusion during the summer and the rains, but continue for a greater part of the year in a climate as of Bangalore.

(D) *Scarlet flowers*

(1) *I. coccinea* (Syn. *I. grandiflora*)*:* Origin East Indies. A medium sized (1.5 m tall) compact shrub, widely cultivated. The brilliant scarlet flowers are borne in profusion during the rains but continue practically throughout the year. The shrub serves very well as ornamental hedge. In Kerala, the flowers are offered in temples.

(2) *I. coccinea* var. *bandhuca:* Origin India. The flowers are deep scarlet.

(3) *I. coccinea* var. *magnifica:* An improved variety of garden origin, with broader leaves and large deep-scarlet flowers.

(4) *I. macrothyrsa* (Syn. *I. duffii*)*:* Origin Sumatra. A large shrub, up to 2.5 m tall, bearing deep scarlet tinged crimson flowers, in 20 cm across corymbs.

(5) *I. splendida:* A shrub of garden origin, bearing orange-scarlet flowers in large corymbs

(E) *Pink flowers*

(1) *I. chinensis* var. *rosea:* Origin Bengal. The flowers (blooming period is summer) are pale pink, but turn reddish with age.

(2) *I.* × *westii:* The flowers open pale rose but become bright rose, or reddish-salmon.

***Jacquinia aurantiaca*; Family: *Myrsinaceae* (some include it under *Theophrastaceae*);**

Origin: Sandwich Islands.

A shrub, 90-180 cm tall, with sharp-pointed obovate-lanceolate leaves. The flowers are borne during April-September in orange racemes. Mostly grown as single specimens. It is propagated from cuttings or seeds. Can be trained into a thick hedge.

Jacquinia ruscifolia; Family: As above; Origin: Havana.

A large, 1.5 to 2 m tall shrub of a beautiful shape. The leaves are sharply pointed. Small orange-red flowers appear during the summer and remain covered by dense foliage. Both the shrubs need some pruning. Propagation as above.

Jasminum: See Chapter XX.

Japarandiba: See *Gustavia insignis.*

Jatropha multifida; Family: *Euphorbiaceae*; Origin: South America; Common Names: Coral Plant, Physic Nut; Vernacular Name: Bengali – *Bherenda.*

A large, 1.5 to 3 m tall, shrub with palmately-lobed foliage. The small scarlet flowers appear in bunches during the summer. The plant is kept bushy by pruning. Propagated easily from cuttings or seeds.

Jatropha pandurifolia (Syn. *J. acuminata, J. hastata*); Family: *Euphorbiaceae*; Origin: West Indies.

A quick-growing, medium-sized, 1-1.5 m tall shrub which bears small scarlet flowers during the summer. Needs light pruning after flowering to keep it in shape. Grows equally well in light shade. Some authors consider *J. hastata* as a separate species.

Jatropha podagrica; Family: *Euphorbiaceae*; Origin: Colombia; Common Name: Guatemala Rhubarb.

A succulent shrub, 50 cm to 1 m tall, with stem gouty at base. The large, long petioled, lobed leaves are arranged at the end of the branches. The orange-scarlet flowers are borne on long-stalked cymes during the rains. It is a good shrub for the rockery. It is also grown in window boxes and pots. No watering should be done in the winter when it sheds the leaves. Propagated by division.

Juniperus chinensis: See Chapter XI, page 203.

J. horizontalis (Syn. *J. prostrata, J. hudsonica, J. sabina* var. *procumbens, J. repens*); Family: *Pinaceae*; Origin: North America; Common Name: Creeping Juniper.

A procumbent shrub, usually with long trailing branches which root to the ground. A suitable shrub for rockery. Propagated from cuttings or by layering. Recently some authors have placed this genus in *Cupressaceae* (*Coniferae*).

Justicia carnea (Syn. *J. magnifica, Jacobinia carnea, J. magnifica, Cyrtanthera magnifica*); Family: *Acanthaceae.*

A sub-shrub with long, sometimes 30 cm long, lanceolate leaves. The carmine-coloured flowers are long, the lower lip recurved. A hardy plant which is easily propagated from cuttings.

Justicia coccinea (*Syn. Jacobinia coccinea*); Family: *Acanthaceae*; Origin: Brazil. A sub-shrub, 60-150 cm tall, which bears crimson-scarlet flowers in the summer.

Justicia gendarussa; Family: *Acanthaceae*; Origin: India, Burma.

A small shrub, 90 cm tall, which bears white with purple spotted flowers. *Justicias* are hardy plants which are planted in the shrubbery and sometimes used as ornamental hedge. These are easily propagated from cuttings.

Kerria japonica (Syn. *Corchorus japonicus*); Family: *Rosaceae*; Origin: China, Japan; Common Names: Japanese Rose, Corchorus.

An extremely bushy shrub, 1.5 m tall. The shrub sheds leaves in the winter and shows new growth of foliage and yellow flowers in the spring. Commonly propagated from young shoot cuttings but layers and root divisions are also alternative methods of propagation. There are double flowering and variegated varieties also.

Kopsia fruticosa: See *Cerbera fruticosa.*

Lagerstroemia indica (Syn. *L. chinensis*); Family: *Lythraceae*; Origin: China, but widely cultivated in India; Common Name: Crape Myrtle.

A pretty deciduous shrub, which never grows taller than 2.5 m if pruned. The flowers are borne in profusion, terminally or subterminally, from the branches. The petals are nicely frilled. The flowering time is during the summer and the rains. The shrub should be pruned to 50 to 60 cm during February. The pruned branches should be used as cuttings for propagation. The common varieties are the 'Bride' with delicate rose flowers, *rosea* with deep pink flowers, and *alba* with white flowers. In the hybrid *L.* × *lancasteri*, the flowers and leaves are much larger. The plant needs pruning during February to keep it dwarf. There are two other cultivars: 'Petite Pinkie' having red-bronze foliage which turns later to rich green, and flowers of rich pink colour; and 'Petite Snow' having white flowers in snowball-like clusters.

Lantana camara (Syn. *L. aculeata*); Family: *Verbenaceae*; Origin: Jamaica.

A widely-cultivated, prickly-stemmed shrub, 50 cm to 2 m tall. The flowers are borne in compact heads. The fruits are round and black. The *Lantanas* are very common and actually have become a weed in many parts of India, but yet are useful as a shrub. But great care should be taken against their spread. They can be grown in the shrubbery, along paths and roads and also for hedging or fencing. The common varieties are: var. *mixta* or *mista* (Syn. *L. mixta*, *L. mista*) – flowers open yellow, change to saffron, and then become brick-red, inner yellow becoming orange; var. *nivea* (Syn. *L. nivea*) – flowers white, the outer ones becoming bluish, in an oval cluster of flowers; var. *sanguinea* (Syn. *L. sanguinea*) – flowers open saffron-yellow but turn to bright red; var. *depressa* – shrub is dwarf, semi-trailing and the flower bright yellow.

Lantana lilacina; Family: *Verbenaceae*; Origin: Brazil.

A non-spiny shrub with oval-lanceolate leaves. Flowers are rosy-lilac borne on large heads.

Lantana sellowiana (*selloviana*) (Syn. *L. delicatissima*); Family: *Verbenaceae*; Origin: South America; Common Names: Weeping or Trailing Lantana.

A trailing shrub with pale mauve flowers on flat heads. The shrub is suitable for rockery, hanging baskets, or for trailing over low walls. All *Lantanas* are hardy shrubs and can be grown without difficulty. All species and varieties are pruned heavily during the rainy season. Propagated by air or ground layering or from seeds. One should be cautious about *L. camara* as it may become a notorious weed.

Lawsonia inermis (Syn. *L. alba*); Family: *Lythraceae*; Origin: North Africa and South West Asia, including India; Common Names: Henna or Tree Mignonette; Vernacular Names: Hindi – *Mehndi*; Bengali – *Mehedi*; Oriya – *Rangota*; Gujarati – *Medi*; Marathi – *Mendi*.

A shrub, up to 2.5 m tall. Grows without care and is mostly used as hedge. The white sweet-scented flowers are borne in profusion, followed by berries. The crushed leaves are used for colouring nails and palms. Propagated from seeds or cuttings. *L. rubra* is a species with rose petals on light green sepals. The shrub needs severe pruning to keep it in shape.

Lemonia spectabilis (Syn. *Ravenia spectabilis*); Family: *Rutaceae*; Origin: Cuba.

A shrub, 2.1 to 2.4 m tall, bearing purplish-red flowers almost throughout the year. The shrub can be grown in the sun as well as in light shade. Propagated by air layering. Another species is *R. rosea*, which is a native of Brazil, and bears rose-red flowers.

Ligustrum japonicum; Family: *Oleaceae*; Origin: Japan; Common Name: Japanese Privet.

A large, vigorous-growing shrub, attaining a height of 2 m. The branches are drooping. The yellowish-white, slightly scented flowers are borne in abundance. The shrub thrives best at medium altitudes. Propagated from cuttings. *L. ovalifolium*, also a native of Japan, is popular as a hedge plant as it can be pruned to any height and shape. The flowers are dull white.

Linum trigynum: See *Reinwardtia*.

Lonicera; Family: *Caprifoliaceae*; Common Name: Honeysuckle.

Loniceras like a good moist soil and full sunshine rather than shade for best blooming and fruiting. The berries, in various colours such as red, yellow, black, purple, blue, and white, give the plant a beautiful look. Propagated from seeds or cuttings. They do well only at medium to high elevations. Some important species are:

(a) *L. fragrantissima* (Syn. *L. caprifolioides*): A native of China. Bears fragrant creamy-white flowers from December to March.

(b) *L. gracilipes* var. *glandulosa:* A native of Japan, bearing beautiful pink flowers in April-May, followed by pendulous scarlet fruits in June.

(c) *L. morrowii:* A shrub from Japan with widespread branches bearing creamy-white flowers which change to yellow. The flowering time is May-June. Fruits are red in colour.

Magnolia fuscata (Syn. *Michelia figo, M. fuscata*); Family: *Magnoliaceae*; Origin: China.

A shrub, 1.5 to 1.8 m tall under tropical conditions. It grows best in porous soil and under light shade. The flowers are yellowish-green stained brown-purple, about 4-5 cm across. The flowering time is the summer and the rains. Propagated by layers.

Magnolia grandiflora: See Chapter XI, page 218.

Magnolia liliflora; Family: *Magnoliaceae*; Origin: Japan, China.

A deciduous, 3 to 4 m tall shrub with straggling branches. The large bell-shaped flowers are purple outside and usually white inside. The flowering season is April to June. It thrives best at medium to high elevations.

Magnolia mutabilis; Family: *Magnoliaceae*.

A shrub, up to 2 m tall, with straight branches arising out of the base. The highly scented buff-coloured flowers appear during the summer and the rains but mostly remain hidden beneath the large leaves. Propagated by ground or air layering.

Magnolia pumila (Syn. *M. coco, Talauma pumila*); Family: *Magnoliaceae*; Origin: China, Java; Vernacular Names: Hindi and Bengali – *Jahuri champa*.

An evergreen shrub attaining a height up to 1.5 m. The drooping cream flowers are borne almost all the year round. Compared to other species, it bears more flowers during the summer and the rains. The foliage gets scorched in direct sunlight. It is propagated by air layering.

Malpighia coccigera; Family: *Malpighiaceae*; Origin: West Indies; Common Name: Miniature Holly.

A dwarf (60-90 cm), bushy shrub with densely-crowded, small, roundish, spiny and shining holly-like leaves. The shrub bears in profusion pale-bluish or pink flowers during the summer and rainy seasons. It is hardy and thrives in any soil under full sunshine. It may be planted in the front rows of the shrubbery, as a specimen in the lawn, or as hedge. Propagated from seeds and by layering. Cuttings also root well.

Malpighia glabra; Family: *Malpighiaceae*; Origin: South America; Common Name: Barbadose Cherry.

A large, 2.5 m tall, much-branched evergreen shrub, having shining, dark green, ovate leaves. The rose or bright purple small flowers are produced during the summer, the rains, and the autumn, followed by beautiful red cherry-like acid fruits, which are used in preparations of jam and jelly and for flavouring. It can be grown as a single specimen in the lawn and in the shrubbery. Propagated from seeds and hardwood cuttings.

Malvaviscus arboreus (Syn. *Achania malvaviscus*); Family: *Malvaceae*; Origin: Mexico.

A densely-foliaged shrub which grows up to 3 m and can be termed as a poor relative of *Hibiscus*. The rich red flowers, similar to those of *Hibiscus*, remain half-open, unlike *Hibiscus*. Flowering is profuse and continues almost throughout the year. The shrub needs drastic pruning in early rains or during the winter. Easily propagated from cuttings.

Malvaviscus conzatii; Family: *Malvaceae*; Origin: Mexico; Common Name: Scarlet Cotton; Vernacular Name: Bengali – *Lanka jaba*.

A popular shrub of eastern India, which is often used as ornamental hedge. The plant produces tubular red flowers throughout the year, the intensity being maximum during the autumn or early winter. The flowers somewhat resemble red chilly and hence the Bengali name. A white flowering type, var. *alba*, and a yellow and green variegated foliage type, var. *variegata*, are also available. Easily propagated from cuttings. The shrub is pruned during early rains to keep it short and bushy.

Manihot utilissima var. *variegata*; Family: *Euphorbiaceae*; Origin: South America; Common Names: Tapioca, Bitter Cassava, Mandioca.

It is an ornamental herbaceous shrub with pale-yellow and green variegated leaves. The shrub is highly susceptible to frost. It is also grown as an indoor and greenhouse plant. Easily propagated from cuttings. Lately some authors are mentioning this as *Manihot esculenta variegata*.

Melaleuca hypericifolia (Syn. *M. oppositifolia*); Family: *Myrtaceae*; Origin: New South Wales.

A tall (2-4 m) shrub of weeping habit, with closely-spaced leaves. The red-stamened flowers are closely set in spikes and look like bottle brush. The flowering time is April to October. Propagated from cuttings.

Memecylon edule; Family: *Melastomaceae*; Origin: Widely distributed in Sri Lanka and South India (exact origin unknown); Common Name: Red Iron Wood; Vernacular Name: Bengali – *Anjan*.

A 2.5 m tall bushy shrub. The sessile, small, blue, fragrant flowers come in round clusters around the axils of leaves and also from the woody stems in early summer. The bright green foliage is also a beauty of this shrub. Difficult to propagate vegetatively but easily raised from seeds. Grows better in partial shade at an early stage. The fruits are red in the beginning, but later turn black.

Memecylon umbellatum; Family: *Melastomaceae*; Origin: As above.

A small (1.5 to 2 m) shrub. The flowers are pinkish blue and the flowering habit is the same as above. The flowers appear in the hot months. The seeds produce 5-8 cm seedlings in a year's time. Vegetative propagation is difficult.

Meyenia: See *Thunbergia*.

Montanoa bipinnatifida (Syn. *Polymnia grandis*); Family: *Compositae*; Origin: Mexico.

A shrub (2.5 m tall) with very large (60-90 cm long) leaves. It branches profusely and bears white-yellow flowers in terminal clusters during the cold months. Propagated from seeds, stem cuttings, and root cuttings.

Muehlenbeckia platyclados (Syn. *M. platyclada*, *Homalocladium platycladum*, *Coccoloba platyclada*); Family: *Polygonaceae*; Origin: Solomon Island.

A bushy, evergreen, glabrous shrub, 120 to 150 cm tall. The stems are flattened, green,

0.5 to 1.5 cm wide. Numerous clusters of white flowers appear on the flat stems during the rains when the shrub produces vegetative growth also. Grows better in light shade. Propagated from cuttings.

Murraya exotica; Family: *Rutaceae*; Origin: India, China, Australia, and Pacific islands; Common Name: Orange Jessamine; Vernacular Names: Bengali and Hindi – *Kamini*; Assamese – *Kaminiphul*; Marathi – *Manchula juti*; Kannada – *Angarakanagida.*

A tall shrub, growing up to 3 m and sometimes even more, with white-greyish bark. The foliage is beautiful, – dark green, glossy and pinnate. The white, strongly sweet-scented flowers appear in great profusion during the rains when the presence of this shrub can be felt from a long distance. It can be grown as a specimen plant, or in the shrubbery, or can be used as hedge. It stands pruning well and can be a subject for topiary. The red fruits are showy. The shrub is propagated from seeds or by air layering.

Mussaenda corymbosa; Family: *Rubiaceae*; Origin: India; Common Name: *Dhoby* Bush; Vernacular Name: Bengali – *Sadapata.*

An erect shrub with slender branches, which thrives better in light shade. The flowers are pale orange and the leafy sepals are white. Flowering time is the rainy season. It is propagated from cuttings or by air layering.

Mussaenda erythrophylla; Family: *Rubiaceae*; Origin: Tropical Africa.

A semi-erect shrub, 1.0-2 m tall, or a climber which may climb up to 9 m at its native place. The shrub thrives well in light shade and in a porous soil. The flowers are sulphur-yellow, but the scarlet sepals are showy. Propagated by layering.

Mussaenda frondosa; Family: *Rubiaceae*; Origin: Tropical Asia.

A shrub, about 90 cm tall, with spreading habit. Needs pruning after flowering to keep it in shape. The leafy sepals are white. Flowering season is the summer and the rains. It is propagated by layering.

Mussaenda luteola; Family: *Rubiaceae*; Origin: Tropical Africa.

A relatively small (1 m) shrub. The flowers are bright yellow and the leafy sepals are light yellow. Flowers during the summer. It is easily propagated from cuttings. Pruning is done during the winter.

Mussaenda philippica; Family: *Rubiaceae*.

This shrub which has emerged from the Philippines in recent years is bushy and very showy. The leaves are large, roundish ovate. The shrub has double, large, showy, leafy sepals which are produced in abundance. The colour of the sepals may be greenish-white, pink and red. A beautiful shrub, it deserves to be grown as a specimen in lawns or in groups, in every garden. *Mussaendas* are difficult to grow in frost-prone areas. They thrive best in humid localities.

Myrtus communis; Family: *Myrtaceae*; Origin: South Europe, West Asia; Common Names: Common Myrtle, Greek Myrtle; Vernacular Names: Hindi – *Vilayati mehandi*; Bengali – *Bilati mehedi.*

A shrub 3-4 m tall. The polished-surfaced lanceolate leaves are fragrant. The white, small, scented flowers appear during February to March, followed by purple-black attractive berries. Propagated from seeds, layers or cuttings.

Nandina domestica; Family: *Berberidaceae*; Origin: China, Japan.

An evergreen handsome shrub, about 2.5 m in height. It is grown for its handsome foliage, which is used in bouquets. Flowers are white and borne in large panicles during June-July followed by purplish-red berries. The shrub grows best in semi-shade. Propagated from seeds or by division of clumps. Cuttings root with difficulty.

Nerium; Family: *Apocynaceae*; Common Names: Sweet-scented Oleander, Rose Bay; Vernacular Names: Hindi – *Kanel, Kaner*; Bengali – *Karabi*; Tamil – *Alari*; Kannada – *Kanagale*; Telugu – *Ganneru.*

One of the most popularly-grown shrubs in India, especially in the northern parts. A spreading, bushy shrub, in which a number of cane-like stems arise from ground level and attain a height of 2.5 to 4 m. The leaves are linear-lanceolate. In India, the most common colours of the flowers are white and pink, though deep red and cream and shades of pink are also available. The flowers are borne in large terminal cymes, the blooming period is from April. The flowers are sweet-scented. The shrub should be pruned to the bare minimum in humid regions as Bengal. The old growths are removed after the new growths form their first cluster of buds. The white juice of the shrub is poisonous and so this is not eaten by cattle. Even the flowers will cause death, if eaten. Propagated from cuttings or by layering.

The common species are: *N. oleander*, *N. oleander* var. *flore pleno*, and *N. odorum* (Syn. *N. indicum*). Semi-double and double varieties are available in either species. There is a variety with beautiful creamy-white variegated leaves (*N. oleander variegatum*). There are several garden varieties. The origin of the former species is Mediterranean region, while that of the latter is north India to Japan.

Nyctanthes arbor-tristis; Family: *Oleaceae*; Origin: India; Common Names: Tree of Sadness (as it blooms at night), Night Jasmine; Vernacular Names: Hindi – *Harisingar*; Bengali – *Seuli*, *Shephalica*; Assamese – *Sewaliphul*; Oriya – *Singarharo*; Marathi – *Parijatak*; Tamil & Kannada – *Parijata.*

A tall (up to 4 m) bushy shrub with rough hairy leaves. Grows without much care in any garden soil in a semi-shaded position. The sweet-scented flowers are pretty with white corolla and an orange-red tube and centre. The flowering season is the autumn and the winter. The flowers open at night and start falling at daybreak. The long woody branches should be pruned after each flowering, otherwise the shrub becomes ugly. Easily propagated from seeds and cuttings during the rains. Several garden varieties, mainly in floral form, are available.

Ochna kirkii; Family: *Ochnaceae*; Origin: Tropical Africa.

A shrub which is 1-1.5 m tall. The flowers are bright yellow and are available almost throughout the year. The berries are scarlet.

Ochna squarrosa; Family: *Ochnaceae*; Origin: Tropical Asia and Africa; Vernacular

Name: Bengali – *Ramdhan champa.*

A tall shrub, at least in the humid climate of eastern India, where it grows up to 2.5 to 3 m. At other places it may attain a height of 2 m or a little less. The yellow flowers appear in profusion during the summer. The red sepals and the black fruits on a red disc-like receptacle or thalamus are perhaps more ornamental than the flowers. Usually propagated from seeds, but semi-hardwood cuttings also root.

Olea fragrans (Syn. *Osmanthus fragrans*); Family: *Oleaceae*; Origin: Himalayas, China, Japan.

A delicate shrub or a small tree, up to 9 m tall, which bears small, sweet-scented, creamy-white flowers several times in a year but mainly during June-August. The plant thrives best in semi-shade. Propagated by layering.

Oncoba spinosa; Family: *Flacourtiaceae*; Origin: Tropical Africa.

A tall (3 m), bushy shrub with very sharp axillary spines 5 cm long. The large (5 cm across), white fragrant flowers are borne singly in terminal and lateral positions on short axillary shoots during the summer. A hardy shrub which is propagated from hardwood cuttings during the rains. It can be placed in a large shrubbery or grown as hedge or screen.

Osbeckia nepalensis; Family: *Melastomaceae* (also spelt *Melastomataceae*); Origin: Nepal.

A dwarf (45 cm), much-branched shrub, having lanceolate leaves which are often spotted brown. The purplish-rose flowers appear during June. Suitable for medium to high elevations.

Paeonia suffruticosa (Syn. *P. moutan*, *P. arborea*); Family: *Ranunculaceae*; Origin: China (Tibet), Bhutan; Common Names: Tree Paeony, Moutan Paeony (also spelt Peony).

A popular shrub of China, Japan, Europe, and parts of America. The shrub is about 2 m tall. The flowers have a close resemblance to rose and may be single, semi-double, and double, in colours varying from white to rose-pink and red. The plants thrive best at medium to high altitudes. Propagated by layering or grafting.

Pentas lanceolata (Syn. *P. carnea*); Family: *Rubiaceae*; Origin: Tropical Africa, Arabia.

A dwarf (45 to 60 cm), erect or decumbent shrub. The arrangement of the flowers has a striking resemblance to Ixoras. The tiny star-shaped pale-purple flowers appear in terminal corymbs during the summer and the rains. The shrub grows well in light shade. The heads of the flowers are excellent for cutting. The shrub can be used for planting in lawns, mixed shrubbery, as ornamental hedge or sometimes planted in pots. The shrub is pruned carefully after flowering. Propagation is easy from softwood cuttings. The var. *alba* produces white flowers; var. *kermesiana* (or *kermesina*) is slightly taller (1.2 m) and bears bright carmine-rose flowers, with violet in the throat.

Petrea arborea; Family: *Verbenaceae*; Origin: West Indies, South America.

An erect, woody shrub, 2-2.5 m tall, which needs rich soil and a sunny situation. The purplish-blue flowers appear in several flushes during the summer. Propagation is difficult but the best method is grafting on *P. volubilis* root-stock.

Philadelphus coronarius (Syn. ***P. pallidus, P. coronarius nivalis***); Family: *Saxifragaceae*; Origin: Southeast Europe, Asia Minor; Common Names: Mock-orange, Syringa.

It is a hardy, tall, spreading shrub, densely covered with leaves and flowers. The flowers are creamy-white, strongly scented and are borne during the spring and early summer.

The other important species are *P.* × *lemoinei*, a hybrid bearing pure white and very fragrant flowers. *P. pubescens* (Syn. *P. latifolius*) also bears white flowers which are rarely scented. This shrub is 3-5 m tall.

The *Philadelphus* species are suitable for growing only at medium and high elevations, either in ground or in pots. They prefer light shade. The shrubs are pruned just after flowering. Propagated from cuttings or suckers or layers.

Phlogacanthus thyrsiflorus (Syn. *Justicia thyrsiflora*); Family: *Acanthaceae*; Origin: India.

A shrub, 1-2 m high, with beautiful laurel-like leaves. The orange flowers are borne in long spikes. The shrub is useful in a shrubbery and is propagated from cuttings during the rains or from seeds.

Phyllanthus nivosus (Syn. *Breynia nivosa*); Family: *Euphorbiaceae;* Origin: Pacific Islands; Common Name: Snow Bush.

A slender, bushy shrub, 1.8 m tall. The variety *P. nivosus roseo-pictus* (Syn. *P. roseo-picta, Breynia nivosa roseo-picta*) has very attractive leaves, mottled with red and pink as well as green and white. The pink colour of the foliage becomes more conspicuous during the winter. It is propagated by separation of suckers and from cuttings.

Plumbago capensis (Syn. *P. auriculata*); Family: *Plumbaginaceae*; Origin: South Africa; Vernacular Names: Hindi – *Chitra*; Kannada – *Chitramula.*

A straggling, dwarf shrub, 75-100 cm tall. Remains in flowering 8-9 months in a year and is one of the most beautiful among the shrubs during this time. The flowers are azure-blue. *Plumbagos* are susceptible to water-logging. A shrub good for edging, or as ornamental hedging, or in the border of a mixed shrubbery. It is propagated by separation of suckers, by layering and from cuttings. The var. *alba* has white flowers.

Plumbago rosea (Syn. *P. indica coccinea, P. indica*); Family: *Plumbaginaceae*; Origin: South Asia; Vernacular Name: Hindi – *Lal chitra.*

A scandent shrub, 1 m tall, which bears brick-red flowers and grows well under light shade. The shrub is slow-growing and flowers only during the winter for 2-3 months. Propagated by division or from cuttings.

Plumbago zeylanica; Family: *Plumbaginaceae*; Origin: South Asia.

A somewhat climbing shrub of dwarf growth. The flowers are white and borne in dense spikes.

Podocarpus macrophyllus; Family: *Taxaceae* (recently, *Podocarpaceae*); Origin: China, Japan.

A tall shrub to a small tree, with linear leathery leaves. The shrub is slow-growing and suitable for planting as single specimens in lawns. Propagated by layering or from cuttings during the rains. The shrub is suitable for growing at a medium or high altitude. *P. taxifolia* is another shrub species.

Pogonopus exsertus (Syn. *P. caracasensis, Howardia caracasensis*); Family: *Rubiaceae*; Origin: Tropical America.

A shrub with teret branches. The flowers are pink and the leafy sepals scarlet. The treatment of this shrub is similar to that of *Mussaenda*.

Poinsettia pulcherrima (*Euphorbia pulcherrima*); Family: *Euphorbiaceae*; Origin: Mexico.

A sun-loving, rapid-growing shrub, attaining a height up to 3 m. The inconspicuous flowers are surrounded by large, coloured, bracteal leaves in a whorl, during the winter. The shrub does not like water-logging. After flowering, the shrub is pruned severely, leaving only about 20-30 cm of the stem. This is suitable for growing in the shrubbery and also in pots. *P. plenissima* has double bracts of vermilion colour. In var. *albida* the bracts are cream-yellow; in 'Rose Queen' the bracts are rose coloured and in var. *alba* the bracts are white. The last-mentioned is not so vigorous. Propagated easily from cuttings.

Polymnia grandis: See *Montanoa.*

Polyscias: See *Aralia.*

Pootia grandiflora; Family: *Apocynaceae.*

A tall (3 m) shrub with long spreading branches, suitable for large shrubberies. Only a few flowers are produced at a time and the flowering time is the summer and the rains. Propagated by air layering.

Portlandia grandiflora; Family: *Rubiaceae*; Origin: West Indies.

A tall, about 2.5 m high, attractive shrub which bears large, 10 cm long and 3.5 to 5 cm across, funnel-shaped white flowers with reddish shade inside throat, all the year round except during the winter. The flowers are scented. The shrub needs protection from the afternoon sun. This can be grown as a solitary specimen in the lawn, in shrubbery, or in scented garden. Propagated easily from cuttings during the rainy season.

Posoqueria latifolia (Syn. *P. revoluta*); Family: *Rubiaceae*; Origin: Brazil, Mexico.

A tall evergreen shrub which bears white fragrant flowers with a long tube, in dense corymbs, during March-April. The shrub is propagated by air layering.

Prunus; Family: *Rosaceae.*

Some of the cherries and plums are very ornamental when in flowering, besides bearing edible fruits. Some of the important species are *P. americana*, *P. maritima*, *P. nigra*, and *P. pumila*. These are propagated from seeds, but since some may not breed true, these may be grafted on *P. avium*. The different *Prunus* species do well only at medium and high altitudes. But some of these can be grown in the northern parts of India. It is possible to grow a few in other parts also, especially those with low chilling requirements.

Punica granatum; Family: *Punicaceae*; Origin: Persia to Northwest India; Common Name: Pomegranate; Vernacular Names: Hindi – *Anar*; Bengali – *Dalim*; Kannada – *Dalambari*; Tamil – *Madalam*.

Punica is a hardy, tall (2.5-3 m) shrub which can be grown in any soil. The weak and interlaced branches are pruned. The flowers are borne in the summer and the rains. The ornamental types do not bear fruits. The different varieties are: *albo-plena* – the flowers are double white; *flore-pleno* (Syn. *rubro-plena*) – the flowers are double red; var. *nana* is dwarf, 50-70 cm tall, with narrow leaves and scarlet, double flowers. This also bears small fruits. Propagated by air layering, grafting and from cuttings.

Quassia amara; Family: *Simarubaceae*; Origin: Tropical America; Common Names: Bitter Quassia; Surinam Quassia.

A shrubby tree which grows up to 4 m in height. In places with hot summer the plant needs protection from the western sun. The bright scarlet flowers, resembling scarlet salvia, are borne during the hot months in terminal clusters. In Bangalore, the blooming period may extend up to September. Propagated from seeds, cuttings, and layerings.

Randia macrantha (Syn. *Gardenia macrantha*); Family: *Rubiaceae*; Origin: Tropical Africa.

A climbing shrub (up to 3 m tall) or a small tree. Bears tubular, yellowish, fragrant flowers in great abundance. Propagated by layering.

Randia maculata (Syn. *Gardenia stanleyana*); Family: *Rubiaceae*; Origin: Tropical Africa.

A much-branched, 3-5 m tall shrub which bears purple flowers with white lobes. Propagation same as above.

Rauwolfia canescens; Family: *Apocynaceae*.

A bushy shrub, growing to a height of 1.5 to 1.8 m and bearing small white flowers. The plant bears, almost always, round berries of green, red, and black colour because of difference in maturity. Suitable for shrubberies and rock gardens. Propagated from seeds and cuttings.

Rauwolfia chinensis; Family: *Apocynaceae*; Origin: China.

A small evergreen shrub, bearing small white flowers and red berries. Propagated from cuttings. *R. serpentina* also bears red berries. The roots and leaves are used as antidote against snake and insect bites.

Ravenia spectabilis: See *Lemonia spectabilis*.

Reinwardtia; Family: *Linaceae*; Origin: North India.

A genus consisting of two species, both being sub-shrubs or near-shrubs. The shrubs bear yellow flowers during November to February. The two species are *R. tetragyna* and *R. trigyna* (Syn. *R. indica*). The difference between the two lies in the shape of the foliage. Propagated from cuttings.

Rhododendron; Family: *Ericaceae*.

Different species of *Rhododendrons* have been dealt under flowering trees and some under azaleas. In the third group are included some species which are shrubs grown at high altitudes only. Propagation and cultivation are the same as of azalea. A few important species are *R. hookeri* with blood-red flowers; *R. keleticum* with purplish-crimson flowers suitable for rock gardens; *R. lindleyi* with white, fragrant flowers; *R. schlippenbachii* with pale pink to soft rose spotted with red-brown flowers; and *R. thomsonii* with blood-red flowers.

Rondeletia speciosa (Syn. *R. odorata*); Family: *Rubiaceae*; Origin: Cuba.

A handsome bushy shrub, 1.5 to 1.75 m in height. The flowers are Ixora-like, orange-red in colour with a yellow eye at the centre. The flowers come in flushes mostly during the hot weather and the rains but continue sparingly almost round the year. The flowers are fragrant. This can be grown in light shade as well as in the sun. Propagated from tip cuttings or layerings.

Ruellia; Family: *Acanthaceae*.

A group of beautiful shrubs and sub-shrubs thriving best under semi-shade. The plants are floriferous, and the flowers appear mostly during the summer and rainy seasons. These should be pinched frequently and after each flowering. These can be grown to cover the vacant spaces in shrubberies and as edges or in hanging baskets. Propagated easily from cuttings. The important species are:

(a) *R. formosa:* Origin: Brazil. Plant 60 cm tall, flowers are scarlet.

(b) *R. macrantha:* Origin: Brazil. Plant 30-180 cm tall, very floriferous. Flowers are rosy with veined throat.

(c) *R. macrophylla:* Origin: Colombia. A sub-shrub, 60-120 cm tall. Flowers are bright scarlet.

(d) *R. rosea:* Origin: Mexico. Hairy sub-shrub. Flowers are rose.

Russelia floribunda (Syn. *R. rotundifolia*); Family: *Scrophulariaceae*; Origin: Mexico.

A pretty shrub, about 120 cm tall, with a stiff habit. The leaves are round and deeply cordate at base. The scarlet flowers appear during early summer. Propagated from cuttings. *R. sarmentosa* is also a straggling shrub, reaching a height of 1.2 m. Bears red flowers.

Russelia juncea (Syn. *R. equisetiformis*); Family: *Scrophulariaceae*; Origin: Mexico; Common Name: Coral Plant.

A twiggy, bushy shrub, 90-120 cm tall, with rush-like pendulous branches. The leaves become minute bracts on the branches. The coral red flowers which take the form of cylindrical tubes appear along the upper portions of the whorled branches. When in flowering the shrub looks most attractive. The flowering time is the summer and the rains. Propagated from cuttings or by division of roots. A quick-growing shrub, suitable for shrubbery and rockery.

Sambucus nigra; Family: *Caprifoliaceae*; Origin: Europe, including Britain; Common Name: European Elder.

A large shrub or a small tree, mainly grown for the handsome foliage. The flowers are yellowish or dull white and are borne almost round the year. Propagated from seeds which germinate easily and also from hard green-wood cuttings and root cuttings. In var. *aurea-variegata*, the leaves are variegated yellow.

Sanchezia nobilis; Family: *Acanthaceae*; Origin: Ecuador.

A very quick-growing sub-shrub, 90-150 cm in height. Grows best in shade or semi-shade. The stems are pinkish-yellow and grows in a zigzag way. The leaves are large, very showy, with light yellow veins against a green surface. Flowers are yellow with red bracts and borne in densely-flowered panicles. Propagated easily from stem cuttings. In variety *glaucophylla* (Syn. *S. noblis variegata*), the leaves are glaucous green, striped white or yellow. A beautiful shrub. Both the types can be grown under the shrubs in a shrubbery to cover vacant spaces.

Serissa foetida (Syn. *S. japonica*); Family: *Rubiaceae*; Origin: Southeast Asia.

A dwarf, bushy shrub which bears numerous white flowers in rounded clusters during the summer and the rains. It can be propagated from cuttings.

Solandra longiflora (Syn. *S. laevis*); Family: *Solanaceae*; Origin: Jamaica.

A shrub, about 180 cm tall, bearing large, 30 cm long, trumpet-shaped, frilled, greenish-white tinged purple flowers. Propagated from seeds or cuttings.

Solanum macranthum (Syn. *S. maroniense*): See page 224.

Solanum rantonnetii (Syn. *S. muticum*, *S. japonicum*); Family: *Solanaceae*; Origin: Paraguay.

An erect shrub, 90-120 cm tall. The violet flowers with yellow centre are borne in profusion during April-May and continues till August. It is propagated from cuttings.

Sophora tomentosa; Family: *Leguminosae*; Origin: West Indies.

A large, 1.8 to 2.5 m tall, bushy shrub with ornamental foliage. Flowers are yellow, pea-shaped and are borne terminally on all branches during the summer and the rains. The shrub is a beauty when in full bloom or even otherwise. This can be grown as a specimen in the lawn or in the shrubbery. Seedlings flower during the third year.

Spiraea betulifolia; Family: *Rosaceae* (Sub-family: *Spiraceae*); Origin: Northeast Asia, Japan.

A group of ornamental shrubs which flower to perfection at medium and high altitudes. The above species is a sun-loving dwarf (0.5-1 m) shrub, but in the scorching sun of the plains the flowers wither in no time. The white flowers are borne in dense panicles. The brown-tipped anthers provide a good contrast against the white corolla. A good plant for shrubbery. There are many other species, some of which are grown as ornamental hedges. Propagated from seeds, cuttings, or layerings.

Stachytarpheta indica (Syn. *S. jamaicensis*, *S. urticifolia*); Family: *Verbenaceae*; Origin: South America.

A dwarf (60 cm), coarse-looking sub-shrub which bears blue flowers during the summer and the rains. Propagated from cuttings.

Stachytarpheta mutabilis (Syn. *Verbena mutabilis*); Family: *Verbenaceae*; Origin: South America.

A hardy shrub or sub-shrub attaining a height of 90-120 cm. The small flowers open crimson and fade to rose, and these are borne in long erect spikes almost round the year but more profusely during the rains. It is propagated from seeds or cuttings.

Streptosolen jamesonii (Syn. *Browallia jamesonii*); Family: *Solanaceae*; Origin: Colombia, Ecuador.

A climbing shrub which attains a height of 1.8-2.4 m, thriving and flowering best at medium-to-high elevations. The bell-shaped orange-coloured flowers are borne in terminal clusters. It is grown as a border plant or in a shrubbery and sometimes in large pots. If the main stem is staked, it can be trained into a standard. Propagated from seeds and cuttings.

Strobilanthes; Family: *Acanthaceae*; Origin: See below.

A group of dwarf herbaceous shrubs (sub-shrubs) which grow without much care. Some are grown for their foliage with metallic lustre, whereas others are grown for the flowers. The more important species are:

(a) *S. anisophyllus* (Syn. *Goldfussia anisophylla*): Origin India. A bushy plant of 90 cm which bears bell-shaped lavender-blue flowers in profusion during the cold season. It thrives at medium-to-high elevations. A good pot plant which is propagated from cuttings.

(b) *S. coloratus* (Syn. *Diflugossa colorata*): Origin Sikkim. The leaves are dark green above and reddish-purple beneath. The flowers, almost funnel-shaped, are large and pale bluish-purple in colour.

(c) *S. dyerianus* (Syn. *Perilepta dyeriana*): Origin Burma. A straggling plant with dark green leaves blotched rich rosy-purple along the middle of lamina. The flowers are long and pale violet.

(d) *S. glomeratus* (Syn. *Goldfussia glomerata*): Origin Sikkim, Burma. It is 1.5 m tall, flowers are funnel-shaped, bluish-purple in colour, with very showy pinkish-purple persistent bracts. Pruning is done during the rains to keep it in shape.

(e) *S. isophyllus* (Syn. *Goldfussia isophylla*): Origin Northeast India. A dwarf, 60-90 cm bushy shrub with swollen joints. The long funnel-shaped flowers are blue and white.

(f) *S. scaber* (Syn. *Sericocalyx scaber*): Origin North India, Burma. A shrub, 30-90 cm tall, which bears thimble-shaped yellow flowers in March.

All the *Strobilanthes* grow better in semi-shade and shade and is propagated generally from cuttings and seeds.

Strophanthus dichotomus; Family: *Apocynaceae*; Origin: Malaysia, India.

An uncommon, erect shrub, 2.5 m in height. Flowers are curious looking with awl-shaped sepals and purple corolla tube having 12-15 cm long white thread-like tails. The flowers appear almost all the year round but remain covered by the foliage. Propagated by ground or air layering.

Syringa persica; Family: *Oleaceae*; Origin: Persia to China; Common Name: Persian Lilac.

A bushy shrub, attaining a height of 1.75 to 2 m. The flowers are lilac, fragrant, and borne in broad panicles during the summer. Grows well at medium-to-high altitudes.

Syringa vulgaris; Family: *Oleaceae*; Origin: Eastern Europe to Caucasus and Afghanistan; Common Name: Common Lilac.

A most popular shrub of the temperate region. The flowers and foliage are most attractive. The flowers are lilac, fragrant, and borne profusely in dense terminal spikes. The heart-shaped leaves are attractive. There are garden varieties in which flowers are tinged blue, violet, purple, or rose. Propagated by grafting and from seeds, hardwood cuttings, and suckers. But the most common method is layering.

Tabernaemontana coronaria (Syn. *Ervatamia coronaria*); Family: *Apocynaceae*; Origin: India; Common Names: Adam's Apple, Crape Jasmine; Vernacular Names: Hindi – *Chandni, Taggar*; Bengali – *Tagar*; Kannada – *Nandi battalu*; Tamil – *Nandiya vattai*.

A very popular handsome shrub especially in eastern India, which bears numerous flowers almost all the year round but more during the summer and the rains. A hardy shrub which thrives best under the full sun. The single types form a good ornamental hedge and can also be grown in a shrubbery. The double flowering variety can be grown as a specimen in lawn, in a shrubbery, or may be formed into an ornamental hedge. Propagated from cuttings. The common garden varieties are: (a) flowers single and white, shrub tall; (b) flowers single and white, shrub dwarf; (c) var. *flore-pleno* (Syn. *plena*), flowers double, shrub tall; and (d) var. *variegata*, flowers white and double, foliage variegated yellow.

The shrubs need regular pruning after flowering to keep them in good shape.

Tamarix; Family: *Tamaricaceae*.

A group of shrubs, sometimes trees, which differ in looks completely from other shrubs. The branches are feathery and whip-like in appearance. The tiny scale-like leaves resemble some of the junipers. The shrubs are extremely suitable for saline soils, sea coasts, and dry localities. It is propagated from its minute seeds or cuttings. The shrubs need careful pruning after flowering. The important species are:

(a) *T. dioica:* Origin India. A tall shrub to a small tree, about 2 m in height, having drooping branches. The flowers are pink and are borne almost round the year.

(b) *T. gallica* var. *indica* (Syn. *T. indica*): Origin Himalayas. Branches are slender and upright. The flowers are pink.

(c) *T. juniperina* (Syn. *T. japonica*, *T. lumosa*): Origin China: Feathery-branched shrub or small tree, bearing bright pink flowers.

Tecoma; Family: *Bignoniaceae*.

A group of shrubs which are most commonly grown in this country. The shrubs are showy and bear terminal clusters of campanulate, brightly coloured flowers. These are suited for shrubbery, or as ornamental hedge, and also as specimens in the lawn. The following species are important.

(a) *T. capensis:* See *Tecomaria capensis*.

(b) *T. gaudichaudi:* A tall (up to 4 m) quick growing, sun-loving shrub which bears clusters of golden-yellow flowers during the summer months, but also continues to bear a few flowers occasionally. It is grown in the back row of a shrubbery. It needs drastic pruning.

(c) *T. smithii:* Possibly an introduction from Australia and is smaller in size than the other species. The bright yellow-tinged orange flowers are produced freely in large panicles. This is supposed to be a hybrid between *T. mollis* and *Tecomaria capensis.*

(d) *T. stans* (Syn. *Stenolobium stans, Bignonia stans*)*:* Origin West Indies. A large, 3 m tall, bushy shrub which grows very quickly and even seedlings flower in the second year. It can be grown in light shade as well as in the sun. The pinnate leaves are handsome. The large golden-yellow flowers are borne in terminal clusters almost throughout the year. The shrub is used to screen a high compound wall or grown in shrubberies or as hedge. The shrub is kept bushy by cutting the branches severely every year.

Tecomaria capensis (Syn. *Tecoma capensis*); Family: *Bignoniaceae*; Origin: South Africa; Common Name: Cape Honeysuckle.

A glabrous climber or near-upright shrub, 4.5 m tall. The tubular honeysuckle-like bright orange-scarlet flowers bloom almost throughout the year. Suited for planting in lawns as a specimen, when it should be trimmed to shape. This can also be planted in large groups or formed into an ornamental hedge or grown as standard.

Tephrosia candida; Family: *Leguminosae*; Origin: India.

A small shrub which bears white or reddish pea-shaped flowers during September-October.

Tephrosia grandiflora (Syn. *Galega grandiflora*); Family: *Leguminosae*; Origin: South Africa.

Tephrosia macrantha; Family: *Leguminosae*; Origin: Mexico.

A shrub, 2-3 m tall, with odd-pinnate leaves comprising 10-12 pairs of leaflets. A handsome shrub which bears purplish-white flowers resembling sweetpeas, in large terminal panicles.

Thespesia lampans; Family: *Malvaceae*; Origin: Asia.

A bushy, 50 to 70 cm tall undershrub, bearing yellow flowers with a glossy brown centre throughout the year. Seedlings bear flowers within a year.

Thevetia nerifolia (Syn. *T. peruviana*); Family: *Apocynaceae*; Origin: Tropical America, Naturalized in India; Common Name: Yellow Oleander: Vernacular Names: Hindi & Punjabi – *Pilakaner*; Bengali – *Kalkephul, Haldekarabi.*

A 3 m tall shrub, but may grow into a small tree if not pruned regularly. The branches are cylindrical and cane-like. The leaves resemble oleander. The greenish-yellow flowers are tubular at base but opening into a funnel-shaped structure. The plant flowers in profusion throughout the year. The foliage is not eaten by cattle and goats. There are varieties with white and orange flowers. The plant can be grown as a specimen in large compounds or can be grown as ornamental hedge. The fruits and seeds are poisonous.

Propagated from seeds and cuttings.

Thunbergia affinis; Family: *Acanthaceae*; Origin: Tropical Africa.

More or less an erect or a rambling shrub, 1.5 m in height. The flowers which are violet with yellow tube are borne during the summer and rainy seasons. It is easily propagated from cuttings. Needs pruning to keep it in shape.

Thunbergia erecta (Syn. *Meyenia erecta*); Family: *Acanthaceae*; Origin: Tropical West Africa.

A compact, bushy shrub, 60-120 cm tall, with erect branches which originate almost from the ground level. In severe winter, the shrub becomes deciduous. The flowers are dark blue with orange throat and yellow tube. The flowering season is during February. The shrub is trimmed during the rainy season to keep it dwarf. A handsome shrub for a mixed border or shrubbery; can also be trimmed into ornamental hedge. Easily propagated from cuttings during the rains, although the seed set is very good. There is a var. *alba* with white flowers.

Thunbergia kirkii (Syn. *T. hookeriana*); Family: *Acanthaceae*.

An erect shrub, 90 cm tall, which bears in the summer and rainy seasons 2.5 cm long, violet-blue flowers. Propagated from cuttings.

Thunbergia natalensis; Family: *Acanthaceae*; Origin: South Africa.

A dwarf, 60 cm tall, bushy shrub with characters similar to those of *T. erecta* except that it is dwarfer and flowers during the summer. The flowers are blue with yellow tube. The shrub grows and flowers well in light shade. Propagated easily from cuttings.

Tibouchina elegans (Syn. *Pleroma elegans*); Family: *Melastomaceae* (also spelt *Melastomataceae*); Origin: South Brazil.

A shrub, 150 cm tall, suitable for growing in hill stations only. The flowers are small, 3-6 cm across, solitary, purple in colour and produced in abundance during May and continues up to October.

Tibouchina semidecandra (Syn. *Lasiandra macrantha*); Family: As in the above species.

It is one of the most attractive, handsome and tall (3-8 m) shrubs of our hill stations. The shrub is easily recognized by its leaves, which are broad at the base and narrowly pointed at the tip and are prominently nerved. The flowers are rich purple, very large, 8-13 cm across in size, and are borne solitary or in clusters of a few flowers throughout the year under glass but mainly outdoors during the autumn. The young plants of this species are suitable for growing in pots while the older ones are better adapted as trellis or pillar plants.

Tibouchinas require very little attention during most part of the year, but should be irrigated regularly during the summer. They prefer a light, well-drained soil. *Tibouchinas* are propagated from cuttings in the spring.

Tithonia tagetiflora (Syn. *T. speciosa*, *T. rotundifolia*); Family: *Compositae*; Origin: Mexico, Central America.

A large, 1.2 to 1.8 m tall, shrubby, herbaceous plant, which bears large, 5-7 cm across, single rich-red or orange flowers during November-January. The other species, *T. diversifolia*, is a spreading and coarse-looking, shrubby, herbaceous plant which bears 6-15 cm across single yellow flowers. Propagated from seeds or cuttings. The plants like full sun and are hardy.

Turnera ulmifolia; Family: *Turneraceae*; Origin: Mexico; Common Names: West Indian Holly, Sage Rose.

A straggling sub-shrub, 60-90 cm tall, which bears solitary, almost sessile, large yellow flowers with maroon centre, almost round the year. The plant flowers in the sun as well as in semi-shade and is enough hardy to survive in the inferior soils of uncared for rock gardens. The plant is propagated from seeds, cuttings, or by division. The var. *elegans* is dwarf and bears almost sessile, pale, or sulphur-yellow flowers.

Uraria crinita; Family: *Leguminosae*; Origin: India, the Philippines.

A sub-shrub, 1-2 m tall, which bears pea-shaped, lavender-coloured flowers, in dense cylindrical racemes about 30 cm long.

Uraria picta; Family: *Leguminosae*; Origin: India, the Philippines.

Almost similar to the above in growth habit and bears purplish flowers in cylindrical racemes 15-30 cm long. *Urarias* are hardy rock garden plants and are rarely grown in the shrubbery. The flowering time is the winter. Propagated from seeds.

Vaccinium corymbosum; Family: *Ericaceae*; Origin: U.S.A.; Common Names: High-bush, Tall or Swamp Blueberry.

A tall (1.2 to 3.5 m) shrub, worthy for its white tinted pink flowers and blue-black globose fruits which are edible. The leaves turn rich-red in the autumn. Propagated by layering or from cuttings. Suitable for medium-to-high elevations.

Viburnum; Family: *Caprifoliaceae*.

A group of deciduous or sometimes evergreen shrubs, rarely small trees, grown for their attractive foliage (often for autumn colour), flowers and fruit. *Viburnums* can be easily cultivated, but they do not like dry situations and poor soils. Propagated from seeds or cuttings. Some important species are:

(a) *V. × burkwoodii:* Evergreen shrub, 1.2 to 1.8 m tall. The flowers which appear during March to May are pinkish, turning white.

(b) *V. × juddii:* A deciduous, bushy and spreading shrub which bears pink, turning almost white, fragrant flowers during April to May.

(c) *V. sieboldii* (Syn. *V. reticulatum*)*:* Origin Japan. A deciduous shrub growing up to 3.5 m. Flowers are creamy-white, fragrant and appear during May to June. The fruits are oval, pink, but turn blue-black on maturity.

Vitex agnus-castus; Family: *Verbenaceae*; Origin: South Europe; Common Names: Chaste-Tree, Hemp-Tree, Monk's Pepper Tree.

A 3 m tall deciduous shrub with aromatic odour. In the summer and the rains, numerous compact terminal racemes of blue or pale violet, fragrant flowers appear on the shrub. Propagated by air layering.

Vitex trifolia; Family: *Verbenaceae*; Origin: Southeast Asia to Japan, the Philippines, Australia.

A deciduous, much-branched, tall (3 m) shrub with tri-foliate leaflets. The flowers are blue or lavender and appear during the summer. The earlier-mentioned species is much more beautiful than this.

Wigandia caracasana (Syn. *W. macrophylla*); Family: *Hydrophyllaceae*; Origin: Mexico to Colombia.

A robust, erect sub-shrub which attains a height of 3 m. The leaves are very large, 45 cm long and 25 cm broad. The lilac-blue flowers, with white tubes, are produced in terminal spikes during the summer. It is propagated from seeds which are minute in size.

Woodfordia floribunda (Syn. *W. tomentosa, Grislea tomentosa*); Family: *Lythraceae*; Origin: India, Sri Lanka.

A shrub, measuring 1-2 m in height. The brick-red tubular flowers are borne along the drooping stem during February-March.

Zenobia pulverulenta (Syn. *Z. speciosa, Andromeda dealbata, A. glauca*); Family: *Ericaceae*; Origin: U.S.A.

A shrub, 60-120 cm tall, with erect or arching branches, sometimes evergreen. The 1.5 cm across, bell-shaped, white flowers are borne in abundance along the young branches during the spring or early summer. The bell-shaped flowers against the blue-green leaves look most ornamental. The old branches should be pruned from time to time. Propagated from seeds, cuttings, and layers or offshoots, if available.

Chapter IX

Bougainvillea

A most popular ornamental shrub, sometimes a climber, bougainvillea was first collected by Commerson, a French Botanist, from Rio-de-Janeiro, Brazil, and was named after the famous French navigator Louis Antoine de Bougainville. It belongs to the family *Nycataginaceae* and has originated in tropical and sub-tropical South America.

Bougainvillea is a quick-growing shrub and varies in height according to different species and cultivars. The leaves are simple, alternate, roundish ovate, or elliptic-lanceolate. Bougainvilleas are armed with stout spines in the stem with the help of which the shrub is able to climb. It is the three brightly coloured petal-like bracts which give beauty to bougainvillea. In recent years, some multi-bracted bougainvilleas have been introduced where the number of bracts are more than three and in such cultivars there is no true flower. The colours of the bracts are innumerable, ranging from white to deep magenta, including other lighter shades such as yellow, orange, pink, mauve, purple, scarlet, crimson, and red. In some cultivars, bracts of two colours are found in the same plant, while in others the colour of the bracts changes with age, thus creating an impression that the plant is bearing two- or three-coloured flowers. The actual flowers in bougainvillea are small, tubular, ridged, and open into a "star" at the apex, the colour of which may be white, light greenish-yellow, cream, yellow, or pink. The flower is usually inconspicuous and remains attached at the base of the bracts.

Bougainvilleas are widely grown in Indian gardens. When in full bloom they present a riot of colours. The peak blooming time in north India is during September to December and again from February to June. However, some cultivars such as 'Snow Queen' and 'Snow White' flower only once during the winter. In places such as Mysore, Bangalore, and Hyderabad the peak flowering season is during February to April and again during July to September. Actually, in the milder climate of these places bougainvilleas flower for a longer period of the year, but the peak flowering seasons are as stated above. The beauty of bougainvillea is that it flowers during the dull summer months, when excepting such trees as *Gulmohar*, Pink Cassias, *Peltophorum*, and Jacaranda very few other ornamental plants produce flowers so freely. As a matter of fact, flowering is best during the dry period. In a wet climate, some of the bougainvilleas flower very little. Bougainvillea does not normally flower in the shade; but when it does, the quality of the flower is poor. In very wet climates bougainvillea flowers very poorly.

Though bougainvillea is a shrub suitable for the plains, a few varieties such as *B. glabra* ('Sanderiana') can be grown at higher altitudes (up to 1,500 to 2,000 m) in some of the hill stations such as Nainital, Almora, and the Nilgiris.

Classification: The garden cultivars of *Bougainvillea* belong to four species. There are about ten species of *Bougainvillea* but only three are of horticultural importance. These are *B. spectabilis*, *B. peruviana*, and *B. glabra*. The last one has three varieties, namely, '*cypheri*', '*sanderiana*', and '*variegata*'. A fourth species, *B.* × *buttiana*, is believed to be a natural hybrid between *B. peruviana* and *B. glabra*.

Uses: Bougainvillea can be used in the garden both as a shrub and as a climber. It is generally grown in the ground but can also be cultivated in pots. It is a very good subject for growing as a specimen in the lawn and is very often grown in that position as a standard, Bougainvillea forms into a colourful ornamental hedge and is impenetrable. Such hedges, once established, need very little attention except occasional pruning to keep them in good shape. The cultivars suitable for hedges are 'Partha', 'Sanderiana', 'R.R. Pal', 'Mary Palmer', 'H.C. Buck', 'Thimma', etc. The first two cultivars form a dwarf hedge. Bougainvilleas are also trained on pergolas or arches and higher gateways. The arches and gateways must be heavily built to withstand the weight.

In Bangalore, it is common practice to train bougainvilleas along the boundary wall. The lateral branches are held in position along the wall by strings and nails. It is also trained vertically against the walls of a building by providing a scaffolding of G.I. wires or wire-nettings for support. Depending upon the cultivar, a bougainvillea can grow into a tree, 10-20 m in height. A dead tree or a stump can also be decorated beautifully by allowing a bougainvillea to climb over it.

It is common practice, especially in Bangalore, Mysore, and some such other places, to grow bougainvilleas on slopes or mounds. To achieve this objective the branches are pegged to the ground at frequent intervals horizontally along the slope or the mound. All the shoots growing vertically are either pegged down, if possible, or removed. If done properly, the whole slope will be covered within a short time and the plant will start flowering. But since such plants are clipped frequently to keep them in a prostrate state, the flowering will obviously be less.

Besides the above uses, bougainvillea is also grown as a shrub in front of a wall or to screen an ugly spot. Bougainvilleas are grown in large pots or wooden and concrete tubs.

Propagation: Bougainvilleas are propagated from cuttings, by ground or air layering and budding. Usually, pencil-thick hardwood leafy cuttings of 15-25 cm length are the best. Some cultivars such as 'Mary Palmer', 'Thimma', 'Louis Wathen', 'Tomato Red', 'Lateritia', etc., are difficult to root from cuttings. In cultivars such as 'Thimma' and 'Cypheri' tender tip cuttings root better than hardwood cuttings. Before the cuttings are inserted they may be treated with growth regulators such as IBA (indolebutyric acid), IAA (indoleacetic acid), or NAA (naphthaleneacetic acid) in solution or talc. The concentration used is generally between 1,000 and 3,000 ppm in the "quick-dip" method. The time for striking cuttings in Delhi and surrounding areas is during July-August or February-March. In eastern India, the cuttings are planted in the nursery during the rains. In a climate as that of Bangalore it is possible to plant cuttings almost throughout the year, but it should preferably be done during June-October. The cultivars which are difficult to propagate from cuttings are raised by layers. For air layers the compost used is leafmould or leafmould with farmyard manure or farmyard manure with soil and sand. Some delicate cultivars can be raised by the inarching method on a vigorous growing rootstock. The cultivars which cannot be grown either from cuttings or by layering are raised by "T" or "Shield" budding. Generally, 'Dr. R.R. Pal', a vigorous cultivar, is used as rootstock. 'Partha' is also a good rootstock and is used at some places. The best budding time is February to March. Plants are also raised from seeds for obtaining new hybrids or cultivars.

Cultural Hints

It is a hardy shrub that can be grown under different types of soil and climatic condition.

Soil and Situation: Being a hardy shrub it thrives well in any type of soil provided there is no water-logging. It grows with varying degrees of success in the sandy soils of Rajasthan, the red soils of Andhra Pradesh and Karnataka, and the black cotton soils of Maharashtra and Karnataka, and even in rocky areas with small pockets of soil. The root system is deep and, once established, the plants grow well, with very little care.

As already stated, bougainvillea does not flower well in shade. A situation which receives the sun, preferably throughout the day or at least up to 12 noon, should be selected for planting bougainvillea. The brighter and more dazzling colours of bougainvillea (e.g., magenta, red, purple) dominate over other colours in the garden. Therefore, bougainvillea of such colours should not be planted near rose or annual beds. They suit very well against the white background of walls or in the corner of a lawn. In north India, young bougainvilleas should be given protection against frost during the winter.

Time of Planting: The best time of planting in northern India is during July to September. Planting in the winter should be avoided as the mortality rate will be high and the growth of plants will be very poor. In eastern India, the planting is done during the rains. In heavy rainfall areas planting may be done just before or after the rains. In Bangalore and at places having similar climate, planting is possible almost throughout the year except during the hotter months of March to May. But the best planting time is between June and October. Pits measuring about 1 m in diameter and 75 cm deep should be dug to receive the plants. About 8-10 kg of well-rotten cowdung manure should be added to each pit at planting time. In the red soils of Bangalore, horse manure may be used in place of cowdung manure, if available.

Manures and Fertilizers: After the basal dose of manure, the bougainvilleas are given 10-12 kg of farmyard manure once a year after the June pruning. At this time, 250 gm of bonemeal should also be applied to each plant. Generally, no fertilizer is applied. But Pal and Swarup (1974) suggest the application of 250 gm of a mixture of ammonium sulphate, superphosphate, and sulphate of potash in the proportion of 1 : 3 : 2 per fully grown bush. This is said to promote profuse flowering. Application of liquid manure by fermenting fresh cowdung or oilcake at fortnightly intervals prior to flowering has been found to be beneficial.

Irrigation: Young plants or those which are newly planted need frequent watering. Once established, the plants need very little or no watering. However, in dry localities, one heavy irrigation per month during summer is beneficial. Frequent watering is harmful. During flowering time watering should be restricted; otherwise, heavy watering will result in the shedding of the bracts.

Pruning: No drastic pruning is needed for bougainvilleas grown in the ground for display. The amount and the method of pruning will depend upon the cultivar and the purpose for which the plant is grown. For example, the specimen bushes growing in the lawn or on the trees and stumps do not require much pruning except pinching or tipping and removal of weak and dead branches. The side branches and erect extra growths are pruned in those plants which grow on a wall. In standards, all growths on the main stem are removed to keep the umbrella form of the plant in shape. Generally speaking, *B. glabra* and *B. peruviana* do not need hard pruning.

In northern and eastern India, the time of pruning is May to June. Tipping or pinching should also be completed by the middle of September. Manures and fertilizers are added at the time of pruning. In southern India also, pruning should be done at this time.

Pests and Diseases

Bougainvillea is a hardy plant and pests and diseases are not a problem. Some leaf spot diseases have been reported by some workers in the U.S.A., but these are not serious in India. Similarly, some leaf-eating caterpillars have been mentioned by some workers in the U.S.A., but these do not cause any serious trouble in India.

Growing Bougainvilleas in Pots

It is also possible to grow bougainvilleas in pots. Pots of 30 cm or more diameter are suitable for this purpose. The pots should be placed over bricks or stone slabs so that the roots do not go into the ground. Large cement and wooden tubs are also very suitable. In an earthen pot a bougainvillea will grow well for about three years, whereas in a large cement pot it may continue to do so for about five years. A rich pot mixture is to be used for the pots and about 1-2 tablespoonfuls of bonemeal should also be added, depending on the size of the pots.

In northern and eastern India, the planting time is July to August, whereas in Bangalore planting can be done any time in the year except during March to May. Potted plants need more watering and during the summer the frequency of irrigation may be once a day or once in two days. Prior to the blooming period, watering should be stopped for about 7-10 days, so that the plants are forced to bloom on leafless branches. During the peak flowering period irrigation is regulated, otherwise the flowering bracts will drop early.

Potted plants need harder pruning compared to the plants in ground. Pruning time is June. In Bangalore this can be done on May also. After pruning cowdung manure should be applied. Soon after flowering starts, the plants should be fed with liquid manure, which is prepared by fermenting fresh cowdung in water and diluting this to tea colour. About one to two litres of this solution should be applied to each pot at 10-15 days' intervals. Now-a-days it is possible to keep the potted plants dwarf by applying chemicals such as maleic hydrazide, B-Nine, phosphon, and cycocel in different concentrations.

The following cultivars do well in pots: 'Blondie', 'Dr. R.R. Pal', 'Lady Mary Baring', 'Mahara', 'Mary Palmer', 'Roseville's Delight', 'Sonnet', 'Spring Festival', 'Summer

Time', 'Thimma', and 'Tomato Red'.

Cultivars

There are innumerable cultivars of *Bougainvillea* including hybrids. Some of the selected cultivars belonging to different species and some hybrids which are commonly grown in India are described below.

(*A*) Cultivars of *B. peruviana*

(1) 'Princess Margaret Rose' (possibly a synonym of 'Lady Hudson'). Pale pink bracts, flower tube greenish purple, large leaves, perpetual blooming habit.

(2) 'Mary Palmer' (patented as 'Surprise' in the U.S.A). It is a sport from 'Mrs. H.C. Buck' which was raised by S. Percy Lancaster in Calcutta and named after Mrs. Palmer in whose garden the sport appeared. White and magenta coloured bracts appear on the same plant. Sometimes, a bract may be partly white and partly magenta. It is a most popular bicoloured cultivar.

(3) 'Dr. B.P. Pal'. A vigorous floriferous white-bracted bud sport of 'Mary Palmer' raised by the Agri-Horticultural Society of India, Calcutta, suitable for growing in pots.

(4) 'Shubhra'. Another vigorous, floriferous, white-bracted sport of 'Mary Palmer' raised at the National Botanical Research Institute, Lucknow.

(5) 'Mrs. H.C. Buck' (Syn. 'Mahatma Gandhi'). Probably seedling of 'Princess Margaret Rose'. Bracts deep rose-magenta. A continuous bloomer. The drooping branches are covered with long sprays of flowers.

(6) 'Partha'. A seedling of 'Princess Margaret Rose'. Bracts change colour from orange or blood-red to purple with age, giving a bicolour effect. A free-flowering vigorous cultivar.

(B) Cultivars of *B. glabra*

(1) 'Cypheri'. A free-flowering large-bracted bougainvillea. Bracts purple, slightly towards red when old. Green cuttings propagate better. A good pot plant.

(2) 'Formosa'. A round-the-year flowering type with pale rosy-mauve bracts. The cultivar is bushy in nature and needs pruning.

(3) 'Magnifica'. Produces a mass effect when in flowering. Colour of bracts is cyclamen-purple which changes with age. A useful cultivar for pergolas.

(4) 'Sanderiana'. Flowers appear all along the branches. Bracts are purple and stars are prominent. A good cultivar for pots, tubs, and as hedge as it stands severe pruning. It is tolerant to moderate frost of northern India. Easily propagated from cuttings.

(5) 'Snow White'. It is possibly a seedling of some *glabra* cultivar. The cultivar has greenish-white bracts which turn to white (with greenish veins) on opening. Flowering is shy in wet soils but better in light sandy soils. Flowers only during the winter in the northern plains. Other white-bracted cultivars are 'Snow Queen', introduced in India by Dr. Homi Bhabha; 'Moonlight Madonna', a climber, 'Alba', and 'Jennifer Fernie', introduced from Africa by Lalbagh, Bangalore.

(6) 'Buddhadas'. Most probally a seedling of 'Snow Queen'. Bracts are white.

(7) 'Dream'. A seedling of *B. glabra*. Branches are drooping, flowering is profuse and bracts are purple.

(8) 'Gagarin'. A seedling of 'Trinidad', named after the Soviet Astronaut. Flowering trusses are terminal and huge. Bracts are mellow-purple.

(9) 'Gopal'. A seedling of 'Formosa'. Bracts are deep-rose, suffused with mauve.

(10) 'Kalyani'. A seedling of 'Gopal'. Bracts are large, brilliant rose-pink to carmine-pink, becoming paler with age. Free-flowering.

(11) 'Sachidananda'. Most probably a seedling of 'Snow Queen'. Bracts are very bright lavender mauve.

(12) 'Stanza'. Most probably a seedling of *B. glabra*. Bracts are chrysanthemum-crimson. Free-flowering.

(13) 'Trinidad'. Probably a seedling of 'Snow Queen'. Bracts are pale mauve.

(C) Cultivars of *B. spectabilis*

(1) 'Aida'. A seedling of *B. spectabilis*. Bracts are carmine-red.

(2) 'Boise de Rose'. Seedling of *B. spectabilis*. Bracts are chamois with green rib and veins. Commonly described as biscuit-coloured. Free-flowering.

(3) 'Jubilee'. A seedling of *B. spectabilis*. Plants are vigorous, bracts are purple.

(4) 'Maharaja of Mysore'. Most probably a natural hybrid between 'Lateritia' and *B. glabra*. Bracts are small, pale rose with a tinge of mauve.

(5) 'Mrs. Lancaster'. A vigorous-growing seedling of 'Maharaja of Mysore'. Bracts are large and magenta-rose in colour.

(6) 'Rosa Catalina'. A compact free-blooming natural hybrid between 'Lateritia' and *B. glabra*. Bracts are red, changing to rose-madder.

(7) 'Tomato Red'. Bracts terracotta-shaded scarlet. Free-flowering.

(8) 'Lateritia' (Syn. 'Dar-es-Salam', 'Brasiliensis'). A shy grower. The bracts change colour with age, first opening as poppy-red, changing to brick-red and finally orange. Stars are prominent. Difficult to propagate from hardwood cuttings but green cuttings are more successful. It thrives well in dry, drought-prone areas.

(9) 'Refulgens'. A good climber, it is always in bloom in long pendulous inflorescences. Bracts are deep purple-mauve and stars are prominent. It is commonly grown as standards and also in pots.

(10) 'Speciosa'. A free-flowering cultivar suitable as weeping standard. Bracts are magenta-rose or orchid-purple.

(D) Cultivars of *B. × buttiana*.

(1) 'Alick Lancaster' (Syn. 'Lilac Queen'). Raised in India by A. Percy Lancaster. Bracts are pale-lilac when young, turning to purple. This is a bud sport of 'Scarlet Queen'.

(2) 'Enid Lancaster'. A bud sport of 'Louise Wathen' raised by A. Percy Lancaster. Bracts are turmeric-yellow shaded orange. Also known as 'Golden Glow' and 'Golden Queen'.

(3) 'Golden Glow'. Possibly a bud sport of 'Mrs. Butt' and very similar to 'Enid Lancaster'. The bracts, when young, are majolica-yellow, then turning to cadmium orange, and lastly to light carmine. Free-flowering in habit.

(4) 'Lady Mary Baring'. A vigorous-growing, free-flowering bud sport of 'Golden Glow'. Bracts are Indian-yellow with green veins. The large cluster of flowers is borne terminally on the branches. It is the best cultivar amongst the yellows.

(5) 'Louise Wathen'. A bud sport of 'Scarlet Queen'. The bracts are golden-salmon, tinged pink, but turn to pale mauve with age. The flowers are malformed.

(6) 'Magenta Queen'. (Syn. 'Purple Queen', 'Mrs. Butt Magenta'). A bud sport of 'Mrs. Butt', raised by B. Rama Rao of Madras. A continuous bloomer with magenta-purple bracts.

(7) 'Mrs. Butt' (Syn. 'Ruby', 'Crimson Lake'). A climbing shrub having crimson bracts when young but changing to purple; stars are prominent.

(8) 'Scarlet Queen'. The bracts are deep rose, very large in size. The tips of the bracts are malformed.

(E) *Some cultivars (seedlings) whose parentage is not known.*

(1) 'Asia'. Introduced by Lalbagh, Bangalore, from Kenya. Blooms profusely all along the branches in compact clusters. The bracts are crimson at opening but change to purple with age.

(2) 'Begum Ali Yavur Jung'. The bracts have some similarity to those of 'Tomato Red' in colour. The leaves are small in size.

(3) 'Eclipse'. Flowering is profuse and the flowers cover the entire length of the branches. The bracts are purple in colour.

(4) 'Kayata'. A free-flowering plant, but the dried bracts persist on the branches. The bracts are rose; sets seed freely.

(5) 'Nawab Ali Yavur Jung'. The bracts are recurved, pink in colour.

(6) 'Pixie'. A dwarf plant which bears flowers continuously. The bracts are soft pink.

(F) *Double-bracted cultivars*

In recent years, several multi-bracted or double-bracted cultivars have been introduced by Dr. J.V. Pancho of the Philippines. Instead of the normal three bracts as found in Bougainvillea, the bracts in these cultivars number between 20 and 40. These are bud sports from cultivars of *B.* × *buttiana*. The different cultivars in this category are listed below:

(1) 'Archana'. A sport of 'Roseville's Delight' with similar bracts, but have variegated leaves.

(2) 'Carmencita'. Lavender-mauve bracts.

(3) 'Cherry Blossom' (Syn. 'Bridal Bouquet'). Multi-bracted. The bracts are white, but suffused with purple all along the margins and the tip. A bud sport of 'Los Banos Beauty'.

(4) 'Double Pink'. The bracts are light pink, similar to those of 'Los Banos Beauty'.

(5) 'Double White'. Bracts are greenish-white with pink margin. In all probability it is 'Cherry Blossom' only.

(6) 'Los Banos Beauty' (Syn. 'Pink Champagne'). A sport of 'Pink Beauty'. The bracts are mellow-purple. The branches are drooping.

(7) 'Mahara' (Syn. 'Million Dollar', 'Manila Magic Red', 'Manila Red'). The bracts, numbering 30-40, are purple. The flowers appear in large clusters. Free-bloomer.

(8) 'Marietta'. A bud sport of 'Mahara' with white and green variegated leaves; margins

pink when young. The bracts are similar to those of 'Mahara'.

(9) 'Roseville's Delight'. A free-flowering vigorous plant with apricot-orange bracts.

(G) *Variegated cultivars*

There are a number of cultivars with variegated leaves that are in cultivation in the garden. These are mostly bud sports. A few of these are described below:

(1) 'Archana'. A double-bracted cultivar, mentioned earlier.

(2) 'Bhabha'. A sport of 'Louise Wathen', arisen at Lalbagh, Bangalore. The leaves are variegated cream and green. The bracts are similar to those of 'Louise Wathen'.

(3) 'Glabra Variegata'. The plant is compact. The leaves are small and variegated cream or white. Suitable as hedge.

(4) 'Jawaharlal Nehru'. Developed at I.I.H.R., Bangalore. The leaves are beautifully variegated yellow and green. The flowers are scarlet.

(5) 'L.N. Birla'. A dwarf variegated (yellow and green) sport of 'Maharaja of Mysore'.

(6) 'Laxminarayan'. A cream and green variegated sport of *B. spectabilis*. Easy to propagate.

(7) 'Louise Wathen Variegata' (Syn. 'Louise Wathen Mediopicta'). The leaves are variegated with a patch of light yellow at the centre of the lamina. It is a bud sport of 'Louise Wathen'.

(8) 'Marietta'. A sport of 'Mahara'. The leaves are variegated white and green. It is multibracted like 'Mahara'.

(9) 'Rao'. A bud sport of 'Mrs. Butt'. The centre of the leaf is green, which is surrounded by white all along. The young shoots and leaves are pink. A very attractive plant. Grows vigorously. The bracts are red.

(10) 'Scarlet Queen Variegata'. The leaves are variegated with cream and green. A sport of 'Scarlet Queen'. The bracts are scarlet like those of 'Scarlet Queen'. A vigorous grower.

(11) 'Sharma'. A sport of 'Mrs. Fraser' with cream and green variegated leaves, with a tinge of red splashed here and there.

(12) 'Spectabilis Variegata'. The leaves are variegated yellow and green.

(13) 'Thimma'. A bud sport of 'Mary Palmer' with double-coloured bracts like those of 'Mary Palmer'. The young shoots are creamish-yellow and pink with some dark spots. A broad yellow variegation at the centre of the large green leaves makes this cultivar one of the best even when not in flower. It performs very well in Bangalore. Propagated from greenwood cuttings or by ground layering.

Chapter X

Climbers

Botanically a climber is defined as a plant which possesses special structures to climb over a support. These special structures may be hook-like thorns (e.g., *Bougainvillea*), tendrils (e.g., *Antigonon leptopus*; *Bignonia*), rootlets (e.g., *Ficus repens*), or modified leaf-stalks. The climbing plants which do not have any such structure but climb over a support or a plant by twining themselves spirally around such supports are called twiners. Such plants belong to the genera *Echitis*, *Hiptage*, *Lonicera*, etc. There are still other plants which fail in their attempt to climb but somehow manage to support themselves over the trunks, stems, or branches of other plants; these are termed ramblers and stragglers. *Quisqualis indica* (Rangoon Creeper) is a good example of such plants. Creepers are those plants which are unable to climb vertically on their own because of their weak stems. Morning Glory is an example of such plants. Trailers are similar to creepers, the only difference being that they do not form any roots at their nodes. In the U.S.A., all types of climbers are termed vine.

Utility of Climbers in a Garden

The importance of climbers is not properly understood by many amateur gardeners. They are important ornamental plants and, if properly selected and planted in suitable situations, enhance the beauty of a garden to a great extent. The bare walls of buildings or boundary walls can be covered beautifully by a colourful climber or a foliage-type of climber. Climbers can be used for the purpose of screening to maintain privacy from the adjacent houses. The climbers are also grown to be trained on trellises, over specially constructed pergolas, arches, arbours, and against pillars or similar structures. Climbers such as Bougainvillea and Morning Glory can be trained along slopes. Such climbers as *Ficus repens* or *Trachelospermum jasminoides* have no difficulty in climbing over walls with the help of their rootlets which attach themselves securely at the points of contact. Other climbers such as Bougainvillea have to be supported by nails driven in the wall and tying with strings or wires. Climbers are also trained to climb over net houses, lath houses or conservatories where shade-loving foliage plants are displayed. Climbers are most suitable for roof gardens, where vertical growth is preferred because of lack of space.

Classification of Climbers

Climbers can be classified according to their growth habit and the purpose for which they are grown.

(A) *Showy flowering climbers:* Climbers of this group generally display a show of colour with their flowers at a certain time of the year and are more appealing to the common man.

(B) *Climbers with scented flowers:* A climber may be attractive because of its beautiful flowers and foliage or the scent of its flowers. But some climbers may possess both these virtues. Some of the climbing jasmines, *Trachelospermum jasminoides*, Honeysuckle, etc., possess sweet scents. These are generally grown near the houses.

(C) *Climbers with attractive foliage:* Certain climbers are mainly grown for their foliage and the flowers of such plants are insignificant or inconspicuous. *Monstera deliciosa*, certain species of *Asparagus*, *Ficus repens*, etc., are grown for their foliage only.

(D) *Shade-loving climbers:* Though most of the climbers thrive best under the full sun there are a few that grow well under light or partial shade. Besides the foliage-type climbers such as *Pothos*, *Asparagus*, and *Hedera helix* which grow in partial shade, the flowering climbers such as *Trachelospermum jasminoides* with their sweet-scented flowers thrive well in full or partial shade. *Clerodendrum splendens*, *Jacquemontia pentantha*, *Thunbergia grandiflora*, etc., can grow well in places receiving partial shade.

Again the climbers are divided into two groups according to the amount of vegetative growth they make and are termed light and heavy climbers. *Quisqualis indica*, *Wisteria sinensis*, and most of the other species of *Wisteria* are heavy climbers. Though under favourable climatic conditions *Pyrostegia venusta* becomes a prolific climber, it is also possible to grow them in limited space by pruning and trimming them. *Petrea volubilis* is another light climber with beautiful flowers.

The climbers discussed in this chapter are all perennials. The annual climbers have been discussed in Chapter VI.

Hints on Cultivation

It is difficult to give any general cultural hints for all the climbers as the requirements vary from genus to genus, species to species, and sometimes even from one variety to another. In general, all climbers need plenty of sunshine and loamy soil with good drainage. For heavy climbers, pits of 75 cm × 75 cm are prepared at the site of planting, whereas for light climbers the pits may be 45 to 60 cm deep and 30 cm in diameter. The quick-growing heavy climbers intended for screening are planted 3 m apart and those with lighter growth are planted 1.2 to 1.5 m apart. *Ficus repens*, planted to cover a wall, are spaced about 25 to 30 cm apart.

Climbers do not grow vertically without support; therefore, they are helped to climb by tying them with strong strings or fine wires or some such support.

Some climbers also need pruning. This will be discussed while describing individual climbers. Manuring once a year just before the monsoon, with F.Y.M. or stable manure, is useful. Application of bonemeal or horn and hoof meal once or twice during the monsoon is also beneficial.

Some Selected Climbers

Adenocalymma alliaceum (sometimes wrongly spelt *Adenocalymna*); Family: *Bignonia-*

ceae; Origin: Brazil.

A heavy but beautiful climber with evergreen shining leathery foliage and large trumpet-shaped pink-mauve flowers blooming during March to June. In the mild climate of Bangalore, the creeper blooms several times in a year. The leaves as well as the flowers when bruised emit a garlic-like smell. The creeper thrives in hot moist climate. Propagated by layering or from cuttings in sand under mist or during the rainy season.

Adenocalymma calycina; Family: *Bignoniaceae*; Origin: Brazil.

The climber is similar to the above but bears yellow flowers which are scented. Flowering is during March to June. Propagation is similar to the above.

Allamanda cathartica var. *grandiflora*; Family: *Apocynaceae*; Origin: Brazil.

An easy-to-grow plant which responds well to manuring during growing season. It requires a lot of water for growth but the drainage should be perfect. The pot-grown plants are re-potted every year. The flowers are large, yellow, and scented, and bloom almost round the year. The var. *nobilis* also bears bright yellow flowers which are scented. For propagation see page 131.

Antigonon guatemalense (Syn. *A. insigne*); Family: *Polygonaceae*; Origin: Guatemala.

The flowers are more numerous and the leaves are broader than those of *A. leptopus*. The flowers are rose coloured and appear to be double.

Antigonon leptopus; Family: *Polygonaceae*; Origin: Mexico; Common Names: Coral Vine, Rosa de Montana, Love's Chain, Sandwich Island Climber.

A tuberous-rooted quick-growing climber, bearing rose-coloured flowers in great abundance throughout the year. The sprays of the flowers are very attractive. The creeper is pruned after flowering, otherwise it will cover the nearby plants. It is commonly grown for the purpose of screening. A rich well-drained soil is suitable for its cultivation. It is propagated from seeds and cuttings and by layering. A variety with white flowers is known as *album* or *albus*. The variety *aliporensis* has deep red flowers.

Aristolochia; Family: *Aristolochiaceae*; Common Names: Birthwort, Calico Flower, Bird's Head.

A group of climbers which bear peculiar duck-shaped flowers, which generally emit a repelling odour. The creepers are quick-growing but bear flowers only after attaining a good length. Propagated from seeds or by layering. Cuttings taken with a heel, root in propagating frame. The common species are:

(a) *A. elegans*; Origin: Brazil; Common Name: 'Calico Flower; Vernacular Name: Bengali – *Pakhilata*.

A slender climber which bears solitary flowers 7.5 cm in diameter. The tube is yellow-green, blotched with white and purple. The outside of the flower is whitish with purple veins. Flowers during the summer. The unpleasant odour of the flower is absent in this species.

(b) *A. grandiflora* (Syn. *A. gigas*); Origin: Guatemala; Common Names: Pelican-flower, Swan-flower, Goose Flower; Vernacular Name: *Hansalata*.

Comparatively a heavy climber. It bears very large swan-shaped flowers which measure 20 to 25 cm across and 50 cm long. The flowers are brown and mottled and are inflated, with a very offensive odour. This creeper should be grown away from the house. Var. *sturtevantii* has the largest flowers and is commonly grown.

(c) *A. ornithocephala* (Syn. *A. brasiliensis* var. *macrophylla*); Origin: Brazil.

An extensive and quick-growing climber for walls and trellises. The foul-smelling flowers are pale purple, netted or spotted with dark purple. Flowering commences during October to November.

Artabotrys odoratissimus; Family: *Annonaceae*; Origin: South China, Burma, the Philippines, India; Vernacular Name: Bengali – *Kanthali champa*.

A large woody climber or half-scandent shrub. The flowers are axillary, solitary, or in clusters of two or three, greenish-yellow in colour when ripe and give a strong scent resembling that of ripened jackfruits. Flowers almost all the year round but more during the summer and the rains. Unsuitable for small gardens because of its huge size. Needs pruning to keep it in shape. Propagation is from seeds or layers.

Asparagus: See Chapter XIII.

Banisteria; Family: *Malpighiaceae.*

A few species of *Banisteria* are climbing shrubs of ornamental look. The leaves are leathery and dark green. The plants thrive in well-drained soil. Propagated from ripe-wood cuttings. The important species are:

(a) *B. chrysophylla* (Syn. *Heteropteris chrysophylla*); Origin: Tropical America.

The flowers are borne in axillary panicles and are orange in colour, turning to reddish.

(b) *B. laevifolia*; Origin: Brazil, Paraguay.

Bears close clusters of golden yellow flowers during the winter when the climber looks charming.

(c) *B. laurifolia*; Origin: Tropical America.

Bears golden-yellow sprays of flowers which look like those of the orchid *Oncidium.*

Bauhinia diphylla (Syn. *Lysiphyllum diphyllum*); Family: *Leguminosae*; Origin: Tropics.

A large evergreen climber with numerous branches. The stem is square and green but coppery when soft. The petals are yellowish-white. The flowers are large and 4-6 flowers are there in each inflorescence. The vine likes moist and lightly shaded places for growth. The flowering season is April to October. *B. vahlii* is another extensive and heavy climber of Himalayan origin. The flowers are white, fading to yellow.

Beaumontia grandiflora (Syn. *Echites grandiflora*); Family: *Apocynaceae*; Origin: India; Common Name: Nepal Trumpet Climber.

A strong, quick-growing ornamental climber which has a woody stem and large shiny foliage. The large, white, trumpet-shaped, lily-like flowers having a faint scent are borne in mass during January to March. The flowers are borne in current season's growth; hence, it is necessary to cut back the branches after flowering. The climber is heavy and is suited for growing over large arches or on large trees. Propagated from cuttings or by layering.

B. jerdoniana; Family: *Apocynaceae*; Origin: India.

A much smaller climber than above, but quite heavy. The flowers are white and scented.

Bignonia; Family: *Bignoniaceae.*

As there are some beautiful flowering trees and shrubs in this genus, so are there some beautiful climbers. Most of these climbers were previously grouped in the genus *Bignonia* but now they have been placed under various genera.

(a) The most common and beautiful among these is *Pyrostegia venusta* (Syn. *Bignonia venusta*). It is a native of Brazil and is commonly known as Golden Shower because of its beautiful golden flowers. The climber is not too heavy and is grown on trellises, and over compound walls or the walls of a house. The climber grows fairly fast in Bangalore but its growth may not be as fast in north India. The plant thrives and flowers best in the full sun and a well-drained soil. The rich orange-yellow tubular flowers are borne during January-February, but in Bangalore and other places having a similar climate several other flushes are also noticed, especially during August-September. The climber is propagated from cuttings or by layering. The other species are:

(b) *Arrabidaea magnifica* (Syn. *Bignonia magnifica, Saritaea magnifica*); Origin: Colombia.

The flowers are delicate mauve to rich purplish-crimson, and appear during the cold season.

(c) *Clytostoma binatum* (Syn. *C. purpureum*, *Bignonia purpurea*); Origin: Uruguay.

A comparatively light creeper which bears purple scented flowers a number of times in the year.

(d) *C. callistegioides* (Syn. *Bignonia speciosa*, *B. callistegioides*, *B. lindleyi*, *B. picta*); Origin: South Brazil, Argentina.

Bears yellow flowers with lilac tinge during the spring to early summer.

(e) *Bignonia unguis-cati* (Syn. *B. tweediana*, *B. gracilis*, *Doxantha unguis-cati*); Origin: Argentina.

A vigorous climber which produces solitary yellow flowers during the summer.

All the above climbers are excellent for being trained over bare stone walls. *B. gracilis* is most commonly trained over pergolas in Bangalore for shading shade-loving foliage plants. The plants are propagated from cuttings or by layering.

Bougainvillea: See Chapter IX.

Camoensia maxima; Family: *Leguminosae*; Origin: Africa.

A heavy, shrubby climber, which bears large, white tinged with gold flowers having frilled petals. The flowering season is the summer and the flowers are scented. The plants are raised from seeds or cuttings.

Campsis grandiflora (Syn. *C. chinensis*, *Tecoma grandiflora*); Family: *Bignoniaceae*; Origin: China, Japan; Common Name: Chinese Trumpet-Creeper.

A climbing shrub with or without aerial rootlets with the help of which the creeper climbs. The deep orange and red flowers are funnel shaped, opening into a trumpet-shaped mouth. The creeper flowers during the rains in 6 to 12 flowered panicles. Propagated easily from root suckers and cuttings. The creeper is pruned during the winter when it is deciduous.

Campsis radicans (Syn. *Tecoma radicans*, *Bignonia radicans*); Family: *Bignoniaceae*; Origin: Southeast U.S.A.; Common Names: Trumpet-Creeper, Trumpet-Vine, Trumpet-Honeysuckle.

The creeper climbs by aerial rootlets and is ideally suited for covering bare walls. The orange trumpet-shaped flowers are borne in 4-12 flowered panicles during the rains. Propagated from cuttings and root suckers.

Chonemorpha macrophylla (Syn. *C. fragrans*); Family: *Apocynaceae*; Origin: India, Malaysia.

It is a large climbing shrub suitable for being trained on strong structures. The leaves are very large, 25 cm long and 10-15 cm wide. The flowers are large (6.5 cm across), creamy-white, scented and are borne profusely in clusters from May to July. It is propagated by layering.

Cissus discolor (Syn. *Vitis discolor*); Family: *Vitaceae*; Origin: Java, Cambodia.

A slender, tendril climber suitable for roof and trellises, and for pot culture. The leaves are velvety-green above, spotted and mottled white; deep reddish-purple beneath. The flowers are greenish-yellow, insignificant. The fruits are globular. The creeper is susceptible to root-rot in poorly-drained soils. The creeper becomes bare of leaves in the winter when it is cut back to 30-45 cm from the ground level, and when watering is restricted to the barest minimum till the new growth starts. Propagated from cuttings in pure sand. The other important species is *C. quadrangularis* (Syn. *Vitis quadrangularis*) which is a native of Tropical Africa. It is quick-growing with very thick fleshy stems which are 4-winged. The kidney-shaped leaves are few in number. The pea-sized fruits are globular and red in colour.

Clematis; Family: *Ranunculaceae*.

This genus consists of a large number of species, of which many are good climbers. Most of the creepers do well in the hills, but *C. flammula* and *C. paniculata* do well in the plains. They like a deep, loamy soil, and partial shade under tropical conditions. The creepers produce flowers only on new growth and hence these are pruned to five or six buds on current season's growth. Propagated by layering or inarching on hardier species.

(a) *C. flammula* (Syn. *C. pallasii*); Origin: South Europe.

This creeper is ideally suited for small gardens, low walls, and wire trellises. The pure white fragrant flowers appear in loose panicles during the rains.

(b) *C.* × *jackmannii*.

It bears violet-purple flowers during July to October.

(c) *C. paniculata*; Origin: Japan.

A hardy creeper suitable for growing on strong trellises. The star-shaped, pure white, scented flowers appear during the early rains.

Clerodendron (also spelt *Clerodendrum*) *splendens*; Family: *Verbenaceae*; Origin: Sierra Leone.

A comparatively dwarf-growing beautiful climber, but in Calcutta and other humid areas it attains considerable height and spread. The broadly ovate, dark green leaves are very ornamental. The creeper bears large clusters of dazzling crimson-coloured flowers of great beauty during the winter. In Bangalore and places having a similar climate the creeper flowers several times. The creeper needs heavy pruning during the rainy season. A most beautiful climber for the Indian plains. Propagated by separation of root suckers or by layering.

Clerodendron thomsonae: See page 140.

Clitoria ternatea: See page 111.

Combretum comosum (Syn. *Poivrea intermedia*); Family: *Combretaceae*; Origin: Sierra Leone.

A woody shrubby climber which needs strong support. The creeper has dense foliage. The orange-red flowers appear in compact terminal spikes during February to June. The plant is pruned back after flowering. Propagated from cuttings or by layering.

Combretum densiflorum (Syn. *Poivrea densiflorum*); Family: *Combretaceae*; Origin: Domesticated in India.

A shrubby woody climber which is difficult to train over pergolas because of its shrubby nature. The flowers are scarlet and are borne during November-April. It is pruned back after flowering.

Congea tomentosa; Family: *Verbenaceae*; Origin: Burma, Assam, Manipur.

A strong, woody, fairly quick-growing climber. The pinkish, large, elliptic and persistent bracts are conspicuous whereas the flowers are insignificant. The flowering season is December to March. Propagated from cuttings and by layering.

Cryptostegia grandiflora (Syn. *Nerium grandiflorum*); Family: *Asclepiadaceae*; Origin: Africa; cultivated in India; Vernacular Name: *Chabuk chari* (Bengali).

A woody climber having long twig-like stems. The large, bell-funnel-shaped, reddish-purple or pale pink flowers appear in the summer and the rains. The flowers are succeeded by triangular fruits. The latex yields an inferior quality of rubber and is poisonous for cattle. Propagated from cuttings or layers during the rainy season.

Delima sarmentosa (Syn. *Tetracera sarmentosa, Trachyptella aspera*); Family: *Dilleniaceae*; Origin: Tropical Asia.

A large, evergreen, woody climber with attractive foliage. The plant likes light shade. The creeper bears terminal, magnolia-like, strong-scented white flowers in panicles during the summer months. Cuttings of young shoots root in frames.

Derris (Syn. *Deguelia*) *scandens*; Family: *Leguminosae*; Origin: Asia to Australia.

A heavy climber requiring strong support. The climber bears numerous pale rose flowers during March to August. It bears numerous pods. Propagated from seeds or by layering.

Echites caryophyllata (Syn. *Aganosma caryophyllata*); Family: *Apocynaceae*; Origin: India; Vernacular Names: Sanskrit, Tamil, Kannada - *Malati*; Bengali - *Malatilata*.

A heavy, twining climber. The leaves are conspicuously marked with reddish veins. The white star-like flowers have irregular twisted petals. The flowers are generally referred to as clove-scented, but the fragrance though agreeable is not exactly clove-like. The numerous flowers are borne during June to September. The plant is quick-growing and suitable for training on tall trees. The climber prefers dry conditions and hence frequent watering should be avoided. Propagated from seeds or by layering.

Ficus repens (Syn. *F. pumila, F. scandens, F. stipulata*); Family: *Moraceae*; Origin: China, Japan; Common Names: Creeping Fig, Indian Ivy.

A beautiful but slow-growing climber, clinging close to conservatory walls with the help of its small rootlets. In course of time the climber completely covers the surface it climbs. *Ficus stipulata* is considered by many as another species with larger leaves, but actually it is a synonym of the above. The var. *minima* (Syn. *F. stipulata minima, F. minima*) has slender growth and smaller leaves. The leaves are small and beautiful. In var. *variegata* the leaves are margined creamy-white. Propagated from cuttings of stems which generally root naturally at each node.

F. radicans (Syn. *F. rostrata*); Family: *Moraceae*; Origin: Unknown, but probably East Indies.

An evergreen trailer which is suitable for growing in hanging baskets. It is propagated from cuttings. In var. *variegata* the foliage are irregularly marked creamy-white.

Gloriosa superba: See pages 238-240.

Gmelina hystrix: See page 146.

Hiptage benghalensis (Syn. *H. madablota*); Family: *Malpighiaceae*; Origin: Tropical Asia; Vernacular Name: *Madhabilata* (Bengali).

A heavy, evergreen climbing shrub cultivated for its sweet-scented white and yellow flowers. Grows without much care. Propagated from seeds which are produced in plenty or by layering. The flowers appear during late winter.

Hoya carnosa (Syn. *H. motoskei*); Family: *Asclepiadaceae*; Origin: South China and Australia; Common Names: Wax Plant, Wax Flower.

A light twiner with shining dark-green oval foliage. Suitable for growing in pots over bamboo trellises or for covering tree trunks and pillars. It grows well in a well-drained shady place. Propagated from stems or leaf cuttings.

Ipomoea; Family: *Convolvulaceae*.

The genus *Ipomoea* consists of about 500 species belonging to the tropics and sub-tropics, of which only a few important ones are discussed below.

(a) *I. acuminata* (Syn. *I. congesta, I. vahliana*); Origin: South America.

A handsome climber of large dimensions bearing large funnel-shaped bluish or bluish-purple flowers with a reddish tinge.

(b) *I. bona-nox* (Syn. *Calonyction aculeatum*); Origin: Tropics of both hemispheres; Common Names: Moon Flower, Good Night Flower.

In sub-tropical climate it flowers within six weeks from planting. The white or greenish-white, mildly fragrant flowers open all of a sudden in the evening and fade by the morning. For rapid germination, the seeds are soaked in water overnight and sown. The creeper is well suited for covering unsightly places and for screening. It flowers during the summer and the rains.

(c) *I. cairica* (Syn. *I. palmata, I. digitata, Batatas senegalensis*); Origin: Tropics; Common Name: Railway Creeper.

A common evergreen climber which is commonly grown on the railway platforms of our country. The leaves are deeply cut into several lobes. The flowers are funnel-shaped, reddish-purple, or purple with a deeper centre and opens in the morning. Flowering is all the year round. A quick grower and is used for screening in the shortest possible time. It is propagated from cuttings or by layering.

(d) *I. carnea*; Origin: Tropics.

A scandent shrub or woody climber suitable for covering waste and marshy lands. Flowers are pink and funnel-shaped and appear between July and November. It is propagated from cuttings.

(e) *I. horsfalliae*; Origin: West Indies.

A handsome twiner with beautiful digitate leaves and bell-shaped rose to crimson flowers which appear in large numbers throughout the year. A good subject for trellises and arches. Propagated from cuttings and by layering.

(f) *I. learii* (Syn. *Pharbitis learii*); *I. indica* (Burman) Merril may be the correct name, but it is not very certain; Origin: Tropics.

A rapid grower attaining a height of 10-12 m. The flowers are beautiful and are lilac or sometimes dark purple in colour and bell-shaped, appearing during July to October. Old plants are to be replaced every two years.

(g) *I. paniculata* (Syn. *I. digitata*, *Batatas paniculata*); Origin: Tropics.

A tuberous-rooted perennial climber, suitable for trellises and for covering pillars. The large purple flowers appear from July onwards. Propagated from cuttings.

(h) *I. purpurea* (Syn. *Pharbitis purpurea*): See page 112.

(i) *I. quamoclit* (Syn. *Quamoclit pennata*): See page 112.

(j) *I. rubro-caerulea* (Syn. *Pharbitis tricolor*): See page 112.

(k) *I. tuberosa*; Origin: India; Common Name: Wood-Rose.

A heavy creeper with digitate leaves and large funnel-shaped yellow flowers. The dry sepals take the form of wooden rose. The hard seeds should be soaked 24 hours before sowing.

Jacquemontia pentantha (Syn. *J. violacea*, *Convolvulus pentanthus*); Family: *Convolvulaceae*; Origin: Tropical America.

A light, neat climber. Grows well in a well-drained semi-shady situation. It is suitable for growing on trellises, arbours, or in pots and tubs with a balloon-shaped overhead framework. The bell-shaped flowers are violet-blue and appear throughout the year in large numbers and thus compensate for the small size of the blooms. It it propagated from seeds, cuttings and by layering.

Jasminum: See Chapter XX.

Lonicera hildebrandiana; Family: *Caprifoliaceae*; Origin: Burma, Thailand, China; Common Name: Giant Honeysuckle.

A heavy climbing shrub which bears flowers in pairs during June-August. The fragrant flowers are creamy-white, deepening to orange. It is propagated from cuttings and by layering.

Lonicera japonica; Family: *Caprifoliaceae*; Origin: Japan to China, Korea; Common Name: Japanese Honeysuckle.

A quick-growing climber, very popular for its sweet-scented flowers which appear in large numbers during the winter. It is a suitable subject for trellis work; can also be trained on small trees. The plants like light shade. The tubular fragrant flowers are white, tinged with purple but turn to yellow with age. It is easily propagated from cuttings and by layering.

Lonicera sempervirens (Syn. *Caprifolium sempervirens*); Family: *Caprifoliaceae*; Origin: U.S.A.; Common Name: Trumpet Honeysuckle.

The brilliant tubular flowers are scarlet outside and yellow within, but without fragrance. Suitable for growing at medium to high altitudes.

Monstera deliciosa: See page 265.

Odontadenia grandiflora (Syn *O. speciosa, Dipladenia harrisii*); Family: *Apcoynaceae*; Origin: Trinidad, Brazil.

A beautiful scandent shrub. The delicately fragrant, velvety, bell-shaped flowers are bright yellow shaded orange in colour, and are borne throughout the year. The young stem cuttings root easily.

Pandorea jasminoides: See *Tecoma jasminoides*.

Passiflora; Family: *Passifloraceae*; Common Name: Passion Flower; Vernacular Names: *Jhumkolata* (Bengali); *Siruppunaikkali* (Tamil); *Tellajumiki* (Telugu).

The genus is famous for its group of creepers which bear beautiful flowers and delicately flavoured fruits. These are tendril climbers which can be trained over trellises or walls. The creepers are easy to grow in a soil rich in lime, with an abundant supply of water. The plants need pruning after flowering which not only keeps them in shape but often help in producing more flowers. In general, *Passifloras* grow better, comparatively, at medium to high elevations. The flowering season is between July and November. The more important species are discussed below. Propagated from suckers or by layering. In most of the species the leaves are digitately lobed.

(a) *P. alba* (Syn. *P. subpeltata, P. atomaria*); Origin: Mexico to Venezuela.

The flowers are large, white and mildly fragrant.

(b) *P. caerulea*; Origin: Central and Western South America; Common Name: Blue Passion Flower.

A vigorous climber. It is one of the hardiest and commonly cultivated *Passifloras*. The leaves are bright green, 5-lobed, and sometimes 9-lobed. The flowering bracts are large, green and the petals are pale pink. The corona is 5.0 cm across, the outer filaments blue at the top, white in the middle and purple at the base. The flowers are slightly scented and flowering is throughout the year. Propagated easily from seeds, but can be raised from suckers which come out around the base.

(c) *P. edulis*; Origin: Tropical South America.

A strong, woody climber. Leaves deeply 3-lobed. Bracts greenish, petals white, tinged with pink. Corona filaments white banded purple. Bears dull purple edible fruits.

(d) *P. kermesina* (Syn. *P. raddiana*); Origin: Brazil.

A slender, glabrous climber with 3-lobed, yellowish-green leaves. The flowers are scarlet, the outer filaments violet-purple, and the inner ones united into a tube. It is not suited for hot and humid zones. The flowers appear in the summer and the rains.

(e) *P. laurifolia* (Syn. *P. tinifolia*); Origin: Tropical America; Common Names: Jamaica Honeysuckle, Water-lemon.

A strong climber with fragrant, violet-purple flowers.

(f) *P. quadrangularis*; Origin: Tropical America; Common Name: Granadilla.

A strong glabrous climber with 4-angled, winged stem and large leaves. The flower is large, fragrant, pinkish, variegated with blue. The fruits are edible. Propagated from seeds and by layering. Best suited for lower elevations.

(g) *P. racemosa* (Syn. *P. princeps, P. amabilis*); Origin: Brazil.

A tall glabrous climber, bearing deep red flowers. The corona filaments are white on the upper side and blue on the lower. The leaves are green and 3-lobed. The plant is best suited for walls and arches. The blooming period is the summer and the rains. It is propagated from seeds and by layering.

Pereskia bleo (Syn. *P. grandiflora*); Family: *Cactaceae*; Origin: Brazil.

(Most of the authors consider *P. bleo* and *P. grandiflora* to be separate species.)

Usually a shrub, but sometimes half-clambering. The cylindrical stem is covered with large spines. The pretty pink flowers appear continuously and resemble single pink rose. It is propagated from cuttings.

Pergularia odoratissima; Family: *Asclepiadaceae*; Origin: China; Common Name: West Coast Creeper; Vernacular Name: *Labanga lata* (Bengali).

An evergreen light climber, producing highly scented greenish-yellow flowers almost throughout the year. It is used by Hindus for worshipping. Propagation is from suckers or layers.

Petrea volubilis; Family: *Verbenaceae*; Origin: Mexico; Common Name: Purple Wreath.

A strong, woody, shrubby climber which bears long wreath-like spikes of star-shaped light mauve flowers during the spring or early summer. The calyx persists for a longer time than the corolla. The climber can be trained as a standard also. When in full bloom the climber is of charming beauty. Propagation is from dormant wood cuttings, layers, or suckers. In var. *albiflora* the flowers are white.

Philodendron: See pages 268-269.

Piper ornatum; Family: *Piperaceae*; Origin: Celebes.

An attractive climber, attaining a height up to 5 m. The young leaves are bright green with pink spots on the upper surface, whereas in the older leaves the spots are dull and become white. It is propagated from cuttings.

Poivrea: See *Combretum*.

Porana paniculata; Family: *Convolvulaceae*; Origin: India, Malaysia; Common Name: Bridal Bouquet; Vernacular Name: *Safed bel* (Hindi).

A strong shrubby climber having large, dark green, heart-shaped leaves. The very small, white, fragrant flowers appear in great abundance, in much-branched panicles, during the colder months. The individual flowers are funnel-shaped.

Porana racemosa; Origin: India, Burma.

It is an annual which has cordate leaves smaller in size than the above. It is commonly called the Snow Creeper. The individual flowers are star-shaped.

Porana volubilis; Origin: Burma, Malaysia.

It is a shrubby twiner, of which the stems are terete and the flowers white.

Pothos: See pages 270 and 271.

Quisqualis indica (Syn. *Q. pubescens*, *Q. glabra*); Family: *Combretaceae*; Origin: Malaysia, Burma, the Philippines; Common Name: Rangoon Creeper; Vernacular Names: *Madhu-malati* (Bengali), *Rangoon-ki-Bel* (Hindi), *Lalchameli* (Marathi), *Irangunmalli* (Tamil).

A widely known garden climber of huge proportions. It is very hardy and quick-growing. This is either trained on large trees or requires strong trellises for support. The curved spines enable it to climb. The flowers are white at the time of opening but gradually turn into pink and reddish and hence white and pinkish flowers can be seen together. The calyx tube is rather long. There are five petals. The flowers are sweet-scented and are produced in large drooping bunches almost throughout the year. There is a variety with the red blooms retained throughout and a white flowering one. The other species, *Q. malabarica*, has smaller sized flowers. Propagated easily from cuttings or layers.

Rhynchospermum jasminoides: See *Trachelospermum.*

Roupellia grata (Syn. *Strophanthus gratus*); Family: *Apocynaceae*; Origin: West Africa.

A heavy, evergreen shrubby climber suitable for strong trellises and pergolas and can also be trained as standard. The flower buds are deep purple, but the fully opened scented flowers are white with a purple tinge and measure 6 cm across; it flowers during the summer months. Propagation is from cuttings. If space is limited, severe pruning is resorted to to keep it in shape.

Solandra grandiflora; Family: *Solanaceae*; Origin: Jamaica; Common Name: Trumpet Flower.

A large shrubby climber which is generally trained over trunks of large trees. It bears large, 20-25 cm long, fragrant, trumpet-shaped flowers which are greenish-white in colour, but turn to brownish-yellow, and appear during the summer months. Propagated from seeds and cuttings. Cuttings from flowering shoots generally produce small plants.

Solanum jasminoides; Family: *Solanaceae*; Origin: South America; Common Name: Potato Vine.

A slender twining shrub suitable for medium-to-high elevations. It needs a porous soil and a shady situation. The small star-shaped, bluish-white flowers appear in 8-12 flowered racemes during the summer. The var. *grandiflorum* is more robust and the flower trusses are larger. It is propagated by layering.

Solanum seaforthianum (Syn. *S. azureum, S. venustum*); Family: *Solanaceae*; Origin: South America.

A beautiful, slender, herbaceous or slightly woody climber which bears lilac flowers throughout the summer. The plant bears crimson berries. Propagation is by layering.

Solanum wendlandii; Family: *Solanaceae*; Origin: Costa Rica.

A coarse-foliaged vine of rapid growth having a few scattered spines on the stems, branches, and petiole. The small flowers are pale lilac and appear in large cymes during April to August. Propagation is by layering.

Stephanotis floribunda; Family: *Asclepiadaceae*; Origin: Madagascar; Common Names: Clustered Wax Flower, Madagascar Jasmine.

A slender and one of the most popular climbers, producing pure white, very fragrant flowers in large bunches. This is suitable for growing over arches or in pots with a balloon structure. The creeper likes a well-drained soil and moist surroundings. The flowers appear during the summer and the rains. Propagation is by layering.

Stigmaphyllon (sometimes spelled *Stigmaphyllum*) *ciliatum*; Family: *Malpighiaceae*; Origin: Brazil; Common Name: Golden Vine.

A slender climber with large bright yellow flowers. The petals are frilled and clawed. The young stems are covered with dense white hairs.

Stigmaphyllon periplocifolium; Family: *Malpighiaceae*; Origin: West Indies.

A scandent shrub with a cylindrical brown stem. The star-shaped yellow flowers are 2.5 cm across. Petals are five: four sub-equal and the fifth smaller in size. Both the above species bloom during the summer months and are propagated by layering or from cuttings.

Strongylodon macrobotrys; Family: *Leguminosae*; Origin: Madagascar, Tropical Asia, Pacific Islands; Common Name: Jadevine.

A large, woody creeper with a twisted stem. Grows well in damp localities. The leaves are trifoliate. The blue-green flowers are 7-10 cm long, half-moon shaped and are crowded in a long flowering stem which may sometimes be as long as 150 cm. The fruits are large, often the size of an infant's head, containing 8-10 large seeds which lose viability in 1-2 weeks' time.

Tecoma grandiflora: See *Campsis grandiflora*.

Tecoma jasminoides (Syn. *Pandorea jasminoides, Bignonia jasminoides*); Family: *Bignoniaceae*; Origin: Australia; Common Name: Bower Plant of Australia.

An evergreen climbing shrub with 5-9 ovate, bright green leaflets. The blossoms are white with pink or red shades on the throat and appear in few-flowered clusters during August to October. Propagation is from seeds or half-matured stem cuttings. In var. *alba* the flowers are larger and white.

Tecomaria capensis: See page 167.

Tetracera sarmentosa: See *Delima sarmentosa.*

Thunbergia; Family: *Acanthaceae.*

In this genus are included a large number of hardy, vigorous, free-flowering climbers with dense foliage. These are useful for growing on arbours, trellises, trees, porches, etc. The leaves of most of the species are coarse. A careful pruning is required to make it flower freely, but vigorous pruning is detrimental. The propagation is from seeds or cuttings and by layering during the rainy season. The important species are described below.

(a) *T. alata:* See page 113.

(b) *T. coccinea* (Syn. *Hexacentris coccinea*); Origin: India, Burma.

A wide-spreading climber with thick heavy leaves, 12 cm long and 8 cm broad. The leaves are variously shaped – lower ovate, upper oblong. The flowers are borne in terminal or axillary racemes which are 12-45 cm long. The flowers are 2.5 cm across, red with orange throat.

(c) *T. gibsonii*; Origin: Africa.

It bears bright orange-coloured flowers. This is actually a species related to *T. gregorii* which itself is a beautiful climber.

(d) *T. grandiflora*; Origin: Bengal (India); Common Names: Heavenly Blue, Clock Vine.

An extensively growing heavy climber with a dense foliage and lovely flowers. The plant grows in any good soil and is useful for covering walls, trellises of tennis courts, and trees. It needs strong support. The flowers are large, 7-9 cm long and broad, corolla tube bell-shaped and of blue colour with yellow shades inside. The flowers are solitary or in short stout racemes. The flowering season is from February till the rains. In var. *alba* the flowers are white. Judicious training and pruning will keep the plant in good shape.

(e) *T. harrisii* (Syn. *T. laurifolia*): Origin: Malaysia.

Similar to *T. grandiflora*, but the leaves are smaller and the flowers are much more deeply blue.

(f) *T. mysorensis* (Syn. *Hexacentris mysorensis*); Origin: Nilgiri Hills (India).

Not as heavy as *T. grandiflora*. The yellow-coloured flowers with purplish tube are borne in chain-like pendulous racemes. Flowering is from January to March.

Most of the *Thunbergias* mentioned above can be grown as winter annuals in the plains of India.

Trachelospermum jasminoides (Syn. *Rhynchospermum jasminoides*); Family: *Apocynaceae*; Origin: China; Common Name: Star Jasmine.

A slow-growing slender shrub which makes considerable growth and bears pure white scented flowers during the summer. The leathery, glossy, dark green leaves are ornamental. The plant thrives well in partial shade. It exudes a white latex when injured. The climber is suitable for growing on trellises and for screening. Propagation is from cuttings and by layering during the rainy season.

Tristellateia australasiae (Syn. *T. australis*); Family: *Malpighiaceae*; Origin: The Philippines, Taiwan, Malaysia, Australia.

A small, handsome climbing shrub which blooms throughout the year. The star-shaped yellow flowers are borne in terminal racemes. This can be grown in pots provided with a balloon-like structure. Propagation is from seeds or by layers.

Vallaris heynei (Syn. *V. dichotoma*); Family: *Apocynaceae*; Origin: Himalayas, Burma, Sri Lanka.

An evergreen, strong-growing climber with pretty foliage. The climber bears white fragrant flowers from February to April. Propagation is by layering.

Vernonia elaegnacfolia (also spelt *elaegnifolia*); Family: *Compositae*; Origin: Domesticated in India.

A quick-growing, evergreen climber, which is the most popular and useful climber for screening. The branches which are pendulous hang close to their support, thus forming a neat screen. The flowers are white in colour but insignificant. It thrives best in places with high humidity. During the rainy season new plants are raised from cuttings.

Wisteria sinensis (Syn. *W. chinensis*, *Glycine sinensis*); Family: *Leguminosae*; Origin: China; Common Name: Chinese Wisteria.

It is considered one of the best climbers for cool dry places. In the plains of West Bengal and places having a similar climate flowering is unsatisfactory. The fragrant mauve or deep lilac flowers appear in their thousands during the summer months (April-May), when they are objects of great beauty. It is a heavy, woody climber, up to 30 m long in extreme cases. The creeper may be trained on walls, porches, or trees. To control growth, severe pruning may be necessary. This may be raised as a small tree and by careful pruning can be trained as a standard or a shrub. Propagation is by layering.

Chapter XI

Shade, Ornamental and Flowering Trees

Trees are an essential feature of a landscape garden, for roadside planting, public parks, along railway lines, in schools and colleges, and some of these can even find a place in small private gardens. A tree serves many purposes in a garden. It provides shade or bears flowers or does both and also plays a significant role in controlling air pollution in cities and towns. It is difficult to draw a line of demarcation and to specify which is an ornamental tree and which is not. The different species of *Acacias*, especially *Acacia arabica* which grows wild in Delhi, Haryana, Rajasthan, and other places and is considered by many as a spiny nuisance is also used for beautification under special circumstances. Fruit trees such as mango, *jamun*, and coconut can also be considered for roadside planting. But, in this chapter only the most common and widely-grown trees in the country, both in the plains and on the hills, are included.

Methods of Planting

Trees are hardier than other ornamental plants, but, even then, proper care is needed in transplanting them if they are to grow vigorously. A pit measuring at least one metre in diameter and of equal depth is prepared for planting a tree. The dug-up soil is mixed with at least 4 to 10 kg of well-rotten F.Y.M. or leaf-mould, depending on the fertility of the soil, and is then refilled into the pit. In places having red soil, such as Bangalore, horse manure may be used instead of F.Y.M. The refilled soil is allowed to settle by drenching thoroughly with water. After 3-4 days, work up the top soil in the pit and plant the tree, keeping the ball of earth, around the roots of the plant, to about 5-7 cm deep from the surface. The ball of earth should not be broken but the roots should be loosened and straightened. The injured roots and branches are cut off. The sapling is planted in the pit at the same depth as it was in the pot or the nursery. Deep burying is harmful. If the soil is alkaline or sandy, the pit is filled up with good soil brought from outside. Only one sapling per pit should be put and casualties, if any, should be replaced. After pressing the soil firmly around the newly-planted tree with both hands, water copiously and stake the plant with a stout bamboo stick. For protection against white ants and chafer grubs it is useful to dust 5 per cent BHC and DDT in equal proportions or to add 5 per cent aldrin dust at the rate of 30 g per square metre to the soil in the pit. Thimet applied at the same rate is also very effective.

Time of Planting

The time of planting will vary from place to place. In the plains of eastern India, if irrigation can be assured, the best time of planting will be in February or early March. But, planting should preferably be deferred to the end of the hot summer and near the onset of monsoon. Sometimes, the planting is also done after heavy rains. In the plains of northern India, planting should preferably be done during the rains in June to August, though March is also a suitable time, if watering is assured during the coming summer. In the hills, planting should be started in spring and continued through the rainy season. In places having a milder climate as Bangalore, planting can be done almost throughout the year except the hotter and the drier months of March, April, and May. In other parts of south India, planting should preferably be undertaken after the rains have started. It is always advisable to plant 2-3 year old plants rather than to start from very young seedlings. The older saplings grow faster than the seedlings. But, pot-bound trees should never be planted as in these plants the tap root gets twisted and the tree never grows properly.

Dwarf trees are planted five metres apart in the rows and the taller and the spreading ones are planted 9-10 metres apart. But a tree with a huge canopy such as the banyan or the rain tree can be planted farther apart.

Management and Care of Transplants

All newly planted trees should be protected from the hot summer winds between March and May. In north India where the winter is severe and the occurrence of frost is common, many of the trees will need protection against cold. During the summer, drenching of the pits at regular intervals is essential, especially during the first five years of growth. It is beneficial to irrigate older trees also during the summer and drought. In heavy rainfall areas, excessive water near the roots of the newly planted trees may cause damage and, therefore, proper drainage should be provided. Delicate young trees may be given protection against hot summer by thatching paddy or wheat straw over bamboo frames. Similarly, in frost-prone areas, susceptible young trees should be protected by covering them during the night from all sides with the help of paddy or wheat straw. Sometimes, the young trees are given wind-screen or shade by planting 4 to 6 seedlings of *Dhanicha* or *Sunn-hemp* or some other leguminous plants such as *Arahar*, 60 cm away from the tree and around it. These protective plants are allowed to grow till the trees become large enough to withstand the vagaries of weather on their own, when the guard plants are removed. Besides protection from adverse climatic conditions, the young trees will need tree guards for protection against grazing by cattle and goats. Tree guards can be prepared with brick and mortar or portable and detachable pre-fabricated reinforced concrete structures or simply by putting coal tar drums after making sufficient perforations on the sides to allow sunlight to reach the young plant.

For obtaining straight trunk, all trees must be given stout staking from seedling stage and trained in an upright manner retaining only one stem. At what height a plant should be allowed to branch will depend on the situation, the type of tree, and the needs of the grower, but, generally, no tree should be allowed to branch below 2.5 metres from ground level. Pruning of branches should be done carefully so as not to cause unnecessary wounds which may invite fungal infection and damage of the main trunk. Also, no stump should be left while pruning a branch. A clean cut should be given with a pruning saw and a fungicide or coal tar applied at the cut end. Any existing hole in a branch or a trunk should be repaired, first by removing all dead wood in the hole and then treating it with some disinfectant such as weak carbolic acid or a solution of mercuric chloride or a suitable fungicide such as Bordeaux mixture or Blitox and then filling the hole with broken bricks and finally smoothing off with cement mortar.

Manuring

All the young trees should receive one dose of F.Y.M. before the rains. The quantity to be applied will depend on the age of the plant. The younger plants are given 8-12 kg and the older plants double the quantity or even more. It is a general practice that once the trees are established they get little or no attention. But to get good performance from them all the flowering trees should be given F.Y.M. once a year. In times of stress and strain (April to June in northern India) the trees should be irrigated. A heavy drenching should be given as mere sprinkling with water is of no use. It is advisable to remove 20-25 year old trees of *gulmohar* and jacaranda and to plant new ones in their places.

Classification of Trees

From the floricultural point of view, trees can be classified into two groups, namely, Shade (or Foliage) and Ornamental Trees and Flowering Trees. The line of demarcation between the two groups may sometimes get blurred. In the first group, only those trees are included which are planted for their showy ornamental foliage or for shade and which are either non-flowering (e.g., cone-bearing *Gymnosperms*) or bear inconspicuous flowers. In the second group are included trees bearing showy and ornamental flowers.

(A) Shade and Ornamental Trees

Abies webbiana (Syn. *A. spectabilis*); Family: *Pinaceae*; Origin: Eastern India.

A handsome tree with beautiful needle-like foliage (up to 7 cm long), growing to a height of 30-45 m and a girth of 6-9 m. It is suitable for growing at high altitudes only and is multiplied from seeds obtained from female cones. It is generally planted singly in parks and botanic gardens.

Adansonia digitata; Family: *Bombacaceae*; Origin: Tropical Africa; Common Names: Baobab, Monkey-bread Tree, African Calabash Tree, Cream of Tartar Tree; Vernacular Names: Hindi - *Gorakh, Gorakh imli, Goram lichora*; Gujarati - *Gorakh chinch*; Tamil - *Paparapulia, Perruka*.

A giant deciduous tree, not very tall (15 m) but with a large swollen trunk having a maximum girth of 10 m at base, tapering upwards. The creamy-white large pendent flowers open at midnight and fade by noon next day. The tree is planted singly in parks and zoological and botanic gardens. It is propagated from seeds.

Adenanthera pavonina; Family: *Leguminosae*; Origin: India, Burma; Common Names: Coral Wood Tree, Bead Tree, Red Sandalwood Tree; Vernacular Names: *Rakta chandan, Ratan gunj.*

A handsome tree, growing to a height of 6-20 m. The green bipinnate foliage is beautiful. The flowers are small, creamish white and insignificant. Thrives best in humid localities. Seeds are bead-like and bright red which germinate with difficulty after prolonged soaking. Can be grown from large cuttings planted during the monsoon.

Ailanthus excelsa (Syn. *Pongelion excelsum*); Family: *Simaroubaceae* (also spelt *Simarubaceae*); Origin: India; Vernacular Names: Hindi - *Maharuk*; Gujarati - *Ardusi*; Malayalam - *Metti gongilyam*; Marathi - *Mahanimb*; Telugu - *Peddamanu.*

A fast-growing, tall (up to 25 m), deciduous tree found commonly in Bihar, Gujarat, and south India. Flowers, in branched panicles, are insignificant. The tree grows fast in poor soil and a dry and hot climate. *A. altissima* (Syn. *A. glandulosa*) is known as Tree of Heaven and is from China. Flowering is same as above, blooming in August. It is planted as a shade tree in north India. Propagation is from seeds.

Albizzia lebbek (spelt differently as *Albizia lebbeck*, *Albizzia lebbeck*); Family: *Leguminosae*; Origin: Tropical Asia, North Australia; Common Names: Siris Tree, Lebbek Tree, Woman's Tongue Tree.

A handsome, deciduous, fast-growing and spreading tree. Grows up to a height of 20-25 m. The tree is somewhat resistant to drought. A good shade tree which is propagated from seeds. *A. moluccana*, *A. procera*, and *A. richardiana* are also grown.

Alstonia scholaris; Family: *Apocynaceae*; Origin: India; Common Names: Devil's Tree, Dita Bark; Vernacular Names: Hindi - *Chatian*; Bengali - *Chatim*; Sanskrit - *Saptaparni*; Marathi - *Satvin*, *Shaitan*; Tamil & Telugu - *Palaigh.*

A tall (16-30 m) elegant tree with whorled branches which provide plenty of shade. The flowers and the long, slender, drumstick-like fruits cover the whole tree during the summer. Flowers are white or greenish-white and strong-smelling during night. Grows well without care and are propagated from seeds.

Anthocephalus cadamba (*A. indicus*, *Nauclea cadamba*); Family: *Rubiaceae*; Origin: India, Nepal, Burma; Common Name: *Kadamba*; Vernacular Names: Hindi, Gujarati & Marathi - *Kadamba*; Bengali - *Kadam*; Kannada - *Kadawala*; Malayalam - *Attutek*; Tamil - *Cadamba*; *Vellai*; Telugu - *Kadambamu.*

The tree is associated with Lord Krishna. A large (10 m tall), handsome tree with spreading crown and drooping branches. The leaves are large, oblong, and shining. The ball-like flowers are sweet-scented. A well-drained soil and a humid climate are best for its growth. Grows rapidly. Planted in large groves or on the roadside. Flowers during the rainy season. Propagated from seeds. It is highly susceptible to frost and even fully grown trees are killed to ground level by frost.

Araucaria cookii (Syn. *A. columnaris*); Family: *Pinaceae Araucariaceae*; Common Name: Araucaria; Origin: New Caledonia (U.S.A.).

A columnar, handsome and very ornamental tree. It is widely used, when young, as a potted plant for indoor or outdoor decoration and is later planted out. It thrives very well in mild climates as of Bangalore and also in eastern India and in the hills. *A. cunninghamii*, a native of Australia, is not as handsome as the above and has sharp pointed leaves. *A. bidwillii* has leaves with stiff points and grows well on the hills. None of the species are suitable for dry hot areas as field plants. *A. cookii* is most commonly grown as a specimen plant. Propagated from seeds. *A. heterophylla* (Syn *A. excelsa*) and its var. *glauca* are also beautiful. Many of the plants grown as *A. excelsa* may be actually *A. cookii*, and rarely the opposite.

Azadirachta indica (*Melia azadirachta*); Family: *Meliaceae*; Origin: India; Common Name: Margosa; Vernacular Names: Hindi & Bengali – *Neem, Nim*; Punjabi – *Neem*; Kannada – *Bevu, Margosa*; Malayalam – *Vepe*; Marathi – *Nimbay, Limba*; Tamil – *Vepa*; Telugu – *Yepa, Vempu.*

A medium-sized or large tree (13-16 m) of elegant appearance, generally evergreen or deciduous in certain climates. Flowers which are small, white and sweet-scented, appear in April. It is hardy and grows well in saline soils and drought conditions. Propagated from seeds which should be sown immediately after ripening. This plant is recommended as an avenue tree for highways.

Brassaia actinophylla (Syn. *Schefflera macrostachya*); Family: *Araliaceae*; Origin: Queensland, New Guinea, Java; Common Name: Umbrella Tree.

A handsome, medium growing (6-7 m high) tree which at young stage serves as a beautiful pot plant also. The tree is evergreen with large radially arranged leaves and has few branches. The spikes (30-60 cm long) are handsome and bright scarlet or coral-red in colour, borne terminally and pointing radially. To improve the appearance and to keep the plant dwarf it is advisable to behead it once in 3/4 years. This tree is planted as a specimen plant in the corner of a fairly large lawn or in public parks. Propagation is from seeds.

Caesalpinia coriaria; Family: *Leguminosae*; Origin: Central America; Common Name: Divi-divi.

A tree with an umbrella-type canopy. The light feathery foliage and the scented yellow flowers borne in July are the plus points in this plant. It is grown more for its ornamental shape than for flowers. However, no grass can survive under this tree. Easily propagated from seeds. The pods are useful for dyeing and tanning.

Calophyllum inophyllum; Family: *Guttiferaceae* (*Guttiferae*); Origin: Old World and American Tropics; Common Names: Alexandrian Laurel, Dilo Oil Tree; Vernacular Names: Hindi – *Sultan champa*; Bengali – *Sultan chanpa*; Marathi – *Undi, Udi, Pumag*; Tamil – *Punnai, Pinnai*; Kannada – *Surabinne.*

The name is derived from Greek – *Kalos* means beautiful, *phyllon* means leaf, and *inophyllum* means 'with strongly nerved leaves'. As the name indicates, the tree is more attractive for its large (10-20 cm long and 7-10 cm broad) shining leaves than for the white, sweet-scented flowers borne in long racemes in May-June. The medium-sized plant has a smooth gray bark and bears round fruits of the size of a lemon. It is a beautiful roadside tree, though slow-growing. The cuttings of half-ripe shoots can be rooted under mist with bottom heat if necessary. But it is normally propagated from seeds. The timber is good for building ships.

Casuarina equisetifolia (*C. indica*, *C. africana*); Family: *Casuarinaceae*; Origin: North Australia; Common Names: Beefwood, She-oak; Vernacular Names: Bengali - *Jhau*; Marathi - *Sura*; Tamil - *Cavukku*; Telugu - *Chowka, Saruku*.

The tree attains a considerable height, 20 m or more, and is not normally planted in a garden. But a plus point of the tree, in the language of Benthall in *Trees of Calcutta*, is "the soft sighing of the air through its innumerable slender twigs is a pleasant sound, reminiscent of the noise of the sea on a distant shore". For unknown reasons birds avoid sitting on the tree and the fallen leaves poison the soil below and affect the grass adversely. It is suitable for growing in dry and hot tracts and is quick-growing. The propagation is from freshly harvested seeds. Trees older than 50 years are affected with fungus and so should be cut down.

Cedrus deodara (*C. indica*, *C. libani deodara*); Family: *Pinaceae*; Origin: Western Himalayas; Common Names: Indian Cedar, Deodar.

A large (up to 60 m tall, with a girth of 11 m), handsome tree suitable for the hills and is planted singly in large public or botanic gardens as a lawn specimen or as a roadside tree. The timber is used for making railway sleepers and the wood yields a fragrant essential oil. Propagation is from seeds.

Cryptomeria japonica; Family: *Pinaceae* (Some authors say *Taxodiaceae*); Origin: Japan, China; Common Name: Japanese Cedar; Vernacular Name: Nepali - *Dhup*.

A tall (35-45 m), handsome tree of conical shape. It thrives well only at 1,500 m or above. Good for avenue planting or for growing singly in large parks and gardens. Propagated from seeds. When planted densely, no vegetation can grow underneath.

Cunninghamia lanceolata (Syn. *C. sinensis*); Family: *Pinaceae* (Some authors say *Taxodiaceae*); Origin: China.

A large tree, attaining a height of 30 m or more. The plant grows well only in temperate climates. Normally propagated from seeds or cuttings of erect shoots. The timber is fragrant. Suitable for roadside planting.

Cupressus funebris (Syn. *C. pendula*) and *C. funebris glauca* (Syn. *C. cashmeriana*); Family: *Pinaceae* (some authors say *Cupressaceae*); Origin: See text below; Common Names: the former is Mourning Cypress or Chinese Weeping Cypress and the latter is Kashmir Cypress.

The first species is of Chinese origin and the latter of Kashmir or Tibet. The Kashmir cypress is up to 20 m tall and is one of the most beautiful of the cypresses; at the same time it is delicate also. *C. funebris* is a little taller than the above.

Cupressus macrocarpa (Syn. *C. hartwegii*); Family: *Pinaceae* (*Cupressaceae*); Origin: Monterey, California, U.S.A.; Common Name: Monterey Cypress; Vernacular Name: Hindi - *Saru*.

A fast-growing tree having a pyramidal shape when young, but spreads with age and may attain a height of 25 m. A hardy species which can be grown as avenue tree in many parts of the country or individually in the lawn. It forms a good hedge also, but needs support at young stage.

Cupressus sempervirens; Family: *Pinaceae* (*Cupressaceae*); Origin: S. Europe, W. Asia, N. India; Common Name: Mediterranean Cypress.

The species *C. sempervirens* var. *indica* (Syn. *C. roylei*) is an erect and stiff tree forming a narrow columnar head and attains a height of 30 to 45 m, whereas *C. sempervirens* var. *horizontalis* (Syn. *C. horizontalis*) has spreading branches and *C. sempervirens* var. *stricta* (Syn. *C. fastigiata*), the Italian Cypress, is similar in habit as *indica*. *C. sempervirens* is difficult to distinguish from *C. macrocarpa* except that the former has relatively small leaves. The plant is extremely popular for formal gardening because of its stiff outline. The fragrant wood is utilized for cabinet making.

Cupressus torulosa (Syn. *C. nepalensis*); Family: *Pinaceae* (*Cupressaceae*); Origin: W. Asia, N. India; Common Name: Bhutan Cypress.

It is a tall tree (up to 45 m), pyramidal in shape at young stage, but forms a wide-spreading crown with age. This plant occurs naturally in limestone formations. The wood is fragrant and resistant to decay. It forms a beautiful formal avenue. The cultivation of different cypresses is almost similar. They prefer a deep sandy-loam soil. Some species may be injured by frost at young stage. The propagation is usually from seeds. Hardwood cuttings can be rooted under mist with bottom heat where necessary. In some species veneer grafting is possible. These plants require several transplantings in the nursery before final planting to form a good earth-ball around the roots as otherwise they do not stand transplanting well.

Dalbergia sissoo; Family: *Leguminosae*; Origin: Bengal; Common Name: Sissoo Tree; Vernacular Names: Hindi – *Shisam*; Bengali – *Sisu*; Kannada – *Biti*; Marathi – *Shisham*; Tamil – *Iti*, *Eriwadi*.

A most common tree in India, attaining a height of 25 m, but more dwarf in dry localities. The plant can stand drought as well as submerged conditions for a fairly long duration. The freshly harvested seeds give a higher percentage of seedlings. Sometimes plants are raised from cuttings also. The plant is occasionally planted along roadside, but remains deciduous once a year. The timber is good for making boats, railway sleepers, gun-carriages, and furniture.

Dillenia indica (Syn. *D. speciosa*); Family: *Dilleniaceae*; Origin: Eastern India; Vernacular Names: Hindi & Bengali – *Chalta*; Kannada – *Bettadakanagal*; Tamil – *Uva*; Gujarati & Marathi – *Karmbal*, *Karmal*; Oriya – *Chalota*; Malayalam – *Chalita*; Telugu – *Kalinga*.

An upright, evergreen, slow-growing, 10-20 m tall tree with rounded crown. The leaves are pretty long as well as broad, with serrated margins. The fragrant flowers are large and white, blooming in July. The fruits are large, acidic, but edible when cooked. Make good *chatni* and pickle. It prefers a humid climate and is susceptible to frost. Stands forest fire well. Plants are raised from seeds and stem-cuttings. It can be grown in small gardens in dry areas and in large public parks as individual trees for shade and decoration.

Diospyros embryopteris; Family: *Ebenaceae*; Origin: Himalayas, South India; Common Names: Persimmon, Ebony; Vernacular Names: Hindi – *Tendook*; Bengali – *Deshi Gab*; Tamil – *Panichchai*.

An evergreen, slow-growing tree of medium height, but spreading to a considerable extent. The branches are shining and blackish-brown. The new leaves are very ornamental because of the brilliant red colour. The fruits are decorative and edible. A good shade tree which is propagated from seeds.

Eucalyptus; Family: *Myrtaceae*; Origin: Australia.

There are many *Eucalyptus* species grown for their ornamental value, as windbreaks, roadside trees, and for fuel and timber. All the species are easy to cultivate and can withstand drought well. There are species suitable for growing in the hills, e.g., *E. globulus* (60 m or more tall and called Blue Gum), the leaves of which are blue-green; *E. ficifolia* (Red Gum, 10 m tall); *E. multiflora* (Syn. *E. robusta*) which is 30 m tall. *E. citriodora* is a moderate tree with lemon-scented foliage, which thrives well in the plains. Many other species also do well in the plains. All the *Eucalyptus* species are said to draw enormous quantities of water from the soil and hence these should not be grown in areas lacking sufficient groundwater. The propagation is from seeds, a high percentage of which are fertile. Almost all the species are brittle and likely to be broken during strong cyclones. The trees mostly grow straight and some are as tall as 100 m in their natural habitat and, thus, give a majestic look. As a windbreak these are planted as close as 2-3 m apart. When grown in an avenue, a spacing of 5-8 m may be allowed depending on the species. All *Eucalyptus* species are fast growing.

Ficus benghalensis (also written *bengalensis*); Family: *Moraceae*; Origin: India, Tropical Africa; Common Name: Banyan; Vernacular Names: Bengali – *Bat, Bar*; Hindi & Punjabi – *Bar, Borh*; Marathi – *Wad*; Tamil – *Ala*.

A huge tree (20 m tall) with a huge crown of foliage occupying a large area. The aerial roots originating from the branches strike into the soil and gives a columnar support to the immense head of foliage. A large number of such roots reach the ground giving the impression of several trees planted side by side and jointed at the top. The leaves are large and leathery and provide a cool shade. The plant bears crimson berries in August-September. The tree is suited for planting in large parks or along national highways (though not very suitable because of its unwieldy size and shape). The plant is propagated from freshly harvested seeds and by planting large cuttings during the rainy season. A large number of birds are attracted during the fruiting season.

Ficus benjamina; Family: *Moraceae*; Origin: Tropical Asia; Common Names: Java Fig Tree, Golden Fig; Vernacular Names: Tamil – *Alamaram*; Telugu – *Marri*.

A moderate to large sized (15-20 m tall) evergreen tree with drooping branches. A slow grower, but of good form and spreading in habit. It is suitable as a shade tree for wide avenues. The globose fruits when ripe are very ornamental. The plant is propagated from seeds.

Ficus elastica; Family: *Moraceae*; Origin: Tropical Asia; Common Name: Indian Rubber Plant; Vernacular Name: Bengali – *Atha Bor*.

A quick-growing, moderately tall (15 m) tree having large (10-30 cm long and 7-15 cm broad), leathery, smooth, and shining leaves. The young leaves remain enclosed in a pinkish stipule giving the plant an ornamental look. It grows best in heavy rainfall areas. A much smaller variety, var. *variegata* (Syn. var. *aurea*), has variegations of creamy-white and yellow on the margins of leaves and some blotches. It is used more as a foliage plant than as a shade tree. The plant can be propagated by air layering. There are several other ornamental varieties (see Chapter XIII).

Ficus infectoria; Family: *Moraceae*; Origin: Tropical Asia; Vernacular Names: *Pakur*, *Pilkhan*.

A spreading, evergreen, low-crowned tree (10-12 m tall), providing a lot of shade and is especially suited for town and city roads. The seedling plants produce straight trunks and high-spreading crowns, whereas those from cuttings produce a low-crowned, ugly-stemmed specimen. Birds flock on the tree in large numbers during the fruiting season.

Ficus krishnae; Family: *Moraceae*; Origin: India; Common Names: Krishna *Bor*, *Makhan Katori*.

A small tree with grey bark. The leaves are folded on the sides in the reverse direction and jointed at the base giving the look of an ice-cream cone. The story goes that Lord Krishna used to store butter in the cup-shaped leaf. Planted as a specimen tree in large lawns.

Ficus parcellii; Family: *Moraceae*; Origin: Pacific Islands.

A shrubby tree with membranous light green leaves mottled creamy-white. Propagation is from cuttings of half-ripened branches.

Ficus religiosa; Family: *Moraceae*; Origin: India; Common Name: Peepul Tree; Vernacular Names: Hindi – *Pipal*; Bengali – *Asvatha*; Tamil – *Drasi*, *Avasi*; Kannada – *Arali-mara*; Malayalam – *Avasai, Arasu*; Telugu – *Ravi, Ragi, Rai, Asvatthamu*.

The peepul is a large tree (30 m tall) with fluttering heart-shaped leaves terminating in a long tapering point. In April, the old leaves are shed and the dark-red-coloured young leaves appear giving the plant an ornamental look. A religious tree for the Hindus and the Buddhists. Suitable, but not ideal, for planting as avenue trees along highways. The plants are raised from seeds.

Ficus retusa (Syn. *F. nitida*); Family: *Moraceae*; Origin: Tropical Asia; Vernacular Name: *Chilkhan*.

A tree, 10-11 m high, with glossy bark and green leaves. The fruits are yellowish when ripe. An excellent evergreen shade tree for wide avenues. Propagation is from seeds. The tree has a few aerial roots.

Ficus roxburghii; Family: *Moraceae*; Origin: India; Common Name: Eve's Apron.

A small tree, 3-6 m tall, with leaves 15-45 cm long and 10-30 cm broad, giving the plant a very ornamental look. It can be grown as a specimen plant in large parks.

Filicium decipiens (Syn. *Pteridophyllum decipiens*); Family: *Sapindaceae*; Origin: India, Sri Lanka; Common Name: Fern-leaved Tree.

A medium evergreen tree (5 to 10 m tall), spreading to about 10 m. The young plant is very ornamental and is used for decoration. The plant is slow-growing, but the grown-up tree is also ornamental specially for its attractive foliage. It is hardy and grows in any good soil. It is propagated from freshly harvested seeds.

Grevillea robusta; Family: *Proteaceae*; Origin: Eastern Australia; Common Names: Silver Oak, Australian Silk, Silk Oak.

A tree growing up to 30 m, but generally about 10-15 m. It is an evergreen tree with finely divided fern-like twice-incised pinnate leaves, the under surface of which is silvery downy. The young seedlings are decorative and can be used as pot plants for indoor decoration. The plant with its golden yellow flowers looks very attractive during April-May and the flowers attract honey-sucking birds and bees. The plant is easily propagated from seeds. The young seedlings need frequent watering. The plant grows quite fast for the first 6-8 years, but may take 20 years to attain a height of 10 m. It can withstand drought remarkably. The older plants need some pruning to keep them in shape. The plant is very ornamental and can be planted along avenues or as a lawn specimen.

Inga dulcis (A synonym of *Pithecolobium dulce*); Family: *Leguminosae*; Origin: Mexico, the Philippines; Common Names: Manila Tamarind, Madras Thorn; Vernacular Names: Hindi – *Bilaiti imli, Jelabi*; Tamil – *Korukapuli*; Bengali – *Dekani Tentul.*

It is a fine thorny hedge when kept pruned, but attains a considerable height if allowed to grow. A hardy tree which can be grown in drought-prone areas. Flowers are white and pods twisted like *jelabi* which are 12-15 cm long.

Juniperus bermudiana (*J. barbadensis*); Family: *Pinaceae* (some authors say *Cupressaceae*); Origin: Bermuda.

A densely branched pyramidal tree, 12-16 m tall, with bluish-green foliage. The leaves are of two distinct types, needle-like and arranged in threes on young trees and scale-like and overlapping in old trees. In transition stages the two kinds may be found on the same tree. A beautiful tree for planting in rows along paths leading to a building or in a park and in formal gardens. Propagation is from seeds and by air layering.

Juniperus chinensis (Syn. *J. sinensis*); Family: *Pinaceae* (*Cupressaceae*); Origin: China, Mongolia, Japan.

A handsome, erect, dense, pyramidal tree (13-20 m tall) or shrub. The branches are short and have leaves of the same kind as above. It is believed that 10 trees of male and female forms differ in shape sometimes, but this is not a constant feature. A hardy tree which is widely planted for decorative and formal gardening. Propagation is the same as above. When planted in a row the distance may vary from 5 to 7 m for both the species. It can be kept in shape by moderate pruning.

Kigellia pinnata; Family: *Bignoniaceae*; Origin: Mozambique (Africa), Sudan, Uganda; Common Names: Fetish Tree, Sausage Tree; Vernacular Name: *Jhar phanoos.*

A medium to large-sized (7-16 m) evergreen and spreading tree having a globular appearance and with a dense branching system. The cup-shaped, mottled, dark purplish flowers with a foetid smell dangle from different parts of the branches in stalks of 1-1.5 m giving the appearance of candelabra. The flowering is in April-May. The fruits are big, heavy, and sausage-shaped. The foliage is not eaten by cattle and can be grown safely without protection. A good foliage tree for parks and also for avenue planting. The plant grows well in arid or humid regions if the soil is good. Propagation is from seeds.

Madhuca indica (Syn. *M. latifolia, Bassia latifolia*); Family: *Sapotaceae*; Origin: India; Vernacular Names: Bengali, Hindi & Marathi – *Mahwa* or *Mohua*; Gujarati – *Mahuda*; Tamil – *Illipe*; Telugu – *Ippi*.

A large deciduous tree (20 m tall) with a spreading habit. The leaves are clustered at the end of branches. When the copper-coloured young leaves appear during the spring, the plant looks most ornamental. The fleshy, creamish-white, globular, scented flowers appear in March-April in dense fascicles near the end of the branches. The flowers are stalked and drooping. The trees are planted along national highways for shade. The flowers and the fruits are a source of food for the tribals. The oil from the seeds is used in soap-making.

Melia azedarach (Syn. *M. japonica*); Family: *Meliaceae*; Origin: North India, China; Common Names: Persian Lilac, Pride of India, Bead Tree, China Tree, China-berry; Vernacular Names: Hindi – *Dake, Drek, Bakain*; Bengali – *Mahanim, Ghoranim*; Tamil – *Mallay vembu, Puvempu*; Malayalam – *Vayambu*; Telugu – *Taraka vepa*; Marathi – *Pejri*; Oriya – *Mahanimbo*.

Melia, a Greek word, means ash, in allusion to the resemblance in the leaves, and the name of the species, *azedarach*, comes from the Persian name *azad-darakht*, meaning an independent tree. A medium-sized (5-8 m tall), fast-growing, deciduous tree similar in appearance to *neem* (*Azadirachta indica*). The leaves which are beautiful and fern-like are shed in the winter and new leaves along with sweet-scented lavender-coloured flowers appear in March-April. The rounded poisonous yellow fruits borne in clusters also look ornamental. The seeds have a central perforation which make them ideal as beads. The saplings are raised from freshly harvested seeds during the rains. The saplings are damaged by frost, but are hardier than *neem*. The plant is not suitable for avenue planting in areas having strong winds as it is shallow-rooted and may be easily blown off. When shade is required to be provided quickly in a garden or park, this tree is ideal.

Mimusops elengi; Family: *Sapotaceae*; Origin: India, Malaysia; Vernacular Names: Hindi – *Mulsari*; Bengali – *Bakul*; Oriya – *Bokulo*; Malayalam – *Bakulam, Elengi*; Tamil – *Vagulam, Lanji maghadam*; Telugu & Kannada – *Pagada*; Marathi – *Ovalli*.

A medium-sized (15 m tall), evergreen tree with a dense crown of dark green shining foliage. The white, very fragrant, drooping flowers are borne in clusters during April-July and September-November and remain hidden in the leaves. A favourite tree of Indian gardens. A tree suitable for planting in large lawns or even as an avenue tree along narrow roads. The tree is propagated from seeds and is slow-growing. There is a variegated-leaved variety which is much dwarf and is a good specimen plant for the lawns.

Picea smithiana (Syn. *P. morinda, P. khutrow*); Family: *Pinaceae*; Origin: West Himalayas; Common Name: Himalayan Spruce.

A tall (50 m or more), majestic tree with wide-spreading branches and long pendulous branchlets. The dark green leaves are more thick than broad. Young plants may be damaged by frost. Planted as a single tree in large parks or botanic gardens at high altitudes. In the high hills this plant looks like a living herbarium as all sorts of orchids and other epiphytes grow on its branches.

Pinus longifolia (*P. roxburghii*); Family: *Pinaceae*; Origin: Himalayas; Common Name:

Indian Pine or *Chir*; Vernacular Names: Hindi and Punjabi – *Chil, Chir*; Bengali – *Saralgach*; Oriya – *Sorolokatho*; Gujarati & Marathi – *Saraladeodora*; Malayalam – *Saralam*; Telugu – *Devadaru*; Tamil – *Suruldevdari*; Kannada – *Saraladevadaru*.

The most commonly grown pine which grows to a height of 30 m or more with a round-topped symmetrical head. The light green, long (up to 35 cm in length), pendulous, needle-like leaves are remarkable. It is grown as a specimen in the lawns and can be found in all important gardens in India, though it thrives best at 500-2,000 m. Yields resin.

Podocarpus macrophylla (*P. longifolia*); Family: *Taxaceae* (recently *Podocarpaceae*); Origin: Japan, China.

A medium-sized, resinous, ornamental, woody plant grown for its foliage. A variety *maki* (Syn. *P. chinensis*) is also grown. Several variegated forms of this variety are in cultivation. *P. nerifolia* (Syn. *P. nerifolius*) with foliage somewhat like *Nerium* is also cultivated.

Polyalthia longifolia; Family: *Anonaceae*; Origin: Sri Lanka, India; Common Name: The Mast Tree; Vernacular Names: Hindi – *Ashoka, Deodar*; Bengali – *Debdaru*; Assamese & Oriya – *Debodaru*; Marathi – *Asoka, Asupala*; Telugu – *Asokamu, Devadaru*; Tamil – *Maraillupai, Asuvattai*.

A tall (20 m), handsome, slow-growing, evergreen tree with lance-shaped, tapering to a fine point and wavy margined shining leaves. Flowers insignificant. This is highly esteemed as an avenue tree for its graceful appearance. Freshly harvested seeds are sown in pots and transplanted at sites without disturbance as these do not transplant well. The var. *pendula*, with drooping branches and the whole plant assuming a pyramidal or conical shape, is very ornamental and widely planted in the gardens along pathways or in front of buildings 2-5 m apart.

Pongamia glabra (Syn. *P. pinnata*); Family: *Leguminosae*; Origin: India and Burma; Vernacular Names: Hindi – *Karanj*; Bengali – *Karanja*; Oriya – *Koronjo*; Tamil – *Punga, Pargam*; Kannada – *Honge*.

A medium tall (10-15 m), deciduous tree with trifoliate, bright-green, glossy leaves, The leaves make very good compost. The plants flower in a leafless condition during April-May. The tree when laden with lilac-coloured flowers looks beautiful. This is a fast-growing tree and is raised from seeds during the rains. This can also be raised from cuttings. Suitable for roadside avenues and railway platforms for shade or ornamental purposes. Also used as a protective hedge.

Populus; Family: *Salicaceae*; Origin: Widely spread over cool and temperate parts of Northern Hemisphere; Common Names: Poplar, Aspen, Cotton Wood.

There are several species of poplar which are very ornamental and are grown in gardens. These are soft-wooded, very fast-growing trees and are planted on the roadside or in groups. The trees are deciduous. They form a very good screen within a very short time.

Putranjiva roxburghii; Family: *Euphorbiaceae*; Origin: India; Common Names: Indian Amulet-Plant, Wild Olive, Child-Life-Tree (as the generic name implies); Vernacular Names: Hindi – *Jiaputa, Jalpitni*; Bengali – *Putranjib*; Punjabi – *Jiyaputra*; Marathi – *Putajan*; Malayalam – *Pongalam*; Tamil – *Karupalli*; Telugu – *Putrajivika*; Kannada –

Amani.

A moderate-sized evergreen tree (10 m tall) with uniform spreading habit. The dense foliage is beautiful, dark green in colour, and gives the plant a weeping appearance. It is a very good avenue tree and can be planted in public places as a shade tree also.

Ravenala madagascariensis (or *madagascarensis*); Family: *Musaceae*; Origin: Madagascar; Common Name: Travellers' Tree; Vernacular Name: Bengali – *Panthapadap*.

The tree collects clear watery sap in the leaf stalk, which is said to be utilized by thirsty travellers in the arid regions. The tree, which has a tall trunk (up to 30 m) resembling that of palm, is evergreen. The leaves are very large, resembling banana leaf, ordinarily 3 m but very often up to 9 m long, crowded in two ranks and spread out in the shape of a fan at the top of the trunk. The tree can be grown singly in the corner of the lawn or a group of 4-5 trees may be planted in a suitable portion of a public garden.

Salix babylonica (Syn. *S. pendula*); Family: *Salicaceae*; Origin: China; Common Names: Weeping Willow, Napoleon's Willow.

A tree, 10-15 m tall, with spreading habit. The branches are olive-green, very pendulous, and slender, giving the plant a weeping look. This beautiful tree of picturesque habit can be grown on the edge of a tank or an artificial pool. The plant thrives better at high altitudes.

Samanea saman (Syn. *Pithecolobium saman*, also spelt *Pithecellobium*, *Enterolobium saman*, *Mimosa saman*); Family: *Leguminosae*; Origin: W. Indies and Central America; Common Names: Rain Tree, Zaman, Saman; Vernacular Names: Bengali – *Bilatisirish*; Oriya – *Bano sirisha*; Malayalam – *Urakkamthongimaram plavu*; Tamil – *Amaivagai*, *Enal-vaka*.

A large, up to 25 m tall, handsome, ornamental shade tree, very spreading in habit. Although the plant bears pink and white silken tufts of flowers covering the whole canopy during March to May and September to November, this is regarded more as a shade and ornamental tree rather than a flowering tree. The tree does not do well in northern India due to cold. This is very suitable for planting along avenues and can also be planted singly in large public gardens. Propagation is from seeds and cuttings. The plant grows rapidly.

Schleichera trijuga (Syn. *S. oleosa*); Family: *Sapindaceae*; Origin: India, Sri Lanka, Burma, Java, Timor; Common Name: Ceylon Oak; Vernacular Names: Hindi – *Gausam*, *Kosum*; Punjabi – *Jamoa*; Gujarati – *Kossame*; Malayalam – *Puvam*; Tamil – *Konji*, *Puvam*; Telugu – *Paparti*, *Pullakaya*; Kannada – *Kakuta*.

A large tree, but the growth is slow. The young leaves, as they first appear, are carmine-red, gradually turning to green. It is an excellent tree to be grown in large public gardens and parks.

Sesbania grandiflora; Family: *Leguminosae*; Origin: India, Mauritius, N. Australia; Vernacular Names: Hindi – *Agasti*, Bengali and Assamese – *Bakphul*; Marathi – *Agathi*; Tamil – *Agathi*; Telugu – *Avisa*; Kannada – *Agase*; Malayalam – *Akathi*.

A very fast growing, upright, handsome tree, attaining a height of 5-9 m. The white or pink flowers are borne during July-August. The tree, when planted closely, form a good screen to hide an unseemly object in the garden. Propagates easily from seeds. Other synonyms are *Aeschynomene grandiflora, Agati grandiflora.*

Sterculia alata; Family: *Sterculiaceae*; Origin: India.

A large foliage tree, often planted along avenues and streets. The leaves are large.

Sterculia diversifolia (Syn. *Brachychiton diversifolia, B. populnea,* or *B. populneum*); Family: *Sterculiaceae*; Origin: Australia.

A very interesting tree, because no two leaves are identical; often called *Pagla* (mad) tree for this reason.

Swietenia mahagoni; Family: *Meliaceae*; Origin: Tropical regions of North and South America, W. Indies; Common Name: Mahogany; Vernacular Names: Hindi – *Abnus*; Bengali – *Mehagoni.*

A large evergreen tree of beautiful look and suitable for avenue planting. Propagation is from seeds.

Terminalia arjuna; Family: *Combretaceae*; Origin: India, Burma, Sri Lanka; Vernacular Names: Hindi and Bengali – *Arjun*; Tamil – *Vellamarda, Kulamaruthu*; Telugu – *Yermaddi.*

A large (20-25 m tall), evergreen shade tree, commonly planted on the roadside. The pale yellowish-white flowers are borne during March to June and attract honey-bees for the honey the flowers contain. The plant is propagated from seeds. The seed is sown half buried in the soil; if buried deep, the seed may fail to germinate.

Terminalia catappa; Family: *Combretaceae*; Origin: Andaman Islands (India); Common Names: Tropical Almond, Indian Almond, Bengal Almond, Myrobalan, Almendro, Amendoeira; Vernacular Names: Hindi – *Badami, Janglilbadam*; Bengali – *Deshibadam, Baksa badam*; Oriya – *Badamo*; Kannada – *Badamalili, Nadu badami*; Marathi – *Janglibadam, Natbadam*; Tamil – *Amandi, Nattuvadumia*; Telugu – *Ingudi, Natubadamu.*

It is a forest tree but is widely grown as an ornamental tree. During the autumn the foliage takes a beautiful reddish tint, making the tree very ornamental. In some places such as Bangalore, this phenomenon may happen twice in a year. The branches are horizontal. It can be planted on roadsides.

Thuja occidentalis (also spelt *Thuya* or *Thuia*); Family: *Pinaceae* (*Cupressaceae*); Origin: Eastern North America; Common Names: American or Common Arbor-vitae, erroneously but commonly called white cedar; Vernacular Names: Hindi – *Morponkhi*; Bengali – *Bilatijhau.*

A handsome tree growing up to 20 m in height. The main trunk is strongly buttressed and divided at ground level into several secondary trunks. The branches and the branchlets are densely and horizontally arranged. Thus the plant gets a pyramidal shape and a rather compact head. The foliage is fern-like and bright green above and yellowish-green below. During the cold season the colour changes to dull brownish-green. The plant stands trimming well and can be given a good shape. The trees are very suitable for planting along pathways in the garden. The pyramidal and majestic look of the plant adds beauty to any garden. The tree can also be planted singly or in pairs in front of buildings or entrance. It can be grown in pots for several years when young for interior decoration or roof garden. Repotting once a year or top dressing with manure is useful. Propagation is from seeds or air-layers. There are several garden forms. The garden form var. *compacta* (Syn. var. *froebeli*, var. *globularis*) is globose with compact, rather thin branches, and light green in colour, while var. *globosa* (Syn. var. *compacta globosa*) is of dwarf globose form and quite similar to var. *compacta* but shorter and smaller.

Thuja orientalis (Syn. *Biota orientalis*); Family: *Pinaceae*; Origin: China; Common Name: Chinese Arbor-vitae.

A pyramidal or bushy tree, up to 12 m in height. A number of erect branches are produced from the base. There are two types – one pyramidal or columnar in habit, and the other dwarfer and rather globular. The branchlets are flattened and arranged at right angles with the branches. The variety *compacta* is dwarf and the best. The use and cultivation are the same as in the case of *T. occidentalis*.

Fruit Trees for Avenues

The following fruit trees are also sometimes planted along roads or avenues and they look quite ornamental.

Averrhoa carambola; Family: *Oxalidaceae*; Origin: India, China; Common Name: Carambola Tree; Vernacular Names; Hindi – *Khamrak*; Bengali – *Kamranga*; Marathi – *Kamarakha*; Tamil – *Sagadam*; Telugu – *Karamonga, Tamarta*.

A medium-sized tree (5-8 m tall) and very ornamental.

Eugenia jambolana (Syn. *Syzygium cumini, S. jambolana*); Family: *Myrtaceae;* Origin: Australia, Tropical Asia; Common Name: Jambolan or Jambolan Plum; Vernacular Names: Hindi & Punjabi – *Jamun*; Bengali – *Kalojam*; Assamese – *Jamu*; Oriya – *Kalajamo*; Urdu – *Jaman, Phalenda*; Marathi – *Jam, Jaman*; Gujarati – *Jambu*; Tamil – *Arugadam, Naval*; Telugu – *Jambuva, Neredu*; Kannada – *Nerale*.

A tall and spreading tree which provides good shade and yields a purple-black oblong to roundish edible fruit which is relished by many people. The falling ripe fruits stain clothing. Propagated from seeds.

Mangifera indica; Family: *Anacardiaceae*; Origin: India, Tropical Asia; Common Name: Mango; Vernacular Names: *Am, Amb*.

Phyllanthus emblica; Family: *Euphorbiaceae*; Origin: Tropical Asia; Common Names: Emblic, Myrobalan; Vernacular Names: Hindi – *Amla, Aonla*; Bengali & Assamese – *Amloki*; Punjabi – *Ambli*; Gujarati – *Amali*; Marathi – *Anvala*; Malayalam – *Amaḷakam*; Tamil – *Amlagam*; Telugu – *Usirika, Amalaki*; Kannada – *Amalaka, Usirikai, Nellikai*.

An ornamental tree suitable for roadside planting. The fruits are highly prized by Indians for their medicinal value.

Tamarindus indica; Family: *Leguminosae*; Origin: Asia, Africa; Common Name: Tamarind; Vernacular Names: Hindi – *Imli*; Bengali – *Tentul*; Marathi – *Ambali*; Tamil – *Ambilam, Puli*; Kannada – *Huli, Amla, Hunise*; Telugu – *Chinta*.

A large tree suitable for roadside planting. An important spice and commonly used in south India and elsewhere.

(B) Flowering Trees

Amherstia nobilis; Family: *Leguminosae*; Origin: India.

This tree, when in flower, is considered to be one of the most beautiful plants in the garden. It is a medium-sized (4-5 m in Bangalore and 10-15 m in Calcutta and Burma) spreading tree. The young leaves remain folded and are of purplish coppery hue with a very ornamental appearance. The flowers appear between February and May in Bangalore and between January and March in Calcutta. About 20 flowers are borne in 50-60 cm long vermilion-coloured hanging racemes, and each vermilion-colour-tipped yellow flower is about 7.5 cm in length. This is a rare tree and worth growing in a garden. The tree requires a humid atmosphere and does not thrive in the drier parts of the country. Propagation is from seeds and by layering.

Barringtonia racemosa; Family: *Lecythidaceae* (placed in *Myrtaceae* by Bailey, 1963); Origin: Malabar (India); Common Name: Bottle Brush Oak; Vernacular Name: *Samundar-ka-phul*.

A medium-sized (10 m tall) tree, bearing small powder-puff like flowers in long pendulous racemes. The flowers are pinkish-white in colour and open at night. This is a moisture-loving plant and is sometimes planted on the roadsides. Propagation is from seeds.

Barringtonia speciosa; Family: As mentioned above; Origin: India (near the sea), Pacific Islands.

A very ornamental, evergreen tree of spreading habit, attaining a height of 15-20 m. The foliage is shining, leathery, and very handsome. The large flowers are showy, the petals are white in colour, and the stamens are tinted purple. Propagation is from seeds and leafy cuttings and by air layering. The tree is planted on the roadsides. There is another species, *B. acutangula*, which is superior to the above species.

Bauhinia

The name is in honour of two botanists, John and Caspar Bauhin, who were twin brothers. The genus comprises shrubs and small trees. Some species are so ornamental that without them a garden will not be complete. The leaves in Bauhinia have the same characteristic, resembling the hoof of a camel. The flowers are numerous and in some species are borne in a leafless condition; many of them are fragrant also. Propagation is from seeds.

(a) *B. purpurea* (Syn. *B. triandra*); Family: *Leguminosae*; Origin: India, Burma, China; Common Names: Purple Bauhinia or Geranium Tree; Vernacular Names: Hindi – *Gulabi kachnar*, *Gairal*, *Kaliar*; Bengali – *Devakanchan*, *Raktakanchan*, *Koiral*; Tamil – *Mandari*, *Kalavilaichi*; Telugu – *Kanchanam*, *Bondanta*; Kannada – *Basavanapada*; Malayalam – *Swarnamandaram*; Punjabi – *Karalli*; Marathi – *Atmatti*; Oriya – *Boroda*.

A medium-sized tree which is almost evergreen, except for a brief period during the winter when the leaves fall. It is branched profusely and has a rounded crown. The bark is ash-grey or brown. The clawed or pointed flowers are large and showy. The flower is rosy purple or lilac or deep pink. Sometimes the base of one petal is white. The plant flowers profusely in November.

(b) *B. tomentosa*; Family: *Leguminosae*; Origin: India, Sri Lanka; Common Name: St. Thomas Tree; Vernacular Names: Kannada – *Vanasampige*; Tamil – *Thiruvatti*; Marathi – *Sona*.

This is a shrub or small tree (4 m tall) and bears showy yellow flowers which are blotched with dull maroon at base. The colour changes to copper yellow on fading.

(c) *B. triandra:* See *B. purpurea.*

(d) *B. variegata*; Family: *Leguminosae*; Origin: India; Common Names: Mountain Ebony, Variegated Bauhinia, Camel's Foot; Vernacular Names: Hindi – *Kachnar* (*Baisakhi*); Bengali – *Kanchan, Kovidara*; Kannada – *Kanchivala*; Tamil – *Mantharai*; Marathi – *Kanchan*; Oriya – *Boroda*; Telugu – *Bodanta*.

A small tree (up to 6 m tall) having an irregular shaped crown. The bark is of brown colour. The flowers are large, fragrant, white in colour, but variegated with rose, purple, and white. The flowers are borne during February to March when the plant remains in a leafless condition. The tree when in bloom looks very beautiful. It is very useful as an avenue tree, and for parks and open places. It is generally propagated from seeds, but cuttings also root, though with some difficulty. It grows fast in the first season and flowers in the second year. The var. *candida* grows to a height of 4 m and bears scented white flowers, beautifully veined green.

Bignonia megapotamica; Family: *Bignoniaceae*; Origin: Brazil; Common Name: Rio Grande Trumpet Flower.

A handsome tree with graceful sweeping branches. Grows to a height of 8-12 m. The leaves consist of one or two pairs with one terminal leaflet and these are bright olive-green in colour. The short-lived, scentless flowers are palest lilac-pink and are produced in numbers during March-April. The tree is good for planting in a lawn or along a small avenue. *B. crispa* (Syn. *Dolichandrone crispa*), called *Padiri* in Tamil, is a common tree in south India. The white flowers are delicately scented. It is propagated by offsets.

Bombax malabaricum (Syn. *B. ceiba*; also known as *Salmalia malabarica*; the name *Salmalia* is derived from the Sanskrit name *Salmali* or *Shalmali*); Family: *Bombacaceae*; Origin: India, Malaysia, China; Common Name: Silk Cotton Tree; Vernacular Names: Hindi – *Simul, shimbal*; Bengali – *Simul*; Oriya – *Simul*; Assamese – *Simalu*; Marathi – *Savane, Sayar*; Tamil – *Ilavu*; Telugu – *Buraga*; Kannada – *Hatti.*

It is a deciduous tree with wide spreading branches and often reaches a height of 25 m. A good shade tree except during the flowering time (January-February) when the tree sheds all the leaves. The large bright crimson or red flowers are borne in clusters on the branches. When in flower, there are very few trees in the world that can compare with *simul* tree. To add to the beauty many birds visit the tree during flowering for the nectar. Occasionally, there may be yellow or white flowered trees also. The egg-shaped fruit pods yield silk cotton, which is used for filling pillows and cushions. The silk cotton tree is useful for planting along national highways and town roads, for shade and beauty. Propagation is from seeds or cuttings.

Brownea ariza (Syn. *B. princeps*); Family: *Leguminosae*; Origin: Tropical America;

An abstract topiary design

Majestic *Araucaria cookii* trees in front of a public building

Bonsai: A pomegranate

Bonsai: A banyan

View of a public park (Cubbon Park, Bangalore). Note the four *Tecoma argentea* in bloom

A lawn (*doob* grass) with stepping stone path

A small home garden — a partial view from the front

The 'Travellers Tree', *Ravenala madagascariensis*

A *Kalanchoe* in flowering

A garden adornment: statue

Landscaping public buildings: gardens of Vidhana Soudha, Bangalore

Common Name: Western Indian Mountain Rose.

This is a small-to-medium-sized (6-10 m) umbrella-shaped tree. The sweeping branches touch the ground. The reddish orange flowers are borne in large bunches during February to April. The seedlings take 10-15 years to come to flowering, but a layer or graft will flower in 5-7 years.

Brownea coccinea; Family: *Leguminosae*; Origin: Venezuela (South America).

It is a shrub rather than a tree, growing to a height of 2-3.5 m. A tree of spreading habit and bears scarlet flowers in bunches.

Brownea grandiceps; Family; *Leguminosae*; Origin: South America.

It is considered to be the best *Brownea*. The tree is very ornamental with its handsome 12 pairs of oblong-lanceolate leaflets. The young leaflets are coppery-pinkish in colour and very handsome. The large, bright scarlet or red flowers are clustered densely, forming a globose head at the end of drooping branches. The flowering season is during April-May.

Bursaria spinosa; Family: *Pittosporaceae*; Origin: Australia.

It is an evergreen, spiny, much-branched shrub to a small tree, up to 5 m tall. The small, scented white flowers are borne in profusion covering the whole plant during November to January. It can be easily grown from seeds.

Butea monosperma (Syn. *B. frondosa*); Family: *Leguminosae*; Origin: India, Burma, China; Common Name: Flame of the Forest; Vernacular Names: Hindi – *Dhak, Palas, Chalcha*; Bengali & Assamese – *Palas*; Gujarati – *Khakda, Khakra*; Marathi – *Kakrucha, Palas*; Punjabi – *Palak chachra*; Oriya – *Polaso*; Tamil – *Palasu*; Kannada – *Muthuga*; Telugu – *Moduga*; Malayalam – *Shamata, Muriku.*

A small-to-medium-sized, 5-10 m tall, untidy deciduous tree with crooked trunks and branches. The leaves are trifoliate, each leaflet being large, rounded, leathery, and rough. The flowers are large and crowded on leafless branches. The flowers are bright flaming scarlet orange with dark olive-green to black calyces, with a shape resembling a parrot's beak. The flowering season is between April and May. It flowers a little earlier in and around Bangalore. The tree is suitable for planting in small groups in a garden or a park. Propagation is from seeds which should be sown when fresh. There is a rare variety, *B. monosperma* var. *lutea*, with yellow flowers. The tree withstands frost and drought very well and does well on saline soils. The commercial shellac is produced on this tree. The bark is used for tanning. When several leaves are stitched together it serves as a plate. The flowers yield a dye.

Callistemon lanceolatus (Syn. *C. citrinus*); Family: *Myrtaceae*; Origin: Australia; Common Name: Bottle Brush.

A beautiful shrub or a small tree, 3 m tall, and of weeping habit. The crowded scarlet flowers with free stamens are borne in spikes 5-10 cm long on the old branches. A branch with flowers looks like a bottle brush. Flowers between February and November, but the peak season is March-April and August-September. The plants are propagated from seeds or layers. A beautiful tree which should find a place in every garden. It is specially suited near ponds or lily pools with the weeping branches falling over the pool.

Camellia japonica (Syn. *Thea japonica*); Family: *Theaceae* (*Ternstroemiaceae* Bailey, 1963); Common Name: Camellia; Origin: Tropical and Sub-tropical Asia.

A shrub or tree, sometimes growing up to 10 m. The leaves are broadly oval, deep shining green above. The tree is compact in growth and does well at high and medium elevations. At a medium elevation it may preferably be grown in semi-shade either in ground or in pots. The plant prefers an acid soil and a lot of humus. The roundish, 7-12 cm across, red or pink flowers are double or semi-double and are very beautiful. The flowering season is between April and June. Propagation is from seeds, cuttings, bud cuttings, layers, and by inarching. In bud cutting, a 2.5 cm stem with a leaf attached to it is taken. About half the thickness of the wood is removed and the leaf with an axillary bud is inserted as cutting. *Camellia* is good for planting singly or in groups in parks and gardens.

Canangium odoratum (Syn. *Cananga odorata, Uvaria odorata, Unona odorata*); Family: *Annonaceae*; Origin: South India, Java, the Philippines, Malaysia and Tropical Pacific Islands; Common Names: Ylang Ylang, Ilangilang, Alangilang.

A quick growing tree with horizontal and arching branches. The greenish-white scented flowers, resembling those of *Artabotrys odoratissimus*, are borne in profusion many times a year. The plant is more suitable for a large garden or a park. It is hardy and grows in any good garden soil. Propagation is from seeds.

Cassia fistula; Family: *Leguminosae*; Origin: India, Sri Lanka; Common Names: Golden Shower, Pudding Pipe Tree, Indian Laburnum; Vernacular Names: Hindi - *Amaltas*; Bengali - *Bandar lathi, Sondal, Sonali*; Assamese - *Sonaru*; Oriya - *Sunari*; Punjabi - *Alash*; Tamil - *Sarak konne*; Kannada - *Kakke*; Telugu - *Relu*; Marathi - *Bahava, Boya*.

The most widely known *Cassia* in India. A medium sized tree, up to 10 m tall, of great beauty when in flower. The leaves are large and pinnate. The young leaves, coppery red in colour, remain folded and pendulous till fully grown. During the hot weather (April-May) the tree becomes most conspicuous with its pendulous racemes of golden yellow, mildly-scented flowers. The tree remains leafless on the commencement of flowering and at the end of the flowering season the leaves start to appear. After flowering is over, the long (30-45 cm) cylindrical pods appear, which become black on ripening. The tree grows well in dry tracts. The tree is of slow growth and may take 10 years to flower. Under favourable climatic conditions it may flower in five years. Propagation is from seeds, but the percentage of germination is poor. Sometimes the tree is defoliated by caterpillars.

Cassia grandis; Family: *Leguminosae*; Origin: Tropical America; Common Names: Pink Shower, Horse Cassia.

The rose-coloured flowers are borne in long drooping racemes in the axils of fallen leaves crowded on old wood during March-April. As the flowers fade, new shoots appear. This is an elegant, spreading and quick-growing tree. Propagated from seeds.

Cassia javanica; Family: *Leguminosae*; Origin: Java.

A beautiful cassia of medium size, with a straight trunk and spreading crown having horizontal branches. The drooping branchlets with feathery leaves are very ornamental. The tree is slightly taller than *C. nodosa*. The leaves are different from those of *C. nodosa*, being bluntly rounded, while the leaves of *C. nodosa* are pointed or lance-shaped. The pink-coloured flowers appear in whorls along with new leaves at the apex of the lateral branches during May-June. The branches also bear clusters of flowers for a short period. While *C. renigera* flowers in a leafless condition, this species is distinguished by its pink flowers being intermingled with foliage. This cassia needs a sheltered position as the wood is brittle and may break due to strong winds. A good tree for avenues and parks.

Cassia marginata (Syn. *C. roxburghii*); Family: *Leguminosae*; Origin: India; Vernacular Name: Tamil – *Vakai*.

A medium-sized tree with drooping and spreading habit. The foliage consists of alternate pinnate leaves of 10-12 pairs of linear-oblong leaflets, the margins of which are coloured. The beautiful terra-cotta red flowers appear during March-June and again in August-September under certain climatic conditions as of Bangalore.

Cassia multijuga (Syn. *C. calliantha*); Family: *Leguminosae*; Origin: Tropical America.

A quick-growing dwarf tree, 4-5 m tall, of spreading habit. The leaves are smaller than those of other *Cassias*. The bright yellow flowers cover the whole tree in August-September, giving it a fantastic look. Propagated with difficulty from cuttings, but can be easily raised from seeds which are produced in dry regions. The tree is suited for moderately dry regions and is planted along streets.

Cassia nodosa; Family: *Leguminosae*; Origin: Burma, Malaysia and a part of India (Assam).

A comparatively large tree with a short gnarled trunk. The tree is similar in general habit to *C. javanica*. The pink flowers appear in bunches from the nodes all along the branches. Hence the name *nodosa*. The tree flowers in a leafless condition between May and June, the leaves first appearing on the lower branches. It is one of the best flowering trees. The tree is recommended for planting in parks, medium-sized gardens, and avenues.

Cassia renigera; Family: *Leguminosae*; Origin: Burma, but found commonly in Maharashtra and Tamil Nadu; Common Name: Burmese Pink Cassia.

A medium-sized deciduous tree. The leaflets are leathery, glossy, and pointed. A large kidney-shaped bract remains attached to the base of the young leaf-stalks, and drops later on. The clusters of bright pink or rose flowers arise from the axils of the fallen leaves all along the branches, during April-June. A quick-growing plant, it flowers in 4-5 years when raised from seeds. It is recommended for planting in large home gardens and parks.

Cassia siamea; Family: *Leguminosae*; Origin: India, Sri Lanka, Malaysia.

A low-growing tree. The leaflets are 6 to 14 pairs, 3-7.5 cm long, oblong, and apex notched and distinctly stalked. The flowers are yellow and crowded in axillary and terminal erect panicles. Used as an avenue or a roadside tree.

Cassia spectabilis; Family: *Leguminosae*; Origin: Tropical America.

A beautiful, medium-sized, umbrella-shaped tree, suitable for roadside planting in towns and cities. The pinnate leaves are long, leaflets lanceolate and bright green in colour. The large (4 cm across), cup-shaped, golden-yellow flowers are borne in 30-60 cm long erect racemes of a beautiful appearance. Flowers during July to August when the whole plant gets covered. Propagation is from seeds.

Hybrids of Cassia.

(a) *C. alipurense:* Cream coloured flowers, resembling *C. fistula.*

(b) *C. atkinsiana:* Intermediate between *C. fistula* and *C. nodosa*. The spikes are pendent but short. Flower colour cream with red sepals.

(c) *C. lancasteri:* A hybrid of *C. nodosa* and *C. marginata*. The flower colour is intermediate. Flowers throughout the rainy season.

All *Cassia* seeds are hard and it is advised to put these in boiling water for five minutes before sowing.

Castanospermum australe; Family: *Leguminosae*; Origin: Australia; Common Name: Moreton Bay Chestnut.

A slow-growing evergreen tree, 13-20 m tall with large, unequally pinnate leaves. The leaflets are 10-15 in number, ovate to oblong in shape, and pointed at the apex. The plant flowers in March in racemes of large orange-crimson flowers. The stout brown pods contain 2-4 large chestnut-like seeds which are edible when roasted.

Chorisia speciosa; Family: *Malvaceae* (*Bombacaceae* by Bailey, 1963); Origin: Brazil.

It is a fast-growing, medium-sized, up to 10 m, deciduous tree. The stems and branches are greenish and thorny. The trunk is bottle-shaped. The tree bears yellow flowers, marked brown at the base of the petals, in a leafless condition, during October. There are other species of *Chorisia* with purple, mauve, and white flowers.

Citharexylum quadrangulare; Family: *Verbenaceae*; Common Name: Fiddle Wood Tree; Origin: W. Indies.

An upright, 10 m tall, deciduous tree with fairly large, dark green foliage. It bears flowers sparingly almost throughout the year. But the drooping racemes of scented white flowers appear mainly during March-May. The branches of old trees may be pruned for vigorous new growth. The tree is propagated from long cuttings and these should be planted at the place where it is intended to be grown. This tree deserves a place in all gardens for its fragrance.

Cochlospermum gossypium (Syn. *Bombax gossypium*); Family: *Bixaceae* (The Royal Horticultural Society *Dictionary of Gardening*, 1956, placed this in the family *Cochlospermaceae* being the only genus of this family); Origin: India; Common Name: Yellow Silk Cotton Tree; Vernacular Names: Hindi – *Gooloc, Galgal, Kumbi*; Bengali – *Golgal*; Marathi – *Galgal*; Gujarati – *Kadachogund*; Oriya – *Konto-palas, Ganiari*; Punjabi – *Kumbi*; Tamil – *Kangilam tannaku*; Kannada – *Arasina burugada mara*; Malayalam – *Appakutakka*; Telugu – *Adaviburaga, Akshotamu.*

A small, 4-5 m tall, branching, deciduous tree with 3-5 lobed leaves. When the tree flowers in February-March in a leafless condition, the golden yellow flowers in terminal clusters look strikingly beautiful. It is grown as an avenue tree, especially in the drier tracts. Easily propagated from seeds.

Colvillea racemosa; Family: *Leguminosae*; Origin: Madagascar (Africa); Common Name: Colville's Glory (in honour of Sir Charles Colville); Vernacular Name: Hindi – *Kilbili.*

A medium-sized tree resembling *gulmohar* in habit when not in flowering. The large foliage is twice-pinnate, the linear leaflets resembling *gulmohar*. During August to September the plant bears long, compact, drooping racemes of orange-red flowers at the end of the branches. The tree is most ornamental when in bloom. It likes moist weather. Propagation is from seeds. Since the tree bears flowers at the end of branches, it should never be pruned.

Cordia sebestena (Syn. *C. speciosa*); Family: *Boraginaceae*; Origin: W. Indies; Common Names: Geiger Tree, Aloes-wood or Scarlet Cordia; Vernacular Names: Hindi – *Bhockar*; Kannada – *Challekendala*; Tamil – *Acchinaruvihli*; Telugu – *Virigi*.

A dwarf tree or a garden shrub of deciduous habit. The bell-shaped orange or scarlet flowers are borne almost throughout the year in terminal clusters, but in more abundance during January to March. The dry wood when burnt emits a good perfume. Propagation is from seeds which should be sown when fresh.

Couroupita guianensis; Family: *Lecythidaceae*; Origin: Guyana; Common Name: Cannon-Ball Tree; Vernacular Names: Hindi – *Tope gola*; Bengali – *Nagalinga*; Tamil – *Nagalingam*.

A huge tree which produces flowers only from the main stem or branches. The flowers are fleshy, large. The outside of the petals are yellow and the inner side deep pink with shades of white. The stamens are united and form into a shape resembling the hood of a cobra. The tree sheds the leaves several times in a year. The fruits are large, round, and resemble cannon balls. The fruits ripen in about a year's time. The tree thrives well in humid localities. Propagation is from seeds or suckers which are produced in abundance. It is suitable for planting only in large gardens.

Crataeva religiosa (Syn. *C. nurvala*); Family: *Capparidaceae*; Origin: Malabar (India) and Society Islands; Common Name: Caper Tree; Vernacular Names: Hindi – *Barna, Bilasi*; Bengali & Assamese – *Barun*; Punjabi – *Barna*; Oriya – *Baruna*; Gujarati – *Varno*; Marathi – *Haravarna*; Kannada – *Mavilinga*; Tamil – *Maralingam, Narvala*.

A small or medium-sized, graceful, deciduous tree. The plant flowers in a leafless condition during April-May, when it looks handsome. The flowers are yellow or reddish-yellow, borne in dense corymbs. The plant stands drought well. The leaves are eaten by cattle; therefore, proper protection should be given. Propagation is from seeds and root-suckers. The seed coat being hard it may take even one year to germinate.

Elaeocarpus ganitrus; Family: *Elaeocarpaceae* (Formerly included in *Tiliaceae*); Origin: India, Malaysia; Common Name: Indian Bead Tree; Vernacular Names: Hindi – *Rudrak*; Bengali – *Rudraksha*; Oriya – *Rudrakhyo*; Marathi – *Rudraksh*; Malayalam – *Rudraksham*; Tamil and Kannada – *Rudrakshi*; Telugu – *Rudraksha*.

An ornamental tree, up to 8 m tall. Bears drooping racemes of white flowers from the axils of the fallen leaves. The seeds are used as beads (*rudraksha*) to be worn round the necks and as Rosary by religious Hindus. Propagation is from seeds or cuttings from ripe leafy shoots. The tree is suitable for planting in lawns and along avenues.

Erythrina blackeii; Family: *Leguminosae*; Origin: India.

A large shrub or a small tree bearing dark scarlet flowers in April. Suitable for planting in small compounds and along town roads.

Erythrina crista-galli (Syn. *E. laurifolia*); Family: *Leguminosae*; Origin: Brazil; Common Name: Common Coral-tree.

It is a dwarf tree, bearing long and dense terminal racemes of scarlet flowers. This is also useful for planting in small compounds. Stems and petioles are sometimes spiny.

Erythrina indica (Syn. *E. carnea*); Family: *Leguminosae*; Origin: India, Polynesia, W. Indies; Common Names: Hindi – *Pangri, Pharad, Dadap*; Bengali – *Palita madar, Rakta madar*; Kannada – *Haluvana*; Tamil – *Kalyana murungai*.

A fast-growing, densely-branched, small tree. The erect stem has black spines. The trifoliate leaves give the plant an ornamental look. The tree flowers in a defoliated condition during February to March and even up to April. The scarlet red flowers appear in racemes in a cluster. The tree is planted for the beauty of its flowers. It can be planted in groups in parks and gardens or along avenues in towns and cities. The trees being small, it is possible to plant them in multiple rows. The var. *alba* bears white flowers. In var. *marmorata* (Syn. *E. marmorata*) the leaves are attractively spotted with white. In var. *parcellii* (Syn. *E. parcellii*) the leaves are variegated yellow along the veins and the midrib.

All *Erythrinas* can be propagated from long stem cuttings. The cuttings are planted *in situ* during the rains. Propagation is also possible from seeds.

Erythropsis colorata (Syn. *Sterculia colorata*); Family: *Sterculiaceae*; Origin: India, Burma, Sri Lanka; Vernacular Names: Tamil – *Marambarutti*; Malayalam – *Maramparatti*.

The tree grows to a height of 10-12 m. The leaves are palmately 3-5 lobed. It bears in a leafless condition numerous orange-red blossoms on the branches during March-April. The bare tree fully covered with orange-red flowers is a sight to see. Recommended for planting in parks, gardens, and along avenues. Birds are attracted by the flowers.

Gliricidia maculata (Syn. *Lonchocarpus maculatus*); Family: *Leguminosae*; Origin: Guatemala to South America; Common Name: Madre Tree of South America (for shading cocoa).

A small, 3-5 m tall, good-looking tree, which bears, during February-March beautiful pale pink flowers all along the leafless branches. The tree is recommended for planting in rows or groups in parks, gardens, and along avenues. The tree is brittle and is often broken by strong winds. The tree should be pollarded annually to keep it dwarf. It can be used as a green manuring crop and the seeds for rat poison. The tree is killed by frost in north India. There is a white flowering species.

Gustavia augusta; Family: *Lecythidaceae*; Origin: Tropical America.

A medium-sized tree with handsome magnolia-like foliage. The plant bears beautiful, scented, white, water-lily-like flowers, during July-August (hence *augusta*), which remain covered by the foliage. There are seedling variations.

Gustavia gracillima; Family: *Lecythidaceae*; Origin: Tropical America.

The tree bears 10 cm across rose-red flowers. *Gustavias* can be propagated from cuttings of ripened wood or layering.

Holarrhena antidysenterica; Family: *Apocynaceae*; Origin: India; Common Name: Easter Tree or Conessi Bark; Vernacular Names: Hindi – *Karra, Kaura*; Bengali – *Kurchi*; Assamese – *Dudcory*; Oriya – *Kherwa, Indraja*; Punjabi – *Kewar*; Kannada – *Kudsalu korchu*; Tamil – *Kulappalai veppalai*; Marathi – *Kodaga, Kurra*.

A small tree having compact branches. During March-May it comes to bloom, when flowers are badly needed in a garden. The creamy-white strong-scented flowers practically cover the whole tree which looks very handsome. The bark yields an anti-dysenteric drug. There is a variety, *pubescens*, which bears in more profusion and is a little dwarf also. Propagated from root-suckers only.

Ixora parviflora; Family: *Rubiaceae*; Origin: India; Common Name: Torch Tree; Vernacular Names: Hindi – *Kota gandal*; Marathi – *Lokandi, Kurat*; Tamil – *Shuludu, Korivi*; Telugu – *Kori, Gorivi*; Bengali – *Shetrangan*.

An evergreen small tree with many branches, which bears fragrant whitish flowers. In Madras, the branches are used as torches by letter carriers. *I. ferrea* is also a small tree (about 6 m tall), which bears a few pink axillary flowers.

Jacaranda mimosaefolia (also spelt *mimosifolia*) (Syn. *J. ovalifolia, J. acutifolia, Bignonia caerulea*); Family: *Bignoniaceae*; Origin: Brazil.

A deciduous, elegant tree with open branching system, attaining a height of 8-9 m. The tree grows very well and flowers profusely in Bangalore. It also flowers well in New Delhi, Dehradun, Chandigarh, and Lucknow. In the plains of West Bengal, the flowering is poor. In Madras and Coimbatore, the plant does not flower. It thrives at an altitude up to 1,600 m. The foliage is fern-like and bipinnate. The bell-shaped blue-mauve flowers are borne in erect panicles during March-May and just at the start of flowering the tree remains leafless for a brief period. In Bangalore flowering starts during February. This beautiful tree is recommended for planting along avenues and in parks and gardens for its handsome foliage and beautiful flowers. It is propagated from seeds and semi-hardwood cuttings. In the U.S.A. grafting is also practised, using seedlings as rootstocks. The grafted plants flower within one year but have a shorter life-span, about 20 years. The branches may be topped once in three years for a bushy growth. The young plants can be grown in pots for their handsome foliage.

Kleinhovia hospita; Family: *Sterculiaceae*; Origin: India.

A small (4 m tall), bushy, evergreen tree with large cordate leaves and small rosy-pink flowers, borne in terminal panicles. Flowers almost all the year round, but more profusely in July-August. The pods are ornamental. Propagation is from layers or seeds.

Lagerstroemia flos-reginae (Syn. *L. speciosa, L. reginae*); Family: *Lythraceae*; Origin: India, Sri Lanka, China, Malaysia; Common Names: Queen's Flower, Pride of India, Crape Flower; Vernacular Names: Hindi – *Jarul*; Bengali – *Jarool*; Assamese – *Ajhar*; Oriya – *Jarulo*; Marathi – *Taman, Mota-bondara*; Tamil – *Kadali*; Kannada – *Maruvachalla*; Telugu – *Varagogu*.

This tree is 8-10 m in height, but in its natural habitat it sometimes grows up to 20 m. The leaves are large (15 cm × 8 cm) and handsome. During February-March the leaves turn somewhat coppery or yellowish in colour. The showy, purplish or purple or mauve flowers are borne in about 30 cm long panicles at the end of all branches during April-May and July-August. Propagation is from seeds. The seedlings are transplanted after one year. The growth is slow during the first couple of years, but afterwards it grows rapidly. The side branches should be pruned to train it into a well-shaped tree, otherwise tends to grow into a bushy shrub. The flowering starts in about five years. The plant grows best in deep alluvial loam. Recommended for growing along avenues, and in parks and gardens.

Lagerstroemia indica (Syn. *L. chinensis*); Family: *Lythraceae*; Origin: China; Common Names: Indian Lilac, Crape Myrtle.

A large shrub or a small tree in suitable climatic conditions. The leaves are smaller than the above. The plant looks most handsome from May to August when the sprays of delicately fringed, 5 cm across flowers of pink, rose, mauve, or white (var. *alba*) cover the whole plant.

Lagerstroemia thorelli; Family: *Lythraceae*; Origin: Tropical Asia (possibly India).

A small bushy tree having white and mauve mottled flowers. The flowering season is from July to September. Easily propagated from seeds. Suitable for planting in towns and cities.

Magnolia campbellii; Family: *Magnoliaceae*; Origin: Sikkim (India).

A large deciduous tree which may grow up to 50 m in the wild, but it is generally much smaller when in cultivation. The tree is not known to flower under 20 years of age. The huge (15-25 cm across) lotus-like fragrant flowers, pink without and deep rose to crimson within, are borne all over the tree in a leafless condition during February and March or May. The tree is most ornamental when in bloom. This plant can be grown at high altitudes only. The flower buds may be damaged by frost. A white flowering variety is also known.

Magnolia grandiflora (Syn. *M. foetida*); Family: *Magnoliaceae*; Origin: Southeast U.S.A.; Common Names: Lily Tree, Bull Bay; Vernacular Names: Hindi - *Himchampa*; Bengali - *Udaypadma*.

An evergreen tree, pyramidal in shape. It may grow up to 25 m, but is generally much smaller in cultivation. The leaves are 15-25 cm long and tapered at both ends. The leaves are thick, glossy dark green above and rusty-felted below. The lotus-shaped flowers are large, 20-25 cm across, and sweet-scented. Propagation is by layering. Veneer grafting or budding may be successful.

Melaleuca leucadendron (*M. cajuputi*); Family: *Myrtaceae*; Origin: Australia, Malaysia; Common Names: Cajeput or Cajuput Tree, Punk Tree, Swamp Tea Tree.

A slow-growing large-sized tree, the foliage resembling that of *Callistemon*. The flowers are borne in slender, cylindrical spikes, 7-20 cm long, and generally white or greenish yellow, but sometimes pink or purple as in Australia. Propagated from cuttings.

Mesua ferrea; Family: *Guttiferae*; Origin: India, Sri Lanka, Malaysia; Common Name: Iron Wood Tree; Vernacular Names: Hindi & Punjabi - *Nagkesar*; Bengali - *Nageswar champa, Nagkesar*; Marathi - *Nagchampa*; Tamil - *Nagasuram*; Kannada - *Nagasampige*;

Malayalam – *Nagachampakam*; Telugu – *Nagakesaramu.*

The wood is reddish brown and very hardy and hence the name Iron Wood Tree. It is a moderate-sized tree with highly ornamental, lanceolate-shaped, shining, leathery leaves. The young leaves are deep red turning to pink, then green as they grow older. The flowers are large, white petalled, with a yellow centre, and are highly scented. The stamens are numerous. The flowering time is February to March. The tree needs partial shade. It likes an abundance of moisture and, hence, is not suitable for dry areas. The seeds should be sown *in situ* as transplantation is difficult.

Michelia alba; Family: *Magnoliaceae*; Origin: Java.

A tree similar to *M. champaca*, but the flowers are white and faintly scented for a day, becoming fetid afterwards.

Michelia champaca; Family: *Magnoliaceae*; Origin: India, Java; Common Name: Golden *Champa*; Vernacular Names: Hindi and Punjabi – *Champa*; Bengali – *Swarnachampa*; Assamese – *Phulchopa*; Oriya – *Chompa*; Gujarati – *Pitochampo*; Marathi – *Sonachampa*; Malayalam – *Champakam*; Tamil – *Shambaga*; Kannada – *Sampige*; Telugu – *Champaka.*

A medium-sized tree, 10-13 m tall, having long leaves with wavy margins and slightly resembling mango leaves. The 5-7 cm long flowers are golden-coloured, highly scented and silky in appearance. The flowering season is between April and May and September and October. The white-flowered species mentioned above and a creamy yellow variety are propagated by grafting on the golden-flowered variety which produces seeds in abundance. The seedlings take about 7-8 years to come to flowering. When a plant tends to bear an abundance of fruits, it should be removed, otherwise the plant may be exhausted and may not bear flowers properly for the next couple of years. The tree likes a humid atmosphere and grows up to medium altitudes. Suitable for planting along avenues and in parks and compounds.

Milletia ovalifolia; Family: *Leguminosae*; Origin: Africa.

A dwarf tree with ornamental ovate-shaped leaflets. It flowers in leafless condition during April. The small lilac-coloured flowers completely cover the tree. It is suitable for compounds of houses, parks, and public places. Propagation is from seeds.

Millingtonia hortensis; Family: *Bignoniaceae*; Origin: Burma; Common Names: Indian Cork Tree, The Jasmine Tree; Vernacular Names: Hindi – *Akasneem, Neem chameli*; Bengali – *Akasneem, Phuljhari*; Oriya – *Bakeni, Reali, Mach-mach, Sita-hara*; Marathi – *Akas nimb, Kavla-nimb*; Tamil – *Mara-malli, Kat-malli*; Kannada – *Beratu*; Telugu – *Kavuki*; Malayalam – *Katesam.*

A quick-growing, straight, tall (20 m) tree with a large foliage divided into leaflets. The foliage is handsome. The branches are drooping. The trumpet-shaped, silvery-white flowers are sweet-scented at night. The tree is likely to be toppled by high winds. It is a very attractive avenue tree. Propagation is from root-suckers which come in abundance or from cuttings. Propagation from seeds is also possible, wherever it is formed.

Nyctanthes arbor-tristis: See page 158.

Oroxylum (also spelt *Oroxylon*) *indicum* (Syn. *Calosanthes indica*); Family: *Big-*

noniaceae; Origin: India, Malaysia.

A small (10-12 m tall), upright tree, which branches vary rarely. The leaves are tripinnate, 90-150 cm long, the leaflets large (5-12 cm by 3-9 cm) and elliptic in shape. The large funnel-shaped, fleshy flowers are arranged in long peduncled racemes. The colour is dirty brown outside and yellowish inside. The flowers are foul smelling. The shoe-shaped capsule is flat, wide, and long (30-90 cm), resembling a sword. The flowers appear during the rains. Propagation is from seeds and also root-suckers.

Parkia biglandulosa: Family: *Leguminosae*; Origin: Malaysia; Vernacular Name: Kannada – *Sivaling mara.*

A large and fast-growing evergreen, upright tree with handsome, feathery, bipinnate foliage, consisting of 45-50 pinnae in a leaf. Each pinna has about 150 leaflets. The flower heads are globular, hanging down the branches suspended by long peduncles. The flower heads are at first brownish and velvety, later turn to white. It is a good avenue and shade tree. Propagation is from seeds.

Parkinsonia aculeata; Family: *Leguminosae*; Origin: Probably Tropical America, but naturalized in many countries; Common Name: Jerusalem Thorn; Vernacular Names: Hindi – *Vilaiti kikkar*; Bengali – *Bilati babla.*

A small (3-4 m tall), thorny tree with drooping branches. The leaves are bipinnate, the leaflets numerous and very small. The plant gets covered with yellow scented flowers during February to March when it looks beautiful. A good avenue tree for dry regions. The tree can be found in some roads of south Delhi. It forms a good hedge. Propagation is from seeds.

Peltophorum inerme (Syn. *P. ferrugineum, Caesalpinia inermis, Baryxylum inerme*); Family: *Leguminosae*; Origin: Australia, the Philippines; Common Names: Yellow *Gulmohar*, Rusty Shield Bearer; Vernacular Names: Tamil – *Iyalvagi, Iya vakai*; Telugu – *Kondachinta.*

A large, handsome tree with a fairly spreading crown. The beautiful foliage are bipinnate and fern-like. The main flowering season is between February and May, but the plant flowers a second time during September to November. The yellow, mildly scented flowers are borne in large panicles, terminally, on the branches. The large copper-coloured pods are conspicuous and ornamental until they become black. It is a beautiful ornamental tree, suited for avenues, parks, and gardens. The tree, if planted with *Gulmohar*, forms a good contrast with its yellow flowers to the scarlet gulmohar. The plant is easily raised from seeds. Cuttings also strike roots. The tree is fairly quick-growing.

Plumeria alba; Family: *Apocynaceae*; Origin: West Indies; Common Names: White Frangipani, Pagoda Tree; Vernacular Names: Tamil – *Peru, Perumallari, Perungalli*; Telugu – *Veyyivarahalu*; Bengali – *Gulancha champa.*

An evergreen tree, 6-7 m tall, having pure white, yellow-eyed fragrant flowers. The leaves are lanceolate to oblong-lanceolate and do not have the conspicuous marginal vein. A good tree for planting in the home garden. It can also be planted along avenues and in groups in a park. The flowering period is during March-April and again during July-September. The tree is easily propagated from cuttings.

Plumeria obtusa; Family: *Apocynaceae*; Origin: Mexico, West Indies.

A small evergreen tree with large dark-green blunted leaves. The margins of the leaves tend to roll under. The flowers are pure white, much larger than the other species and less funnel-shaped than the other *Plumerias*. The broad petals are conspicuously overlapping and slightly incurved so as to give them a rounded shape. A good tree for planting in home gardens and in parks. Propagation is from cuttings.

Plumeria rubra (Syn. *P. incarnata*); Family: *Apocynaceae*; Origin: Central America (From Mexico to Panama); Common Names: Frangipani, Pagoda Tree, Temple Tree, Tree of Life, West Indian Red Jasmine; Vernacular Names: Hindi – *Chameli, Gul-e-chin*; Bengali – *Gulancha, Kat champa*; Assamese – *Golanchi*; Oriya – *Golochi*; Gujarati – *Dholo champo, Radha-champo*; Marathi – *Khairchampa*; Kannada – *Kadu-sampige*; Tamil – *Ilattalari, Perungalli*; Telugu – *Arhataganneru*; Malayalam – *Arali.*

Possibly the most widely-grown tree in our gardens. It is a low-growing, 6-7 m tall, spreading tree of deciduous habit, shedding the leaves between the winter and the rainy season. The young trees generally do not lose their leaves. When injured, the tree exudes a milky latex. The leaves are large, broadly lanceolate, tapering at both ends, with conspicuous parallel veins. The leaves are crowded in spirals at the apex of the long branches. The flowers are very variable in colour, generally white with a yellow throat, rose, or yellow. Sometimes the petals are tinged with pink below. The flowers are very fragrant. There is another type, var. *acutifolia* (Syns. *P. acutifolia, P. acuminata, P. mexicana*), where the flower is white with a yellow eye. The flowering period is the same as in *P. alba.* If the plant is uprooted and left with exposed roots, it will continue to grow and bloom for some time and if replanted will resume its normal life as if nothing had happened. For this character, the plant is sometimes called "Tree of Life". The plant can be easily raised from cuttings during the rains. The cuttings should be allowed to wilt for a few days before planting. There are many hybrids of *Plumeria* with variations in flower colour, including a predominantly red-flowered variety.

Poinciana regia (Syn. *Delonix regia*); Family: *Leguminosae*; Origin: Madagascar (Africa); Common Names: *Gulmohar*, Peacock Flower, Flamboyant, Royal Poinciana; Vernacular Names: Hindi – *Gulmohar*; Bengali – *Radhachura, Krishnachura, Raktachura*; Oriya & Assamese – *Radhachura*; Marathi – *Gulmohr*; Tamil – *Mayarum*; Telugu – *Shima sankesula*; Malayalam – *Alasippu.*

It is a large, spreading, and somewhat umbrella-shaped tree, with light, feathery, bipinnate foliage. In March, the tree generally sheds the leaves in northern and eastern India, whereas in Bangalore it may or may not shed all the leaves. The flowers are borne in large masses covering the whole tree. The colour and the size of the flowers vary from tree to tree. The flower colour may be light orange or deep scarlet and various intermediate shades. The markings on the standard petals also vary greatly. The flowering time is between April and May and sometimes continues till July. In Bangalore, the tree may flower as early as in March and continue to bloom till August. The plants also remain comparatively small in Bangalore. It is a beautiful tree when in full bloom and is suitable for planting along avenues and in parks. It can be alternated along avenues with *Cassia fistula* in north India and with *Jacaranda* in Bangalore, though in some years the flowering season of the two may not synchronize. The tree is easily propagated from seeds and is quick-growing. The seed may be soaked in hot water for about ten minutes before sowing for better germination. The plants are also raised from cuttings. In areas susceptible to frost, the young plants should be protected by providing thatches in the winter. Lawn and other plants cannot be grown under this tree.

Prunus campanulata (*P. cerasoides* var. *campanulata*); Family: *Rosaceae*; Origin: Formosa, but cultivated in S. Japan; Common Name: Japanese Cherry.

Some of the *Prunus* species of Japan bear very ornamental flowers in a leafless condition. *P. campanulata* is a glabrous tree, 7-10 m tall, and bears pendulous, bell-shaped (campanulate) flowers which are deepest red in colour among the Japanese cherries.

Prunus sargentii (Syn. *P. serrulata* var. *sachalinensis*, *P. serrulata sargentii*); Family: *Rosaceae*; Origin: Japan, Korea; Common Name: Japanese Flowering Cherry.

A large tree with smooth shoots. The flowers are rose-pink, borne during March to April.

Prunus sieboldii (Syn. *P. pseudo-cerasus* var. *sieboldii*, *Cerasus sieboldii*, *C. watereri*); Family: *Rosaceae*; Origin: Japan, China; Common Name: Japanese Flowering Cherry.

A slow-growing strong tree producing showy pink double flowers.

Prunus subhirtella var. *pendula* (Syn. *P. pendula*, *P. itosakura*, *Cerasus pendula*, *C. itosakura*); Family: *Rosaceae*; Origin: Japan; Common Names: Rose-Bud Cherry; Japanese Weeping Rose-Flowered Cherry.

A small tree with drooping and crooked branches. The white to rose-pink flowers are borne during March-April.

Most of the ornamental cherries do well when grafted on an easy-to-root cherry rootstock such as *Prunus domestica* or peach. The cherries flower better in places having a cool winter. Suitable for planting in home gardens or in groups in public parks.

Pterocarpus indicus (Syn. *P. dalbergioides*, *P. flavus*); Family: *Leguminosae*; Origin: E. Indies; Common Names: Burmese Padouk, Burmese Rosewood, Gold Mohur of Singapur.

A tall tree (10-12 m), with spreading rounded top and drooping branchlets. The sweet-scented golden flowers are borne several times in a year. It is suitable for large gardens. The wood is a valuable timber.

Pterospermum acerifolium; Family: *Sterculiaceae*; Origin: E. Indies; Vernacular Names: Hindi & Marathi – *Kanak-champa;* Bengali – *Moochkunda*, *Kanak-champa*.

In north and east India, it grows into a slender tall tree, but in the south it does not grow tall. The leaves are large, heart-shaped, downy and hairy below and green above. The young stems also have a downy cover. The flower is large, the calyces are also downy but of *khaki* shade. The calyces, five in number, split and roll back gracefully releasing the creamy-white petals. The flowers are very fragrant and open between February and May. It is good for planting in parks. Propagation is from seeds.

Rhododendron; Family: *Ericaceae*; Origin: See below.

Rhododendrons are mostly shrubs and a few Himalayan species are trees which bear beautiful flowers. They do well in temperate regions or at high altitudes. The plants are easily raised from seeds in seed pans containing sandy peat. The seeds can be sown in fresh sphagnum moss also, but in this case the seedlings should be transplanted as soon as they can be handled. The other methods of propagation are layers and grafts and occasionally cuttings. *Rhododendrons* are suitable for planting as specimens in lawns, rock gardens, or in groups in parks backed by dark green foliage of the conifers. The important Himalayan "tree" *Rhododendrons* are as follows:

(a) *R. arboreum* is an evergreen tree which grows up to 13 m. The blood-red, pink, or white, usually spotted flowers are borne in trusses of 20.

(b) *R. barbatum* is a large, evergreen tree about 20 m in height. The flowers are packed in a hemispherical truss, 10-12 cm across. The flowers are bell-shaped and blood-red, with five black-crimson blotches at the base.

(c) *R. falconeri* is an evergreen tree, up to 10 m tall, bearing bell-shaped flowers of creamy-white to pale yellow with purple blotches at the base.

(d) *R. grande* is another evergreen species, 12-13 m tall. About 20-30 flowers are borne in a rounded truss, 16-19 cm across. The flowers are bell-shaped, ivory-white, blotched purple at the base. It is one of the best Himalayan species.

All the *Rhododendrons* mentioned above flower during March-April in the hills.

Rhododendrons are affected by a number of pests and diseases but these are not serious in India. For pests see Chapter XXVI.

(Here only actual *Rhododendrons* are described and *Azalea*, which also come under the same genus, are not included. For *Azalea* see page 133.)

Saraca cauliflora; Family: *Leguminosae*; Origin: India.

A medium-sized tree having 10-12 leaflets, the upper leaflets oblanceolate and about 30 cm long, while the lower ones are shorter and more oblong. The scarlet flowers are borne in dense, nearly sessile corymbs.

Saraca declinata; Family: *Leguminosae*; Origin: Java, Sumatra.

A tree closely related to *S. indica*. The leaflets are 6-8 in pairs and reddish when young. The flowers are yellowish-orange, borne in clustered panicles.

Saraca indica; Family: *Leguminosae*; Origin: India, Malaysia; Vernacular Names: Hindi – *Ashok*, *Sita-ashoka*; Bengali – *Asok*, *Asoka*; Gujarati – *Ashopalava*; Marathi – *Ashoka*, *Jasundi*; Oriya – *Osoko*; Kannada – *Asoka*; Tamil – *Asogam*; Telugu – *Asokamu*; Malayalam – *Asoka*, *Hemapushpam*.

A small-sized, 7-10 m tall, handsome, evergreen tree, with branches spreading in all directions and forming a shapely crown. The leaflets are 3-7 pairs, opposite, lanceolate and 12-15 cm long. The young leaflets come out in drooping bunches, are greyish-white to pinkish in colour, and remain folded. The rich orange, slightly fragrant flowers are borne in roundish panicles on the branches and stems. The flowering season is the summer.

All the *Ashokas* are easily propagated from seeds. *S. indica* is suitable for planting as a specimen in lawns or in groves in a park. The tree is widely planted in and around Calcutta. This is also found in Bangalore and Lucknow.

Sesbania grandiflora (Syn. *Agati grandiflora*, *Aeschynomene grandiflora*); Family:

Leguminosae; Origin: Mauritius, India to North Australia; Common Name: The Sesban or Agastya; Vernacular Names: Hindi - *Agasti*; Bengali & Assamese - *Bakphul*; Oriya - *Buko, ogosti*; Gujarati - *Agathio*; Marathi - *Agathi*; Tamil - *Agathi*; Kannada - *Agase*; Telugu - *Avise*; Malayalam - *Akathi*.

This is a very quick-growing small tree having ornamental value. The flowers are very large and pea-shaped. There is a creamy-white flower variety and another variety in which the flowers open greenish-white in the evening, but turn pink in the morning. The flowers are edible. The flowering time is December to January. The plant can be utilized for developing into a quick hedge, or as a shade tree for foliage plants in which case it should be headed back. It is also used as a green manure plant. The plant is easily propagated from seeds and attains almost full height in six months and flowers the same year.

Solanum macranthum (Syn. *S. maroniense*); Family: *Solanaceae*; Origin: Brazil; Common Name: Potato Tree.

A quick-growing tree which attains a height of 4.5 m. The branches are herbaceous and sparsely prickly. The flowers, bluish mauve in colour, are borne in clusters above the large and nicely lobed leaves. The flowers are 2.5 cm across and resemble potato flowers. The flowering season is from March to November. The fruits are as large as a billiard ball. The plant is easily propagated from seeds and cuttings. The plant needs replacement every 4-5 years.

Spathodea campanulata; Family: *Bignoniaceae*; Origin: Tropical Africa; Common names: Flame or Tulip Tree, Fountain Tree, Scarlet Bell Tree, Scarlet Fountain Tree, Syringe Tree or Uganda Flame Tree, Bell Flambeau Tree, Squirt Tree; Vernacular Names: Hindi - *Rugtoora*; Tamil - *Patadi*; Kannada - *Neerukayi mara*.

One of the best trees introduced in India which is useful for shade, ornamental look, and beauty of its flowers. In some parts of the country such as Bombay and Bangalore, it behaves as an evergreen tree but in drier parts this plant sheds all the leaves for a short time during the summer. A tree with upright branches, it may attain a height of 30 m. The leaflets are oval in shape and dark green in colour. The large, crumpled, bell-shaped, bright scarlet-orange, crimson or almost vermilion flowers are borne in slightly branched racemes or panicles. The flowering time varies from place to place. In Bombay, it flowers throughout the cold season and the early part of summer. In north India, the flowering is during February to March. But in Bangalore, the main flowering season is between October and December, though sporadic flowering is seen in July also. The capsules point upwards like fingers. The tree is good for planting along avenues and in groups in large parks or especially in undulated landscape. It is not suitable for small compounds. The young flower buds are hollow and contain a clear liquid. Children use this as water-squirts, by squeezing. The tree likes a rich and well-drained soil. Propagation is from seeds, root-suckers, and sometimes cuttings. *Spathodea* can grow at altitudes up to 1,200 m.

Spathodea nilotica; Family: *Bignoniaceae*; Origin: Central Tropical Africa.

A much smaller tree than the above but similar in appearance. The leaflets are ovate-oblong, whereas in the above these are ovate-lanceolate. The flowers are almost like the above species but a little bright.

Sterculia colorata: See *Erythropsis colorata*.

Stereospermum suaveolens (Syn. *Tecoma suaveolens*); Family: *Bignoniaceae*; Origin: Asia (including India) and Africa; Vernacular Names: Tamil & Kannada – *Padiri*; Bengali – *Parul*; Hindi – *Paral.*

A large and spreading tree, popular in south India, generally planted in temples. The leaves are pinnate, the leaflets 12 cm long and broadly elliptic. The scented pale or dark purple flowers are bell-shaped and are borne in large, loose, terminal panicles. The flowers are used for *puja*. Propagation is from seeds, cuttings, and by air layering.

Tabebuia avalandi; Family: *Bignoniaceae*; Origin: Brazil.

An umbrella-shaped, small-sized (5-7 m tall) tree of unsurpassable beauty when in bloom during November to January in Bangalore. The plant has a good shape and looks ornamental. The mauve flowers are borne terminally in round bunches. It is very suitable for planting as a specimen tree or for avenues, especially in cities and towns in view of its dwarf nature. This beautiful tree must find a place in the garden. This is susceptible to frost, especially when young.

Tabebuia heterophylla (Syn. *T. triphylla, Bignonia leucoxylon, Tecoma leucoxylon*); Family: *Bignoniaceae*; Origin: West Indies.

A tree with digitate leaves. The leaflets are oblong lanceolate. The flowers are rose-pink or near-white, suffused pink, borne in loose few-flowered panicles.

Tabebuia rosea (*T. pentaphylla, Tecoma rosea*); Family: *Bignoniaceae*; Origin: Mexico, Central America, Guatemala.

A small-to-medium-sized tree. The leaves digitate, simple in young plants, leaflets 5, ovate to oblong and acuminate. The flowers are pale mauve and are borne in large terminal clusters. A good tree for avenues.

Tabebuia spectabilis (Syn. *Bignonia spectabilis, Tecoma spectabilis*); Family: *Bignoniaceae*; Origin: Columbia.

A small (up to 6 m tall) tree with ornamental, digitate foliage. The leaflets are ovate-oblong. The plant sheds leaves during March to April, or even earlier in Bangalore, when the plant is covered with bright yellow funnel-shaped flowers, 7·cm across, in dense, short, terminal panicles. The plant looks most beautiful at this time, but only for 10-15 days. A good tree for parks and home gardens.

Tecoma argentea; Family: *Bignoniaceae*; Origin: South America.

A most striking tree when in flowering. A small (5-8 m tall) tree with thick, narrow, lanceolate leaves. The plant bears dense, terminal clusters of bright yellow, funnel-shaped flowers in a leafless condition. The tree in full flowering looks most beautiful. It is good for planting along avenues and in parks and home gardens. The tree at young stage is likely to be damaged by frost. A popular avenue tree in Bangalore.

Tecomella undulata (Syn. *Tecoma undulata, Bignonia undulata*); Family: *Bignoniaceae*; Origin: India, Arabia; Common Name: Wavy-leaved Tecoma; Vernacular Names: Hindi – *Rugtrora*; Punjabi – *Lahura, Luar, Regdau, Rohira, Roir.*

A large shrub to a small tree. It has drooping branches and greyish-green foliage. The leaves have wavy edges. The orange-yellow flowers are borne in bunches at the apex of lateral branches. The tree looks beautiful when in full bloom from March to April. It is a drought-tolerant plant and suitable for landscaping and avenue planting in dry regions. It is propagated from seeds and cuttings.

Thespesia populnea; Family: *Malvaceae*; Origin: Tropical Asia (including India) and Africa, Pacific Islands; Common Names: Portia, Oil-nut, Mahoe, *Bhendi* Tree; Vernacular Names: Hindi - *Bhendi, Pipal*; Bengali - *Parash, Porash*; Gujarati - *Bendi, Parasapiplo*; Marathi - *Bendi*; Tamil - *Kallal, Pursa*; Telugu - *Galgaiavi*.

A medium-sized, 10-13 m tall, quick-growing, evergreen tree. The leaves are poplar-like and form into a head of foliage, providing enough shade. The large bell-shaped flowers are yellow in colour with reddish blotches at the base. The old flowers turn rosy-violet. The flowers resemble those of the vegetable *bhendi* (okra, lady's finger) or cotton. Easily propagated from seeds or large cuttings. The tree flowers almost throughout the year, the peak time being the cold season, and hence very suitable as an avenue tree. Also provides sufficient shade. The plant is salt-resistant and hardy.

Wrightia coccinea; Family: *Apocynaceae*; Origin: India.

It is a small tree (4-5 m), with elliptic or elliptic-lanceolate leaves 7-12 cm long. The tree bears numerous red flowers in the hot season when it looks beautiful. The plants are raised from seeds.

Wrightia tinctoria; Family: *Apocynaceae*; Origin: India.

A tree similar to the above. Its wood is ivory-white and is used in the manufacture of toys. The tree bears star-shaped, jasmine-like, scented, white flowers, 1-2 cm across, during March to May.

Chapter XII

Ornamental Bulbous Plants

The word "bulb" in commercial horticulture means much more than the strict botanical interpretation. Besides bulbs, this includes corms, rhizomes, tubers, fascicled roots, and pips (as in lily-of-the-valley, *Convallaria majalis*) also. In this chapter the culture of some commonly-grown bulbous plants, as the term means horticulturally, is discussed. The bulbous plants are valued for their flowers and foliage (e.g., *Caladium*).

Outdoor Culture of Bulbs

The primary and the most important prerequisite for growing bulbs is to have a perfect drainage. If good drainage is assured, it is not very difficult to grow bulbs in any type of soil. From the foregoing sentence one should not conclude that left to themselves bulbs will start flowering on their own. But compared to many other flowering plants, it is much easier to grow bulbs provided the basic cultural norms are followed. For example, before planting bulbs in a heavy clay soil it will be necessary to add leaf-mould or coarse sand or both to improve the drainage.

Sites for bulbs: The bulbs can be planted in ground at many places. There is an advantage of growing bulbs in the natural way as these can be left in the ground for several years to flower there without lifting, dividing, and replanting them, provided sufficient spacing is allowed initially. But only hardy bulbs such as football lily, hardy species of *Amaryllis*, *Zephyranthes*, *Crocus*, *Iris* (on the hills), *Oxalis* species, some lilies (e.g., Tiger lily) on the hills, tuberose, and others should be grown in this manner.

Bulbs are commonly grown in borders, including herbaceous borders, without lifting them frequently. It is best to arrange bulbs in borders in an informal manner putting them in groups to create a most natural effect. How many bulbs in a group have to be planted will depend upon the length and the breadth of the border. A big group of 30 or 40 small flowering bulbs in a large border is quite effective.

Planting of bulbs in formal or informal beds is a common practice for plants such as canna, *Hyacinthus* (in hills), *Amaryllis*, gladiolus, *Crocus*, *Iris* (in hills), dahlia, and others. It should be borne in mind that while planting in beds only one colour of the same cultivar in one bed should be accommodated to ensure blooming at one time, harmony in appearance and uniform height.

The scope of planting bulbs in different places in the open is almost unlimited. In the rock garden one can include *Crocus*, *Zephyranthes*, *Oxalis*, different species of *Narcissus* in the hills, and some species of this during the winter in north India.

It is a good idea to grow bulbs in the shrubbery as a ground cover and for beauty. In the plains of India bulbs such as football lily (*Haemanthus multiflorus*), *Zephyranthes*, different species of *Amaryllis* or *Hippeastrum* and *Crocus* can be utilized for this purpose. In the hills many bulbs such as *Lilium*, *Anemone nemorosa*, and *Ornithogalum umbellatum* are grown in a woodland garden or a shrubbery. It is also possible to grow some bulbs under spreading trees such as *Scilla campanulata*, *S. sibirica*, and *Leucojum muscari* in the hills. In the plains of India, one may try such bulbs as *Zephyranthes* and football lily.

When naturalized in lawns, bulbs present a glorious picture. But the choice of bulbs is limited: *Cooperanthes* and *Zephyranthes* are suitable for the plains and Narcissi and daffodils for the hills. Not all grasses are suitable for bulbs. For example, *Imperata* (Miscanthus) *arundinacea* tends to pierce the bulbs with the new-growing shoots. But *doob* (*Cynodon dactylon*) is a suitable grass for growing bulbs. Generally, the grasses in the meadow, orchard, cemetery or woodland garden which can be left uncut till the bulbs complete flowering and the leaves die off are suitable for such planting.

An exclusive garden may also be raised for bulbs in the kitchen garden or some other convenient places for cut flowers and propagation. Bulbs can also be grown near architectural features (sundial, statue, etc.) of a garden. Bulbs such as tuberous begonias, achimenes (both for hills), *Oxalis* and *Zephyranthes* are suitable for hanging baskets.

Planting season: There is no fixed time of planting as this will vary according to the type of bulb and the climate of the place. In India, having varied climatic conditions, the time of planting will vary widely, from place to place. For example, gladiolus can be planted almost throughout the year under the climatic conditions of Bangalore, though the best planting seasons are June and October-November, whereas it is planted in the middle of September or by mid-October in Delhi. The planting season for different bulbs will be given for each type when these are dealt individually later in this chapter.

In the hills in India the bulbs should be planted in February when the chances of frost are over. But it is also a practice to plant early in the autumn, so that the bulbs get well established before the winter to flower in the spring.

Planting: The ground should be thoroughly prepared and well-rotted cow or horse manure is added at the rate of a handful per bulb or 2-4 kg per square metre area. Bonemeal at the rate of 80-100 gm per square metre should also be added. The newly-planted bulbs should not come in contact with fresh manure.

The depth of planting will vary according to the type of bulb, and to some extent on the nature of soil also. For example, in a sandy soil if a tall bulb such as gladiolus is planted at shallow depth it is likely to be toppled at the time of flowering with even a moderate wind pressure. As a general rule, the bulbs are planted at about twice or thrice their own depth. The uniformity in depth of planting should be maintained for the same kind of bulb or a deeply planted bulb will flower later than a shallow planted one even if both are planted on the same day. The base of the bulb should rest firmly at the bottom of the hole made and no airspace should be left below (Fig. XII. 1).

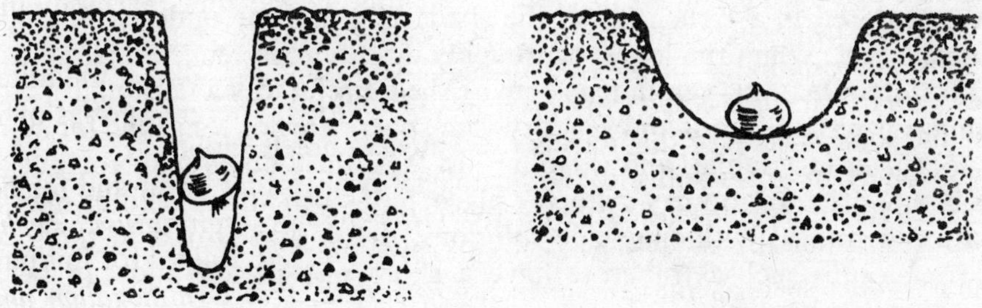

Fig. XII.1

A. Incorrect way of planting a bulb B. Correct way of planting a bulb

In the bulbous plants, which form a long neck-like *Crinum* the bulbs should be planted with the neck remaining above the ground. It is a good practice to sprinkle a little sand at the base where the bulb rests, especially for the choicest *Liliums*.

Dormant bulbs should never be planted as they may rot in the ground. Before planting, the bulbs should be soaked in a 0.1 to 0.2 per cent fungicide solution such as Captan, Benlate, or Brassicol for prevention against some bulb-borne diseases. Before planting, the soil should be damped sufficiently and the bulbs intended for planting kept in shade for a few days prior to planting. The bulbs in rectangular or square beds are arranged in parallel lines, staggering the bulbs in the alternate lines.

Cultural hints: Watering should be very sparse till the bulbs sprout. Once sprouted, many of the bulbous plants such as gladiolus and *Liliums* will need staking when these are about 15 cm in height. The bulbs benefit from liquid manure feeding as suggested for gladiolus at fortnightly intervals till the flower buds expand. If sufficient organic manure in the form of well-rotten cowdung or leaf-mould has been added there will be hardly any requirement for chemical fertilizers. However, it has been observed that many bulbs respond well to the application of fertilizers. Nitrogen at the rate of 5-15 g, phosphate (P_2O_5) 20-40 g and Potash (K_2o) 20-40 g per square metre are found to be beneficial. All the fertilizers should be applied a few days before planting. Half of the nitrogen dose may be applied as basal dose and the other half before the appearance of flower buds. Watering has to be done carefully and the quantity will depend upon the type of bulb. It is suggested by some experts to remove the anthers with pollens. If the flowers are not fertilized they will last longer. But this practice is feasible only on a small scale.

Lifting and storing: It is important to treat the bulbs carefully after flowering. A bulb has three phases in its life history, namely, the resting phase, the growing period, and the flowering season. After flowering most bulbs enter the rest period at the end of which they start growth and flower. But in certain bulbs such as football lily or certain amaryllis, the bulbs straightway produce flowers after the rest period and then develop foliage. In gladiolus, tuberose, etc., first the foliage develops followed by flowering. After flowering is over the foliage in most bulbs starts to die. The bulb should not be lifted till the foliage turns yellow. It should be borne in mind that food manufactured by the foliage of this year will have its impact on the capacity of the bulb to produce good flowers next season. In some bulbs such as gladiolus and football lily the leaves fall off naturally, while in others they have to be cut off only when turned yellow. The bulbs are lifted carefully and the leaves are cut off leaving 2.5 cm attached to the bulbs. Similarly, roots are also cut off. In such bulbs where the foliage have not turned yellow they are not disturbed till turned yellow and are stored in a trench under the shade of a tree or under a dry airy shelter with the roots intact. When the leaves thoroughly dry they are removed along with roots, cleaned, and stored in a well-ventilated room or shed, preferably a dark place.

Both damp storing conditions and very dry atmosphere are not congenial for storing bulbs especially for gladiolus, dahlia, and caladium. In fact, their bulbs may rot due to fungal attack in a damp place. In too dry a storing place bulbs will shrivel. Bulbs may be dusted with B.H.C. and treated with fungicide solutions, dried, and stored in trays containing dry sand. Some bulbs such as tuberose have a short rest period and can be replanted almost immediately after lifting.

It is not necessary to lift all bulbs every year as some do very well when left on their own in the ground for 3-4 years. *Amaryllis belladonna*, *Crinum*, football lily, and many *Lilium* species are generally not lifted from the ground every year, whereas gladiolus is generally lifted every year. Similarly, hyacinths and tulips on the hills do much better when lifted annually. The bulbs which are left in the ground are not watered till the new growth starts.

Some bulbs such as *Alpinia*, *Crocus*, *Crinum*, *Hippeastrum*, and tuberose do not get completely devoid of foliage and look quite ornamental even without flowers.

Bulbs in Pots and Bowls

Growing of bulbs in pots and bowls has its own charm and glamour as they can be used for indoor or outdoor decoration. The pot is filled with a good and porous compost consisting of equal parts of loam, leaf-mould and well-rotted cowdung along with a little sand leaving a gap of about 2-3 cm from the brim. The bulb is planted with the crown just above the compost. After planting, the soil is soaked thoroughly and the pots are placed on board or slate to prevent worms getting in. To preserve moisture the pots are sometimes surrounded and covered with 10-12 cm thick fibre or ash till the bulbs make a growth of about 2-3 cm. After growth starts watering will also start liberally but too much saturation is not good. Keeping the pots in dark encourages better root growth. After the bulbs start growth they are gradually removed to sun and should never be exposed all of a sudden to too much light.

The size of pots varies according to the type of bulb and similarly the number of bulbs to be planted per pot will also depend upon the size of the bulb as well as the pot. For example, only one tuber of dahlia is planted in a 25 cm pot, whereas at least 3 gladiolus corms can be accommodated in such a pot.

Almost any bulb can be grown in pots. But before planting each pot should be washed thoroughly, especially the used ones, as they may be contaminated with diseases.

Bulbs such as daffodils, hyacinths, *Freesia*, *Narcissus*, *Ornithogalum*, Anemones, *Scilla*, etc., are suitable for growing in bowls in a temperate climate, whereas *Zephyranthes*, *Cooperanthes*, and *Crocus* are suitable for the tropical plains of India.

Bulbs are grown in glazed bowls of 12 cm depth containing some fibre as growing media. Since coconut fibre is the easily available material in India, it can be used for growing bulbs in bowls. The growing medium is prepared by mixing three parts of shredded coconut fibre, one part of coarse leaf-mould, and half part each of small grit and small pieces of broken charcoal. To this about one-fourth part of coarse sand should be added. This medium is then soaked nicely in water, squeezed, and then filled in the bowl and pressed gently. The bowl is filled up to one-half its depth for larger bulbs such as hyacinth and to three-quarters for smaller bulbs such as *Scilla* and *Fritillaria*. The tips of bulbs should remain above the level of the fibre. The bowls are watered just enough to keep the fibre moist and any extra water is drained off by tilting the bowls to one side. The bowls are kept in a dark airy place till the crowns grow to a few centimetres. At this time these are gradually exposed to light till the leaves are dark green when they can be placed on the window sill or elsewhere. Any sudden exposure to light at the initial stage will prove disastrous. The bowl should be turned occasionally or the plants will become lopsided as they tend to grow towards light. Some people are of the opinion that no manure should be used, but we find no reason why a weak clear liquid manure should not be used after the flower spikes appear.

After flowering is over, the bulbs are taken out and planted in the ground as these will not flower again in the bowl due to depletion of stored food. Every year fresh bulbs are needed for bowl culture.

Some Important Bulbous Plants

Achimenes longiflora (Syn. *A. kleei, A. jaureguia*); Common Names: Achimenes, Magic Flower, Widow's Tears, Nut Orchids; Family: *Gesneriaceae* (also spelt *Gesneraceae*); Origin: Guatemala, Mexico.

It is rather a dwarf (15-30 cm tall) plant with somewhat tubular petunia-like flowers of violet-blue and whitish beneath. There are many horticultural types with a wide range of flower colours. The peak flowering season is during the rains, but continues to flower during June to November. Achimenes do better at medium-to-high elevations than in the plains.

These are suitable for growing in hanging baskets, bowls and pots. Three tubers are planted in a small-sized pot. The propagation is by tuber or terminal stem cuttings raised in sand during the rains. Adequate drainage, plenty of moisture, and partial shade are a few criteria in the successful cultivation of this bulb.

In *A. grandiflora*, the flower is crimson or violet-purple and somewhat smaller than *A. longiflora*.

Acidanthera bicolor; Common Names: Scented Gladiolus, Peacock Orchid; Family: *Iridaceae*; Origin: Abyssinia.

A tall (90 cm) plant resembling gladiolus and bearing white flowers with star-shaped crimson-maroon blotch in the throat, which are sweet-scented and prized as cut flowers.

The corms are small and planting starts from October in the plains for flowering during December to March. In the hills, planting is done during March-April for flowering during the autumn.

A well-drained, richly manured (preferably cowdung) soil and sunny sheltered location are the criteria for the successful cultivation of this plant. The plant is suitable for rockeries, pots and herbaceous borders.

There is a variety *murielae* which is stronger in growth.

Agapanthus umbellatus (Syn. *A. africanus*); Common Names: Blue African Lily, Lily of the Nile; Family: *Liliaceae*; Origin: Africa.

A bulb with long (50 cm) graceful evergreen leaves suitable for cultivation between 1,000 and 2,000 m altitude. The pale blue or blue flowers are funnel-shaped, 5 cm long, and are borne in umbels of 10-30 flowers. A white flowering type is also sometimes seen.

This is suitable for growing in tubs or pots, near water pools, or in beds and borders. Planting is done is spring for flowering during July to September. The plants are lifted once in 3-4 years. An open situation, rich soil, and plenty of water are good for its cultivation.

Allium; Family; *Liliaceae*; Origin: Northern Hemisphere (Temperate zones).

Some species of the onion group bear attractive flowers for growing in rock gardens, shrubberies, borders, for naturalizing and for cut blooms.

The bulbs are planted in light soils, 5-7 cm deep and 7-10 cm apart. They do well only at medium-to-high elevations where these are planted during May to October, depending upon the variety. But most of the types bloom between May and June.

(a) *A. albopilosum* (Syn. *A. christophii*): 61 cm tall, star-shaped lilac flowers.
(b) *A. azureum* (Syn. *A. caeruleum*): 30-60 cm tall, sky-blue flowers.
(c) *A. moly:* 15-30 cm tall, yellow flowers.
(d) *A. neapolitanum:* 20-30 cm tall, flowers white with coloured stamen.
(e) *A. ostrowskianum:* 20 cm tall, flowers rose coloured, star-shaped.

In all the species, the flowers are mostly borne in many-flowered umbels on long stems.

Amaryllis belladonna; Common Name: Belladonna Lily; Family: *Amaryllidaceae*; Origin: South Africa.

A very common bulb for both the plains and the hills. The plant is 60-90 cm tall having long strap-shaped leaves. The flowers are rose-coloured, large, trumpet-shaped and are borne in umbels of 2-4 flowers at the end of a long, hollow but stout stem.

The planting time is September-October in the plains and March to April on the hills. Propagation is by bulbs which may be left in the ground for 3-4 years or lifted annually.

This is an attractive flower for cut blooms, beds, borders and pots. The bulb is hardy and blooms in early summer in the plains and from winter to summer on the hills. There are several botanical varieties mainly differing in flower colour.

Anemone coronaria; Common Names: Anemones are generally known as Windflower Anemone, but this particular species is known as Poppy-flowered Anemone; Family: *Ranunculaceae*; Origin: South Europe to Central Asia.

There are two types of anemone flowers, one group resembling chrysanthemum, while the other, poppy. It is a dwarf (15-25 cm) species with lobed and irregularly cut foliage. The solitary large flowers (single or semi-double) may be of crimson, scarlet, purple, pink, blue, mauve, or white colours.

Anemones do best on the high hills but can be grown in the northern parts of the country having a long cool winter. The planting time is October in the northern plains and autumn or spring on the hills.

The plants are suitable for growing in rock gardens, pots and borders and are raised from seeds, division, and cuttings, taken from rootstock. The seeds are rubbed with sand before sowing and the tubers are soaked in water for 24-36 hours before planting. The plants look best when planted in clumps of at least a dozen or more. A well-drained, richly-manured soil and a sunny or semi-shaded situation are best for its cultivation. The tubers are lifted after flowering.

A species, *A. japonica*, with rosy purple or carmine flowers is also grown.

Arisaema; Common Names: Cobra or Snake Lilies, Green Dragons; Family: *Araceae*; Origin: Mostly Asiatic, but some are of tropical African origin.

Hardy tuberous rooted (rarely rhizomatous) plants which have curious looks. Generally do well on the hills and are planted in the spring for flowering during the rains. The flowering spathe looks like the hood of a cobra.

(a) *A. flavum:* Has two pedately divided leaves per plant. The leaves have a 30-40 cm long stalk. The spathe is greenish yellow with purple on the inside.

(b) *A. nepenthoides:* The spathe is whitish inside and yellow outside. The spadix is whitish.

(c) *A. speciosum:* Commonly known as 'Snake lily', it blooms at higher altitudes only during March. The plant has only a single leaf and the green stalk is sometimes specked with purple or brown. The spathe is striped with purple. The colour of the spadix is dark brown or purple with a pinkish yellow tail.

Begonia; Common Names: Begonia, Elephant's Ear, Beef-steak Geranium; Family: *Begoniaceae*; Origin: Subtropical and tropical countries except Australia.

Begonias can be classified into four groups: (a) Bulbous or almost so, (b) Tuberous, (c) Rhyzomatous, and (d) Fibrous-rooted. Of the four, the last three are more common. Though fibrous-rooted, Begonia is not considered as a bulbous plant; it will be included in this chapter for the sake of convenience.

(a) *Tuberous-rooted Begonias*

These are exceedingly beautiful plants which can be grown with success only on the high hills under glass, on the window or in a sunny veranda. Flowers are very beautiful resembling roses, but the texture is more fine and in some modern florist begonia the flowers are larger than roses. The margin of the petals may be plain, fringed or fimbriated. The flowers are generally double but some may be single especially if raised from seeds. The flower colour has a wide range from purest white to purple, pink, yellow, primrose, orange shades, scarlet, and crimson. The plants are generally 45 cm tall.

Tuberous-rooted begonias are generally grown in pots but can also be grown in hanging baskets (especially the pendulous ones), bowls, and beds in a shaded border.

These are generally raised from imported tubers. These should not be exposed to direct sunlight. In high rainfall areas the plants should generally be grown under glass. The tubers are first planted in small pots and then transferred to a 25-cm pot containing a mixture of equal parts of soil and leaf-mould with a good sprinkling of old building material. On the surface of the soil broken crocks should be placed so that the leaves do not touch the soil. Watering is done sparingly at the beginning but increased when growth starts vigorously. Liquid cowdung manure (weak solution) may be applied once a fortnight during vigorous growing period till flowering starts. The tuberous-rooted begonias cannot stand the exposure of their roots which are delicate and surface-rooted. The flowering is during August to October. The planting starts in February on the hills and continues till March to get a continuous blooming period. Watering is stopped after flowering is over. The leaves die off soon and the tubers are lifted and stored in sand or dry fibre in a cool place.

These can be raised from seeds also which are very fine in texture and are mixed with fine sand before sowing (1 seed : 10 sands). The seed pans are covered with glass sheets and watering is done by placing the pan in a larger pan containing water. The water enters the pan through the drainage hole and moistens the compost.

Tuberous-rooted begonia flowers in Bangalore, where it is planted in April-May to get flowers in September-October.

Some of the important species are:

(i) *B. bicolor:* The flowers are light-rose in colour and the leaves are spotted white.

(ii) *B. davisii:* The inflorescence is an umbel bearing six flowers of bright red colour.

(iii) *B. froebelii:* Flowers large, scarlet-coloured; var. *nana* is ideal for conservatories on the hills.

There are numerous beautiful cultivars.

(b) *The Semi-tuberous or Bulbous Group*

B. socotrana and Gloire de Sceaux are good examples of this group. These are difficult to cultivate, especially the former, bearing bright rose flowers in the winter in temperate countries but will flower in May-June in the high hills in India. The compost should have more leaf-mould compared to the fibrous-rooted begonias.

(c) *The Fibrous-rooted Begonias*

B. haageana, B. metallica, B. nitida (Syn. *B. minor, B. obliqua*), and *B. semperflorens* var. *gigantea* and others belong to this group. The Gloire de Lorraine group also belongs to this category. The begonias belonging to this group generally grow well in the plains of India, provided these are grown in humid, cool, and semi-shaded situations.

Most of the species can be multiplied by seeds or healthy stem cuttings, while the Lorraine group multiplies best through tip cuttings. The Semperflorens begonias bear beautiful small clusters of shell-like flowers in shades of light to deep pink, red, salmon, white and orange. Some of the species in this group are also liked for their attractive foliage. The plants of this group can be cultivated in open beds between November and February. During the summer these should be taken out and planted in pots and transferred to shade. A compost consisting of two parts loam, one part leaf-mould, one part well-rotted manure, and one part sand is recommended for these plants.

(d) *The Rex Begonias*

These begonias are cultivated mainly for their large, handsome and ornamental foliage. Most of the varieties and cultivars in this group have been derived from the species *B. rex*.

Rex begonia can be grown in the plains in a humid, shaded conservatory with care. To provide proper drainage in the compost recommended generally for other begonias, some broken pieces of charcoal and brick should be added.

Propagation is from seeds, leaf cuttings, or by division. For large-scale multiplication propagation from leaf cuttings is best. A well-matured but not old leaf is taken and the main nerves under the surface are cut at several places and this is pegged or weighted down on the surface of a properly drained propagation bed. Roots will develop at each cut and a tiny foliage will develop at these places. A triangular leaf-cutting with a portion of the main petiole attached to it can be inserted with the petiole end downwards in the propagating bed. A single leaf with the petiole inserted in propagating bed also roots easily. The young plants thus obtained are transferred to small pots containing sifted well-drained soil. Watering has to be done carefully and both over- and under-watering have to be avoided. The grown-up plants are not allowed to flower and seed as these may weaken the plants.

Belamcanda chinensis; Common Names: Blackberry Lily, Leopard Flower; Family: *Iridaceae*; Origin: China.

A rhizomatous plant, about 60-90 cm high, with iris-like leaves. Very commonly grown all over India. The flowers are 45 cm across and orange in colour with crimson spots here and there and are borne in long stems. There are hybrids of different colours including yellow, red, and apricot and some are unspotted.

Propagation is from seeds or by division. Planting is done during February-March in the plains and October-November on the hills for flowering during August-September.

The plants are suitable for pots, shrubberies and borders, and are planted in clumps of 3-4. The spikes are also cut for table decoration.

Caladium; Family: *Araceae*; Origin: Tropical America.

A popularly grown plant, valued for its highly ornamental foliage. The leaves are broadly arrow-shaped, membranous in texture and the size varies from a few to 30 cm across. It is almost impossible to describe the colour of the leaves which may vary from white to purple, bronze, pink, and deep crimson and blotched, splashed, and spotted with conspicuous colours.

The caladiums are deciduous and are in their best during the rains in the plains of India. The plants should be grown in well-drained compost consisting of equal proportions of garden loam, leaf-mould, well-rotted manure and sand. Some pieces of charcoal are added to keep the compost sweet. From practical experience it is known that addition of a little lime improves the colour of the foliage. The plant thrives best under moist semi-shaded conditions. Under too much shade the colour of the leaves will fail to develop properly. The fully grown plants need proper staking.

Propagation is from seeds or the tuberous rhizomes. The rhizome is sprouted in moist sand first before planting. After the leaves die off, the tubers are lifted and stored in a cool dry place.

The plants can be grown in beds, pots, and hanging baskets. The potted plants can be used for display in conservatories, interior decoration, and in the veranda.

Canna species and hybrids; Common Name: Canna; Family: *Cannaceae*; Origin: Tropical America, Asia.

The modern cannas of horticulture are hybrids derived from crosses between different species and hybrids. The height of modern hybrid plants varies between 45 and 180 cm.

There is a very dwarf hybrid measuring about 30 cm in height. The leaves resemble those of banana and may be green, bronzy-green, or bronze coloured and there are cultivars having streaked variegated foliage. Cannas are hardy and beautiful plants bearing flowers in different colours such as white, cream, yellow, orange, pink, scarlet, deep red, and various shades of these colours.

This is one of the best plants for growing in beds, especially in the midst of lawns. Canna beds should preferably be set against the background of a hedge. Generally, each bed is allotted for one colour, but a combination of different coloured cultivars of the same growth habit and flowering season can also be accommodated in one bed.

Propagation: Cannas are generally propaged by the division of rhizomes. The best season for planting the rhizomes is just before the rains, i.e., during June-July. For raising new hybrids, propagation by seed becomes essential.

Cultural hints: A sunny situation and rich soil are required for the cultivation of canna. The beds should be prepared to a depth of 60 cm. A layer of 10-12 cm well-rotten manure, preferably stable manure, is spread on the surface and worked with a spade. The rhizomes are planted 7-10 cm deep in the soil. The spacing from row to row and plant to plant in the bed will vary from 60 cm to 1 m depending upon whether the cultivar is dwarf, medium, or tall. Within about 6-7 weeks of planting, flowers will appear, but are removed to encourage vegetative growth. The stalks which have finished flowering should be cut down to ground level to have continuity in flowering. The plants are fed with liquid manure, prepared by decomposing oil-cake, during flowering time. An application of a handful of bonemeal per plant at the time of planting is recommended. Cannas generally flower in two flushes, the first during September-December and the second between February and April. The beds should be mulched with leaf-mould or manure after the first flush of flowering. The beds need digging at least once a month. Heavy watering at longer intervals are preferred to frequent light waterings. The cannas are replanted every year. The rhizomes are lifted in May with a portion of 15 cm stem attached and stored in a shady place before planting in June or July.

Cannas can also be cultivated in pots of 30 cm containing a compost of stable or cowdung manure and garden soil in the proportion of 2 : 1. The plants need repotting every 6 to 9 months.

Cannas are susceptible to frost damage. The beds should be flooded the previous evening when danger of frost becomes imminent. Leaves damaged by frost should be removed.

Clivia miniata (Syn. *Imantophyllum miniatum*); Common Name: Kaffir Lily; Family: *Amaryllidaceae*; Origin: Natal, Africa.

An evergreen flowering bulbous plant suitable for the hills only. These are planted in the sun or in semi-shaded places in February-March. The orange-yellow or scarlet coloured, funnel-shaped flowers are borne in umbels. This is a useful houseplant.

Convallaria majalis; Common Name: Lily-of-the-Valley; Family: *Liliaceae;* Origin: Asia and Europe.

A dwarf, 15 cm tall, plant with creeping rhizomes and the bell-shaped flowers are very popular for their fine fragrance. The flowers are generally white, but a pink-flowered cultivar is also available. A rare variety named *prolificans* has double flowers. This grows best only on the hills but can be tried for one or two seasons in the northern plains during the winter. The creeping rootstock is planted horizontally about 3 cm below the surface. Feeding with liquid manure is helpful. The plant flowers in May-June on the hills.

Cooperanthes; Family: *Amaryllidaceae*; Origin: Calcutta (India).

This genus comprises hybrids between species of *Cooperia* and *Zephyranthes*. It was first raised by Percy Lancaster in Calcutta in 1900 and subsequently further developed by his son S. Percy Lancaster. The hybrids are intermediate between the parents but more floriferous. The plants are very showy when in flowering and can be grown in pots and in clumps. Cultivation is similar to that of *Zephyranthes*. Individual flowers last only one or two days.

The best known hybrid is *C.* × 'Alipore Beauty' (*Cooperia oberwettii* × *Zephyranthes robusta*), with flowers of soft lilac rose on white within and deeper without.

Crinum; Common Names: Crinum, Cape Lily, St. John's Lily; Family: *Amaryllidaceae*; Origin: Tropics, especially Asia and Africa.

These are tall growing plants with strap-shaped leaves. The plants are grown in borders, pots, and near pools. The plants like swampy or marshy and semi-shady places. Propagation is from bulbs. The flowering is profuse during the rains. The plants can remain in pots or borders for several years. It is a hardy plant needing very little care. The spacing allowed is 30-40 cm if not replanted.

(a) *C. asiaticum:* Bears white flowers in umbels of 20 flowers. As the name implies, it is of Asiatic origin.

(b) *C. bulbispermum* (Syn. *C. longifolium*, *C. capense*): It is a plant of South African origin and has white flowers flushed red on the outer side which are borne in a stout stalk in a cluster of 8-12 blooms in a scape.

(c) *C. moorei* (Syn. *C. mackenii*, *C. makoyanum*, *C. ornatum*): An African species. The flowers are white but flushed red and are borne in 8-10 flowered umbels.

(d) *C* × *powellii:* A hybrid with reddish flowers and greenish tubes. The umbel is about 8-flowered.

Crocus sativus; Common Names: Crocus, Saffron Crocus, *Kesar*; Family: *Iridaceae*; Origin: Asia Minor (Italy to Kurdistan).

In India, crocus can be grown well only on the hills. The common Saffron of commerce cultivated popularly in Kashmir is referred to by some authors as *C. sativus kashmiriana*. This is a winter-flowering species of dwarf habit and grass-like foliage which bears small (2.5-5 cm across), rounded, fragrant flowers on long slender stalks. The flowering is during September to December. There are three types of crocus: one group flowering in the autumn, the second in the winter, and the third in the spring. There are a large number of species and large Dutch Crocus hybrids. Some other important autumn-flowering species are *C. longiflorus* (flowers bright lilac with orange throat), *C. medius* (lilac-blue), and *C. speciosus* (bright lilac with purple veins). The winter and the spring flowering species are *C. aureus* (Syn. *C. maesicus* or *maesiacus*) or Dutch Yellow Crocus (orange-yellow), *C. biflorus* or Scotch Crocus which has flowers from white to lilac and combinations of these basic colours, *C. sieberi* (lavender-blue), and *C. stellaris* (orange-yellow with black lines on the outside), which is probably a hybrid.

Crocuses are grown in rockeries, beds, pots, in small clumps under trees, shrubberies, and lawns.

Corms are planted quite close (8-12 cm) and the time of planting is July-August for the autumn flowering and the autumn for the spring flowering. The plants can be grown in sunny or semi-shady places. To overcome damage of corms by rats after planting, these are dipped in water for a minute or two and rolled over red lead powder before planting. The corms are not lifted from the ground after flowering but to avoid overcrowding these are divided once in five years or so.

Cyclamen; Common Names: Cyclamen, Sowbroad; Family: *Primulaceae*; Origin: Mediterranean region.

Cyclamens are herbaceous low-growing plants with tubers or corms. The cyclamens can be divided into two classes – the hardy outdoor classes or the greenshouse types. Cyclamens can be grown with success only on the high hills of India. Under great care this can be grown in the north but the plants last only one season.

The most commonly grown species is the *C. persicum* var. *giganteum*, in which the flowers are very large and the colours may be white, pink, purple, red, and various other colours. The reflexed petals give the flower the appearance of a butterfly. In many cultivars the petals are fringed, frilled, and margined. Some cultivars have fragrance also.

Cyclamens are generally cultivated in pots and some hardy types are grown in beds under large trees. Propagation is from seeds but corms or tubers are more convenient. The seeds are sown in well-drained seed pans in sifted compost. The seed pan is kept just moist after sowing and is covered with a glass pane and kept in dark until the seeds germinate. The sowing season is August to September. The seedlings are pricked when sufficiently large to handle and then transplanted in 25 cm pots. The seedlings take 15-18 months to flower. The seedlings are normally true to form. The corms are planted in October in the hills or during August for indoor display during winter to early summer. The corms or tubers may be cut into small pieces, each containing an eye. The corms are never allowed to remain buried more than half covered. A good potting compost consisting of 2 parts leaf-mould, 1 part soil, 1 part sand is needed for its growth. The plants like humidity and this can be maintained by frequent syringing of the foliage. But over-watering may cause rotting of the corms. A single 2-4 cm diameter corm may produce about a dozen or more flowers, but it is not very rare for a corm to bloom 30-40 flowers at a time in the high hills. After the flowering is over the corms should be kept watered till the leaves wither away. Then these are lifted, allowed to rest, and repotted for the next season. The plants need partial shade, a moist condition and a temperature ranging from 10° to 15°C for producing best quality flowers.

Some more important species flowering during the summer and the autumn are *C. africanum* (rose pink), *C. europaeum* (carmine), and *C. neapolitanum* (rose and white).

The following species flower during the winter and the spring: *C.* × *atkinsii* (rose pink or white lined and spotted red), *C. coum* (carmine with deep carmine spot), and *C. repandum* (bright carmine).

Dahlia: See Chapter XX.

Eucharis grandiflora (Syn. *E. amazonica*); Common Name: Star of Bethelhem, Amazon

Lily; Family: *Amaryllidaceae*; Origin: Colombia.

The popular name, Star of Bethelhem, is also applied to *Ornithogalum*. The plants are dwarf (20-40 cm) with long (30-45 cm) ovate-lanceolate foliage. About 6-7 sweet-scented white flowers are borne in an umbel over a 30 cm long scape. The flowers are smaller in size than *Crinum* and the flowering season is the winter.

The plant is suitable for pot culture, indoor decoration, borders, and cutting.

The plants like semi-shaded positions and warm humid conditions. So it thrives better in the plains than on the hills. The plants are raised from bulbs which are planted during September or October. Frequent repotting is not needed.

Freesia; Family: *Iridaceae*; Origin: South Africa.

The present-day large-flowered cultivars of freesias are nothing but hybrids between the different species such as *F. armstrongii*, *F. leichtlinii*, *F. refracta* and others. Many of the hybrid forms are included as *F. hybrida*. The freesias have tubular and funnel-shaped flowers in various colours such as white, yellow, orange, pink, blue, purple, rose, and carmine.

Freesias are suitable for pots, shallow pans, beds, and are also used for indoor decoration and for cutting. The flowering raceme is cut just when the first flower opens and kept in vase. The other buds will open in succession.

Freesias are propagated from seeds or corms, but it is best to propagate from corms to get the true form. The seedlings take about two years to come to flowering. The planting of corms in the plains is staggered from September to October to have an extended flowering period from February to March. Corms are planted in 15-20 cm pots containing a compost consisting of 2 parts good garden loam, 1 part leaf-mould, and 1 part well-decomposed cowdung manure. In heavy soils sand is also added. On the hills the planting is done during August to November and removed under glass to protect from frost. The corms are watered sparingly and kept in shade till the time of sprouting. The plants respond well to liquid manuring. The plants should be staked. After flowering is over, irrigation is stopped gradually and when the foliage dies off, the pots are removed in shade for replanting in the next season.

Fritillaria imperialis; Common Name: Crown Imperial; Family: *Liliaceae*; Origin: Asia, Europe, and North America.

Tall (90-100 cm) plants with narrow, whorled foliage. They grow well only on the hills. The flowers are brick-red, coppery-red, yellow, and orange, but have an obnoxious foxy odour.

Fritillaria is grown in beds, borders, rock gardens and for naturalization. The flower is suitable for cutting.

Propagation is through bulbs which are planted between August and November. The requirements for successful cultivation are partial shade, rich and moist soil. The bulbs are not lifted for several years.

Gladiolus: See Chapter XX.

Gloriosa superba; Common Name: Climbing Lily, Glory Lily; Family: *Liliaceae*; **Origin**: Asia (including India) and Africa.

A native of India and is found growing in the wild in many parts of India. It is a creeping

plant which climbs with the tendrils. The long-lasting flowers open primrose-yellow and later turn to orange-red or dark red. The petals are long, twisted or wavy, and heavily corrugated. It is suitable for growing in pots or in ground. The flowers last long in vase.

The tubers are planted horizontally in March-April and the flowering season extends from July to September. The plant needs a light soil and plenty of watering. The plant is susceptible to frost.

G. rothschildiana having crimson and yellow large flowers are also commonly cultivated. *G. carsonii* bears many flowers in a cyme and are maroon or dark chocolate with yellow towards the centre.

Gloxinia: See *Sinningia speciosa*.

Haemanthus multiflorus (Syn. *H. kalbreyeri, H. tenuiflorus*); Common Names: Football Lily, Blood Flower, Red Cape Lily; Family: *Amaryllidaceae*; Origin: Tropical Africa.

A relatively tall (45-50 cm) plant with large oblong leaves. The inflorescence is a football-like globular head which is actually a many-flowered dense umbel, blood-red in colour. The flowers are borne in a 30 to 90 cm long scape.

Football lily can also be grown in pots, besides shrubberies and borders. The bulbs are planted in February-March. The flowers precede the foliage. During resting period in the winter these are not watered. A light, well-drained soil is best suited for its culture. The bulbs may be left in the ground for several years.

Hedychium; Common Names: Garland Flower, Ginger Lily, Butterfly Lily; Family: *Zingiberaceae*; Origin: South East Asia.

A herbaceous perennial with rhizomatous stems and 1-2 m tall. The white, scarlet, or yellow flowers are borne in large showy spikes during the rains. Some species bear fragrant flowers.

The propagation is by division of rhizomes. They look nice when grown in clumps. Semi-shady positions and plenty of moisture help *Hedychiums* to grow properly.

(a) *H. coccineum:* Flowers are scarlet.

(b) *H. coronarium:* Plants are 1 to 2 m tall. Flowers are snow-white and fragrant. Flowering is during August to September in the plains and May in the hills. The flowers are produced in long spikes which look beautiful. The species is best suited for the lower elevations.

(c) *H. flavum:* Bears yellow or orange flowers in dense spikes, which are strongly perfumed.

(d) *H. gardnerianum:* Lemon-yellow flowers are borne in long spikes, and are perfumed.

All species except (b) do well in the hills only.

Hemerocallis fulva; Common Name: Day-Lily; Family: *Liliaceae*; Origin: Asia and Europe.

A hardy perennial with grass-like leaves. The roots are fleshy and tuberous. The plants are tall (60-90 cm).

H. fulva has yellow-orange flowers, while in *H. flava* it is clear lemon-yellow, in *H. aurantiaca* it is orange with a purple flush. There are cultivars having double flowers but singles are more common. The flowers are borne in long and branched stems, which are large and trumpet-shaped.

The day-lily is suitable for growing against the background of a herbaceous border or in front of a shrubbery and near pools and streams. This can also be grown in large tubs.

The plants are multiplied by division during March and can also be propagated from seeds. The plants prefer a rich soil, a sunny situation, and an abundance of moisture. Day-lily flowers more freely at medium elevations than in the plains. The flowering season is between April and June.

Hippeastrum; Common Name: Amaryllis; Family: *Amaryllidaceae*; Origin: South America.

A number of plants which are now constituents of this genus used to be included in *Amaryllis* formerly. The *Hippeastrums* with their large beautiful flowers in colours of white, blood-red, and crimson bloom during the summer months. The modern hybrids, especially the Dutch hybrids, have an array of flower colours including beautiful striping and mottling. The large trumpet-shaped flowers are borne in umbels of 2-4 flowers on a stout scape 25-50 cm long.

Hippeastrum is grown in beds, borders, and pots and is an important flower for cutting. *Hippeastrum* can also be used for edging paths and walks.

Propagation is from seeds or offsets (bulblets). Seeds of good strains generally breed true to the type. The offset should be removed only when it comes off easily. The bulbs are planted during September-October or preferably during December-January when they are dormant. In the hills, planting should start from October and may continue up to March-April. The plants flower during March-April in the plains and in the winter or summer in the hills depending upon the time of planting. In Bangalore, flowering may start even earlier. Once planted, the bulbs need hardly any special attention and would continue to flower every summer. The repotting or replanting is usually done at intervals of 4-5 years. But it is always beneficial to remove the top few centimetres of soil and replace this with a good compost when the growth starts. Feeding with liquid manure once a week is also beneficial during the period of active growth and flowering. *Hippeastrum* goes to rest in the winter in places having a cool temperature. But in places having a mild climate such as Bangalore the plants remain evergreen. There is a veiw that pot-bound *Hippeastrum* flowers better and hence people adhering to this view advise not to use pots more than 5 cm in diameter than that of the bulb planted.

Hippeastrums thrive well in a sunny or semi-shaded position, especially in places like Delhi (otherwise the flowers may not last long in the hot sun), a well-manured soil, and plenty of moisture.

Diseases and pests: For pests and diseasse see Chapter XXVI.

Some important species of Hippeastrum are listed below:

(a) *H. aulicum:* Flowers are large and rich crimson. It is known as 'Lily-of-the-Palace'.

(b) *H. equestre:* Commonly called 'Barbadose Lily'. Its flowers are bright red with green at the base.

(c) *H. reginae:* Commonly called the 'Mexican Lily'. It bears bright red flowers with a large greenish-white star in the throat.

(d) *H. reticulatum:* Flowers are soft pink and white with darker veins. Some authors describe the flower colour as bright mauve or purple-red with crimson veins. It may be possible that the flower colour varies with climate.

Hosta (Syn. *Funkia*); Common Name: Plantain Lily; Family: *Liliaceae*; Origin: China and Japan.

The plants are perennial on the hills. The tufted-looking plants flower during the rainy season in the plains of India. The foliage is also ornamental.

The plant is extremely suitable for planting in a wild garden. They are also planted in shrubberies, on slopes of land, in rockeries and pots. They produce flowers even in shade, and like moisture.

Propagation is by the division of rhizomes. Once planted, the plants are not disturbed unnecessarily. The most common species is *Hosta plantaginea*. The flowers are trumpet-shaped, white and fragrant. The variety *grandiflora* (Syn. *Funkia grandiflora*) has larger flowers than the above.

Hyacinthus orientalis; Common Name: Common Hyacinth; Family: *Liliaceae*; Origin: Syria, Lebanon, Turkey, Palestine, and Asia Minor.

This species is the original stock of the present-day beautiful hybrid hyacinths. This species produces slender flower stems on which racemes of small blue florets are borne. But the present-day hybrids have stout stems with racemes of 30-40 florets of bell-shape. The colour of the hybrid flowers varies from pure white to yellow, purple, mauve, dark and light blue, pink, and red.

Hyacinths flower well only on the hills, but plants grown from imported bulbs or those obtained from the hills flower satisfactorily in Delhi and other parts in the north. Hyacinths are excellent for growing in beds, pots, and rockeries. These are also grown indoors in bowls, pots, shallow pans, and special hyacinth glasses. This is also suitable for window gardens. The flowering is during February-March in the plains and May to June in the hills. Before planting in September-October in the plains and February in the hills the bulbs are sprouted by keeping them in a cool, dark place in soil or peat for about 8-10 weeks.

For multiplication 1.5 cm deep incission is given to the mother bulb across the base and planted. Small bulblets will be formed at the cut end. The first flower from the bulbs meant for propagation should be removed. Propagation by this method is successful only in the hills. Seeds can be sown in a sandy soil for producing new plants. At the end of the season, the bulbs should be lifted, dried, and stored for planting in the next season.

Hymenocallis littoralis (Syn. *H. adnata*); Common Name: Spider Lily; Family: *Amaryllidaceae*; Origin: South America.

The plants are medium to tall (30-60 cm) and have strap-shaped leaves and white fragrant flowers which resemble spiders.

The plant can be grown in clumps near pools or along paths and in pots. A good soil, proper drainage, and a sunny or semi-shaded situation are all that is required for its cultivation. Propagation is by offsets. The bulbs should be rested after flowering.

Iris; Common Names: Iris and Flag; Family: *Iridaceae*: Origin: Southeast Asia and Southern Europe.

There are numerous species of Iris and most of them thrive best in the hills. However, some of these, e.g., *I. fimbriata* (Syn. *I. japonica*) and *I. germanica*, can be grown in semi-shade in places such as Delhi, Punjab, and U.P. The irises have two forms of growth. One group has fleshy rhizomes or roots and the other has bulbs. These are again subdivided into 10 or 11 groups. The largest and the most popular group is the bearded irises. These are again divided into tall bearded irises (*I. germanica*) and the dwarf bearded irises which have been developed from three species, namely, *I. chamaeiris* (Syn. *I. lutescens*), *L. flavescens*, and

I. pumila.

The tall bearded irises grow best in a sunny situation and a well-drained soil containing lime. These are propagated from seeds and by division. But seeds do not come true. The dwarf bearded irises grow about 15 to 25 cm high and are good for rock gardens.

The English Irises (*I. xiphioides*, Syn. *I. anglica*), the Spanish Irises (*I. xiphium*, Syn. *I. hispanica*) and the Dutch Irises (developed by crossing *I. tingitana* with an early flowering variation of *I. xiphium*) are bulbous plants and are available in shades of white, blue, and yellow. The culture is similar to other bulbous plants. The following Himalayan irises are of rhizomatous type: *I. nepalensis* have pale lilac flowers and *I. kashmiriana* and *I. kamaonensis* (Syn. *I. duthiei*) have white and purple mottled lilac flowers. Only the Japanese irises do not like lime in the soil. The English irises like moisture and a heavy soil, while the other bulbous irises prefer a cool and dry weather.

Kniphofia (Syn. *Tritoma*) *uvaria* (Syn. *K. aloides*); **Common Names: Red-Hot Poker, Torch Lily, Flame-flower; Family: *Liliaceae*; Origin: South Africa.**

The plant is tall (60-120 cm) and has clusters of reed-like foliage. The flowers are borne at the end of a leafless stem in densely-packed racemes about 15 cm long. The individual flowers are tubular and coral-red in colour. But the older flowers in the lower portion of the raceme first turn to orange and then greenish yellow.

Torch-lily thrives best in the hills but can be grown in Delhi for a season or two after which it perishes. Propagation is from seeds or rhizomes. In the hills, it flowers during the rains. A fertile light soil and full or partial sunlight are the conditions needed for its culture. These plants should not be disturbed for several years. It needs protection from frost.

Lilium: species and hybrids; Common Name: Lily; Family: *Liliaceae*; Origin: Japan, China, Korea, U.S.A., and some other countries.

In India, most of the lilies thrive best on the hills and in places having a mild climate such as Bangalore. But the Easter Lily (*Lilium longiflorum*) can be grown with success in the plains in many parts of the country.

Besides various *Lilium* species and varieties there are several beautiful hybrid cultivars.

The cultural requirements for various *Lilium* species will vary. But it is possible to give a general guideline. Lilies need a well-drained fertile soil of at least 60-90 cm depth. A south-facing bed with a little slope will be ideal for most lilies. Some favour lime in the soil while others may be injured in a calcareous soil. As a general rule, lilies are planted two and half times the depth of the bulb. Similarly, bulbs with stem roots (roots arising from the stem above the bulb which act as support to the plant) are planted deeper than the bulb with basal roots only. In a heavier soil shallow planting is advocated. If at any time a lily throws up roots above the ground level these are to be covered immediately by earthing up. The planting distance will vary with the shape of the plant as well as the size of the bulb. Small lilies are planted about 15 cm apart, the medium-sized 20-25 cm apart, and the large ones 30-45 cm apart. All lilies are staked when in bloom and the dead flowers are removed, as soon as they dry up, without cutting the stems.

Lilies are propagated from seeds, bulbs, scales, bulblets, and bulbils. Lily seedlings are more vigorous in growth and more disease-free than plants grown vegetatively. But seedlings take a long time to flower, sometimes as long as 5-6 years. From mature bulbs a few scales may be removed and planted in a mixture of fine soil and sand for propagation. Each scale will give rise to a plant, but will take 2-3 years to come to flowering. Some *Lilium* species such as *L. bulbiferum*, *L. sargentiae*, and *L. tigrinum* produce in the upper leaf axils green or purple button-like growths called bulbils (see Fig. III.9, page 23) which are also used for propagation. Many of the stem-rooting lilies produce bulblets which are also used for propagation.

Lilies can be planted in various ways. Different types of lilies can be accommodated in a lily border just as a herbaceous border or planted in mixed herbaceous border and shrubbery. Some lilies are also quite suitable for planting in rock gardens (e.g., *L. auratum*, *L. bulbiferum croceum*), near water gardens (e.g., *L. pardalinum*) and pots (e.g., *L. hansonii*, *L. longiflorum*, *L. tigrinum*). Before planting lily bulbs, a little sand should be placed at the bottom and no fresh manure should come in contact with the bulbs. In the northern plains the lilies may be planted in September-October. In the hills, the early summer flowering types are planted during the autumn and those flowering during July-August may be planted in early spring. The lily bulbs dry out easily and it is not advisable to lift them for 3-4 years. Unnecessary disturbances may delay flowering and also affect the quality of the flowers.

The following *Lilium* species are cultivated more popularly:

(a) *L. amabile* (Korea): 45-120 cm, flowers grenadine red spotted black; stem-rooting.

(b) *L. auratum:* 'Golden-rayed lily' or 'Queen of lilies' of Japan, 1.5-2 m, flowers bowl-shaped, ivory-white or pale yellow with crimson-purple or brown bands and spots; stem-rooting.

(c) *L. brownii* (Korea): 90-120 cm, flowers large, trumpet-shaped, creamy-white within, thoroughly tinged purple or green outside.

(d) *L. bulbiferum croceum* (Syn. *L. aurantiacum*, *L. croceum*): Orange lily of Europe, 1-1.5 m, flowers cup-shaped and tangerine-orange; stem-rooting.

(e) *L. candidum* (East Mediterranean): 'Madonna Lily', 30-150 cm, flowers pearl-white.

(f) *L. chalcedonicum* (Syn. *L. rubrum*): Greek lily, 60-120 cm, flowers bright scarlet, olive-brown at base.

(g) *L. davidii* (Syn. *L. pseudo-tigrinum* of gardens, *L. thayerae*) (China): 90-120 cm, flowers reflexed, orange or scarlet with black spots. *L. d. willmottiae* is a popular variety.

(h) *L. hansonii* (Korea): 90-120 cm, orange-yellow spotted brown; stem-rooting.

(i) *L. henryi* (China): 180-240 cm, late flowering, flowers large, orange-yellow dotted reddish-brown; stem-rooting.

(j) *L. longiflorum* (Formosa): 'Easter lily', 60-90 cm, trumpet-shaped, pure white fragrant flowers, early flowering; stem-rooting. Raised easily from seeds also.

(k) *L. martagon* (Europe, Asia): 'Turks cap' or 'Martagon Lily', 90-150 cm, flowers recurved, dull purplish-red spotted black and purple, scent unpleasant.

(l) *L. nepalense* (Nepal): 60-90 cm, flowers funnel-shaped, recurved, fragrant, greenish-yellow with deep purple in throat; stem-rooting.

(m) *L. pardalinum* (California, U.S.A.): 'Panther' or 'Leopard Lily', 150-210 cm, reflexed flowers, red with yellow centre and maroon spots.

(n) *L. pyrenaicum* (Pyrenees Mountains of France and Spain): 60-90 cm, greenish yellow flowers with dark spots, odour unpleasant.

(o) *L. regale* (Syn. *L. myriophyllum*) (China): 90-180 cm, free-flowering, pure white trumpet lily with a golden yellow throat and the backs of the petals shaded wine-crimson, tips recurved, quickly raised from seed; stem-rooting.

(p) *L. sargentiae* (China): 120-150 cm, trumpet-shaped flowers somewhat similar to *L. regale* but suffused brown outside and more beautiful; stem-rooting produces bulbils in the stem.

(q) *L. speciosum* (Syn. *L. lancifolium*) (Japan): 30-120 cm, pure white spotted red flowers are bowl-shaped, fragrant, late flowering, stem-rooting. The Syn. *L. lancifolium* is claimed as erroneous by some authors.

(r) *L. tigrinum* (China, Korea, Japan): 'Tiger Lily', 90-120 cm, flowers reflexed, soft reddish-orange (fire-red) with numerous purplish spots; bulbils in the stem are black, stem-rooting.

Diseases and pests: See Chapter XXVI.

Narcissus; Common Names: Daffodil, Narcissus, *Nargis*; Family: *Amaryllidaceae*; Origin: Northern Hemisphere, Europe including British Isles, Asia including Japan, Canary Islands, North Africa.

To clear a misconception in the nomenclature it may be explained that while daffodil is the popular name, narcissus is the Latin name, but both are the same. But often the large trumpet types only are referred to as daffodils, while the short-cupped types are called narcissus. The plants are not very tall (45 to 65 cm); in some species it is only 7.5 cm. The foliage is narrow grass-like. The flower has two distinct portions, the central portion is referred to as corona which is trumpet-shaped or cup-shaped or almost flat, the outer surrounding part being composed of the petals or perianth segments. Narcissus has been divided into separate sections for the sake of convenience and the basis for this classification is in reference to the colour, size, and general shape (specifically the corona) of the flower. These are Trumpet, Double, Large cupped, Small cupped, Cyclamineus, Triandrus, Jonquilla, Tazetta, Poeticus, and Miscellaneous. Some of the above groups have their own hybrids. The general colours of the flowers are yellow, white, pink, or red and combinations of these colours. The flowers may be borne singly or in clusters of ten or more depending upon the sections to which the species or hybrids belong.

Narcissus can be planted in formal or informal beds especially along a water pool or a channel. Naturalization in grass is also very effective. Planting in groups and clumps are most effective, when put in borders or in lawns. These are grown indoors in pots and bowls. The flowers are cut for decoration. These should be placed in water immediately after cutting early in the morning.

A well-drained loamy soil is suitable for most narcissi, although the larger types prefer a deep and stiff soil. Well-rotted leaf-mould or F.Y.M. is best for narcissi growing. Bone-meal is added at the rate of 100 g per square metre area. Some plants can withstand partial shade, while some others actually grow well in shade.

The planting time is October in the hills and September or October in the northern plains. The flowering is during December to April in the hills and December to February in the plains.

After flowering is over water is withheld, but the stem is never cut when it is still green. Afterwards the bulbs are lifted, dried in the sun, and stored in a cool shady place for the next planting. Potted bulbs can be shifted along with the pots in the shade of a large tree. On the hills, the bulbs are lifted only once in 3-4 years for obtaining offsets for multiplication.

Indoor cultivation: Bulbs are cultured in pots or bowls for display indoors. The potting compost consists of one part each of good soil, leaf-mould, and sand and half part of broken charcoal. In a 15 cm pot about 6-8 bulbs can be accommodated. The bulbs are planted with their necks remaining above the surface of the compost. The pots with the bulbs are kept in a dark, cool but airy room till the bulbs sprout and attain a height of 5-7 cm. Then these are brought indoors and placed on the inner-side of a sunny window only at the time of the emergence of the flower buds. For culture in bowls see pages 230-231.

(a) *N. cyclamineus:* Flowers are drooping, and lemon-yellow; corona is orange-yellow and crenate at the mouth.

(b) *N. jonquilla:* 2-6 flowered, flowers are bright yellow and very fragrant. Corona is cup-shaped and bright yellow.

(c) *N. poeticus:* 'Pheasant's eye', 'Poet's narcissus'; flowers are solitary, pure white and fragrant. Corona is flat with a red-crisped edge.

(d) *N. pseudo-narcissus:* 'Trumpet Daffodil' or 'Common Daffodil'; flowers are small to rather large, scented, and drooping. Flower is light yellow and the corona bright yellow which is straight but cut irregularly at the mouth.

(e) *N. tazetta:* 'Polyanthus Narcissus'; 4-8 flowered, corona is cup-shaped and lemon-yellow; flower colour is variable, scented.

(f) *N. triandrus:* 'Angel's Tears'; flowers are white, drooping, and are borne in 6-flowered inflorescence, corona is cup-shaped.

Ornithogalum umbellatum; Common Name: 'Star of Bethelhem'; Family: *Liliaceae*; Origin: Europe including Britain and North Africa.

A bulbous plant best grown on the hills. The star-shaped flowers are white inside and green striped white outside. The flowers are borne at the end of a 15 cm scape in racemes of 12-20 flowers.

The bulb is suitable for planting in pots, borders, shrubberies, rockeries and for naturalization in lawns. Closer planting is advisable for natural look. It is valued as a cut flower.

The plant is propagated from offsets or bulblets. Planting is done during October in the northern plains or hills. In the northern plains, the flowering time is March-April, whereas on the hills it is April-June. A comparatively rich and well-drained soil and an open situation are ideal for its growth, but it can tolerate partial shade especially in the plains. At higher altitudes the plants may suffer from frost. Heavy watering is avoided until initial growth starts and watering is withheld after flowering. After flowering, bulbs are lifted and stored in cool and dry places. In *O. arabicum* the flowers are fragrant, white with dark centre and in *O. nutans* the flowers are white inside and greenish outside.

Oxalis; Common Name: Oxalis; Family: *Oxalidaceae.*

A genus comprising 800 species which are dwarf in nature, having trifoliate weed-like leaves. The tiny flowers have a variety of colours such as white, yellow, pink, and purple-red. A rich compost is not desirable for oxalis culture, the suitable compost being one part each of garden soil and leaf-mould and half part course sand. Oxalis grows best in semi-shaded locations.

Oxalis is suitable for growing in rockeries, shrubberies, pots, and hanging baskets under tall trees.

Propagation is mainly from bulbs but some species propagate well from seeds. Sowing of seeds or planting of bulbs is undertaken in October in the plains and in September on the hills. The seeds are sown in March-April.

(a) *Oxalis bowiei* (Syn. *O. purpurea*): A Mexican species with large rosy-pink flowers.

(b) *O. corymbosa* (Syn. *O. martiana*): A plant of South American and East Asian origin, it is suitable for hanging baskets. The bluish-violet flowers are borne in groups of 5-12.

(c) *O. deppei:* A Mexican weed; flowers are small and deep pink. The leaves are clover-like.

(d) *O. magellanica* (Syn. *O. lactea*)*:* Originated in Bolivia and South Australia. The flowers are white and suitable for stony places and walls.

(e) *O. rosea* (Syn. *O. floribunda*)*:* An annual which is propagated from seeds. The flowers are rose-coloured. Originated in Chile.

Polianthes tuberosa: See Chapter XX.

Ranunculus asiaticus; Common Names: 'Common Garden Ranunculus', 'Turban Buttercup', 'French Ranunculus'; Family: *Ranunculaceae*; Origin: South East Europe, Syria, Iran.

A dwarf (30-45 cm) plant with beautiful cut foliage. The flowers are double, yellow in colour and about 4 cm across. The modern garden cultivars have a wide range of flower colours such as white, yellow, orange, purple, and red. A new garden strain has Paeony-flowered blooms.

Propagation is from seeds and claw-like tubers. The claws are firmed downwards. Planting time is October in the northern plains and in the spring in the hills after the danger of frost is over.

It thrives best on the hills. The best flowering season is from July to September/October. The plant prefers a well-drained moist soil. In the plains, the tubers are lifted after the plants die and are stored in cool places. In the hills, unless the weather is severe, the tubers may be left in the ground.

Richardia: See *Zantedeschia.*

Sinningia speciosa: (Syn. *Gloxinia speciosa*); Common Name: 'Gloxinia'; Family: *Gesneriaceae*; Origin: Brazil.

A most ornamental plant which bears bell-shaped flowers in great abundance and has ornamental, oblong, obtuse, velvety-green foliage. The flowers may be white, pink, blue, violet, purple, scarlet, bright red, or blue with a white border or scarlet with white border. The petals in some cultivars are fringed and frilled. There are double flowering cultivars also.

Gloxinia is best suited for growing in pots in the porch, window gardens, indoors, and in a greenhouse.

Propagation is from seeds, tuberous rhizomes, and leaf and stem cuttings. The plants grow best only in the hill stations. In Bangalore, the planting is done towards the end of April/May to get flowering during August-September. The tubers can be planted in the northern plains in December-February or June which flower during August-September. In the hills, planting is done in the spring to flower during the summer. The plants like moist soil and sunshine, but not direct light. Drainage should be perfect. There are large flowering varieties which produce enormous blooms. Some F_1 hybrids such as 'Atlas' and 'Discoverer' are also well known.

Tulipa; Common Name: Tulip; Family: *Liliaceae*; Origin: Europe, Central and Western Asia, North Africa.

Tulip is an important flower in the western countries, and Holland is the chief producer. Tulip does well only on the hills and in Kashmir and Kulu valleys in India. Some odd specimens may flower in Delhi and around from imported bulbs or bulbs brought from the hills.

Tulips are low-growing plants with about four oppositely placed leaves of lanceolate, linear or broad and ovate shape. The flowers are generally borne singly but often up to four in a stalk, which are cup- or egg-shaped. The tulip flowers are self-coloured or a combination of two or more colours. All the natural colours from white to black can be found in modern tulips.

Tulips have various uses in the garden. These are grown in beds, borders, and for naturalization in grass, under trees and shrubs. Tulips are also grown in pots, bowls, and window boxes. Some dwarf growing species are suitable for rock gardens also. Tulip is a valuable cut flower.

Propagation is done from seeds, bulbs, and also bulblets. The bulbs are planted during November-December in the hills and October in Delhi. But if the winter is severe and there is danger of frost, the bulbs should be planted in February for flowering in April to June. The bulbs are planted about 7-10 cm deep. Tulips thrive best in a well-drained, rich soil and a sunny situation. But they do quite well under semi-shade where the flower colour lasts longer. Fresh manure should never come in contact with bulbs. Bonemeal can be added at the rate of 100 g per square metre area. To obtain better quality flowers, the bulbs should be lifted from the ground, after the leaves die off, every two or three years. About six bulbs are grown in a 20 cm pot.

The following species are suitable for growing in rock gardens:

(a) *T. biflora:* Pale yellow or white inside and tinged with green or red outside.
(b) *T. clusiana:* White inside and rosy-red outside with dark blotch at base.
(c) *T. kaufmanniana:* Creamy-yellow within, carmine outside.
(d) *T. linifolia:* Bright crimson with a bluish block blotch at base.
(e) *T. praestans:* Pale to dark red flowers.

The species *T. aitchisonii* and *T. stellata* are natives of the Himalayas. Besides, there are numerous cultivars some of which are double flowering.

Zantedeschia (Syn. *Richardia*) *aethiopica*; Common Name: Arum Lily, Calla Lily, Lily of the Nile; Family: *Araceae*; Origin: South Africa.

The species is also known as *Richardia africana.* The leaves are arrow-shaped, twice as long as broad and dark green in colour. The spathe is white with creamy at base and 7-25 cm long. The tip of the spathe is cuspidate. The leaves somewhat resemble arum.

The plants grow well in the Nilgiri hills as well as in the other hill stations. In the mild climate of Bangalore also this lily thrives well. But it is difficult to cultivate it in the plains.

A well-manured soil, full sunlight, and cool and moist conditions are suitable for its growth. It is propagated from rhizomes. The species *Z. elliottiana* (Syn. *Richardia elliottiana*) has bright yellow flowering spathes and *Z. rehmannii* (Syn. *Richardia rehmannii*) has pink spathe.

Zephyranthes; Common Names: Zephyr Lily, Fairy Lily; Family: *Amaryllidaceae*: Origin:

South America.

A very dwarf (15-20 cm) growing plant having grass-like foliage and crocus-like flowers borne on long slender stalks. A garden hybrid *Z. ajax* has pale yellow flowers. The other important species are *Z. candida* with white flowers, *Z. grandiflora* (Syn. *Z. carinata*, *Z. tsouii*) with large rose-pink flowers, and *Z. rosea* with rose-pink flowers.

Zephyranthes is frequently used as edging for paths and beds. When grown in mass in grass or under trees it gives very good effect. This is also grown in front of shrubberies, pots, and rock gardens.

The plants are propagated from bulbs or seeds. A good garden soil and a sunny or semi-shady position are required for this hardy plant. The bulbs are generally not lifted for 3-4 years.

Chapter XIII

Ornamental Foliage Plants

These are also often called greenhouse plants, but the latter name may become confusing under our climatic conditions. In the temperate greenhouse those plants are grown which are otherwise difficult to grow in the adverse climatic conditions outside. Therefore, many plants are included as greenhouse plants in the temperate countries which may not be a greenhouse plant in our country. In this chapter only those plants are included which need a cool, shady, and humid atmosphere to grow and the foliage is ornamental in nature. Some plants are also included which bear ornamental flowers, but only under cool and shady conditions in India. Ferns, palms, *Selaginellas*, etc., also belong to this group but these have been dealt with in separate chapters.

In most parts of India these foliage plants are grown in a greenhouse or lath-house with partial shade and lot of humidity. But in a climate as that of Bangalore many foliage plants such as croton, *Dieffenbachia*, etc., grow quite well in the open. The construction of greenhouses and lath-houses have been dealt with in Chapter XXIII. A greenhouse should be so situated that the foliage plants within receive only the eastern and north-western sun but not the afternoon sun. If needed, artificial shade should be given from the south and the western sides. The lath-house, often known as fern-house, as these are mostly used by amateur gardeners to grow ferns only, should never be constructed under a heavy-foliaged tree as the drips from the tree will damage the delicate foliage plants.

Cultivation and Care

Besides shade or partial shade the next important thing is humidity, especially in places having a long dry summer. It is advisable not to have a *pucca* (cemented) floor as this is not congenial for retaining moisture. If the whole floor is covered with boiler ash or cinder this will hold enough moisture and will not allow the floor to become muddy after each watering or rain. Pebbles of miniature size can also be used for preserving moisture and giving a firm floor. Flooring with kiln-burnt bricks is also very good, though somewhat costly. If the flooring is made in any of the ways suggested, watering can be freely done, if the plants are in pots, without making the floor muddy. But in a conservatory watering has to be done with a little more care as the plants are planted in the ground. In a conservatory, watering is done more freely if the soil is porous, but cautiously when the soil is heavy. The paths which are made of gravel, pebbles, cinder, or bricks are also given a dressing with water to increase the atmospheric humidity. It is also advised to syringe the plants with water early in the morning and also once more as the day advances, in places having a dry summer. This will not only provide additional moisture but will also free the foliage from dirt. Lily pools with fountains or running streams may also be constructed inside the lath-house which helps maintain a certain degree of humidity.

A wise greenhouse gardener should not place too many deciduous plants in a group in one place, especially in a conservatory (in a conservatory the plants are planted in the ground) as that place will create a gap during the winter when the leaves drop. In a conservatory it is a good idea to bury a few 25 cm pots here and there in the rockery, which can be utilized to put some plants in flower such as geraniums and African violets. When such plants complete flowering they are replaced with some other flowering plants. If no flowering plant is at hand the gaps can be filled by ornamental foliage plants. The greenhouse should be kept clean by removing dead and yellowing leaves.

The foliage plants are propagated from stem cuttings (croton, geranium), leaf cuttings (Begonia rex, African violet), stem node cuttings (*Aglaonema, Dieffenbachia*), by air layering (croton, *Dracaena*), division of suckers (*Anthuriums, Aspidistra, Dracaena*), division of rhizomes (*Anthurium, Alpinia*), and from corms (*Caladium*), tubers (*Alocasia, Colocasia*), crowns of fruits (*Ananas*), and seeds (*Asparagus plumosus*, geraniums, coleus).

The pests and diseases of foliage plants are almost similar to those of house plants.

Select List of Ornamental Foliage Plants

Achimenes: See page 231.

Aechmea (*Bromeliaceae*).

Plants similar to *Billbergia*. The leaves are ornamental with cross band. There are many beautiful species. Treatment is similar to that of *Billbergia*.

Aglaonema (*Araceae*).

Perennial ornamental plants belonging to the tropics. The leaves are green, blotched with greenish-grey or variegated. A compost consisting of garden loam, leaf-mould, broken charcoal, sand and old mortar are good for their growth. Syringing of the leaves with clear water everyday helps improve their appearance. Propagation is by division of the basal shoots and also from terminal or node cuttings of the stem.

(a) *A. commutatum:* The leaves are leathery, oblong-lanceolate, deep green in colour with markings of silver-grey. There are many beautiful horticultural cultivars.

(b) *A. costatum:* A very attractive dwarf species; the leaves are stiff, broadly ovate, dark-green with white spots all over and a band of white through the mid-rib.

(c) *A. hospitum* (Syn. *A. brevispathum*): The leaves are spear-shaped, leathery, glossy and ovate with creamy-white spots.

(d) *A. hospitum variegatum:* A richly marbled white variety of the above.

(e) *A. modestum* (Syn. *A. sinensis*): The somewhat pendent leaves are leathery, waxy-green, and ovate-acuminate in shape.

(f) *A. modestum variegatum:* A variety of the above with creamy variegation.

(g) *Aglaonema*, 'Parrot Jungle': A hardy cultivar with lanceolate leaves which are painted silver-green on deep green.

(h) *A. roebelenii* (Syn. *Schismatoglottis roebelenii*): A showy plant having leathery leaves which are largely variegated with silver against the greyish-green background of the basal colour.

(i) *A. rotundum:* A dwarf variety with leathery broad-ovate or rhombic leaf, dark metallic glossy green in colour, mid-rib and lateral veins pink; the reverse side is wine-red with pink veins.

(j) *A. treubii:* The leaves are leathery, narrow, bluish-green in colour, blotched with silver-grey.

Alocasia (*Araceae*).

The plants of this genus have beautiful showy foliage. Some hardy species have origins in India and others are mostly from Asia. The plants are usually rhizomatous or tuberous rooted. Most of the species have rest periods. They love a moist and shady situation. A very porous compost consisting of one part each of good soil, leaf-mould, cowdung or horse manure in case the soil is red earth, one part sand, half part mortar rubbish, and one-fourth part of broken charcoal is good for *Alocasias*. Propagation is from one-bud node cuttings of stem or tubers or rhizomes.

(a) *A.* × *amazonica* (*A. sanderiana* × *A. lowii grandis*): A beautiful bushy hybrid. The leaves are dark green, the veins are white and the margins scalloped.

(b) *A.* × *argyraea* (*A. longiloba* × *A. pucciana*): The leaves are large sagittate-peltate, leathery, silvery-shining and dull underneath.

(c) *A. cuprea* (Syn. *A. metallica*): A beautiful and delicate species. The leaf blade peltate and ovate with deeply depressed veins, bronzy-purple with metallic sheen above and deep violet underside.

(d) A. 'Hilo Beauty': A somewhat small-growing species, 75-90 cm tall, stalks bluish-black, leaves medium-sized, green, irregularly marked with blotches of light pale-yellow.

(e) *A. korthalsii* (Syn. *A. thibautiana*): The leaves are sagittate, olive greyish-green, many grey veinlets radiating from the greyish-white mid-rib, undersurface purple.

(f) *A. lindenii:* The leaves are long-pointed, cordate, green above with yellowish-white mid-rib and veins; the stalks white or greenish-white.

(g) *A. longiloba* (Syn. *A. gigantea, A. eminens*): The leaves are bluish-green, broadly arrow-shaped, margins and veins silver-grey, petioles greenish-white mottled purple.

(h) *A. lowii:* The leaves are cordate-sagittate, leathery, peltate, olive-green with thick white ribs, purple beneath. Var. *grandis* has broader leaves, which are deep metallic brownish-green with greyish-green veins. In var. *veitchii*, the leaves are smaller, main veins with wide white margins, smaller veins whitish, stalks green-striped, variegated rose and green.

(i) *A. macrorhiza* (also spelt *macrorrhiza*): 'Giant Elephant's Ear'. A very large plant up to 4.5 m or more, the leaves sagittate-ovate, large (60 cm), somewhat leathery, waxy-green, ribs prominent, margins wavy. In variety *variegata* the leaves are blotched and mottled with white.

(j) *A. portei* (Syn. *Schizocasia portei*): The leaves are deeply lobed, triangular, waxy and fresh green.

(k) *A. watsoniana:* The leaves are large, corrugated, bluish-green above with silver-white veins and margins and purplish beneath.

(l) *A. zebrina:* Arrow-shaped large green leaves, stalk slender with zebra-like grown bands.

Alpinia (*Zingiberaceae*).

The plants have ginger-like roots, effective for mass planting in swampy low grounds. *A. sanderae* is an attractive ornamental species. The somewhat 2-ranked leaves are pale green, edged, and obliquely banded from the centre to the margin with pure white. In *A. vittata* (Syn. *Renealmia ventricosa*), the leaves are variegated with stripes of creamy-white. In rich soil variegations fail to develop properly. Propagation is by division of clumps.

Amorphophallus (*Araceae*).

The common edible tuber-yielding plants are sometimes grown as ornamentals. Some of the species are *A. bulbifer*, *A. bulbifer marmoratus*, and *A. campanulatus*. It is called *Zimikand* in Hindi, *Ol* in Bengali, and *Suran* in Marathi.

Ananas (*Bromeliaceae*).

The foliage even in the orchard type of *Ananas* (pineapple) is quite ornamental. But there are certain ornamental types also.

Propagation is from the crowns of the fruits or suckers from the base of the plants or slips. Single leaf-cuttings from the crown with a portion of stem attached is very economical. Some beautiful species are *A. comosus*, *A. comosus porteanus*, and *A. comosus variegatus*.

Anthericum: See *Chlorophytum*.

Anthuriums (*Araceae*).

A beautiful group of tropical aroids. These are horticulturally divided into two sections: foliage types and flowering types. Though all *Anthuriums* bear flowers, in the flowering types the flowering spathe is large and very ornamental but the foliage is generally not so attractive as those in the foliage section. The flowering spathe is generally red, rose, cream, or white in colour. Recently, cultivars with double flowering spathes have been introduced.

Anthuriums thrive best in well-drained soil but like a warm, humid and shady situation. In pots, as the plants grow, the roots tend to get exposed which should not be allowed to happen and these should be covered with compost or moss or repotted after cutting back and starting on fresh-roots. Propagation is generally from seeds which of course do not breed true-to-type. Other propagation methods are node cuttings and pieces of rhizomes or division. A compost similar to that for *Alocasias* will suit *Anthuriums* also.

The flowering spathes of *Anthuriums* are really beautiful and stand on the plants as well as in vase for quite a long time. The spathes have a great export potential and they stand very well in transport.

The important flowering type species are *A. andraeanum* (also spelt *andreanum*), *A. andraeanum giganteum*, *A. andraeanum rubrum*, and *A. scherzerianum*. There are many beautiful hybrids also. The important foliage types are *A. clarinervium*, *A. cordatum*, *A. crystallinum*, *A. magnificum*, *A. veitchii*, and *A. warocqueanum*. There are many hybrids also.

Aphelandra (*Acanthaceae*).

These are beautiful ornamental foliage plants from South America. The base colour of the leaves is generally shiny green, while the veins are variously coloured prominently. The important species are *A. aurantiaca roezlii* (Syn. *A. roezlii*), *A. squarrosa*, *A. squarrosa* var. *louisae*. For cultur see page 132.

Aralia (*Araliaceae*).

Many of the Aralias, as this is commonly known to the amateur gardeners, actually belong to other genera such as *Dizygotheca*, *Polyscias*, *Fatsia*, *Panax*, *Pseudopanax*, etc., but are also synonymous. For convenience sake most of these will be dealt here. Many of these species are quite hardy and can be grown in the open where the climate is not very extreme. Most of the species are suitable pot plants, grown in conservatories and also used for indoor gardening. The delicate types do not thrive well in the plains of India. The plants are multiplied by stem cuttings, air layering, and occasionally from seeds. These ornamental foliage plants prefer to be grown in under-sized pots. It is necessary to protect them from hot sun. These are natives of North America, Asia, and Australia.

(a) *A. balfouriana* (Syn. *Polyscias balfouriana*, *Panax balfourii*): A leafy bushy shrub; leaflets usually 3, roundish or kidney shaped, coarsely-toothed.

(b) *A. balfouriana marginata:* Leaflets are variegated with irregular white borders.

(c) *A. balfouriana* 'Pennockii': An attractive cultivar with large, leathery, cordate-oval leaves, variegated and tinted with creamy-white or pale green.

(d) *A. elegantissima* (Syn. *Dizygotheca elegantissima*): An excellent plant for table decoration; the leaves are very graceful, palmately compound, leathery, leaflets 7-10, filiform (thread-like); the narrow segments are lobed, have a metallic sheen, and are red-brown in colour.

(e) *A. filicifolia* (Syn. *Polyscias filicifolia*, *Panax filicifolius*, and *P. filicifolium*): A beautiful plant with bipinnate fern-like leaves cut into narrow lobes, bright green in colour, mid-rib purplish.

(f) *A. guilfoylei* (Syn. *Polyscias guilfoylei*, *Nothopanax guilfoylei*): A tall shrub of 4.5 m; the leaves pinnate, about 40 cm long; leaflets often lobed, irregularly spine-toothed, margins white and often splashed grey. Useful as an ornamental hedge. Var. *victoriae* is smaller in size, the leaves much divided into small segments of variable sizes and margined white.

(g) *A. sieboldii:* See *Fatsia japonica*.

(h) *A. veitchii* (Syn. *Dizygotheca veitchii*): The leaves are digitate on slender and rather long stalks; leaflets about 11, thread-like, coppery green above, dark-red below, margins wavy.

See also *Polyscias* and *Pseudopanax*.

Asparagus (*Liliaceae*).

A group of beautiful foliage plants of which the following species are worth growing.

(a) *Asparagus africanas:* There is a little confusion about this species as in some latest books this has been mentioned as *A. cooperi*, whereas in other established books *A. cooperi* is a separate species.

The species is quite ornamental and resembles *A. plumosus*; phylloclades in fern-like sprays, branches thread-like but not as fine as in *A. plumosus*, leaves awl-shaped, spines red-brown, flowers white, sweet-scented, berries red. A native of Africa.

(b) *A. densiflorus* cv. 'Sprengeri' (Syn. *A. sprengeri*): True leaves are reduced to thorns, branchlets fluffy because of the soft green needle-like leaflets; flowers white, small and fragrant; berries red. A most ornamental species. A native of Natal (Africa).

(c) *A. plumosus* (Syn. *A. setaceus*): A beautiful species from South Africa, commonly called 'Asparagus Fern'. A climber with numerous horizontally spreading branches; phylloclades bristle-like, green, finely pointed, prickles sharp on thin wiry stems. There is a cultivar 'Pyramidalis' (Syn. *cupressoides*) which is of dense habit; in this the bristle-like cladodes and also the stems grow erect, resembling *Cupressus*. In var. *nanus*, the phylloclades are shorter and more numerous.

Aspidistra lurida (Syn. *A. elatior, A. punctata*) (*Liliaceae*).

A native of China, a popular old-fashioned indoor plant; the leaves basal, oblong-lanceolate, 40-50 cm long and quite broad and leathery. In var. *variegata* the leaves are ornamentally and alternatively striped green and white. It likes a rich soil and a humid atmosphere, but the variegated form fades in rich soil. Propagation is by division.

Beaucarnea: See pages 300-301.

Begonia.

The ornamental foliaged begonias have been dealt with in pages 234-235.

Bertolonia (*Melastomaceae*, also spelt *Melastomataceae*).

A group of tiny herbaceous plants having very ornamental foliage, requiring semi-shaded location and well-drained soil for their growth. Propagated from seeds or stem cuttings and also from leaves by cutting across the veins and placed flatly face upwards in sharp moist sand.

(a) *B. maculata* (Brazil): The leaves are cordate, ovate, densely hairy all over, light green in colour, margins reddish, lightly silver-green along the centre, undersurface deep red.

(b) *B. marmorata* (Syn. *Eriocnema marmoratum*) (Ecuador): Has very ornamental, vivid green, velvety leaves, parallel veins streaked white on the upper surface, lower surface uniformly rich purple.

(c) *Bertolonia* cv. 'Mosaica': It is a German cultivar, having velvety-green leaves with white longitudinal bands, iridescent in many colours.

Billbergia (*Bromeliaceae*).

Herbaceous epihytic plants of dwarf nature; leaves harsh, rigid, somewhat like those of pineapple and generally have cross bands; flowering in erect panicles. The suckers producing flowers usually die after flowering but new suckers arise from the base. The flowering bracts are usually colourful. The plants need good drainage and a shady location. These are usually grown in pots or like orchids on tree-trunks or in rockeries. A compost consisting of equal parts of garden loam, leaf-mould, manure, and sand plus good quantities of pieces of bricks and charcoal suits the plants. Propagation is from suckers. For quick multiplication, plants are also raised from seeds but the process is slow. Some important species are *B. decora, B. nutans, B. pyramidalis, B. rosea, B. vittata* (Syn. *B. leopoldii*), and *B. zebrina* (Syn. *Bromelia zebrina*).

Breynia: See *Phyllanthus nivosus roseopicta.*

Caladium: See page 235.

Calathea (*Marantaceae*).

It is difficult for a layman to distinguish between *Calatheas* and *Marantas* and hence in horticulture both are treated in the same way. Other related genera in the family are *Ctenanthe* and *Stromanthe*. These plants are somewhat delicate but not difficult to grow with care. The plants like plenty of moisture but dislike stagnation and hence a perfect drainage should be ensured. It is best to grow them in a compost consisting of equal parts of loam, sand, and leaf-mould, but the more delicate among them are best grown in sand and leaf-mould. They also need a moist atmosphere for which constant syringing with soft water is beneficial. Hard water containing minerals causes leaf injury. A moist atmosphere helps control the red-spider mite, the chief enemy. *Calatheas* should be repotted every year or the leaves become very crowded and do not grow properly. At the time of repotting care is taken not to damage the roots. The roots are cleaned with water to remove the soil and then the plants are divided, each portion having a few roots, by sharp cuts and planted separately. The operation is done generally during February, but can also be done during the rains in the plains.

Some of the important species of both *Calatheas* and *Marantas* are listed here. The plants belong to South America, some extending up to Central America.

Calathea insignis, *C. lindeniana*, *C. magnifica*, *C. makoyana*, *C. medio-picta*, *C. metallica*, *C. ornata* 'Sanderiana', *C. rotundifolia fasciata*, and *C. zebrina* (Syn. *Maranta zebrina*).

Maranta arundinaceae var. *variegata*, *M. bicolor*, *M. leuconeura* var. *kerchoveana*, *M. leuconeura* var. *massangeana*, *M. repens*, and *M. zebrina* (Syn. *Calathea zebrina*).

Callisia elegans Syn. *Setcreasea striata* (*Commelinaceae*).

A low-growing, vigorously branching, trailing plant. The leaves are fleshy, clasping, triangular, olive-green above with white stripes and dark purple underneath. A most ornamental plant.

Chlorophytum (Syn. *Anthericum*) *comosum picturatum* (*Liliaceae*).

An ornamental rosette type succulent; the leaves are 30 cm long, strap-shaped, green in colour with yellow central longitudinal bands; flowers appear in slender racemes which also give rise to new plantlets afterwards.

C. comosum variegatum.

Margins of leaves variegated white.

C. comosum vittatum (Syn. *C. elatum vittatum*).

Commonly called 'Spider plant' because small plantlets with roots, developing from flowering panicles, remain hanging and look like spiders. Plants are, more or less, similar to the above except that the recurving leaves are channelled and narrow-linear.

Codiaeum (*Euphorbiaceae*).

Many people do not know that *Codiaeum* is the botanical name of crotons and are more familiar with the latter name. The garden crotons differ in many ways from the true crotons and are much more attractive. Crotons are popular ornamental shrubs with colourful and attractively-shaped leaves, which are much more commonly grown than plants of similar type. The cultivation of crotons is also easy as the plants are comparatively hardy. Crotons can be used in the garden as ornamental hedges for screening, as specimen plants in the lawn, and as a member of the shrubbery. They are also widely grown in pots or tubs for outdoor as well as indoor decorations.

For obtaining new varieties crotons are raised from seeds. The male and the female flowers in crotons are borne in different clusters which are easily recognized. In the garden, crotons are multiplied by tip cuttings during the rainy season and one hundred per cent success may be obtained. Cuttings of old shoots also root but not so easily. Large healthy plants can be obtained straightway by air layering. While taking the cutting or air layering a bright shoot should be selected.

Crotons like a warm and moist climate. They cannot withstand frost; as a result it is very difficult to grow them in northern India which not only experiences frosty weather but has a dry and hot long summer, detrimental to the proper growth of these plants. In the hill stations also the plants can hardly be grown except in the warm greenhouse. To get the best colour for the foliage the plants should get enough of the cool morning sun and remain protected from the scorching sun late in the day. Unless the sun is really hot most of the types can be grown in the open also, though not at their best, as the plants can adjust to a great extent in new environments. In too much of shade the colour of the foliage does not develop properly. Since crotons are grown for their attractive foliage the aim should be to have bushy plants and to achieve this the main stem should be pinched when the plant is about 15 cm tall. The side shoots should also be pinched when these are 20 to 30 cm or more long and the procedure is repeated until a reasonably good bushy plant is obtained. As the old plants lose the bottom leaves they look leggy, when the plants should be cut back to about 15 cm from ground level during the rains, to obtain good bushy plants in about 6 months. The plants should be syringed regularly which not only improves the appearance of the foliage but also keeps in check thrips and red spider mites which are the two main enemies of crotons.

Crotons like a comparatively rich but well-drained compost consisting of two parts of good soil, two parts of cowdung manure (horse manure for red soils), one part leaf-mould, one part of sand, and half part of old mortar rubbish. Addition of bonemeal (50 to 100 g, depending upon plant size) keeps the plant in good shape for a long time. It is advisable to repot crotons once a year. Top dressing with equal quantities of good soil and cowdung or horse manure once in 3 or 6 months after scraping out a portion of the old soil is good for the growth of the plants.

In crotons the shape of the foliage is generally of three types and they are classified accordingly. These are: broad-leaved, narrow-leaved, and trilobed. In some cultivars, a portion of the leaf lamina, generally in the mid portion or at the tip, turns into a thread-like shape.

There are innumerable cultivars and most of these an amateur gardener cannot distinguish. Therefore, it is advised that while purchasing one should select plants which are attractive and appealing. Some of the cultivars developed at the Indian Institute of Horticultural Research are listed below:

'General Kumaramangalam', 'Mohan Kumaramangalam', 'Gnat'', ''Pokhran', 'S.B. Lal Singh', 'Sonar Bangla', 'Vijayanta', and 'Vikrant'.

Coleus (*Labiatae*).

A perennial (sometimes annual) herbaceous plant very commonly grown in the garden for its colourful foliage. The plants grow to a height of 30-50 cm. The plants are more commonly grown as pot plants, but are a great favourite for the window box garden and a few dwarf types are also suitable for carpet bedding.

Coleus propagates easily by tip cuttings of slips from the leaf axils. The rooted cuttings are potted singly in 15 cm pots containing a mixture of three parts good garden loam, one part each of leaf-mould, sand, and cowdung manure and horse manure. Plants are also raised from seeds, and seedling plants have better foliage colour than plants from cuttings. To obtain a bushy growth plants are stooped once or twice. As the plants grow these are transplanted in 20, 25, or even in 30 cm pots containing some compost if a large specimen bush is desired. Liquid horse or cowdung manure is applied once in ten days or once in a fortnight to obtain a rapid and healthy growth. It is generally observed that pot-bound plants develop better leaf colour. Flowers should be removed constantly. Older plants become lanky and are frequently attacked by mealy bugs and a white scale. Health cuttings from affected plants are taken and then these are destroyed by burning.

Most of the garden cultivars of *Coleus* are developed from *C. blumei*.

Colocasia (*Araceae*).

These are tuberous-rooted herbs which are somewhat similar to *Alocasias* requiring treatments as in the case of *Caladiums*. Some attractive species are *C. affinis jenningsii*, *C. antiquorum fontanesii*, *C. antiquorum illustiris*, *C. esculenta* (Syn. *C. antiquorum esculenta*, *Caladium esculentum*), and *C. fallax*.

Commelina benghalensis variegata (*Commelinaceae*).

A trailer with watery succulent stems, leaves pale green striped and edged white. A native of Tropical Asia and Africa.

C. communis aureo-striata.

The bright green leaves are banded and striped with cream.

Cordyline (*Liliaceae*).

Cordyline is closely related to *Dracaena* and differs from the latter only in the structure of the ovary. An amateur grower can distinguish between the two only after long acquaintance. Several of the species of *Cordyline* are commonly known as *Dracaena*. Some species previously included in *Dracaena* are now placed in another genus *Pleomele*.

Cordylines are evergreen shrubs or trees and are prized for their richly coloured foliage, including variegated forms. *Cordylines* as well as *Dracaenas* are excellent pot plants at their juvenile state some of which are also suitable for growing in borders and shrubberies under a mild climate, for shade gardens and greenhouses. At young stage of growth most of the species are widely used for table decoration.

The plants like a rich compost mixed with a little lime in the form of old mortar, which promotes colour development. The plants prefer pot-bound condition, plenty of diffused light, sufficient water supply, and proper drainage for a luxuriant growth. Syringing with clear water is beneficial for the plants. The plants should be top-dressed with a rich compost once in six months. The leggy leafless plants should be cut back or air-layered near the bottom to encourage new growth.

Cordylines and *Dracaenas* are propagated by division of suckers in species producing suckers or more frequently by air layering and from node or terminal cuttings of the stem. The old stems are cut into bits, each containing one node or joint and then placed horizontally in moist sand in seed pans with the bud pointing upwards. The sand is kept just moist and the cuttings soon develop roots and start growing. Terminal cuttings also root, but the plants are not as good as those from node cuttings.

Some of the important species of *Cordyline*, *Dracaena*, and *Pleomele* are mentioned below:

(A) *Cordyline.*

(a) *C. australis* (New Zealand): It has a dense cluster of leathery arching leaves, 30-90 cm long, 2.5-6 cm broad, very ornamental and suitable for indoor decoration at young stage. This also goes by the name of *Dracaena indivisa* in horticulture.

(b) *C. australis aureo-striata:* The leaves are marked with longitudinal creamy-yellow lines.

(c) *C. banksii* (New Zealand): It has strap-shaped recurving dark green leaves, with pale yellow mid-rib. Sometimes the veins are red or yellowish, measuring 90 cm long and 7-8 cm wide.

(d) *C. rubra* 'Bruantii' (Syn. *Dracaena bruantii*): It has coppery-red gracefully recurved foliage, leaves oblanceolate and densely arranged.

(e) *C. terminalis* (India, Malaysia): The 'Red Dracaena' of commerce; leaves 30-60 cm long, 5-10 cm wide, normally green but developed by cultivation into various shades of red. Some beautiful varieties and cultivars of this are:

(i) 'Firebrand': A compact rosette; leaves recurving, purplish-red with a satiny texture, young leaves mahogany-red, mid-rib crimson.

(ii) 'Madame Eugene Andre': Broad leaves green to coppery with red border, new foliage shocking pink.

(iii) 'Tricolor': A colourful rosette; leaves variegated cream, pink and red over a base colour of fresh green.

(B) *Dracaena.*

(a) *Dracaena deremensis* 'Bausei': It has sword-shaped leathery leaves having two broad white marginal bands separated by a narrow milky-green central stripe.

(b) *D. deremensis* 'Longii': The leaves are pointed, corrugated lengthwise, abruptly drooping, having a white central band.

(c) *D. fragrans lindenii* (Syn. *D. lindenii*): It has oblanceolate recurved leaves, deep green with yellowish green bands.

(d) *D. fragrans massangeana* (Syn. *D. massangeana*): The leaves are striped and banded light green and yellow along the centre.

(e) *D. fragrans victoriae* (Syn. *D. victoria*): A beautifully variegated species with graceful pendent leaves; streaks of silvery-grey at the centre with broad margins of cream to clear golden yellow.

(f) *D. godseffiana* (Congo): A small plant; leaves are small (7-10 cm long, 3-5 cm broad), firm, green with numerous bright-cream spots.

(g) *D. godseffiana* 'Florida Beauty': A seedling of 'Kelleri', covered almost entirely with creamy-white blotches.

(h) *D. godseffiana* 'Kelleri': A cultivar which is blotched and marbled more numerously with ivory-white.

(i) *D. goldieana* (South Nigeria): The most ornamental *Dracaena* requiring a moist warm climate for its growth. The cordate ovate, rather large (18-20 cm long, 10-12 cm broad) leaves are marked with cross bands of pale green against the deep green basal colour, maturing almost to white.

(j) *D. sanderiana* (Cameroons, Congo): The leaves are narrow lanceolate, elegantly twisted, deep green with broad white margins. There are several beautiful cultivars such as 'Borinquensis', 'Margaret Berkery', etc.

(C) *Pleomele*.

(a) *Pleomele* (*Dracaena*) *reflexa* (Africa, India): The leaves are leathery, densely clustering, short and narrow, glossy-green, wavy and reflexed. The stem is branching type. It is a good potted plant; in var. *angustifolia*, the long, narrow, strap-shaped leathery leaves are crowded around the stem and spreading.

(b) *P. reflexa variegata* 'Song of India' (India, Sri Lanka): It is a variegated form of the above, the leathery leaves are exquisitely margined by two broad bands of yellow or cream. A strikingly beautiful plant.

Cryptanthus: Plants similar to *Billbergia*.

The important species are:

(a) *Cryptanthus beuckeri* (*Bromeliaceae*): Epiphytic, similar to *Billbergia*, leaves in open rosette, ellipitic with slender point, marbled rich-green over pale green. A native of Brazil.

(b) *C. zonatus* (Syn. *Pholidophyllum zonatum*, *Tillandsia zonata*, *T. zebrina*): An attractive plant with oblong lanceolate leaves, 15-22 cm long, soft, leathery; margins wavy, densely spiny, and coloured with alternating bands of green and ashen on the upper surface. A native of Brazil.

(c) *C. zonatus zebrinus* (Syn. *C. zonatus fuscus*): A very attractive plant commonly known as 'Zebra Plant'. The long wavy leaves are bronze-purple with prominent silver cross-bands, resembling a zebra.

Ctenanthe (*Marantaceae*).

A Brazilian plant closely related to *Calathea*. Some important species are *C. compressa* (Syn. *C. pilosa*), *C. lubbersiana*, and *C. oppenheimiana* (Syn. *Stromanthe porteana oppenheimiana*, *Calathea oppenheimiana*).

Cuphea: See pages 141 and 142.

Curculigo recurvata (Syn. *C. capitulata*) (*Amaryllidaceae*).

A native of Java. Plants stemless; leaves long, lanceolate, palm-like and recurved; the small flowers in dense heads appear near the base, which are yellow in colour.

Cycas.

Some cycases are very good greenhouse and indoor plants. For details see Chapter XV.

Cyclanthus bipartitus (*Cyclanthaceae*).

The plants resemble palms, stemless; leaves appear in clusters, milky-juiced, large, quilted and forked at top; flowering spathe yellow, scented. Another important species is *C. cristatus*.

Cyperus alternifolius (*Cyperaceae*).

The 'Umbrella Plant' is a native of Madagascar. A perennial clustering bog plant with ribbed stems which are up to 1.2 m high, each stem is crowned with numerous grass-like green leaves in an umbrella-like fashion; in var. *gracilis*, stems are thin wiry, leaves narrower than type; in var. *variegatus*, stems and leaves are banded lengthwise with cream-white, sometimes wholly white.

C. papyrus (Syn. *Papyrus antiquorum*).

The 'Egyptian Paper Plant' is a good plant for the pools. The stems are 2.5 m tall and at the apex are borne brush-like umbels of pendent thread-like foliage.

Dasylirion glaucophyllum (Syn. *D. glaucum*) (*Liliaceae*).

A beautiful plant with hundreds of rosette of linear leaves, 90 cm long and 5-7 cm broad, slightly convex, margins having small hookspines, tips splitting.

Dianella tasmanica (*Liliaceae*).

The two-ranked leaves are broadly sword-shaped, up to 1.21 m long, turned downwards, margins spine-toothed, the small pale blue flowers are borne on long stalks, followed by most ornamental blue berries. In var. *variegata*, the leaves are striped yellow. A native of Tasmania and Australia.

Dieffenbachia (*Araceae*).

These tropical American plants are very handsome with their ornamental broad foliage and are cultivated widely in the garden. Unfortunately, the plants are poisonous, the acrid sap causing intense pain if it goes into the mouth. No part of this plant should be placed in the mouth.

The plants need shady and warm humid conditions. These are generally grown as potted plants and often for indoor decorations also. Frequent syringing with clean water is beneficial. The plants like a rich, well-drained compost mixed with little quantities of small broken bricks and charcoal. Propagation is from terminal and node cuttings. The procedure for striking node cuttings is the same as for *Pleomele*. *Dieffenbachias* should not be allowed to flower as the flowers weaken the plants. *Dieffenbachias* differ from *Aglaonemas* in floral characters and an amateur may find it difficult to distinguish between the two. Some important species are described below:

(a) *D.* × *bausei* (Syn. *D. picta bausei*): A hybrid between *D. picta* and *D. weirii*. The pointed leaves are beautifully coloured with yellowish-green and dark green with white and dark green spots at the edges.

(b) *D. bowmannii:* The leaves are very large, up to 75 cm long and 30 cm broad, the colour of the leaf is deep rich green, with lighter lateral parallel markings.

(c) *D. exotica* (Syn. *D. arvida*): A mutant possibly from *D. hoffmannii*. Very attractive. The leaves are ovate, matt-green and highly variegated creamy-white.

(d) *D. hoffmannii:* The leaves are deep satiny green, much marbled and spotted white including the petioles, mid-rib white.

(e) *D. picta* 'Rudolph Roehrs': A most attractive mutant; the leaves almost yellow throughout or chartreuse, with blotches of ivory-white; the mid-rib and margins dark green.

(f) *D. picta* 'Superba': An attractive mutant of *D. picta*, creamy-white variegation on the glossy green leaf lamina.

(g) 'Janet Weidner': A sport of the above but with more variegation and marbling.

Doryanthes (*Amaryllidaceae*).

Rosette of numerous succulent sword-shaped leaves up to 1.8 m long, inflorescence giant. The arching leaves resemble some *Dracaenas*. Flowering is delayed in our country. Useful for growing in greenhouses or shade gardens. Propagation is from suckers.

(a) *D. excelsa:* Inflorescence 4.5 m long with heavy heads 10 cm in diameter, crimson-red with paler inside flowers, bracts green. Commonly called 'Globe Spear-lily'.

(b) *D. palmeri:* Inflorescence 3-4.5 m long, spike pyramidal with 6 cm diameter, funnel-shaped scarlet-red flowers, bracts coloured. Commonly known as 'Spear-lily'. Both the above species are natives of Queensland.

Dracaena: See *Cordyline*.

Episcia (*Gesneriaceae*).

Herbaceous perennial plants from Tropical America having fleshy showy foliage and often bearing showy flowers. They are good for hanging baskets. They need a porous compost with plenty of sand, broken bricks, and pieces of charcoal. A shady, cool, moist atmosphere suit them best. Propagation is by division of clumps and from cuttings which are rooted in sand. There are many good species and cultivars.

Eranthemum: See page 144.

Eupatorium: See Chapter XXIII under Edgings.

Excoecaria bicolor.

This is a beautiful shrub and can also be treated as an inmate of a greenhouse. For details see page 145.

Fatsia japonica, Syn. *Aralia sieboldii* (*Araliaceae*).

An attractive foliage plant from Japan. The leaves are leathery, palmately lobed, shining green in colour, the broad lobes terminate in points and are toothed. The var. *variegata* is a mutant where the leaves are margined or variegated creamy-white especially towards the tip.

Ficus elastica (*Moraceae*).

This is commonly called 'India Rubber Plant'. A good houseplant but grows into a tree up to 30 m tall. The leaves are showy, large, leathery, oblong in shape and deep green in colour; the young leaves are enclosed in a rosy sheath. In var. *variegata* the leaves are variegated grey and the edges are creamy-yellow. Other important and beautiful cultivars are *F. elastica* 'Decora', in which the leaves are larger and broader with prominent lateral veins, at right angle with the ivory mid-rib which is reddish beneath, the growing point is covered with a red sheath. The variegated forms are *F. elastica* 'Decora Schrijvereana' and *F. elastica* 'Doescheri'. In *Ficus elastica* 'Rubra' the young foliage is maroon-coloured and the mid-rib is red. *Ficus parcellii* is a small shrub with wildly variegated and marbled leaves.

Among the creeping types the variegated forms of *Ficus pumila* (Syn. *F. repens*) var. *variegata* and *F. radicans* var. *variegata* are grown as greenhouse plants.

These plants need a semi-shaded humid atmosphere but in a climate as that of Bangalore they grow very well in the open also. Propagation is by air layering or from cuttings. The creeping types generally develop roots on the nodes.

Fittonia (*Acanthaceae*).

These are trailing herbs with beautiful foliage. The important species are:

(a) *Fittonia argyroneura* (Syn. *F. verschaffeltii argyroneura*): A dwarf, creeping herbaceous plant with bright-green oval-shaped leaves netted with white veins.

(b) *F. verschaffeltii*: A dwarf, trailing herb, with sturdy oval leaves, dark green in colour netted with deep red veins. In var. *pearcei*, the veins are light rose-pink. Fittonias require ample water, good drainage, and semi-shaded positions. A compost consisting of garden loam, leaf-mould and silver sand suits them most. The flowers should be removed as they appear. They propagate easily from cuttings.

Furcraea: See page 298.

Gynura aurantiaca (*Compositae*).

A native of Java, commonly referred to as' Velvet Plant'. This herbaceous perennial grows to a height of 60-90 cm, has velvety fleshy leaves covered with violet or purple hairs and have purple veins. The flower heads are brilliant orange. Propagation is from cuttings.

G. sarmentosa.

It is possibly a hybrid between *G. aurantiaca* and *G. bicolor*. A loosely twining herb with reddish stem. The leaves have shallowly-lobed margins and are broadly lanceolate, covered with purple hairs. Lower surface is coloured wine-red. Flowers are small and orange-coloured. A good houseplant. Propagation is from cuttings.

Heliconia (*Musaceae*).

Dwarf plantain-like plants belonging to South America. The plants are grown for their colourful foliage and in some species these have beautiful markings. Some species bear boat-shaped flowering bracts which are useful for indoor decoration. The plants do not tolerate direct sun. These are suitable for growing in large (30-40 cm) pots, and for growing in beds in a shade garden. They need a rich compost and plenty of water supply during active growing period, but during November to February watering should be done sparingly when the plants go to rest without shedding the foliage in tropical climate. Propagation is by division of root-stock as in canna or from seeds. Small bits of roots are rooted in small individual pots which are transferred later to large pots. *Heliconias* attain their best look in the rains. Some important species are: *H. angustifolia* (Syn. *H. bicolor*, *H. braziliensis*), *H. aurantiaca*, *H. bihai* (Syn. *H. aurio-striata*), *H. humilis*, *H. metallica*, and *H. psittacorum*.

Hoffmannia (*Rubiaceae*).

Tropical American herbs or shrubs grown for their beautiful foliage. The plants like a shady situation, a well-drained rich soil containing small quantities of lime which augments leaf colour. Propagation is from cuttings.

(a) *H.* × *Fantasia*: A hybrid of *H. ghiesbreghtii* × *H. roezlii*. The leaves are quilted, dark coppery olive-green with lighter veins, undersurface bronzy-red.

(b) *H. ghiesbreghtii*: The leaves are lanceolate, soft velvety quilted, bronzy to moss green, rosy-red on the reverse. In var. *variegata*, the leaves are blotched with creamy-white, yellow, and red. *H. refulgens*, *H. refulgens* 'Vittata', and *H. roezlii* are also very ornamental. The latter is considered to be a hybrid by many.

Homalomena (also spelt *Homalonema*) (*Araceae*).

Herbaceous or shrubby plants from Asia, tropical Africa, and tropical America. The cultural requirements are the same as for *Dieffenbachia* and *Caladiums*. They are grown for their handsome foliage.

(a) *H. picturata* (Syn. *Curmeria picturata*): The heart-shaped large leaves are variegated silvery-white on either side of the mid-rib. The leaves are borne on long slender petioles.

(b) *H. rubescens* (Syn. *H. rubrum*): The leaves are heart-arrow-shaped reddish-green with brownish edge tinged red, purplish beneath. Petiole slender and red in colour.

(c) *H. wallisii* (Syn. *Curmeria wallisii*): The leaves are broad, oval, reflexed, dark olive-green in colour with beautiful blotches of yellowish-silver.

Impatiens: See page 115.

Iresine: See Chapter XXIII under Edgings.

Manihot (*Euphorbiaceae*).

Commonly known as 'Tapioca'. It is a shrub but is often grown as a greenhouse and indoor plant for its ornamental foliage. For details see page 156.

Maranta: See *Calathea*.

Miconia (*Melstomaceae* also spelt *Melastomataceae*).

A native of Tropical America. The genus comprises trees or shrubs with large striking leaves having strong veins. The plants are useful for potting and for planting in beds in the shade garden. They need a fibrous compost, perfect drainage, abundant water, and semi-shaded humid situations. Propagation is from cuttings.

(a) *H. hookeriana* (Syn. *Cyanophyllum hookeriana*): The leaves are elliptic, 30 cm long, surface velvety, wrinkled, deep olive-green with silvery lines running laterally from base to apex, underneath reddish-purple.

(b) *M. magnifica* (Syn. *Cyanophyllum magnificum*): The thin ovate leaves may be up to 75 cm long, velvety green, the sunken veins are ivory-white, with a reddish-purple underside.

Monstera (*Araceae*).

Mostly tropical root climbers possessing exquisitely beautiful foliage and are very suitable for growing in a greenhouse or as houseplants for indoor decoration. The plants may be used for covering the naked stems of large trees, can be grown on pillars in shade gardens or as tub plants. *Monsteras* need a rich well-drained soil, a humid atmosphere, and a shaded position for proper growth. Syringing the foliage with water, especially during the hot months, is useful. Propagation is by division of the growing stems. The stem is cut into bits, each containing three nodes, and inserted in moist sand as in the case of node cuttings.

(a) *M. acuminata* (Guatemala): The stem is flattened, climbing, leaves oval or ovate-oblong, 20-25 cm long, perforated in old leaves.

(b) *M. deliciosa* (Syn. *Philodendron pertusum*) (Mexico, Guatemala): The showiest grower among the *Monsteras.* Stem 6 m long or more. A heavy root-climber, forms long hanging cord-like aerial roots. The leaves are large, 90 cm across, leathery, perforated with oblong holes and pinnatifid (clefted) at margin. The juvenile stage with less perforations is known as *Philodendron pertusum.* In var. *variegata* (Syn. *Philodendron pertusum variegatum*) the foliage is irregularly marbled greenish-yellow or cream. *M. deliciosa* bears an edible fruit with a taste and flavour in between pineapple and banana.

(c) *M. obliqua* (Brazil): Elliptic tapering leaves develop irregular holes, more or less of oval shape when old.

(d) *M. obliqua expilata* (Syn. *M. leichtlinii*)*:* A form of the above where the leaf lamina is perforated, so much so that it looks nothing more than an ovate skeleton of veins only.

M. friedrichsthalii, *M. perforoides*, and *M. pertusa* are also ornamental.

Musa (*Musaceae*).

Some of the bananas have ornamental foliage and are grown occasionally as greenhouse plants or used in landscape gardening. The plants need plenty of water and a rich soil. Propagation is from suckers.

(a) *M.* × *paradisiaca vittata:* A strikingly beautiful plant. Leaves light green with variegation of milky-green and white, the mid-rib is also white and the margins are red.

(b) *M. sumatrana:* Deep greyish-green leaves are blotched with wine-red having a reddish undersurface but only in young leaves.

(c) *M. zebrina:* A native of India and Java. The blackish blood-red and purple blotches are scattered irregularly over the dark green leaves, the channelled mid-rib is brownish-red and the undersurface is wine-red.

Neoregelia (*Bromeliaceae*).

The plants are similar to *Billbergias* and require similar treatments. Grown mainly for the ornamental foliage, but the flowering bracts are also very ornamental. Some important species are *N. carolinae* 'Tricolor', *N. marmorata*, *N. sarmentosa chlorosticta*, *N. spectabilis* (tip of otherwise green leaves are red), and *N. zonata* (olive-green leaves marbled and banded wine-red).

Nepenthes khasiana (*Nepenthaceae*).

A curiosity plant from the Khasia Hills in India. Commonly known as 'Pitcher Plant'. It has a tubular (7-20 cm long, 2.5-4 cm wide), green to reddish-green pitcher with a lid, at the end of the mid-rib of the leaf. Grows in a boggy shaded position and is carnivorous in nature, trapping insects through the pitcher. Propagation is from cuttings as well as seeds. Suitable for growing in hanging baskets also.

Nidularium (*Bromeliaceae*).

The plants are related to *Billbergias* and need similar treatment. There are many ornamental species.

Ophiopogon (*Liliaceae*).

A lovely plant with tufted, linear, grass-like leathery leaves. It needs a sandy-soil. Propagation is by division. Some important species are *O. intermedius argenteo-marginatus*, *O. jaburan variegatus*, and *O. japonicus*.

Oplismenus: See Chapter XVII.

Oxalis (*Oxalidaceae*).

Oxalis is grown as a greenhouse and indoor plant, mainly for its foliage, but it also bears pretty flowers. For details see pages 246-247.

Panax: See *Aralia*, *Pseudopanax*, and *Polyscias*.

Pandanus (*Pandanaceae*).

The common 'Screw Pine' has several vernacular names, viz., *Keya* and *Ketaki* in Bengali, *Keora* in Hindi, *Thasai* in Tamil, and *Gedege* in Kanarese. These are tropical shrubby plants grown for the highly scented flowers or the ornamental foliage. The ornamental foliage types are generally cultivated in the greenhouse, protected from direct sun. The variegated types lose the colour in a rich soil and hence should be grown in a rather poor soil containing just sufficient nutrients and moisture. Propagation is by offsets or division of suckers. *Pandanus* are good for potting and for indoor decoration. The common screw pine is used as a hedge also.

(a) *P. baptistii* (New British Isles near New Guinea): The leaves are spirally arranged, arching beautifully and tapering to a long point. The leaves are blue-green in colour with 1-4 longitudinal yellowish-white stripes and have no spines.

(b) *P. graminifolius:* Another attractive species, which is sometimes confused with *P. pygmaeus*.

(c) *P. sanderi* 'Roehrsianus' (Timor, Malaysian Archipelago): The leaves are sword-like, 120 cm long, shining green with length-wise stripes and bands of light and golden yellow. The young foliage is golden orange in colour. Leaves are armed with fine spines.

(d) *P. utilis* (Syn. *P. elegantissimus*, *P. odoratissimus*) (Madagascar): The common 'Screw Pine'. Leaves are strap-like, leathery, long, and spirally arranged, spiny at the margins. The plant becomes a tree about 18 m tall with age and bears highly scented flowers liked by Hindu women. The scent of the flower, it is alleged, attracts cobras which remain hidden in the flowers and hence one should not handle the flowers without proper inspection. The name *P. odoratissimus* is also ascribed to the species *P. tectorius*.

(e) *P. veitchii* (Polynesia): A beautiful houseplant. The leaves are 60 cm long, thin-leathery, somewhat recurving, spiny-toothed, narrowing to a long point, deep green in colour and broadly margined with creamy-white.

(f) *P. veitchii compactus* (*nelsonii*): More compact than the above, leaves rather rigid and dense, the variation is pure white and more clear than the above.

Panicum: See *Oplismenus*.

Pedilanthus tithymaloides: See Chapter XXIII under Hedge and page 301.

P. tithymaloides cucullatus (*Euphorbiaceae*).

A native of Central America. A herbaceous succulent plant with erect, fleshy, black-green stems; leaves lanceolate, pale green tinged red and dark green near the mid-rib. A hardy plant which is propagated from cuttings.

Pelargonium (*Geraniaceae*).

In tropical climate, the geraniums are grown in greenhouse conditions. For details see page 116.

Pellionia (*Urticaceae*).

The genus represents about 15 species most of which are creeping in habit and have showy ornamental foliage. The plants prefer a rich sandy-loam soil and a moist shaded position. The plants are propagated by division or from cuttings. These are good greenhouse subjects and can also be grown in hanging baskets or in shaded rockeries or as undergrowths in shade gardens. As the plant trails on the ground it develops roots at each joint, thus facilitating propagation.

(a) *P. daveauana*: A succulent creeper; stem pinkish; leaves 2-ranked, flattened, oval when young and lanceolate with age and up to 6.5 cm long. The colour of the leaves is dark bronzy olive-green with a slight violet tinge, the central area is pale green or grey. A native of Vietnam, Malaysia, and Burma.

(b) *P. pulchra*: A fleshy creeper, having a pinkish stem and stipules; leaves are obliquely oblong, blunt and obliquely heart-shaped at base, light green to greyish in colour but thoroughly covered with a blackish network of veins, the lower surface is pale purplish.

Peperomia (*Piperaceae*).

Small, succulent, herbaceous, mostly perennial plants with very ornamental foliage, often prettily marked. The plants make ideal subjects for the greenhouse but some species are also suitable for hanging baskets and for table decoration. A compost consisting of two parts of garden loam, one part of cowdung manure, one part leaf-mould, one part of sand or old rubbish is good for growing *Peperomias*. Composts meant for ferns also suit them. They can also be grown in raised beds but the drainage must be perfect. The plants need protection from direct sun and a humid atmosphere, but over-watering will result in rotting. The plants are raised from seeds but more conveniently from short stem cuttings or one-node stem cutting with a leaf attached.

Leaf cuttings are also used for propagation, with the long petiole inserted in pure sand and the leaf blade spreading horizontally on the surface of the sand. The sand is kept just moist, and the roots appear soon. The rooted cuttings are planted at first singly in 7.5 cm pots and then shifted to 15 cm pots before finally potting them in 20-25 cm pots. *Peperomias* are usually fleshy and creeping in habit. The flowers appearing like slender catkins are also very attractive.

There are innumerable attractive species and cultivars and it will not be possible to discuss them in detail. But the following are recommended.

(a) *P. argyreia* (Syn. *P. sandersii*) (Brazil): A rosette almost without stem; leaves orbicular or ovate, fleshy; petioles red; leaf blade green to bluish with radiating silver lines.

(b) *P. caperata* 'Emerald Ripple' (Brazil): The leaves appear in dense clusters, heart-shaped, heavily corrugated and quilted, fresh green in colour, while the valleys are tinted brown and ridges grey.

(c) *P. caperata* 'Tricolor': Leaf corrugated, milky-green, margined with creamy-white with tinges of red around the base.

(d) *P. griseo-argentea* (Brazil): In Europe this is commonly known as *P. hederaefolia.* A very attractive species. Leaves round cordate, looking like shield, quilted, painted with glossy silver; the sunken veins have a purplish tinge. 'Blackie' and 'Nigra' are two attractive cultivars.

(e) *P. magnoliaefolia*: The leaves are like those of *Magnolia grandiflora.* A native of W. Indies.

(f) *P. metallica* (Peru): A most attractive species with red stem and narrow waxy leaves which are of coppery colour having a metallic lustre and silvery-green band along the centre; the undersurface has reddish veins. A delicate and rare species.

(g) *P. obtusifolia* (Syn. *P. clusiifolia*) (Venezuela): An attractive species well suited for dish garden. Leaves obovate, fleshy, leathery. The var. *alba* and the var. *variegata* are beautifully variegated; in the former the young leaves are albino. The albino is delicate in nature.

(h) *P. prostrata* (Colombia): It is the garden synonym of *P. brevipes*. A tiny beautiful creeper, with very small (0.8-1 cm) diameter, circular leaves which are bluish grey in colour, variegated like filigree work of silver on the upper surface. It is suitable for hanging baskets.

(i) *P. rotundifolia* (Syn. *P. nummularifolia*) (Puerto Rico to Jamaica): Another small rambling creeper with very tiny (0.8-1 cm), round, fleshy and waxy leaves, pale grey in colour. Can be grown in hanging baskets. The following species are also very attractive:

P. caperata variegata, *P. clusiaefolia variegata*, *P. elongata*, *P. incana*, *P. marmorata*, *P. polybotrya* and *P. verschaffeltii*.

Philodendron (*Araceae*).

Mostly shrubby creeping plants (rarely tree-like), or stemless herbs of great ornamental value for their graceful foliage which are leathery, heart-shaped, egg-shaped, arrow-shaped or oblong or ovate, either entire or lobed. In some species, the foliage is perforated and pinnatifid. The *Philodendrons* have the same uses in the garden as *Monstera*. The cultural requirements and propagation are the same as for *Monstera*. The *Philodendrons* comprise 120 species many of which are ornamental especially at juvenile stage but may become unwieldy with age. A few species are discussed below.

(a) *P. andreanum* (Colombia): The leaves are velvety in texture, oblong sagittate, i.e., like an arrowhead, dark green with coppery reflection and ivory-white veins.

(b) *P. bipinnatifidum* (Brazil): The leaves are very large (90 × 45 cm) and bipinnatifid. There are 10-12 segments on each side of mid-rib, lobes narrow and lobed twice.

(c) *P.* × 'Burgundy': A commercial hybrid. Leaves leathery, arrow-shaped, about 30 cm long, deep green flushed reddish; stem wine-red; petioles winged. The young leaves are burgundy-red.

(d) *P. dubia* or *P. dubium* (South Mexico, Guatemala): The juvenile stage of *P. radiatum* is actually cultivated in the garden as *P. dubium*, a beautiful climber with broad, bright green, deeply-lobed leaves but at juvenile stage the incision is less pronounced.

(e) *P. gloriosum* (Colombia): A slow creeper; leaves large, ovate-heart-shaped, deep green with snow-white veins; margins have thin pink edge; young leaves have a silky sheen.

(f) *P. mamei* (Ecuador): Stemless; leaves arrow-shaped or rather cordate (heart-shaped), handsomely marbled white.

(g) *P. melanochrysum* (Colombia, Costa Rica): Called 'Black Gold' as the foliage is almost velvety black.

(h) *P. micans* (possibly the juvenile state of *P. oxycardium*): A native of Tobago and Dominica. A weak vine. Leaves heart-shaped, bronzy with silken sheen above and reddish beneath.

(i) *P. pertusum:* See *Monstera deliciosa.*

(j) *P. pertusum variegatum:* See *Monstera deliciosa variegata.*

(k) *P. selloum* (Brazil): Stem is stout tree-like with age, produces thick aerial roots. The foliage is strikingly ornamental, the leaves are pendent, 60 cm long, blade pinnatisect (cut down to the mid-rib in a pinnate way), the segments themselves are again lobed or pinnatifid, but the juvenile leaves are only lobed. There are many cultivars and variegated species.

(l) *P. squamiferum* (Guyana): A twisting creeper, leaves 25 cm long and 12-25 cm broad, deeply 5-lobed, but young leaves are either entire or 3-lobed, petioles densely covered with green to red bristles.

(m) *P. wendlandii* (Costa Rica, Panama): Stem very short, leaves 30 cm or more long, obovate with thick mid-rib, arranged like a bird's-nest, green.

The *Philodendrons* have showy flowering spathes.

Phyllanthus angustifolius (Syn. *Xylophylla elongata, Xylophylla montana*).

An interesting plant from Jamaica with flattened leaf-like branches on the edges of which are borne tiny reddish flowers.

P. nivosus roseo-picta (Syn. *Breynia nivosa roseo-picta*): See page 160.

Pilea cadierei (*Urticaceae*).

A native of Vietnam. It is commonly called 'Aluminium Plant'. It is a rapid-growing succulent, suitable as an edge plant in the shade, as a rock plant in shaded portion, and also as ground cover in a shade garden or in a shrubbery. The obovate, quilted leaves are about 7-8 cm long, vivid green or bluish green in colour and painted shining silver aluminium on the surface. It can be propagated easily from small stem cuttings.

P. muscosa (Syn. *P. microphylla*).

A native of West Indies variously known as 'Artillery Plant', 'Gunpowder Plant', and 'Pistol Plant', because of the fact that the pollens are ejected with force which can be observed if the plants are kept in the sun just after the foliage has been watered. This is a small trailing densely branched plant with handsome minute fern-like foliage. The plants are useful as ground cover under trees and shrubs, in the shade garden, in rockeries under semi-shade and also for growing in pans, hanging baskets, and for carpet bedding. Propagation from cuttings is very easy and often the plants are self-sown. *P. cadierei* 'Minima' and the cultivar 'Silver Tree' are also very ornamental.

Pleomele: See *Cordyline*.

Polyscias fruticosa (Syn. *Aralia fruticosa, Panax fruticosum, Nothopanax fruticosum*) (*Araliaceae*).

An erect shrub from India to Polynesia (1.8-2.5 m tall). Leaves leathery, irregularly 3-pinnate; margins of leaflets cut into segments, and often edged white; the rachis and stalk spotted.

P. fruticosa 'Elegans' (Syn. *Aralia, Panax*).

The leaves are leathery, green, tripinnately cut and lobed.

P. paniculata variegata.

A handsomely variegated species. Leaflets pinnate, leathery, serrated, green splashed with creamy-white and greenish-white. Treatment and propagation of *Polyscias* are the same as for *Aralias.*

Portulacaria afra: See page 302.

Pothos: See *Scindapsus*.

Pseudopanax crassifolius (Syn. *Aralia crassifolia*) (*Araliaceae*).

A native of New Zealand and commonly known as 'Lance-wood'. While young, the leaves are ovate-lanceolate, finely toothed, dark green with yellow blotches above and purple beneath; older leaves 3-5 foliate.

(a) *P. ferox* (New Zealand): Leaves leathery. stiff, very variable but always single, linear at juvenile state but later sword-shaped; chocolate-brown in colour and having toothed margins.

The plants should be treated in the same way as *Aralias.*

Rhoeo spathacea (Syn. *R. discolor*) (*Commelinaceae*).

A native of Mexico. Leaves are lanceolate, stiff and waxy in rosettes; the colour of the yellow leaves are dark green above and purple below. The minute white flowers peep through the boat-shaped bracts.

(a) *R. spathacea vittata:* A variegated form of the above. Leaves are variegated, pale yellow with tints of red in longitudinal stripes. The plants need a moist soil and are propagated by division.

Ruscus hypoglossum (Syn. *R. hypophyllum hypoglossum*) (*Liliaceae*).

A native of Europe and Asia Minor. An interesting tufted, compact, dwarf (45 cm tall) shrub. The tiny greenish flowers are to be found in the mid-rib of the cladodes (leaf-like flattened branches); berries very rare, red in colour. Propagation by division of creeping rootstock. A useful plant for planting in shade gardens in the vicinity of tree-roots.

Sansevieria: See page 302.

Schismatoglottis (*Araceae*).

The plants are somewhat similar to *Aglaonema* and *Alocasia*. The foliage is very ornamental, oblong or ovate-cordate, often marbled or spotted. The plants need a shady moist atmosphere and a well-drained rich soil. Propagation is by division of rhizomes. Some of the important species are *S. neo-guineensis* (Syn. *S. novoguinensis*), *S. neo-guineensis variegata*, and *S. picta*.

Scindapsus aureus (*Araceae*).

A native of Solomon Islands. Commercially known as *Pothos* and is commonly called 'Money Plant'. A fleshy root-climbing vine with thick long aerial roots. The leaves at young state are broadly ovate, waxy, green, having yellow variegation. The mature leaves become 60 cm long and the blades become lobed or slashed. This is also known as *Raphidophora aurea*.

(a) *S. aureus* 'Marble Queen': It is a beautiful mutant. The leaves are richly variegated and streaked and may become pure marble-white with few or no green streaks when grown in bright light.

(b) *S. aureus* 'Tricolor': The leaves are marbled and spotted pale green, deep yellow, and creamy.

(c) *S. pictus* (Indonesia, the Philippines): A tree climber, attaching closely to the bark; leaves leathery; obliquely ovate-oblong, irregularly spotted silver over dark green.

(d) *S. pictus argyraeus* (Syn. *S. argyraeus*, *Pothos argyraeus*): A juvenile state of the above with cordate bluish-green leaves, marked and edged with silver.

These plants need shade, plenty of moisture, and a rich porous soil. The plants may be allowed to grow over a tree or a pillar, or can be grown in pots as greenhouse or indoor plants. Some people grow them in bowls containing water. Propagation is easy from cuttings.

Setcreasea purpurea (*Commelinaceae*).

A native of Mexico, commonly called 'Purple Heart' because the plant is purple-coloured in bright light. The plants are trailing; leaves clasping, lance-shaped, purple; surface of leaves covered with pale hairs; the blue flowers resemble orchid blooms. This is easily propagated from cuttings. The plant is suitable for growing in hanging baskets, rockeries and as a greenhouse plant. It also serves well as an edging plant.

Strelitzia reginae (*Musaceae*).

'Bird-of-Paradise' is a native of South Africa. The flower resembles the above-named bird. A stemless, compact plant, about 1.5 m tall. The stiff leathery banana-like bluish-grey leaves are borne in long peduncles. The flowers are exceedingly beautiful which are long stalked and emerge from the green boat-shaped bracts which are bordered red; the numerous petals are pointed and brilliant orange in colour; there is an arrow-shaped tongue of contrasting vivid blue. The fully-opened flower looks like a flying bird.

The plant is rather difficult to grow in the plains of India. It needs a shady, humid atmosphere and a well-drained gritty soil. Propagated from seeds or by division. *S. alba* (Syn. *S. augusta*) is a white flowering hardy species.

Syngonium (*Araceae*).

Quick-growing, ornamental foliage climber suitable for small pots and trellises. The plants are quite hardy and not fastidious about soil. Leaf-mould with coarse sand is quite suitable for growing these plants. Propagation is easy from cuttings. The plants root from nodes wherever these touch the ground. Leaves are generally sagittate (like arrowhead) or 3 cut; many are variegated. Some important species, varieties and cultivars are mentioned below:

S. podophyllum (Syn. *Nephthytis liberica*), *S. podophyllum albolineatum*; 'Emerald Gem', 'Imperial White', and 'Tricolor' are some of the beautiful cultivars of *S. podophyllum*, and *S. wendlandii* creeper with tri-lobed velvety green leaves having white veins in juvenile state.

Tacca chantrieri (*Taccaceae*).

A native of Malaysia. Commonly called 'Bat-Flower' because of the bat-like inflorescence in both shape and colour. There are two broad wing-like maroon-black bracts, from which trailing filaments or 'whiskers' come out. The black flowers are followed by heavy berries. *T. aspera* (Syn. *T. cristata*) is another interesting species. They need a porous soil and are propagated by the division of rhizomes.

Tradescantia (*Commelinaceae*).

Members of this genus are low-growing herbs many of which are trailing or of creeping habit; some have beautiful variegated foliage when grown in bright but not scorching light.

The plants need a lot of moisture and a good soil for proper growth. They are useful for covering rockeries, for edging or carpet bedding. The trailing types look beautiful when grown in hanging baskets. Propagation is very easy from cuttings. Actually most of the trailing species produce rootlets in the nodes wherever they touch the ground.

(a) *T. albiflora* 'Albo-vittata' (Central America): A freely branching plant having comparatively large and lanceolate fleshy leaves which are bluish-green, striped as well as bordered white.

(b) *T. blossfeldiana* (Brazil): Creeping or trailing plant branching profusely. Stem purple; leaves olive green above with purplish underside and covered with silvery hairs. There is a variegated form with cream to yellowish stripes.

(c) *T. fuscata* (Brazil): The latest synonym is *Siderasis fuscata* but is also known as *Pyrrheima loddigesii*. A clustering rosette; leaves broad and oblong, olive-green in colour with a central silvery band and covered with brown hairs, the reverse is purple. The large blue flowers appear at the base.

(d) *T. navicularis:* See page 304.

(e) *T. sillamontana* (Syn. *T. pexata*, *T. villosa*): A beautiful fleshy trailing plant commonly called in the trade as 'White Velvet' or 'White Gossamer'. The leaves are fleshy, ovate, clasping, arranged in ranks and completely covered with fluffy white wool.

(f) *T. zebrina:* See *Zebrina pendula*.

Xanthosoma (*Araceae*).

A group of ornamental foliage plants with arrow-shaped leaves which are often variegated. The compost for growing should mainly consist of leaf-mould. They need a shaded moist atmosphere and good drainage. Propagation is by division, or by cutting the rhizome or rootstock into pieces and planting them in light soil.

(a) *X. lindenii* (Syn. *Phyllotaenium lindenii*): The leaves are thin-leathery, large, arrow-shaped, green, shiny with the central and main veins silver-white. A horticultural cultivar, 'Magnificum', has larger leaves and more prominent cream to white veins. *X. atrovirens* and its varieties, *X. roseum variegatum* are some other important types of *Xanthosoma*.

Yucca (*Liliaceae*).

These are bold, stiff-leaved plants, somewhat resembling in between *Aloe* and *Agave*. Suitable for planting in the lawn as specimens or for mass grouping in large gardens, or as a rock plant in large rock gardens. These are generally grown in the succulent collection. A well-drained sandy-loam soil suits them most. Propagation is from seeds, offsets, stems, or rhizome-cuttings. Flowers are borne in a column-like inflorescence arising from the centre of the plant.

(a) *Y. aloifolia* (Syn. *Y. serrulata, Y. crenulata*) (U.S.A., West Indies): Leaves dagger-like, sharply pointed, arranged in crowded rosette, and galucous-green in colour. Flowers cup-shaped, creamy white tinged purple at the base. In variety *marginata* the margins of the leaves are creamy yellow; in var. *tricolor* the pungent dagger-like leaves have ivory margin, with partly yellow centre; the young leaves are tinged red.

(b) *Y. gloriosa* (Syn. *Y. acuminata, Y. gloriosa acuminata, Y. integerrima*) (U.S.A.): The most commonly cultivated species are variously called "Spanish Dagger", "Adam's Needle" for the needle-pointed leaves. The leaves are sword-shaped, long, flat, grey-green with reddish margins. Flowers are bell-shaped, white and striped purple on the outer side. The plant suckers freely.

Zebrina pendula (Syn. *Tradescantia zebrina, T. tricolor, Cyanotis vittata*) (*Commelinaceae*).

A trailing and popular plant for hanging baskets, rockeries, edging and carpet bedding. The ovate leaves are succulent, deep green or purple with two silver bands and purple beneath. Flowers are rosy purple.

The plants root at the nodes and can be propagated from cuttings. A semi-shady moist atmosphere suits the plant most. There are several horticultural varieties with variously variegated leaves such as 'Discolor', 'Discolor multicolor', 'Quadricolor', etc.

(a) *Z. purpusii* (Syn. *Tradescantia purpurea*): A rampant creeper. The foliage is purplish brown with light stripes, purple beneath. Flowers are lavender-coloured.

Chapter XIV

Cacti and Other Succulents

The group of plants belonging to the family *Cactaceae* and many other plants belonging to the genera *Agave*, *Aloe*, *Cotyledon*, *Echeveria*, *Euphorbia*, *Haworthia*, *Kalanchoe*, *Sansevieria*, *Sedum*, *Sempervivum*, *Stapelia*, etc., are all succulent plants. The plants of the cacti group have areoles while no other succulents (or other plants) have these structures. There are many other differences between cacti and other succulents and hence these are dealt with separately.

(A) CACTI

Cactus (plural cacti) is the name of one genus, but this has become the common name for all the plants belonging to this family. Botanically each plant in the family *Cactaceae* must have all the following five characteristics in common:

(a) Areoles or spine-cushions are invariably present in all the members. With a few exceptions, these areoles are usually armed with spines or with fine barbed hairs (glochids) or with both.
(b) All are perennial plants.
(c) The fruit is one-celled berry.
(d) All belong to dicotyledones group.
(e) The flower petals arise from the top of the ovary.

Besides, the plants of the cacti group usually are almost leafless, with only vestiges of leaves present which are not visible to the naked eye, except in the genus *Pereskia*.

No other group in the whole plant kingdom has such strange variation in plant shape and form, size and colour of areoles and thorns, the beauty of the flowers with a remarkable metallic sheen, as the cactus group of plants. The range of colour of flowers may vary from white to pink, purple, yellow, crimson, dark crimson-red, violet, and combinations of various colours with spots and streaks. Unfortunately, the flowers are short-lived. For example, in genera *Astrophytum*, *Chamaecereus*, *Echinopsis*, *Lobivia*,

Parodia, *Rebutia* the flowers gradually open up with the increase in intensity of sunlight and close down by dusk, while in *Selenicereus* sp. the flowers open after dusk and close down by dawn. In some species such as *Haageocereus bicolor* and *Gymnocalycium* sp., the flowers last for a couple of days. The size of flowers also varies from 1 cm in some *Mammillarias* to about 35 cm across in a few of the night-flowering types.

In their natural habitat, most of the cacti grow in adverse climatic conditions and to adapt themselves to these trying conditions they have developed strange shapes and sizes, with fleshy, juicy, and jointed stems with the virtual absence of leaves. The fleshy green stems are used as reservoirs for storage of water to be used during drought conditions, and they serve such physiological functions as manufacture of food, as a proxy to the missing leaves. The general assumption that cacti are desert plants is not quite correct. Many are found growing in the prairie type of lands where some other vegetation such as dwarf trees, shrubs, and coarse grasses also grow. There are species which grow in a good loamy soil, while others inhabit rocks and mountain slopes. Beautiful flowering cacti such as *Epiphyllums* grow as epiphytes on trees in the forests of Brazil.

Housing of cacti: It is quite likely that due to the process of domestication many of the cacti species have lost their natural hardiness as a result of favourable growing conditions available in cultivation. It is often observed that if a cactus growing in semi-shade is suddenly removed to a more sunny situation it develops burns and scorching due to bright sun. It is for such reasons that the domesticated cacti should be housed in suitable locations. All cacti like plenty of sunshine, but it is the morning sun which should be preferred. In regions where the sun becomes very hot during the summer, it is advisable to exclude midday and afternoon sun. Cacti grown in shade become lanky and do not flower or flower sparsely. Excess of water is harmful for most of the species and hence all cacti should be housed under glass or polythene roof, although a few hardy species of *Cereus*, *Echinopsis*, and *Opuntia* can be grown in the open also, under our climatic conditions. For the production of plenty of flowers and for development of good texture of plants, abundant circulation of fresh air is essential. Hence, a cactus house should have only the roof made of glass, fibre-glass or polythene, supported on pillars, while the sides remain open. Such houses should run in the north-south direction facing the east; the west may be blocked with a brick wall in extremely hot places.

Compost: Though cacti are quite hardy and rather easy to cultivate, the success depends, perhaps, more on the use of the correct type of compost and watering. Although the aerial portion of the cacti, i.e., the stems, are fleshy and sometimes quite large, the underground roots are generally very few and small in length. As a result their capacity to absorb moisture rapidly is limited. Moreover, the cacti may not also need large quantities of moisture. For these reasons it is essential to have a porous compost. A compost consisting of one part loamy or sandy loam soil, one part of well-rotten cowdung manure, two parts of leaf-mould of medium coarseness, one part of sharp sand (not silver sand) and small quantities of broken pieces of bonemeal is quite suitable for growing most of the species, whether globular or columnar in shape. To this compost 10 per cent BHC should be dusted to keep away the ants and to some extent root mealy bugs. People having a lawn can prepare a good compost which has been found to be very suitable for growing most of the cacti. When the lawn is scraped, generally annually or every alternate year, the scrapings containing the green blades of grass, a portion of the roots along with soils and residues of organic matters are dumped in a pit. While the exposed grass on the top dries up the lower layers in the bottom decompose slowly. The dried-up grass is burnt and mixed up with the rest of the partially decomposed compost and covered up to facilitate full decomposition. A fully decomposed compost of this type is porous and quite good for growing cacti after sieving and addition of bonemeal. A freshly prepared compost should not be used for planting, but a time gap of 15 days should be allowed between mixing and planting. Cacti, especially the very thorny types, do not require heavy feeding with fertilizers, if they are repotted at least every alternate year. Generous feeding with fertilizers makes the plants lush green and these become liable to rotting.

Containers: Earthen pots should always be preferred to glazed pots and metallic or plastic containers, as these have more pore spaces for exchange of air. The pots should be cleaned, or preferably sterilized, and dried before potting. The correct size of the pot should be used which is just large enough to hold the plant, as over-sized pots will hold more moisture which will be harmful for the cacti growing in these pots. Plants grown in small pots become pot-bound and stunted in growth.

Time and method of planting and potting: Planting of cacti, either in ground or in pot, can be undertaken any time of the year by experts who know the precautions to be taken while planting. But the amateur growers are advised to do the planting or potting when growth starts slowly in March or April.

One-third of the pot at the bottom is filled with bold pieces of crocks with the concave face down to facilitate drainage. Then the compost is filled, leaving a gap of about 2 cm on the top to facilitate watering. Before planting all dead roots, roots attacked by disease and affected by nematodes should be removed. The nematode-affected roots are generally swollen, forming tubercles and hence can be easily identified. The cut end of roots should be treated with sulphur dust or blitox paste. Often root mealy bugs may be present which should be picked up and the roots washed with a mixture of methylated spirit (5-10%) and nicotine (2.5%). At the bottom layer of the compost a few crystals of paradichlorobenzene may be put to drive away the mealy bugs. The roots of the treated plants should be dried before planting. If all the roots are in a bad condition they should be cut off and the plant re-rooted.

While repotting, the same principle is followed as in the case of planting or potting. It is not necessary to repot every year, especially the young ones, as long as they appear healthy and vigorous. The large specimens should be repotted every alternate year or once in three years.

Practical experience has shown that seedlings, small plants, or even cuttings of smaller sizes do well when they are planted close to the sides of the pot rather than at the centre. This practice lessens the chances of the young plants rotting due to excess of water.

Unless the compost is really in a dry state the newly planted cacti, especially those with injured roots, should not be watered for a few days after planting as these might rot easily because of the presence of an abundance of moisture.

While planting or repotting the spines should not be damaged, because once damaged they will never grow again. Since most of the cacti are spiny it is difficult to handle the plants during these operations. Cactus growers are advised to wear rubber gloves and to use a table spoon while filling the compost. The compost around the plant may be firmed with an old table knife handle.

Watering: In their natural habitat the cacti are adapted to severe drought conditions. In cultivation, correct watering is the most important single factor in the success of growing these plants. Over-watering is more dangerous than under-watering as the former practice may kill a plant while the latter may only hamper proper growth.

The frequency of irrigation and the amount of water needed depend upon the time of the year, the type of compost, whether a potted or a ground plant, and finally upon the type of the cactus. The water used should be neutral or nearly so in reaction. A plant in a small pot will need more frequent irrigation than a plant in the ground. The principle of watering a cactus may be summed up as follows: never water a cactus again if the soil is still damp. One way of knowing whether the compost in a pot is dry or wet is to hit the side of the pot gently with a wooden hammer or batten and if the resultant sound is a ringing one the compost is dry, while a dull sound will indicate that it is still wet. For a plant planted in the ground, the top soil may be forked up a little to inspect the moisture content in the inner layers. Cacti should not be watered during the winter. The watering should gradually be minimized from the autumn to bring the plants to rest. To prevent any drying-up of roots the plants may be watered a little, once in a month during the winter. Watering should be commenced again from March to April. But in places where the winter is not so severe irrigation should not be delayed so much.

In a large collection, the plants may be watered conveniently with the help of a hose fitted with a fine spray nozzle. During the hot summer months the plants may be sprayed or syringed with water. Water dripping from roof is bad as this causes rotting. Any leak in the roof should be rectified promptly. Water should not also collect in the depression of the plants as this also causes rotting.

Staking: Most of the cacti do not need any staking as they can stand firmly on their own roots. But some of the tall columnar types may need some kind of staking. Painted split-bamboo stakes are quite useful.

Propagation: The propagation of cacti is done through seeds, cuttings, offsets, and grafting; seedlings take a longer time to come to specimen size and the small seedlings need much more care than cuttings or graftings.

(a) *Raising from seeds:* For rapid multiplication nurserymen and enthusiastic amateurs specialized in cactus growing raise the plants from seeds. The seeds of most of the cacti germinate between a week and a month. The *Opuntias* may be rather slow in germination. The best time of sowing in the plains is between March and September. In a cold climate, if bottom heat is provided, sowing can be done at any time. The optimum temperature is 21°C.

Seeds are best sown in shallow seed pans, preferably new ones, otherwise clean and sterilized pans should be used. A good seed-compost consists of equal parts of well-decomposed sifted leaf-mould and sand. Some people advocate adding of small quantities of loam, powdered brick, and charcoal to this compost. The compost is filled in the pans in which a few layers of crocks should be placed at the bottom for drainage. The compost is firmed gently. It is advisable to use sterilized compost to prevent damping-off of seedlings. A little quantity of BHC is added to the compost to keep off ants as they carry away the seeds. The seeds are sown thinly and as evenly as possible. The smaller seeds are not covered generally, while the larger ones should be covered with a fine layer of sand. A layer of sand on the top prevents the growth of algae to some extent. The seed pan should be watered from bottom by placing it to half of its depth in a container of water. Then the pans are covered with a glass sheet and shifted to shade. Alternatively, these can be covered with newspaper sheets. Moisture will sometimes accumulate on the inner-side of the glass which should be wiped off once or twice every day, as falling drops of water may injure new seedlings or may wash off the compost covering the seeds. When a reasonable number of seedlings have germinated the glass should be removed and the seedlings shifted to a semi-shaded place as young plants cannot withstand direct light. If the seedlings are crowded or grow rapidly these should be pricked in a seed pan or trough in the very first season, but normally they are not disturbed for about 12 to 15 months. Afterwards, when sufficiently grown, these are transplanted in small individual pots. Planting of young seedlings in individual pots should be discouraged, as the tender roots are likely to be damaged and it becomes a problem to keep each individual pot moist. It is easier to water a seed pan full of seedlings than individual pots. The compost for pricking and for small seedlings is the same as seed-compost.

(b) *Offsets and cuttings:* The raising of plants through cuttings poses no problem. But there is a limitation as the large globular and solitary stem type plants cannot be multiplied by this method. Often many cacti produce offsets from the base of the mother plants, which may root while still attached to the plant. Such rooted offsets can be separated and planted immediately while unrooted offsets are used as cuttings. The following genera produce offsets freely: *Echinopsis*, *Lobivia*, *Mammillaria*, and *Rebutia*. If specimen plants are wanted quickly offsets should not be separated for propagation, unless it is absolutely necessary to multiply the species rapidly.

The plants from which the cuttings are to be taken should not be watered for a few days to enable the plants to use up the excess of water. This procedure reduces the chance of rotting of the cuttings in the propagation bench. The cuttings are taken with the help of a sharp knife and the cut should be smooth and transverse. The cut end should immediately be smeared with either fine sterilized sand or charcoal powder and kept in a shady place for 4 to 5 days, depending upon the size of the cutting and species, for drying. This is done in order that the cut surface is hardened and forms a callus as also to prevent rotting. The compost for raising the cuttings consists of equal parts of sharp sand and leaf-mould. The base of the cutting should just penetrate the compost and these are staked if necessary. Deep planting should always be avoided. During the process of rooting the cuttings are watered sparingly even during hot weather. In tropical climate, it is better to keep the cuttings in semi-shade till they root. It will be better if the cuttings are raised under glass or polythene cover. New growth emerging from the cuttings is the best indication of rooting, which may take 15 days to one month depending upon climatic conditions. After the cuttings have established well, these are shifted to pots or beds containing the usual compost. It is a useful method for rapid multiplication of rootstocks.

(c) *Grafting:* Grafting is an interesting and common feature in the cultivation of cactus. Moreover, this has got certain advantages over the other methods. For example, grafting helps raise specimen size plants quickly; a few delicate species which do not perform well on their own roots can be grafted on hardier rootstocks; and cactus with chlorophyl-poor body such as *Gymnocalycium mihanovichii friedrichii* 'Rubra' and cv. 'Multicolor' can thrive only when grafted on a green rootstock. Grafting is also useful for multiplication of hybrids, cristate, and monstrous forms. Another advantage is that if a portion of a rare species is damaged by insects, diseases, or any other cause, the species can be saved by grafting a healthier portion from the same plant on a suitable rootstock. Pendent forms of cacti such as *Aprorocactus flagelliformis*, *Chamaecereus silvestrii* (also spelt *silvestri*), *Echinocereus pentalophus*, and *Zygocactus* look better when grafted as half standards on vertical stocks. However, one major disadvantage of grafting is that the majority of the scions lose the characteristics typical of the species. But generally grafted plants grow better and flower more profusely.

In the process of grafting, the first and foremost thing is to select a healthy and right type of rootstock. There are three types of grafts, namely, *flat*, *cleft*, and *splice*.

In flat graft, both the scion and the stock should be of more or less the same width at the place of graft union. After selecting the scion and the stock, a smooth transverse cut is made on both, and the cut end of the scion is placed firmly on the severed top of the stock immediately. These can be held together firmly by passing two large rubber bands, in a criss-cross manner, over the top of the scion and fixing them underneath the pot. A more simple method is to bend over criss-crossly two (or one) pliable but strong split bamboo stakes in a hairpin fashion over the scion and fixing the ends in the ground along the two sides of the stock. To facilitate easier bending the portion of the stake passing over the scion should be scraped with a knife to remove the wood portion and leaving back only the cane-like skin of the bamboo. This latter method is suitable especially when grafting is done in beds. To avoid injury to the scion a piece of cloth may be placed over it before passing the rubber band or bamboo stake. After about a month the pressure is released by which time the union should have taken place. The stock and the scion may also be given an inclined cut at 45° angle, as in the case of tongue graft, instead of the customary transverse (horizontal) cut.

The cleft graft is performed with such pendent species as *Aporocactus flagelliformis*, *Chamaecereus silvestrii*, etc. First the stock has to be beheaded at the desired height and a desired V-shaped notch is given on the top of the stock just long enough to accommodate the scion and never longer than that. Then two slanting cuts are given on both sides at the bottom of the scion to form a wedge. To obtain a perfect union all the cuts should be even and not irregular. Then the wedge-shaped cut of the scion is inserted in the cleft of the stock and held in position by pinning two long cactus spines through the stock and the scion. In addition to this, it may be necessary to tie some string or raffia around the graft to prevent the notch from spreading. The string or raffia should not be tied too strongly as it will cut into the stock. The spines and the strings are to be removed carefully once the union is complete.

In splice graft one side of both the scion and the stock is spliced at a 45° angle and both held in position by wrapping with raffia. With the globular-shaped species it is possible to graft by making a cone-shaped groove at the base of the scion and inserting it into a stock in which a wedge-shaped cut is inflicted to just fit in the cone of the scion. Then these are kept in position as in the case of flat graft. But, in general, flat graft should be preferred to this. (For illustrations see page 28.)

Grafting is generally done in the summer as the climate is dry at that time and the possibility of rotting is less. In and around Delhi, September is found to be a good period. As soon as grafting operations are over, the newly grafted plants are removed to a warm shady place, so that the freshly cut surface does not dry up rapidly. Normal watering of the stock has to be done, taking care that water does not enter the graft joint.

Rootstock-scion incompatibility may lead to various problems such as stunted growth of the scion, the scion sending roots into the stock, bulging of the stock below the joint, etc., and hence it is imperative to select the right stock for each scion. A list of suitable stocks for different scions are given below, based on personal experience.

Name of rootstock	Name of Scion
A. *Cephalocereus albispinus*	1. *Cephalocereus senilis*
	2. *C. senilis* 'cristatus'
	3. *Cleistocactus strausii* 'Cristate'
	4. *Espostoa plumosa*
	5. *E. plumosa* 'Cristate'
	6. *Haageocereus bicolor*
	7. *Mammillaria angularis* 'Cristate'
	8. *Notocactus leninghausii*
	9. *Opuntia tunicata*
B. *Cephalocereus* sp.	1. *Cleistocactus strausii* 'Cristate'
	2. *Haageocereus bicolor*
C. *Cleistocactus baumannii*	1. *Aporocactus flagelliformis*
	2. *Chamaecereus silvestrii*
	3. *Mammillaria elongata*
	4. *M. gracilis* (*fragilis*)
D. *Myrtillocactus geometrizans*	1. *Cephalocereus senilis cristatus*
	2. *Gymnocalycium mihan ovichii rubra* and hybrid
E. *Opuntia tomentosa*	1. *Opuntia microdasys*

Pests and diseases: See Chapter XXVI.

Classification

The family *Cactaceae* is divided into three tribes:

Tribe I. *Pereskieae* – Consisting of one genus *Pereskia*
Tribe II. *Opuntieae* – Some genera of this tribe are: *Pereskiopsis, Pterocactus, Nopalea, Opuntia* and *Grusonia*
Tribe III. *Cereeae* – This large group is again subdivided into 8 subtribes.

Subtribe I

Cereanae: Plants of this subtribe are terrestrial, erect, bushy, or occasionally diffuse, slender, or stout; the stems and the branches are with few joints or many-jointed; generally very spiny. Flowers are borne singly, rarely many, on the upper part of matured areoles. Flowers may either be day-blooming or night-blooming. A few important genera of this subtribe are: *Cephalocereus, Cereus, Cleistocactus, Espostoa, Haageocereus, Myrtillocactus, Oreocereus, Pachycereus, Pseudoespostoa,* and *Trichocereus.*

Subtribe II

Hylocereanae: Some important genera are, *Aporocactus, Hylocereus* and *Selenicereus.* The plants of this subtribe are mostly epiphytic, elongated, vine-like climbing, pendent, or trailing. The stems and branches are ribbed or angled, rarely flat. These plants are mostly night-blooming with large flowers.

Subtribe III

Echinocereanae: Low-growing terrestrial cacti, solitary or occasionally forming clumps; stems ribbed and spiny. Flowers are borne from lateral old areoles and not on the top of the plant. Some important genera are: *Chamaecereus*, *Echinocereus*, *Echinopsis*, *Lobivia*, and *Rebutia*. The cacti of this subtribe are commonly known as 'Hedgehogs'.

Subtribe IV

Echinocactaneae: Terrestrial plants, usually low and small, but occasionally several metres tall, solitary, sometimes in clumps; stems are ribbed, spherical or short cylindrical. Flowers are solitary, borne from young areoles on the apex of the plant and are among the showiest of all. The important genera are: *Ariocarpus*, *Astrophytum*, *Echinocactus*, *Ferocactus*, *Gymnocalycium*, *Hamatocactus*, *Lophophora*, *Notocactus*, and *Parodia*.

Subtribe V

Cactanae: The characteristic of this terrestrial subtribe is the curious wooly or bristle-like structure formed through the crowding of areoles on the top of the plant. The genera included are *Cactus* (*Melocactus*) and *Discocactus*. The plants are commonly known as '*Melon Cactus*' or Turk's cap.

Subtribe VI

Coryphanthanae: Low-growing terrestrial plants, mostly globose, occasionally cylindric and rarely elongated, solitary but frequently forming clusters; stem tuberculate and the tubercles are usually arranged spirally. Flowers are borne solitary at lateral or apical areoles but never at spine-areoles. Some important genera are: *Coryphantha*, *Mammillaria*, and *Thelocactus*.

Subtribe VII

Epiphyllanae: Mostly tropical epiphytic plants, generally growing on trees, but also in ground which is rich in humus. The stems are flattened leaf-like, many-branched and usually spineless. Flowers are large and showy, usually night-blooming. Some important genera are: *Epiphyllanthus*, *Epiphyllum* (*Phyllocactus*), and *Zygocactus*.

Subtribe VIII

Rhipsalidanae: These are epiphytic plants, generally growing on trees, occasionally on rocks and are commonly called 'Chain Cacti'. Stems are much-branched, slender, cylindric (teret) or flat, spineless; except in two genera flowers are small and without tube. Some important genera are: *Erythrorhipsalis*, *Pseudorhipsalis*, *Rhipsalidopsis*, and *Rhipsalis*.

Some Important Cacti

Aporocactus flagelliformis (Mexico).

The 'Rat-Tail-Cactus'. Stems are slender, creeping or long pendent, closely ribbed having brownish spines; flowers are 7-8 cm long and crimson in colour.

Ariocarpus fissuratus (Syn. *Anhalonium fissuratum, A. engelmannii*) (Mexico, W. Texas, U.S.A.).

A star-shaped rosette, the surface horny and irregularly fissured; the flowers are lovely pink. To a novice onlooker the plant will appear to be shrivelled and dead.

A. retusus (Syn. *Anhalonium retusum, A. furfuraceum, A. prismaticum*) (Mexico).

The plants are globular or more or less depressed; the numerous erect leaf-like tubercles (rounded protruding body) form a rosette, greenish-grey in colour; flowers are pale-yellow. It is known as *Ariocarpus trigonus* in horticulture.

Astrophytum asterias (Syn. *Echinocactus asterias*) (North Mexico).

The plants are low, circular, rather dome-shaped, with a star-shaped depression, green but tinged coppery with white scales and areoles are woolly, spineless. The flowers are yellow with red markings in throat.

A. capricorne (Syn. *Echinocactus capricornis*) (North Mexico).

Sub-globose; up to 25 cm high; ribs 7-8; spines curved; flowers are lemon-yellow with red throat, large, widely spreading.

A. myriostigma (Syn. *Echinocactus myriostigma*) (Central Mexico).

Generally 5-ribbed, which are very broad, acute, covered with small white spots; areoles spineless. Flowers are yellow, with brown tips on the outer petal. Commonly called 'Bishop's Cap'.

A. ornatum (Mexico).

Somewhat similar to the above but with 8 ribs which are sharper and wavy; spines few and curved. Flowers are lemon-yellow.

Cactus melocactoides: See *Melocactus melocactoides.*

Cephalocereus senilis (Mexico).

A beautiful cactus of slender column, growing up to 12 m in natural habitat, closely ribbed, covered with long gray hairs; flowers are rose in colour. Commonly called 'Old Man Cactus'.

C. senilis cristatus.

A curious cristate form of the above; fan-shaped, thickly covered with snow-white hairs.

Cereus peruvianus monstrosus.

A garden form commonly called 'Curiosity Plant'. The monstrose branches grow continuously in an irregular fashion; bluish-green and spiny.

C. peruvianus monstrosus minor.

A dwarf plant somewhat similar to the above; deep green, brown spines are fine needle-like.

Chamaecereus silvestrii (Syn. *Cereus silvestrii*) (Argentina).

The short cylindric green stems come in clusters, which are covered with soft white spines. The flowers are orange-scarlet and borne very freely, especially when grafted. There are two other forms: (a) crested, and (b) variegated yellow.

Cleistocactus baumannii (Syn. *Cereus baumannii*, *C. tweediei*) (Argentina, Paraguay, and Uruguay).

The columnar stems are thin, ribbed and branching at base, covered with whitish spines, sometimes yellow or brown. Flowers are tubular and orange-scarlet.

C. strausii (Syn. *Cereus strausii*, *Borzicactus strausii*) (Bolivia).

The slender columnar stem is green, clustering in habit and covered with bristle-like white spines with yellow tinge. Flowers are tubular and carmine in colour. There is a cristate form, which is fan-shaped and beautiful.

Coryphantha echinus (Syn. *Mammillaria echinus*) (Western Texas, U.S.A.).

A small, ball-like globular plant up to 5 cm diameter, solitary; numerous radial white spines pressed against the plant. The flowers are yellow. Other species are *C. compacta* and *C. radians*.

Echinocactus grusonii (Mexico).

'Golden Barrel'. Sometimes jokingly referred to as 'Mother-in-Law's Chair', because of the cushion-shape and numerous spines. This develops into a giant globe of about one metre diameter, numerously (21-37) ribbed, which are thin and high, light green, but covered with numerous golden-yellow spines which fade with age. Flowers are yellow and are borne on the apex during day time. It is generally solitary but sometimes develops two to three heads from the base; var. *cristatus*, a crested form of the above, looks like a large cock's comb which is covered with golden spines.

Echinocereus dasyacanthus (Texas, U.S.A., Mexico).

'Rainbow Cactus'. The plants are small, cylindric; spines are radial, spreading, pinkish when young but changing to gray. Spines of intermediate ages are variously coloured, hence the popular name. Flowers are yellow and tinged red.

E. longispinus (Probably Mexico).

The plants are elongated; spines long, cream coloured but red-tipped. *C. purpureus* and *E. roetteri* are also recommended.

Echinopsis.

The generic name is Greek, meaning hedgehog, referring to the armament of the plant. Plants are generally low, globular or short cylindric. The flowers are large and often fragrant (e.g., *E. multiplex*). Some species such as *E. hamatacantha* are night-flowering.

Epiphyllum.

The plants of this genus were previously known as *Phyllocactus*. Plants are generally epiphytic, branches are flat, leaf-like, sometimes 3-angled, spines numerous when young, but at adult stage wanting. Flowers are very large. Though generally day-blooming, some are nocturnal of which the 'Queen of the Nights' *E. oxypetalum* (Syn. *Phyllocactus oxypetalus*, *P. grandis*, *P. latifrons*) bears large fragrant white flowers tinged red outside. There are a few beautiful hybrids.

Espostoa lanata (Syn. *Pilocereus lanatus*) (Peru, Ecuador).

The plants are columnar, 2-4 m high. Stem sometimes branching, 4-10 cm in diameter, covered with cotton-like white hairs through which the numerous spines protrude out; flowers are pinkish. A beautiful species commonly called "Peruvian Old Man".

E. plumosa.

Another beautiful species similar to the above but more hairy, with smaller thorns. There is a cristate variety which is very ornamental, fan-shaped, covered with white hairs, the hair colour changing when grafted, depending upon rootstock.

Ferocactus horridus (California, U.S.A.).

Globular large plant; broadly-ribbed, numbering 8-13; spreading radial spines 8-12, all central spines are straight, reddish except one which is long, flattened, and hooked. Flowers and fruits not known.

F. latispinus (Mexico).

The plants are large, globular, somewhat depressed; ribs 15-23; radial spines 8-10 in number, white to pink, central spines are 4 or more, stouter and highly coloured compared to radials, straight except one which is flat and hooked. Flowers are rose to purple.

F. pringlei (Central Mexico).

Stems are at first globose, later becoming cylindric; one of the largest (height up to 3 m) species, grows in clumps; ribs 16-18 compressed; spines are hard, sharp, curved and red in colour. Flowers are yellow and outside red.

Gymnocalycium.

Low, globular, strongly ribbed plants; ribs tubercled, tubercles often protruding at base forming a knob resembling a chin and hence the common name 'Chin Cactus'. The flowers are bell-shaped, usually large in comparison to plant size; white, pink or rarely yellow.

(a) *G. mihanovichii* (Paraguay): The plants are hemispherical; greyish-green, banded red; ribs 8 in number, prominent, acute; spines spreading, radial, and straw-coloured. The flowers are yellowish-green tinged with red, free-flowering.

(b) *G. mihanovichii friedrichii* (Uruguay, Argentina): Similar to the above in shape, but with triangular ribs and coppery-green in colour; spines banded cream and yellow. The flowers are pink when fresh.

(c) *G. mihanovichii friedrichii* 'Rubra': A small globular cactus strikingly ornamental, being bright red (or yellow) in colour. Should be grafted for better growth as chlorophyll content is poor.

(d) *G. multiflorum* (South Brazil to Argentina): Small globular or sometimes short columnar; 10-15 ribbed; spines radial, flat. The flowers are light pink. A variable species with many listed varieties.

(e) *G. quehlianum* (Argentina): A depressed globe; 8-13 ribs, dark olive; tubercles rounded; spines radial, recurved. The flowers are white with red at the centre.

(f) *G. saglione* (Argentina, Bolivia): The plants are globular, often very large; ribs 13-32 depending on size, low and broad; spines are curved, dark brown to black; the flowers are white or slightly tinged with pink.

Haageocereus bicolor (South America).

A columnar cactus of light green colour; ribs dense and 14 in number; areoles thickly covered with creamy, woolly tufts; spines are radial, fine, tipped brown.

Hamatocactus setispinus (North Mexico, Texas, U.S.A.).

The plants are globose, 15 cm tall; ribs thick, 13 in number; spines radial, grey-brown, the central spines with fish hook. The flowers are yellow with red in the throat.

Hylocereus lemairei (Syn. *Cereus lemairei*) (Trinidad, Tobago).

A climber with 3-angled dark-green stems; rooting on one side; flowers are very large (25 cm long); linear outer segments yellowish-green, tipped bronze; the rather broad inner segments are white, tinged pink; stamens are numerous; night-blooming.

H. undatus (Syn. *Cereus undatus*) (Brazil).

Epiphytic; stem 3-angled; among the largest night-blooming *Cereus*. The 30 cm long white flowers bloom only for one night.

Lobivia famatimensis (Syn. *Echinocactus famatimensis*) (Argentina).

The plants are small, cluster-forming; stems short cylindric; spines many, short, yellowish; very free-flowering, especially when grafted; flowers borne at the top of the plants, colour varying from pale yellow to red.

Some other species are *L. ferox*, *L. hertrichiana*, *L. jajoiana*.

Mammillaria bocasana (*Mamillaria*, as sometimes written, is not correct).

The *Mammillarias*, according to old American use, is referred to as *Neomammillaria* but not by International Rules of Botanical Nomenclature. The *M. bocasana* is commonly called the 'Powder Puff', because of its appearance like a cotton ball as the small round plant is covered with silky white hairs; the fish-hook spines are brown in colour. The flower buds are rose-coloured, but flowers are described as white.

M. bravoae (Mexico).

A small (5-7 cm high) globose plant almost covered with spines and wools; radial spines are white and the central one is stiff. The flowers are pink.

M. camptotricha (Mexico).

Popularly known as 'Bird's Nest', as the small globes form into clusters; bristle-like spines are yellow, often twisted; flowers are white within and greenish without.

M. candida (Central Mexico).

Popularly called 'Snowball'. Plant globose, grows in tufts, almost hidden by white spines; flowers are 2 cm long and rose-coloured.

M. elongata (Central Mexico).

The plants are small, cylindric in shape, densely clustering, erect, ascending or prostrate; the interlacing radial yellow spines are arranged in a most artistic manner almost covering the plant; flowers are white.

M. fragilis (Syn. *M. gracilis*) (Mexico).

Appropriately known as 'Thimble Cactus' because of the like shape; stem branching near the top breaking off easily; radial spines white; flowers are small, lasting several days, cream-coloured but outer side pinkish.

M. geminispina (Syn. *M. bicolour*, *M. nivea*) (Mexico).

The plants are club-shaped becoming cylindric, clustering; radial spines many, white, central few (2 to 4), stouter, longer and black-tipped. The flowers are red.

M. hahniana (Mexico).

Popularly called 'Old Lady'. A small globe somewhat flattened on the top, either solitary or sometimes in clusters. Covered with white spines and long white hairs. Flowers are tiny and crimson in colour.

Mammillaria is a very large genus comprising about 150 species, many of which not mentioned above are also quite ornamental.

Melocactus melocactoides (Syn. *M. violaceus*, *Cactus melocactoides*) (South East Brazil).

The characteristic feature of this genus is the formation of a woolly cap or cephalium on the apex when the plants attain flowering size. It contains wool and bristles but no spines. At this stage the main stem stops growing but the cephalium grows annually.

This species is small (8 cm high, 15 cm broad), conical, light green, with 10 broad ribs; radial spines white; cephalium white, with brown bristles; flowers are pink.

Notocactus haselbergii (Syn. *Malacocarpus haselbergii*) (South Brazil).

A beautiful, small (7-8 cm in diameter), globular species having 30 low ribs; the silvery-white spines with yellowish tip are soft. Flowers are orange-red in colour.

N. leninghausii (Syn. *Malacocarpus leninghausii*) (South Brazil).

Very attractive cylindrical column, up to 1 m tall; ribs closely arranged, covered with soft yellow spines. Flowers are citron-yellow.

N. scopa (Syn. *Malacocarpus scopa*) (Brazil, Paraguay).

Globular when young but becoming cylindric with age. Almost covered with white radial and long brownish central needle-like spines; flowers are satiny and canary-yellow in colour. Var. *cristata* is really ornamental, fasciated and coiling into a serpentine form, covered all over with white hairs. All these *Notocactus* species also go by the generic name *Echinocactus*.

Nyctocereus serpentinus (Syn. *Cereu serpentinus*) (Mexico).

Stems are cylindric, many-ribbed, growing in clumps, initially erect, then clambering against bushes or walls or, if no support available, creeping or hanging up to 3 m long; areoles woolly; spines white to brownish; funnel-shaped flowers are large and white; a night-bloomer.

Opuntia clavariodes (Chile)

Grows in a fasciated or cristate form looking like 'Sea Coral'; very interesting.

O. microdasys (North Mexico).

Called 'Bunny-Ears', because of the young pad-like growths coming on top of the older ones and appearing like ears. Joints are flat, oblong or roundish, velvety; areoles close and conspicuous because of neat arrangement and light brown tufts of bristles which stick easily if touched and painful. Flowers are yellow. In var. *albispina* the areoles have soft white glochids; in cv. 'Lutea' the glochids (tufted barbed hairs) are pale honey-coloured.

O. tetracantha (South Arizona, U.S.A.).

A low, branched bush (1.5 m tall), slender joints cylindrical in shape; spines needle-like; flowers are greenish-purple.

O. tomentosa (Mexico).

Grows to a tree (3.5 m tall); joints very fleshy and thick; green spines few; flowers are yellow; the carmine-red fruits at the apex look ornamental.

O. tunicata (Central Mexico, Ecuador to Chile).

A species with whorled branches, growing to 60 cm, joints globular or short cylindrical. The plant is covered thickly with vicious white sheathed and barbed spines; flowers are yellow.

O. vestita cristata.

A mutation. Joints twisted into fan-like shape and covered densely with white hairs.

O. vulgaris (Syn. *O. monacantha*) (Brazil, Argentina, Uruguay).

Joints flattened, fleshy, green, practically spineless; flowers are yellow; the unripened fruits root and form little ears. In var. *variegata*, the joints are patterned green and white.

Some other species are *O. cylindrica*, *O. gosseliniana*, *O. pachypus*, *O. rufida*, and *O. subulata*.

Oreocereus celsianus (Syn. *Pilocereus celsianus*) (Bolivia, Peru, Chile).

Grows in clumps to about 1 m high; the new branches are either prostrate or ascending at basal portion, erect above, 8 cm thick; ribs 10; areoles have long hairs and reddish spines; flowers are dark red.

O. maximus (South Bolivia).

Stout columnar; spines colourful and body entirely covered with white hairs.

O. ritteri.

A beautiful columnar species; body almost entirely hidden with long silvery silken hairs.

O. trollii (Bolivia, Peru).

Columnar, branching at base; body covered with long, silky creamy hairs; flowers are rose in colour.

Parodia aureispina (Argentina).

Commonly called 'Tom Thumb', a small bluish-green round plant; ribs spiral; spines white, centrals bright yellow. The flowers at early stage are golden-yellow in colour.

P. maassii (*maasii*) (Syn. *Malacocarpus maasii*) (Bolivia)

A small round plant, yellowish-green in colour, ribs spiral and prominent, 13 in number, spines radial on knobs, white in colour, central spine hooked.

Pereskia aculeata (also spelt *Peirescia, Peireskia* and *Perescia*) (Tropical America).

A shrubby, leafy plant with woody stems, branches clambering, behaving as vines; spines brown, needle-like; flowers are creamy, lemon-scented. Var. *godseffiana* is a beautiful sport with broader leaves, which are yellow-green with tints of copper and red, undersurface rosy-red.

P. bleo (Colombia, Panama).

Grows to a small tree up to 6 m in height. Young branches are reddish. Leaves 20 cm long; spines black; flowers are pink.

Pseudoespostoa melanostele (Peru).

Similar to *Espostoa lanata*, having rows of brown-tipped needle-like spines; densely covered with white hair, the density increasing with age.

Rebutia fiebrigii (Bolivia).

A small plant, globose to cylindrical, clustering; tubercles spiralled; spines white; flowers are vermilion.

R. haagei (North Argentina).

A green to reddish clustering cactus; tubercles spirally arranged; radials bristle-like; flowers are pink.

R. kupperiana (Bolivia).

A globe with depressed top; tubercles in 20 rows covered with light spines; very free-flowering even when about 2.5 cm high; flowers are scarlet, attractive.

Rhipsalis.

A genus of *Cactaceae*, quite different from the desert types and needs much moist atmosphere. These are epiphytes, the stems are much-branched, either cylindrical, angled or flattened. The areoles are hairy, woolly or with bristles, and are found irregularly scattered on the cylindrical forms or on the margins of the flattened forms; the small flowers are borne singly at an areole during day or night; fruits are round, white or coloured, and juicy. Some important species are *R. capilliformis*, *R. cassutha*, *R. clavata*, *R. crispata*, *R. micrantha*, *R. oblonga*, *R. robusta*, *R. roseana*, *R. teres*, and *R. virgata*.

Selenicereus grandiflorus (Syn. *Cereus grandiflorus*) (Jamaica, Cuba).

The stems are clambering, 2.5 cm in diameter, night bloomer; flowers about 30 cm long, white inside, salmon outside, with a powerful vanilla smell. It is an epiphyte.

S. urbanianus (Syn. *Cereus urbanianus*) (Cuba).

An epiphyte possessing aerial roots, clambering, 4-5-angled joints, stems usually light-green but occasionally turn purple throughout. Spines about 1 cm long, brownish. A night bloomer; flowers are large (20-30 cm long), white within, tan without. *S. vagans* (Mexico) is another night bloomer.

Trichocereus.

The plants are columnar with ribs more or less branched; night bloomer, flowers large, largely white, fragrant with scales and hairs on the flower tube.

(a) *T. candicans* (Syn. *Cereus candicans*) (Northern Argentina): Flowers are very large and highly sweet-scented.

(b) *T. schickendantzii* (*Echinopsis schickendantzii*) (Argentina): Flowers are white, lasts for several days, some reports say it is scented while others disagree; night bloomer.

(c) *T. spachianus* (Syn. *Cereus spachianus*) (West Argentina): Flowers are 30 cm long and 15 cm in diameter, white.

Zygocactus.

An epiphyte, stems are flattened, jointed, dichotomously branched, margins serrated. There are two prominent teeth or claws at the apex. The only species is *Z. truncatus* (Syn. *Epiphyllum truncatum*) (Brazil), which is commonly known as 'Crab Cactus' or 'Thanksgiving Cactus'. The flowers are 6-7 cm long, last for several days. Flowers are irregular from scarlet to white. This is generally grafted to give the pendulous plant some height. *Pereskia* or *Hylocereus* can be used as stock. There are several beautiful hybrids.

(B) SUCCULENTS

Succulents which form a large cultural group in the garden are fleshy plants. The main characteristics of succulent plants are very fleshy foliage or stem or both. These plants inhabit mostly dry desert localities in open situations, and are capable of withstanding long hot spells of drought as they store sufficient moisture in their succulent body parts. The atmosphere in alpine regions at high altitudes is also physiologically dry and the vegetation of such places is also succulent in nature which enables them to withstand a dry climate. The salinity of sea coasts also influences the growth of succulents.

Succulents show a great diversity in form, many having very ornamental appearance, while others bear beautiful flowers. Cacti also come under the cultural group of succulents but these have been dealt with separatey above.

The cacti differ from the succulents in that they have areoles which are minute vertical pores occurring as incomplete canals and not perforating the wall completely. In rare cases areoles are perforated by minute pores through which a mucilaginous secretion oozes out. The succulents have leaves, whereas except in one or two genera these are absent in the cacti. The succulents are in demand in the garden primarily for their hardiness and modesty in their requirements in cultivation.

The succulents have various uses in the garden. They can be used in the same way as foliage and flowering plants for decorating rooms, balconies, and verandas. Many succulents are suitable for window-box gardening, bowl gardening, trough gardening. hanging baskets, etc. Succulents are grown as specimen plants in lawns (e.g., variegated *Agave* and *Furcraea*), in pots and rock gardens. Many succulents such as *Agave* form an impregnable hedge, while others such as *Echeveria* are very good as edging or as carpet bedding plants.

Climatic and soil requirements: (a) *Climate:* Succulents inhabiting the arid regions will do well in most of the places in India. But the types which are natives of the alpine regions can be cultivated with success only under a sub-tropical climate of medium to high elevation. The succulents do very well in the mild climatic conditions of Bangalore, Kalimpong, and other places with a similar climate. The climatic and other requirements of most of the succulents, with a few exceptions, are almost similar and hence these can be grown together under one greenhouse. The plants of the mesembryanthemum group such as *Lithops* need special treatment regarding watering and these may be grown separately from others. As a general rule, all succulents need plenty of air, sunshine, a porous soil with good drainage, and a relatively dry atmosphere. Some of the succulents such as *Haworthia* and *Gasteria* dislike direct sunlight.

(b) *Soil:* The succulents are not at all fastidious about the soil as long as it is porous. Although a few can grow well even in pure sand and others in a comparatively rich compost, it is possible to grow a majority of them in a common compost. The following compost may be used for all the succulents.

Leaf-mould	1 part by volume
Coarse sand	2 parts by volume
Well-rotted cowdung	$\frac{1}{3}$ part by volume
Garden loam	$\frac{1}{3}$ part by volume
Crushed brick and charcoal	$\frac{1}{3}$ part by volume

For *Agaves*, *Aloes*, and *Euphorbias*, the compost can be made a little richer by increasing the quantities of leaf-mould and cowdung manure. *Aeonium* and *Echeveria* like a sandy porous soil, but not very sandy. Many succulents like the presence of lime in the compost while others also do not resent this and hence it is advisable to put traces of ground chalk or limestone or old mortar in the general compost.

Housing: In the tropical climate of India, the succulents can be housed in a greenhouse, with all the sides, except the west, open. The west side is covered by raising a brick or stone wall just short by 45-60 cm from the roof and this gap is covered with glasspanes. The width of such roofs is generally between 4.5 m and 6 m. The succulents are potted and displayed in a gallery. It is also possible to grow them directly in soil, but this practice should preferably be avoided, especially by the amateurs.

Propagation: For a majority of the succulents, propagation from seeds is possible. The ideal climatic conditions required for proper germination are a temperature of 30-35°C and a relative humidity of 90 per cent. After germination, the seedlings should be shifted to a drier place.

Vegetative propagation by means of cuttings is often practised. The cuttings are taken from a mature and woody growth. The cutting is taken with a sharp knife and the cut end is placed in powdered charcoal and allowed to dry thoroughly. The cutting is then rooted in sterilized coarse sand. *Adenium*, *Euphorbias*, *Pedilanthus*, etc., can be propagated by this method. The succulents such as *Cotyledon*, *Echeveria* and *Sedum* can be multiplied from leafy cuttings. Well-developed and matured leaves are detached and treated the same way as stem cuttings and are just laid over coarse sand for rooting. But many people suggest a compost consisting of two parts of coarse sand and one part of dry leaf-mould for rooting of stem and leaf cuttings. The leaves or stem cuttings should never be inserted deep as it may cause rotting. There are a few succulents such as *Kalanchoes*, which bear on the margins of the leaves small adventitious buds, which are detached and planted like cuttings. *Agaves* and *Furcraeas* are increased by the division of suckers. In many species of *Agave* and *Furcraea* bulblets are produced in the inflorescence in quantities and are used for propagation.

Cultural hints: Of all the cultural practices, watering is the most important. A perfect drainage is absolutely necessary. For this purpose, one-fourth of the pot is filled with broken bricks and charcoal and then the compost is filled. The best time to water a succulent is when the soil is sufficiently dry, but not completely dry. Succulents have a very small and limited root system and hence over-watering is harmful. Watering by immersion is not advisable as it leaches out plant food. However, succulents may be watered by placing them over a receptacle containing just enough water the pot can absorb.

Repotting is an important cultural practice in succulents. The time for repotting depends upon the type of succulents. Those having a rest period are repotted at the commencement of growth during the spring, while others having no rest period can be repotted any time but preferably, for convenience sake, during the spring. The general belief is that all succulents benefit from repotting every alternate year. But the vigorous growing ones whose growth covers the whole pot surface within one year should be repotted every year. This repotting is essential since before each watering the compost has to be checked to make sure that the plant needs watering. But if the succulent covers the pot surface completely it is not possible to check the compost. At the time of repotting the old compost including the crocks should be replaced completely. The compost should be filled in a "crumbly moist" (just sufficiently moist to form a ball when moulded) condition. When this method is followed the repotted plant need not be watered for about 4 to 7 days, depending upon how long the compost takes to dry.

In the summer, some succulents need shading, and many will not thrive well at all inside a glasshouse because of excessive heat. Such plants should be kept out of doors during the summer. If the heat is not excessive, partial shade may be provided by a lime wash of the glass roof.

For pests and diseases, see Chapter XXVI.

Some Important Succulents

Adenium obesum: See page 131.

Aeonium (*Crassulaceae*).

Succulent herbaceous plants with woody and more or less branched stems. The leaves are borne in rosettes at the end of the branches. Plants like plenty of water in the summer and a rich sandy soil. Propagation is from seeds and cuttings. A few important species are:

(a) *A. arboreum* (Syn. *Sempervivum arboreum*): A native of Morocco. Branches few; leaves are in dense rosettes at the end of branches.

(b) *A. canariense* (Syn. *Sempervivum canariense*): A native of Canaries Islands and commonly known as 'Giant Velvet Rose'. Leaves are in large, loose rosettes.

(c) *A. sedifolium* (Syn. *Aichryson sedifolium, Sempervivum sedifolium*): A native of Canaries Islands, a twiggy herbaceous shrub with tiny succulent obovate leaves.

(d) *A. tabulaeforme* (Syn. *Sempervivum tabulaeforme*): A circular-shaped plant.

(e) *A. undulatum* (*Sempervivum youngianum, S. undulatum, S. urbicum*): A native of Canaries Islands. A beautiful species.

(f) Another popular species is *A. haworthi.*

Agave (*Amaryllidaceae*).

The plants are sometimes commonly known as 'Century Plants' because of the popular belief that they flower once in 100 years, but actually, different species flower between 10 and 40 years. Agaves are very common ornamental succulent plants with typical xerophytic looks. The leaves are borne in rosettes which are thick, leathery, and sword-like. The margins of the leaves are usually teethed, the apex ending in a stout spine. Sometimes, the margins may be smooth but the tip will have a spine. The flowers are borne in a large column-like flower-stalk containing innumerable florets. The flowers last from several days to months and form seeds or bulbils which are used for propagation. With the maturity of seeds, the plant dies gradually. Plants produce offsets which are often rooted and are used for multiplication. Agaves are planted in rockeries, as specimens in lawns, tubs, large pots, and also for fencing.

Some important species are listed below:

(a) *A. americana:* A native of Mexico, with large rosettes; margins of leaves are toothed.

(b) *A. americana* var. *marginata:* Margins of leaves are broadly yellow.

(c) *A. americana* var. *medio-picta:* In this variety, the broad yellow line is along the centre of the leaves.

(d) *A. angustifolia* var. *woodrowii* (Syn. *A. woodrowii, A. cookei*): It is a very ornamental species having dwarf compact growth. The leaves are creamy-white with silver-grey stripes along the centre of the leaves. Difficult to cultivate.

(e) *A. filifera:* A native of Mexico. Many-leaved rosettes. Forms offsets. The leaves are stiff, bright-green, with white lines along the margin which split into loose filaments.

(f) *A. filifera compacta* (*A. pumila* of hort.): It is much more compact than the above.

(g) *A. lophantha poselgeri* (Syn. *A. lechuguilla*): A native of Mexico. The plant produces suckers. Rosettes are loose. The leaves are stiff, undulated, channelled, and snake-like.

(h) *A. parviflora*: A native of Arizona. Rosettes are small. Most ornamental. The leaves are dark green with white margins on the upper surface. Margins split into white threads.

(i) *A. sisalana:* A native of Mexico, commonly known as 'Sisal' in India and produces fibre for making ropes.

(j) *A. striata* (Mexico): A short-stemmed rosette plant. Leaves are dense, flaccid, angled, linear, up to 45 cm long, and narrowing to the tip to about 1 cm. Leaves are glaucous-grey with a dark green shade.

(k) *A. stricta* (Syn. *A. striata stricta*) (Mexico): Commonly called 'Hedgehog-agave'. Roundish rosette; leaves stiff, dense, spreading, incurving, narrow and linear, green in colour, 35 cm long.

(l) *A. verschaffeltii* (Mexico): An attractive species with a small rosette.

(m) *A. victoriae-reginae* (Mexico): A most attractive species which looks like a lotus flower.

(n) *A. victoriae-reginae laxior* (Syn. *A. ferdinandi-regis*): Similar to the above but the leaves are fewer and less compact.

Aloe (*Liliaceae*).

A group of hardy succulent plants, mostly stemless or with short stems but some species are tree-like. Leaves are in rosettes, fleshy, margins entire or toothed. In some species, the surface is variegated or spotted or striped. Many species bear attractive tubular flowers in long stalks. Aloes should not be exposed to rains as water will collect between the leaves and the plants may rot. Propagation is from seeds or suckers.

Some important species are listed below:

(a) *A. aculeata* (Transvaal): Leaves are very fleshy, broad lanceolate, bluish in colour, up to 60 cm long and 12 cm broad; flowers are lemon-yellow.

(b) *A. aristata* (South Africa): A rosette; leaves dense, slender, dark green covered with white warts.

(c) *A. brevifolia* (Cape Province, Africa): Leaves are triangular-oblong, thick, pale-green, flat above, rounded to keeled below; flowers are red in dense racemes; var. *variegata* is highly variegated and streaked with yellow.

(d) *A. excelsa* (South Rhodesia): Becomes tree-like, up to 6 m high.

(e) *A. ferox* (South Africa): Develops a 4.5 m stem; leaves are fleshy, bronze-green, and margins have brown-red teeth. Flowers are orange-red.

(f) *A. glauca* (Cape Province): Margins of leaves having red-brown spines. Flowering is in tall racemes of salmon-red colour.

(g) *A. pietersii* (Transvaal): Leaves are grey-glaucous, concave, armed with strong grey to brown spines on the margins. Flowers are orange-red.

(h) *A. saponaria* (Natal, Transvaal): Leaves are fleshy, broad lanceolate, bluish-green blotched with yellow-green bands; margins are toothed with large cream to brown thorns; flowers are red-yellow.

(i) *A. sladeniana* (South West Africa): Stemless rosettes are attractive and form into culsters. Leaves are moss-green, flushed with pink and covered all over the surface with translucent, white, oblong spots and margins are white-edged with small white teeth. Flowers are of rose colour.

(j) *A. transvaalensis* (Transvaal): Leaves are milky-green with many white-transparent oval spots arranged in cross bands and margins are armed with light-brown teeth. Flowers are salmon in colour.

(k) *A. variegata* (Cape of Good Hope): Leaves are blue-green with white spots arranged in irregular cross-bands. Flowers are tubular and salmon-coloured.

(l) *A. variegata ausana* (South West Africa): Similar to the above but the white spots are not arranged in cross-bands as in the above species.

(m) *A. zebrina* (Transvaal, Angola): Leaves are green but sometimes purplish, having numerous white spots which form almost straight lines across the leaves; racemes are branched and flowers are red.

Apicra (*Liliaceae*).

Closely related to Aloe but resemblance is much closer to *Haworthia* being differentiated from the latter by the regular form of the flowers. These are ornamental small plants useful for indoor decoration. The cultivation is like that of *Haworthia* and *Gasteria*. The important species are: *A. deltoidea*, *A. dodsoniana*, *A. egregia*, *A. pentagona* (Syn. *Aloe pentagona*, *Haworthia pentagona*), *A. spiralis* (Syn. *Aloe spiralis*, *Haworthia imbricata*). All are natives of Africa (South Africa, Cape Province, and Cape of Good Hope).

Beaucarnea: See *Nolina*.

Bowiea volubilis (*Liliaceae*).

A native of South Africa and often known as 'Sea Onion' or 'Climbing Onion'. It is grown more as a curiosity than for its decorative value. The light-green bulb, 20 cm, sometimes 45 cm, in diameter, remains above the ground from which a twining branched stem of green colour, with a few fleshy basal leaves, climb upward if supported by a trellis. The greenish-white flowers are insignificant and are borne at the end of the vine. It likes a sandy but rich soil and should be kept dry during the resting period.

Bryophyllum (*Crassulaceae*).

A small group of succulents similar to *Kalanchoes* and often included in it, but being distinguished from the latter by the drooping and not erect flowers, the sepals jointed, not free, and the stamens inserted at the base of the corolla tube, not above. Propagation is easy as plantlets appear in the notches of the leaves sometimes even on the plants or when placed in moist sand. The plants like a rich, well-drained soil and plenty of water. Flowering is during the winter. Some important species are: *B. crenatum*, *B. daigremontianum* (Syn. *Kalanchoe daigremontianum* or *daigremontiana*), *B. tubiflorum* (Syn. *Kalanchoe tubiflora*). Leaves are cylindrical in the last mentioned species. Flowers are reddish. Plantlets develop at the tips of the leaves. A native of Madagascar and the tropics.

Caralluma (*Asclepiadaceae*).

The plants resemble *Stapelias* and require similar cultural treatments. The stem is usually 4-angled and branching is generally at ground level. Some important species are *C. burchardii* (Canary Islands). *C. europaea* (Syn. *Stapelia europae*, *S. gussoneana*) (Mediterranean Island), *C. lutea* (South Africa) and *C. mammillaris* (Syn. *Stapelia pulla*, *Boucerosia mammillaris*, *Piaranthus pullus*) (South Africa).

Ceropegia (*Asclepiadaceae*).

Succulent plants which are often without leaves, but mostly twining with opposite fleshy leaves. The roots are generally tuberous or corm-like. Flowers are borne in small clusters which are beautiful and somewhat resemble *Stapelias*. A warm temperature, a well-drained soil and plenty of moisture suit the plants well. Propagation is from cuttings. Some of the important species are: *C. dichotoma* (Canary Islands), *C. elegans* (India), *C. fusca* (Canary Islands), and *C. woodii* (South Africa).

Cissus cactiformis (*Vitaceae*).

A native of South and tropical Africa. The species is more or less similar to *C. quadrangularis* but the internodes are much thicker, about 4 cm across.

C. quadrangularis (Syn. *Vitis quadrangularis*).

A very thick fleshy 4-angled succulent which is much contracted at nodes. It is a quick-growing climber and is generally leafless. A sandy-loam soil and a moist situation suit the plant well. This and the above species are propagated from cuttings. Flowers are green.

Cotyledon (*Crassulaceae*).

Ornamental succulent herbs, shrubs or sub-shrubs. They like a rich sandy soil. Some grow compactly and form clumps. Plants are propagated from seeds, cuttings, and sometimes from leaf cuttings with difficulty. Mainly grown for their ornamental look. Some important species are:

(a) *C. barbeyi:* A native of Eritrea and South Arabia. A succulent with large, fleshy shovel-like leaves which are whitish-green in colour.

(b) *C. flanaganii* (Syn. *C. decussata*)*:* A native of Cape Province to South Africa. A succulent with loose rosettes of leaves the margins of which are irregular. Leaves have a spatula shape and are grey-green in colour with white glaucescence.

(c) *C. orbiculata:* A native of Cape Province and shrubby in habit. Leaves are rounded above but narrowed at the base, whitish-grey with red margins.

(d) *C. undulata:* A native of Cape Province with shrubby habit. Leaves are broad, wedge-shaped, thickly covered with a silver-grey powdery substance and wavy at margins.

Crassula (*Crassulaceae*).

This genus represents a large number of succulent shrubs and herbs. Excepting a few, all *Crassulas* are easy to cultivate, but they perform better on the hills. Propagation is usually from cuttings including leaf cuttings. Seeds are also used for raising plants. In species like *C. falcata* enough material for cutting is not available and hence the top is cut and used as a cutting. The cuttings are dried for a few days in shade before inserting them in beds. A rich well-drained soil containing enough sand is suitable for the cultivation of *Crassulas*. They are natives of South Africa, tropical East Africa, Madagascar; and Arabia. Most of the plants have attractive and handsome forms. Some of the important species are: (a) *C. arborescens* (Syn. *C. cotyledon*) – the species rarely flowers; (b) *C. argentea* (Syn. *C. portulacea, Cotyledon ovata*); (c) *C. argentea* var. *variegata* – leaves variegated cream to orange-yellow with an orange-red edge; (d) *C. columnaris*; (e) *C. cornuta*; (f) *C. deceptrix*; (g) *C. densa* (Syn. *C. cephalophora*); (h) *C. falcata* (Syn. *Rochea falcata*); (i) *C. hemisphaerica*; (j) *C. imperialis* – a hybrid between *C. pyramidalis* and *C. lycopodioides*; (k) *C. lycopodioides*; (l) *C. obvallata*; (m) *C. orbicularis*; (n) *C. perfossa*; and (o) *C. pyramidalis* – a beautiful succulent with a conical shape, resembling pyramid.

Dasylirion (*Liliaceae*).

Evergreen herbs which are highly ornamental. Suitable for rockeries, terraces, staircases, etc. Propagation is from seeds or cuttings when branches are produced. There is a doubt whether it is a true succulent.

Duvalia (*Asclepiadaceae*).

These succulents are allied to *Stapelias*. *D. elegans* (Syn. *Stapelia elegans*) and *D. pillansii* are two important species.

Dyckia (*Bromeliaceae*).

They somewhat resemble *Agaves* but the spines are smaller and they flower regularly. A perfectly drained soil is essential. Propagation is from suckers.

Echeveria (*Crassulaceae*).

A much-branched succulent plant which forms clumps or small bushes with short stems. The leaves are ornamental, in various shapes and sizes, very fleshy and spirally arranged into rosettes. Most of the species are beautiful and suitable for growing indoors. Propagation is easy by means of cuttings. Pieces of stems, leaves and sometimes even the flower stalks strike root if dried well before planting. Seeds also germinate easily. Other cultural requirements are the same as for *Cotyledons*. *Echeverias* are suitable for carpet bedding. In many species the rosettes of leaves look like a fully open rose bloom. Some of the beautiful species are: (a) *E. agavoides* (Syn. *Cotyledon agavoides*, *Urbinia agavoides*); (b) *E. crenulata* (Syn. *E. hoffmannii*); (c) *E.* × *dasyphylla*; (d) *E. elegans* (Syn. *Cotyledon elegans*); (e) *E. glauca* (Syn. *E. secunda glauca*); (f) *E.* × *imbricata*; (g) *E. pallida*; (h) *E. peacockii* (Syn. *E. desmetiana*); (i) *E. secunda* (Syn. *Cotyledon secunda*); and (j) *E. setosa*.

Euphorbia (*Euphorbiaceae*).

This genus represents a large group of plants but only the succulent types will be dealt with here. The succulent *Euphorbias* are of varied habit. The form in *Euphorbias* may vary from thorny fleshy shrubs to plants having slender branches with the thickness of a pencil, some even with columnar stems, while others having spherical stems are also there. Besides these, various intermediate forms are also available. In most of the species the leaves are very much reduced or are small and soon drop off. But in a few species the leaves are conspicuous. Though the majority of *Euphorbias* have sturdy thorns from the leaf base, some are thornless as well. Some resemble the cacti in form and are often mistaken as such. *Euphorbias* exude a milky latex when injured.

Most of the *Euphorbias* are quite hardy under our climatic conditions and can be cultivated with ease. They need plenty of light and ventilation, a well-drained soil with plenty of coarse sand, broken pieces of charcoal and bricks. The foliage types can be watered quite freely while the more succulent types should be watered very carefully. The quick-growing types need a rich compost more than the slow-growing ones. The compost consists of a mixture of old cow manure and good loam, the proportion varying according to the type of the plant. The succulent types propagate easily from cuttings, provided the cuttings are dried well for about a couple of days before putting them in the beds. A mixture of peat (or leaf-mould), powdered charcoal and sand is suitable for planting cuttings. Many species can be raised from seeds also.

Except *E. splendens* and *E. bojeri*, which produce attractive flowers, most of the succulent types produce insignificant flowers. The milky juice in *Euphorbias* is poisonous although the latex of some species is used as medicine by natives.

A few handsome species are discussed below:

(a) *E. antiquorum:* A native of India. A thorny succulent with many branches; the stem is 4 to 5-angled, while the branches are generally 3-angled, constricted at joints. Leaves are small, roundish. Grows up to 3.5 m.

(b) *E. canariensis:* A native of Canary Islands. The plant is cactus-like, branching at ground level and growing up to 9 m. The stems are 4, 5 to 6-angled and brownish-green in colour.

(c) *E. decepta:* A spineless low-growing succulent of the Cape Province. The stem is globose, 7.5 cm in height and greyish to olive-green in colour. Numerous brownish-green branches arise from the top of the stem bearing deciduous small leaves.

(d) *E. globosa:* It is also a native of Cape Province. A dwarf succulent, devoid of spines and consisting of top-shaped main stem with globose tuberculate branches arranged haphazardly.

(e) *E. horrida:* A native of Cape Province. A cactus-like thorny succulent. The columnar stem has 8-14 sharp ridges with deep furrows. The plant is unisexual and is propagated from seeds.

(f) *E. inermis* (Syn. *E. serpentina*, *E. viperina*): A native of Cape Province. The main stem is thick with a series of circular, tuberculated branches 2 cm thick and 30 cm long. The plant appears like cactus.

(g) *E. mammillaris:* It is a native of Cape Province. A unisexual plant which is commonly known as 'Corncob Cactus'. The main stem is cylindrical and 8-17 angled. The angles are tubercled resembling a corncob. Usually there are many club-like lateral branches and spines are occasionally present. In var. *variegata*, the column is greenish-white with fresh green markings and pink tinge.

(h) *E. neriifolia:* A native of south India. A succulent tree with whorled branches and oleander-like leaves. The spines are short and dark-coloured. The stem is 5-angled.

(i) *E. neriifolia cristata:* A monstrose form from India with dissolved ribs.

(j) *E. obesa:* A unisexual plant from Cape Province. The stem is unbranched, more or less spherical or more tall than broad. Stems are usually 8-angled and not very conspicuous except at the apex. The surface is pale-green with dull-purple longitudinal and transverse stripes. There are rows of small knobs along the ridges. Propagation is from seeds only. The plant looks exactly like a globose cactus.

(k) *E. pseudocactus:* A thorny shrubby succulent from Natal. The branches are arranged in spray and are 4-angled.

(l) *Euphorbia splendens prostrata:* A shrub with spreading, somewhat climbing branches. The plant is full of grey spines and persistent small dark-green leaves. The plant flowers almost round the year in terminal clusters; the rounded bracts are red in colour. In var. *lutea* the bracts are cone-shaped, lemon-yellow in colour, tinged red at the edges. The plants are natives of Madagascar and are often referred to as 'Crown of Thorns'. To make the plants bushy, the stem is cut back.

(m) *E. symmetrica*: A thornless, dwarf succulent from Cape Province. The plant is a spherical globe like *E. obesa* but much wider. There are 8 broad ribs, and transverse purple bands.

(n) *E. tetragona:* A succulent attaining the size of a tree. The branches are arranged in whorls and are 4-angled.

(o) *E. tirucalli:* Though a native of South Africa, it was introduced very early in cultivation in India. A succulent attaining the height of a tree. The slender, brush-like branches are round. Leaves are small and soon fall off. A very common succulent in south India which is used as a hedge. Plants are unisexual. The milky juice is poisonous.

(p) *E. tuberculata:* A native of Cape Province. The stout stem is underground from the apex of which arise many prostrate, spineless branches arranged in a circular fashion. The plant resembles a cactus.

(q) *E. valida*: A unisexual plant from Cape Province. The plant is more spherical when young, but turning to cylindrical with age.

Some other important species are: *E. abyssinica, E. bupleurifolia, E. caput-medusae, E. caput-medusae minor, E. clandestina, E. lactea, E. lactea cristata, E. polygona*, and *E. trigona.*

Furcraea (Syn. *Fourcroya, Fourcraea, Furcroya, Furcroea*).

A group of succulent desert plants from tropical America resembling agave. It belongs to the family of *Amaryllidaceae*. Some of the species are very suitable for planting as specimens in the lawn, especially the variegated ones. They are also useful for rockeries, greenhouses and for growing in pots and tubs for display. The cultivation is the same as for agave. The flowering stem bears many bulbils which are used for propagation. Some important species are listed below which are really very ornamental and deserve a place in every garden.

(a) *F. gigantea* (Syn. *F. foetida*): It has up to about 50 fleshy, sword-shaped leaves which are 1.2 to 1.8 m long, borne in a rosette. The leaves are shining green and have negligible spines. The stem is about 90-120 cm. The flowering stalk is 9-12 m high. The flowers are milk-white within and greenish without.

(b) *F. gigantea striata:* A very showy variegated-leaved variety. The variegation is in stripes and broad bands and are ivory and whitish-green over rich-green.

(c) *F. selloa:* An almost stemless species with a rosette of 30-40 leaves measuring 0.90-1.2 cm in length with upcurved brown spines in the margins.

(d) *F. selloa marginata:* The leaves are glossy-green with broad creamy margins which may finally turn yellow.

Gasterhaworthia (*Liliaceae*).

A bigeneric hybrid between *Gasteria verrucosa* and *Haworthia margaritifera*, commonly known as 'Royal Highness'.

Gasteria (*Liliaceae*).

A group of very showy plants with fleshy thick leaves somewhat resembling *Aloes.* The leaves are usually arranged in 2 ranks (distichous) and sometimes in rosettes also. The tongue-shaped leaves are spotted with purple or white. They are popular plants for growing in rooms. Propagation is usually from leaf cuttings or offsets. Seeds generally do not breed true to type. *Gasterias* are native of South Africa. Some important species of *Gasteria* are: (a) *G. croucheri*, (b) *G. hybrida* (a hybrid), (c) *G. lingua* (Syn. *G. disticha, Aloe lingua*), (d) *G. maculata* (Syn. *Aloe maculata*), (e) *G. trigona*, (f) *G. verrucosa* (Syn. *Aloe verrucosa*) – the two-ranked dark green leaves are densely covered with whitish-grey tubercles.

Gastrolea.

It is a new genus formed as a result of intergeneric crosses between *Aloe* and *Gasteria*. The plants are intermediate in look. The following are the different types: (a) *G. beguinii* (*Aloe aristata* × *Gasteria* sp.), (b) *G. smaragdina* (*Aloe variegata* × *Gasteria candicans*), and (c) 'Spotted Beauty' (*Aloe* × *Gasteria*).

Graptopetalum paraguayense (Syn. *G. weinbergii, Cotyledon paraguayensis, Echeveria weinbergii, Sedum weinbergii*).

A succulent with a rosette of thick fleshy leaves resembling *Echeveria*. It belongs to the family *Crassulaceae* and is a native of Mexico.

Greenovia (*Crassulaceae*).

The plants are closely related to *Sempervivum* and also resemble *Aeonium* without the tall stem. The plants produce offsets. Propagation is from seeds. *G. aurea* (Syn. *Sempervivum aurem*) and *G. dodrantalis* (Syn. *G. gracilis*) are important species.

Haworthia (*Liliaceae*).

A group of small succulent plants, which are natives of South Africa. These were formerly included in *Aloe* but are now grouped separately because of the 2-lipped flower. The plants may or may not have a short stem. The leaves are fleshy, arranged in rosettes or closely overlapping, often covered with pearly tubercles. Flowers are insignificant horticulturally.

Haworthias are very attractive plants suitable for room decoration. They are easy to cultivate and a compost consisting of good loam, sharp sand and leaf-mould or peat, in equal proportions, suits them well. Propagation is by means of offshoots which are produced freely in most species. Leaf cuttings also root but with difficulty. Propagation from seed is not reliable as *Haworthias* hybridize easily. *Haworthias* should be repotted every two years.

While the majority of the species are in the form of dwarf rosettes, and some are star-shaped, there are others which are columnar and elongated in form. The species *H. truncata* is a horticultural novelty and is known as 'Window Plant'. The leaves are arranged in 2-ranks, oval-shaped in cross-section and rough warty in texture. The end of each leaf is flat, as if cut by a sharp knife, and is translucent. The important species are listed below:

(A) Rosette Forms.

(a) *H. arachnoidea* (Syn. *arachnoides*); (b) *H. attenuata clariperla* – a striking species, the leaves are covered with large white tubercles; (c) *H. browniana*; (d) *H. cuspidata*; (e) *H. fasciata* (Syn. *Aloe fasciata, Apicra fasciata*); (f) *H. mirabilis* (Syn. *Apicra mirabilis*); (g) *H. papillosa*; (h) *H. retusa* (Syn. *Aloe retusa, Apicra retusa*); (i) *H. setata*; (j) *H. subfasciata*; (k) *H. tessellata* (Syn. *Aloe tessellata* also spelt *tesselata*); and (l) *H. turgida* (Syn. *Aloe turgida*).

(B) Columnar or Elongated Forms.

(a) *H. coarctata* (Syn. *Aloe coarctata*); (b) *H. glauca* (Syn. *Aloe glauca*); (c) *H. reinwardtii* (Syn. *Aloe reinwardtii*); (d) *H. reinwardtii* var, *chalwinii*; and (e) *H. viscosa* (Syn. *Aloe viscosa, Apicra viscosa*).

Kalanchoe (*Crassulaceae*).

Ornamental succulent plants similar to *Bryophyllum*. The plants are natives of tropical Asia, Africa, and America. The leaves are fleshy and in many species spotted. The plants like a rich porous soil. Propagation is easy from cuttings and seeds. Leaf cuttings also root in many species. Many species produce ornamental flowers which last long in bouquets. The branches producing flowers die off after flowering or seed formation but new growths appear from base. The seeds are very minute and should not be covered after sowing. Some important species are described below:

(a) *K. daigremontiana* (Syn. *Bryophyllum daigremontianum*): Leaves are arched, fleshy, tricornered, and brownish-green in colour. Plantlets are produced at the margins of leaves. Flowers are grey-violet in colour.

(b) *K. eriophylla:* Leaves are thick, fleshy, ovate, green in colour but shaded coppery-purple at the margins.

(c) *K. fedtschenkoi* 'Marginata': An attractive sport from *K. fedtschenkoi*. The pale bluish-grey leaves are blotched with creamy-white at the margins.

(d) *K. marmorata* (Syn. *K. grandiflora*): A beautiful species with obovate leaves notched at the edges. On both surfaces of the leaves there are brown markings. Flowers are white.

(e) *K. quartiniana:* Leaves are smooth, broadly elliptic and toothed. It flowers in white terminal clusters.

(f) *K. tomentosa* (Syn. *K. pilosa*): A most beautiful succulent, known as the 'Panda Plant'. Stem is erect and branching. The leaves are spoon-shaped, soft, fleshy, and entirely covered with a dense white felt. The teethed margins of the leaves are tinged brown The flowers are white marked with brown stripes.

(g) *K. tubiflora* (*Bryophyllum tubiflorum*): See *Bryophyllum tubiflorum*.

Mesembryanthemum (Sometimes spelt *Mesembrianthemum*).

It belongs to the family *Aizoaceae* and is chiefly a native of South Africa but a few are from Australia, New Zealand, Canary Islands, and Arabia also. This genus includes the most interesting plants among the succulents, some are strikingly beautiful in plant form. Most of them bear beautiful flowers also. Formerly, a large number of species were included in this genus but based on recent studies the genus has been divided into many genera including *Mesembryanthemum*. In many species, the plants are in the form of tufts; the shape of bodies are obconical and the upper surface or the apex is truncated, as if cut across by a knife. The upper surfaces are often marked with artistic dots and stripes and resemble mimicry forms of coloured stones.

The shrubby types among the various genera have no soil preference provided it is porous. The delicate succulent types need a proper compost mixture, the chief criterion of which is good drainage. The compost in addition to other ingredients like leaf-mould and garden loam should also have quantities of old mortar, fine brick dust and plenty of sharp clean sand. Propagation is generally from seeds or cuttings. Unfortunately, a large majority of these plants do not do well in the plains of India.

Some of the important genera in this group are: *Aloinopsis*, *Argyroderma*, *Conophyllum*, *Conophytum*, *Faucaria*, *Gibbaeum*, *Glottiphyllum*, *Lithops*, *Mesembryanthemum*, *Ophthalmophyllum*, *Pleiospilos* and *Stomatium*.

Nolina recurvata (Syn. *N. tuberculata, Beaucarnea recurvata, B. tuberculata*).

It belongs to the family *Liliaceae* and is a native of South East Mexico. The plant attains a height of 9 m; the stem is slender, swollen at the base and branching only at the top. The leaves are green, slender, recurving, smooth-edged, 2 cm wide and 90-180 cm long. Propagation is from seeds, cuttings, or from offsets, if formed.

Pachyphytum (*Crassulaceae*).

A native of Mexico and closely allied to *Echeveria* and hybridizes easily with the latter. The cultivation is similar to that of *Cotyledon*. In *P. bracteosum* (Syn. *Cotyledon pachyphytum*), the leaves are obovate, swollen, club-shaped and curved. *P. brevifolium*, *P. compactum*, *P. hookeri* (Syn. *P. aduncum*, *Echeveria aducna*), and *P. oviferum* are some other important species.

Pachyveria (*Crossulaceae*).

A genus created as a result of hybrids developed involving crosses between *Pachyphytum* and *Echeveria*. The plants are intermediate in look. × *P.* 'Curtis' and × *P. haagei* are hybrids in this group.

Pedilanthus (*Euphorbiaceae*).

Variously known as 'Slipper Plant', 'Slipper Spurge', "Bird Cactus', 'Red Bird Cactus', and 'Jew bush'. These are cactus-like shrubby succulents. The flower resembles a bird's head and hence the name 'Bird Cactus'. The plants are hardy and suitable for greenhouse collection, indoor display and rock garden planting. Propagation is easy from cuttings.

(a) *P. tithymaloides* (Syn. *Euphorbia tithymaloides*, *E. carinata*, *E. canaliculata* and *Pedilanthus carniatus*): A shrub attaining a height of 90 cm when fully grown. Leaves are lanceolate, leathery and dark green in colour. In var. *variegatus*, the leaves are variegated white and tinged carmine-red. In var. *cucullatus*, the leaves are pale green but dark along the mid-rib and tinged red. Flowers are red. *Pedilanthus* is a native of West Indies.

Pelargonium (*Geraniacae*).

Some of the *Pelargonium* species have succulent stems. Most of these are suitable for growing in a milder climate. They prefer a rich sandy soil. Propagation is from seeds and cuttings. For details of cultivation see page 117.

(a) *P. carnosum:* A native of South Africa. A weird-looking succulent with a thick stem. Leaves are pinnately lobed and pinnatifidly cut. Flowers are in 6 to 8 flowered umbels and are creamy-rose in colour.

(b) *P. gibbosum:* Commonly called '*Gouty Geranium*' or '*Knotted Geranium*'. It is a native of South Africa. A climbing, intricately branched shrub, the branches being thin. Leaves are pinnately lobed and ruffled, while the greenish-yellow flowers emit scent in the evening.

(c) *P. juttae:* A native of South West Africa. A curious-looking low-growing succulent with short, tuberously thickened stem. There are a limited number of short branches on which the long-stalked, bisected leaves are borne.

(d) *P. tetragonum* (Syn. *Geranium tetragonum*): It is another important species, commonly called 'Square-stalked cranesbill'.

Portulacaria afra (*Portulacaceae* or *Portulaceae*).

A native of South Africa. It is commonly called 'Elephant bush'. A succulent shrub which may attain a height of 3 metres, with thick, fleshy, and browny stems. Leaves are obovate to roundish, thick, and juicy. Flowers are inconspicuous and pink in colour. In var. *variegata* the leaves are creamy-white with a flush of carmine-red at the margins. It is propagated from cuttings. A good succulent for growing in semi-shade in a rock garden and also for indoor decoration.

Sansevieria (*Liliacae*).

Herbaceous perennials which are succulent in growth and are usually grown for the elegant, erect, stiff, and sword-like leaves which are usually variegated. These are quite hardy plants and are frequently used for indoor decorations as they need very little sunlight. These are also suitable for growing in rock gardens. A sandy loam to rather heavy soil suit them quite well. They do not need frequent watering. Propagation is easy by division of suckers. They may be raised also from leaf cuttings about 7-8 cm long, placed in a sandy soil.

(a) *S. cylindrica:* A native of South Tropical Africa. Leaves are cylindrical, rigid. leathery and arching in tufts of 8-10 and often 1.5 m long.

(b) *S. stuckyi:* A native of Rhodesia. A creeping rosette in which the leaves are cylindrical and up to 90 cm in height.

(c) *S. trifasciata:* It is a native of Natal, Africa, and is most popularly grown *Sansevieria*. It is commonly known as 'Snake Plant' because of like' appearence. Leaves are linear-lanceolate, concave, fleshy, up to 120 cm long, deep green in colour with cross bands of grey white. Flowers are greenish-white emitting fragrance during night.

(d) *S. trifasciata* 'Golden Hahnii': A very showy, lowly rosette plant with firm leathery leaves which are having golden-yellow bands along the margins. A sport from 'Hahnii'. Some other variegated forms of this species are 'Goldiana', a beautiful plant, *S. trifasciata laurentii* 'Compacta' is also a very beautiful mutant.

(e) *S. trifasciata 'Hahnii':* Commonly called 'Bird's nest'. It is a small rosette which is a sport from *S. trifasciata laurentii.*

(f) *S. trifasciata laurentii:* It is the most popular variety in the garden, favoured for its sword-shaped leathery leaves, banded on either side with a colour lighter than the deep-green of the leaf blade proper.

Some other species are: *S. aubrytiana*, *S. ehrenbergii*, *S. guineensis* (Syn. *S. thyrsiflora*), *S. zanzibarica*, and *S. zeylanica.*

Sedeveria: See *Sedum* below.

Sedum (*Crassulaceae*).

These are perennial herbs, rarely sub-shrubs, which are succulent in nature. Most of the species are very showy and popular with the garden lovers. A few species are useful for carpeting poor and waste places (e.g., *Sedum acre*) such as cemeteries. Some species are also very popular for carpet bedding. Sedums are also useful as rock plants especially in a mild climate or in the shady portion of a tropical rock garden. Sedums are excellent for decorating greenhouses and the best species is *S. spectabile*, which is possibly the showiest among the sedums. This species and some others are also suitable for window gardens, roof gardens, hanging baskets, dry walls, and for indoor decoration. Some species are also used for edging. Sedums propagate easily from cuttings, by division, or sometimes from leaf cuttings. Seeds are also used for propagation. The compost should generally be porous loam but species like *S. acre* prefer a damp soil.

Sedums are fleshy, erect, and decumbent plants, sometimes even moss-like in nature. The leaves are very variable, flat to cylindrical in form, entirely or nearly so, generally alternate, but sometimes opposite or whorled. Sometimes the rosettes are bulb- or bead-like. Flowers are generally white or yellow or rose, sometimes red and very rarely blue.

There are innumerable ornamental species, but mention may be made of the following:

(a) *S. adolphii:* A small rosette, leaves yellowish-green with a reddish margin.

(b) *S. bellum:* A beautiful clustering succulent with rosettes of small, obovate, fleshy, light green, waxy leaves rounded above.

(c) *S. craigii:* A clustering plant, with obovate leaves, which are yellowish-green, glaucous, from silver-grey to pink and bluish.

(d) *S. hintonii* (of horticulture): It is a beautiful cultivar of clustering habit; the swollen, grey, obovate leaves are covered with bristle-like silvery hairs.

(e) *S. morganianum:* Commonly called 'Burro's Tail'. It is a beautiful hanging plant with tassels of short leaves which are spindle-shaped and yellowish-green in colour. It is suitable for hanging baskets.

(f) *S. pachyphyllum:* Commonly called 'Jelly Beans'. It has fleshy club-shaped leaves.

(g) *S. sieboldii:* A very ornamental creeping sedum suitable for hanging baskets. Leaves are in whorls of three, slightly serrated, blue green in colour with pinkish margins. In var. *medio-variegatum*, the centre of the leaves is cream-yellow and the stems are pink in colour.

(h) *S. stahlii:* Commonly called 'Coral Beads'. It is a dwarf succulent with opposite leaves which are almost bead-like. There are certain intergeneric hybrids between *Sedum* and *Echeveria* which are placed in the genus *Sedeveria*. Two of the hybrids are × *Sedeveria derenbergii* and × *S.* 'Green Rose'. *Sedums* are natives of northern temperate and cool regions and also Mexico.

Sempervivum (*Crassulaceae*).

The different species of this genus are found in the mountain regions of central and southern Europe, and also Caucasus to northern Russia. They are commonly known as 'House-Leek'. These are low plants forming clumps or cushions and are generally stemless. Leaves are alternate and arranged in spiral rosettes in a crowded fashion, sessile, lanceolate, obovate or acute, thickly fleshy, and green, reddish, or bluish in colour. The flowers are borne in terminal many-flowered inflorescence and are greenish, white, yellow, purple, or pink in colour. The rosettes bearing flowers perish after flowering or fruiting. *Sempervivums* are comparatively hardy plants requiring a porous soil consisting mostly of finely broken brick rubble; they can be grown even in humus if the drainage is perfect. Propagation is by division of old clumps or from cuttings of rosettes. Propagation from seed is not advised as the species cross easily between themselves. *Sempervivums* are used for carpet bedding, rock gardens, dry walls, and also for growing indoors in pans.

There are a large number of species but mention may be made of the following:

(a) *S. arachnoideum*: A beautiful plant with a slightly convex rosette about 5 cm across and densely arranged. Very fine cobweb-like hairs connect the leaf tips. Flowers are reddish.

(b) *S. arachnoideum* var. *glabrescens:* Rosettes are larger but loose and leaves are spreading with sparse cobweb.

(c) *S. soboliferum: Sempervivums* in general, and this species in particular, send out new rosettes from the parent plant and hence are referred to as "Hen and Chicken House-Leeks". Rosettes in this species are crowded and flattened-globose.

(d) *S. tecotorum:* Also known as 'Hen-and-Chicken House-Leek' and 'St. Patrick's Cabbage'. Miniature rosettes sprout from the mother rosettes; leaves are obovate, cuspidate, grey-green in colour and tipped brown.

(e) *S. tectorum* var. *calcareum:* Leaves are broadly painted reddish-brown at the apex.

Senecio (*Compositae*).

This genus is distributed in Cape Province and mostly in South Africa. This being a widely variable and widely distributed genus, it is difficult to give any concise definition. Leaves are flat or roundish, spatulate, or even more or less cylindrical, linear, or in some species reduced to scales. Most of the *Senecios* have flowers with yellow ray florets. *Senecios* are not very difficult to cultivate and grow well in a loamy soil. The plants are propagated from greenwood cuttings, seeds and the hardy species are even multiplied by division.

The following succulent species are worth collecting: *S. crassissimus*, *S. haworthii* (Syn. *Kleinia tomentosa*), *S. herreianus* (Syn. *Kleinia gomphophylla*), *S. jacobsenii* (Syn. *Notonia petraeus*), *S. pendulus* (Syn. *Kleinia pendula*), *S. scaposus*, and *S.* (*Kleinia*) *stapeliaeformis*.

Stapelia (*Asclepiadaceae*).

These are fleshy cactus-like succulents, grown for their odd shapes and for the singular and often large beautiful flowers. The stems are fleshy, sometimes decumbent (reclining) at the base, but the numerous branches are erect. The branches are leafless, coarsely 4-angled but at young stage rudimentary leaves may be found in the teeth of the branches. The flowers are often large, deeply lobed, generally star-shaped and mostly reddish-brown or purplish in colour often marked with yellow. The flowers have a dirty smell of rotten fish or meat.

Stapelias are not difficult to cultivate. To prevent rotting of the plants watering should be reduced to a point so that the plants do not shrivel. On very hot days the plants should be sprayed with water. The compost consists of loam, compost from lawn cuttings, cowdung, broken bricks and old mortar and fine chips of charcoal. Repotting should be done every year when old stumps are removed and used as cuttings. Propagation is also easy from seeds which germinate in a few days' time. *Stapelias* cross with each other easily and the species found in the garden are not true to the type and hence difficult to identify. They are especially suitable for growing indoors in pots. They are found mostly in South and South West Africa and a few species are also found in tropical Africa and East Indies. Some important species are: (a) *S. asterias* ('Starfish Flower'); (b) *S. bella* – probably a hybrid; (c) *S. gigantea* – flowers gigantic (up to 40 cm across) and yellow in colour; (d) *S. grandiflora* – flowers are purple-brown with whitish or purplish whiskers on the edges of petals; and (e) *S. variegata* – flowers are very attractive, greenish-yellow in colour with purple-brown spots.

Tradescantia navicularis (*Commelinaceae*).

While most of the *Tradescantias* are treated as greenhouse foliage plants, this may be included in the succulent collection. It is a half-creeping, tufted plant with short-jointed stems rooting at the nodes. The imbricated (overlapping) leaves are arranged in two rows which are shaped, fleshy and grey-green in colour. It is easily propagated from cuttings. It is a suitable ovate, boatplant for cultivation indoors and in a greenhouse. It is commonly known as 'Chain-Plant'.

Chapter XV

Palms and Cycads

PALMS

There are about 150 genera and 1,100 species of palms, mostly found in the tropics, belonging to the family *Palmaceae*. Plants in this group are most striking, usually having straight unbranched cylindrical or columnar trunks at the end of which there is a spreading canopy of huge pinnate or digitate leaves distinguishing the group from almost all other forms of vegetation. The great botanist, Linnaeus, called them "princes of the vegetable kingdom". Some palms throw suckers from ground level, while a few others are huge climbers (*Calamus*). Many of the palms are of great economic value. Ornamentally, many of the palms are very useful for decorative purposes and no modern garden is complete without their collection. The easy cultivation of palms and their delightful appearance make them great favourites in the gardens. It is a pity that many of the ornamental species can be used for decorative purposes only up to a certain age when they are somewhat small in size. Afterwards they become unwieldy and unmanageable. But the beauty they add to the garden when they are young is worth the trouble to cultivate them.

Utility: Palms are useful for decorating the entrance of a house, the veranda, the landings of staircases, for display in shade gardens, and for growing in the conservatory. Some of the species, though slow-growing, are excellent for avenue planting (e.g., the royal palm). Palms are very useful as potted plants and are widely used for decoration purposes.

Economically, palms are of great importance and the members of this family supply almost all the necessities of life from food of various kinds (e.g., dates from date palm), materials for building construction (e.g., trunks of palmyra palm used as poles or beams in house construction), thatching of roofs (e.g., leaves of palmyra and coconut palms), basket- and mat-making, fibre for ropes, oils (e.g., coconut oil), betel nuts, intoxicating liquors from tapping of saps and by fermentation (e.g., palmyra and date palm), and many other things.

Most of the palms are propagated from seeds. A few others such as *Rhapis* sp. produce suckers from the ground level and can be propagated by division of the clumps. The size of the seed may vary from the size of a pea to that of a coconut or even larger in some kinds. It is difficult to generalize the rate of germination and growth, but many seeds take a long time to germinate and it may not be a surprise if some germinate one year after sowing. Therefore, pans in which palm seeds have been sown should not be disturbed for a long time even if there is no immediate germination. The hardest seeds may be soaked for a couple of hours in water a little below the boiling point to soften the seed coat before germination. The stratification method suggested for rose seeds may be followed for some hardcoated seeds if these do not germinate properly. The compost for sowing consists of 2 parts garden loam and 1 part silver sand. The pans should have a good drainage as the seeds will remain in the pans for quite a long time. The seeds are sown thickly and are covered with their own depth of soil. In the hill stations, the pans may have to be kept under bottom heat for proper germination. Seeds can be sown any time of the year, but some people prefer to sow during the spring while others during the rainy season. Sowing during the rains appears to be more reasonable in the plains of India. As soon as the first pair of leaves develop fully, the seedlings should be pricked in small pots of a size sufficient to hold the roots and the seed and never in over-sized pots. As the palms grow they are transplanted to slightly bigger pots holding about 2-3 cm more soil around the plant than the previous ones. Palms develop long roots; therefore, seedlings should not be delayed for transplanting.

Cultural hints: Palms are more or less easy to cultivate. In the plains of India, most of the palms will grow better inside conservatories than in the open. These can be kept under the shade of trees also. In the hills, these should preferably be grown under glass. In the extreme climate of Delhi and other places in the north the palms may suffer on account of heat or frost, but more on account of frost. Proper precautionary measures, such as providing shade and humidity during the summer and shelter during the winter, should be taken.

The most important point to be remembered in pot culture of palms is that they like restricted root-room, i.e., they prefer to be pot-bound. The palms should never be repotted until the roots are overcrowded and matted near the base and when it would appear as if the roots are about to break the pot. At this time the palm is repotted without disturbing the ball of earth. The crocks are removed and the palm, with the ball of earth intact, is inserted in the next bigger size of pot which will hold about 5 cm more of soil around the ball of earth. The base of plant (collar) from where the roots emerge should just rest on the surface of the soil. A deeper planting would be harmful. Care should be taken not to injure any roots at transplanting as this may be fatal. Once repotted in this manner it will not be necessary to transplant again, at least for two years. If repotting is not needed, top dressing with a good compost may be given once or twice a year. The compost for potting consists of 2 parts good garden soil, 2 parts leaf-mould, 1 part well-decayed cowdung manure, and half to one part sand. Proper drainage material should be provided at the bottom. Transplanting should preferably be done during the rainy season. The foliage remains shining and green if the plants are fed with liquid manure prepared by fermenting oil cake and diluting the same to tea colour and applied once in fifteen days. The feeding may be alternated with a solution prepared by dissolving about 30 gm of ammonium sulphate in one litre of water. The quantity needed will depend upon the size of the plant.

Palms make quick growth in a warm humid atmosphere as of Calcutta or the coastal regions. In mild climatic conditions, the growth is slow. In hot places syringing the foliage with water is beneficial. Palms like a liberal supply of water provided the drainage is good. It is difficult to revive a palm which has suffered injury due to lack of water supply. Many palms are likely to be injured by frost while some are quite hardy.

The palms are classified into (a) feather-leaved and (b) fan-leaved. The more important species are listed below.

(A) Feather-leaved Palms

Acanthophoenix crinita (Syn. *A. herbstii, Areca crinita*).

It is 12-15 m tall. The trunk is cylindrical and thickly covered with black needle-shaped spines. The leaves are 3-4 m long and the stalk has dense bristles. Propagation is from seeds which may take up to two years to germinate.

Areca catechu (Tropical Asia).

The betel nut palm is not grown for ornamental purposes.

A. lutescens (Syn. *Chrysalidocarpus lutescens, Hyophorbe indica, H. commersoniana*).

A native of Bourbon. Forms an attractive clump and has a beautiful yellowish stem, up to 8 m tall. The arching leaves, yellowish green in colour, are very graceful. The species is very ornamental and is grown in pots and is planted in lawns and in shade gardens. Propagation is from suckers or seeds.

A. triandra.

It is a native of Assam and is similar to *A. catechu* but more ornamental.

Arenga saccharifera (Syn. *A. pinnata*).

A native of Malaysia, it is commonly known as 'Sugar Palm' or 'Sago Palm'. It is a beautiful tall (6-12 m) species with long (1-1.8 m) leaves which are silvery or white beneath. The leaf stalk is smooth. It is good for avenue planting.

Bentinckia nicobarica (Syn. *Orania nicobarica*).

A graceful palm from Nicobar, 20-30 m tall. The leaves are 1.5-2.5 m long and the leaflets are 25-60 cm long.

Calamus lindeni.

Commonly called the 'Rattan Palm' or 'Cane Palm'. It is thorny and rambling in habit and is a native of Malaysia and Indonesia. When young it can be grown in pots. *C. ciliaris* is a native of India (also Indonesia), and is excellent for table decoration when young. The stems and leafstalks are spiny.

Calyptrocalyx spicatus (Syn. *Pinanga globosa*).

It is a native of Moluccas and is good for pot culture.

Caryota mitis (Syn. *C. furfuracea*).

It is a native of Burma, Malaysia and Indonesia and is commonly known as 'Clustered Fishtail Palm', because of the leaflets which are delta- or fish tail-shaped. The palm is quick-growing and attains a height up to 7 m. A spineless palm with a stout, ringed stem, it is excellent for decoration when young. The leaves are 1-3 m long and the wedge-shaped leaflets are 10-20 cm long. It produces suckers freely.

C. urens.

It is commonly called 'Wine Palm'. When mature, it yields sap (toddy or wine) up to 13 litres a day. The pith is considered to be the best sago. A very large species (20-25 m tall) with a thick (50 cm) stem. The leaves are 3-6 m long, and the leaflets are 15-20 cm long, 10 cm wide, obliquely wedge-shaped, leathery and irregularly toothed. It produces no suckers. It is a native of India and Malaysia.

Cocos nucifera.

Coconut palm. For ornamental purposes the king coconuts with yellow-green or red fruits are useful.

Cyrtostachys lakka.

A stately palm from Borneo, it is ornamental when young. The leaf sheath is bright red and has numerous spineless stems. Another species is *C. renda*, which is commonly known as 'Sealing Wax Palm'.

Dictyosperma alba (Syn. *D. album, Areca alba*).

A genus of one species of slender unarmed palm, up to 10 m tall; leaves 1.2 to 2.5 m long and pinnate. The leaflets are 60 cm long, bright green on both sides, and the stalk is whitish. It is a native of Mauritius. In var. *rubrum* (Syn. *Areca rubra*) the main veins and leaf margins are dark red, more so in young plants. The palms are suitable for house and table decoration.

Dypsis (Syn. *Chrysalidocarpus*) *madagascariensis.*

A clump forming, slender-trunked, feather-leaved palm from Madagascar. The plant is somewhat similar to *Chrysalidocarpus lutescens* but with weaker leaflets with drooping tips. A branching palm.

Elaeis guineensis.

A palm from Guinea, it is commonly known as 'Oil Palm'. The plant is about 10 m tall with stout deeply ringed stems which are not spiny. The leaflets are 3-5 m long, dark green in colour, and the stalk is spiny-toothed; leaflets are 50-60, linear lanceolate, and acute. A lovely plant suitable for pots.

Geonoma gracilis (Syn. *G. riedeliana*).

A palm from Brazil. Somewhat short in growth (2-3 m tall), and useful for table decoration. Flowers are very showy and in long drooping spadices in yellow colour. Propagation is by seeds or suckers.

Hyophorbe amaricaulis (Syn. *Areca speciosa, Mascarena lagenicaulis*).

A native of Mauritius, bottle-shaped near the base. The plants may attain a height of 20 m. The leaves are pinnate, spreading 1.2-1.8 m. The pinnae are stout and in pairs of 40-60. The stalk is deep maroon in colour, glaucous, with an orange line along outer edges of midrib.

H. verschaffeltii (Syn. *Areca verschaffeltii, Mascarena verschaffeltii*).

It is 8 to 10 m tall. The trunk is swollen at about half-way mark. The leaves are 1.2 to 1.8 m long, gracefully arching at the top; pinnae in pairs of 30-50, about 60 cm long and 3 cm wide. The leafstalk is about 7 cm long with a yellow band from the upper part of the leaf sheath to the end of the blade. The flowers are orange-yellow in colour.

Kentia.

A genus consisting of very handsome species of palms, somewhat similar in look to the *Areca* but differing in growth habit. While the *Kentias* are decumbent in growth when young, the *Arecas* are upright. The genus also includes two species of the genus *Howea*. This genus is most popular for general decorative work and for growing indoors or in a greenhouse. The stems and leafstalks are spineless.

(a) *K. belmoreana* (Syn. *Howea belmoriana*) : The stems grow up to 10 m in natural habitat. A graceful palm with long, large, arching pinnate leaves.

(b) *K. forsteriana* (Syn. *Howea forsteriana*) : A beautiful species similar to the above. A native of Howe Island near Australia.

(c) *K. macarthurii* (*Actinophloeus macarthurii*, *Ptychosperma macarthurii*): A species from New Guinea, and is very popular. A hardy palm, 3.5-5 m tall. A number of suckers arise from the base, making the plant bushy.

Martinezia caryotaefolia (Syn. *Aiphanes caryotaefolia*).

It is 10-15 m tall and a native of Columbia, Ecuador, and Venezuela. The stem, the leafstalk and the back of the leaves are densely covered with long spines. The species is very ornamental.

Oreodoxa: See *Roystonea.*

Phoenix dactylifera.

The 'Date Palm'. Not generally included in small gardens.

P. reclinata (Syn *P. leonensis*, *P. spinosa*).

It is 7-9 m tall. The leaves are 1.8-2.5 m long, and the leaflets 30-60 cm, linear and bright green.

P. roebelenii (also spelt *roebelinii*).

A very handsome palm, suitable for pot and decoration. The graceful arching dark green pinnate leaves, when grown in pots, almost conceal the pots. The leaves are shining dark green. The plant is dwarf and does not exceed 60-90 cm in height. In pot-grown plants the suckers are removed from the base and only one stem is retained.

P. rupicola.

It is considered to be the most handsome species in this genus. It grows to a height of 5-6 m. The trunk is about 20 cm thick. The leaves are 3 m long, bright green, and flaccid; leaflets 45 cm long, 2 ranked. It is a native of north India and Nepal.

P. sylvestris.

The wild date or toddy palm is quite ornamental when young.

Pinanga kuhlii (Syn. *Ptychosperma kuhlii*, *Seaforthia kuhlii*).

A clustering feather palm; bamboo-like; forms several slender yellowish-green stems up to

10 m in height. When young, it is very well-suited for table decoration.

Ptychoraphis augusta.

A native of Nicobar Islands. When young, it is quite ornamental like the *Cocos weddeliana* or *Geonoma gracilis*. The other species, *P. singaporensis* (Syn. *Ptychosperma singaporense*), is a many-stemmed palm. The stems are slender, case-like, ringed, black in colour and feather-leaved.

Ptychosperma: See *Kentia, Ptychoraphis*, and *Pinanga*.

Roystonea oleracea (Syn. *Oreodoxa oleracea*).

Commonly known as 'Cabbage Palm', it is a native of West Indies. It is 30-40 m tall, swollen at base. The leaves are 3-4.5 m long and gracefully arched. Good for avenue planting.

Roystonea regia (Syn. *Oreodoxa regia*).

Commonly known as the 'Royal Palm', sometimes also called, though not correctly, 'Bottle Palm'. The plant grows to a height of up to 20 m or a little more, usually with a bulge on the trunk in the middle portion, giving the shape of a bottle or a barrel. The plant is unequalled among the palms in its beauty, majesty, and grace. The palm is very slow-growing and is at its best when well-grown, that is, after 10 or more years of growth. It is generally planted along avenues.

Seaforthia elegans.

It is more appropriately known as *Archontophoenix cunninghamii* (other synonyms *Ptychosperma cunninghamianum, P. elegans*). This species is well known to the florists and decorators by the first stated name and is considered to be very graceful and stately. It is suitable for planting in ground. It is propagated from seeds. A stout and straight palm.

Wallichia disticha.

A low (3-5 m tall) palm with 2·3 m long graceful leaves. The leaflets are 30-60 cm long, 3 cm wide. A native of the Himalayas. Propagated from seeds or by separating the suckers gradually so that they get rooted before being completely detached.

B. Fan-leaved Palms

Borassus flabellifer.

The common 'Palmyra Palm' (Vernacular names: *tal, tatninge*). Not generally grown in gardens.

Chamaerops excelsa (Syn. *Trachycarpus excelsa, T. excelsus, T. fortunei, Chamaeropsis fortunei*).

Variously known as 'Windmill Palm' and 'Fortunes Palm'. It is a native of China and Japan. A slow-growing, medium palm (3-12 m tall) which is very ornamental. The stem is covered with a mat of long, dark-brown fibres. The fan-shaped leaves may become as large as 1.2-1.5 m across.

Chamaerops humilis.

A native of the Mediterranean countries, South Europe, and Algeria. The only palm which is native to Europe. Attains a height of 1.8-3 m and forms a cluster of stems at the base. The stalk is about 1.2 m long, armed with stiff straight or hooked spines. The leaves are fan-shaped and are about 45 cm long and split near the base into narrow segments.

(a) Var. *arborescens:* Tree-like, with a stem 1.8 to 2.4 m in height.
(b) Var. *argentea:* With silvery-green leaves.
(c) Var. *elegans:* Of slender form and elegant habit.

Coccothrinax: See *Thrinax*.

Corypha australis: See *Livistona australis*.

Corypha umbraculifera.

'Talipot Palm'. It is a native of South India and Sri Lanka. Grows to a height of 30 m. The leaves are fan-shaped, plaited, forming a complete circle of 3.5 m, bright green in colour. The stalks are spiny.

Hyphaene thebaica.

Commonly called 'Doum Palm', is a native of Egypt. Attains a height of 15 m. This may be called as a botanical wonder with its repeatedly branching habit. The branches are smooth. The leaves are fan-shaped, 60-75 cm long, and the petiole is spiny. The orange-yellow fruits are edible.

Latania commersonii (Syn. *L. rubra*).

Often called *L. borbonica*, but the name is erroneous. It is known as 'Bourbon Palm'. The stem is smooth, slender, about 2 m tall. The stalk is deep chocolate-red, smooth. The leaves are wedge-fan-shaped, deeply cut, and recurved gracefully. The veins and segments are tinged with red, margins entire but spiny in young plants. A most handsome species. It is rare and costly. It is a native of Mauritius. Propagated from seeds.

L. verschaffeltii (Syn. *L. aurea*).

Another handsome species. The leaves are fan-shaped, roundish, deeply cut, pale green; the ribs golden especially when young; the stalks are 60-120 cm long, orange-margined, and spiny in young plants.

Licuala grandis (Syn. *Pritchardia grandis*).

A tropical, very ornamental fan-leaved palm, 1.8 to 2.5 m tall. The leaves are orbicular or almost round, bright green, plaited, margins lobed and toothed; the stalk slender and spiny. Best as an indoor pot plant. A native of New Britain Island near New Guinea. It is propagated from seeds.

L. peltalta.

A fan-leaved palm of Indian origin. The palm is 4-5 m tall. The leaves are peltate, and the lobes of the leaves droop gracefully. It is propagated from seeds.

L. spinosa (Syn. *L. horrida*).

It is a native of Malaysia to Java. The leaves are large, dark green, and plaited. The stalks are rigid and armed with black curved thorns. It suckers densely.

Livistona.

This genus of fan-leaved palms is most popular and is grown very extensively in the gardens. The propagation is from seeds. The trunk of this genus is ringed and clothed with the remains of dead leaf-sheaths. The commonly grown species are:

(a) *L. australis:* It is about 25 m tall. The leaves are 90-120 cm across, divided into 40-50 segments up to the middle or further below. The margins of the brown stalks are spiny. It is a native of Australia.

(b) *L. chinensis* (often erroneously called *Latania borbonica*): Commonly known as 'Chinese Fan Palm'. There is confusion regarding the height of the plant, many people describing it as between 1.8 and 2 m tall, but possibly it grows up to 9 m in the natural habitat. The leaves are many and kidney-shaped, 1.5 m across, cut into segments which are deeply forked, measuring 30-60 cm long and 2.5-5 cm wide in the lower portion while the middle ones are 90 cm long. In between the segments there are threads. The stalk is as long as the leaf-blade and the lower half is covered with brown spines in young plants but these usually disappear afterwards. It is the most hardy species in the genus.

(c) *L. humilis:* A decorative species when young.

(d) *L. rotundifolia:* It is a native of Malaysia. When young, it is one of the most ornamental species among palms. The stems are 12-15 m tall. The leaves are nearly round, 90-150 cm across, deeply divided into segments of 60-90, cut only one-third to base when young and are glossy green in colour in both surfaces. The stalks are 1.8-2 m long, the margins are armed with sharp, spiny teeth. A very good pot plant for indoor and outdoor decoration.

Pritchardia filifera: See *Washingtonia filifera.*

Pritchardia grandis: See *Licuala grandis.*

Pritchardia pacifica.

It is a native of Fiji, Tonga, and is sometimes known as 'Fiji Fan Palm'. The fan-leaved palm has a smooth, attractive trunk, 9 m tall, and 25-30 cm thick. The leaves are many, shortstalked, 1.2 m long and 100 cm broad, and are densely covered with brownish-white tomentum at young stage but later glabrous, bright olive-green, and leathery. The leaves, having about 90 segments, are slightly cut at the apex. A most ornamental species suitable for pot culture.

Rhapis excelsa (Syn. *R. aspera*, *R. flabelliformis*, *R. kwaniwortsik*).

A native of China and Japan, often known as 'Ground Rattan Cane or Palm' and 'Large Lady Palm'. The stem is bamboo-like, 45 to 120 cm tall, and as thick as the thumb, and forms clumps from underground suckers. The stem is persistently covered with fibrous base of the leaves. The leaves are 5-7, glossy green, deeply cut into several segments, stalks obscurely toothed in the margins. It is propagated from suckers. A good palm for pots and shade gardens. In var. *variegata* the leaves are variegated.

R. humilis.

It is more or less similar to the above but less vigorous. The leaves are larger than the above and the stalks are smooth. A native of south China, it is commonly called 'Slender Lady Palm'.

Sabal adansonii (Syn. *S. minor*, *S. glabra*).

A native of southern United States, commonly known as 'Dwarf Palmetto' and 'Blue Palm'. The stem is so short that it remains buried in the ground and appears stemless. The leaves are orbicular, fan-shaped, 60-90 cm long, the segments 20-30, slightly cleft at the apex. The stalk is smooth and shorter than the leaf. A noble looking and a most ornamental palm.

The other species are *S. blackburniana* or *blackburnia*. This includes *S. umbraculifera* of gardens. It is called 'Fan or Thatch Palm'. A native of West Indies, it is a good window palm, when young. *S. palmetto* is the 'Cabbage Palm' or 'Palmetto' and is a native of southern United States. The leaves are segmented and hanging with numerous threadlike fibres. The foliage is bluish-green on the upper surface and grey below.

Stevensonia grandiflora (*S. borsigniana*, *S. grandifolia*, *Phoenicophorum borsigianum*, *P. sechellarum*, *Areca sechellarum*).

It is a native of Seychelles, about 12 m tall. The stem is very spiny when young but less so when old. The stalk is 25-45 cm long, spiny, copper-coloured when young; the leaf blade wedgeshaped, obovate, shortly bifid, about 1.8-2 m long and nearly as broad, lightly cut and toothed at the margin. When young, the leaves are very ornamental, being rich cinnamon brown in colour. A rare and very handsome species but rather difficult to grow.

Thrinax argentea (Syn. *Coccothrinax argentea*).

It is a native of West Indies. The stems are 3.5 to 4.5 m tall and the foliage is very ornamental. The leaves are deeply divided, star shaped; the segments are united at the base, glossy green above and silvery below. The stalk is slender and long.

T. parviflora.

Another beautiful species which is a native of Jamaica, Cuba, and Florida. The petiole is reddish at the base. The leaves are 90 to 120 cm across, having about 40-50 segments, cut half-way to base, with an almost circular shape and of green colour on both sides with yellow ribs.

T. radiata (Syn. *Coccothrinax radiata*).

This is another ornamental species, which is a native of West Indies.

Verschaffeltia splendida.

A very handsome palm with aerial stilt roots. The top of the stem and the petioles are spiny at first. The foliage 90-150 cm wide, bright green veined and shaded with yellow. A very rare species. A native of Seychelles.

Washingtonia filifera (Syn. *Pritchardia filifera*).

A cylindrical robust-stemmed palm, 15-25 m tall. The petioles are thorny and long. The leaves are orbicular, fan-shaped, 90-150 cm across, and very filiferous and deeply cleft.

The nomenclature of palms presents some confusion because of various reasons. Often a genus or species is substituted by others. It is argued that most of the palms are natives of the tropics of Asia and America, and to some extent Africa, whereas most of the noted authors on palms are from Europe and as a result discrepancies in nomenclature have crept in. Another practical difficulty is that in palms, because of their large leaves, it is almost impossible to make herbarium sheets for identification. A basic defect in the nomenclature of plams is that many of the species were named from gardern specimens even before they came to flowering and fruiting. In this chapter we have tried to include all the synonyms wherever possible.

CYCADS

In the tropics of the Indian plains it is not possible to grow most of the conifers, excepting a few species. Most of these grow well only in the hills while some others do quite well in a mild climate as that of Bangalore. For example, the tree *Araucaria* grows quite well in Bangalore and also in many other places such as Calcutta and even in Delhi, but in the latter place mostly as a pot plant. A few of such trees belonging to the conifers have been dealt with in Chapter XI. Here we shall deal with some ornamental plants which are allied to *Coniferae* in their fructification. Some of these plants which are very beautiful ornamentals are grown in the gardens of the tropics and subtropics. They look like palms but have no botanical relationship with them. These plants belong to *Cycadaceae* and hence are commonly referred to as Cycads.

The general cultivation of cycads is similar to that of palms and similar types of soil and climate will suit them. Many people commonly refer to them as palms of various kinds but as already stated these plants have nothing in common with palms except the general habit of growth.

Most of the plants in this group are slow in growth and prefer partial shade for their growth in the tropics, while in the hills these may have to be grown under glass. They are suitable for planting in ground or as pot plants. The long lasting ornamental leaves are used for table decoration and in many countries these are used for religious ceremonies and for decorating temples and altars. Many plants of this group are injured by frost but most often recover from such shock.

The propagation is chiefly from seeds, which remain viable for a month or so after ripening but should preferably be sown within a month. Suckers are also formed occasionally and these are separated and planted.

Some insects attack the foliage for the control of which see Chapter XXVI.

The *Cycadaceae* has nine genera. Some species of the following genera, namely, *Cycas*, *Dioon*, *Encephalartos*, *Macrozamia* and *Zamia* are commonly cultivated in the garden.

Some Cycads

Cycus cercinalis (*C. thouarsii*).

It is a native of India, Madagascar and New Guinea and is commonly known as 'Fern Palm'. The stem is erect, cylindrical, marked with leaf scars, and usually grows to a height

of 4.5 m but may attain a height up to 12 m. The stem is usually unbranched. The appearance of the tree is palm-like, with the crown having a graceful rosette of fern-like, stiff, glossy, gracefully curved pinnate leaves, the leaflets being flat on the margins. The young leaves are covered with reddish-brown hair. The male and female inflorescences are borne in separate plants. Fertilization is best done by hand. The large seeds are poisonous but after several soakings and changing of water these are ground up and made into cakes for consumption. It forms a good pot plant when young.

C. revoluta (Syn. *C. inermis*).

It is commonly called the 'Sago Palm'. It is about 3 m tall, sometimes branched above. A native of south Japan to Java. The leaves are 60 cm to 3 m long, recurved, and the leaflets are many, stiff in nature, and rolled downwards.

C. rumphii.

Another handsome species which is a native of Burma, Sri Lanka, and Andaman and Nicobar Islands.

Dioon edule.

An ornamental palm-like foliage plant which is very ornamental and more or less similar in appearance to cycas. The stem is about 90 cm tall and of similar dimension. The pinnate leaves have spiny tips. The petioles are covered with white wools at young stage. The seeds are large and are made into a kind of arrowroot. The cones are 30 cm long.

D. spinulosum.

The plants grow up to 15 m and have a slender trunk, crowned by a noble rosette of spreading pinnate leaves up to 1.8 m long. But the margins have 5-8 spines. Both species are natives of Mexico.

Encephalartos caffer (*E. caffra*).

It is commonly called 'Kaffir Bread'. The plants are handsome, palm-like with a stout stem which may grow up to 5.5 m. A crown may consist of 14 leaves, each about 60 to 120 cm long. It is a native of South Africa.

E. hildebrandtii.

A beautiful plant growing up to 6 m. The leaves are up to 2.7 m long, and the main stalk is woolly when young. Both male and female cones are quite long, between 45 cm and 60 cm. It is a native of tropical East Africa.

Macrozamia spiralis (Syn. *M. tridentata*).

Of the various species, this is the most commonly cultivated. The trunk is short and usually underground. The leaves are 80 to 100, up to 1.8 m long, and the flat leaflets are spiny-tipped.

Zamia.

This genus differs very little from cycas but the leaves are more leathery and fern-like. The plants are tropical and subtropical and resemble palms and to some extent ferns. They are propagated by division when the stems are clustered and also from seeds and offsets. But in our environmental conditions they produce hardly any seeds and suckers very rarely. Hence the propagating materials or plants are to be imported. Zamias are generally stocky and short plants and sometimes the trunk is subterranean and more or less tuberous. Some of the important species are *Z. furfuracea* (Mexico, Columbia) and *Z. integrifolia* (Syn. *Z. media commeliniana*).

Chapter XVI

Ferns and Selaginellas

FERNS

In the vegetable kingdom, ferns are one of the oldest members along with *Cycads* and others. To give the readers a glimpse of their great age it may be stated that long before such gigantic but now extinct animals as dinosaurs lived on this good earth, the members of ferns were already firmly established, living, producing offspring, and decomposing from prehistoric times. The fossil remains from coal and rocks confirm their great age. The ferns are diminutive relatives of the tree ferns which have descended from the Carboniferous Age about 350,000,000 years ago. Among the non-flowering group of plants, ferns are undoubtedly the most ornamental and the most beautiful. Ferns have a great capacity to adapt themselves to a new place and thus add to the beauty of that place with their elegant and graceful foliage. There are about 12,000 species and varieties of ferns. India has over 1,000 species, but only a very small percentage of this is under cultivation.

Botany of Ferns: The ferns and some fern-allies belong to the class *Pteridophyta* in the classification of the plant kingdom. The entire fern plant may be only a small hair-like prostrate stem consisting of a few moss-like leaves, or it may also be a tall tree of 24 m. Irrespective of whether the plant is a tiny or a large specimen, it is made of three general organs, namely, the root, the rhizome or the stem, and the leaf which is generally referred to as frond. Most of the ferns are *terrestrial* in habit while a large number are *epiphytic*, besides others which are *aquatic*, found in ponds, lakes, and stagnant pools. The rhizome in the ferns is in reality a modified stem. The shape of fronds or leaves in ferns may be bipinnate (doubly divided), pinnate, or with uniquely-shaped cordate pinnules or may be entire (undivided). The size also varies, from very small to very large.

In the life-cycle of an individual fern plant there are two distinct phases, known as alternation of generation. These two distinct phases are represented by two separate and unlike plants. The ordinary fern plant is asexual and is botanically a *sporophyte*, which produces spores generally in spore cases, called *sporangia*, which are found in masses, referred to as *sori*, on the under-surface or margin of the leaf. But in some cases the *sori* are grouped in spikes or panicles, and in a few extreme cases these are spread in a layer on the entire under-surface of the leaf. The sexual phase or *gametophyte* in the fern develops when the spore starts germinating and is represented by a tiny, usually scale-like heart-shaped green *prothallus*. This organ bears the sexes – *antheridia* representing the male, and *archegonia* the female – on the under-surface. The fertilization takes place in the *arehegonium* and the resulting egg develops directly into a fern plant. All these organs, except the young plant, are extremely minute and are visible only under a microscope.

Utility of Ferns: As has already been stated, though ferns are non-flowering plants, they are extremely handsome because of the elegance of their foliage and hence are widely useful for different purposes in the garden. It is a common practice to allot a whole conservatory, usually called a "fernery", for growing ferns. Ferns are very suitable for a shade garden as most of the ferns like a semi-shaded situation for growth. Many ferns find their place in rock gardens. Ferns are also grown extensively in pots and pans for indoor decoration or for display in the veranda. Some of the elegant species are very good for wire or hanging baskets. Some epiphytics such as staghorn ferns (*Platycerium*) are grown on tree trunks or on a ball of coconut fibre, suspended from strings. Ferns are also very good and most natural subjects for the wild garden. In ornamenting a "dry wall" ferns play an important role in imparting an old and natural look to the wall.

Ferns can be used with advantage in decorating roof and terrace gardens. The fern leaves, because of their elegance, are highly prized for their use in vase and table decoration, in preparing bouquets, buttonholes, and wreaths. The dried leaves, specially of the various maiden-hair ferns, are very popular for pasting on greeting cards. A fern enthusiast can have an exclusive fern border in semi-shade or can even include ferns in mixed borders. Ferns are also grown in bottle garden, wardian case, or terrarium and window garden. The moisture-loving ferns should find a place in a marsh garden.

Propagation of Ferns: It has been stated earlier that spores of ferns produce new plants. But an ordinary gardener seldom uses spores for propagation. Use of spores is necessary to obtain new hybrids as well as for rapid propagation. The simplest method of collecting the spores is to obtain some mature *sori*-bearing leaves and laying them on a hard-smooth white paper with the *sori* face (that is the under-surface) down. Then the leaf is covered with a blotting paper to prevent the spores from blowing off as well as for absorbing the excess moisture in the leaf. Within a couple of days the *sporangia* will empty their contents on the white paper, when the blotting paper is removed carefully. The chaff on the paper is removed by gentle winnowing and the paper is folded along the centre. With some vigorous tapping of the paper the spores will accumulate in the central fold, and these are then collected. It is desirable to sow freshly collected spores, although some spores may remain viable for years together.

Spores are sown in shallow pots or pans. Various pot mixtures may be used, but the following has been found to be very satisfactory by some experts. A 10 cm deep pot should first be filled with coarse sphagnum moss to a depth of about 6 cm. Then a compost, consisting of equal parts of humus, top soil, building sand, and very finely chopped sphagnum moss, is prepared by mixing the constituents thoroughly, which is also screened. The pots should be sterilized along with the compost by heat. The clay pots are then submerged in a pan of water up to the neck and then left long enough to make the contents as well as the pots saturated with water. The excess water should be drained off. The coarse sphagnum at the bottom serves as a reservoir of water in between the two waterings. Sowing the microscopic spores uniformly needs some experience. One way of obtaining fairly uniform sowing is to spread the spores on a board paper and tapping this over the sowing pan. After sowing, the pans should be covered with a glass-pane and removed to a warm, shady place. The subsequent watering is done by placing the pots erect on a saucer containing water with the drainage hole submerged. The germination time varies with the species. The spores may either be chlorophyllous (green-coloured types) or non-chlorophyllous, though the majority of ferns produce the latter types. The chlorophyllous types may germinate within a few hours to a couple of days from the time of sowing, while the non-chlorophyllous types take about 5-15 days. When the small plants are sufficiently large to handle, these are pricked 2-3 cm apart in seed pans containing a mixture of equal parts of sifted leaf-mould and sand. Till the fronds are well developed these should not be watered from above but the method suggested above should be followed. Propagation from spores can be done any time in the year but preferably during a warm and moist period.

The easiest and the most common method of propagation is, of course, propagation by division of the clump or by division of the creeping rhizome. Any part of the rhizome can be used for this purpose. Another reason for propagating by rhyzomes is that some kinds are sterile, while others do not breed true to type from spores. The best time for propagation by division is the spring when repotting is done. But in most parts of the plains this can conveniently be done during the monsoon also.

The staghorn ferns (*Platycerium*) produce suckers from adventitious buds on their roots as they grow older, and these are easily separated. Some ferns such as *Asplenium viviparum*, *A. bulbiferum*, and *Polystichum angulare* produce a number of bulbils or plantlets on the surface of their fronds, while *Woodwardia radicans* or *W. orientalis* produce a solitary bud at the apex of the fronds. In such species, the fronds should be pegged down flatly in sandy leaf-mould or the tip should be pegged as in *Woodwardia* to encourage the plantlets to form roots. Once rooted, these are separated and planted individually. The tip cuttings from the trunks of certain tree ferns, e.g. *Cibotiums* and *Alsophila australis*, root in a moist porous soil or when kept in a container having coarse gravel in shallow water.

Cultural hints: Most of the ferns are quite hardy by nature, and given proper condition and environment for growth, are not very difficult to cultivate. The first thing that should be kept in mind is that the ferns dislike bright sunshine except in a few cases. For example, *Nephrolepsis exaltata* adapts well in the open sun. The temperature requirements of ferns depend upon the origin of the species. Ferns of the tropics like a warm humid temperature and those from the high hills or temperate regions grow best under a cool climate.

It has been stated that ferns like a shady situation for their growth. But there are ferns, especially in the temperate regions, growing on cliffs, mountains or walls, which prefer to send their head (foliage) in the sun and the roots in the shade. In the tropics, ferns are always grown in shade or semi-shade. The natural choice of site for ferns will be under the shade of trees. But care must be taken to keep the ferns relatively free from the competition of the roots of large trees. This is ensured by digging around the trunk of the concerned tree to a depth and finding out relatively root-free areas where the ferns can be planted in pockets. Of course, over a period of time, the tree roots will also reach that pocket for the plant nutrients but by that time the ferns will have established themselves. When ferns are grown under trees it is essential to mulch them with leaf-mould twice a year – once during the spring and again during or after the rains. This will give the ferns the much needed additional nutrients, replenishing the loss due to competition from the adjoining trees.

There are two major disadvantages in growing ferns under large trees. Firstly, in the summer the trees evaporate enormous amounts of soil moisture and the ground around the trees quickly become dry. In the summer, additional arrangements for water should be made preferably by sprinklers. The other disadvantage is that if the tree is infested with greenfly, the ferns growing below them get covered with a sticky honey exuded by the greenfly. In the long run, a sooty mould may develop on the fern leaves because of this honey.

The soil where ferns are to be grown should be light, porous, well drained in nature, and should contain enough lime. The moisture-loving ferns should be planted in the marsh garden or around a lily pool.

For growing ferns in pots and pans the same compost as suggested for germination of spores can be used, but with a little addition of bone meal. The compost should be sterilized by heat. The same compost should also be used in the pockets when grown in the ground. In the above compost, one half part of broken pieces of charcoal may be added. The bottom of the pan or pot should be filled with broken crocks. Oversized pots should never be used and the size of the pot should be just enough to hold the roots of the fern.

Ferns are generally repotted when they fill up the pot and at this time the division of clumps or rhizomes is also done. It is difficult to give a general suggestion as to how often a fern should be repotted. If the fern has covered the whole pot, it is time to repot. Again, if the compost has turned sour or exhausted, which will be evident from the poor health of the plant, repotting is necessary. But it should be kept in mind that poor health could also be due to such other factors as diseases, water shortage, etc. In case of repotting, only the next size pot, that is, about 2-3 cm more in diameter, should be used. Many people advise feeding with liquid manure, which can be prepared by fermenting oil cake or fresh cowdung. Top dressing with leaf-mould can also be done once or twice a year.

The success of growing ferns depends much on watering. More ferns die of over-watering and bad drainage than due to under-watering. The requirement of water differs from one kind of fern to another, although all of them like moisture. The general recommendation is to supply enough water to penetrate the soil thoroughly and to permit the excess water to drain off. Over-watering accentuated by poor drainage may result in the death of the plants. A general sign of lack of moisture in the compost is wilting of foliage and the sign of over-watering is yellowing of foliage when it is not due to old age or disease. A grower will gain by experience about the amount of water a particular species needs.

When grown indoors, ferns will like to have moist, cool air, around their foliage. Rooms heated with steam are not often suitable for growing the ferns.

In the north Indian plains, it is very difficult to grow most of the ferns because of excessive dry heat during the summer and frost during the winter. But with a little effort it is possible to grow many ferns under such conditions. The best thing is to construct a conservatory and allot the same exclusively to ferns so that it becomes a "Fernery". The conservatory is a semi-shaded place which is suitable for growing ferns. The inside humidity can be increased by providing a shallow pool and also a stream running around the fernery. During hot and dry weather the ferns should be protected against draughts of hot winds and provided with humidity by occasional syringing. It is also possible to acclimatize ferns of the Himalayas in the hot climate of north India. Some good work in this direction has been done at the National Botanical Research Institute (formerly National Botanic Gardens), Lucknow.

Pests and Diseases: There are a few pests and diseases in ferns which have been included in Chapter XXVI.

Some Common Ferns for the Garden

As has already been stated, there are about 12,000 species and varieties of ferns and a large number of these are already in cultivation in different parts of the world. Ferns of almost any type are very handsome and deserve inclusion in an ornamental garden. But it is not possible even to describe a small fraction of these huge species in a book like this and hence we will limit our list only to a very few common and handsome species.

Adiantum (*Polypodiaceae*).

Ferns of this group are most popular in the Indian gardens, and are commonly known as 'Maiden-Hair Ferns'. The Greek word *adiantos* means "unwetted", as the fronds in this genus do not get wet even when plunged in water.

Adiantums are not difficult to cultivate and grow well in the general compost recommended earlier. They enjoy plenty of light but not direct sunlight. Overhead watering and syringing are disliked by the plants in this genus. Propagation is usually by division. The beautiful foliage is usually borne at the end of a relatively long, slender polished black or purplish stalks.

Adiantums are good as potted plants and are widely used for veranda or indoor decoration and for hanging baskets. The fresh leaves are used in preparing bouquets and buttonholes along with flowers and the dried leaves are very handsome in preparing greeting cards. Some of the common species are: *A. aethiopicum* (Syn. *A. assimile*), *A. bausei* (a horticultural form), *A. bellum*, *A. capillus-veneris* (common maiden-hair), *A. cuneatum* and its varieties, *A. decorum* (Syn. *A. tinctum* var. *wagneri*, *A. wagneri*), *A. formosum*, *A. macrophyllum*, *A. moorei* (Syn. *A. amabile*), *A. peruvianum*, *A. tenerum*, *A. tenerum* var. *farleyense*, *A. trapeziforme*, *A. victoriae*, and *A. villosum* (Syn. *A. rhomboideum*).

Alsophila (*Cyatheaceae*).

Variously known as 'Grove Fern' and 'Norfolk Island Fern', these moisture-loving ferns, belonging to the order *Cyatheaceae*, are generally called 'Tree Ferns'. The plants like shade and plenty of moisture. The plants of tree fern groups have a straight, tall stem or trunk similar to that of a palm. The large leaves are borne at the apex of the trunk giving the plants a palm-like appearance. The ferns are suitable for cultivation at medium-to-high altitudes. At a lower altitude, one may attempt to grow these plants by providing them with shade and plenty of moisture. The trunk should be wrapped with wet moss and water applied by syringing. Some of the ornamental species are *A. australis*, *A. cooperi*, *A. glauca* (Syn. *A. contaminans*), and *A. latebrosa*. The first two species are natives of Australia, while the others belong to India and tropical Asia.

Aneimia (or *Anemia*) *rotundifolia* (*Schizaeaceae*).

Ferns of this Brazilian genus are known as 'Flowering Ferns'. In this fern, the portion of the leaf which bears spores does not have any expanded leaf lamina and hence looks like an inflorescence. The sterile fronds consist of pairs of oval leaflets (pinnae) along the rachis and have the normal look of leaves. The ferns are suitable for pot culture and hanging baskets. Propagation is normally from spores but the barren portion of the frond roots at the tip.

Angiopteris evecta (*Marattiaceae*).

A swamp-dwelling fern from Japan to Australia and Madagascar. The plants develop a stout stem. The bipinnate leaves are borne in a stalk as long as 4.5 m. The pinnae are linear-oblong in shape, succulent, dark green in colour, and finely toothed. Propagation is from the fleshy scales at the base of each frond, which has two dormant buds. These are laid in sand without dividing and are covered with moist sphagnum moss.

Asplenium (*Polypodiaceae*).

This genus consists of some attractive species of ornamental ferns, either with entire leaves or with finely divided fronds. Many species produce plantlets on the surface and the edge of pinnae and are multiplied from these. Other methods of propagation are by division of the crown and from spores.

The most common species is the 'Bird's-Nest Fern', *A. nidus* (Syn. *Thamnopteris nidus*, *T. australasicum*, *A.* and *T. nidus-avis*). A native of the tropics of the old world, it is a graceful species with long (up to 1 m) undivided fronds, each about 7-20 cm broad. The fronds in a large specimen plant form a cup-like cluster, in the shape of a bird's nest. Some other ornamental species are *A. bulbiferum* (India, Australia, New Zealand), *A. cardiophyllum* (Formosa), the fronds in which are anthurium-like, and *A. myriophyllum* (Syn. *A. rhizophyllum*), a native of tropical North America. Most of the *Asplenium* like plenty of moisture.

Blechnum (*Polypodiaceae*).

Many species of this genus are good for indoor decoration. The genus resembles *Lomaria* and the plants have short stems and many have palm-like leaves. Propagation is from spores. The leaves turn black and brown if watered from overhead. Some ornamental species are *B. auriculatum*, *B. brasiliense*, and *B. occidentale*.

Ceropteris: See *Gymnogramma*.

Davallia (*Polypodiaceae*).

This genus consists of some of the most handsome ferns, with scaly creeping rhizomes which in some species resemble a hare's or a squirrel's feet. There are species which form excellent pot plants, while others are most suitable for hanging baskets. *D. bullata* ('Squirrel's Foot Fern') is a suitable species for hanging baskets and for training on frames resembling birds, animals, etc. *D. mariesii* (Syn. *D. bullata mariesii*) or the 'Ball Fern' is also suitable for this purpose. *D. fijiensis* (also spelt *fejeensis*), the 'Rabbit's Foot' fern, is also a handsome species. Some botanists consider this as a variety of *D. solida*. There is a variety *plumosa* in this species which too is handsome. *D. canariensis* is the 'Hare's Foot' fern.

Drynaria (*Polypodiaceae*).

A genus mostly of epiphytic habit and suitable for outdoor decoration. *D. quercifolia* (Syn. *Polypodium quercifolium*) has fronds which are large and have some similarity to oak leaves. A native of India, Queensland, Fiji.

Dryopteris (*Polypodiaceae*).

A large genus of ferns with handsome dissected foliage. Most of the species are easy to cultivate. Propagation is generally from spores. Some botanists, especially in England, refer some species of *Dryopteris* as *Nephrodium*. The cultivation of both is the same as there is hardly any difference between the two.

Gymnogramma (*Polypodiaceae*).

Previously this group included the attractive ferns known as 'Gold and Silver' ferns. The under-surface of finely cut leaves of some species are covered with yellow or white powder or farina and hence are so called. Presently the ferns of this group are placed in *Pityrogramma*, *Ceropteris* (Syn. *Ceratopteris*), and *Gymnopteris*. The ferns in this group should not be watered from overhead. The important species are as follows:

(a) *Pityrogramma chrysophylla* (Syn. *Gymnogramma chrysophylla*, *Ceropteris chrysophylla*): This is a beautiful 'Gold Fern' of graceful habit.

(b) *P. pulchella* (Syn. *Ceropteris pulchella*, *Gymnogramma pulchella*): A 'Silver Fern' of upright habit. The fronds are densely covered beneath with white powder.

(c) *P. sulphurea* (Syn. *Gymnogramma sulphurea*, *Ceropteris sulphurea*).

(d) *P. tartarea* (Syn. *Ceropteris tartarea*, *Gymnogramma tartarica*): A 'Silver Fern'.

Lomaria (*Polypodiaceae*).

This genus is allied to *Blechnum* and is generally included therein. *L. gibba* (Syn. *Blechnum gibbum*) is a handsome species.

Lygodium (*Schizaeaceae*).

A group of twining or climbing ferns belong to this genus. *Lygodiums* are suitable for growing in pots over balloon-like trellises made of G.I. wire fitted to the pots. Some important species are:

(a) *L. circinatum* (Syn. *L. dichotomum*, *L. pedatum*).

(b) *L. japonicum* (Syn. *Ophioglossum japonicum*): Leaflets papery, pretty.

(c) *L. scandens* var. *microphyllum*: A native of the Himalayas, Africa, and Australia. A very handsome twining fern.

Nephrodium: See *Dryopteris*.

Nephrolepsis (*Polypodiaceae*).

The genus consists of about 35 species of hardy, handsome tropical and subtropical ferns. Most of the species are ideally suited for growing in hanging baskets because of their drooping nature, while others are good for conservatories. Most of the species produce slender runners and are easily propagated from these. These ferns are popularly known as 'Sword Ferns' or sometimes 'Ladder Ferns'. Among the various attractive species the following are recommended: *N. acuminata* (Syn. *N. davallioidies*), *N. cordifolia* (Syn. *N. tuberosa*, *N. cordata*), *N. exaltata*, *N. exaltata bostoniensis compacta*, *N. exaltata* 'Elegantissima', and *N. muscosa* (Syn. *N. exaltata* 'Muscosa'), a cultivar.

Onychium japonicum (Syn. *O. capense*) (*Polypodiaceae*).

A fern from Japan, China, and Sikkim. A small fern whose waxy-green fronds are lacily sub-divided into small segments. Suitable for indoor decoration.

Osmunda regalis (*Osmundaceae*).

The best species in the genus, it is popularly called the 'Royal Fern'. This is well adapted for open moist localities and hence are often recommended for planting in marsh or bog gardens. The fronds are feather-like, plain or sometimes crested; the fertile portions are contracted. The fern may grow as tall as 150-180 cm. Osmunda root-fibres make good compost, especially for growing certain orchids. It can be multiplied by the division of the clump. The fertile top portion of the fronds takes the appearance of an inflorescence and hence the other colloquial name, 'Flowering Fern'.

Pellaea (*Polypodiaceae*).

These are small ferns suitable for growing in limestone rocks and hanging baskets. The compost for them should be light and porous with minimum soil. The roots should never go dry and at the same time the foliage should not be syringed. These are grown in semi-shade. Most of the species can be propagated from spores. The recommended species are *P. cordata*, *P. falcata* (Syn. *Platyloma falcatum*), *P. geraniifolia* (Syn. *Doryopteris concolor*, *Pteris geraniifolia*).

Pityrogramma: See *Gymnogramma*.

Platycerium (*Polypodiaceae*).

These ferns are very handsome and epiphytic in nature. The fronds are divided at the margin like the horns of stags and hence the ferns are popularly known as 'Stag-Horn Ferns' or 'Elk's-Horn Ferns'. These can be grown like orchids in logs, preferably supported by fibres such as coconut and sphagnum moss. These ferns can also be attached to free fern blocks. These are also suitable subjects for hanging baskets. They are propagated from young plants which are developed from adventitious buds from their roots. All species like a moist atmosphere, with perfect drainage. *P. bifurcatum* is the common 'Stag-Horn Fern'. Other species are *P. alcicorne* (many consider it as a Syn. of *P. bifurcatum*), *P. grande* (*superbum*), *P. hillii*, etc.

Polypodium (*Polypodiaceae*).

A large genus of ferns, many species of which are very handsome. The rhizomes should not be buried deep; only the roots should be covered. Only well-established plants produce handsome fronds and hence these should not be disturbed for a long time, once planted. Most of the species produce spores freely and plants raised from spores have better shapes compared to those obtained by the division of rhizomes. The plants dislike continuous dripping of water. Most of the species are hardy and suitable for outdoor culture. In many species the fronds are strap- or tongue-shaped and are somewhat leathery in texture. The following species are recommended for growing: *P. aureum* (*Phlebodium aureum*), *P. quercifolium* (see *Drynaria quercifolia*), *P. subauriculatum* (Syn. *Schellolepsis subauriculata*), suitable for hanging baskets, and *P. verrucosum* (Syn. *Schellolepsis verrucosa*).

Polystichum (*Polypodiaceae*).

These are moisture-loving hardy ferns which grow better in semi-shade or shade, though tolerate enough sunlight. These ferns dislike poor drainage. The propagation is best from spores, as division of the crowns is a delicate operation. The fronds are very useful in flower arrangements. Some recommended handsome species are *P. adiantiformis* (*adiantiforme*) (Syn. *P. capense, Aspidium coriaceum, Rumohra adiantiformis*), *P. aristatum* var. *variegatum*, a handsome variegated species, and *P. setosum* (Syn. *P. discretum*).

Pteris (*Polypodiaceae*).

A large genus which consists of several handsome species, many of which possess beautifully variegated fronds. The ferns in this genus are excellent for pot culture and dish gardening meant for table decoration. The species with creeping rhizomes offer propagation through division, but others are easily propagated from spores. Some recommended handsome species are: *P. cretica, P. cretica* var. *albo-lineata*, with broad central white band, *P. ensiformis* var. *victoriae, P. longifolia, P. multifida* (Syn. *P. serrulata*), *P. quadriaurita* var. *argyraea, P. tremula*, and *P. tripartita*.

Some other ornamental ferns not mentioned above are *Tectaria fuscipes, T. trifoliata* (*Aspidium trifoliatum*), *Woodwardia chamissoi, W. orientalis, W. radicans*.

SELAGINELLA

Selaginella belongs to the family *Selaginellaceae* and is commonly known as 'Club Moss'. The plants have small scale-like leaves and may be a minute prostrate annual to an erect or even climbing perennial. The fern-like foliage is much-branched and very ornamental. The colour of the leaves is in various shades of green and sometimes are bronze or bluish. Selaginellas like shade and moisture for growth. A light soil suits most of them. The soil is kept open by mixing charcoal pieces. The creeping stems root almost at every joint and stem cuttings are taken for propagation. The propagation from cuttings can be done in quantity. The fronds can be layered in damp, mossy soil for rapid propagation. Selaginellas are very handsome and should be grown in large numbers wherever possible. Some Selaginellas can be grown in pans as semi-aquatics, some are suitable for rockeries. Selaginellas also form good edging for beds in a shade garden. It can form a carpet bedding of emerald gold. The fronds can be used in preparing bouquets in combination with flowers. In our country the Selaginellas grow best during the rainy season. Most of the species do much better in a mild climate as that of Bangalore and at medium altitudes. Some of the handsome species are: *S. canaliculata* (Syn. *S. caudata, S. chinensis, S. sinensis*), *S. caulescens, S. cuspidata, S. denticulata, S. emmeliana* (also spelt *emiliana*), used for table decoration, *S. erythropus, S. martensii* var. *variegata, S. pilifera* and *S. pulcherrima*. The species *S. laevigata* is an old favourite.

Lycopodiums (*Lycopodiaceae*).

A group of *Pteridophytes* which resemble selaginellas and have erect or trailing stems. Lycopodiums thrive well only in mild moist climates. They are useful for table decoration, wardian cases and conservatories. The most ornamental species are *L. clavatum, L. phlegmaria, L. squarrosum*, and *L. taxifolium*.

Chapter XVII

Ornamental Grasses, Bamboos, and Reeds

There are a number of ornamental grasses, bamboos, and reeds, which are handsome in look and quite easy to cultivate, but, with the lone exception of the lawn grass, they do not generally find any place in our gardens. The main prejudice against the ornamental bamboos or grasses is that they harbour snakes. It is a fact that uncared for bamboo bushes with their overgrowths and hollow cut-stumps provide good shelter for the reptiles. For that matter, even a unweeded shrubbery or an untrimmed hedge can also be the hiding place for snakes. A well-kept clean bamboo clump or a neat clump of ornamental grasses cannot have any room for the reptiles.

Bamboos and grasses can be grown in any kind of soil in a moist but well-drained situation. There are annual ornamental grasses as well as perennial grasses including the ornamental bamboos. The annual grasses do not thrive well under the climatic conditions of the Indian plains but they do well in the hills. In this chapter our emphasis will be on the perennials. Some of the ornamental annual grasses produce flowers of everlasting nature which are good for vase decoration. Many of these annuals are good as pot plants, while some are fit for the rockeries. In the list of perennial types, the ornamental bamboos are generally grown in clumps. They look particularly beautiful when planted in sloping lands above the waterline of a pool. Sometimes, bamboos are also planted along an avenue or a garden path. A few dwarf types (e.g., *Lagurus ovatus*) grow well in pots. The ornamental bamboos or perennial grasses are generally hardy and can be grown in full sun or by partial shade. The propagation of these ornamentals is either from seeds or cuttings or division of the rootstocks from a clump. The planting is best done during the rains. The bamboos and perennial grasses are deep-rooted and hence a deep cultivation is beneficial. The clumps, especially of bamboos, should be divided every 2-3 years, all dead stumps should be removed and only a portion of the clump can be retained to keep them within bounds. For potted plants it will be a wise move to repot them every year and in the process all dead stumps should be removed and new compost added.

Many of the bamboos, grasses, and reeds, which are naturally suited for regions around sea level, also thrive quite well, though not as good, up to an elevation of 1,500 m.

There are a large number of annuals and perennial types of bamboos, grasses, and reeds which are very ornamental for growing in the garden but in this chapter we shall confine our discussion to a few types only,

The plants described below belong to the family *Gramineae*, unless mentioned otherwise.

Agrostis.

This genus consists of about 100 species of annual or perennial grasses. The flowers are generally borne in many-branched spreading panicles with numerous small spikelets. The seeds are sown in moist weather. Many of these are very pretty and grow easily in an ordinary garden soil, but do well only at higher elevations in India. The inflorescence is dried and preserved for decoration during the winter. These are also good for pot decoration.

(a) *A. elegans:* A tufted annual grass up to 30 cm tall. A native of western Mediterranean. Panicles are loose with 2-5 basal branches and the spikelets are very small.

(b) *A. nebulosa:* It is called 'Cloud Grass' and is a native of Spain. It is a small, 30 cm tall, tufted annual grass. The flowers are used for winter decoration. Seeds are sown *in situ.*

(c) *A. pulchella:* An annual, 15-30 cm tall, more rigid in habit than the above, but equally handsome.

Andropogon (Syn. *Cymbopogon*) *citratus.*

Commonly known as 'Lemon Grass', it is a bushy, densely tufted grass not taller than 50 cm. The leaves are long, narrow, green and lemon-scented. Propagation is by division.

Apluda aristata.

A small miniature bamboo-like plant growing to a height of 12-50 cm. It is suitable for growing in pots or in rockery.

Arundinaria auricoma (Syn. *Bambusa fortunei aurea*, *Pleioblastus viridistriatus*).

A bamboo with a slender round stem, suckering and growing up to a height of 90-180 cm. The stem is dark-purplish, the stem-sheaths are persistent, hairy at the base and margins. The leaves are green, striped yellow, but often more yellow than green, and the lower surface downy. A native of Japan.

A. chrysantha (Syn. *Bambusa chrysantha*, *Sasa chrysantha*).

About 90-180 cm tall, with reed-like culms and branches several at one joint. Leaves bright green, occasionally striped with yellow. Origin is Japan.

A. fortunei (Syn. *A. variegata*, *Bambusa fortunei* var. *variegata*, *Sasa fortunei*, *Pleioblastus variegatus*).

It is considered to be the most attractive dwarf bamboo in cultivation. A low-growing plant with gracefully arching branches. Leaves are 7-15 cm long and up to 2.5 cm broad, strikingly banded with white, on both surfaces, especially the lower, hairy. Origin is Japan.

A. hindsii (Syn. *Bambusa erecta*, *Pleioblastus hindsii*, *Phyllostachys hindsii*).

A Japanese bamboo, growing up to 3.5 m. The stem is olive-green. The branches are many, forming clusters at the upper joints. The leaves are 15-25 cm long, dark green, glaucous beneath.

A. japonica (*Bambusa metake*, *Pseudosasa japonica*).

Commonly called the Femde Arrow Bamboo of Japan. The stems are 3-4.5 m tall, very hollow. The leaves are dark green, 10-20 cm long and 2.5-5 cm wide; the upper surface of the leaves is dark glossy-green while the lower surface is white and finely pubescent.

Arundo donax var. *versicolor* (Syn. *A. donax* var. *variegata*, var. *picta*).

It is the variegated-leaved 'Giant Reed' of the Mediterranean. A tall (3.5-4 m) perennial grass, with a knotty rootstock and a stout woody stem. The graceful arching leaves are arranged alternately, about 60 cm long, and grey green in colour ribboned with creamy white. The showy plume like panicles, 30-60 cm long, are reddish, later turning white. It is propagated from seeds or by division. A stem placed in water produces shoots from joints, which in turn produce roots, and is then removed for planting. The plant grows well by waterside in ordinary soil. *A. conspicua* is another tufted species but less ornamental and hardy than the above.

Bambusa (Bamboo).

Most of the ornamental bamboos cultivated in the garden are now included in the following genera as per modern classification: *Arundinaria*, *Phyllostachys*, *Sasa*, and *Shibataea*. But there are still some ornamental species which are known by the genus *Bambusa* even now. Bamboos like a lot of moisture for their growth and hence these may be planted in groups or clumps along the edges of ponds or streams. The dwarf ornamental species which have greater demand in home gardens, can also be planted in pots or tubs. Some of the common species are described below.

(a) *B. eutuldoides:* It is a large-clump bamboo from China, about 9 m tall. The stiff erect hollow canes are 2-2.5 cm in diameter, olive-green in colour but invariably having handsome greenish-white or ivory stripes longitudinally.

(b) *B. gracillima:* It is a Chinese bamboo, 1.8 m tall, with the 1-cm-thick stem arching outwardly and uniformly. The young stems are green and the older ones orange.

(c) *B. multiplex* var. 'Alphonse Karr' (Syn. *B. nana* '*Suochiku*'): The dense branches and culms are bright yellow, irregularly striped with green, whereas the culm sheaths bear yellow stripes. It is 7.5 to 12 m high. Origin is Japan.

(d) *B. nana* (Syn. *B. disticha*): It is 60-90 cm high, and the stems are slender, zigzag and green in colour or sometimes tinged with purple. A beautiful species. The green leaves produced in two vertical ranks give the plant a distinct and unusual appearance. Origin is uncertain.

(e) *B. pygmaea* (Syn. *Sasa*, *Arundinaria*, *Pleioblastus*): A very dwarf (15-30 cm) Japanese bamboo, in which the stems are much-branched, slender with prominent nodes and purple in colour. A very hardy species, it is used as undergrowth but outgrows very quickly forming a carpet.

(f) *B. ventricosa:* A Chinese bamboo which is commonly called 'Buddha's Belly Bamboo'. It is so called because the species, especially the pot-grown ones, have swollen internodes.

(g) *B. vulgaris:* A tall (15-24 m) bamboo which forms open clumps and are arching in nature. The stem is at first green but later turns yellow. Origin is India, West Indies,

Africa, Java, South and Central America.

Briza media.

'Common Quaking Grass'. It is a perennial grass with creeping rhizome, 30-60 cm high. The leaves are long, grass-like, and 0.5 cm wide. Flowers in panicles, 15 cm long, usually purplish, nodding gracefully with wind. It is a native of Europe, Asia Minor, and Siberia.

The inflorescence is suitable for cutting and is dried for winter decoration. The grass thrives best at medium elevations. It is propagated from seeds and is grown in a good loamy or humus soil in the open. *B. maxima* is an annual, more or less similar in habit.

Cortaderia argentea (Syn. *C. selloana*, *Gynerium argenteum*).

Commonly called 'Pampas Grass'. Has long ribbon-like leaves and large terminal pyramidal panicles of silver white, sometimes tinged rose, resembling a feather. The grass thrives well only on the hills.

Cyperus alternifolius: See page 261.

C. diffusus (Syn. *C. elegans*, *C. laxus*).

Similar to the above. It is used for table decoration. Belongs to *Cyperaceae.*

Eragrostis elegans (Syn. *E. interrupta*).

An erect or spreading annual, 30-60 cm tall, having small, about 39 cm, feathery, rather narrow, panicles with ascending branches. Culture is similar to that of *Briza.* The seeds are sown during the monsoon.

Eulalia japonica (Syn. *Miscanthus sinensis*).

A handsome and most popular perennial ornamental grass, of Japan and China, in cultivation. It is a hardy grass which is propagated by division and the inflorescence is used for dry bouquets. It is cultivated in the hills and as a pot plant in conservatories at medium elevations. The var. *variegatus* (Syn. var. *fol. striatis*) have leaves striped yellow; var. *gracillimus* is of slender habit with narrower leaf, and in the form *univittatus* (Syn. *Eulalia gracillima univittata*) there is a median yellow stripe; var. *zebrinus* is banded with transverse yellow streaks.

Festuca ovina var. *glauca* (Syn. *F. glauca*).

It is commonly called 'Blue Fescue' and is about 25 cm long. The leaves are all filiform, about 0.1 cm thick, silvery blue; the panicles are somewhat one-sided and the spikelets are purple. It is suited for cultivation only in a cool weather. Used as borders.

Gynerium argenteum: See *Cortaderia argentea.*

Mariscus umbilensis: (Syn. *Cyperus natalensis*, *C. umbilensis*).

Belongs to *Cyperaceae* and is a native of South Africa. It is a rush- or grass-like herb which is suitable for planting on the edges of water gardens.

Oplismenus hirtellus variegatus (Syn. *O. burmanni variegatus, Panicum variegatum* of florists).

A perennial, weak, more or less creeping grass, rooting at nodes. The leaves are lanceolate, rather broad, striped daintily with white and pink. It is commonly used as edging in conservatories and for growing in hanging baskets.

Papyrus antiquorum (Syn. *P. sicula, Cyperus papyrus, C. syriacus*).

Commonly called 'Papyrus' or 'Egyptian Paper Reed'. It belongs to the family *Cyperaceae* and is a native of Egypt. It is perennial, dark green, 3-angled and the plants are about 3 m tall. The leaves are in umbel at the crown and are gracefully drooping. Suited for growing in shallow water or moist soil.

Pennisetum ruppelii (Syn. *P. ruppelianum, P. orientale*).

A native of Abyssinia. An annual, 60-180 cm tall. The leaves are long and slender. The slender, spike-like inflorescence is graceful, about 15-25 cm long, plumose and often purple.

Phalaris arundinacea var. *picta* (var. *variegata* of Horticulture); also known as *P. arundinacea* var. *elegantissima*.

It is commonly called 'Ribbon Grass' and 'Gardener's Garters'. It is planted along the banks of streams or pools and is also used for pot culture. The plants thrive well under partial shade. The grass is 60-180 cm tall. The leaves are long, 1.5 cm broad, and striped white longitudinally.

Phormium tenax (*Liliaceae*).

It has long (2.4-3 m), stiff, reed-like leaves, which are 10-12 cm broad, with a narrow orange or reddish margin. In var. *variegatum* the leaves are striped yellow and green. It is a handsome plant suitable for growing in moist soil.

Phyllostachys aurea (Syn. *Bambusa aurea*).

It is known as 'Golden Bamboo' or 'Fish Pole Bamboo'. The stems are 5 m tall, about 4 cm thick, and the internodes at the base are very short and bright yellow in colour. It is a native of China.

P. bambusoides.

A giant (21 m tall) bamboo which is 12 cm in diameter. The stems are green but are eventually striped or yellow.

P. heterocycla.

'Tortoise-shell Bamboo'. It is very curious-looking and ornamental. The knots formed in old stems resemble tortoise shell.

P. nigra (Syn. *Bambusa nigra*).

The 'Black Bamboo' of Japan and China. It is about 7.5 m tall. The stems are green at first, then speckled and ultimately completely black.

P. ruscifolia (Syn. *P. kumasaca, Phyllostachys kumasaca, Bambusa ruscifolia, B. viminalis, Shibataea kumasasa, S. kumasaca*).

The stems are usually 90 cm tall, tufted, slender, zigzag and flattened between joints which are short. There are 3-4 short branches at each joint having 2-4 narrowly ovate, slender, pointed and bright green leaves.

Sasa fortunei: See *Arundinaria fortunei*.

Sasa pygmaea: See *Bambusa pygmaea.*

Stenotaphrum secundatum variegatum.

It is called '*St. Augustine Grass*'. It is very ornamental. A creeping or erect (15 cm tall) type of stoloniferous grass in which the stems are flattened, the linear leaves are flat, firm and handsomely banded creamy white. The leaf tip is blunt.

Shibataea kumasasa: See *Phyllostachys ruscifolia.*

Tricholaena rosea (Syn. *T. repens*, *T. grandiflora*, *T. violacea*, *Panicum tonsum*).

'Natal Grass'. It is an annual or short-lived perennial and one of the most ornamental grasses. Often also known as 'Rubby Grass'. The panicles are 7-20 cm long, the spikelets are silky, wine-coloured when freshly opened, turning purplish on drying. It is propagated from seeds and is used for borders and dry bouquets. Grows in cool climates.

Uniola paniculata.

Commonly called 'Sea Oats'. The culms are 120-240 cm tall, arising from creeping rhizomes. The spikelets are narrower, forming long drooping panicles.

Chapter XVIII

Cultivation of Plants in Pots

Cultivation of plants in pots is useful in several ways, but it is most advantageous to those who live in flats and have no space for growing plants in the ground. For them this is the cheapest method of gardening. Another advantage it provides is the easy mobility, i.e., a potted plant can be handled easily and can be moved to any part of the garden or even inside the house for the purpose of beautification. Some of the delicate plants such as tuberous-rooted begonias which seldom survive in the ground can be grown successfully in pots. Other such beautiful flowering plants as chrysanthemum and dahlia are grown in pots and when they bloom they are used for decorating various parts of the garden and the house-front. There are several occasions when growing plants in pots becomes essential. For example, if delicate seedlings are first pricked singly in 5-7 cm pots and then transferred to ground at a later date there will be very few casualties. The main disadvantage of pot culture perhaps is that if the right size of pot is not used or if a plant is left too long in a pot, it gets pot-bound and does not grow properly. One may also argue that this method is against nature, though this argument may not hold much water against the arguments in favour of pot cultivation.

Pots: Pots are generally made of burnt porous clay. A porous pot is important as it allows exchange of gas and moisture which are essential for healthy root growth. Earthen pots come in various sizes, from the tiny 5 cm pot to the giant 45 cm pot. A pot is generally conical in shape, tapering towards the base, thus providing more soil for the feeding roots at the top. The small 5 cm pots are used for pricking seedlings. Grafts or seedlings having long tap roots are first transplanted in long nursery pots. For sowing seeds, seed pans are used which are of a larger diameter (say 30 cm) but have less depth (say 7 cm). Plants having shallow root systems, such as certain categories of ferns, need pots having less depth than usual. For orchids, special pots are used with perforations all over the sides to allow plenty of aeration and drainage needed by them. It is even possible to grow some bog plants in specially-made pots. These consist of one pot placed inside the other with a narrow space in between, meant for filling it with water, damp moss or sand, so that the soil in the inner pot growing the bog plant receives moisture constantly. The top diameter of a pot is almost equal to the depth of the pot, but in larger pots the depth may be a little less than the diameter. Now-a-days pots made of special unbreakable polythene are used, which prove economical compared to earthen pots.

Potting: Before potting is done it is essential to clean the pot thoroughly, especially those which were used earlier, with clean water or, preferably, with hot soda water. The idea is to get rid of any disease or fungus that may be hibernating. Moreover, a dirty pot is not very porous and will not allow exchange of air properly. The new pots are washed because a dry pot draws away the moisture from the compost within.

At places having very cold winter the pots should be stored in a dry frost-proof shelter. Dry pots are not affected by frost, whereas wet pots are likely to crack when the moisture inside is frozen and expanded.

Before filling the pot with compost, large crocks with the concave side down are placed over the drainage hole so as to ensure that the hole is not blocked by the compost. Over this several small pieces of crocks are placed. Then a layer of coarse leaf-mould or half-decayed leaves is spread. The compost is then filled in, leaving about 1.5-2.5 cm space on top, depending upon the size of the pot, to hold sufficient water. Small pieces of charcoal, cinder, or bricks may be placed at the surface of the compost to prevent it from being washed out and to help retain moisture. It is very difficult to keep these particles in place, as during forking the soil, these may get into the compost. The best way is to water the pots gently with a hose or a water-can fitted with a hose. To preserve moisture, watering should be followed by forking. The ball of earth is buried about 3-5 cm deep in the soil and the compost is rammed firmly. The ball of earth around the roots of the plants being clayey or sticky, is broken by soaking, before potting.

Potting Compost: The usual constituents of a potting compost are a good garden soil, preferably loam, which generally forms the greater part of the compost, leaf-mould or peat-moss, coarse sand, old mortar rubble, well-rotted manure, charcoal, and cinder ash. The proportion of the constituents will vary according to the type of the plant. For plants which need rich potting compost, the amount of soil in the compost is reduced and other constituents such as leaf-mould and manure are increased. Some greenhouse plants such as *Anthurium*, ferns, *Alocasia*, and *Dieffenbachia* need rich compost. Actually *Anthurium* or *Alocasia* will do well even in a bed consisting of leaf-mould and manure. In clayey soils, the proportion of sand in the compost will be more than in the alluvial soils of eastern or northern India. It has already been stated that the proportion of the various ingredients in a soil will vary according to the type of the plant grown and the nature of the loam used, but, nevertheless, the following compost with a little variation, depending on the situation, will be suitable for most of the general type of soft-wooded plants. The special pot mixtures required for other plants have been mentioned in the respective chapters.

Good garden loam (preferably fibrous)	:	2 parts
Leaf-mould or peat-moss	:	1 part
Well-rotted manure	:	$\frac{1}{2}$ part
Coarse river sand	:	$\frac{1}{2}$ part

The compost should be stored in potting sheds. The potting compost at the time of use should not be bone-dry, but should be moderately moist. If the compost is held firmly in the hand, it should get moulded into the shape of the hand, but not sticky and should crumble down when disturbed.

Method of Mixing the Ingredients: The ingredients of the compost are generally not sieved except to remove grass roots and large stones. In the case of sowing of seeds or planting cuttings the compost is generally sieved. The ingredients are mixed thoroughly by placing them one above the other in layers. For example, to prepare the above compost two baskets of garden loam is first spread on the ground, over this one basket of leaf-mould is spread, followed by half a basket of well-rotted manure, and finally half a basket of coarse sand. The process is repeated till a good heap is formed. Before putting the subsequent layers all the spread-out layers are moistened lightly. After the heap is made it is turned and the ingredients are mixed thoroughly.

Repotting: Repotting is an important operation. The time when repotting is to be done depends on several factors. When a potted plant is to be repotted with the ball of earth intact in a larger pot or in ground, this operation can be undertaken almost any time of the year as the roots do not get disturbed during the process. When the ball of earth is to be disturbed or broken because of its being old or getting sour, the transplanting should be undertaken in a season when the roots are in active growth. However, orchids and some other plants, especially those which are transplanted bare root, are transplanted at the "resting" period.

To ascertain if transplanting is essential, a plant or two of the same batch of similar species are turned out of the pot and observed. If the roots are seen to wind round the side of the pot, it is time for repotting, provided the plant is not in "resting" season. Under the conditions in the Indian plains, the rainy season will be the best time for transplanting. Plants which have started new growth should preferably be not transplanted. For repotting a potted plant in a larger pot the plant is tossed out of the pot gently. It is not always necessary to repot the plant in a larger pot. If the winding roots around the ball of earth are not many, the plant is repotted in the same sized pot after necessary root pruning. If the soil has become sour or old a portion of it is removed and new compost is filled in between the ball of earth and the pot and then rammed firmly. When a larger pot is used, it should be just one size bigger than the previous one. Plants having sufficient root growth can be put in a pot slightly larger than the one needed.

Do's and Don'ts in Potting and Repotting: It should be kept in mind that a right-sized pot be used for potting and never an over-sized pot. While potting, a plant should be planted neither too deep nor very shallow as both are harmful. A plant potted at shallow depth will not have sufficient hold and will be shaken by even a little disturbance. A safe method is to plant the old ball of earth at the same depth as it was in the previous pot.

Immediately after potting, the pots are removed to shade to enable the roots to get established. After the plant is established, it is brought out in the sun, gradually increasing the exposure time. Nutrients in pots are quickly used up by the plant and some of these are also washed out. To replenish them, potted plants need to be fed by liquid manure or by "top-dressing" at regular intervals. In the latter method, the top 2-7 cm compost is removed from the pot by scraping and the gap is filled with fresh compost. Top-dressing can be done once in 3-4 months, whereas liquid manure can be applied once in a fortnight. The hard crust on the surface of the compost is forked occasionally to a depth of 1-2 cm to facilitate aeration and penetration of water.

Termites and earthworms should not be allowed to enter the pots. The pots should be placed over a couple of bricks placed side by side.

Watering: It is often due to mismanagement in watering that potted plants suffer. The common mistake is to water either very little or too much. Over-watering will make the compost sour very soon as the compost is deprived of aeration. Watering is to be so adjusted that the compost in the pot is allowed to grow dry after each watering, but not too dry to cause wilting. One way of knowing whether a pot needs watering is to gently tap the outside of the pot with a wooden hammer or rafter. If the resulting sound is metallic, the pot needs watering; and if the sound is dull, watering will not be needed. The mode of watering will vary from plant to plant. Leafy and succulent plants need more frequent watering than the woody plants. Certain plants may need watering even twice a day during the summer. The pots are watered sufficiently to soak the compost and water should not run down the drainage hole. Some potted plants enjoy syringing especially during hotter months. Freshly planted pot plants may also be syringed a couple of times during the day in addition to regular watering.

Chapter XIX

Bonsai

Bonsai is a fascinating art which has thrived in Japan for hundreds of years and has become an integral part of Japanese culture. In Japanese homes, there are plants which are known to be hundred years old and in one instance a specimen is said to be about one thousand years old. This treasured art is passed from generation to generation in Japanese society.

The word *Bonsai* is a combination of two Japanese words *Bon* (meaning shallow pan) and *Sai* (meaning plant), which can be translated as "tray planting". By growing the trees in shallow containers the growth is checked or dwarfed but the plant is not starved. The great Japanese expert on *Bonsai*, Kyozo Morata, is on record to have said, "Like a pet animal, it needs water, sunshine and nourishment." These fascinating trees, in addition to being grown outdoors, are great favourites for growing indoors in such places as living rooms, verandas, terraces, window-sills, and conservatories. The city dwellers can with advantage take to this culture. The *Bonsai* differs from a pot plant where either the foliage or the flower is the main criterion, but here the artistic look of the plant in a miniature form maintained for many years is all that is wanted.

Criterion for Selecting Plants: In Japan, the suitability of various plants for the different types or shapes of *Bonsai* has been tested and established beyond any controversy because of the long tradition of cultivating this art. There is a comprehensive list of available trees, shrubs, and creepers suitable for different objectives. But in India we are at a loss because there are hardly any data available on this account, except the great personal contribution of the late V.P. Agnihotri of New Delhi. He perfected a technique for developing *Bonsai* from common trees grown in India and built up a collection of a few hundred beautiful *Bonsai* specimens.

Among the various factors that determine the suitability of a plant as a *Bonsai* the foremost is its hardiness. It should stand the rigours of growing in a shallow container, withstand the operations of pruning of roots, branches and tough training but still manifest itself as a lively replica of a normal plant. All *Bonsai* specimens should have a trunk which looks natural. The growth of such plants should be in harmony with the container where it stands and the branches should also reflect a rhythm consistent with the artistic value. The trees whose trunks are strong and thick at the ground level give a good appearance. The plants which show seasonal variations in growth pattern and flowering are greatly favoured. For example, some trees which change their leaf colour in one season or others bearing fruits in another season, all add colour to the collection.

Though not exhaustive, but a fairly long list of tropical plants are cited here as a guide for using as subjects for *Bonsai.*

Adansonia digitata, Adenanthera pavonina, Bombax malabaricum, Brassaia actinophylla, Bamboos (e.g., *Phyllostachys nigra* Syn. *Bambusa nigra* commonly called Black Bamboo. *P. bambusoides, Arundinaria simonii,* etc.), *Berberis thunbergii, Butea monosperma, Callistemon lanceolatus, Chorisia speciosa, Erythrina crista-galli, E. parcellii, Ficus religiosa, F. benghalensis, F. infectoria, F. mysorensis, F. retusa, Jacaranda mimosifolia. Kigellia pinnata, Mangifera indica, Melia azedarach, M. azadirachta, Milletia ovalifolia, Punica granatum* var. *nana, Putranjiva roxburghii,* and *Thespesia populnea.*

Several conifers can also be grown as *Bonsai* under tropical conditions. These are *Juniperus chinensis, J. prostrata* (Syn. *J. horizontalis*), and *Pinus khasiana.* Other conifers suitable for temperate climate are *Cedrus deodara, Cryptomeria japonica, Ginkgo biloba, Picea excelsa, Pinus densiflora, P. sylvestris, Taxus baccata,* etc.

There are certain shrubs listed below which are also trained as *Bonsai: Adenium obesum, Azalea* (temperate), *Brya ebenus, Berberis asiatica, Cotoneaster horizontalis* (temperate), *Fortunella japonica, Hamelia patens, Jatropha podagrica,* and *Murraya exotica.*

The following woody climbers are also used as subjects for *Bonsai:* Species of *Combretum, Derris scandens, Hiptage madablota, Roupellia grata* (Syn. *Strophanthus gratus*) and *Wisteria sinensis.*

One can experiment with many other trees and shrubs not listed above for *Bonsai.*

Classification of Bonsai: Bonsai are classified into two groups, firstly by the shape of the trunk and, secondly by the number of trees grown in a single pot (see Fig. XIX.1):

(1) *Upright or Chokkan style:* In this style a single specimen is grown with a straight and upright stem.

(2) *Winding or Kyokkuk style:* In this a single trunk is twisted once or several times, showing as if it has struggled for its existence in the wild.

(3) *Oblique or Shakan style:* This is also referred to as windswept style. The plant is grown in an oblique position as if swept by wind and consequently sometimes having more branches on one side.

(4) *Gnarled or Hankan style:* This differs a little from the winding style. Here the trunk is twisted near the ground once or twice to form a knob or a loop.

(5) *Cascade or Kengai style:* Here the trunk hangs over the edge of the container as if it was hanging over the ledge of a mountain.

(6) *Ikadi-Buki style:* A tree is often laid or trained horizontally and several branches are allowed to grow vertically at intervals resembling individual trees. Alternatively, 2 or 3 plants of the same kind may be planted in a single container. Another method is the clustered *Bonsai* where several branches are allowed to grow from ground level in a clustered manner.

(7) *Clasped to Stone style:* A single tree is grown with its roots trained over a stone or some blocks of boiler ash. Plants grown in this way presents the look of age and tenacity.

How to Start: The first thing is to collect the suitable type of plant. A plant for *Bonsai* may be raised by seeds, seedlings, cuttings, layering, or plants collected from the wild. Plants raised from seeds or seedlings will take some time to train as *Bonsai.* Hence, many people start with layering of a large branch, cutting and grafting. Some wild "old" plants, if selected properly, will need very little training. Large seedling plants can also be obtained from

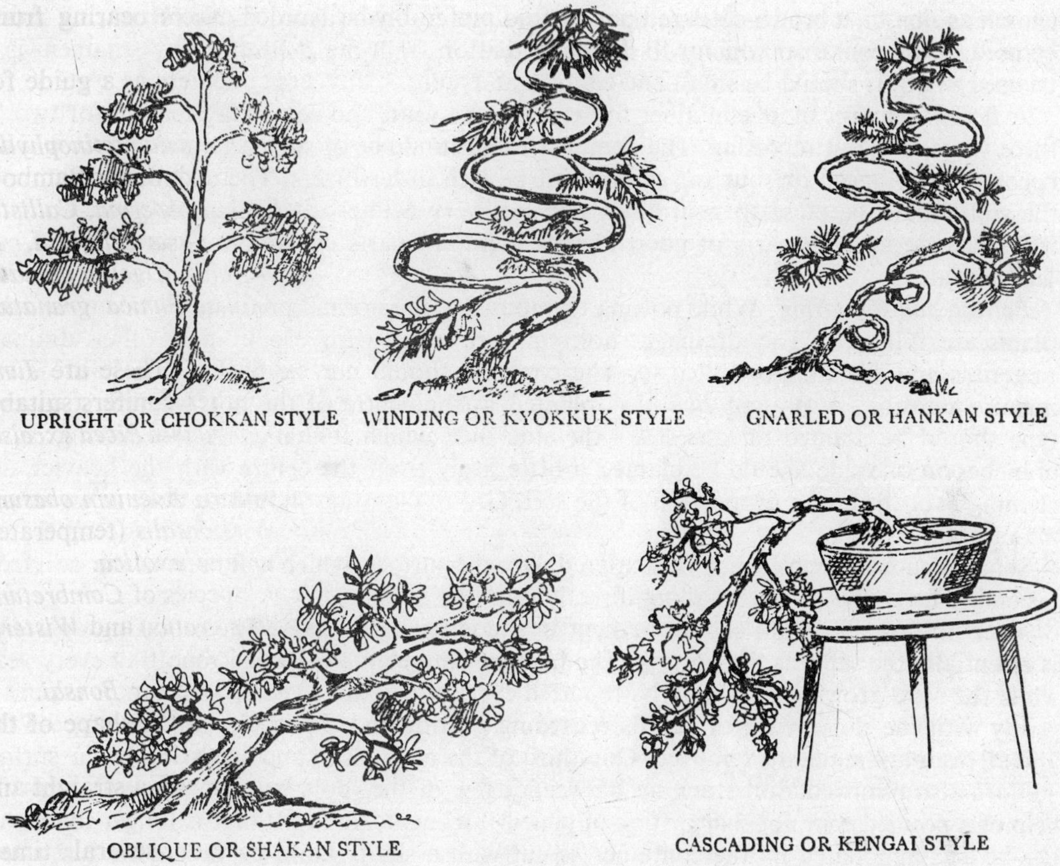

Fig. XIX.1

nurseries. The wild plants and plants grown in a nursery should be uprooted with care and grown in an ordinary pot for a year or two and then transferred to the shallow container. For other type of plants also these should preferably be grown in the ground for one or two years to encourage strong root growth and then transferred to a suitable container after necessary root and branch pruning. Pot-grown plants can be transferred to a container directly.

Containers and Soil: While selecting a container it must be kept in mind the meaning of *Bonsai* (*Bon* meaning shallow). Although the choice of a container depends to a certain extent on the personal choice of the grower, we cannot ignore certain guidelines. Whatever may be the shape, the container must be shallow. The shape may be round, oval, or rectangular. For a single upright stem an oval or a round container will suit better, while for the cascade or *Ikadi-buki* style a rectangular container looks more artistic. An unglazed container is preferred to a glazed one as the former allows aeration of the soil. A container should also have one or two drainage holes, depending on size. The colour of the container should be as natural as possible and is generally black, green, or brown. For certain specimens, even a white-coloured pot is quite effective. When a flowering tree having red or pink flowers is chosen as *Bonsai*, a brown-coloured pot should preferably be avoided. No ornaments should be used in the container which will detract attention from the *Bonsai*. If any ornament is to be used at all, it should be small and of antique type.

A *Bonsai* remains in its container for at least one year, and some may remain for two to three years, without repotting. The compost used should be of such a nature that it will not become waterlogged or sour and should not be rich in fertilizers. The chemical reaction of the soil should be close to neutrality, i.e., not very acidic or alkaline. A good compost will be made up of 2 parts of good fibrous loam, 1½ parts coarse river sand and 1 part leaf-mould.

Potting and Repotting: While potting or repotting, the normal precautions taken for other plants are followed. The drainage holes are covered with crocks and other drainage materials and the compost filled in. The compost should not be packed too hard as this impairs aeration. A straight *Bonsai* is planted in the centre of the pot, whereas a cascade type should be planted on one side – the side over which it hangs. A tree having more branches on one side should be planted a little away from the centre with the heavier side running over the larger proportion of the soil. Group plantings should be done in an artistic manner.

Some attractive pebbles may be scattered over the surface, which will prevent the soil from becoming hard due to water falling directly over the soil.

After potting, the plant should be kept in shade for a few days. The repotting operation is essentially the same as for potting. The fast-growing plants will need repotting every year, while the slow-growing ones may be repotted every two or three years. The plant is taken out gently with the earth ball (for details regarding potting and repotting, see Chapter XVIII) and all drainage materials removed. One-third of the old subsoil and a portion of the surface soil are also removed. Soil stuck up between a few of the roots may be removed, with the help of a pointed, but not sharp, stick of pencil-thickness. Any root that is longer than what can be accommodated in the container is cut with a sharp knife or secateurs just about 2 cm away from the side of the container. The plant is then placed in position and new compost is filled in. The plant is immediately watered. No wiring or pruning should be done up to at least six months or a year after repotting.

Training: The bending and training should be done as soon as it is possible to tie a wire. But some of the plants with a soft bark will be damaged if wired very early. For training and bending, copper wire is best as it is pliable. The gauge (thickness) of the wire will depend on the specimen, but this should be soft enough to work with. Before using the wire this should be heated to make it pliable, then cooled before using. But once used for winding after heating, the wire gradually regains its original strength. When the wire is to be reused it should be heated again. When wiring a branch, first wind the wire round the trunk several times and then around the branch. While winding a branch the union with the trunk is to be held with one hand or otherwise the joint may snap. The wire coils should be spaced evenly by about 0.5 to 0.6 cm apart. To give the specific shape the wire has to be in place for several years, but rewinding is to be done every six months or so to avoid injury to the bark. For slow-growing plants the wire can be kept in position for about a year. A lapse of a few months should be allowed between two rewindings. It is always advisable, especially for tender specimens, to wind the branch first with cotton tape or raffia and then wind the wire over this to avoid injury. For upright and oblique styles very little wiring is needed. Stakes should be placed in the desired angle and the stems are tied to it with raffia. Some branches

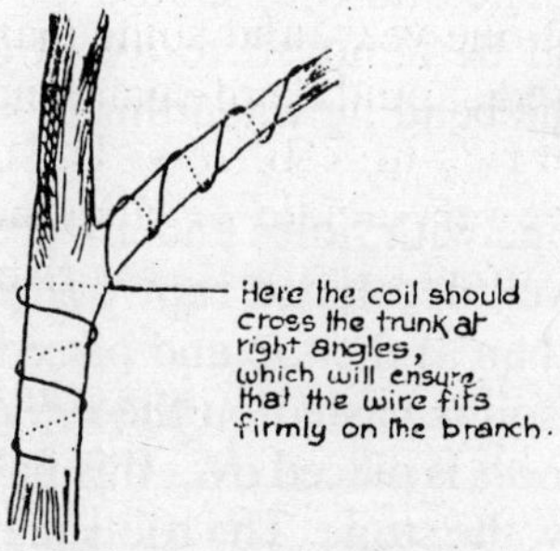

METHOD OF WIRING BRANCH
The wire should be evenly spaced.

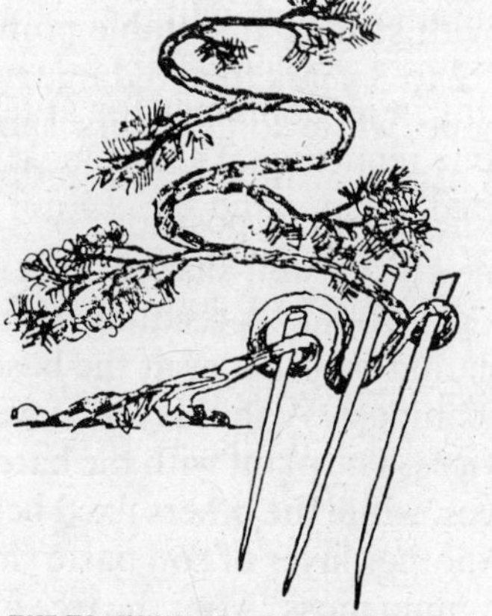

THE TRAINING OF A GNARLED BONSAI
Demonstrating the use of pegs.

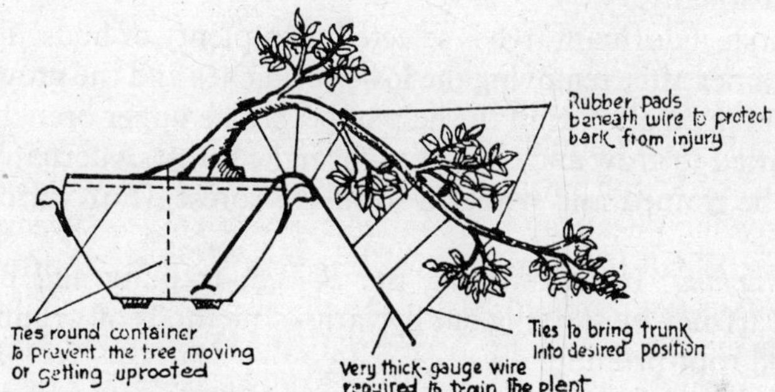

THE TRAINING OF A CASCADING BONSAI

Fig. XIX.2.

may need a little wiring to give them the artistic shape. All the knots in the ties should be redone periodically to avoid damage to the bark.

For the cascading style, a stout wire should first be pushed to the base of the container through the soil, and then bent over the side at an angle in which the plant is to be shaped. The trunk that is to hang below is then tied to this wire with strong strings at several points to bend it in the desired shape. These strings can be tightened further at regular intervals. Wherever the strings are tied on the trunk a protective band of rubber sheet (from tubes of pneumatic tyres) is to be placed round the trunk to avoid injury. The branches will need wire coiling to give them artistic shape.

In the winding trunk style the winding is done round the stem to bend it on one side in the first season. After this is achieved, it is bent in the opposite direction. Simultaneous backward and forward bending is not practised; it is practised when the plant is only one or two years old, although an expert may start with 4 or 5 year-old plants also.

The most difficult style is the gnarled style. For this a soft-trunked plant is needed. It will take a long time to gnarl and twist a trunk with the help of both wire and stakes. The stakes are planted firmly in the soil and to keep them in position they may be supported by stones.

The first twist round the stake should preferably be given near the ground level. The trunk is tied to the stake with strings at suitable points. The ties should be remade several times in a year for the first 2 to 3 years to avoid injury. The branches should be wired to bring them into an artistic shape, while some others should be removed to avoid overcrowding. While gnarling an older tree it will be better to do the bending when this is still in the ground as the stakes can be held firm in position.

In clasp-to-stone style a well-shaped, soft stone with holes and crevices is to be selected so that the roots can get a hold. A seedling with well-developed root system is selected for this style. The stone should be broader at the base than at the top and placed over a shallow container filled with compost. A thin layer of wet soil is pasted on the top of the stone and into the holes and crevices. The plant with the bare roots is placed over this paste, a few roots going inside some crevices, while the others fixed below the stone. The roots may be tied in position with soft twine. Another layer of soil paste should be applied over the roots. Then this layer is covered with sphagnum moss. After the tree is established the moss as well as the ties should be removed. The roots will gradually take the shape of the stone. In dry weather, the plant should be syringed occasionally.

In *Ikadi buki* style a dormant tree is selected with plenty of buds. The trunk is laid horizontally on the container after removing the lower branches and the growing tip, and kept buried partly. This is held in position with the help of pegs. The upper branches may also be removed and the buds allowed to grow and trained to form the forest. Alternatively, the upper branches themselves may be trained and wired to form the forest. (For various methods of training, see Fig. XIX.2.)

Pruning and Pinching: In establishing the *Bonsai*, training and pruning go together. A *Bonsai* is kept dwarf not by starving but by various methods of pruning, viz., shoot pinching, leaf-pinching, and root-pruning.

Pruning or pinching of shoots is an important operation in *Bonsai* growing. This helps maintain the shape of the plant and reduces total leaf area which in turn keeps the plant dwarf as it is in the leaf where the food is manufactured. The amount, the mode, and the time of pinching depend upon growth habit, response to pinching, and rate of shoot growth of the species. Pinching of the branches is also not done at the same length or direction, but at different lengths to give an artistic shape. If the growth is very vigorous, pinching may be resorted to more than once a year. It should not be done frequently as such pruning will give rise to weak shoot growth. Some branches may be just pinched at the apical bud level while others may be removed from further below. If some secondary shoot growth is not wanted these should be removed when quite young or such removal leaves a scar on the branch. Conifers having needles should not be pinched through the needles as the half-cut needles look ugly.

A further method of reducing leaf area is done by leaf pinching. This is done with vigorous-growing deciduous trees at the time of active growth. Leaf pinching is done only once a year in addition to the normal fall of leaves. This can be done either by removing all the leaves from just above the petiole or by reducing the leaf area. The latter is achieved by either cutting half of the leaf lamina or removing the alternate leaves. After leaf pinching supply of water should be drastically reduced since leafless plants are unable to draw enough water. But the plant should receive enough sunshine. During all these pinching operations the plant should not be shaken at the base and roots disturbed.

It is also necessary to prune a certain amount of roots keeping in view with dwarfed shoot growth. During transplanting and repotting time the root pruning is done, as has been described earlier.

Watering and Manuring: As *Bonsai* grows on shallow containers, watering is an important aspect. In places having severe winter, watering once a week during that period may be sufficient.

During the summer months watering three times a day may be necessary. During this period syringing of the leaves with fine spray of water two or three times a day will be beneficial. During the rains watering should be reduced depending on the weather. Over-watering is harmful. The compost should be filled in a container leaving about 2-3 cm space from the brim to facilitate watering. One way of ensuring whether or not a plant needs watering is to tap the container with a wooden hammer on the sides. If a dull sound comes out the plant should not be watered, but if there is a ringing tone the plant needs watering.

Since there is limited soil in a *Bonsai* container manuring is absolutely needed for proper vegetative growth and to encourage flowering or fruiting. For the layman, it will be difficult to decide the amount of fertilizer that is to be added. The best thing will be to ferment one kilogram of mustard cake or groundnut cake in four to five litres of water for about a week, and then dilute this solution 5-10 times and apply to each plant. Alternatively, two table spoonfuls of powdered cake may be applied on the surface of the soil and then worked in slowly and watered. Generally, manuring is done twice a year, once during the spring and again during the rainy season. For vigorous-growing plants, manuring once a month during the growing season, followed by watering is practised. Too much manuring should be avoided as this may cause soft and sappy growth. One will learn by experience the amount of manure or fertilizer to be added.

Pests and Diseases: A *Bonsai* grower should look out for diseases and pests and arrange for spraying of insecticide and pesticide at regular intervals.

Mame Bonsai

The growing of very dwarf plants not exceeding a height of 5-20 cm is known as miniature or *Mame Bonsai.* To keep the plants in miniature shape they are grown in very shallow containers containing not more than 2-3 cm compost. Plants of *Mame Bonsai* are raised *in situ* from seeds. Because of its small size, *Mame Bonsai* can be shaped only in the upright, oblique, and cascading styles. Group planting is most effective in raising attractive *Mame Bonsai.* The method of training and pruning will be similar as in *Bonsai.* Leaf pinching is particularly very important as this keeps the plant dwarf. More frequent watering is needed since the container is shallow.

There is no knowledge as to which tropical plants are suitable for *Mame Bonsai.* But the following plants of the temperate regions are suitable subjects for *Mame Bonsai* – species of *Acer, Azalea, Berberis, Cotoneaster, Salix, Pinus, Picea,* and *Cryptomeria japonica.*

Chapter XX

Commercial Floriculture

A. FACTORS AFFECTING FLOWER PRODUCTION AND HOW TO OVERCOME THEM

The cut flowers produced for sale must be of superior quality so that these have consumers' appeal. Various factors responsible for the quality production of cut flowers and possible remedial measures are discussed below.

1. *Quality of Seeds and Planting Material:* The biggest snag in India is the non-availability of genuine and good quality planting material. Nurseries producing genuine and quality seeds are very few and even the seeds available from some leading nurseries are not always up to the mark. Sometimes, the percentage of germination is also very poor. Unless the seed is of good quality, the flowers produced from such stock will also be of poor quality and fetch lower price in the market. Nursery stocks such as corms of gladiolus or budded roses are often below standard or not genuine. The authors have found that if a particular cultivar of gladiolus is ordered, the supply received is often a mixture of different cultivars or of a different cultivar. This omission may be deliberate, while on many occasions this may be due to various other factors. Sometimes, the corms of different cultivars get mixed up due to improper storage or mishandling by the labour. One major factor that is responsible for this mixing-up is that different cultivars of gladiolus are grown on the same piece of land alternately. As a result, a portion of the cormels of one cultivar remains unharvested due to neglect and often a new cultivar is planted on the same plot. Thus, the corms of two or more cultivars get mixed up. Some nurseries bud roses 15 cm above the ground level which is definitely not good. Moreover, many of the grafts supplied are not genuine material. If a rose plant is purchased from north India which is budded over Edouard rose, it will not do so well in places like Bangalore or Calcutta, where plants budded on *Rosa multiflora* perform better.

A commercial grower must ensure first that the seeds or plant materials are genuine and of superior quality. These should be purchased from nurseries of repute which he or his neighbours have once tested and found to be reliable. First a small quantity of seeds may be purchased to test the germination percentage and quality. Similarly, for other plant materials such as corms and budded plants only nurseries of good reputation should be contacted. Rose grafts purchased should be on the right type of stock suitable for the locality.

Unfortunately, we do not have any seed certification and quality control organization on floricultural products in India. The government should take some action on this aspect to save the customer from being cheated.

2. *Know-how:* Before venturing on any cut-flower business the grower should make sure that he has sufficient knowledge about the cultivation of the commercial flower he wants to grow. Any deficiency on this score will ruin the whole show. Growing of flowers for cutting purpose is a specialized job which calls for experience and talent on the part of the grower.

3. *Water:* Water is the most vital constituent of a living plant and constitutes up to 90 per cent or more of the total fresh weight. Water is needed for nutrient uptake, photosynthesis, cell formation, cell elongation, transpiration, etc. Enormous amount of water is released to the air by plants through transpiration. Therefore, the amount of soil moisture present in any soil is of vital importance for plant growth. The amount of moisture in the air (humidity) also is a limiting factor in plant growth. For example, tuberoses or orchids do better in a humid climate, whereas most of the rose cultivars prefer dry climate. The depth of water table is also important for plant growth. Plants need a water table at an optimum depth and too close or too deep water tables are bad for plant growth.

The shortage of soil moisture can be supplemented by timely irrigation, mulching, and proper crop planning. For example, gladiolus can be planted during May in Bangalore, but it may be advisable to postpone the planting till June, otherwise water may become a limiting factor. It is difficult to control atmospheric humidity under field conditions. Therefore, it will be better to avoid growing certain crops in particular regions having unfavourable humidity conditions. If the water table is too close to the surface, proper drainage has to be arranged.

4. *Soil:* The type of soil, soil profile, colloidal property of the soil, pore space, and available plant nutrients are all important to a lesser or a greater extent for plant growth. The different types of soil are: clay, which is sticky; sandy, which is gritty; silt, which has a silky nature; and when none of these characteristics are perceptible and the soil can be moulded in the palm, that soil must be a loam. There may be intermediate types of soil also which are mixtures of two or more of the above. For example, clayey-loam or sandy-loam. Different floricultural plants need different types of soil for their optimum growth and development. Only a few bog plants prefer a clay soil, while most of the commercially grown flowers cannot be grown on clay soil because of poor drainage. Most of the crops grown commercially will do best in a loam. A perfect garden soil is one which has the following components in the proportions mentioned: $\frac{1}{4}$ clay, $\frac{1}{6}$ humus, $\frac{1}{2}$ sand, and $\frac{1}{12}$ lime. But there is hardly any soil which is ideal or perfect to this extent, but many of the loamy soils come very near to this characteristic. A sandy soil is not suitable for growing most of the commercial flowers, firstly because it has got a very poor water holding capacity and, secondly, these soils lack sufficient nutrients. It is possible to some extent to improve the texture of the soils. For example, the water holding capacity of a sandy soil can be improved considerably by adding organic-decayed matter or green manure crops. Similarly, the texture of a clayey soil can be corrected by improving the drainage which again can be done by the addition of either organic matter or coarse sand or both. If the nature of a sandy soil cannot be improved up to the desired level, the crops grown in it must be supplied with extra quantities of nutrients and by frequent and heavy watering.

The soil pH is an important factor in determining plant growth. The soils may be neutral or acidic (pH below 7) or alkaline (pH above 7). The soil preference of plants may vary.

Some require slightly acidic soils, others prefer slightly alkaline conditions, while still others favour almost a neutral range. For example, chrysanthemum requires a sandy loam soil with neutral or slightly acidic reaction. In acidic or alkaline soils many of the plant nutrients get bound and are not available to the plants for utilization. For example, in chalky soil the iron in the soil becomes soil-bound. Before undertaking to grow any crop it is absolutely necessary to test the soil reaction and know the pH preference of the particular crop to be grown and correct the soil pH accordingly. If the soil preference of any particular crop is not known, it is safe to start with a near-neutral soil. Soil pH can be corrected by various methods. Alkaline soils can be corrected by the addition of acid-forming chemicals such as aluminium sulphate, sulphur, and gypsum (calcium sulphate). Alkaline and saline soils are generally formed as a result of bad drainage. Improving the drainage by addition of sand and organic matter also helps lower the pH. In saline soils, the salts get deposited on the surface which are generally soluble in nature. To get rid of these salts the land is levelled first, then enough drainage is arranged, if required even by pipe drainage, and finally the land is flooded several times to leach out the salts. Unless the drainage is perfect, these salts will reappear. The acidic soils are formed in high rainfall areas due to the constant leaching of many soluble carbonates. Acidic soils may also result due to the presence of parent acidic rock material. Constant addition of ammonium sulphate can also make a soil acidic. An acidic soil can be made more alkaline by the addition of hydrated lime (calcium hydroxide) or limestone (calcium carbonate). Even quick-lime (calcium oxide) may serve this purpose. The lime requirement for improving the soil pH for certain crops may be considerable. For example, to raise the pH of the upper 15 cm of loamy soil from 5.5 to 6.5 about $1\frac{1}{2}$ tons of quick-lime or 2 tons of hydrated lime or 3 tons of pulvarized limestone per acre will be required. If the soil is sandy in nature or leaching is of habitual occurrence the operation has to be repeated every 3-5 years. For correcting alkaline soils, gypsum at the rate of 2-4 tons per acre may be needed. Cultivation of alkaline and salt-tolerant crops such as rice, sugarcane, barley, etc., also helps to reduce the salts gradually.

The water holding capacity and the nutrient levels of the soil are also important for the growth of the plants. In light soils the water holding capacity can be improved by the addition of organic matter. The amount and type of soil moisture is also important for plant growth. Out of three types of groundwater, namely, hygroscopic water, gravitational water, and capillary water, only the last is available for plant growth. The amount of capillary water depends on the structure, texture, and organic matter content of the soil. Fine textured soils which are granular in nature and contain organic matter can retain the maximum amount of capillary water.

The soil nutrients are depleted by the process of removal by (1) crops, (2) erosion, (3) leaching, and (4) volatilization. The depletion of soil nutrients can be checked by proper soil management which includes (a) fertilization, (b) liming, (c) proper cultivation, (d) crop rotation, (e) inter-cropping with leguminous crops, and (f) mulching.

A sound cultivation includes weed control, aeration, water penetration, and incorporation of organic matter. Depth of cultivation is very important as it facilitates proper root penetration, weed control, etc. Cultivation also checks evaporation of soil moisture.

The importance of crop rotation is well known and need not be discussed in detail. If the same crop is planted year after year in the same plot of land, the diseases and pests of the particular crop get multiplied manifold. If rotation is followed with another crop, which is not an alternate host, the diseases and pests are gradually eliminated from the plot.

Mulching helps, among other things, conserve soil moisture. Erosion of soil can be checked by terrace cultivation wherever land is slopy. The terraces are made along the contour of the land. The purpose of terracing is to check the flow of water, as the force of the running water washes away the top soil. Another method of checking erosion, especially in high rainfall areas, is not to allow any land to remain fallow, but to grow some cover crops which will prevent soil erosion. Strip cropping on slopy or sandy soils is another method of preventing erosion. This consists of growing alternate rows of cultivated crops and non-cultivated crops on the contour. The non-cultivated crop is usually a legume or a grass, which plays the role of holding the soil and at the same time improves the soil nutrient especially if it is a legume. These provide organic matter, materials for mulching, and fodder also.

Proper drainage of the soil is very important as most commercial crops including those of tuberose which require a substantial amount of water, will resent prolonged water-logging. The drainage of a soil can be improved by addition of organic matter or coarse sand and if necessary by providing underground drainage.

Soil air and soil temperatures are also important for plant life. Soil air provides aeration for plant roots. In a water-logged soil the plants die as the roots do not get air for breathing. The soil air content can be improved by drainage. Certain seeds will not germinate if soil temperature goes down or above a certain limit.

5. *Effect of Temperature:* Temperature affects the vital process of plant life like manufacture of carbohydrates or protoplasm which is the physical basis of life, transpiration and/or respiration, nutrient absorption, and translocation. Sometimes the temperature may be so low that the plants are unable to manufacture carbohydrates or make protoplasm. At other times, the temperature may be too high resulting in very high rates of respiration and/or transpiration. At high temperature, stomata may close down to check high rate of transpiration and in the process cause reduction in the manufacture of carbohydrates. Higher rates of respiration use up the available carbohydrates which otherwise would have been used for growth and production.

According to their temperature requirements plants are grouped as cool season crops and warm season crops. Tulip and dahlia need cool temperatures for proper flowering. On the other hand, balsam and gomphrenum do best under warm climatic conditions. Within each group, different crops need some specific temperature ranges for their proper growth and flowering, which is termed as the optimum temperature for the plant. An optimum temperature is that with which a particular plant is able to have maximum photosynthesis to manufacture carbohydrates and have normal respiration rate so that the excess carbohydrates are available for obtaining the highest marketable yields. Temperatures below or above the optimum level thus may cause several abnormalities.

The optimum night temperature is very important for crop plants as new cells and the protoplasm for these cells are made only during night. The manufacture of new cells is essentially a biochemical reaction and as in any other biochemical reactions here also temperature either retards or enhances the process. The night temperatures below or above the optimum range result in low yield. This happens because either the rate of respiration increases to a considerable extent at high temperature or the rate of photosynthesis decreases at a faster rate than the corresponding reduction in respiration rate.

Lower temperature is also responsible for dormant periods in plants in the temperate zones as the plant parts above ground level cease to grow because of unfavourable weather conditions. Some tropical plants also have periods of dormancy because of high temperatures and dry seasons. Some plants have a specific rest period which is controlled by some internal conditions of the plant. The principal plant structure which has a rest period is the vegetative bud, although some seeds also have rest periods. Sometimes, the period of dormancy and rest may overlap.

The best way to avoid the effect of unfavourable temperature is to grow the right type of flowers at the proper place. A plant such as tulip which needs cool weather cannot be cultivated successfully in places as Bangalore or Poona. Another point is to grow the crop at the right season. For example, *Antirrhinum* grows best under cool weather and hence this should be grown only in the winter season in north India. The rest period in plants can be broken by vernalization, which is the method of artificial application of cold treatment to buds and seeds to induce flowering. The application of gibberellin has similar effects as of vernalization in certain plants. Certain chemicals, such as ethylene chlorohydrin, break the rest period in a few hours' time in certain plants, whereas the vernalization requirement may be of a few hundred hours of cold treatment. Similarly, dormancy in certain bulbous plants can also be broken by treating with chemicals such as thiourea, ethylene chlorohydrin, etc. It is possible to provide the plants the right amount of temperature under a greenhouse. This method is adopted for growing many flowering crops commercially in the western countries under glass in controlled environmental conditions.

Plants are also physically injured because of low temperatures. Some plants are winter-hardy while others are not. Moreover, immaturity of tissues also causes winter injury. Water deficiency in woody plants may make them susceptible to winter injury. Protecting the more susceptible plants from frost injury during the winter is one of the methods to avoid this type of injury. Of course, the best thing is not to grow any crop commercially which is likely to be injured or killed by frost.

High temperatures also cause physical injuries to plants. Leaf scorching and scalding are some of the effects of high temperature.

Temperature also influences the occurrence of diseases and insect pests. More favourable the temperature is for the survival of the pests, generally more will be the damage caused by insects, pests, and diseases.

6. *Effect of Light:* Light is another factor which is of vital importance to plant life as it is the source of energy for them. Effect of light on plant life depends upon its quantity or intensity, the quality or kind of light, and the number of day length hours or *photoperiod.*

(A) *Intensity:* Plants vary in their need for light. There are plants like rose and carnation which require light of high intensity. Some plants such as ferns, *Anthurium*, and African violets need light of low intensity, i.e., partial shade. There are some plants such as *Nandina domestica* and *Lemonia spectabilis* which can grow well under a wide range of light intensity. The intensity of light also affects the rate of photosynthesis, respiration, and transpiration. A plant requiring full sun will have its stomata closed in a cloudy weather resulting in less photosynthesis. The transpiration rate also varies within a species as a result of partial shade or full sun. The newly planted young growing plants are thus shaded to check transpiration. In some plants, the flower colour may fade due to intense light. In many variegated foliage plants, the variegation will not manifest if plants are kept in too much of shade. In many bright-coloured foliage plants the colour of the foliage becomes dull in too much of shade. Again direct sunlight may scorch the bright colour of the foliage. A sun-loving plant growing in shade or partial shade will have lanky and sickly growth. It may, however, be noted that most of the important floricultural crops need the morning sun for the production of satisfactory blooms. Therefore, if a flower has to be grown successfully commercially it must receive the right amount of light.

(B) *Quality of Light:* The composition of light is made up of the visible spectrum containing violet, blue, green, yellow, orange, and red, which may be further divided into far red and invisible spectrum, e.g., ultraviolet and infrared. In certain plants "white" light (all the visible spectrum) is much better for growth than if certain portions are eliminated. But certain plants when grown in red or blue lights produce greater dry weight compared to plants grown in white light. It is reported that green light inhibits plant growth. Red light promotes seed germination in certain plants and is suitable for optimum growth of seedlings in many plants.

The role of invisible spectrum (ultraviolet and infrared) of light possibly has no useful role in plant growth and development. The glass of the greenhouse absorbs most of the ultraviolet and infrared light but even then the crops grown inside are of excellent quality. Experiments conducted to see the beneficial effects of these lights did not produce any tangible results.

(C) *Photoperiod:* The duration of light during the course of a day is termed as photoperiod. The duration of photoperiod has a profound effect on the vegetative growth, flowering, and sometimes also on the bulb formation in a large number of plants. Plants can be sub-divided into three groups according to their daily requirement of light and dark period.

(a) *Long-day plants:* These plants require long light periods combined with short dark periods to form flower buds. Examples are *Delphinium*, stock, China aster, gardenia, calendula, etc.

(b) *Short-day plants:* The plants in this group need short light periods combined with long dark periods. Plants such as cosmos, poinsettia, chrysanthemum, and *Kalanchoe* belong to this group.

(c) *Day-neutral plants:* These plants flower any time irrespective of the number of dark or light periods and are known as day-light neutral plants or simply day-neutral plants. Examples are tuberose, carnation, dianthus (pinks) and African violet. These are known to be photo-insensitive plants having no specific light or dark requirements.

If short-day plants such as chrysanthemum are allowed to grow in long-days they will continue to make vegetative growth and fail to form flower buds. Similarly, long-day plants will not flower unless they get the minimum number of hours of daily light. Generally speaking, within any specific 24-hour day, the majority of long-day plants would need 8-10 hours of "continuous" dark period and most of the short-day plants will need 10-14 hours of "continuous" dark period to produce flower buds. Even day-neutral plants which are not dependent on day length for flower formation are benefited by the way of improvement in flower quality if they receive a specific number of day-light hours.

It is not very clear how photoperiod is responsible, in short-day and long-day plants, in controlling flower formation. One opinion is that the length of the dark and light periods determines the period of protein synthesis. The short days combined with long nights possibly hinder the synthesis of protein as a result of which protoplasm formation and consequently new cell development is greatly impaired. The direct consequence of this phenomenon is the accumulation of carbohydrates which are utilized for the formation of flower-bud primordia. But the more tangible and widely accepted theory is that the number of hours of day-length is responsible for the formation and translocation of flower – including hormones which are formed in the leaves.

The amount of day-light hours to control flowering in commercial flowering crops has been wisely exploited by growers especially in chrysanthemum. If left to themselves under natural day-length conditions the chrysanthemums generally form flower buds around October and complete flowering by the first week of January, under northern Indian conditions. But in western countries, the growers have been able to market flowers all the year round by exposing the plants during the winter by additional light and shortening the light period by placing the plants in dark chambers during the summer. This can be achieved by complete screening with thick black cloth or black polythene sheets during a portion of the day-light hours. Additional light delays flowering in chrysanthemum, while dark periods hasten flowering. Thus, by this method it is possible to extend the blooming periods of chrysanthemum from the traditional one-and-half months to several months if not all the year round. Actually, under north Indian conditions it has been possible to produce chrysanthemum blooms every week right from October to April, by successive planting and treating the plants with short days and long days. For creating long days, illumination by the combination of fluorescent tubes (40 watts) and incandescent lamps (60 watts) suspended 120 cm above the plants in alternate rows is effective. The general advice is to provide 20-30 watts of light for every 30 × 30 cm of space. By prolonging the flowering period it is possible to avoid market glut and to produce flowers in off-season which are a novelty and which fetch higher returns.

But while emphasizing the importance of photoperiod in plant life it must be borne in mind that recent investigations have pointed out that in a large number of plants including chrysanthemums, which are affected by photoperiod, temperature is equally important in controlling plant responses. In all probability, a combination of both temperature and photoperiod determines the various plant responses, although there are varietal differences, especially in chrysanthemum.

The role of growth regulators in breaking the day-length effect has been studied. Gibberellic acid sprays induce long-day plants to flower in non-inductive short photoperiods. But it has not been possible to have flowering on short-day plants under long-days with gibberellic acid spray.

In a poor country like India, the growers may not be in a position to adopt the various methods of altering the day-length hours and, hence, it is advisable to grow each plant at the proper season of the year.

Experimental evidence shows that the amount of photoperiod has a direct bearing on the production of bulbs in plants. For example, it has been observed that photoperiod influences the formation of corms and cormels in gladiolus.

7. *Site:* The importance of site in growing commercial flowers cannot be overlooked. Site includes topography, i.e., the contour of the land, its elevation, depth, and the quality of soil. If the contour is too slopy one has to spend an enormous amount of money and effort to make it cultivable which will add to the cost of production. As far as practicable, only levelled lands should be selected for cultivation of commercial flowers. Elevation also is important as this determines the temperature of the place and which in turn influences the type of flower which could be grown. Depth of soil is also important for root penetration, amount of nutrient available, and so on. A shallow depth of soil will be unsuitable for growing a crop such as rose which is a fairly deep-rooted crop and requires about 75 cm

deep soil. A site receiving full sun or at least the morning sun up to noon in the day is essential for growing flowers. Moreover, a site should be nearer to a large city or near a rail-head or an airport for efficient marketing of the produce.

8. *Nutrients:* The plants need certain nutrients for their growth. Among the non-mineral nutrients *carbon*, *hydrogen*, and *oxygen* are needed by plants. Carbon is obtained from the carbon dioxide of the atmosphere. Decaying organic matter also produces a considerable amount of carbon dioxide for the atmosphere and growing plants. Hydrogen is obtained from water. Carbon is a constituent of all organic compounds whether in plants or in animals and constitutes about 50 per cent of the dry weight of the majority of the crop plants. Similarly, hydrogen also is a constituent of almost all the organic compounds made by plants. The oxygen requirement, in the process of photosynthesis, is obtained from carbon dioxide. Water also is the source of free oxygen liberated at the time of photosynthesis. Oxygen is needed for the formation of all carbohydrates, lipides, proteins, and other related substances. The oxygen needs for respiration of plant cells is obtained from the atmosphere.

Among the mineral nutrients *nitrogen*, *phosphorus*, and *potash* are required by plants in large quantities and, hence, these affect plant growth more compared to other mineral nutrients. However, in addition to these, three elements, *calcium*, *sulphur*, and *magnesium* are also needed in fairly large quantities than the other essential elements such as *iron*, *manganese*, *zinc*, *copper*, *boron*, and *molybdenum* which are needed in traces. These essential elements required for plant growth are found naturally in soils. But sometimes these elements get depleted in the soil thus limiting plant growth. The causes of depletion may be due to removal by crops as a result of continuous cropping without replenishing the used element, leaching, and erosion. Another cause may be that though an element is present it is not present in the available form as the nutrient gets bound and one of the causes for which may be the soil pH. Some are soluble in water and are easily leached out of the vicinity of the plant roots as a result of heavy irrigation or rain. Sometimes, erosion of the top soil which is rich in plant nutrients causes the deficeincy of plant nutrients. To overcome all these problems it is advisable to follow crop rotation, add enough of plant nutrients to the soil after each cropping, avoid leaching, and stop soil erosion by the methods suggested earlier in this chapter.

(a) *Nitrogen:* The deficiency of nitrogen is manifested by the yellowing of the middle portion of the leaf blades while the margins remain green in monocot plants and uniform yellowing of the leaf blade in dicot plants. In both the groups, the deficiency symptoms are first observed in older leaves. Since nitrogen-deficient plants have less chlorophyll the manufacture of carbohydrates is impaired as a result of which there is poor growth and production. Excess of nitrogen results in rapid vegetative growth, dark green leaves, soft and succulent growth, and delayed flowering. The deficiency should be corrected by replenishing with organic manures such as well-rotted cowdung, horse dung, etc., and by the addition of fertilizers such as ammonium sulphate, urea, etc.

(b) *Phosphorus:* Phosphorus has an important role in photosynthesis, respiration, cell division, and for sugar-starch transformation in plants. It functions as a part of the enzyme system which participates in many vital chemical reactions. Phosphorus is required for the formation of nuclear proteins. It is needed for early maturity and for seed and fruit production. Its deficiency causes stunted growth, late maturity, shrunken seeds, reddish or purplish areas or reddish and purpling of the older veins in the leaves. The deficiency is corrected by the addition of phosphatic fertilizers. Sometimes, in very acidic or alkaline soils phosphorus becomes unavailable and, hence, the maintenance of correct soil pH which is generally between 5.5 and 6.8 is essential in phosphate management.

(c) *Potassium:* Potassium is not a constituent of any plant compound or tissues. Probably potassium plays a role in the synthesis of amino acids. Potassium is believed to give resistance to plants against diseases. Its deficiency causes weak stem, underdeveloped roots, and discolouration of leaves. The margins of leaves first show yellowing which gradually increases in size and turns brown. The older leaves show the symptoms first.

(d) *Sulphur:* Sulphur is a necessary part of glycine molecule and for vitamin B. This is also needed in the formation of chlorophyll.

(e) *Calcium:* It is an important constituent of the cell wall. This element is responsible for the absorption of nitrate nitrogen. A typical symptom of deficiency is that the upper leaves become yellowish-green while the lower leaves remain green.

(f) *Magnesium:* Magnesium is a part of the centre of molecule of both chlorophyll a and b and is also constituent of the cell wall. Magnesium deficiency causes loss of chlorophyll and the symptom manifests on the leaves where the areas in between the veins become pale in colour while the veins themselves remain green.

(g) *Trace elements:* As a result of deficiency of both manganese and iron, chlorophyll fails to develop. In boron deficiency, cell division is slowed down and the plants fail to produce vegetative and flower buds. Zinc is necessary for cell elongation as it is necessary for the formation of precursor of indolacetic acid, the latter-mentioned plant hormone giving the property of elasticity to the cell wall. Zinc-deficient branches have small, narrow, mottled leaves and such branches generally die back. Copper is an essential constituent of certain oxidizing enzymes such as tyrosinase and ascorbic acid oxidase. This element is also needed for the formation of chlorophyll. It is known that molybdenum is needed for the reduction of nitrates in the leaves and possibly in the growing points. Besides this, the element has some other important functions in plant life.

As has already been stated, only nitrogen, phosphorus, and potash are needed by plants in large quantities, while the others are needed only in small amounts. Generally, the trace elements are present as impurities in commercial fertilizers but in case deficiency symptoms are noticed, minute quantities of these elements may be given as soil application or foliar sprays. Some commercial preparations such as "Navras" containing these elements can be used for this purpose.

Recent investigations have revealed that a few other elements such as chlorine, iodine, silicon, and aluminium are also essential for growth in certain plants.

9. *Mechanical Support:* This is also an important aspect in the commercial production of flowers. Many of the commercial flowering plants are herbaceous in nature and need mechanical support either for the plant itself or for the flower stem or for both. For example, carnation or gladiolus when grown for flower production needs some mechanical support so that the flowering stems do not break or lodge with the velocity of wind or heavy rains. One way of doing this is to provide split bamboo stakes for each flowering stem and tying these to the stake loosely with gunny twines or *sutli*. But this involves a tremendous amount of labour. Alternatively, sturdy bamboo stakes or stout bamboo poles may be planted at regular distances closely along the rows of the plants on both sides and then run pliable wire strings or gunny threads through these poles so that the plants get support from both sides. In the case of taller plants, two rows of threads on each side should be given to make the

support more firm.

10. *Disease and Pests:* Regular measures to control insect pests and diseases should be undertaken as insect or disease infested flowers fetch less price or may be discarded outright by the traders. The measures include crop rotation, clean cultivation, use of disease-free stock, and regular spraying of insecticides and fungicides.

11. *Marketing:* This is an important aspect of floriculture. Before venturing to grow any flowers it is necessary to make a market study in the region ascertaining the local preferences and fascinations. For example, flowers of *Barleria* and *Crossandra* are in great demand for hair-do in the southern States but there is hardly any demand for these flowers in other parts of the country. Hence, cultivation of these plants anywhere other than near Bangalore, Madras or Hyderabad will be unwise and uneconomical. Flowers such as gladiolus fetch better values in cities as Bombay or Delhi rather than Bangalore as in the latter-mentioned city the local population prefers flowers for hair decoration rather than for table decoration.

Flower being a highly perishable commodity, the centre of production should normally be very near the centre of consumption. But in modern times with fast transport facilities, including air transport, it is possible to grow flowers even at reasonably distant places from the centre of production without much hampering the quality of the product during transportation. Land being costly near big cities there is a need for growing the flowers farther away where land may be available at a cheaper price. But while selecting the site fast transport should be ensured.

B. PRODUCTION OF CUT FLOWERS

Cut flowers are needed in India mainly for hair decoration, garlands, *veni*, worship, table decoration, and various other uses during religious ceremonies and social functions. Some flowers, especially roses, are also produced near Delhi mainly around Kanauj for preparing *attar* (scent). But commercial production of flowers for export has not made much headway although there is vast scope to do that. The most important commercial flowers cultivated on a large scale for cut flower trade are rose, jasmine, china aster, chrysanthemum, tuberose, gerbera, marigold, gladiolous, carnation, crossandra, and barleria. The detailed cultivation aspects of these flowers are dealt with in the following pages.

C. FACTORS AFFECTING KEEPING QUALITY OF CUT FLOWERS AND REMEDIES

In the production of cut flowers one of the most important aspects is to deliver the flowers in garden-fresh condition to the market. But this is rather a ticklish problem, and more so with the comparatively delicate flowers.

Two sets of factors are responsible for the keeping quality of cut blooms. The first is the internal mechanism of the plant body. This includes (1) the rate of water absorption and (2) transpiration. The principal plant factors on which these processes depend are (a) the relative area of absorption and (b) the total water-holding capacity of the tissues. After the flower is detached the area of absorption is reduced drastically, whereas the proportionate area for transpiration is much higher. Therefore, unless something is done to reduce transpiration the cut flower will wither in no time. Respiration is the second internal factor which affects the life of the cut flower. Besides, there are some environmental factors that affect cut flower life. These are: (1) temperature; if the temperature is high there will be faster rate of respiration and burning of tissue and as a consequence the life of the flower will be shortened; (2) relative humidity is another factor which has a bearing on the transpiration loss, higher the humidity the less is the transpiration rate and *vice versa*; (3) wind velocity is also very important with regard to cut flower life; strong hot breeze will cause higher rate of transpiration and shorten the life of the cut flowers.

The above discussion will indicate that to prolong the cut flower life one has to control the environmental factors to reduce the transpiration and respiration rates. The rate of respiration depends on the amount of sugar available in the plant tissues, temperature, and use of certain chemicals to regulate it. During the process of respiration the sugars stored in the plant tissues are burnt and consequently the cut flower life will depend much on the potential availability of sugar at the time of cutting and on the lowering of the rate of respiration after cutting. It should be ensured that at the time of cutting the plant tissues contain a good amount of sugar. One way, and possibly the best way, of achieving this is to grow the flower at the optimum temperature range needed by that particular species. Proper feeding and disbudding also help in some way to store extra sugar in the tissues. Wherever possible the low atmospheric temperature should be taken advantage of in prolonging the life of flowers. It might have been noticed that cut flowers or even blooms on the plant itself last longer on cooler days compared to hot seasons. Similarly, cut blooms kept in an air-conditioned room, have more life than those kept at a normal room temperature.

Certain chemicals also help reduce respiration rate and thus prolong the life of cut flowers. These chemicals also control the bacteria and fungi in the vase water which may otherwise cause rot of the stem. The following two compositions are commonly placed in the vase water containing the flowers for the above purposes: (1) a solution composed of potassium aluminium sulphate, ferric oxide, sodium hypochlorite, and sugar, and (2) a solution composed of manganese sulphate, hydrazine sulphate, and sugar. In western countries the florists themselves supply commercial preparations of these in the form of a tablet along with the flowers sold and the buyer has simply to drop the tablet in the vase water. Some recent research findings on prolonging the life of cut flower is reviewed below which will be quite useful to commercial growers.

Cathey and Heggestad (1973) suggest that SADH can reduce ozone damage to petunias. Marousky (1973) and Rogers (1973) have done elaborate studies and reviewed work in the field of causes of wilting of cut flowers and how to prolong vase life. Among many other causes, xylem tissue in the cut flower stem may pull up air bubble under certain conditions and thus block the vessels causing impediment to waterflow and consequent wilting.

There are a number of physiological factors which may cause stem plugging. It is reported that the cut flower life of roses is shortened due to plugging in the stem caused by complex gum-like materials having the characteristics of lignins and tannins. But this view is contested by many. It has also been observed that phenolic compounds cause stem plugging in mimosa. The lignification of the cut stem in chrysanthemum causes wilting of the flower. The harvest cut in chrysanthemum should be made at a place where the stem is reasonably soft to overcome the problem of early wilting. Some investigations definitely point out that this cause of physiological stem blocking is at least partially enzymatically mediated. Experimental evidence is available to show that enzyme inhibitors or conditions unfavourable to enzyme activity such as pH of the vase solution, have helped to prolong the vase life of flowers. For example, low pH range (3.0 to 4.0) of vase solution reasonably improved the keeping quality of vase flowers. Similarly, enzyme inhibitors such as azide and DNP, and chelating agents such as cupferron and 8-HQ (8-hydroxyquinoline) when placed in vase solution improved the keeping quality.

The plugging of stem may also be caused by microbial organisms such as bacteria, yeast, and fungi. There are many chemicals which are used in the vase solution to prolong vase life, mainly because of their bacteriocidal property. The substances evaluated are amphyl, sodium hypochlorite, 8-hydroxyquinoline sulphate (8-HQS), copper sulphate, zinc acetate, silver nitrate, and aluminium nitrate. Among these, 8-HQS was found to be the best, although hydroxyquinoline citrate (8-HQC) is also very effective. The growth of micro organisms was reduced at 10 ppm 8-HQC solution, almost eliminated at 100 ppm, and completely inhibited at 300 ppm and above. It is also suggested to add succinic acid-2, 2-dimethyl-hydrazide (SADH) to the holding solution of 8-HQC and sucrose. Chrysanthemum buds harvested one week prior to the normal harvesting time and kept in a vase solution of 8-HQC and Sucrose opened normally. Gladiolus spikes harvested as small as 150 mm and held in 8-HQC + sucrose increased in size comparable to those remaining intact on the plant. Results of similar nature were obtained with snapdragon and gypsophylla also.

Another recent method in storing is the hypobaric method. This method may also be called as Low Pressure Storage (LPS). The method lays emphasis on the control of native growth regulators ethylene and CO_2. The hypobaric equipment provides for storing the flowers or nursery stock in a controlled chamber where the pressure and temperature are low and the humidity is relatively high. Under such conditions, ethylene and CO_2 levels are maintained at the most beneficial level.

To understand the hypobaric method of storage it is necessary to know the roles of ethylene and CO_2 on the cut flower life. Ethylene causes the fading of many, but not all, species of flowers. Very low concentrations of ethylene in the atmosphere may cause "sleepiness" in carnation (30-125 ppb, i.e., parts per billion), increased branching of roses, decreased keeping quality of cut roses, carnations, snapdragons, stocks, narcissus, and daffodils (50 ppb).

It has been observed that many excised flowers produce ethylene which causes fading of the flower. Investigations have showed that carnation, narcissus, rose, larkspur, *Cattleya, Vanda, Dendrobium, Phalaenopsis, Calceolaria*, snapdragon, and daffodil produce ethylene and are also affected by the gas. The quantity of ethylene produced by flowers and other plant tissues is greatly accentuated by tissue damage, air pressure, chilling, and disease. Any of these factors could cause premature fading during storage and transit.

But the process of fading can be delayed or retarded by a low temperature and air pressure, permanganated or brominated carbon, low concentration of oxygen or high concentration of carbon dioxide (CO_2). At low temperature respiration is reduced and so are the other metabolic activities including production of ethylene. Low temperature also makes ineffective any ethylene already present. The cut blooms produce ethylene which accumulate round the flowers thus causing fading. Brominated carbon or permanganate reacts with the evolved ethylene and thus prevents fading. Higher concentration of CO_2 not only reduces respiration rate but also appears to reverse completely almost all known effects of ethylene. On the other hand, low oxygen concentration limits the ability of ethylene to function.

The hypobaric storage has shown that most cut flowers store best at 40 to 60 mm Hg.

Carnation buds can be stored in hypobaric condition for at least 9 weeks with no loss in ability to open or subsequent shelf life. Gladioli unopened flowers stored at 60 mm Hg at 1.7°C for 4 weeks in hypobaric condition opened normally and completely, whereas buds stored in ordinary cold storage for only one week did not open properly. Even fully opened flowers of carnation, China aster, and chrysanthemum can be stored at 60 mm Hg for longer time without much adverse effect.

D. SOME HOME REMEDIES FOR PROLONGING VASE LIFE

Some home remedies may be tried to increase the vase life of flowers which may bring marginal improvement. The vase containing the flowers should not be placed near a fire or under an electric fan circulating hot air. To improve relative humidity the blooms may be misted with clean and cool water. A barber's sprayer is most suitable for this purpose. The cut end of the flower stem should be immersed immediately after harvest in clean water. Some also suggest charring of the cut end before putting in the vase. Everyday a portion of the cut stem end should be cut under water and replaced without exposing this again to air. This prevents the atmospheric air from entering the water absorbing (xylem) vessels in the stem and thus keeping them free for water absorption. A little amount of sugar added to the vase water may improve the keeping quality. A very dilute solution of copper sulphate solution added to the water may check rot due to fungi.

E. PACKAGING OF CUT FLOWERS

Packaging is another important aspect in the flower trade and much depends on the proper method of packing to ensure the garden-fresh quality of flowers for the consumer. If packing is not proper, the flowers may wither or suffer mechanical injury due to bad handling during transit. But, unfortunately, in our country the ordinary commercial growers pay scant respect to packaging. The major factor for this neglect is possibly the poor economic condition of the grower who cannot afford the luxury of modern packaging technique. Another reason may be that most of the flowers produced in India such as tuberose, China aster, marigold, etc., can withstand certain amount of rough handling in transport. But it is a common sight to see that flowers arrive on the markets in a badly mutilated condition as a result of which a substantial portion of the produce is lost. The common method of packing is to send by bamboo baskets lined on the sides and top with hessian cloth. Jasmine buds are even sent in gunny bags. But it is high time that our growers and flower traders paid some attention towards proper packaging.

The main principles of packaging aimed towards long storage life and keeping quality are to lower the rate of respiration, transpiration, and cell division during transportation. An ideal package should have the following qualities to fulfil the above requirements: (i) small volume, (ii) airtight, (iii) waterproof or non-water absorbent, and (iv) strong enough to withstand handling. If the package is oversized for the contents, there will be lot of empty space and flowers will tend to transpire more. Moreover, too much space will give the contents enough room for movement during transportation which may damage the flowers. If the packge material is of non-absorbent type, with the help of a little misting of water it will be possible to maintain high humidity which will prevent wilting. When cellophane or polythene wrap is used, besides maintaining relative humidity because of its gas-proof nature, it helps maintain a high level of carbon dioxide concentration, say around 5 to 15 per cent, which in turn keeps the respiration rate low. Even though the package may have been designed to maintain high relative humidity, since the stems are unable to draw water directly during transit, a relative water deficiency will result in the tissues which will slow down cell division, as water is one of the most vital factors in the process of cell division, thus preventing ageing.

The most suitable package material for cut blooms is hard cardboard boxes which are not only cheap costwise, but also light in weight. Because of the light weight the cost of transporation will be less and it will be easy to handle. Before packing the flowers a cellophane or polythene wrap should be provided all round for the purpose mentioned in the foregoing paragraph. Some of the Southeast Asian countries such as Thailand tuck a wet cotton at the cut end of the flower stem (especially for orchid blooms) which is wrapped inside a polythene wrapper and tie the loose upper end of the wrapper with a rubber band or thread to maintain humidity.

F. PRODUCTION OF POTTED PLANTS

With the oil boom in West Asia a vast market has been opened up for India and other countries to sell potted plants especially foliage plants and some shrubs and trees. Already many nurseries in the country are exporting foliage plants such as *Monstera, Philoderndron, Dracaena, Pleomele reflexa variegata*, etc., to these countries.

Besides the foreign market there is a vast scope for selling these plants in the internal market also. Already a number of nurseries spread all over the country are doing brisk business in selling potted plants such as dahlia and chrysanthemum and different flowering annuals and sell these with a premium when the blooms start opening. For detailed cultivation practices of potted plants, please consult Chapter XVIII. The details of packing live plant material has been discussed in pages 34-35. The production of cacti, especially the grafted ones, and succulents is also a very profitable business.

G. PRODUCTION OF BULBS AND SEEDS

Production of underground propagation materials such as bulbs, corms, and tubers offers a good avenue of income. There is a great potential for developing an export market for bulbs such as amaryllis, tuberose, canna, football lily, etc., if a multiplication programme is embarked upon in a big but systematic way. There is a shortage of genuine planting materials of these kinds in the country and there is a scope for exploiting these aspects commercially.

There is a genuine shortage of reliable plant seeds in the country. Except few nurseries, seeds marketed by all others are not dependable. Sometimes, even the seeds from the large seedsmen do not germinate properly or are of poor quality. If commercial floriculture is to be given a good footing in the country there is an urgent need for starting a chain of reliable seed production centres. Production of seeds of ornamental plants is a lucrative trade as the return per capita is much more than any other branch of agriculture. Moreover, the F_1 hybrid seed multiplication has a good export potential. The production of F_1 seed is a costly affair as a lot of manual labour is involved. F_1 seed production becomes costly in advanced countries because of high cost of manual labour. India with its cheap labour force can compete easily and capture a portion of the foreign market. But to make this business commercially successful a strict quality control agency has to be established to certify the seeds. The potential of hybrid seeds as foreign exchange earners had been amply demonstrated by the enterprising young scientist proprietors of Indo-American Hybrid Seeds at Bangalore.

CULTIVATION OF SOME IMPORTANT CUT FLOWER PRODUCING ORNAMENTALS

There are a number of plants which produce cut flowers for marketing. The consumer preference of a cut flower varies from country to country or even from one place to another within the same country. For example, orchid cut flower is very widely sold in the U.S.A., but in the Indian flower market we hardly see any orchid bloom. Similarly, the cut flowers of barleria or crossandra are so popularly traded in some of the southern States of India whereas these flowers are never marketed in eastern or northern India.

In the following pages we shall deal with the cultivation of China Aster, Carnation, Chrysanthenum, Dahlia, Gerbera, Gladiolus, Jasmine, Marigold, Orchids, Rose, and Tuberose.

Besides these plants, other ornamentals such as Anthurium, Barleria, Cyclamen, Crossandra, many flowering annuals such as Sweet Pea and others, Lilies, etc., also produce flowers suitable for cutting. For their cultivation, the respective chapters may be consulted.

China Aster

Callistephus chinensis; Family: *Compositae* (Now *Asteraceae*); Origin: China and Japan; Other Common Name: Aster.

The China aster was introduced from China in the early eighteenth century. The original plant was a single flowering and branching type having a height of about 60 cm. Since its introduction to Europe the plant has undergone a remarkable change in form and size and also colour of flowers. Now the plants range in height from 15 cm to about 1 m with pompon flowers about the size of a button to large flower heads having single, double, anemone-flowered, peony-flowered, incurved, quilled or shaggy flower types. The colour range is so great that today China aster is one of the most valuable garden flowers. In all the different types the colours include pure white, many shades of pink, primrose, pale blue, mauve, purple, dark blue, and scarlet. The pure yellow colour is not found in aster.

Use: China aster is excellent as a cut flower. It is widely cultivated in many parts of the country and especially around Bangalore for this purpose all the year around. The cut flowers last long in water. In Bangalore, the plant as a whole is cut and sold as cut flower. It is grown for bedding and potting. The dwarf Pompon and Lilliput types can be grown in window boxes and in mixed borders.

Classification: There are several types of asters which are mainly classified according to the form of flowers. The different types are as follows:

(1) *Chrysanthemum flowered* (both tall and dwarf): The flowers look like Chrysanthemum.

(2) *Mammoth Peony flowered:* The height of plants vary between 60 and 75 cm. The huge

double flowers are 10-12 cm across and gracefully incurved like peonies. Flowering stems are between 35 and 50 cm and without much branching.

(3) *Giant Californian Asters:* The plants are about 80 cm tall, beautiful, large, and curled. The interlaced flowers are borne on long erect stems of about 50 cm. Some cultivars such as 'Los Angeles' and 'Elmonte' bear flowers about 15 cm in diameter.

(4) *Comet:* They bear double feathery flowers, long narrow petals incurving. Both dwarf and tall varieties are available. The Crego and Ostrich Feather belong to this group.

(5) *Victoria:* These are dwarf to intermediate types. The flowers cover the entire plant.

(6) *Pompon:* The plants are about 35 cm tall and bear beautiful round pompon flowers.

(7) *Branching Asters:* They are pyramidal in form with many lateral branches. 'America Beauty', 'Queen of the Market', and 'Invincible' belong to this group.

Another aster with needle-shaped petals and growing up to 25 cm is known as Aster Unicum.

Propagation: Propagation is usually done by seeds. Seedlings are raised in seed pans or sometimes sown directly in the beds. Seeds are sown during August to October and even earlier in June or July in areas where rainfall is low. In the northern plains, earlier sowing is useful for obtaining better blooms. If sowing is delayed, flowers are affected by high temperature prevailing in the summer. In mild climate, as that of Bangalore, China Aster is grown almost throughout the year, but June sowing is definitely better. In the hills, sowing should be done in April. Water should be given very carefully and it is advisable not to water unless the seed pans become fairly dry. Seeds should be thinly covered with soil. When seedlings are in 4-leafed stage they should be transplanted to 7.5 cm pots, before final re-potting or transplanting in beds. If seeds are sown directly in beds they should be thinned to the desired spacing. For dwarf type spacing of 20 cm and for tall type 30 cm spacing is recommended both between rows and between plants.

Culture: All the types mentioned above are quite hardy and easy to grow. China Aster needs rich, porous, and well-drained soil. It grows best in a sunny situation. They bloom in about 3½ to 4 months from the date of sowing. The late flowering cultivars such as Giants of California take 4-5 months. At the time of bed preparation, sufficient organic manure, about 4-5 kg per square metre, in the form of cowdung or horse-dung is added to the soil. Application of liquid manure once in a fortnight, prepared by fermenting fresh cowdung along with a handful of ammonium sulphate, after the flower buds appear is beneficial. The surface of the soil should be hoed occasionally to keep down weed growth. Hoeing also helps conserve soil moisture. The taller cultivars may need staking but this is not necessary for the dwarf types. Watering should be done early in the morning or in the afternoon.

To obtain a continuous flowering over a longer period, the sowing should be done at 7-10 days' intervals, especially for cut flower trade. The faded flowers should be removed regularly to encourage more flowering.

Pests and Diseases: See Chapter XXVI.

Carnation

Dianthus caryophyllus; Family: *Caryophyllaceae*.

Carnation, a native of Southern France, is a flowering annual but in reality it is treated as biennial and succeeds well as such. The word *Dianthus* is a derivative from a Greek word meaning the Divine Flower. According to Theophratus it was in cultivation in Greece as early as in 300 B.C. There are several distinct types of carnations and each of these can be grown successfully in cooler parts of the plains and at medium-to-high elevations. In the plains of other places only Marguerites do better.

Classification: The carnations are classified as follows:

(a) *Border Carnations and Picotees:* These types are distinguished by their well-frilled flowers having smooth-edged broad petals. They are dwarf in habit and branch mainly at the base. Based on the colour pattern of the petals, they are sub-divided into, (i) Self, having a single even colour in the petals without spots or stripes; (ii) Bizarres and Flakes, having petals with white ground colour and stripes and blends of different colours. In Bizarres, every petal has stripes of two or more different colours but in Flakes the stripes are of one and the same colour distributed evenly; (iii) Picotees have narrow or broad rounded and smooth petals of either yellow, white, or buff background, bordered with a band of colour at the margin; (iv) Fancies are those which cannot be placed in any of the above groups, but are distinguished by large well-shaped brilliant-coloured petals. Fancies generally have irregular markings or blended colours.

(b) *Perpetual Flowering:* These comprise the American Tree Carnations and the perpetual Malmaisons. Since they bear flowers throughout the year, the name perpetual is given and they are hybrids involving several *Dianthus* species. The first perpetual flowering carnation was sold by Peter Fisher, a Scotsman, in the middle of the nineteenth century to the American Copper King, T.W. Lawson, and named after him for 30,000 dollars. This created a sensation and these varieties became popular instantly. The flowers are generally self-coloured and the stems are longer and branching. The petals have serrated or fringed edges. In some cultivars the petals are only frilled. The perpetual flowering types withstand rigours of transport well and hence preferred by florists. In temperate regions, having severe winter, the perpetuals should be grown in glasshouses, but in the plains of India these could be grown in the open during the winter. Malmaisons have generally pink-coloured flowers which are sometimes as large as rose flowers.

(c) *Maguerite Carnations:* This is a group of Carnations having clove-scented flowers of all shades of colour with fringed petals. They produce flowers freely for a prolonged period but the flowers do not stand for a long time when cut. They are also perennial but are treated as annuals and can be easily raised from seeds. The Chabaud carnations are also quite famous and can be raised from seeds.

Cultivars: In India, it is difficult to get good seeds of Carnations. Some seedsmen are marketing Maguerite and Chabaud Carnations. The Chabaud cultivars are 'Aurora' (salmon-rose), 'Jeanne Dionis' (white), 'Marie' (yellow), 'Nero' (dark red), etc. Some cultivars of border carnations are Madonna (white), 'King Cup' (yellow), 'Crimson Model' (crimson), etc. The important perpetual flowering cultivars are 'Allwood's Crimson', 'Ancient Rose', and 'Canadian Pink'. Modern American Sim Carnations are also available in several colours.

Propagation: (a) *Sexual:* Carnations, especially the Maguerite and Chabaud types, are easily raised from seeds. In the plains of India, seeds should not be sown earlier than July or August and their sowing may extend up to October. Seedlings raised from seeds sown earlier than this are weak. In Bangalore, it is possible to raise seedlings during May and June or even earlier, but the best sowing season is between October to November. In the hills, sowing should be done during August to October and

March to April depending upon temperature. The seeds are sown thinly in seed pans or seed beds. Watering should be done judiciously to avoid damping-off. The seeds germinate in about 5-8 days. At four-leaf stage, the seedlings are pricked or transplanted singly in 7.5 cm nursery pots using a compost of equal parts of leaf-mould and good garden soil till final potting or transplanting. The seedlings are hardened by gradually exposing them to more and more sunlight hours. When these grow to a height of 15 cm, they are transferred to 25 cm pots or in beds.

(b) *Cuttings and layers:* Terminal cuttings measuring about 10 cm with four leaves are taken from the side shoots which have not yet developed any flower buds. The lower cut is given below a node and the leaves at the base are removed before inserting the cuttings in the rooting media, consisting of sand or vermiculite. The leaves should not come in touch with the sand. The cuttings may be raised in a mist-chamber, a small plastic house, or in the open if humidity is sufficient. In about a month's time, the cuttings develop sufficient roots, when these are transferred singly to 7.5 cm pots containing a potting mixture of equal parts of soil and leaf-mould and half part sand.

The stout lateral shoots not having any flower bud are selected for layering. Layering is done in pots or directly in the ground. The lower leaves on the selected branches are removed. A sharp cut is given with a knife about half the thickness of the branch, between two nodes starting just above the lower node and ending a little below the upper node to form a "tongue". The "tongue" is kept open with the help of a wedge and inserted in the soil and held in position by a small hair-pin or layering pin made out of split bamboo. The terminal end of the shoot is kept above ground. The layers generally root earlier than the stem cuttings. After rooting, these are separated from the mother plants. The young cuttings and layers are treated in the same way as young seedlings.

Cultural hints: Carnations do well in a well-drained soil, rich in lime and sunny situation. Carnations are great favourites for pot cultivation also. Deep planting, water-logging, and mulching with manure are harmful as these may cause stem or collar rot.

The manurial mixture for beds should consist of one part each of bonemeal, sulphate of potash, and calcium nitrate or half of it as ammonium sulphate. Sixty grams of this mixture is added per square metre area. Alternatively, it can be used in liquid form by dissolving 30 g of mixture in 15 litres of water and applied at the rate of 5 litres per square metre at fortnightly intervals. The potting mixture should consist of one part of cowdung manure or horse manure, one part red earth (for south) or good garden soil, one part leaf-mould, and half coarse sand. Sprinkling of small amounts of common salt in the soil is good for carnation. To prevent water-logging small pieces of charcoal may be added in the potting mixture. The plants should be potted in 25 or 30 cm pots. The cowdung manure is applied deep into the soil. Too much of cowdung or fertilizers will delay flowering and may also cause disease or splitting of the calyces. When the plants are 25 cm long they are staked with split bamboo sticks painted green, and inserted in the soil near the plant. A wire ring may be placed around the plant near the top supported by stakes. The spacing in ground should be 40 cm either way. But still closer spacing is also followed.

Stopping and Disbudding: When the plants are about 20-25 cm long, pinch (or stop) about 7.5 cm of the growing point just by bending near a node or with the help of a shaving blade, as this encourages development of side shoots. Half of the lateral growths arising from the leaf axils are also stopped when they are 15 cm long. About 3-6 shoots are retained for flowering. De-shooting is also practised to remove unwanted weak shoots. In border carnations stopping is not practised as these produce one stem only.

Disbudding is practised to obtain long stalks and larger blooms. The axillary or secondary flowers borne on the main stem or terminal shoots are disbudded as soon as they are large enough (pea size) to handle. Only one apical bud is kept in each stem.

Calyx splitting: The calyces, in some cultivars, split when the flowers start to open. This is a bad trait in the carnation and the cause is attributed to various factors such as overfeeding, varietal character (hereditary), high temperature, a sharp drop in night temperature, a low nitrogen level, or boron deficiency. Sometimes, to prevent the splitting of the calyx, the adjoining bud is retained till the top bud shows colour. Another method followed for preventing the calyces from splitting is to tie a string or put a rubber band round the calyces at one-third the distance from the base of the calyx. This is possible only on a limited scale for plants meant for exhibition.

Carnations can be grown in the hills as perennials. After flowering, the plants are beheaded just above the ground level and shifted into pots which are kept in protected places during the winter. Although the plants can last for 3-4 years, it is advisable to discard these after two years as after that the vigour declines to a great extent. The carnation comes to flowering in about 4-6 months after sowing.

Chrysanthemum

Family: *Compositae* (*Asteraceae*); Origin: Europe, Asia, America, and Africa; Vernacular names: Hindi - *Guldaudi*, *guldawri*; Bengali - *Chandramallika*; Punjabi - *Bagauri*; Gujarati - *Guldaudi*; Marathi - *Gulsevati*; Tamil - *Gavanthi*; Kannada - *Savantige*; Telugu - *Chamanti*.

Chrysanthemum is also commonly known as 'Glory of the East' or as 'Mum' in short in the U.S.A. It is one of the most widely cultivated garden flowers and ranks probably next to the rose in popularity. There is hardly any other garden flower which has such diverse and beautiful range of colour shades, widely different flower shapes, and height range as chrysanthemum.

Broadly speaking, the florist's Chrysanthemum is a short-day plant, which under normal circumstances fails to form the flower buds if the day length exceeds 14½ hours, nor develops the buds if the day length is beyond 13½ hours, with the exception of certain early flowering cultivars. The plant needs longer days for proper vegetative growth.

History and Development: Chrysanthemums were in cultivation in China at least 3,000 years ago. The plant was brought into England from Japan in the seventeenth century and to North America about 150 years later. The modern florist's Chrysanthemum (*Chrysanthemum morifolium* and *C. indicum*) evolved as a result of endless variation and hybridization from *Chrysanthemum morifolium* (Syn. *C. sinense*), *C. indicum*, and *C. ornatum* (this, according to Bailey, 1963, is perhaps a form of *C. morifolium*). The florist's or glasshouse Chrysanthemums have come to form a group by themselves, and Bailey (1963) suggests that these may be distinguished as *C. hortorum*.

Classification: The chrysanthemum can be classified into several groups according to the size and shape of the flowers. The major groups are described below.

(1) *Incurved:* The ray florets curve upwards and inwards towards the centre of the bloom to form a globular shape.

(2) *Incurving:* In this the ray florets incurve loosely and irregularly and do not form a

ball as in above.

(3) *Reflexed:* The florets are drooping in this type.

(4) *Anemones:* The disc in this class consists of disc florets and is hemispherical in form. The ray florets are either lingulate or quilled.

(5) *Pompon:* Flowers are very small sized. The disc is concealed or absent.

(6) *Singles:* The petals are arranged in one or not more than five rows with a prominent central disc.

(7) *Korean:* These are small single or double flowers with a conspicuous central disc.

(8) *Spider:* The ray florets are elongated and tubular, curved, and have hooks at the tip of the petals.

(9) *Spoon:* Ray florets are tubular with a spatula or spoon-like opening at the tips.

(10) *Rayonnaetes:* The petals are quilled in this type.

(11) *Exhibition:* The outer florets are reflexed and the inner incurved. The ray florets are generally twisted and irregularly overlap each other. A very attractive group.

(12) *Decorative:* Fully double flowers with flat petals and the central disc is generally absent.

Propagation: The most common methods of propagation in chrysanthemum are by suckers and cuttings.

(a) *Suckers:* After the flowering is over the plants are beheaded to a height of about 15-20 cm from the ground level. During January-February, new suckers start appearing round the base of the stool and these are separated and planted in 10 cm pots. The suckers have their own root system. The pots should be kept in light shade to allow more rooting.

(b) *Cuttings:* After flowering is over the plants are cut to ground level as stated above. New shoots arise from the base as well as from axils of leaves. Young tip cuttings measuring 5-8 cm are taken just above a node. The lower leaves are removed and the basal portion of the cuttings are treated with some root promoting hormones like seradix before planting. The rooting media consists of 2 parts of loam, 1 part coarse sand, and 1 part of leaf-mould. The compost should be sterilized by steam or by 2 per cent solution of formalin.

Recent investigations have indicated that the practice of removing cuttings by knife from a node and subsequent removal of basal leaves are laborious processes and this possibly causes spread of diseases. The best method is to snap the cuttings from the stock plants and directly inserting them in the rooting media without removing the lower leaves. If the stock plants are treated with long days and fed with a weak nitrogenous liquid fertilizer, the shoots from such plants snap cleanly. It has been observed that if optimum conditions for rooting are provided there is no difference in rooting performance between a nodal cutting and an internodal cutting. The time for raising cuttings is June-July in northern and eastern India. In Calcutta, the time is extended up to August. In Bangalore, Mysore, Madras and other places in south India Chrysanthemum is prepared for flowering during July to August and January to February and, hence, the cuttings are prepared six months ahead of these dates. Cuttings raised during July in north and east India reduce the growing time by 5-6 months compared to plants raised from suckers during February. Moreover, plants raised from cuttings are free from unwanted branching. After inserting the cuttings these are kept in partial shade for two weeks and watered 2-3 times a day. If the weather is very dry the cuttings may be covered with an inserted pot or polythene sheet to prevent desiccation. The best method is to raise the cuttings under intermittent mist. It is also possible to raise Chrysanthemums by air or ground layering on a limited scale. The single and Korean type can be raised from seeds also. For raising new hybrids propagation by seed is essential.

Compost: The suckers and cuttings are transferred to 10 cm pots immediately after rooting in a compost consisting of one part each of coarse sand, garden soil, and leaf-mould and traces of wood ash. The second potting is done two months after the first in the case of suckers raised during February, in a bigger pot of 15 cm. The potting mixture consists of a richer mixture containing one part sand, one part garden soil, two parts of leaf-mould, a quarter part of wood ashes, and two table spoonfuls of bonemeal. The final repotting is done in August in 25-30 cm pots. The potting mixture should be very porous and rich. The following mixture is recommended:

Garden soil	1 part
Leaf-mould	2 parts
F.Y.M.	1 part
Broken pieces of charcoal and bricks (rubbish) and wood ashes	½ part
Bonemeal	2 table spoonfuls/pot

In the red soils of Bangalore horse manure may be used in place of F.Y.M. About one to two table spoonfuls of superphosphate should also be added to this compost. In south India, first repotting is done immediately after rooting and the second and final two months after the preceding repotting.

Feeding: The cuttings make vegetative growth till September in north and east India. Feeding of suckers with liquid manure should start from July and for cuttings from August. In other places, the feeding may be started 15 days after the final repotting. The liquid manure is prepared by fermenting fresh cowdung and oil cake in a drum for 5-6 days (about 1-2 kg of each in 10 litres of water). It is diluted to tea colour and applied at the intervals of 7-10 days at the rate of 500 ml to 1 litre per pot. Feeding of weak fertilizer mixtures is also useful and various formulae have been recommended. It is suggested that a fertilizer solution containing 10 g of potassium nitrate, 30 g of ammonium sulphate and 30 g of superphosphate dissolved in 10 litres of water should be applied twice during September at fortnightly intervals. About 500 ml of this solution is applied to each plant. In places other than north and east India this solution can be applied one month after final potting. The application of liquid manure prepared from cowdung should be continued till the buds are half open. Overfeeding is harmful and this can be judged by snapping a leaf in two or by folding at the centre. If the leaf breaks into two clean halves and if the colour of the leaf is also dark green, feeding should be stopped immediately. If the pot is too much vacant from the brim a top dressing covering 2.5 cm of the vacant space with the following compost is suggested during September:

1. Mustard/*Neem* cake	¼ part
2. Cowdung/Horse dung	½ part
3. Wood ash	1 part
4. Garden loam	1 part

Watering: Over-watering of chrysanthemums is very harmful. If the compost is porous enough, watering on alternate days during the hot summer of north India will not pose any problem. During rainy season watering should be regulated. During this period or at any other time, if water accumulates in the pot, it should be drained off by tilting the pots. It will be better to repot such plants along with the original compost intact in new pots with sufficient drainage. Over-watering causes yellowing of leaves and mortality of plants due to rotting of roots. During the hot summer and after the rainy season syringing of the foliage with water twice or thrice a week to maintain turgidity and to clean the foliage of dirt is recommended.

Staking, Stopping, and Disbudding: The methods of stopping (pinching) depends upon the nature of bloom one intends to obtain. If only one bloom per plant is envisaged, no stopping is needed. But if three or six stems are needed for the plant, stopping is resorted to. The tip of the main stem measuring 3-5 cm is removed (stopped or pinched) when the plants are 8-10 cm tall. This stopping will encourage the lateral shoots (or breaks) to develop from the leaf axils. Three strong laterals are retained and others removed. The laterals retained for flowering should preferably consist of one central stem and two on either side of it. The laterals are staked with strong split bamboo stakes inserted in the compost with a view to giving support and also to see that these are spread out from each other. The first crown bud develops at the end of each lateral which contains maximum number of ray florets and will give the largest bloom, though may not be the best bloom. This is retained and all other growth arising from the leaf axils is deshooted (removed). Sometimes, the crown bud in the laterals is stopped to obtain the second crown bud which arises from the leaf axils. In many cultivars, the second crown buds produce flowers of more intense colour, harder in texture, and more symmetrical in growth. However, in most cultivars the first crown bud produces the largest bloom.

Disbudding starts in October or as soon as the flower buds appear. In chrysanthemum, if all the buds in one stem are allowed to bloom, the flowers become smaller in size. Therefore, in large flowering cultivars only one bud per stem is allowed to bloom and others are removed (disbudded). The ideal time for disbudding is when the buds surrounding the central one have developed 0.5 cm long pedicels. Too early disbudding may cause injury to the central bud which is retained to bloom. Deshooting is also practised from time to time by removing all side shoots before they attain the size of 2.5 cm. The aim is to divert all the energy of the plant towards the bud which has been retained for flowering. However, in Singles, Koreans, and Sprays no disbudding is done. In these types, even deshooting is restricted to prevent the plant from being too much crowded. When the buds start showing colour the bamboo stake is cut just below the basal level of the bud so that it does not obstruct the bud in developing into a perfectly shaped flower.

For obtaining a single bloom per plant no pinching is done, but disbudding and deshooting are done as mentioned above. All suckers should also be removed from the base of the plant, otherwise they will take away the nutrients which should have gone to the flower bud.

Sen Rin Tsukuri Method of Growing: The literary meaning of the Japanese words *Sen Rin Tsukuri* means "growing thousand blooms". In this style the plant is trained by repeated pinching in such a way that one single plant will bloom a large number cf medium sized flowers. The newly planted suckers, for growing such plants, are kept in full sunshine till February. In March, these are planted in the ground on raised beds 90 cm apart. The main stem is pinched when it attains a growth of 10 cm. The laterals arising from the leaf axils are also pinched to encourage side-growths. Pinching is done several times to allow numerous side branches to grow. After June no pinching is done, as side branches developing after this date are too small and pose a problem in arranging the blooms on such short branches. All the laterals are staked to spread them in an artistic manner so that the plant takes a circular shape at the top. The plants remain in the ground till September or October when these are transplanted in 30 cm pots. During the end of October or nearer that period the plants are removed from the ground with utmost care with a ball of earth round the roots. Three persons are needed for transplanting - two for holding the ball of earth and one for supporting the portion of the plant above ground. The plant is then placed in the pot. During July to October, when the plant is in the ground, it should receive liquid manure at 10 days' intervals. Disbudding is done allowing only one bud per spray. Each lateral is supported by stakes and the plant is given an artistic shape, usually a circular shape, at the top. Two to three days prior to transplanting watering should be withheld and immediately after transplanting the plant is irrigated thoroughly.

Field cultivation for cut flowers: The small-flowered double Korean type of chrysanthemums are also cultivated in the ground for making garlands and *veni* in parts of Tamil Nadu, Karnataka, Rajasthan, Madhya Pradesh, and Bihar. Yellow and white coloured cultivars are preferred. The propagation is done from suckers. In south India, the flowering season is from July to January, whereas in other parts it is only during November to December or at the most up to the first week of January.

The land is ploughed two or three times and 15-20 tons of F.Y.M. is added per acre. Fertilizers at the rate of 25 kg of nitrogen, and 40 kg each of P_2O_5 and K_2O per acre may also be added. Some growers also apply a top dressing of 25 kg of urea per acre at the time when flower bud appears. No pinching or disbudding is followed. Irrigation at intervals of 8-15 days depending on weather and weeding at regular intervals are practised. The yield of flowers per acre is about 3,000 to 4,000 kg.

Some large flowering cultivars are also grown on a single stem or three stems to produce cut flowers for table decoration on a limited scale around cities such as Calcutta and Bombay.

Controlling Growth and Flowering in Chrysanthemum: It is possible to regulate the growth and flowering of chrysanthemum by manipulating the day length and by chemicals. Recent investigations have shown that in many plants, including chrysanthemum, which are affected by day length, temperature is as important as day length in determining plant response. In chrysanthemum, probably both temperature and day length play an equal part in determining plant response but in certain cultivars the importance of one over the other factor may be more. The rate of photosynthesis, respiration, absorption of water and minerals, and transpiration are affected by temperature.

(i) *Controlling growth by day length:* If short-day conditions are created by keeping the whole plant in a dark chamber or covering it effectively by black polythene sheets for 14 hours a day (from 5 p.m. to 7 a.m.), flowering time can be hastened by up to 30 days in many cultivars.

Similarly, blooming can be delayed till April in north India by treating the plants with long days. It is done by illuminating the plants with a combination of fluorescent tubes (40 W) and incadescent lamps (60 W) by hanging them alternatively 120 cm above the plants, so that the plants do not get more than 9 hours of continuous dark period. Illumination during the middle of the night has been found to be more effective. There is another benefit of long-day treatment, besides delaying blooming period. When plants are given additional lighting in the early stages of vegetative growth, they grow more vigorously resulting in increased size of plants and higher yield of flowers. Such long-day treated plants need frequent pinching for branching. By giving treatments of long and short days according to a schedule, which is determined by practical experiments for different cultivars, it is possible to obtain flowering every week starting from October and extending till April, provided the temperature is optimum. It has been observed that temperatures between 10 and 30°C are tolerable, whereas temperatures lower than 10°C inhibit flower bud initiation and above 30°C affect flower bud development.

To obtain planting material for successive planting it is necessary to maintain mother stocks under artificial long days which promote active growth.

(ii) *Growth regulation by chemicals:* The chemical B-Nine helps to reduce the height of the plants. The concentrations are 0.25 per cent, if the spraying is done twice at 3-week intervals just after pinching, and 0.5 per cent if only one spraying is resorted to. To ensure maximum absorption of the spray the plant should be dry and turgid. Besides imparting compactness to the plant, B-Nine also improves the colour of the foliage and the proportion of the different plant parts. Recent researches have opened up the possibilities of obtaining earlier flowering and higher yield by application of gibberellin sprays.

(iii) *Autumn blooming cultivars:* The operations mentioned above to regulate blooming involve some expenditure and precision in timing of different operations, which is beyond the capacity of amateur growers. The best method would be to breed early blooming cultivars. In the National Botanical Research Institute at Lucknow a beginning has been made towards this end. Some very early blooming cultivars were crossed with local cultivars, which were forced to bloom early by manipulating the day length. Some early flowering hybrids have been raised which are under observation. These hybrids open up the possibility of doubling the blooming time from 2 to 4 months.

Dahlia

Dahlia is a tuberous rooted half hardy herbaceous perennial belonging to the family *Compositae* having its origin in Mexico. Dahlias with their most gorgeously coloured flowers are very popular in the Indian gardens and are widely used for garden display and indoor decoration. Dahlias have a wide range of flower colours and diversity in the form of the flowers and so can cater to the tastes of a large number of garden lovers. Dahlia is not difficult to grow and even an amateur can raise the prize dahlias.

The height of dahlia plants varies from 30 to 180 cm depending upon the cultivar. The dahlia flower consists of a certain number of outer ray florets in which the male organs are modified into a strap-shaped petal, arranged round a central disc of bisexual florets. Actually the ray florets in dahlia have all the flower colours, whereas the disc florets are generally yellow. In the double flowered cultivars more of the male organs get converted into ray petals with proportionate reduction in the number of discs.

The modern garden cultivars of dahlias have arisen as a result of crossing between various species and varieties mostly consisting of the following: *D. variabilis*, *D. coccinea* (Syn. *D. cervantesii*), *D. crocea*, *D. imperialis*, *D. juarezii*, and *D. merckii*.

Classification: A joint Dahlia Committee formed by drawing equal number of members from the National Dahlia Society and the Royal Horticultural Society in England has classified the modern dahlias into the following main types:

(1) ***Single-flowered dahlias:*** Flowers in this group consist of a single row of ray florets

round the central disc. The blooms hardly exceed 10 cm in diameter.

(2) *Star dahlias:* The blooms are small, consisting of 2 or 3 rows of somewhat pointed ray florets which are slightly incurved forming a cup round the central disc.

(3) *Anemone-flowered dahlias:* In this group, the flowers have a dense dome-shaped central disc of tubular florets surrounded by an outer ring of ray florets.

(4) *Collarette dahlias:* The flowers in this group somewhat resemble the singles, but in addition to one layer of ray florets surrounding the central disc there is an additional layer of inner ring of narrow small florets (the collar), which are about half the length of the outer florets.

(5) *Paeony-flowered dahlias:* This consists of semi-double blooms of a few rows of ray florets surrounding the central disc. This class is again sub-divided into large, medium, and small, depending upon the size of the bloom.

(6) *Decorative dahlias:* The blooms in this class are fully double and the central disc is not visible until the blooms are aged. The ray florets are flat and broad with a blunt point. The petals are generally not much twisted.

(7) *Cactus dahlias:* These also have fully double blooms without any central disc showing at least until the bloom is too old. The ray florets in this group are narrower and more pointed than in the decorative dahlias and part of the petal is revolute (i.e., tend to curve backwards), or incurving, or straight.

Both the classes, decorative and cactus, are sub-divided according to size into giant flowered (over 25 cm diameter), large flowered (above 20 cm but less than 25 cm), medium flowered (over 15 cm but less than 25 cm), small flowered (over 10 cm but usually below 15 cm), and miniature flowered (normally less than 10 cm).

(8) *Double show and fancy dahlias:* The blooms in this group are also fully double, almost globular, and the central florets are comparatively smaller than the outer. The margins of the petals are incurved, tubular, and blunt at the mouths. The flowers are over 10 cm in diameter.

(9) *Pompon dahlias:* These resemble exactly the double show and fancy dahlias in shape of bloom except that these are much smaller in size. The class is sub-divided, depending upon the size of bloom, into large (over 7.5 cm but below 10 cm), medium (over 5 cm but below 7.5 cm), and small (less than 5 cm).

(10) *Miscellaneous dahlias:* A small group of dahlias which are not covered by any one of the above classes. The orchid flowered dahlias resembling somewhat the cactus types except that the petals here are involute, i.e., curved forward across the width forming a tube and showing the reverse side instead of revolute, belong to this class. The chrysanthemum-flowered types are also included in this class.

(11) *Dwarf bedding dahlias:* The height of the plants in this group is between 30 and 60 cm but the formation of the flower may belong to any of the above groups.

Uses: Dahlia is one of the most popularly grown garden flowers and its popularity can even be compared with that of rose. Because of its popularity the plant is grown in various places in the garden. It is widely grown in exclusive dahlia beds, each bed generally comprising one colour and type, and in annual, herbaceous, or mixed borders. Dahlias are also suitable for planting in front of the shrubberies or in between newly planted shrubbery to fill up the interspaces and bring more colour. Dahlia is a very good potted plant. The blooms are also prized as cut flowers.

Propagation: Dahlia is a perennial plant and can be propagated by various methods other than the normal method of raising from seeds.

The propagation from seed is never followed for the modern dahlia hybrids except the single types, as seedlings do not breed true to type. The double types if raised from seeds throw a large number of semi-doubles and singles. The seeds are sown in September-October in the plains and March-April on the hills. In Bangalore, the best blooming period is from August to September and, hence, sowing should be done in May. Places having similar climatic conditions should follow this rule. The seedlings are raised in nursery and transplanted in beds after pricking.

The other alternative is to propagate by the vegetative method which will ensure a progeny true to type, that is, without variation in size or colour of the blooms and other plant characters, except of course in cases of rare mutations or "sports".

One of the vegetative propagation methods is by the division of the tuber. Generally, the old clumps will have three or more tubers attached to each stem depending upon the cultivar. The tuber should be separated with the help of a sharp knife, with at least one "eye" attached to it. The so-called eye in the tuber is situated at the stem end. The tubers are planted during June in the plains, but this should be delayed for some time in heavy rainfall areas as Culcutta. In Bangalore, the planting of tubers is undertaken in May. In the hills, this can be done in March-April. The tubers are planted 15 cm deep and 30-90 cm apart depending upon the class of dahlias. The large decorative types need more spacing than the dwarf bedding types. The cut end of the tubers should be treated with 0.2 per cent captan or 0.1 per cent benlate to prevent rotting. Plants raised from tubers start flowering during November-December in the northern plains and in August in Bangalore. In eastern India, these come into bloom in December-January. Propagation by tuber is considered best when only a few plants are required.

For raising a large number of plants, true-to-type, the best method of propagation is by cuttings. The cuttings are taken from the young green shoots, produced from the crown of the tubers, when these are about 7.5-10 cm in length. The shoots should be cut as close to the crown as possible with a sharp razor blade just below the first node above the tuber. This will encourage more growths from the secondary eyes which are as good as cuttings. If cuttings are taken too high from the tuber, fresh shoots arise from the leaf axils left behind, rather than the secondary eyes of the tuber. The quality of these shoots as cuttings are not good as most of them are hollow.

The lower leaves in the cutting may be removed by clipping and the cut end should be treated with a root promoting hormone such as Seradix B before planting. The cuttings are planted 5 cm apart in shallow boxes or pans containing a porous mixture of coarse sand and leaf-mould or peat or vermiculite alone. Under tropical conditions, the cuttings raised in shade and sheltered from heavy rains will root easily. After rooting, the cuttings are pricked singly in 7.5 cm pots containing one part each of soil and leaf-mould before final planting in pots or beds.

In Delhi and places around, the cuttings can be taken periodically from July to middle of September. Cuttings taken later than this date generally remain stunted in growth with the onset of severe winter cold and start growing again only in February and flowering later. Cuttings raised early may flower in December-January in Delhi. It has been observed by the authors that cuttings raised under protection during December in Delhi flower at the end of February very satisfactorily, but the plants never grow more than 45 cm in height. Cuttings taken earlier and grown under the protection of a pergola or similar structure but

TUBER BENCHED UP FOR PRODUCTION OF CUTTINGS

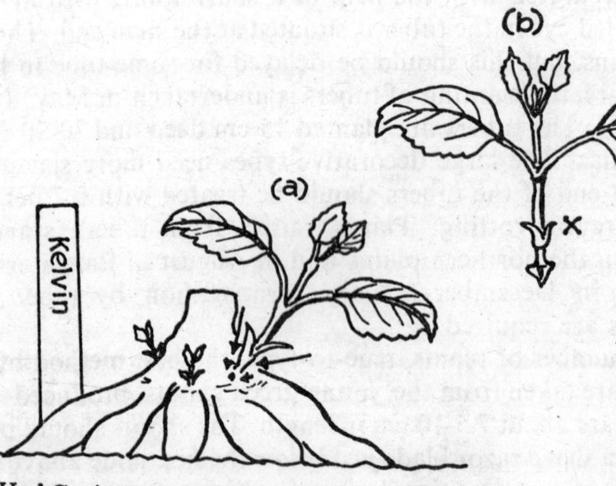

Heel Cutting

(a) Removing and cutting from tuber (Cut downwards at arrowed points)

(b) Cutting after removal—The small pairs of leaves (x) should be cut off.

receiving sufficient sunlight grow normally and bear flowers from February to April in and around Delhi. In heavy rainfall areas as Calcutta, good cutting material from the tubers is available from mid-September and October only. The shoots sprouting during the rains are generally woody and thin and are not quite suitable as cuttings. In Bangalore, cuttings are taken in May or early June. In Madras, the planting is generally undertaken during October or November for flowering in January or February. In the hill stations the tubers may be sprouted under glass in December-January for striking cuttings in March-April, making arrangement for bottom heating if necessary. All cuttings should be labelled properly for future identification.

Some people claim that heel cuttings root easily and root better than ordinary cuttings. Heel cutting is taken with a small portion of the crown attached to the base. This can be done easily with the help of a sharp knife having a thin point (rose budding knife). Heel cutting should be taken with minimum destruction to the secondary eyes.

Cultural hints: Growing of dahlia does not pose any problem and this can be grown with comparative ease in any type of soil but a medium textured soil of neutral pH or little acidic (pH 6.5) in reaction should be preferred. The site selected should be in the open, receiving ample sunshine. Dahlias grown under shades of trees become lanky and weak. If the soil is too acidic this will require an addition of lime. Similarly, alkaline soils will have to be treated with gypsum. Hydrated lime or gypsum may be added at the rate of 300 – 400 g per square metre depending upon the soil reaction. Dahlia is not a deep-rooted plant and, hence, very deep digging is not essential. A general digging up to 45 cm will be quite sufficient for light and medium soils. In heavy clay soil, deep digging and drainage are essential.

Manuring: It is generally agreed that if the soil is properly prepared and plenty of manure is incorporated as basic dose, there will be hardly any need for artificial feeding during the later stages of growth. About 4 kg of well-rotten cowdung manure or leaf-mould per square metre of soil should be sufficient for most types of soil except for very poor and sandy soils, where the quantity should be a little more. Well-rotten stable manure is considered to be more suitable for growing dahlias. Compost made from lawn mowings is also very good. Bonemeal at the rate of about 100 g per square metre area should also be added as basal dose. Potash is an important element for dahlia culture as it promotes healthy plant and tuber growth and improves flower quality. About a basketful of wood ash should be added per square metre at the time of bed preparation. Too much application of nitrogen impairs the keeping quality of the tubers. One top dressing with cowdung or leaf-mould at the time of the appearance of flower buds will be quite beneficial especially in poor soils. Feeding with liquid manure as suggested for chrysanthemum once at fortnightly intervals can be followed especially in poor soils till the flower buds open. But overfeeding should be avoided. Before feeding with liquid manure the ground should be soaked thoroughly to ensure proper penetration of the liquid manure.

Stopping and thinning: When the plants have attained a height of about 25-30 cm and developed 4-6 pairs of healthy leaves, the plants are stopped, i.e., the growing point is removed. Very weak-growing plants and recently planted dahlias should not be stopped. Though the main object of stopping is to obtain early bloom it is also aimed at getting side branches, making the plant bushy and increase the size of blooms. The giant and large decorative dahlias are allowed to have three side branches on light soil and five in rich soil. The alternate branches in the successive internodes are removed.

The same procedure is followed in the case of giant and large cactus dahlias and here the branches left are five and seven in light and rich soils respectively. The medium decorative and cactus are allowed to develop six and up to nine branches respectively in the above soil types. The cultivars vary in their growth habit and this method has to be adjusted keeping in view the vigorous and weak-growing cultivars.

The process of deshooting is not followed for small decorative and cactus types unless these develop too many branches. In such events the weak branches should be removed. Here again the operation will depend upon the cultivars and the nature of soil.

For double show and fancy dahlias the technique followed for giant cactus should be adopted. The medium and small pompons receive the same treatment as for the small decorative and cactus types.

The stopping schedule given above is for raising blooms for exhibition. Same procedure

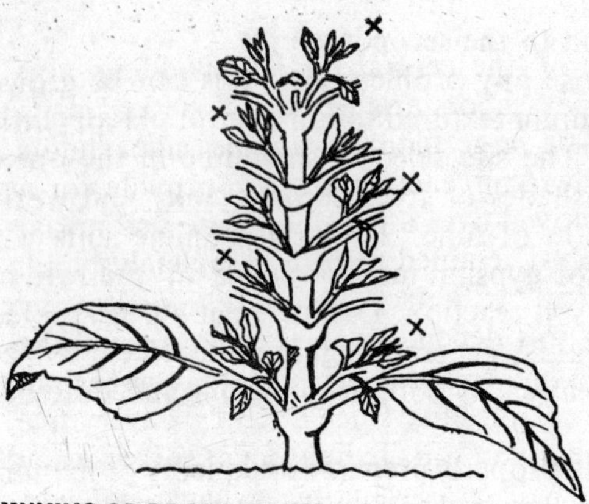

THINNING OR DE-BRANCHING
Side shoots marked x are removed. This method is suitable for giant decoratives or cactus.

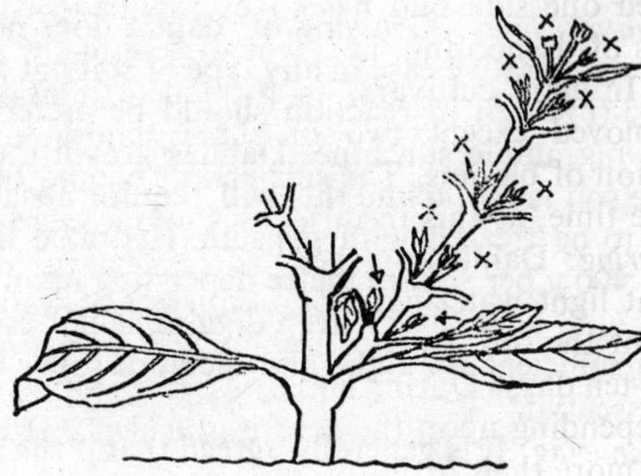

DISBUDDING GIANT FLOWERED VARIETIES
Only one branch depicted for clarity. Buds and side shoots marked x removed. The two arrowed shoots retained. At a later date the weaker out of the two shoots is removed

should be followed in the case of dahlias for garden decoration, but the restrictions may not be of the same degree.

It is also possible to some extent to control the date of flowering by stopping. Approximately it can be said that giant decorative dahlia will bloom six weeks after stopping, provided all other factors, such as climate, soil, health of plant, remain constant. Above all, this time varies also from cultivar to cultivar to some extent.

The side shoots should not be removed until these are about 10-15 cm in length, because before that stage it will not be possible to judge the quality of the branches. The endeavour will be to retain the most healthy branches.

Staking: Once the dahlias start growing vigorously it will be necessary to provide strong stakes and to tie the plants to the stakes. It may be necessary even to stake the side branches and as such very often one stake may not be sufficient to tie all the branches. The giant cultivars may need one stake for each branch, fixed in the ground in a triangular fashion around the plant. This method though very useful may prove to be quite laborious and expensive when a large number of plants are to be staked. An alternative method may be tried. Two rows of strong stakes may be fixed on either side of the plant rows and these should be connected from one end to the other by passing one or two lines of firm G.I. wires to which the branches are tied. To keep the wires stretched, the ends of the wires should be fastened to wedges driven to the ground. The branches should be tied to the stakes with soft strings or *sutli*, and the knots around the branches should be loose enough to allow the branches to increase in girth.

While staking and tying, no branch should be forced too far apart from its natural position. For one or two such odd branches one additional stake should be provided. Though the operation of staking appears to be cumbersome, it is absolutely necessary or at a later stage when the branches are about to bloom it may break off with the strong breeze and cause a great disappointment.

Disbudding: The main objective of disbudding is to obtain quality bloom. Dahlia will generally have three flower buds at the end of each branch. The central or "Crown" bud is retained for blooming while the other two are removed at "pea" stage. If the crown bud is damaged one side bud has to be retained in place of the central bud. Disbudding also consists of deshooting, i.e., removing some of the side shoots arising out of the leaf axils below. In giant cultivars, to produce exhibition type large blooms, all the side shoots are removed except two or three shoots low down on each branch to provide for a succession of blooms. The side shoots should be removed over a period of time and not at any one time, as this in rare cases may cause the shoot retained to turn completely blind.

Watering: Dahlias are benefited by heavy watering at longer spells rather than by frequent light watering. The frequency of watering will depend upon the climate and the nature of the soil. In cool and humid weather it may not be necessary to water more than once in ten days. During hot spells the plants will need watering twice a week or every five days depending upon the severity of heat and the nature of soil. In light soils the frequency will be more than in heavy soils. The young plants will need watering at more frequent intervals than the grown-up and established plants.

Pot culture: There is hardly any difference in the method of cultivation of dahlia in pots than in ground. The pot mixture consists of 1 part garden soil, 1 part sand, 1 part leaf-mould, 1 part manure (well-rotten cowdung or stable manure), and ¼ part of broken pieces of charcoal. About 2 table spoonfuls of bonemeal should be added to each pot. Pots of 25-30 cm diameter are suitable for growing large dahlias. The frequency of irrigation in pots will be more than in ground. Liquid manure feeding is also helpful.

Frost: Dahlias are killed by frost. The danger of damage from frost is quite serious in and around Delhi. The plants may be grown under a pergola or under the protection of a shrubbery or a tree but should receive sufficient amount of sun at least in the morning. Naturally the plants are to be protected more on the northern and western side keeping in view that the plants at least receive sufficient sun during the morning hours.

Shading: Dahlia blooms raised for exhibition may be damaged by hot sun or dry winds and pouring rains. In dry and hot places the blooms may be shaded by cheese cloth, brown paper, or cartridge paper to prevent them from drying. The papers are fixed over frames supported by stakes. Shading of blooms have two disadvantages. The first thing is that the beauty of the garden is impaired by the assorted shading materials fixed all round the dahlia beds. Secondly, the process of shading causes general loss of colour in a number of cultivars. So unless absolutely essential this procedure should not be adopted for general purposes.

Lifting and storage of tubers: After the flowering is over and the plants turn yellow the plants are headed back to 15-20 cm above ground level. The tubers are then lifted carefully with the help of a garden fork taking care not to injure the crown. No tuber should be lifted from the ground unless it is fully matured. In the hills, after the flowering is over in September, the plants may be automatically killed by frost. If this does not happen, the plants are headed back and tubers lifted in November.

After lifting, the tubers are cleaned by removing the soil attached to it. Some experts also recommend partial drying of the tubers outdoors for a few days before storing if the weather is clear, otherwise this can be dried indoors. The tubers are then stored in cool airy rooms on shelves containing layers of sand.

The tubers should be dusted with 5 per cent DDT or BHC and flowers of sulphur to protect them from insects and diseases. The tubers should not be allowed to shrink from heat and to prevent this occasional sprinkling with water is necessary. The ideal storage temperature is 4-7°C with free supply of air and low humidity. The potted plants, after heading back, are removed to the shade of a large tree and the tubers are left in the pot itself. Occasional sprinkling with water is essential to prevent the tubers from drying. The thin or very small tubers should be potted in slightly moist soil shortly after lifting.

Pot tubers: Dahlia cuttings are raised in pans or boxes which are later pricked in 7 cm pots before transplanting. Sometimes, some of these cuttings become surplus and are not transplanted. These surplus cuttings generally do not produce any flowers but they form small tubers which are referred to as "pot tubers". To encourage formation of pot tubers such plants are pinched frequently to keep the growth checked. The pot tubers are very good for further propagation as the plants grown from them have the same quality of blooms as the mother plants.

A few useful hints: In some giant and large cultivars the blooms are borne on very long stalks which may snap from the first joint in the event of strong winds. Such blooms may be secured by placing green-painted, rounded, smooth, thin but strong bamboo supports extending from just below the bloom to a point 25-30 cm below the first joint. In cultivars having weak neck the flowers droop down. Such blooms may be forced to remain at right angle if some soft padding material is tied at the end of the bamboo stake supporting the stalk.

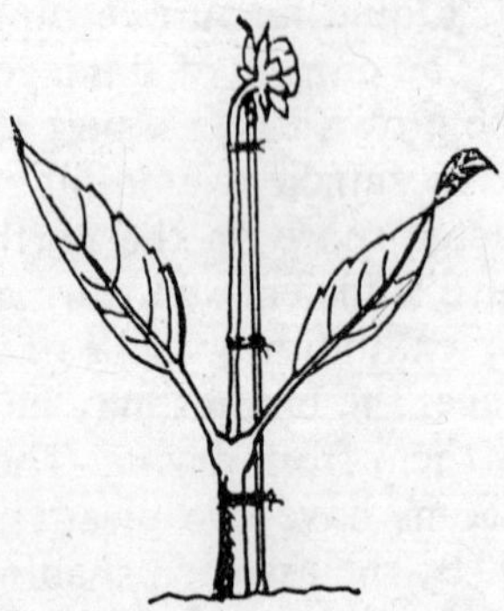

SUPPORTING A DAHLIA BLOOM
The stake (green painted) tied at 3 places, one tie should be in the main stem. Soft padding is provided below the bud.

Gerbera

Family: *Compositae*; Origin: South Africa; Other Common Names: 'Transvaal Daisy', 'Barberton Daisy', 'African Daisy'.

Gerbera is a dwarf perennial herbaceous plant which is named after Gerber, a German Naturalist. The genus consists of about 50 species of Asiatic and African origin. The best species which is commonly grown in the garden along with different hybrids is *Gerbera jamesonii.* The plants are stemless, dwarf, 30-45 cm tall, and are hairy throughout. The leaves are 12-20 cm long, 5-7 cm broad, deeply lobed, with a fairly long (15 cm) petiole. The foliage is arranged in the form of a rosette at the base. The flowers are daisy-like, 7-10 cm across, but in certain hybrids these may be as large as 15 cm across. The flowers may be single or double and are available in various self-coloured cultivars as well as in bicolours. The colour may be white, cream, lemon-yellow, brick-red, orange, pink, salmon, scarlet, maroon as well as in many other shades. The flowers are borne in long and slender stalks.

Uses: The long stalked star-like flowers are pretty and long lasting both in field conditions and in vase and hence the flowers are very popular for garden decoration as well as for the vase. Beds exclusively meant for growing gerberas should be located in a sunny situation for a succession of blooms, almost throughout the year, as the plants generally bloom perpetually. Besides, gerbera is also suitable for growing in mixed borders, rock garden, and pots. Gerbera is also a valuable flower for cutting and hence is most suitable for commercial floriculture. The flowers are hardy and stand the rigours of transportation admirably. The long keeping quality fetches a good market price.

Propagation: Propagation of gerbera is from seeds and by division. Plants raised from seeds do not breed true to the type unless self-pollinated.

The surest and simplest method of multiplying a desired cultivar is to propagate it by the division of the mother plant. The division of a clump is also necessary after every 2 or 3 years to keep the plants vigorous and free-flowering. The clumps may be cut into small segments, each possessing a few well-developed leaves and roots with the help of a sharp knife or a secateur. The very large clumps may initially be halved and quartered with the help of a sharp spade. Further divisions are done with a sharp knife. After the division is complete the foliage and the roots are trimmed back to 7-10 cm in length. The best time for division, for most part of the country, is during the rainy season. In a mild climate as that of Bangalore the operation is possible throughout the year. In the hills, it may be done in the spring.

The gerbera seeds lose viability within about three months from maturity and, therefore, must be sown immediately on maturity. The month of June is good for sowing as the seeds need warm weather for germination. The seeds need a friable well-drained soil for sowing. The seeds are sown in pans or boxes and covered thinly with sieved compost or well-rotted cowdung manure. After completion of sowing, the seed pans should be moved into a warm and sheltered place. The pans should not be allowed to dry up. Freshly harvested seeds germinate in about 10 days. The seedlings are pricked at two-leaf stage in individual 7 cm pots or in boxes about 3 cm apart. These are planted in permanent positions when grown large enough.

Cultural Hints: Gerbera is a fairly hardy plant and can be grown both in the plains and in the hills. The climate of Bangalore is very suitable for growing this valuable commercial flower. Hill stations, with climate as that of Kalimpong (1,200 m), are also suitable for gerbera. Gerbera needs a well-drained soil with plenty of moisture in it. To ensure good drainage in high rainfall area, growers prefer growing gerbera in raised beds. Gerberas suffer from crown and root rots in poorly drained soils.

Gerbera grows in a wide range of soil types except heavy clayey soils. But the best soil is a well-drained loam or a sandy loam. Gerbera requires a soil with plenty of organic matter. If soils are deficient in organic matter a layer of well-rotted cowdung manure (about 5-7 cm thick), leaf-mould, or humus should be spread over the surface and then incorporated with the soil at the time of digging. Gerbera is a deep-rooted plant and hence dislikes transplanting. Digging up to a depth of 45-60 cm should be done. The digging and manuring is done about a fortnight ahead of the planting time.

The planting is done 35-45 cm apart within the rows, and rows are spaced 30-35 cm apart. Plants grown for the commercial flower production are given lesser spacing of 35 cm. The plants should never be planted deep which may cause the crown rot to develop. The crown of the plant, i.e., the meeting point of the leaf-stalks and roots, should remain above ground level. Immediately after planting, water is given copiously, but then it is restricted until the new sprouts start coming. Gerbera needs an open sunny place and will not thrive in shade. But protection from the west is beneficial, especially in hot climate.

The fertilizer requirement of gerbera is not perfectly known. However, if a good basic dose of organic manure is applied at the time of planting, the plants may not need any other additional fertilizer doses until the time of flower bud appearance. At this stage, a complete balanced fertilizer dose made up of one part of ammonium sulphate, two parts of super phosphate, and one part of muriate of potash should be applied at the rate of 60-80 g of this mixture per 3 m of plant row. Fertilizers high in nitrogen are not good for gerberas. Mulching with compost, leaf-mould, or straw during the warmer months is good for conserving moisture and checking weed growth. Thick mulching placed closer to the plants should be avoided as this may promote crown rot. Top dressing with fertilizer should be given only at intervals of 8 weeks depending upon the health of the plants.

Gerbera flowers only at the end of the second year when raised from seeds. The plants flower best only in the third year. Vegetatively propagated plants flower much earlier.

Recent research has demonstrated that gerbera can be multiplied very rapidly by tissue culture.

Disease and Pests: See Chapter XXVI.

Gladiolus

Family: *Iridaceae*; Origin: South Africa; Other Common Names: Sword Silly, Corn Flag.

The word gladiolus is derived from the Latin word *gladius*, meaning a sword, because of the sword-like leaves of this plant. The gladiolus is popular for its attractive spikes having florets of huge form, dazzling colours, varying sizes, and long keeping quality. Possibly this was first brought into cultivation from its native home in South Africa during the Greek period. The development of gladiolus started only at the beginning of the eighteenth century when William Herbert (1806) in England produced inter-specific hybrids. Later, in 1837, Hermann Josef Bedinghaus developed Gandavensis Hybrids in Belgium which proved to be very helpful in the production of modern gladioli. The species *Gladiolus primulinus* was found in 1902 near the Victoria Falls of the Zambesi river. It was used in hybridization to produce large hooded flowers of light colours. The present-day cultivars have arisen as a result of crossing between the following species: *Gladiolvs aurantiacus*, *G. blandus*, *G. byzantinus*, *G. cardinalis*, *G. cruentus*, *G. dracocephalus*, *G. psittacinus*, *G. primulinus*, *G. trimaculatus*, *G. tristis*, and *G. tristis* var. *concolor*.

Uses: Gladiolus is by far the best suited bulbous plant in India, but unfortunately this is not so commonly grown as tuberose. In Holland and other European countries it is most popular and ranks next to Tulip only. Gladiolus is ideal as cut flower. It is also very good for beds, herbaceous borders, and does quite well in pots.

Classification and Cultivars: There are two types of gladiolus: one having large flowers and the other butterfly and miniatures. Both of these groups have early, mid-season and late flowering cultivars. In butterfly types, the spikes are small, but the various coloured flowers, sometimes with darker throats, look beautiful and are most suitable for small gardens. The primulinus (*G. primulinus*) is another type of gladiolus where the spike is slender and the florets have a hooded upper petal.

The colour range in gladiolus is fantastic and almost any colour from near-black to white, pink, violet, lilac or mauve, greenish, "smoky", and combinations of these colours is available.

Some of the important cultivars, viz., 'Bis Bis', 'Cherry Blossom', 'Cornus', 'Donald Buds', 'Dusk', 'Friendship', 'Geliber Herald', 'High Fashion', 'Jo Wagenear', 'Lily Blue', 'Melody', 'Oscar', 'Peter Pears', 'Snow Princes', 'Stormy Weather', 'Tintorente', 'Tropic Seas', 'Vink's Glory', 'Watermelon Pink', and 'Wild Rose' have been found suitable for cultivation around Bangalore for cut flower purposes by the Indian Institute of Horticultural Research, Hessaraghatta, Bangalore.

A number of new hybrids with attractive colours have been developed by breeders of the Indian Institute of Horticultural Research, Hessaraghatta. So far, 6 hybrids and one mutant gladioli have been released by the Institute.

Culture: Gladiolus can be grown in the plains and up to an altitude of about 2,500 m. It should be grown in well-drained soil and in a sunny situation sheltered from strong winds. The soil should be dug thoroughly and powdered finely. At the time of final land preparation, 5 to 6 kg of well-rotten cowdung manure or leaf-mould and 60 g of bonemeal should be added per square metre area. Adding of too much manure should be avoided as this tends to make the flower spikes too tall and slender.

For cultivation of gladiolus in pots, the pot mixture should contain one part of good soil, two parts of cowdung manure or leaf-mould, and half part of coarse sand. About two table spoonfuls of bonemeal and a handful of wood ashes should also be mixed in the potting mixture.

Planting Time: The best planting time in the plains is between September and October. Planting by the middle of August can also be tried if the weather is cool. In places with mild climate as that of Bangalore, gladiolus is planted in June to obtain flowers in August. Planting in October and November may be done to obtain flowers in the winter. Planting can also be practised at other times of the year with varying degrees of success. Planting during January and February is not advisable as the flowers appearing in the hot weather of March to May will wither away very quickly. Recent research conducted at the Indian Institute of Horticultural Research, Bangalore, indicated that the best planting seasons were June and October to November for flower production in Bangalore. However, it was noticed that flowering could be obtained round the year with varying degrees of success. In the hills, the gladiolus flowers during May-June or August-September. Hence, planting should be undertaken during March-April. In eastern India planting should be delayed till the monsoon rains are over. Successive planting, at an interval of 7-10 days, should be done to get continued flowering over a longer period.

Propagation and Planting Procedure: Gladiolus is planted from corms, which are often wrongly called bulbs. It can also be grown from seeds, but plants thus raised do not come true to type. However, sexual propagation is followed by breeders only to evolve new varieties by hybridization. For quick multiplication, small corms, called "cormels", obtained in various sizes along with the large corms are planted after soaking in water for 24 hours. Before soaking, the brown scales over the cormels are removed carefully. The cormels take about 2-3 repeated plantings to produce proper bloom. In Bangalore, the cormels come to flowering earlier and in the second planting they produce very large blooms.

To obtain flowers of good size, corms of at least 4-5 cm across should be planted. To obtain exhibition-type large spikes corms of at least 7.5 to 10 cm diameter should be planted. A medium-sized corm, with high crown (top sized), is better than a flat large corm. Before planting, the brown scales covering the corms should be removed and the corms soaked in 0.2 per cent Captan or 0.1 per cent Benlate solution for 30 minutes. It is better to get the corms sprouted before planting by keeping them in a dark warm place covered with black polythene or by burying the corms in 5 cm to 7.5 cm layer of moist sand. After sprouting, planting may be done in beds or pots.

The corms are planted 15-25 cm apart in a row and rows are spaced 30-45 cm apart. A good spacing between rows is necessary to facilitate inter-culture operations. If more cormel production is required the corms should be planted shallow (3 cm deep), but such shallow planting may cause lodging in high wind areas. Planting at 5 cm depth should suit most of the localities. Where high winds prevail a planting depth of 7 cm is recommended. The depth of planting is measured from the tip of the corm.

Feeding and Watering: It has already been pointed out that heavy feeding is not desirable. It is best to manure with well-rotten cowdung, leaf-mould, or compost while chemical fertilizers are not used generally. But, recent researches show that nitrogenous and phosphatic fertilizers used in judicious quantities give good results. Liquid manure, prepared by fermenting fresh cowdung in water and then diluting it to tea colour, may be applied once a week. Application of light soot water once a month or fortnight is recommended. A light mulching with leaf-mould, when the flower spikes start appearing, is useful.

A good soaking with water once a week will be quite sufficient. In the winter, irrigation should be given only once in 10 days. During the rains the gap between two irrigations can be adjusted depending on the frequency and amount of rain received.

Staking: When the spikes start emerging, staking with strong bamboo stakes should be provided, so that the spikes do not break or fall down during winds. Staking is not needed if the plants are grown closely or in clumps. The miniatures and butterfly types generally do not need staking. When the plants grow to a height of about 25 cm earthing up should be done once.

Cutting flowers for vase: After the first floret (the lowermost) shows colour, the spike is ready for cutting. The spike is cut with a sharp knife as close to the base as possible, causing least injury to the foliage. Actually at no stage of growth should the gladiolus foliage be injured as this hampers proper growth of corms as well as flowers.

Immediately after cutting, the spikes should be placed in buckets containing water. When kept in vase, all the upper florets will open in succession and the spikes will last for a long time. To prolong the vase life, a little portion (1.5 cm) of the spike from the base is cut under water on alternate days and fresh water is put daily in the vase. Cut flower life can also be extended if the cut flowers are kept in a solution of 600 ppm of 8-HQC (8-hydroxyquinoline citrate) plus 4 per cent of sucrose.

Harvesting and storage of corms: If the spikes are not cut, the gladiolus should not be allowed to go to seed as this robs the developing corm of its vitality. When the leaves turn yellow and start drying, the corms along with the cormels are lifted from the ground, preferably with the help of a fork, with proper care, avoiding injury. Before storing, the corms are dried in semi-shade and then dusted with 5 per cent D.D.T. and stored in a cool room in shallow trays or polythene bags. Sufficient perforations should be made in the polythene bags. The best storage method is to store in a refrigerator or cold storage. The cormels of each cultivar should be stored separately for further planting. To minimize rotting in storage, it is advisable to soak them in Benlate (0.1 per cent) or Captan (0.2 per cent) for half-an-hour and then dried and stored.

Breeding of Gladiolus

Historical review of breeding: The improvement of gladiolus through breeding started at the beginning of the eighteenth century when William Herbert in England (1806) crossed different species and produced numerous hybrids, but all proved to be sterile and were never commercialized. The first hybrids to enter commerce were the work of James Colville, in England, who used *Gladiolus tristis* var. *concolor* and *G. cardinalis* to create his Colvillei hybrids in 1823. Three of his original hybrids are still being grown, namely, 'Roseus', 'Rubra', and 'Albus'. In 1838, Schneevogt, in Holland, produced Ramosus hybrids by crossing *G. cardinalis* with *G. blandus*. Hermann Josef Bedinghaus of Belgium made successful crosses which proved to be very much helpful for the development of present-day garden gladiolus. In 1837, he crossed *G. psittacinus* with *G. cardinalis* and produced Gandiensis (Gandavensis) hybrids. They were very much popular and were having enormous demand from growers and nurserymen. Later, nurserymen developed as many as 140 cultivars from the original hybrids. William Hooker, in England, crossed *G. blandus* with *G. psittacinus* in 1846 and named his hybrids as Brenchleyensis hybrids. In 1948, Carolus crossed *G. blandus* with Ramosus hybrids and produced Leopoldi hybrids. Shortly after, Cole, in England, crossed Gandavensis hybrids with *G. blandus* naming the off-spring Willmorecanus hybrids. Eugene Souchet in 1852 combined Gandavensis and Ramosus hybrids with *G. blandus* and produced hybrids which were admired in the U.S.A. for the first time. Nanus hybrids were produced in the Channel Islands around 1855 by using *G. cardinalis* and *G. venustus* as parents. John Standish, in England, crossed the Brenchleyensis hybrids with *G. cruentus* and introduced over 100 cultivars. In 1877, Max Leichtlin of Germany hybridized *G. cruentus* on Gandavensis hybrids and produced Leichtlin hybrids and later, in 1891, John Lewis Childs (U.S.A.) named them as Childsi hybrids. Victor Lemoine, in France, crossed *G. purpureo-auratus* with Gandavensis hybrids, thus developing his Lemoine hybrids and from them 152 named cultivars were introduced in 1878. Sander in 1889 crossed *G. dracocephalus* with the gandavensis strain. In 1900, Lemoine crossed *G. aurantiacus* with some Leichtlin hybrids, thus resulting in the production of Precoces hybrids. In 1902, *G. primulinus* was discovered near the Victoria Falls. It has hooded petal to protect the reproductive organs of the flower. James Kelway, in England, crossed Gandavensis with Lemoine hybrids to produce Kelwayi hybrids. Between 1898 and 1910 some 20 such cultivars were introduced. The first ruffled gladiolus was produced by Kundred in 1911. Shortly afterwards he introduced the lacinated type, then the recurved and snapdragon types. Van Tubergen, in Holland, introduced the Herald hybrids from Nanus × Childsi strains. Philip O. Buch, New Jersey, crossed Herald hybrids with *G. primulinus* and the resulting hybrids were named Heralinus hybrids.

Around 1935, Collingwood Ingram produced the first bigeneric hybrids by crossing *G. concolor* × *Homoglossum revolutum.* These were fragrant in evenings but were never commercialized. In 1956, Mrs. John Wright, in New Zealand, attempted the second recorded bigeneric cross using the cultivar 'Filigree' × *Acidanthere murieliae.* Later, one of her hybrids was crossed with *A. murieliae* and a really fragrant Gladanthera hybrid, named 'Lucky Star', was developed. It is being used by other breeders for breeding fragrant varieties.

Most gladiolus breeders use the modern hybrid cultivars whose complex genetic background provides an excellent source of genetic variation. Floret colours range from white to dark purple encompassing most shades of yellow, pink, and red. Only dark blue, bright green, and black are conspicuously missing from the modern cultivars. Petal structures are varying from plain to heavy ruffling; substance varies from light to extremely heavy; and floret types vary from round to triangular; cupped to wide open. More than 10,000 cultivars have been recorded during the past 200 years. As mentioned earlier, from a gene pool of 250 known species, only 12 have been used in the development of modern garden gladiolus. The potential genetic variability would have been more if the 238 unused species could have been used in the breeding programme. With the introduction of bigeneric hybrids, there is an additional source of genetic variability now.

Breeding Objectives: For breeding new varieties, one should be very clear in his objectives. The two most important aims in today's hybridization are health and beauty in the new seedling. Good health means ability to grow well in most of the soils and climatic conditions, to produce good corm and cormels, to be resistant to *Fusarium, Botrytis* and other diseases, and to develop suitable flower spikes year after year for cut flower or exhibition use. For beauty, the breeder should consider the numbeı of florets, placement, texture, ruffling, colour clarity, floret size, spike length, symmetry and vigour. There should be an eye-pleasing blend of all the features. There is less demand for novelty types such as doubles and multi-branched spikes. Recently, there is some fancy to breed fragrant gladiolus also. Although improvements are needed in many colour classes for quality, the major need of gladiolus is increased resistance to serious diseases such as *Fusarium* and *Botrytis*.

Floral biology: Before proceeding further, it will be worth while to describe gladiolus flower. The particular type of inflorescence characteristic of gladiolus is the spike. As such, the individual florets are attached directly to the axis. As in the case of many monocotyledons, gladiolus florets also possess floral parts in threes. The outermost three segments make up the calyx and the next whorl of three segments comprise the corolla. Sepals and petals collectively make up the perianth of the flower. The perianth surrounds three stamens and a tricarpeled pistil with a three-forked stigma. The ovary contains between 75 and 150 ovules. Each flower bud is enclosed separately within its own spathe which consists of two green bracts. The anthers dehisce within 3-4 hours after urfurling of the perianth (tepals), with most of the pollen falling on the ground or landing on the lower tepals or lip of the floret. The stigma stands some distance above the anthers and is not receptive until it becomes feathery. The fruit of gladiolus is termed as capsule.

Hybridization technique and seed setting: The aim of the mechanism followed in hybridization is the production of seed of known parentage. Keeping this in view, one must adopt the procedure that will assure seed set without contamination by pollen from unknown sources. In many cultivars the style begins to curl downward on the third day of opening the floret, bringing the receptive stigma in contact with the anthers and any adhering pollen. This provides an opportunity for self-pollination. Natural cross-pollination is rare as the pollen is heavy and is not carried away by wind and bees rarely contact stigma. Sometimes stigma comes in contact with the wings of visiting bees which carry foreign pollen and consequently contamination takes place. The hybridization in gladiolus is done by hand emasculation and hand pollination. To protect the pollen parent from any contamination is of prime importance. It is done by covering the florets with butter paper bags. Each floret which is to be used as seed parent must be emasculated prior to its opening. All the anthers should be removed and the tips of the tepals (perianth) of the floret should be tied with soft thread. Emasculation is done preferably in the afternoon. It can be done in the morning also but before 7.30 a.m. Stigma is receptive between 10 a.m. and 1.00 p.m. when it becomes feathery. Pollination is more successful in the morning before the stigmatic surface dries completely. Pollen from the male parent is dusted on the sticky stigmatic surface. This process is done on the following day of emasculation to ensure pollination. The easiest way of pollination is by brushing the stigma with an anther which has pollen adhering to it. The best period for hybridization under Bangalore conditions is October to February. Gladiolus pollen grains do not lose viability for 3 months if they are stored in a desiccator at low temperature. They remain viable for one week at room temperature. Four or five lower florets should be used for hybridization and the remaining part of the spike should be removed as soon as the pollination of the 4th or 5th bud is completed. After pollination, the florets are again tied at the top with thread and tags mentioning the parentage and date of crossing should be tied with the floret. If the pollination is successful, the style will begin to wilt within 48 hours. It is advisable to use two spikes for each combination to ensure against loss of one due to diseases or breakage in the fields. The pods generally mature after 4-5 weeks of pollination. They turn brown in colour and begin to split. If left too long in the field, the seed may be damaged by insects and diseases or may be shattered by wind. As soon as the pods become brown and start to burst, they should be picked off from the spike and put in paper bags for complete drying. The number of seeds in a pod may vary considerably, depending upon the compatibility of the cross. A maximum of 90-100 seeds can be expected per pod from a successful cross.

Sowing of hybrid seed and aftercare: If the weather is favourable, the seeds can be sown just after complete drying as gladiolus seeds do not have any dormancy. Before sowing, the seeds should be rubbed between two layers of cloth to remove the waxy covering. It has been reported that the waxy covering contains some substances that retard the germination process. Removal of "wings" or the waxy covering allows better contact of the seed with the available soil moisture and increases germination. The seeds should be sown in seedpans in 2 cm deep furrows with individual seed not closer than 2.5 cm to allow development of large cormels. It is suggested that the soil mixture should be sterilized with either steam sterilizer or 2 per cent formalin. After sowing the seeds, the furrows are covered with saw dust as it helps retention of moisture. Watering should be done with a water can having a fine rose. The seeds will start germinating after 15 to 20 days of sowing. If the seedlings are given good care, many of the cormels will be 1 cm to 1.5 cm in diameter within six months. When the foliage of seedlings turns brown, the cormels are dug, cleaned, dried, and stored. It will be better if just after 3 months of storing the tunic (scale covering) is removed from the cormels because tunic contains inhibitors and prevents sprouting. After sprouting, the cormels are planted in well-prepared beds at a distance of 10 cm each way. Generally, in the first season the hybrid cormels do not produce spike. It is interesting to note that under Bangalore climatic conditions, the hybrid cormels flower in the first season itself. When the leaves dry up, the newly formed corms are lifted and stored in the usual way and planted again as soon as sprouting starts. They are evaluated for various desirable characteristics as and when they produce spikes. Those which are found to be useless from all points of view are rejected and the selected ones are carried forward for further evaluation. After final selection, they should be sent to different places for location tests.

Mutation breeding: Mutations in gladiolus are common especially in floret colour. Generally, the mutants are lighter in colour than the originals and a few have made outstanding contribution to the gladiolus world, such as 'White Friendship'. This cultivar was developed as a bud sport from the cultivar 'Friendship'. Most colour mutations are simple chimaeras with no effect on germ cells. Attempts at inducing mutations with irradiation or chemicals have had limited success, with most of the mutations being horticultural freaks only. Recently at the Indian Institute of Horticultural Research, Bangalore, a chimaeral variation in floret colour has been induced by gamma rays in the cutivar 'Wild Rose'. After thorough evaluation this mutant has been released.

Genetic studies: In gladiolus, relatively little has been known with respect to the specific genetic details. Like other crops, the concept of dominance and recessiveness applies in the case of many gladiolus traits but it is not known initially which particular gene in gladiolus exhibits dominance or recessiveness and which traits show a blending mode of inheritance. The basic chromosome number in the genus is 30 but the number present in the typical modern gladiolus is 60. Certain Gladiolus species as well as hybrids derived from them are characterized by having a different chromosome number, i.e., 30, 45, 75, 90, or 120. Therefore, garden gladiolus may be referred to as tetraploid. Due to its polyploid nature, many features or characteristics of gladiolus exhibit a gradual variation from one extreme to another. Accordingly, it is not merely a matter of being tall or short, small-flowered or large-flowered, heavy-substanced or thin-substanced, dark-pigmented or light-pigmented, completely susceptible to disease organisms or fully resistant, etc. Rather, because of lack of dominance and/or the participation of more than a single quartet or alleles, these and many other traits may be expressed to varying degrees.

In so far as the specific hereditary situation underlying a quantitative trait in terms of number of genes involved, status of dominance, etc., may not yet have been determined; and in so far as exact knowledge concerning the degree of homo- or heterozygosity of the parents making up a given cross may well not be available, one is not able to predict with any certainty how the offspring will turn out with respect to the trait in question. There is, however, very definite advantage in being aware of such a mode of inheritance as happens to be available. For example, one might know much more readily what to do when working towards such goals that involve intensifying or toning down a trait – be it colour, size, etc. Knowledge of quantitative inheritance helps determine that how large a seedling population must be grown in order to produce the desired traits. The extreme types (AA BB CC, etc., or aa bb cc) occur with least probability in crosses involving heterozygous parents, the expected probabilities being 1/16, 1/64, and 1/128 in case of two, three, and four heterozygous pairs of alleles characterizing each parent respectively. Some breeders have reported close linkage in heavy substance of petals and sepals and heavy ruffling; heavy ruffling and short flower heads; blue-violet flower colour and susceptibility towards certain diseases; large floret size and loose floret placement; flower colour and corm pigmentation. These apparent cases of linkage may well be explainable on the basis of the phenomenon of pleiotropy, i.e., multiple expressions (one major and one or more minor) of a single gene. Thus, perhaps the action of a single gene might influence cell acidity-alkalinity which then in turn influences such diverse characteristics as flower colour and susceptibility to certain diseases. Likewise, the degree of ruffling and flower-head length both might perhaps in some way be geared to hormonal levels within the plant which in turn might be under the control of a single gene.

It appears that various flower colour patterns (coloured varieties with white throats, coloured throat blotches on a lighter coloured or white background, picotee patterns and self-coloured types) might be individual expressions of multiple genes. Parentage studies have shown that most deep violets, yellows, heavily ruffled types, miniatures, and giant-flowered types have been derived from deep violets, yellows, and the like. Thus, in breeding for whites, one might well have a better chance of producing what he desires with respect to floral form, floret number, and floret size from a given number of offspring all of which happen to be whites, than when their colour is variable. Broadly speaking, this can be accomplished by inter-crossing two recessive-type whites. Unlike recessive-type whites, cream and yellows, the inhibitor-type varieties possess extraneous water-soluble pigments of the lavender-red purple category (usually anthocyanins) in their stamens and/or inner portions of petals and sepals. Inhibitor-type individuals, even though they are basically white (or cream or yellow) possess genes underlying these other pigments. The action of these (colour) genes is, however, prevented in large part as a consequence of the simultaneous presence of one or more inhibitor (suppressor) genes. On the basis of Mendelian Law of Segregation and Independent Assortment, inhibitor-type varieties may be expected to transmit colour genes to their offspring without they themselves expressing such genes provided such offspring do not inherit an inhibitor gene along with colour gene(s). The crossing of a recessive white with a non-inhibitor type yellow may prove to be a very efficient means of attaining cream or light yellows.

Likewise, outcross between blends which are due to the simultaneous presence of both water-soluble (vacuolar) and water-insoluble yellow (plastid) pigments and recessive-type whites or yellows; or to varieties possessing intense water-soluble pigmentation may in turn be made for the purpose of diluting or intensifying one or both of the pigment types making up the blend. In another case of outcrossing, a probable way to improve the floret form of deep violets would be to outcross one of the plain-petal type deep violets with a light violet that has especially good form. Although a backcross to a light violet would be preferable, the use of intensely ruffled recessive-type white may also be considered. Subsequent use of such offspring in a series of backcrosses to both parental forms and/or by inter-crossing sister seedlings (sib-crossing), one may achieve his aim after three or four generations. Outcrossing is the principal method for the breeder for increasing the depth of the "gene pool".

Jasmine

Family: *Oleaceae*; Origin: Mostly of Indian origin and also of China, Burma, Australia, and S.E. Europe.

Jasmines are the most common flowers in Indian gardens and also important commercially as flowers of some species are valued as cut flowers and a few yield essential oil for the perfumery industry. The flower is cultivated in India from olden times as is evidenced in the book *Kamasutra* written by Vatsayana (A.D. 300-400). In this book it is written that a virtuous wife should plant in the garden, amongst other shrubs, jasmine also. In the epic *Mahabharata* also, which was compiled around 500 B.C., we find the mention of jasmine.

From another record it can be found that jasmine was introduced in Iran from China probably during the period 250 B.C. to A.D. 226. Evidence gathered from the mural paintings from Pompeii (100 B.C. to A.D. 50) which was buried under the lava of Vesuvius, indicated that Romans also cultivated jasmine. The Arabs introduced this flower in Spain probably in A.D. 756.

From the foregoing narration it will be evident that jasmine is one of the oldest cultivated flowering shrubs. It is estimated that at least 12,500 kg of cut flowers of different kinds of jasmine are sold every day in the important cities of Madras, Bangalore, Bombay, Delhi, and Calcutta. The above figures are from an earlier report and the present-day consumption must be much higher. Most of the jasmine flowers are used for preparing garlands, *venis* for hair decoration especially in south India, and also for the purpose of worship. A large quantity of some species is also consumed for extraction of essential oils.

Classification: The classification of *Jasminum* is rather confusing, apparently because of the existence of a number of synonyms. Thus, while according to Baker (1877), Cooke (1905), and Duthie (1911), the *Jasminum* consisted of about 90 species, Bailey (1958) and Rendle (1925) maintained that the genus consisted of 200 or more species. A careful study undertaken by Veluswamy *et al.* (1973) reveals that there are only about 89 species in this genus, thus approximately confirming the estimates of Baker, Cooke, and Duthie. Of the various species only a few are cultivated in the garden. Out of the species with scented flowers only three are commercially cultivated for their cut flowers and for the extraction of perfume. These are *J. auriculatum*, *J. grandiflorum*, and *J. sambac*. But some other species are also cultivated for their ornamental value. Among these the following are important: *J. arborescens*, *J. humile*, and *J. pubscens* var. *rubescens*.

J. grandiflorum: It is a native of India and is popularly called 'Royal Spanish', 'Catalonian', or 'Italian Jasmine'. It is also known as *Chameli* and *Janti* in vernacular. The leaves are shining dark green with 5-7 leaflets. This is a twining shrub with pendulous branches. Flowers are white with a purplish tinge underneath and highly scented. It is one of the best white flowering jasmines and is cultivated for extraction of perfume. The scientists at the Indian Institute of Horticultural Research, Hessaraghatta (Bangalore), have observed two types of clone in this species. These have been differentiated as 'Pin' and 'Thrum' types. In the former type, the style points out of the corolla tube, while in the latter the style remains hidden inside the flowers. The 'Pin' type is much more free-flowering and heavy yielder than the 'Thrum' type.

J. sambac, the 'Arabian jasmine', is a native of India. It is a dwarf-growing shrub with almost sessile leaves having wavy margins. It is also possible to grow it as a short climber. The highly scented white flowers are borne in clusters of 3-12. The flowers may be single, semi-double, or perfectly double. The double flowering types are known as *motia* or *mogra* in north India. In south India, the large double or the 'Tuscan' or the 'Grand Duke of Tuscany' is known as *boddu malle*, the double but small-flowered type as *gundu malle*, and the semi-double as *donthara malle*. In Bengal, the different names for these types are *bela*, *khoya*, *motia*, or *mogra* for the large double, *moturia* for semi-double, and *rai* for the largest double flowers. This species is widely cultivated in Karnataka, Andhra Pradesh, Tamil Nadu, and also to some extent in West Bengal for use as cut flowers, especially for the unopen flower buds. The species flowers mainly during the summer and rainy seasons but several other flushes during other parts of the year is quite common.

Another creeping species, *J. auriculatum*, commonly called *jooee*, has smooth foliage with three leaflets. The flowers are white, single, or double and are borne in clusters having many flowers. It is a native of India, Malaysia, and Africa. A very high-yielding cultivar of this species named *Pari Mullai* has been evolved at the Tamil Nadu Agricultural University, Coimbatore.

J. arborescens (Syn. *J. roxburghianum*) is a dwarf shrub with white flowers, blooming mainly during the winter months but flowering is also extended up to the beginning of summer. This is popularly known as *naba mallika*. It is a native of India and Sri Lanka.

J. pubescens var. *rubescense* is the pink-budded jasmine and is commonly known as *kunda* in West Bengal. The open flowers are white on the upper surface and pinkish underneath. It is a native of China and India.

J. humile, commonly known as 'Italian jasmine' or *swarna chameli*, is a shrub native to Tropical Asia but also found in southeastern Europe. The leaflets are 1-3 paired. The yellow, faintly scented flowers are borne in clusters of 2-4. It is a winter flowering species.

J. angustifolium is a native of India. It is commonly known as *mallika* and bears star-shaped, white, fragrant flowers during the winter.

Propagation: Jasmines are generally propagated by cuttings of almost mature wood and also by layers, commonly ground layers. The best time for propagation for either method is the rainy season. The cuttings are raised in the nursery beds and planted in the field after one year. Seeds are hardly used for propagation but are used for raising new hybrids.

Climate and Soil: Jasmines are generally hardy plants and with irrigation can be grown in a wide range of soils and climate. The best soil type suitable for their growth is sandy loam or clayey loam. In clayey soil though the vegetative growth is quite fibrous, flowering is poor. Jasmines cannot withstand water-logging and most of the types will be killed in a water-logged soil in no time. *G. grandiflorum*, as well as *J. sambac*, is sensitive to frost. A warm climate with a mild winter is suitable for most of the cut flower yielding *Jasminum* species.

Cultural hints: Once planted the jasmine remains in the field for 10 to 15 years. This fact makes it obligatory to plough the land deeply and prepare it thoroughly by ploughing and harrowing. The best time for planting in most parts of India is during the monsoon. But one can plant jasmine almost round the year in climates as of Bangalore, except during March to May, if irrigation is available.

J. sambac is planted 75 cm apart in rows spaced 1 m apart. Some people recommend a spacing of 1 to 1.5 m either way. The correct spacing depends upon the nature of growth which again depends upon the climatic factors, soil and whether or not irrigated. The creeping types need a wider spacing, which are generally trained on arbours. Here again, there are divergent views on the spacing to be followed. While some people recommend a spacing of 1 m from plant to plant in the row which are spaced 1.5 m apart, others recommend spacing of 2.5 to 3.5 m either way. Here also the spacing will depend upon the factors mentioned earlier. In commercial cultivation the plants are not grown on arbours.

Irrigation is stopped during the month of November to enable the plants to rest. At this time the plants shed the leaves and if this does not happen often the leaves are removed by hand. Irrigation should commence slowly and the frequency increased after the appearance of flower buds. Irrigation at a fortnight's interval is recommended during April to June depending on rainfall. But in dry climate weekly irrigation is not uncommon. It is advisable that watering should be stopped between two flushes until new flower buds appear.

The plants are pruned to half their length during the last week of January or the first week of February. In Bangalore much earlier pruning in late November or mid-December is also advocated for some types. A light pruning is also advocated during the month of July, especially for the vigorous growers. *J. sambac* needs light pruning. Chemical defoliants such as Pentachlorophenol were found to be effective substitute for normal pruning on *J. auriculatum* and the application also increased yield by 65 per cent. Chemicals like Ethrel, Boll's eye, etc. are also used as defoliants.

A deep earthing-up is done at the time of pruning and the roots are allowed to weather in sun for a few days. At this time well-rotten F.Y.M. is applied at the rate of 10-15 kg per bush. The same dose may be repeated after the first flush of flowering. In France, light doses of ammonium sulphate are added with irrigation water. Heavy doses of fertilizer are not good for jasmine as this tends to promote vegetative growth and reduces yield. But Muthuswamy & Madhava Rao (1972) have recommended high doses of fertilizer for the high-yielding cultivar *Pari Mullai.* They recommend 300 g of ammonium sulphate, 750 g of super-phosphate, and 200 g of muriate of potash per bush at the time of pruning in addition to 15 kg of F.Y.M. The dose is repeated in July. The dose has to be halved for plants between the ages of 1 and 2 years and quartered for plants of about six months. The dose is recommended for commercial growers, but it is doubtful how many growers can afford to follow this routine. However, one thing is certain that Indian growers hardly supply any nutrient to jasmines which will be evident if compared with yields (see below) of other countries. It is high time that appropriate doses of nutrients were applied to improve yield and quality.

Yield of flowers: The yield of fresh flowers varies depending upon the species. Most of the species start commercial production only from the second year and reach maximum production in about the fourth year. The production starts declining from the tenth year onwards. *J. sambac* produces, on average, about 1,000-1,600 kg of flowers per hectare. In India, *J. grandiflorum* and other climbing species give higher average yield of around 1,500-2,000 kg per hectare. The yield of *J. grandiflorum* is about 3,000 kg in France and 4,500 kg per hectare or more in Sicily, Italy, and Morocco. Muthuswamy & Madhava Rao (1972) claim that *Pari Mullai* has given a yield of about 6,700 kg per hectare per year. The yield appears to be on the very high side and may be obtained only under perfect cultural practices in an experimental trial and is unlikely to yield as much in commercial cultivation. But certainly there is great scope for improvement of the average yield by adopting improved cultural practices including the use of plant nutrients.

In India, the flowers are harvested in bud stage in the afternoon by handpicking which involves lot of labour and consequently higher costs. In France, an electrical harvester with a suction device is used which sucks the flower bud from the base of the joint at calyx.

Yield of essential oil: Jasmine yields a very low percentage (0.07) of volatile cil on water distillation. But on solvent extraction a concrete yield of between 0.28 and 0.3 per cent w/w is obtained. The concrete of jasmine is not used as such in perfumery but the perfumery ingredients are further purified by 95 per cent alcohol whereby jasmine absolute is obtained. It is estimated that one ton of fresh jasmine flowers yield 2.8-3 kg of concrete which in turn yields 1.3 to 1.5 kg of absolute (otto) depending upon season, time of collection, method of extraction, and species (Gupta & Sharma 1972). The best time of harvest for the extraction of concrete is eaily morning.

Pests and diseases: See Chapter XXVI.

Marigold

(i) *Tagetes erecta* (African Marigold);
(ii) *Tagetes patula* (French Marigold); Family: *Compositae;* Origin: African Marigold - Mexico; French Marigold - Mexico and South America.

The Marigolds are one of the easiest annual flowers to cultivate and have wide adaptability to different soil and climatic conditions. The plants, with their attractive flower colours bloom for a considerably long period and the flowers keep remarkably well when cut. All these favourable points make Marigold one of the most popular annual flowers in India, for garden display as well as for commercial cultivation.

Uses: Marigolds are ideal for cut flowers, especially for making garlands. In popularity as cut flower, Marigolds probably rank next only to Jasmines in India. Sometimes, the whole plant is cut and used for decoration. They can be planted in beds for mass display, in mixed borders and can also be grown in pots. The dwarf types, especially French Marigold, are very suitable for hanging baskets, window boxes, rockeries, and edgings. When grown in newly-planted shrubbery, they help fill the gaps and add colour. The French Marigolds are also suitable for planting in drifts and along driveways or paths.

Classification: Marigolds are broadly divided into two groups, namely, African Marigold and French Marigold. The former generally grows taller and the latter is a dwarf type.

(a) *African Marigold:* These are vigorous tall-growing (up to 90 cm) plants having large globular flowers measuring up to 15 cm or more across and in shades of lemon yellow, bright yellow, golden yellow, orange, and nearest to white. Very recently, a white flowered cultivar has also been developed in the U.S.A. There are also dwarf-growing types (20-30 cm) with similar flower heads. The doubles always throw a smaller percentage of single flowers. One of the most beautiful variations in this group is the Carnation-Flowered Marigold, with medium-sized double flowers resembling that of a carnation. Some of the cultivars in this group are 'Guinea Gold' (throws no singles), 'Apricot', 'Sun Giants', 'Primrose', 'Fiesta', and 'Golden Yellow'. The other important variation is the "Chrysanthemum-Flowered" types having double, shaggy, incurved Chrysanthemum-like flowers including some dwarf strains. Some of the cultivars in this group are 'Glitters', 'Mammoth Mum', and 'Crown of Gold' (tall), and 'Happiness', 'Spun Gold', and 'Cupid' (dwarf). There is a type called 'Honeycomb' in which the orange coloured tubular petals are arranged in a honeycomb fashion.

Many people do not like to grow the Marigolds because of the characteristic marigold odour of the foliage. Some cultivars with odourless foliage in both Carnation and Chrysanthemum-Flowered types have been developed recently. These include 'Hawaii', 'Crown of Gold', 'Honeycomb', and 'Cupid'. Besides these, the F_1 hybrids such as 'Toreador' and 'Climax' are grown widely. These exhibit vigorous growth and uniformly giant globular blooms which are borne in great abundance for garden display and cutting.

(b) *French Marigold:* These are compact in habit, 30-37.5 cm tall and flower profusely in singles or doubles. The recent extra-dwarf types (15 cm) which just look like cushions remain fully covered with blooms (e.g. 'Petite'). The flowers in this group may be self-coloured, striped, spotted, and blotched. The colour range is deep scarlet, mahogany and rusty reds, primrose, yellow, golden-yellow, orange and combinations of these colours.

The dwarf double cultivars include 'Rusty Red', 'Flame', and 'Spry'. In 'Naughty Marietta' and 'Dainty Marietta', the flowers are single but blotched maroon in yellow base. The other attractive singles are (a) 'Star of India', with maroon flowers marked and striped with yellow and (b) 'Harmony' having deep orange-centred flowers, edged crimson. There is a tetraploid cultivar also.

(c) *Single Signet:* It is another dwarf (30-37 cm) marigold, botanically known as *Tagetes tenuifolia* (Syn. *T. signata*). The single flowers may be orange, lemon, or yellow. There is a more compact and dwarf (22 cm) variety, *pumila* in this species. Some of the cultivars in this group, namely, 'Ursula', 'Golden Gem', and 'Lulu', are suitable for edging or rock gardens.

(d) *Interspecific hybrids:* There is an interesting group of hybrids between African and French Marigolds, developed in America. These are known as Red and Gold Hybrids. The plants are intermediate in character, that is, medium tall (about 60 cm) with double flowers measuring 5-7 cm across and in colour combinations of red to orange. 'Nugget', with yellow flowers, is the most recent interspecific triploid hybrid developed in the U.S.A.

Cultural hints: Marigolds are one of the most easily grown annuals. The requirements for the cultivation of Marigold are well-drained soil and a sunny position. They can be grown in various types of climate and almost any part of the year except in very cold winter when the plants are likely to get damaged by frost. French Marigolds perform well on a light soil. The African types need well-manured and moist soil. The soil should be cultivated well and a good dose of cowdung manure (8-10 kg per square metre area) should be incorporated in the soil to obtain the largest and most double flowers. The French Marigolds should not be given too much of manure as they will produce more vegetative growth and less blooms. The seeds are sown in May-July in the nursery beds or seed pans during August-October and February-March in the plains. In the hills, sowing is done during March-April. The seedlings should be pricked when 2-5 cm tall and after about a month these are transplanted in permanent beds at 30 cm apart for French and 30-45 cm apart for African Marigolds or in 25 cm pots. Seeds can also be sown directly in permanent beds and thinned out after germination. Plants can be raised from cuttings, especially during the rains. The seedlings tend to produce flower buds at a comparatively young stage when these should be pinched to obtain bushy and compact growth. The French Marigolds start to flower in about 2-2½ months and the African Marigolds in 2½-3 months after sowing of seeds.

Orchids

Orchids are flowering plants, all of which belong to a single botanical family *Orchidaceae*. The Chinese are known to have grown a few *Cymbidium* species of orchids for centuries, but it was not until 1731 that the first orchid *Bletia purpurea* (Syn. *B. verecunda*) was grown by an Englishman, Peter Collinson, who was the first among the westerners to grow an orchid.

It has not yet been estimated, with accuracy, the number of different kinds of orchids which grow in the wild in different parts of the world. According to Alex D. Hawkes (1965), approximately 24,000 valid species is a safe and reasonable estimate. In addition, about 32,000 hybrid forms produced by the efforts of man are in existence. The number of hybrids are increasing at the rate of about 1,000 per annum.

Orchids are perennial herbs. *Zeuxine strateumatica* (L) Schltr, however, appears to be an annual orchid, though further research may prove this to be wrong. Under special conditions certain other imperfectly known terrestrial orchids may prove to be annual in nature, having different growth habits. Those orchids growing on trees and shrubs are called *epiphytes* or "air-plants", and some others growing on rocks are termed *lithophytes*. The ground-dwelling ones are referred to as *terrestrial*, while some other relatively little known ones growing on decaying organic matter in the ground or rotting logs are known as *saprophytes*. The saprophytes which consist of only a few genera such as *Corallorrhiza* and *Galeola* are difficult to introduce in cultivation because of the delicate relationship between the roots of the orchids and the growing medium. A few orchids grow as semi-aquatic, under special conditions and two Australian genera, *Cryptanthemis* and *Rhizanthela*, are subterranian in habit with only the inflorescences reaching the surface and in certain extreme cases even flowers remain underground during their complete life-cycle.

Taeniophyllum khasianum is a tiny orchid which has no stem or leaves but only roots of green colour. *Galeola falconeri* is the largest Indian orchid with 40-45 cm rhizomes and the plant attains a height of 2.2 to 2.5 m, whereas the species *Eria pusilla* is the smallest orchid, the entire plant measuring only about 1 cm.

The common belief is that orchids are found only in dark damp spots in a jungle. This is not at all true. They are found in all parts of the globe from the sandy deserts of Australia and Africa, as semi-aquatic plants, in the snow-covered peaks of the Himalayas and in many other such odd places. In temperate regions, mostly terrestrial orchids are found, whereas in the tropics the epiphytes growing on tree-tops are more frequent compared to lithophytes inhabiting on rocks, which are less common. Contrary to popular belief that orchids are parasites, so far no true parasite is known to exist in the orchid kingdom. The terrestrial orchids have cylindrical, thread-like, either branching or not, and often quite long roots. Great many of the terrestrial species have numerous root-hairs. The epiphytic and lithophytic orchids have aerial roots consisting of a central axis, surrounded by a covering of loose and spongy tissue, known as *Velamen*, the cells of which are filled with air only during dry season. The velamen with its thin epidermis can absorb moisture very rapidly and stores it for an appreciable length in time. The roots in epiphytic orchids are adventive roots and produced from the stem nodes, which also help them climb up trees.

The growth habit of some orchids may reverse under special circumstances. A epiphyte or a lithophyte if falls by chance to the ground by breaking of a branch or otherwise, may be forced to grow on the ground as a terrestrial plant. Many such orchids would develop root system complete with root-hairs, like those of the true terrestrial types. Certain species of *Epidendrums*, *Cymbidiums* and others do not show any preference either as epiphytic or terrestrial and the same species can be found to thrive well under both the conditions. Many of the terrestrial types may take to epiphytic living especially in a pocket of decaying organic matter in a branch-crotch or similar other position. Some orchids, especially *Epidendrums*, *Vanillas* and some others begin their life as true terrestrials and later develop into epiphytes.

The orchids are placed into three groups, according to the style of their vegetative growth: sympodial, monopodial, and pseudomonopodial. In the sympodial group, genus such as in *Cattleya*, the growth of the axis ceases at the end of each flowering season and continues in the next season by a newly-grown axis. A single plant will thus produce a series of annual axes in course of time. The pseudomonopodials lying between the sympodial and monopo-

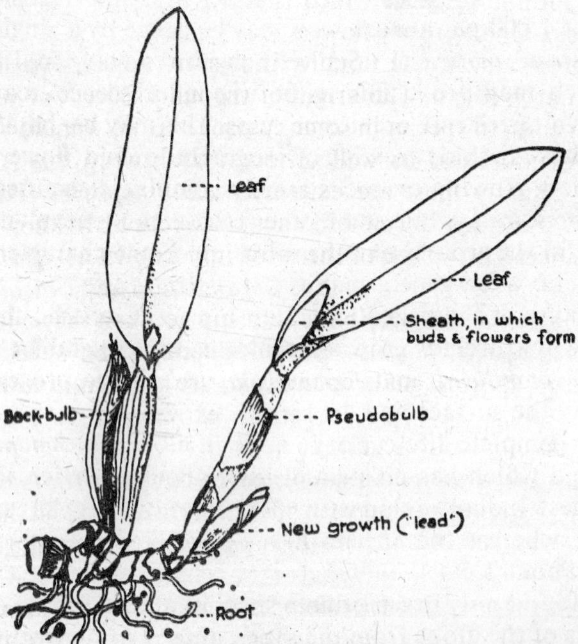

SYMPODIAL HABIT OF GROWTH
The parts of a typical *Cattleya* orchid plant.

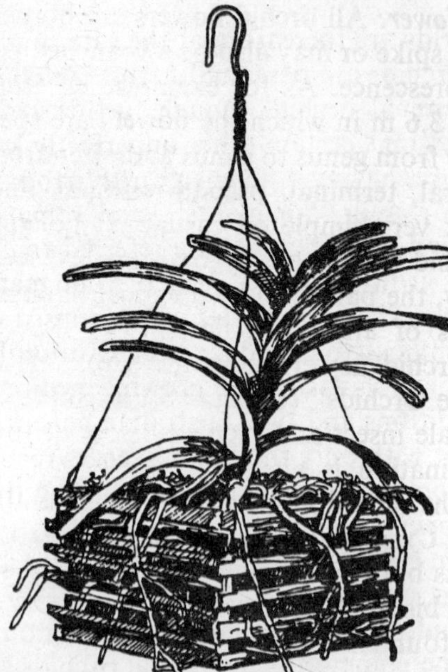

MONOPODIAL HABIT OF GROWTH
An epiphyte orchid growing in slatted baskets. Note the aerial root system.

dial, grow simultaneously in the apical direction season after season, as well as in the horizontal direction. The monopodials grow continuously in one direction only and their stems lengthen from one season to another, and produce aerial (adventitious) roots generally along their whole length.

Morphology of Orchids

A rhizome is found only in the sympodial types of orchids, with few exceptions as in pseudomonopodials, which is actually a primary stem, from which arises the secondary stems. The rhizomes vary in shape, size and form from genus to genus and from species to species within the same genus. In some orchids, the rhizome may be so shortened as to be scarcely visible. Most *Cattleyas* and *Laelias*, as well as many other orchids, have distinct rhizomes and in many cases these are conspicuously elongated as in *Bulbophyllum* and *Coelogyne*.

Pseudobulbs, which are not true bulbs, are found in a large number of epiphytic orchids with sympodial growth-habit. These are actually secondary stems which are highly specialized, to some extent thickened and consisting of one or more internodes acting as storage organs for moisture and food. The shape of pseudobulbs vary greatly and these are generally globular, pear-shaped or spindle-like. In orchids possessing pseudobulbs, the leaves are borne either all along the length of the pseudobulb or at the apices, which may be solitary or numerous. The leaves in many epiphytic orchids are very thick or fleshy and the surface is made up of a hard cuticle or "skin" to prevent loss of water. In monopodial orchids, the leaves are generally longer than broad and may be paper-thin in texture to rigid or fleshy.

Flower: All orchid flowers are borne in an inflorescence which is either a simple raceme or a spike or may also be a branched panicle. One to many flowers may be borne by a single inflorescence. As for example, in *Arachnanthe lowii*, the floral axis measures between 1.8 and 3.6 m in which the flowers are spaced at 7-10 cm. The origin of the inflorescence may vary from genus to genus and often from species to species in some cases. This may be basal, lateral, terminal, pseudo-terminal, and pseudo-basal as well. Though the orchid flowers look very simple on casual examination, actually these are extremely complex structures. Orchid flowers are zygomorphic, that is, if these are bisected in equal halves on one plane only, the parts will be identical in appearance. This is one of the most important characteristics of an orchid flower.

Orchid flowers show incredible ability to mimic insects, birds, etc. As for example, the "Bee Orchids" (*Ophrys* sp.) resemble insects as bee or wasp. One such species resembles a female insect and even emits odour to attract the real male counterpart and in the process pollination is effected.

The orchid flowers are bisexual and much more rarely unisexual, as in most *Catasetums* and *Cycnoches*. The unisexual flowers have the staminate or male and pistillate or female parts borne in separate blossoms and even sometimes in separate inflorescences altogether. But bisexual flowers are the general rule in *Orchidaceae* and here the male and the female parts are found in the same flower. The ovary is inferior, consisting of 1 or 3 cells (called locules) and is pedicellate, that is, the ovary is combined with the pedicel or flower stalk. Many of the orchids are not scented while others such as *Aerides odoratum* and *Coelogyne ochracea* are highly fragrant. In a few orchids the odour may be repulsive (e.g., *Masdevallia velifera*).

The flower consists of seven parts – three sepals, three petals, and the column or *gynostemium*. The lip or *labellum* is actually a highly modified petal and is the most attractive part of the flower. The lip may be of any conceivable form varying from tubular, lobed, sac-like to indistinguishable from the other petals. The lip, irrespective of its form, serves as a platform for visiting insects. In bud stage, the lip is actually the uppermost petal, but in almost all orchids this becomes the lowermost one by a peculiar twisting of the pedicellate ovary. The phenomenon is known as resupination. In orchids, the sepals and petals are both typically coloured, which is not the case in most other flowers. For example in Rose (genus *Rosa*) the sepals are green and petals are coloured.

The sexual parts are situated in the column. The stigma or stigmatic surface is somewhat convex in shape and sticky. The pollen in orchids is generally not separated in tiny grains (actually the grains are indistinguishable), but more or less compact and waxy masses known as *pollinia*. The pollinia are borne in a single anther, with few exceptions, at close to the apex of the column. The number of pollinia may vary from two to eight and is often a critical determining factor in the genera of this family.

Importance and Economic value of Orchids

The orchids are fancied primarily for their highly decorative flowers. The flower on the plant may last from a few days to a few months. Generally the flower will not wither unless it is fertilized by an external agent. The cut flower life of orchids is also quite good. A large number of hybrids have been developed and they are cultivated for sale of flowers. The commercial hybrids and species generally belong to the genera *Vanda*, *Dendrobium* *Paphiopedilum*, *Cattleya*, *Cymbidium*, etc. In the U.S.A. alone the production of flowers and plants for sale is a multimillion dollar business. The other countries which produce orchid plants and flowers for sale on a large scale are Singapore, Thailand, Malaysia, and the Philippines. In India, only some nurseries in Kalimpong export orchid plants on a limited scale. But, unfortunately, they do not generally multiply them, but just collect the plants from the jungles and export, with the result that a large number of highly valued species such as *Coelogyne cristata*, *Cymbidium elegans*, *Dendrobium densiflorum*, *Thunia alba*, *Vanda coerulea* etc., are almost at the verge of being extinct. Though our orchid wealth is vast, so far we have failed to exploit these commercially on a scientific basis. It is possible to establish a large industry for sale of flowers and plants. Breeding of new hybrids should also be taken up on a large scale.

Besides the sale of flowers and plants, there are some other uses of orchids. The original sources of commercial vanilla is the orchid species *Vanilla planifolia*. The vanilla is obtained from the dried and cured seed capsule. The medicine salep is also obtained from some orchids, especially in some species of the genus *Orchis*. The drug *chin shih hu* obtained from the dried stem of *Dendrobium nobile* is being used in China for a long time as an antipyretic and tonic. Besides, a large number of alkaloids including the famous "Dendrobine" have been isolated from various orchid species.

Leaves of some *Anoectochilus* are used as vegetable in Malaysia. The dried pieces of leaves of *Dendrobium salaccense* when cooked with rice imparts a delicate flavour. Dried and cured leaves and pseudobulbs are sometimes used to prepare beverage and tea and also for medicinal purposes. In the Philippines, Indonesia and elsewhere the flexible stems of certain *Dendrobiums* are used for weaving baskets. The fleshy tubers or pseudobulbs in *Geodorum nutans* in the Philippines and some members of *Cyrtopodium* in Brazil yield a glue which is widely used in the manufacture of musical instruments such as guitar.

There are many other minor economic uses of orchids.

General Cultural Hints

Most of the commonly cultivated orchids are air plants; therefore, wherever these are grown, it is of prime importance to see that they get sufficient air at all times. For this reason, most of tropical orchids belonging to the genera such as *Vanda*, *Aerides*, etc., are grown in hanging baskets rather than in pots or benches, as in this way they receive more air current. Light is another important factor for the proper growth of orchids. In general, orchids like diffused sunlight. Though in nature one may find some orchids growing in full sun, when brought into cultivation care must be taken to see that the leaves are not scorched by direct sunlight. It is always advisable to expose most of the varieties to diffused sunlight at least during a part of the day.

The orchids are categorized into three groups depending upon their heat requirements. The optimum night temperatures for these groups are, cool loving 45-50°F, intermediate 55°-65°F, and hot or warm 65°F and above. Moisture or humidity is one of the most important factors for the proper growth of orchids. There is a tendency to over-water orchids which is harmful. Most of the orchids would like moderate watering and then allowed to dry relatively between two waterings. When these are grown in moss, osmunda fibre and bark preparations, the media retains a lot of moisture and care should be taken to ascertain the humidity in the medium before another watering. Plants grown in baskets may need more frequent watering than those grown in pots as water is likely to evaporate more quickly from baskets. In cloudy weather, especially during the winter months, watering should be less frequent than during the summer months.

Rain water is best for orchids, but this is not easily available to the grower. Alternatively, tap water is quite suitable. While watering the orchids with a hose a fine spray should be used, instead of applying water with a force. To keep humidity in the greenhouse (glass-house) or lath-house, moisture retaining gravels or pumice may be spread on the floor. The benches, the gravels, and the air may be kept humid by spraying water once or twice in the day depending upon weather.

The orchids, having a rest period, should receive less amount of water, more air circulation, and somewhat cooler temperature than normal during the rest period.

Under natural conditions the orchids receive their nutrient requirements through bird droppings, rain water, and decay of organic matter. While in greenhouse or lath-house cultivation, the plants need to be fertilized artificially. Some bonemeal may be applied to the compost. Besides bonemeal, the orchids may be given other nutrients in the form of N, P, and K, especially N. A fertilizer mixture where the proportion of N, P, and K are equal (a formula like 10 : 10 : 10) should be used. About half to one teaspoonful of this mixture is added to about 4.5 L of water and applied to the plants at fortnightly intervals. The nutrient solution is either sprayed or applied to the compost. Too much feeding is harmful. Fertilizer mixture is to be given once a fortnight.

Composts, method of planting, and containers: For epiphytic orchids, a large number of growing media are in use. Osmunda fibre, the dried roots of the fern Osmunda, is a widely used medium. Sphagnum moss, tree fern fibre and blocks, are also widely used. A mixture of broken bricks and moss is used often under Indian conditions. Some sterile media such as charcoal pieces and vermiculite are also good for many epiphytic orchids. Small pieces of porous rocks also serve as a potting medium.

While potting the orchid in a pot or in a hanging basket the first thing is to ensure drainage by putting large pieces of the potting material or broken bricks near the drainage hole or below the baskets. Then the plant is placed and a layer of sphagnum moss is placed around the plant. Then the potting mixture should be firmly placed round the plant till the basket or pot is filled up. A stake may be necessary to keep the plant in position. The growers are given a special note for growing the genera such as *Stanhopea*, *Acineta*, and *Gongora* which should be grown in hanging baskets having sufficient gap at the bottom because the blossoms of these genera are borne straight downward. When growing in tree fern rafts, the plant is tied to the raft with a soft copper wire and gunny twine along with sphagnum moss. All newly planted orchids should be watered judiciously till they develop new roots and get established.

A general potting mixture for the terrestrial orchids is suggested and the grower may vary this a little depending upon climatic conditions and requirement of individual plants. The mixture consists of one part of rich humus, one part of well-decayed leaf-mould, half a part of decomposed and dried cow or sheep manure, and one part of chopped sphagnum moss and osmunda fibre (or sphagnum moss if osmunda is not available) and chopped tree-fern fibre. For all terrestrial orchids, perfect drainage should be ensured. These orchids love more sunlight compared to many epiphytic orchids.

Orchids are grown in earthen pots, baskets or tree-fern blocks. The terrestrials are generally grown in earthen pots or in beds. In both the cases sufficient drainage materials should be provided at the bottom. A special type of earthen pot is made with perforations on the sides for aeration and drainage, which is suitable for growing the epiphytes with sympodial growth habit. Various types of hanging baskets are used for growing the epiphytes. The slatted baskets made of moisture resistant wood such as teak are specially suited for epiphytes. The size of the basket will depend upon the type of the orchid. Monopodial orchids such as *Aerides odoratum* need smaller baskets as their growth is vertical. But sympodials such as Cattleya which spread horizontally will need larger baskets. Such plants once established should not be repotted normally, but the stale compost or fibre replaced once in two years. The ground orchids are grown in earthen or plastic pots.

Propagation: The orchids are grown by seeds, tissue culture and vegetative propagation. For the amateur grower, it will be difficult to propagate by seeds or tissue culture. Though vegetative propagation is a slow process, it will be easier for an amateur grower.

Cattleyas, *Dendrobiums*, and *Cymbidiums* can be propagated by division of pseudobulbs. The pseudobulbs may be divided by severing the rhizome with a sharp knife. Plants selected for division should have several pseudobulbs or stem or otherwise flowering will be lessened after division. One must remember that the bigger the pseudobulb, the quicker the plant will flower. Offshoots, called *Keiki* in Hawaii, are the plantlets produced from the upper or middle portion of the stems, or pseudobulbs from some *Dendrobium* and *Epidendrum* are also separated with 3-4 roots for propagation.

In certain genera such as *Phaius* and some species of *Phalaenopsis*, the flower stalks can be cut after flowers are faded and laid in moist sand or moss for rooting in a warm dark place. One *Keiki* is produced from each internode.

It is a very slow process to propagate the monopodial types such as *Vanda* vegetatively. So it will be best to propagate these by seed or tissue culture. Some orchids such as *Vanilla planifolia* can be grown by stem cuttings.

The orchid seeds are produced in their thousands and sometimes in millions inside each capsule. The seeds do not have any stored food substances or endosperm, as seeds of other plants contain, so the seeds are grown in cultural flasks containing sterilized nutrient-agar media (Knudson's Formula 'C' or some other formula) in an aseptic condition. When the seeds germinate and seedlings are sufficiently large in about 4-6 months' time they are put to community pots containing finely chopped osmunda fibre or tree-fern fibre along with a portion of the agar round its roots. Then these are transferred individually into bigger pots as they grow, beginning from a 7.5 cm pot. But some people suggest to remove all traces of agar from the roots by thorough washing with distilled water. For details of raising orchids from seed, consult the next section on "Breeding of Orchids".

Housing the Orchids

The orchids may be housed in a greenhouse (glasshouse) or lath-house or wardian cases. The first and last mentioned structures are meant for temperate regions, where the temperatures are low and orchids need protection from low temperature and adverse weather conditions. A lath-house is suitable for warm or tropical regions. But if a greenhouse with glass (or fibre-glass) roof is needed for breeding or other research purposes in a tropical climate, only the roof should be made of glass, while the sides are kept open. For security reasons the sides may be enclosed with thick wire mesh or expanded metal or welded mesh or insect-prof mesh. A greenhouse with glass panes on roof and all sides will get heated too much in a tropical climate and become useless, unless air-conditioned.

A greenhouse is a costly structure of cement, steel or aluminium and glass and not many amateur growers may afford to construct one. Luckily in India most part of the country is having tropical climate. It is not, therefore, necessary to construct a glasshouse except in the high hills and some other places, with fairly cool climate during the winter months. While constructing the greenhouse, adequate ventilators should be provided for proper air circulation, as this is the most important factor for the growth of orchids. The ventilators should be controllable, so that during storms they could be closed. Humidity is another imporant factor and this can be provided by having a built-in spray system inside the greenhouse. Some absorbent materials such as porous gravel, boiler ash or similar other material may be used in the pathways to keep himidity high. The greenhouse should receive adequate light. During hot months, if needed, shade may be provided by gunny cloth, or lime painting the glass roof or by growing seasonal creepers over the roof. If only a particular portion of the greenhouse needs shade, it is provided by spreading a gunny or cheese cloth on the roof only on that portion.

For tropical regions a lath- or slat-house is best for growing orchids. The size of the house will depend on the number of orchids to be grown. The lath-house should be constructed firmly to withstand the vagaries of weather. For details regarding the construction of all types of greenhouses, consult Chapter XXIII.

In advanced countries people living in metropolitan cities, where space is a problem, the wardian cases with thermostatically controlled heating system, are available for growing orchids. These are small glass-walled cases, most suitable for small collections.

Orchids can also be grown in open beds in temperate regions. Many of the epiphytic types can be grown on the trees, by attaching them with the help of copper wire, nails, coconut fibre, and sphagnum moss. Some lithophytes can be grown on rocks. These are placed on some depression of the rocks and fixed there with the help of sphagnum moss and stone chips till their roots attach to the stones. It is obvious that trees having rough barks and stones with rough and porous surface will be more suitable for growing the orchids as these surfaces hold more moisture.

Some Selected Orchids

Aerides

The word *Aerides* has been derived from a Greek word meaning air plant. The genus consists of about 60 species of aerial or lithophytic orchids. These are natives of China, Japan, Indonesia, India, and Burma. The culture of this genus is similar to that of *Vanda*.

(a) *A. multiflorum* Roxb. (*A. affine* Wall.): Native of the Himalayas, Burma, Mergui Archipelago, S.E. Asia. Inflorescence up to 30 cm long, densely-flowered; flowers white with purplish spots.

(b) *A. odoratum* (Poir.) Lour.: Old specimens up to 2 m tall and often branched. Leaves 15-25 cm long and 5 cm broad, the racemes of flowers are up to 60 cm long and densely-flowered. Flowers 2.5 to 5 cm long, very fragrant, white tipped with magenta; flowers during June to September.

Cultural hints: A. multiflorum is an easy-to-grow orchid, in hanging baskets or fern blocks and recommended for beginners. A compost made of well-burnt crocks, fern fibre, and moss is suitable for this species. The top layer should have sufficient moss which will maintain humidity.

Anoectochilus roxburghii (Wall.) Ldl.

This is a terrestrial orchid and native of the Himalayas. Plants of this genus are collectively known as "Jewel Orchids". Leaves are beautiful, up to 5 cm long and 3 cm broad, soft-textured, dark velvety-green with branching network of golden veins, median part having a distinct golden zone, margins flushed reddish. Inflorescence about 30 cm long, sparsely-flowered, flowers greenish, lip white.

Cultural hints: A difficult to grow orchid which should have a perfectly drained porous and rich compost consisting of osmunda and tree-fern fibre, chopped sphagnum, pea sized gravel (porous) and leaf-mould. It requires high temperature and humidity at all times and likes semi-shade condition. The root system of this orchid should never be disturbed.

Arundina bambusifolia Ldl. *A. graminifolia* (D. Don) Hochr.

Terrestrial orchid is grown outdoors and native of Southern China, the Himalayas, Malaysia and Indonesia; plants up to 2.5 m; leaves like grass up to 30 cm long. Flowers are fragrant, 7 cm long and pale rose-mauve in colour.

Cultural hints: Plants are grown in specially prepared beds with rich compost. It needs tropical conditions for growth. Flowering is seriously hampered sometimes, if grown in pots.

Calanthe

The genus consists of about 150 species of terrestrial or lithophytic orchids. A native of South Africa, Madagascar, China, Japan, the Himalayas, Fiji and West Indies.

(a) *C. masuca* (D. Don) Ldl. : It has deep violet flowers which fades to lilac.

(b) *C. veratrifolia* R. Br. : Evergreen orchid, sometimes 2 m tall when in bloom. Inflorescence is up to 2 m long, flowers 2.5 cm long, white in colour except some red or yellowish warts at base of lip.

(c) *C. vestita* Ldl. : The flowers come in racemes, about 7.5 cm across and whitish in colour.

Cultural hints: Grown in pots containing a compost consisting of rich loam, well-rotted manure and chopped osmunda and tree-fern fibres. Should be grown in half shade with plenty of moisture.

Cattleya

Cattleyas are widely known to the orchid lovers. They are distributed from Mexico to Bolivia, Paraguay, Argentina, and Brazil. Flowers are borne singly or in few-flowered racemes from the apex of the pseudobulbs. There are a few thousand hybrids within this genus and similar number of crosses between *Cattleya* and allied groups (e.g., *Laelia*, *Brassavola*).

Cultural hints: Cattleyas are easy to grow and thrive well indoors in pots or baskets. In warm climate, these can be grown outdoors on trees or rocks. Potting mixture consists of osmunda or chopped tree-fern fibres, with broken crocks at bottom. These are also grown in sterile media such as charcoal. When the plants outgrow they should be repotted or divided, if necessary.

Most of the *Cattleyas* are delightfully fragrant. *Cattleyas* are divided into two sections depending upon certain plant characteristics. Some of the important species belonging to each of these sections are listed below.

(A) *Labiata Group:* (a) *C. dowiana* Batem; (b) *C. eldorado* Lind; (c) *C. gigas* Lind & Andre; (d) *C. labiata* Ldl.; (e) *C. labiata* var. *gaskelliana* (Sand.) Veitch: (f) *C. lawrenceana Rchb.* f.; (g) *C. mossiae* Hook; (h) *C. trianaei* Lind. & Rchb. f.

(B) *Diphyllus Group:* (a) *C. aurantiaca* (Batem ex Ldl.) P.N. Don; (b) *C. bicolor* Ldl.; (c) *C. bowringiana* Veitch; (d) *C. citrina* Ldl.; (e) *C. guttata* Ldl.; (f) *C. skinneri* Batem.

Coelogyne

A genus of pseudobulbous epiphytic orchids mostly found in higher elevations. Native of Nepal, Burma, India, Malaysia, and China.

Cultural hints: They are best grown in pots tightly packed with osmunda or tree-fern fibre, with crocks sprinkled liberally for proper drainage. They are never exposed to full sun and like plenty of humidity.

(a) *C. cristata* Ldl. : This is a cool loving species. Pseudobulbs are almost globular, clustered and about 10 cm tall with 2 leaves. Leaves are up to 25 cm long. Inflorescence is drooping, fragrant flowers 10 cm across and pure white in colour.

(b) *C. odoratissima* Ldl.: A native of the Nilgiris in India and Sri Lanka. The flowers are sweet scented, white with yellow stains on the lips.

Cymbidium

There are about 70 species. The orchids in this genus are terrestrial, lithophytic or semi-epiphytic in nature mostly belonging to the tropics and sub-tropics of Asia. Plants have short and stout pseudobulbs.

Cultural hints: The majority of *Cymbidiums* in cultivation are terrestrial or semi-epiphytic. They take to pot culture well, preferring a certain amount of pot-binding. The pots are filled with chopped osmunda or tree-fern fibre, sphagnum moss, rich leafy humus, with sufficient broken crock at the bottom for good drainage. They like a cool spot with diffused sunlight and relatively high humidity except during flowering when watering should be restricted.

(a) *C. aloifolium* (L) Sw.: The species is almost without pseudobulbs. Inflorescence is pendulous and flowers brownish-yellow. This is a native of South and Northeast India as well as Sri Lanka.

(b) *C. eburneum* Ldl.: The Pseudobulbs are obscure. Inflorescence is about 30 cm long, erect, usually 1-2 flowered; flowers ivory-white, fragrant, 7 cm in diametre.

(c) *C. ensifolium* (L) SW, is also found in India, which bears fragrant greenish-yellow flowers having purple veins.

(d) *C. giganteum* Ldl.: The Pseudobulbs are ovoid and compressed. Leaves are linear-ligulate, acute, to 1 m long. Inflorescence is stout, as long or longer than the leaves. Flowers are spaced rather distantly, which are 7.5 to 10 cm in diameter, lasting long and faintly scented. The colour of flower is light yellow-green, striped longitudinally with red. Flowers appear in September-October.

(e) *C. grandiflorum* Griff.: This is somewhat similar in habit to the above. Leaves are 60 cm long, ligulate and acute. Inflorescence is stout, more than 75 cm long and with 7-12 flowered apical raceme. The flowers are long-lasting, green in colour but lip light yellow spotted with red-purple. The plant flowers in November.

(f) *C. lowianum* Rchb. f.: The habit is similar to *C. giganteum* Ldl. The inflorescence is stout, arching to about 1 m in length with 15-25 flowers in ihe raceme. The flowers are long lasting, fragrant or not, greenish-yellow in colour with prominent reddish veins. Flowering time is late winter or early spring.

Dendrobium

This is the largest genus in *Orchidaceae* comprising over 1600 species, distributed in the tropics of Asia, Australia, and New Zealand. They are found growing from hot sea level to snow-covered mountains. Numerous hybrids have been developed. They bear most showy and sometimes highly fragrant flowers many of which are prized as cut flowers. Generally, they possess pseudobulbs which are thickened bearing alternate and somewhat succulent leaves. The lips of the flowers are highly coloured than the other parts and generally clawed. The column possesses a characteristic foot. Dendrobiums are mostly epiphytes and few are lithophytes.

Cultural hints: Since they grow from tropical to cool climate, it is difficult to suggest one set of cultural requirements. *Dendrobiums* do not like stale conditions near root zone and hence proper drainage is essential. The cool-growing *Dendrobiums* such as *D. nobile* are deciduous, while species growing in tropics such as *D. bigibbum* and *D. anosmum* are evergreen. The former groups need cooler temperature and a rest period, while the latter has no rest period. They are best grown in well-drained pots or baskets, containing osmunda or tree-fern fibre and broken crocks. They can be grown on tree-fern blocks and the lithophytes on rocks.

(a) *D. chrysotoxum* Ldl.: The pseudobulbs are clustered and spindle-shaped but get ribbed with maturity. The inflorescence is loose, a few to many-flowered; fragrant, golden or orange yellow in colour and the disc of the lip has a curved orange band. The flowers are long lasting and bloom in the autumn-spring.

(b) *D. densiflorum* Wall.: Pseudobulbs are club-shaped, 4-angled and green in colour. Inflorescence is borne from the upper nodes of pseudobulbs, which is pendulous and many-flowered. The flowers are 5 cm across faintly-scented, bright butter-yellow, lip hairy and orange-yellow in colour. Flowers generally appear in the spring.

(c) *D. nobile* Ldl.: The species is extremely variable in floral colour and several horticultural forms exist. The fragrant flowers are white with rose tips and throat of lip is deep crimson or crimson purple. Flowering is during spring or early summer.

(d) *D. pierardii* Roxb.: Pseudobulbs are very slender, stem-like, drooping and 2 m in length. Leaves are deciduous especially during blooming. Inflorescence is 1-3 flowered, 5 cm across, fragrant, semi-transparent, pale blush white or rosy-blush and lip pale yellow. Flowering is in the spring or early summer.

Other important species are: *D. aggregatum* Roxb., *D. amoenum* Wall. ex Ldl., *D. chrysanthum* Wall., *D. farmeri* Paxt., *D. fimbriatum* Hook., *D. formosum* Roxb., *D. longicornu* Ldl., and *D. thyrsiflorum* Rchb. f.

Paphiopedilum

Over 50 species of this genus are known. It is sometimes wrongly called as *Cypripedium* in horticulture. Distributed from China, the Himalayas, Southeast Asia, and Indonesia. Thousands of hybrids are known. The flowers, often called as "Lady's Slippers" because of the like-shape of flowers, are produced singly or in few-flowered racemes on short to elongated inflorescences.

Cultural hints: They are terrestrial and best grown in pots containing mixture of about equal parts of chopped osmunda and sphagnum moss and small amounts of fibrous loam and/or coarse white sand. They like lot of humidity especially at rooting zone and grown in a semi-shaded situation, but too much of shade retards flower production.

Many of the species in this genus are famous for their flowers which are mainly used as cut flowers.

(a) *P. fairrieanum* (or *fairieanum*) (Ldl.) Pfitz.: Flowers are about 5-7 cm across, very ornamental and handsome. Dorsal sepal large compared to flower, whitish streaked and netted with violet with wavy margins. Petals are also wavy margined and of similar colour. Lip is greenish-violet with purple vein. Flowering is in October.

(b) *P. hirsutissimum* Pfitz.: A very free-flowering handsome species, the flowers being 10 cm across. Lateral petals are very ornamental with green and deep purple blotches. Flowering time is April to May.

(c) *P. insigne* (Wall.) Pfitz.: Another beautiful species which has got flowers having 10-12 cm diameter. Dorsal sepal apple-green with numerous brownish-purple spots. Lip is helmet-shaped, yellowish-green, shaded with brown. Flowering season is autumn.

(d) *P. venustum* (Wall.) Pfitz.: A species with very attractive leaves which are marbled and blotched with pale greyish-green above and undersurface is velvety, heavil mottled with dull purple. Flowers are rather small compared to other species (7 cm in diameter). lowe ing is in the winter or spring.

Some other species are *P. druryi* (Beddome) Pfitz. an *P. villosum* (Ldl.) Pfitz.

The flowers of *Paphiopedilum* are large and showy and valued as cut flowers. The flowers are borne singly in long stalks and last long.

Phaius

About 30 species of showy terrestrial orchids known and found in East Africa, Tropical Asia, Himalayas and Fiji Islands.

Cultural hints: They do best in large well-drained pots having a compost made up of rich loam, well-rotten manure and small amounts of chopped osmunda and tree-fern fibres. Though they like a brightly lighted spot, care should be taken against burning of foliage by direct sun. They also like high humidity and temperature.

P. wallichii Hk. f. (Syn. *P. tankervilliae* (Ait). Bl.): Pseudobulbs clustered closely and dull green in colour. Leaves are 3-4 and to about 1 m long. Inflorescence is borne from pseudobulb base and up to 1.25 m long. Flowers are fragrant, long-lived, 10 cm across, powdery white colour outside and dark wine-red inside. This species flowers in the spring or summer.

Stanhopea

A most interesting group of epiphytic orchids bearing pendulous inflorescence which are produced from the base of the pseudobulbs and hang downwards. These are grown in hanging baskets with sufficient gaps at the base to allow the inflorescence to come through. Flowers of most species highly fragrant, but flowers do not persist for more than 3 days.

Vanda

There are more than 70 species of *Vanda* and there are thousands of hybrids made within this genus and with allied groups (viz. *Ascocentrum*, *Renanthera*, *Aeriedes*). *Vandas* are very famous for their cut flowers. Some of the naturally occurring *Vandas* produce flowers suitable for the vase.

Cultural hints: These are comparatively easy to grow and are highly recommended for the amateur growers. They are recommended for pots and baskets and also in tree-fern blocks. One can grow them on trees in the gardens also. The potting mixture varies from chopped osmunda to tree-fern fibres and even porous rocks. They have a continuous growth period and should never be allowed to go dry. Heavy watering, plenty of light, high humidity and good amounts of fertilizer are essential for their growth.

(a) *V. coerulea* Griff. ex. Ldl.: This is the famous blue vanda. The stem is robust, growing to over 1 m, densely leafy and leaves are leathery. Inflorescence is erect or arching to about 60 cm long which is 5-15 flowered, flowers 10 cm across, pale-blue with darker markings and lip deep purple-blue. This is mostly autumn flowering.

(b) *V. roxburghii* R. Br. (Syn. *V. tessellata* (Roxb.) Hk.: Inflorescence ascending or sub-erect, about 20 cm long and 5-8 flowered, flowers 5 cm across, fragrant, white outside, light green inside tessalated with light brown. Commonly found in the plains of India growing on mango and other trees. Autumn flowering.

(c) *V. teres* (Roxb.) Ldl.: The stems and leaves of this species are terete. Inflorescence is 15-30 cm long, bearing 3-6 flowers. Flowers are 7-10 cm across, fragrant, rose-purple and flushed with white, lip yellow with red spots. Autumn flowering.

Some other important orchids are *Bletia hyacinthina* R. Br., *Coelogyne flaccida* Ldl., *Epidendrum radicans* Pav. (Syn. *E. ibaguense* HBK.), *Laelia grandis* Ldl. and *Paxt.*, *L. purpurata* Ldl. and *Paxt.*, *L. superbiens* Ldl., *Oncidium papilio* Ldl., *O. sphacelatum* Ldl., *Phalaenopsis amabilis* (*L.*) Bl. (*Phalaenopsis* consists of an important group of epiphytic orchids producing beautiful and showy flowers; there are many hybrids with allied groups and the flowers are in demand for cut flower trade.), *P. schilleriana* Rchb. f., *Rhynchostylis retusa* (L.) BL., *Saccolabium giganteum* Ldl., *Sapthoglottis aurea* Ldl., *S. plicata* Bl., *Thunia alba* (Wall.) Rchb. f., *T. marshalliana* Rchb. f. and *Vanilla planifolia* Andrews.

There are numerous inter-generic and interspecific hybrids grown for commercial cut flower production.

BREEDING OF ORCHIDS

History of Breeding

In comparison to other ornamentals, the history of breeding of orchids is very well documented and is of very recent origin. Till 1852 the orchids were new to the Europeans and people were learning to know and grow them. In 1852 J. Dominy with the help of his doctor friend Harris of Exeter made the first orchid hybrid between the two *Calanthe* species, *C. masuca* × *C. furcata*. The hybrid flowered in 1856 and was named as *C. dominyi*. Dominy devoted much of his time in the breeding of orchids and in the next twenty years produced about 25 hybrids including the first bigeneric hybrid Laelio-cattleya in 1863 (*Cattleya mossiae* × *Laelia crispa*).

After the pioneer work of Dominy, a large number of hobbyists and growers picked up the hybridization technique and tried all sorts of combinations and hybrids were made in the genera such as *Cattleya*, *Calanthe*, *Dendrobium*, *Aerides*, *Paphiopedilum* *Cymbidium*, *Lycaste-Zygopetalum*, *Miltonia*, *Laelia*, *Phaius*, including many bigeneric and trigeneric hybrids and about 1,500 hybrids were registered by 1960. Considering the difficulty in the germination of orchid seed this was a phenomenal record. By intercrossing the best of the hybrids and again selecting the best, the orchid hybridizers succeeded in gradually producing hybrid groups which were much superior in appearance to the original species.

The success of the earlier hybridizers was mostly dependent on the ease with which different orchid genera cross with each other and give fertile hybrids. It must be mentioned that most of the orchids still seem to be in an active state of evolution and the so-called orchid genera are in reality artificial constructs and in most cases these groups inter-breed very easily, to cite such an example, × *Potinara* contains four groups, *Brassavola*, *Cattleya*, *Laelia*, and *Sophronitis*.

The other major factors which have played a major role in the orchid breeding is polyploidy. In the genera such as *Cymbidium*, *Paphiopedilum*, *Phalaenopsis* and the members of the *Cattleya* tribe (*Brassavola*, *Cattleya*, *Laelia*, and *Sophronitis*), polyploidy has been of extraordinary effect. This coupled with inter-generic compatibility has culminated in the formation of hybrid groups which show both greater size and hybrid vigour compared to parental species.

Lastly, it may be mentioned that the man-made hybrids in orchids far exceed that of described species. To date there are more than 45,000 recorded hybrids with 577 established hybrid generic names such as *Aeridopsis* (*Aerides* × *Phalaenopsis*), *Aeridovanda* (*Aerides Vanda*), *Brassolaelia* (*Brassavola Laelia*), *Laeliocattleya* (*Cattleya* × *Laelia*), *Potinara* (*Brassavola* × *Cattleya* × *Laelia* × *Sophronitis*), *Vandacostylis* (*Rhyncostylis* × *Vanda*), etc.

These bigeneric and trigeneric hybrids have been obtained by fertilizing a species of one genus with the pollen of a species of the other genus. The first bigeneric hybrid was obtained by crossing *Phaius grandifolius* with *Calanthe vestita*. Hybrids have also been obtained from *Cattleya intermedia*, *C. loddigesii*, and *Laelia elegans* each crossed with *Sophronitis grandiflora* and its hybrids. This started a series of hybrids named as *Sophrocattleya*.

In addition to the man-made hybrids, a large number of natural hybrids have also been recorded in most of the geographically combined species or through introgressive hybridization. Conclusive proof for the existence of such hybrids has now been further confirmed with the help of identical forms raised from the same pair of species, e.g., the hybrids obtained from *Phalaenopsis aphrodite* and *P. rosea* was totally similar to the earlier-described *P. intermedia* of Lindley, who had also suspected it to be a natural hybrid between these two species. Similarly, artificially raised hybrids between *Odontoglossum pescatorei* and *O. triumphans* raised by Veitch proved identical to *O. excellens* of Reichenbach who also suspected this species to be of hybrid origin.

Hybridization Techniques

An orchid flower has three outer and three inner floral parts (tepals). One of the inner three parts is highly modified in various shapes and is called labellum or lip. The lip is the most striking part of an orchid flower, and is not only shaped differently but has its own markings of colour and is fantastically decorated with crests, horns, tails, and other appendages.

The diagnostic part of an orchid is the column which is a fleshy structure that consists of fused reproductive parts. At the tip of the column is the anther which bears the pollen.

Just below the anther, separated by a partition(Rostellum), is the female receptive organ - the stigma. It is a small cavity-like depression filled with highly sticky stigmatic fluid.

The orchids can be divided into two main groups on the basis of pollinia. In the *Paphiopedilum* group, the pollen is contained in two sticky pollen masses, one on each side of the sterile staminodium. In the *Cattleya* group, a solitary dehiscent anther is located on the top of the column. This anther is readily removed and reveals 2-8 pollinia.

In the nature most of the orchids are cross-pollinated by insects. Pollination of orchids under controlled conditions in green houses is a very simple procedure and involves removing ofpollinia from one orchid flower and placing the same in the stigmatic pouch of the desired species with the help of a blunt forceps. Care should be taken to pollinate the flowers at their prime and pollination should be done as soon as the flower is fully open. Early pollinations are frequently more successful than later ones. Sometimes it becomes necessary to make crosses between species which do not flower at the same time. In such cases the sticky pollinia from the male parent is removed shortly after the flower is open and put in a small vial which is stoppered with a cork or cotton plug and stored in refrigerator at 4°C. Pollen stored in such a way remains viable for several months and in some genera even for a year or two.

When crosses are made between two genera having flowers which differ greatly in size it is always better to use the smaller-flowered plant as the female parent. It is also more rewarding to use proven parents in a new cross combination than to try two new ones.

If the cross is successful, the flower will start to wither within 24 hours of pollination and within a week the ovary will start to swell. During its further development one can see the characteristic striations or markings on the developing capsule. A capsule generally ripens within 6-12 months depending upon the species.

The ideal time to collect the seed is when it starts to dry and cracks begin to show on the capsule. However, it is possible to harvest capsules much earlier than full maturity and still get viable seeds. On average, there are 500,000 to 600,000 seeds per capsule. After the seed is removed from capsule it should immediately be sown on culture media for germination or stored in the manner described above.

Germination of Seeds

Seeds of orchid are extremely small and in many cases as fine as talcum powder, and close to one million are found in one capsule. Due to their smaller size, the seeds contain very little or no stored food to nourish the embryo during germination. It must find a ready supply of food when it starts to grow under natural conditions. The decaying vegetable matter is decomposed by fungus and bacteria thus releasing minerals and sugars which help the germination of orchid seeds in the nature.

Under controlled conditions and in orchid houses the seeds are germinated in different culture media containing inorganic nutrients, agar, and sugars. The most widely used medium at present is Knudson's C medium. The composition of this medium is:

Calcium nitrate $(CaNO_3)_2$, $4H_2O$	1.00 g
Monobasic potassium phosphate (KH_2PO_4)	0.25 g
Magnesium sulphate $(MgSO_4$, $7H_2O)$	0.25 g
Ammonium sulphate $(NH_4)_2$ SO_4	0.50 g

Ferrous sulphate ($FeSO_4\ 7H_2O$)	0.025 g
Manganese sulphate ($MnSO_4$, $4H_2O$)	0.0075 g
Sucrose ($C_{12}H_{22}O_{11}$)	20.00 g
Agar	15.00 g
Water (distilled)	1 litre pH = 5.0-5.2

The growing of orchid seed with the help of culture media has revolutionized the commercial orchid growing as every viable orchid seed can become a plant. This has put orchid growing at par with other commercial greenhouse crops. A large number of growers have taken up the aseptic culture of orchid seed on a commercial scale and even sell the flasks with orchid seeds in various stages of germination. The period from the germination of the seed to the first flowering of the plant varies in different genera. In *Cattleya* and *Laelia* the shortest recorded period is 6 years while in *Dendrobiums* the period is 4-5 years and in *Calanthe* it is 3 years. It is generally agreed that the period of terrestrial orchids from the germination of the seed to the first flowering of the plant is much shorter compared to epiphytic species.

Rose

According to the classification of Rehder, about 120 species of roses are in existence. These are distributed in temperate and tropical zones of New Mexico, North America, Abyssinia, north-east Africa, the Himalayan belts of India, and in mountains of Japan and China. The species of roses found growing wild in India are *Rosa gigantea* Collet. [Syn. *Rosa odorata* (Andrews) Sweet var. *gigantea* (Collet ex Crepin) Rehd. and Wils.], *Rosa leschenaultiana* Wight and Arn., and *Rosa foetida* Herrm. (Syn. *Rosa eglanteria* Linn., not Mill.), *Rosa clinophylla* Thory (Syn. *Rosa involucrata* Roxb. ex Lindl.), *Rosa longicuspis* Betrol, *Rosa moschata* Herrm, Muskrose, *Rosa macrophylla* Lindl., *Rosa webbiana* Wall. ex Royle, and *Rosa sericea* Lindl.

So far as garden roses are concerned, only about eight species have played a major role in their development as ancestors. These are: *Rosa damascena*, *Rosa foetida*, *Rosa chinensis*, *Rosa gigantea*, *Rosa gallica*, *Rosa wichuraiana*, *Rosa moschata*, and *Rosa multiflora*.

Development of Modern Roses

The development of the modern roses began with the crosses made between the Chinese and the European roses. As a result of these crosses, several types of roses, such as the Portland, Bourbon, Noisette, and the Hybrid Perpetual were produced. The first Hybrid Tea rose, 'La France', was developed by Guillot in France in 1867, by crossing Hybrid Perpetual with Chinese Tea rose. The Hybrid Tea (H.T.) roses are renowned for their large and well-shaped blooms. The first yellow Pernet rose was developed by another French rose breeder Pernet-Drucher, by crossing the Hybrid Perpetual with *Rosa lutea* (*Rosa foetida*), commonly known as Pernet rose. The first H.T. yellow rose was named 'Soleil d' Or'.

The first polyantha rose was produced possibly as a result of a cross between *Rosa multiflora* and a variety named 'Dwarf Pink China'. The polyanthas produce numerous flowers at a time and are in bloom for a considerable length in time. But during the late thirties the breeders lost interest in this group when a new class, the Hybrid Polyanthas, now known as Floribundas, was developed.

The first Floribunda was developed in 1912 by Danish breeder Poulsen and was named 'Rodhatte'. But the Floribundas actually created sensation during 1924 only, when 'Else Poulsen' and 'Kirsten Poulsen' were released. The main reason for the popularity of this class is that they produce more number of blooms than H.T. and of better form than the Dwarf Polyanthas. In the U.S.A., another type of rose has been developed by crossing the Floribundas with H.T. and are grouped as Grandiflora. But this class is not recognized in the U.K. and some other countries. An example of Grandiflora class is the famous cultivar 'Queen Elizabeth'.

Among the other important types of roses there are the Miniatures, which are dwarf plants producing miniature flowers and are suitable for edging, potting as well as growing out-of-doors. Some of the Miniature roses are 'Pixie', 'Baby Masquerade' and 'Baby Gold Star'. Other important types are the Climbers, Ramblers, and Shrub roses. The climbers are of upright habit requiring support for their proper growth. In this group, both types of blooms multi-petalled like H.T. as well as single-petalled in clusters such as Floribunda, are found. The Ramblers, as the name implies, are of rambling habit and are suitable for covering fences, walls, and trellis. The Modern Shrub roses are the result of innumerable cross-breedings and they cannot be classified in any other group. Some of the well-known shrub roses are 'Sparrieshoop' and 'Fritz Nobis'. Besides the above-mentioned types, there are some other types of roses such as Cabbage Roses, Moss Roses, Musk Roses, and Austrian Briars.

Requirements for the Rose Garden

While planning a rose garden the following factors are to be taken into consideration:

1. Since roses grow better in full sun, the site selected should receive full sun at least for half the day, preferably in the first half. A little shade in the afternoon in hot climate is in fact helpful.
2. Roses should be grown away from the proximity of the buildings and trees, to avoid shade and competition from the roots.
3. Roses are likely to be damaged by strong winds and they should be protected from direct gusts of wind.
4. Roses are also susceptible to water-logging and, hence, proper drainage should be ensured.
5. If grown in shade, roses become susceptible to diseases such as mildew, rust, etc.
6. Roses should be grown separately from other plants and provision of space should be provided to add new promising cultivars in the future.
7. Some rose blooms are also injured by frost, and hence such cultivars should preferably be not grown in frost-prone areas or protected against frost.

Plans and Designs

A rose garden may be formal or informal. If roses are to be grown in a mixed garden, these can be grown in separate beds or as standard roses along a pathway. The spacing of the plants in the bed should be carefully planned to facilitate cultural operations such as hoeing, weeding, and pruning.

It is better to plant only one cultivar in one bed and if this is not possible, one may grow more than one cultivar in one bed but of the same colour. The Hybrid Teas, Floribunda, and Polyanthas should be planted in separate beds.

Planting

The beds or individual pits may be dug before the rains, so that the soil may settle down during the rains. When roses are planted individually, circular pits measuring 75 cm to 90 cm in diameter and 60 to 75 cm deep are prepared. The beds are also dug to the same depth as above. While digging the pits or beds, care should be taken to return the top soil at the top of the beds or pits being prepared.

Care of plants on arrival: As soon as plants are received they should be kept in shade. The roots of plants supplied bare-rooted should be soaked in water for 3-6 hours before planting. Some nurseries, especially from the eastern part of the country, supply plants with a stiff clay ball. Before planting, this ball should be broken and removed carefully.

Method of planting: At the time of planting, each pit should receive about 4-8 kg of well-rotten cowdung manure and a handful of bonemeal. A small hole is dug proportionate to the size of the root system of the plant at the centre of the site and the plant with earth ball or bare-rooted is lowered in the pit. In the case of bare-rooted plants, the roots are to be spread out properly. The point of bud union should be kept above the ground level. The soil round the stem must be made firm by treading with feet or otherwise. The planting distance varies from 60 to 75 cm from plant to plant and row to row depending upon the cultivar. Cultivars such as 'Queen Elizabeth', 'Happiness', and 'Super Giant' grow vigorously and need wider spacings. The end rows in a bed should be about 30 to 40 cm away from the border. The planting distances are 90-120 cm in Standards, 3 m in Climbers, and 45 cm in Polyanthas.

Time of planting: The best planting time is between September and October in the northern plains and most other parts of the country. The planting can be extended up to February, but early planting ensures good flowering in the cold season. In Bangalore and other similar climatic zones, early planting from July is also feasible. It is better to avoid planting during heavy rains. In temperate zones, planting is advised in open weather at any time between October and April. The plants should be labelled with plastic or metal labels after planting. The standard roses are provided support with firm and stout bamboo stakes.

Transplanting: Transplanting should be done during the cooler months only. Before transplanting, the soil around the plant is watered one or two days before lifitng. On lifting, all foliage should be removed, excess roots trimmed, and twiggy and tender growth are removed.

After-care: Immediately after planting, the plants are watered thoroughly. Newly planted roses need watering frequently just to keep the soil moist. The canes (branches) may also be sprinkled one or more times a day. Roses like one thorough soaking once in five to seven days during the summer and once in 8-10 days during the winter. Watering may not be necessary at all in a well-spread rainy season. Frequency of watering depends on the nature of the soil. Sandy soil would need light irrigation more frequently than heavy soils. The soil should be hoed lightly between two irrigations with a fork when the surface begins to dry up.

The suckers arising from the rootstocks are removed immediately. The suckers are distinguishable from the growth of the scion by their number of leaflets, colour, and size of the leaves. The suckers should be removed at the point of origin and not just at the ground level. If needed, the soil is to be dug up a little to facilitate this operation. In Standards, the sucker may also appear in the above ground portion of the stem and these should be removed promptly.

Disbudding or removal of axillary (side) buds is practised to obtain exhibition-type blooms, especially in the large-flowered types. Only one (the apical) or two buds are allowed to flower in each stem. The faded flowers and hips (fruits) are removed to obtain better quality blooms in subsequent flushes.

Pruning and manuring should be done at approriate time. Proper care and preventive steps are to be taken against pests and diseases.

Propagation

Roses can be grown by seeds, cuttings, layers, budding, and grafting.

Seeds: To produce new hybrids, through breeding, the seeds formed in the fruits, known as hips, are sown for obtaining hybrid seedlings. In temperate regions, the rootstocks needed for budding can be raised through seeds also.

Cuttings: Some of the vigorous cultivars can be grown from cuttings. Some of the Climbers, Ramblers, and Polyanthas are raised by this method. Miniatures are more widely propagated by cuttings rather than by budding. The most important use of cutting is to raise rootstocks for budding. Cuttings root better if a growth hormone like "Seradix" is used.

Layering: For the multiplication of Climbers and Ramblers, this method is quite useful. While raising layers also, hormones such as "Seradix", may be used for better results as in case of cuttings.

Budding and Grafting: Grafting or budding is an operation in which a portion of a tissue (one or more buds) of one plant is transferred to another by a proper technique. The object is to enable one to utilize the vigorous root system of the other for a proper combination using the best characters of both. The variety which gives the root system is called "Stock" or "Understock" or "Rootstock" while the cultivar grafted upon stock is known as "Scion".

Several stocks are suitable for grafting. In northern India, Edouard rose (*Rosa borboniana*) is most popular, whereas in the south, Bihar, West Bengal, and northern hills *Rosa multiflora* is used for budding or grafting. Recently, a new stock, *Rosa indica*, is being used for budding in northern India which is resistant to powdery mildew. Very recently, at the Indian Institute of Horticultural Research, Bangalore, the second author has located a thornless mutant which has been found to be very suitable as rootstock.

There are three essential steps in budding; (a) preparing the stock, (b) selection of the right bud, and (c) insertion of the bud into the stock. A strong stem of the rootstock of about the thickness of an ordinary pencil is selected for budding and all other branches are removed. All the thorns are removed by pulling a gunny cloth along the length of the branch. The bud is inserted at about 5-7 cm above the ground level. The standards are budded at 1 m, while the weeping and half standards are budded at 2 m and 45 cm respectively. For the latter three categories, Edouard rose is most suitable as rootstock.

The next step is to select the "eyes" or "buds" (bud-wood). These "buds" should not be confused with flower buds; these are small vegetative swellings at the axils of leaves from which new shoots can develop. The best bud for budding is the one which is quite plump but has not burst or elongated. The first few buds in a branch, below the faded flower, are quite old and elongated and the lower ones are too young for budding. Only the few buds in the middle are suitable for budding. The thorns are removed from the bud sticks by gentle pressure of the hand and the leaves are cut leaving about 2 cm of the stalk in each bud, which serves as a handle at a later stage of the budding operation. The bud with a little wood is taken out from the stem in the shape of a shield, measuring 2.5 cm in length, with the help of a sharp budding knife. The portion of the wood remaining attached with the bark is removed gently. A T-shaped cut about 2.5 cm long is made in the stock and the bud inserted in the incision, after trimming it to size, if necessary, with the help of the flat end of the budding knife. The bud should point upwards. The small stalk left on the bud earlier will be of help during insertion of the bud. After the operation is over, no part of the "Shield" should protrude outside, and if any tissue is doing so, it should be trimmed neatly. The bud is then tied around firmly with a 1 cm wide and 45 cm long alkanthene tape, keeping the eye uncovered. The union will take place within 3-4 weeks, when the stem of the rootstocks is cut a little above the point of bud union and the tape is removed. The budded plant will be ready for transplanting in about six months. Sometimes, budding is also done on stocks grown in 6-inch nursery pots, in which case the transplanting operation becomes easy.

1. A portion of the scion along with a 'bud' and leaf; 2. The same 'budwood' from which thorn is removed and leaf cut off, leaving a portion of the petiole; 3. The shield-shaped bud removed along with a chip of wood; 4. The chip of wood removed from 'budding eye'; the leaf petiole acts as a handle; 5. The T-shaped incision made on the stock; 6. The incision is being opened up to facilitate the insertion of the budding eye; 7. The budding eye inserted in the T cut; 8. Securing the 'eye' by wrapping polythene tape; 9. The bud securely tied. The polythene should not be wrapped over the 'eye'.

T-BUDDING IN ROSE

In some parts of West Bengal and Bihar, grafting by the inarching method is followed using *Rosa multiflora* as rootstock. In this process, a lot of budwood is wasted, because whereas budding is done with one "eye" from the stem, a branch required for inarching will have a number of "eyes" upon it.

The best time of budding in north India is during February to March, although budding can be done during October also. In Bihar and West Bengal the grafting by inarching is done during September-October and budding during October-November and February to March. In places having climate like that of Bangalore it is possible to bud any time during the year, but the best time is between February and April. In the hills, budding may be done during the spring.

Pruning

Time of pruning: In most parts of India, pruning is done during October. In Europe and America, the pruning is done after the dormant winter season and the same may be followed for some of our hill stations. In Bangalore, pruning is done twice a year just before the two flower shows, which are held during Republic Day (January) and Independence Day (August). The Hybrid Teas are pruned 42 days before the flower show, compared to 45 days in Floribundas. Three extra days are given for their clusturing habit of flowering.

In a climate as that of Delhi, pruning should start by October 8 and be completed before October 20. The same time will be suitable for most of the other areas with little variation according to climate and local needs.

The general principle of pruning is that all dead, diseased, or insect-infested wood and the weaker branches are to be removed. A sharp cut at an angle of 45 degrees, about 5 mm above a healthy bud, is given.

Pruning of Hybrid Tea Roses: Pruning of Hybrid Teas should start around the end of first week of October in and around Delhi. The cultivars, whose flowers are susceptible to cold, should be pruned earliest followed by those whose flowers show better in the cold temperature. Personal experience will help in identifying the cultivars belonging to these categories but generally it is observed that 'McGredy's Ivory' and 'Kardinal' belong to the first category and 'McGredy's Sunset' and 'Ena Harkness' to the latter category.

Only the strong and healthy canes are retained and all other twiggy, and diseased shoots are removed. The new "maiden" plants are not pruned but only tipped before planting. In the first season after planting, pruning should be moderately hard to encourage strong and healthy growth from nearer the bud union. While consulting American books on roses one should remember that nurseries there mainly deliver 2-year-old rose bushes and hence pruning recommended there is for 2-year-old plants. The canes in older plants are pruned by shortening the last season's growth by half their length. To obtain larger exhibition blooms, a harder pruning by retaining only 3-4 buds on each stem is followed. Many cultivars do not stand harder pruning and experience will help to recognize such plants. For example, The 'Doctor' and 'Crimson Glory' cannot stand hard pruning. There is no hard and fast rule regarding the amount of pruning to be given and experience will show the variations to be adopted for different cultivars. Hard pruning is not advocated for hot climate. In general, yellow and bicolour roses require moderate pruning. It is to be recommended that roses are damaged more due to too severe or wrong pruning than due to want of pruning. It is advisable to prune above the buds which are pointing outwards, so that shoots emerging from these do not crowd at the centre or cross each other.

Pruning in Floribundas: In the first season, pruning will be similar to that of Hybrid Teas to encourage more growth from the base. In the consecutive seasons, pruning should be light and only weak and diseased branches and tips of branches criss-crossing one another and forming lanky growth should be removed. In the established plants, sometimes it may be found necessary to remove the old canes to encourage new growth.

Pruning in Climbers and Ramblers: The Ramblers produce annually from the base slender stems, while Climbers are less prolific. In Ramblers, the older canes do not produce flowers and should be removed to encourage new flowering shoots. The climbers are actually a heterogeneous group and combines several types of plants. In Hybrid Tea climbers the best flowers are borne by the short branches emerging from the old canes. It is advisable to remove the tops of such old canes above these branches and sometimes even some of the canes may be removed to accommodate new growth.

Pruning of Standard Roses: The pruning of Standard roses is somewhat similar to that of bush roses. The aim should be to have a well-shaped plant with the crown growing symmetrically in all directions. The Standards grow less vigorously as the sap travels from the roots to the crown through the 1 m stem of the rootstock and hence the pruning should be generally less vigorous.

The pruning of Polyanthas and Miniatures should involve only tipping of branches that have just flowered and removal of dead and spindly branches.

Thinning, Removal of Suckers, and Disbudding

The removal of damaged and surplus new growth in the following weeks after pruning should be followed as a routine. Any sucker appearing from the stock should be removed promptly.

It has already been pointed out that in Hybrid Tea Roses only one or at best two buds should be allowed to flower upon each shoot so as to have large-sized blooms. All other buds should be removed or disbudded.

Wintering of Roses: This is an operation before pruning, the aim of which is to withhold or curtail the supply of water to the bush for a period of 3-7 days depending on weather and age of the plant. By doing so, the sap flow is reversed to the roots from the weak shoots which are to be pruned. The soil round the roots is removed to a depth of about 15 cm to expose them to sunshine up to three days. Consequent to this the leaves turn yellow and drop and some of the weak branches also dry up. Proper care should be taken to prevent the plant from drying up due to any excessive exposure. Some experts, however, do not agree that wintering is an essential operation.

Nutrition

Roses are gross feeders. At the time of pruning, well-rotten cowdung manure is applied. The doses would vary with the soil type and its fertility status. Generally, 4-8 kg cowdung manure per bush will be quite adequate. A handful of bonemeal should also be added along with the manure. A little BHC or chlordane or aldrin should be mixed with the organic manure to prevent the attack of white ants.

Inorganic manures or fertilizers are more quick-acting and small quantities of these are very effective. Nitrogenous fertilizers are more important. Phosphates also help in the production of more and better quality blooms. Indian soils are quite rich in potash, but potash fertilizers in combination with others give good results in some regions.

Complete rose fertilizers such as "Rose-Mix" are now available in the market. For those who want to prepare their own mixture, the following formula recommended by the Royal National Rose Society of England may be followed:

Nitrate of Potash	6 parts by weight
Sulphate of Ammonia	3 parts by weight
Superphosphate (Single)	16 parts by weight
Sulphate of Potash	8 parts by weight
Sulphate of Magnesium	2 parts by weight
Sulphate of Iron	½ part by weight

B.P. Pal, the well-known Indian rose expert, has recommended the following mixture for different regions of the country.

Groundnut cake	5 kg
Bonemeal	5 kg
Ammophos (11 : 48)	2 kg
Superphosphate (single)	2 kg
Potassium Sulphate	1 kg

For areas such as Bangalore, the mixture suggested by the Royal Society is good. A quantity of 100-g per bush of the mixture are incorporated into the soil after pruning.

Liquid manure prepared by fermenting gingelly (sesamum, *neem* or mustard) oilcake is also given when flower buds have formed and till their opening. The liquid should be diluted to a pale straw colour before applying at the rate of 2-5 per square metre area. Liquid manure can also be prepared by dissolving quick-acting chemicals such as Nitrate of Potash and Phosphate of Potash at the rate of 30 g each in 15 L of water. About five litres of this liquid is added to each square metre at 5-7 days' intervals during the period when buds begin to form until they bloom.

Foliar Feeding: A foliar spray is prepared by dissolving 14 g of the following mixture in 10 L of water and sprayed at fortnightly intervals.

Urea	2 parts
Dihydrogen Ammonium phosphate	1 part
Potassium Nitrate	1 part
Potassium Phosphate (diabasic)	1 part

About 14 g of liquid soap may be added to the mixture as a spreader. The concentration should not be exceeded as otherwise scorching of leaves may take place.

To supply trace elements, the following mixture may be used as foliar spray:

Magnesium Sulphate	7 g
Ferrous Sulphate	7 g
Borax	3.5 g

Seven grams of the above mixture may be dissolved in five litres of water and sprayed along with liquid soap once a month or three months depending upon the circumstances.

This spray will help obtain bright-coloured flowers. The above two mixtures should not be mixed together but sprayed separately. It is often useful to spray insecticides or fungicides along with any of these sprays. All foliar sprays should be done before 8 a.m. on a sunny day and spraying should be avoided when the flowers have opened. Some readymade mixtures are also available for foliar spray.

Irrigation

Roses do not like water-logging but need plenty of water for their optimum growth and development. The frequency of watering will depend upon the weather and the nature of soil. Sandy soils need more frequent watering than clayey soils. In loamy and alluvial soils such as that of Delhi, thorough watering once in seven days in the summer and once in 10-15 days in the winter is considered sufficient. In rainy season, necessary adjustment will have to be made. In between each watering, when the top soil becomes dry, it is useful to stir it up with a *khurpi* or a hand-fork. Frequent light irrigation is harmful; hence, heavy drenching should be given at each irrigation. In eastern India, where rainfall is heavy, no irrigation may be required at all during the rains. In a climate like that of Bangalore, having red soil, one heavy irrigation at five days' interval is recommended throughout the year, except during the rains when this has to be adjusted depending on the rainfall.

Growing of Roses in Pots

There is no denying the fact that roses grow best in ground. But they can certainly be grown in pots also especially for those who live in sunny flats. Many of the Miniature Roses also do better in pots than in ground. The size of the pots will be larger for H.T.s and Floribundas (30 to 35 cm) and smaller for Polyanthas and Miniatures. The large flowering types can also be grown in tubs. The potting mixture consists of 3 parts of garden soil, 1 part of well-rotten cowdung manure, 1 part of leaf-mould, and 1 part of wood-ash. It is advisable to remove 7.5 to 10 cm of the top layer of the compost every year and replace this with a fresh mixture.

Potted roses may be watered on alternate days in the summer or even every day and discretion should be used to water in other seasons. If the leaves of potted roses turn yellow, the drainage hole at the bottom should be checked to ensure drainage.

Accoring to B.P. Pal, the following rose cultivars grow well in pots.

Hybrid Teas	*Floribundas*
'Flaming Sunset'	'Allgold'
'Fragrant Cloud'	'Delhi Princes'
'McGredy's Sunset'	'Flamenco'
'Michele Meilland'	'Rumba'
'Montezuma'	'Zambra'

In addition, most of the Polyanthas and Miniature roses also do well in pots.

Hybridization

The modern rose has a complex ancestry, incorporating into them enormous heterozygosity. This coupled with polyploidy is the basis for the large variability which results in the production of thousands of kinds of seedlings from which a breeder could select the desired types. New cultivars are produced by spontaneous mutation, induced mutation and hybridization. Breeders all over the world have produced thousands of rose cultivars. Due to the polyploidy and complex heterozygosity breeders have not yet exhausted the genetic potential in roses. It is, therefore, possible to get cultivars superior than the existing ones through controlled breeding. It is also possible to produce the blue rose which does not exist in the present colour spectrum of roses.

Before taking up hybridization work in roses, the objective should be clearly outlined and upon this will depend the selection of parents. The cultivars which produce hips are selected as female parents and pollen fertile cultivars are used as male parents. If the breeder is not having information on these two aspects it will not be possible to obtain hybrid seeds from all the parental combinations.

After the selection of the parents, male and female parents are prepared for crossing. Preparation of the female parent consists of emasculation. For this a flower bud which has opened three-fourths of the petals is selected. The calyx is removed with sharp scissors. This helps in identifying the crossed hips even if the tag is lost after making the cross. Then the petals are plucked out exposing the stamens. The emasculation is done by cutting away the anther filaments, which carry the anther, with scissors. While doing this the anthers should not be cut. In case anther is cut, the scissors should be sterilized in rectified spirit before emasculating the next flower. After emasculation, the stigma is exposed and it is protected by covering it with a small butter paper bag which is securely tied to the flower stalk. Next day, the butter paper bag is removed and the stigmatic surface examined. If it is moist and sticky then it is ready for pollination. For pollination parents which produce fertile pollen are selected. Flowers are removed in the morning before the anthers dehise. Their petals are removed and the hip with undehised anthers are kept in a clean petri dish and kept in the sun. In doing so, the anthers dehise. The anthers are rubbed on the stigma of the female flower to effect pollination. One male flower could be used for pollinating three to four female flowers. After pollination, the female flower is covered with the butter paper bag and it is tied securely to the stalk of the flower. The names of the female and male flowers are written on a label indicating also the date on which the cross is made and tied to the flower stalk. After pollination, the hip is allowed to develop. It takes about three to four months for the hip to mature. When it turns pink it is ready for harvest. It has been observed by scientists of I.I.H.R., Bangalore, that in some cultivars the hips can be harvested much before full maturity, and seeds from such hips germinate better.

Stratification: In order to get good germination the rose seeds are subjected to cold treatment. This process is called stratification. The seed coat in rose is very hard and this has to be softened by proper treatment before stratification. The procedure followed at the Indian Institute of Horticultural Research, Bangalore, for scarification and stratification (Alexander, 1977) is given below:

When the hips turn pink they are harvested and allowed to dry for 2-3 days. Then hips are cut open and the seeds removed. The seeds are then immersed in concentrated H_2SO_4 for 30 minutes, after which they are washed in running water for 2-3 hours. After scarification, the seeds are mixed with moist fine sand and kept in a small plastic bag whose mouth is sealed properly after keeping the proper identification labels. These packets are kept in a refrigerator for three months and the temperature maintained at 3°C during this period.

Seed Sowing: The seeds are sown 5 cm apart in seed pans. One part garden soil, one part sand, one part decomposed farm yard manure, and one part leaf-mould is the suitable compost for sowing the seeds. After sowing, the seedling pan should be kept moist. Most of the seeds germinate in about one month's time. When the seedlings are a few centimetres high, they are transplanted into larger pots. The seedling pan from where the seedlings are transplanted should be kept moist for six to seven months more since some more seeds might germinate later.

Handling of Seedlings: Rose seedlings generally flower when they are 7 to 10 cm tall. The seedlings normally attain this height in three to four months. The flower buds should be pinched off as soon as they appear until the plants are reasonably well established and sturdy. Only then they should be allowed to flower. At this stage, primary selection is made. The promising ones are selected for budding on *Rosa multiflora* or other rootstocks. Second selection is made when the budded plants are two to two-and-a-half years old. Budded plants which show promise as good cultivars and which have performed well consecutively for three to four seasons are selected and multiplied. These are shown to different people to get their opinion. The final selection is made from these and released as new cultivar.

Rose Breeders in India: Rose breeding is rather of recent origin in India. The pioneer among the breeders is B.S. Bhattacharji. Among the rose breeders, B.P. Pal has bred a large number of beautiful new cultivars. The other rose breeders who have contributed considerably are M/s. K.S.G. and Sons, Bangalore; Vishnu Swarup from I.A.R.I., New Delhi; Surendra Singh of Nalagarh; and M/s. Laveena Roses.

Indian Rose Cultivars

Pal (1972) has described in his book *Rose in India* 242 rose cultivars released in India. All of these have not been registered with the National Registration Centre of India, but most of them have been listed in the catalogues of rose nurseries in India.

Recently, some rose cultivars bred in India were recognized in the U.S.A. and the U.K. 'Mohini' is patented in the U.S.A. by M/s. Jackson and Perkins (Swarup *et al.*, 1973b, 1974). 'Banjaran' evolved by B.P. Pal received very good reports from the U.S.A. and it was awarded the first and second prizes in the rose shows held there. Swarup *et al.* (1972) has given a list of some new Indian roses registered.

The Division of Vegetable Crops and Floriculture, I.A.R.I., New Delhi, has released 33 cultivars as a result of intensive research for more than one decade.

Some of the promising Indian rose cultivars are briefly described below.

Cultivars Bred by B.P. Pal

H.T. Roses

'ARUNA' (1968)

Seedling of 'Independence', bud small, pointed; flower medium (7.5 cm), double (35 petals) cupped, slightly fragrant, brilliant orange scarlet; foliage normal green, glossy, bushy, free flowering.

'BELLE OF PUNJAB' (1965)

'Montezuma' 'Flamenco'; bud ovoid; flower double, high centred, slightly fragrant, pink; foliage dark glossy, very vigorous, compact, abundant bloom, pink blend.

'DARK-BOY' (1965)
Seedling of 'Nigrette'; bud ovoid; flower medium (7 cm), double (25 petals), full, slightly fragrant, velvety dark, maroon red; foliage soft green, moderate bloomer.

'DELHI APRICOT' (1964)
Bud pointed; flower medium (6.5 cm), double (60 petals), slightly fragrant, apricot-yellow; foliage light green, soft, bushy.

'DELHI SUNSHINE' (1963)
Seedling of 'Mme Charles Sauvage', bud pointed; flower medium (5.5 cm), double (40 petals) full, deep cream, flushed pink; foliage abundant, light green, glossy. Bushy; profuse blooming.

'DR. HOMI BHABHA' (1968)
Seedling of 'Virgo'. Bud large, long pointed; flower very large (11 cm), double (90 petals) high centred, slightly fragrant, white but sometimes with faintest shading of cream or amber in the depths; foliage abundant, leathery, plants vigorous and upright habit with abundant blooming.

'GOLCONDA' (1968)
Seedling of 'Mme Charles Sauvage'. Bud small, ovoid; flower large (9.5 cm), double (55 petals), cupped, pale yellow with deep apricot in the centre. Strong tea type fragrance; foliage abundant, leathery. Bushy, compact, moderate blooming.

'KANAKANGI' (1968)
Seedling of 'Mme Charles Sauvage'. Bud small, globular; flower medium (7.5 cm), semi-double (33 petals), open, strongly fragrant, gold and apricot coloured; foliage leathery; bushy, open, blooms abundantly.

'NAẒNEEN' (1969)
Seedling of 'Queen Elizabeth'. Bud large, ovoid; flower large, double (30 petals), high centred, moderately fragrant, pale soft pink; foliage light green, glossy. Vigorous, free flowering.

'PALE HANDS' (1965)
Seedling of 'McGredy's Ivory'. Bud pointed; flower very large, double, fragrant, ivory white to buff and peach; foliage leathery; vigorous, upright, free blooming.

'SCENTED BOWL' (1965)
Seedling of 'General McArthur'. Bud ovoid; flower medium (6 cm), double (20 petals), strongly fragrant, glowing red; free blooming.

'WHITE NUN' (1968)
Seedling of 'Virgo'. Bud long, pointed; flower medium (10 cm), double (30 petals),

open, white; foliage glossy; vigorous, upright; blooms very freely.

Floribunda Roses

'APSARA' (1966)
Bud ovoid; flower medium (6 cm), semi-double (22 petals), open, salmon pink; foliage normal green, glossy; vigorous, upright; blooms freely.

'AZEEZ' (1965)
Bud pointed; flower medium size, double, high centred, coral pink, reverse lighter; foliage leathery; upright, compact, bears moderately.

'BANJARAN' (1969)
Bud small, pointed; flower small, double (25 petals), cupped, slightly fragrant, gold with orange red blend; foliage medium, leathery. Vigorous, upright, compact; blooms abundantly.

'CELESTIAL STAR' (1965)
Bud ovoid; flower double, orange vermillion; foliage glossy; vigorous, compact; blooms freely.

'CHAMBA PRINCESS' (1967)
Bud pointed; flower medium (6 cm), semi-double (8 petals), open, slightly fragrant, clear pink; foliage normal green, leathery and plants bushy. One of the best Indian Floribundas.

'DELHI BRIGHTNESS' (1963)
Bud pointed; flower medium (7 cm), semi-double (14 petals), open, orange pink; foliage glossy.

'DELHI DAINTINESS' (1963)
Bud pointed, flower medium (6 cm), semi-double (11 petals), open, light-pink with deep pink at back of petals; foliage glossy. A vigorous, upright, compact, and free blooming cultivar.

'DELHI MAID' (1963)
Bud pointed; flower medium (6 cm), single (5 petals), open, flame orange colour with gold towards the base, very free flowering.

'DELHI PINK POWDER PUFF' (1965)
Bud ovoid; flower medium (5 cm), double (56 petals), open, soft pink; free blooming.

'DELHI PRINCE' (1963)
Bud pointed; flower semi-double (20 petals), open, slight tea type fragrance; glowing deep pink; vigorous, bushy; free blooming.

'DELHI PRINCES' (1963)

Bud ovoid, cerise-red but deep-pink on opening, slightly fragrant. A vigorous plant having compact habit and glossy bronzy foliage. Profuse blooming.

'DELHI ROSETTE' (1965)

Bud small, ovoid; flower medium (4 cm), semi-double (28 petals), brilliant orange scarlet; free blooming.

'DELHI SHERBET' (1963)

Seedling of 'Gruss an Teplitz'. Bud ovoid; flower medium (5.5 cm), double (25 petals), full, strong fragrance, deep rose pink; blooms freely.

'FUGITIVE' (1965)

Seedling of 'Mrs. Oakley Fisher'. Bud pointed; flower semi-double, open, apricot-yellow, becoming light, blooms freely in clusters.

'LOREE' (1969)

Seedling of 'Frolic'. Bud small, long pointed, flower small, double (25 petals), cupped, slightly fragrant, white with faint pink flush having very good keeping quality; blooms profusely.

'NUTKHUT' (1969)

'Rumba' × 'Cocorico'. Bud small, pointed; flower small, very double (95 petals), pompon-like, slightly fragrant, coral red with very good keeping quality; blooms freely in large clusters.

'ORANGE CUP' (1965)

Seedling of 'Cocorico'. Bud pointed; flower medium, single, open, very fragrant, orange-scarlet, clustering, profuse blooming.

'PANCHU' (1966)

Flower medium (6 cm), semi-double (13 petals), ruby red, blooms freely.

'ROSE SHERBET' (1962)

Seedling of 'Gruss an Teplitz'. Flower medium (5.5 cm), double (21 petals), fragrant, deep rose pink; blooms freely.

'STANZA' (1969)

Flower medium (6 cm), semi-double (16 petals), slightly fragrant (tea scent), light pink; very vigorous, bushy and compact; blooms freely.

'TEMPLE FLAME' (1965)

Seedling of 'Orangeade'. Flower medium (5 cm), semi-double (24 petals), brilliant or orange flushed with red, thorns many.

Miscellaneous

'DELHI PINK PEARL' (1962). Climbing Polyantha or Rambler

Sport of 'Echo'. Flower medium (5 cm), semi-double (31 petals), open, pearly pink; blooming profuse.

'DELHI STARLET' (1963), Miniature

Seedling of 'Goudvlinder'. Bud pointed; flower small, semi-double, slightly fragrant (Musk), light yellow.

'DELHI WHITE PEARL' (1963), Hybrid Musk

Seedling of 'Prosperity'. Bud ovoid; flower medium size, double, slightly fragrant, pearl-white, clustering.

Roses Released by the Division of Horticulture and the Division of Vegetable Crops and Floriculture, I.A.R.I., New Delhi

Hybrid Teas

'BHIM' (1970)

Scarlet red, huge well-formed blooms are held majestically on strong stems. Flowers have long lasting quality and stand well to severe cold winters. Fragrant.

'CHARUGANDHA' (1972)

Crimson red with velvety sheen on petals, very large well-formed blooms are borne singly on strong shoots in the winter. Highly fragrant.

'CHITRALEKHA' (1972)

Ruby red colour of flowers giving an effect of smoky colour in cold winter. Large size flowers are borne freely on long and strong shoots . Vigorous.

'CHITWAN' (1972)

Beautiful, greenish-yellow buds open to large and shapely primrose-yellow flowers. Good for bedding.

'GANGA' (1970)

Golden yellow, pointed buds look very charming and develop into very shapely flowers. Flowering is almost continuous throughout the season. Ideal for bedding.

'GULZAR' (1971)

Deep magneta red, exquisitely formed blooms are of exhibition type. Fragrant.

'HANS' (1970)

White, shapely blooms are borne singly on long shoots. The flowers are not affected by cold injury.

'MRIDULA' (1975)

Very long, pinkish white and shapely buds open to soft white, fragrant blooms of large size.

'MRINALINI' (1972)

Phlox pink, long pointed buds open to large flowers of shining white petals recalling the silky texture of a poppy flower.

'PUSA SONIA' (1968)

Bright golden yellow flowers of very attractive shape. During hot months the colour of flowers becomes light yellow.

'RAJ KUMARI' (1975)

Long slender buds open to large and well-formed deep pink flowers.

'RAKTAGANDHA' (1975)

Shining red, large pointed buds open to large high centred flowers. The flowers are tolerant to winter injury. Slightly fragrant.

'SUGANDHINI' (1969)

Light rhodamine pink, exquisitely-formed, well-filled blooms, profuse flowering, fragrant; good as standard.

'SURABHI' (1975)

Phlox pink, long tapering buds, open slowly. Flowers are of lasting quality.

'SUJATA' (1971)

Deep crimson red, well-formed high centred large blooms are borne freely on medium shoots. Fragrant.

'SUREKHA' (1969)

Buds well-formed blooms, medium-sized, coral-red; a good vase rose.

Floribundas

'ARUNIMA' (1975)

Deep pink long-lasting flowers are produced in small bunches. Suitable for flower arrangement.

'DEEPIKA' (1975)

Very showy flowers of manderine red with lighter reverse and yellow base.

'DEEPSHIKHA' (1975)

Dutch vermilion double and open flowers are borne in small and big clusters. Flowering is good in the summer and rainy season also.

'HIMANGINI' (1968)
Ivory white, medium-sized, long lasting flowers are produced in huge trusses covering the whole plant.

'KAVITA' (1972)
The open flowers come in attractive shades of orient pink with light yellow base.

'MOHINI' (1970)
An unusual chocolate brown colour with yellow tinge at base of petals at low temperature and dazzling orange at high temperature. Floriferous.

'NAVNEET' (1971)
Creamy white, beautiful, semi-double flowers are produced in small and big bunches on long and strong shoots. Good for bedding.

'NEELAMBARI' (1975)
Beautifully formed pointed buds develop to a full bloom of classic rose form and magenta rose colour.

'PREMA' (1970)
Soft pink with deep pink edges, very shapely and long-lasting flowers are borne profusely in big bunches. Resistant to most diseases and pests.

'RUPALI' (1971)
Deep rose madder; large flowers come in bunches of amazing size. Slightly fragrant.

'SADABAHAR' (1969)
Deep pink buds. Flowers medium-sized. The flowers are produced in clusters in such a great number that the plant remains covered with mass of flowers almost throughout the season. Ideal for bedding and border. Makes beautiful half-standards. Very vigorous. Resistant to most of the diseases.

'SHABNAM' (1975)
White compact small flowers are produced in enormous clusters.

'SHRINGAR' (1972)
The flowers are deep camellia rose with lighter base and reverse.

'SHUCHITRA' (1972)
Rhodamine pink with Mimosa Yellow base and silvery white reverse, beautifully formed double flowers are produced in singles and large clusters. Good for bedding.

'SURYODAYA' (1968)
Dazzling orange, double, well-shaped flowers borne in single and clusters of 4 to 6 flowers, free-flowering.

'USHA' (1975)

The small long-lasting flowers of deep camellia rose colours with lighter reverse are produced in big clusters.

Selected List of Cultivars Released by M/s. K.S.G.'s Roses, Bangalore

'AHALYA' (Floribunda, 1969)

A sport from the famous 'Summer Snow' having all its characteristics. The colour is very pale alluring pink.

'ANUPAMA' (Hybrid Tea, 1971)

Progeny of 'World's Fair Salute' and 'Happiness'. A superb non-fading crimson red with petals prominently veined darker crimson.

'ASWINI' (Hybrid Tea, 1972)

A startling brilliant ruby red. Blooms freely.

'CHAITRA' (Hybrid Tea, 1972)

Free flowering, fiery red in colour.

'CHITRANGINI' (Hybrid Tea, 1970)

A multi-coloured rose of high promise, producing perfect blooms of light yellow suffused pink turning to orange and red as flower ages, with striking yellow heart and reverse. Free flowering and slightly fragrant.

'CLIMBING KRONENBURG' (1973)

A climbing sport of the famous Hybrid Tea. Free-flowering and constant bloomer.

'CITY OF PANJIM' (Hybrid Tea, 1972)

A delightful bicolour of a luminous off-white on the reverse of petals. Flowers are of excellent shape and form, large, double, opening out from tight shapely buds. It has an alluring fragrance.

'DEVADASI' (Floribunda, 1967)

A vigorous plant producing attractive flowers of darkest blackish velvety-crimson flowers singly and in clusters, having fragrance.

'GAYATHRI' (1973)

Flowers deep ruby-red having slight magenta touch with a paler and cream reverse and base; a novel colour. Fragrant.

'JWALA' (1973)

Seedling of 'Firecrescent' X 'Gartenfreude'. Scarlet vermilion buds open out to dazzling scarlet flowers.

'KALPANA' (Hybrid Tea, 1968)

Buds on long stalks, open to carmine-red with more than 70 thick petals. Flowers long lasting and produced freely on healthy bush.

'KAMINI' (Floribunda, 1968)
Flowers are of medium size and has H.T. shape and form. The colour is a delightful blending of light vermilion and gold, overcast salmon.

'KANCHANI' (Hybrid Tea, 1968)
A sport of 'Lemon Chiffon'. Yellow heavily flushed and splashed red giving a wonderful colour effect. Free flowering. Plant vigorous.

'KOLAR PRINCESS' (Floribunda, 1972)
The most attractive quality of the flower is that in the morning and evening the petals remain unfolded. The colour is camellia rose, heavily veined in darker shading. Fragrant.

'KOMALA' (Hybrid Tea, 1971)
Flowers rose-pink, large, double and fragrant.

'PADMAVATHI' (Hybrid Tea, 1969)
Double flowers are pastel pink inside and white outside.

'PRAKASH' (Floribunda, 1969)
The flowers are most glittering and dazzling vermilion-scarlet. Very floriferous.

'PRITI' (Polyantha, 1971)
A sport of 'Margo Koster'. Masses of soft pink lovely ball-like flowers appear in huge clusters.

'PRIYA' (Floribunda, 1972)
Flowers are Hybrid Tea type. The colour is creamy-white to white sparkingly edged apricot and crimson.

'PUSHKALA' (Miniature, 1973)
Tiny pure white flowers appear in clusters. Very free-flowering.

'SHYAMALA' (Hybrid Tea, 1972)
Flowers are innocent soft pale pink.

'SRINIVASA' (Hybrid Tea, 1969)
Flowers are fragrant, deep rose-red with white reverse.

'SUSHMA' (1973)
Very full, double flowers of a charmingly frilled lilac and lavender pink. Flowers exude fruity fragrance. Very free-flowering.

'SWAPNA' (Floribunda, 1968)
Buds creamy-white opening to soft pure-white flowers.

'VEENA' (Floribunda, 1972)
A cross between 'Europeana' and 'Dearest'. A very floriferous cultivar with deep coral pink large blooms. Flowers slightly scented.

Roses Released by Surendra Singh

'CHANDERI' (Hybrid Tea, 1968)
Seedling of 'Peace'. Bud long, pointed, pink blend; flower medium, double (60-70 petals), high centred, slightly fragrant with very good keeping quality.

'HEER' (Hybrid Tea, 1969)
Seedling of 'Picture'. Flower medium, double (60 petals), high centred, slightly fragrant, rose pink. Resistant to mildews and black spots.

'NAZR-e-NAZAR' (Hybrid Tea, 1968)
Large, well-shaped blooms of satin-pink. Seedling of 'Clovelly'.

'RUKHSAAR' (Hybrid Tea, 1969)
Seedling of 'Virgo', bud greenish long, pointed, flower medium, double (over 40 petals), high centred, cream coloured with shell pink centre.

'YAMINI KRISHNAMURTHY' (Hybrid Tea, 1968)
Seedling of climbing 'Sterling Silver'. Blooms are shapely, fragrant and lilac-mauve in colour. Resistant to mildews and black-spots.

Roses Released by M/s. Laveena Roses

'EIFFEL TOWER' (Climbing Hybrid Tea, 1967)
Sport of 'Eiffel Tower'. Flower large (11 cm), double (35 petals), high centred, strongly fragrant, light pink, vigorous, abundant bloom.

'GOLDEN GIANT' (Climbing Hybrid Tea, 1967)
Sport of 'Golden Giant'. Bud pointed. Flower large (8 cm), double (38 petals), high centred, slightly fragrant, yellow.

'LAVEENA' (Hybrid Tea, 1969)
Seedling of 'Kiss of Fire'. Bud long pointed. Flower large (9 cm), double (57 petals), high centred, yellow with pinkish tinge.

Since 1972, rose breeders at the Indian Institute of Horticultural Research, Bangalore, are carrying on intensive research to produce new rose cultivars suitable for cultivation in southern India, especially for export of cut flowers. The climate of Bangalore and of the adjoining areas makes it feasible to grow the roses for cut flowers for most part of the year. Already the Institute has developed a number of promising rose hybrids which are currently under testing and may be released for cultivation in due course. Bhattacharji's of Deoghor,

Bihar, have to their credit of developing a large number of cultivars, many of which are popular in the garden.

List of Some Rose Cultivars Classified According to Colour of Flowers

I. *Hybrid Teas*

(a) **White**: 'Grand Mughal', 'June Bride', 'John F. Kennedy', 'Dr. Homi Bhabha', 'Matterhorn', 'Mount Shasta', 'Pascali', 'Tushar', and 'Virgo'.
(b) **Yellow**: 'Buccaneer', 'Ganga', 'Golconda', 'Golden Giant', 'Landora', 'McGredy's Yellow', and 'McGredy's Sunset'.
(c) **Apricot, Buff, Amber, and Brown Shades**: 'Cannes Festival', 'Diamond Jubilee', 'Helen Traubel', 'Tahiti', 'Thais', and 'Whisky'.
(d) **Pink**: 'Confidence', 'Eiffel Tower', 'First Prize', 'Fleet Street', 'La France', 'Maria Callas', 'Picture', and 'South Seas'.
(e) **Orange, Orange Scarlet, and Coral**: 'Aruna', 'Command Performance', and 'Montezuma'.
(f) **Vermilion, Red to Dark Crimson**: 'Avon', 'Bhim', 'Christian Dior', 'Crimson Glory', 'Kalima', 'Mister Lincoln', 'Papa Meilland', 'Oklahoma', and 'Super Star'.
(g) **Lavender**: 'Africa Star', 'Blue Moon', 'Clair de Lune', 'Lady X', and 'Sterling-Silver'.
(h) **Bicolours and Colour Blends**: 'American Heritage', 'Bajazzo', 'Caprice', 'Cleopatra', 'Dr Valois', 'Inge Horstamann', 'Kiss of Fire', 'Kronenburg', 'President Herbert Hoover', and 'Tahiti'.
(i) **Novel Colours**: 'Careless Love' (pink with white streaks) and 'Anvil Sparks' (coral red with yellow stripes).

II. *Floribundas*

(a) **White**: 'Himangini', 'Iceberg', 'Moon Raker', and 'Saratoga'.
(b) **Pink**: 'Dearest', 'Delhi Princess', 'Else Poulsen', 'Junior Miss', 'Queen Elizabeth', 'Stanza', and 'Tiki'.
(c) **Yellow**: 'Allgold', 'Arthur Bell', and 'Fugitive'.
(d) **Orange-Scarlet, Coral, and Vermilion**: 'Anne Marie', 'Celestial Star', 'Flamenco', 'Orangeade', 'Orange Sensation', 'Shola', 'Spartan', 'Zambra', and 'Zorina'.
(e) **Red to Dark Crimson**: 'Alain', 'Devdasi', 'Dorothy Wheatcroft', 'Marlena', 'Valentine', and 'Zizi'.
(f) **Mauve and Lavender**: 'Angel Face', 'Lavender Princess', and 'Lilac Charm'.
(g) **Bicolours and Colour Blends**: 'Daily Sketch', 'Junior Miss', 'Redgold', and 'Sea Pearl'.

III. *Polyanthas*

'Katharina Zeimet' (white), 'Baby Faurax' (lavender), 'China Doll' (pink), 'Chattillon Rose' (pink), 'Orange Triumph' (orange), 'Paul Crampel' (orange), 'Ideal' (dark crimson), 'Border King' (red), and 'Flamboyant' (red).

IV. *Miniature*

'Cricri' (coral, shaded salmon), 'Josephine Wheatcroft' (yellow), 'Baby Masquerade' (lemon chrome), 'Little Flirt' (bright red, with yellow reverse), 'Cinderella' (white), 'Dwarf King' (red), 'Mimi' (clear pink), 'Twinkles' (white), 'Starina' (red with gold reverse).

V. *Climbers and Ramblers*

'Clg. Show Girl' (pink), 'Clg. Virgo' (white), 'Clg. Crimson Glory' (dark red), 'Marechal Niel' (lemon), 'Delhi White Pearl' (white).

The above cultivars are recommended for the plains. The following cultivars are suitable for the hills. 'Golden Showers' (yellow), 'Mermaid' (primrose-yellow), 'Sanders White Rambler' (white), 'Dorothy Perkins' (pink), 'American Pillars' (pink blend), 'Blaze' (red), 'Violet' (mauve), and 'Excelsa' (rose to light crimson).

VI. *Scented Roses*

(a) *Hybrid Teas:* 'Avon' (red), 'Blue Moon' (lavender), 'Crimson Glory' (crimson), 'Chrysler Imperial' (red), 'Clair de Lune' (lavender), 'Eiffel Tower' (pink), 'Kanakangi' (golden), 'La France' (pink), 'Oklahoma' (red), 'Sugandha' (red).

(b) *Floribundas:* 'Elizabeth of Glamis' (salmon), 'Magenta' (magenta), 'Pink Parfait' (bicolour), 'Spartan' (orange-red, shaded coral).

(c) *Miniatures:* 'Midget' (carmine-red), 'Pixie' (white), 'Twinkles' (white).

(d) *Climbers and Ramblers:* 'Albertine' (coppery-chamois), 'Clg. New Yorker' (velvety scarlet), 'Coral Dawn' (pink), 'Delhi White Pearl' (white), 'Dorothy Perkins' (rose-pink), 'Golden Showers' (yellow), 'Honeymoon' (golden yellow), 'Marechal Niel' (bright yellow), 'Penelope' (shell-pink), and 'Sympathie' (dark red).

VII. *Exhibition Types*

The exhibition type cultivars bear large and shapely blooms. Important cultivars are listed below.

'Anvil Sparks' (coral-red, with yellow stripes and splashes), 'Avon' (red), 'Ballet' (deep pink), 'Christian Dior' (scarlet), 'Dr. Homi Bhabha' (white), 'Eiffel Tower' (pastel pink), 'First Prize' (deep rose), 'Golden Giant' (yellow), 'John F. Kennedy' (white), 'King's Ransom' (golden yellow), 'Maria Callas' (rose pink), 'Montezuma' (salmon), 'Papa Meilland' (crimson-velvet), 'Pusa Sonia' (golden yellow), 'Rose Gaujard' (bicoloured), 'Sterling Silver' (lilac), 'Super Star' (light vermillion), 'Thais' (strawberry pink), and 'Virgo' (white).

TUBEROSE

Tuberose (*Polianthes tuberosa*); (Family: *Amaryllidaceae;* Origin: Mexico; Common Names: *Rajanigandha* (Bengali, Hindi), *Gul-e-Shabu* (Urdu), *Sugandharaja* (Kannada and Tamil).

The grass-like leaves in tuberose are long and narrow and the plant is dwarf, hardly exceeding 15-20 cm in height. The leaves are light green in colour and crowded in an arching form. The flower buds are either single or double with tubular shape. The flowers are pure white in the single cultivar and tinged with red in the double (known as 'Pearl') cultivar. A semi-double cultivar somewhat resembling double also exists. There is another cultivar 'Variegated' where the leaves are variegated yellow on the margins. The flower is single with little tinge of red in bud stage but turning to white when opened fully. The flowering stalk (or spikes) arise from the centre of the cluster of leaves and are usually from 50 cm to 90 cm high or sometimes a little more in case of variegated and single cultivars. The flowers are highly fragrant in 'Single' tuberose compared to 'Double' cultivar where the fragrance is less.

The spikes are useful as cut flowers in vase decoration and bouquets, while individual florets are used for making *veni*, garland, button-holes or crown, the last named ornament is used during marriage or other religious ceremonies such as birthday and *Annaprasana* in Bengal. The spikes generally last for 7-12 days in vases depending upon the room temperature.

Tuberose is planted in beds and borders, but can also be grown as potted plant.

Propagation

The plants are grown from bulbs. Bailey (1951) in his book *Manual of Cultivated Plants* describes the rootstock as a bulb-shaped tuber and places this genus in the family *Agavaceae*. These are planted before or during the rains in the plains and in May-June in the hills. The old roots are cut away at the base of the tubers before planting. The larger tubers in the centre of the cluster are best for planting. The smaller ones take 2 to 3 years to come to flowering depending on their size. In high rainfall areas with rich soil, or in places with assured irrigation, the tubers are planted 7-10 cm deep and at least 30 cm apart from plant to plant and row to row, especially when these are to be left undisturbed for two to three seasons. In dry areas, comparatively closer spacing is recommended.

Manurial and Fertilizer Requirements

Sufficient F.Y.M. or town compost should be added to the soil (2-4 kg per square metre) before planting. Nitrogen is highly beneficial for increasing leaf numbers, number of spikes, and number of florets. Phosphate is also beneficial for good quality flowers. A fertilizer dose of 10-20 g of Nitrogen, 20-40 g of P_2O_5, and 20-40 g of K_2O per square metre can be used. Half the dose of N and the full doses of P_2O_5 and K_2O are supplied at the time of planting and other half of N is applied as top-dressing when the flower spikes start to appear. A heavy irrigation once in 5-10 days, depending upon weather conditions, is necessary. The tuberose is voracious feeder. Application of weak liquid manure prepared by fermenting fresh cowdung or oilcake mixed with small quantities or urea or ammonium sulphate during the period of vigorous growth and when the flower spikes start to appearing is recommended. Around Bangalore, application of potash is not effective in influencing growth and flowering.

Cultural Hints

Tuberose thrives best in fertile and well-drained soil and sunny places, though sometimes it can also be grown in partial shade. To have a succession of flowering the tubers should be planted at 15 days' interval throughout the year especially under Bangalore conditions. In places as Calcutta also, with high rainfall and mild winter, it is possible to have flowers almost throughout the year. The peak flowering season, in the plains, is between June and October. In the northern plains, the plants are damaged by cold temperature and frost and hence its commercial cultivation is not so profitable. Amateurs may grow it on a small scale in protected places such as under large trees or in shrubbery borders during the cold season. Since the main flowering season is during the summer and the rains, the growers of north Indian plains may cut the leaves at ground level before the winter and allow the tubers to rest in the ground during the winter. After flowering is over, it is advised to cut down the spikes to encourage more flowering. The tubers are left undisturbed in the ground during the winter. The tubers are left undisturbed in the ground for about 2 to 3 years when these are separated and replanted. If the tubers are not uprooted every 2 to 3 years the spikes tend to become smaller from the third year. The tubers do not have any prolonged rest period and can be replanted four to five weeks after harvesting. During flowering, the unhealthy and dried leaves should be removed occasionally.

Tuberose can also be grown in 25-30 cm pots. The pot mixture consists of 2 parts of garden soil, 1 part of each of leaf-mould, F.Y.M., and coarse sand, and ½ part of coarse charcoal pieces.

Pests and Diseases

See Chapter XXVI.

Some Research Findings

Militu *et al.* (1970) have found that N is the most important factor in controlling flower quality. Jana *et al.* (1974) working with tuberose have found that high N and P proved very effective in increasing the number of leaves and bulbs. These two elements also increase the number of spikes and flowers. Deficiency of either of these elements suppresses flowering and low K reduces the number of flowers in the rachis. According to Motial (1973), high doses of N increase the yield of fresh flowers. Allincai (1960) has reported that an application of granular superphosphate at 60 g per square metre combined with phospho-bacterin, increases the number of flowering plants, the number of flowers per inflorescence, and also the life of the flowers. Mukhopadhyay & Bankar (1978) have observed that N is the most important factor in controlling vegetative and floral characters.

Chapter XXI

Principles of Garden Designs

The planning of a garden is an art. It is said that a poet is born, and in that sense a garden architect is also born. But that does not mean that a garden architect has nothing to learn. On the contrary he should learn enough of geology, geography, garden history, styles of gardening, and above all should have a profound knowledge about plants.

It is imperative to know at the outset what is meant by landscape gardening. A landscape may be defined as any area, either big or small, on which it is possible or desirable to mould a view or a design, whereas landscape gardening may be described as the application of garden forms, methods, and materials with a view to improving the landscape. The art of designing is known as "landscape architecture," although the older term "landscape gardening" is also popular. To help the garden designer, the different symbols used in a garden plan are given in Fig. XXI.1.

Before discussing the principles of gardening in detail, it will be worthwhile to make some useful suggestions. The first and the foremost thing is not to imitate another garden which has secured a prize in a competition. One has to develop one's own design giving due consideration to the local conditions. One more mistake which is commonly made is to plant many more specimens than a garden can accommodate causing overcrowding. In a landscape garden any difference in levels has to be taken advantage of, but in a perfectly flat land it will be costly to create artificial undulations. In each garden there should be at least one feature or if there be a second these two should harmonize with each other. Before planning a design one must be sure for what purpose the garden is - utility or beauty or both.

Initial Approach: In theory, everyone would like to have a perfect plot of land, but in actual practice the plot available for gardening, in three out of five cases, either will not be in a good site or the shape and size will not be ideal. Whatever may be the case, one should not throw one's hands up in despair even if the site appears to be hopeless. A good designer is one who will make best use of such a site. As has already been stated, land with natural undulations should never be levelled, but rather the differences in levels should be utilized with advantage. Fencing, though may not look artistic, is essential in any garden. But the fencing should be such that it looks natural as far as practicable and it should not obstruct any natural view. For example, if there is a natural forest scenery or a hillock just outside the boundary it should be incorporated in the garden design in a thoughtful manner so that it appears to be a part of the garden.

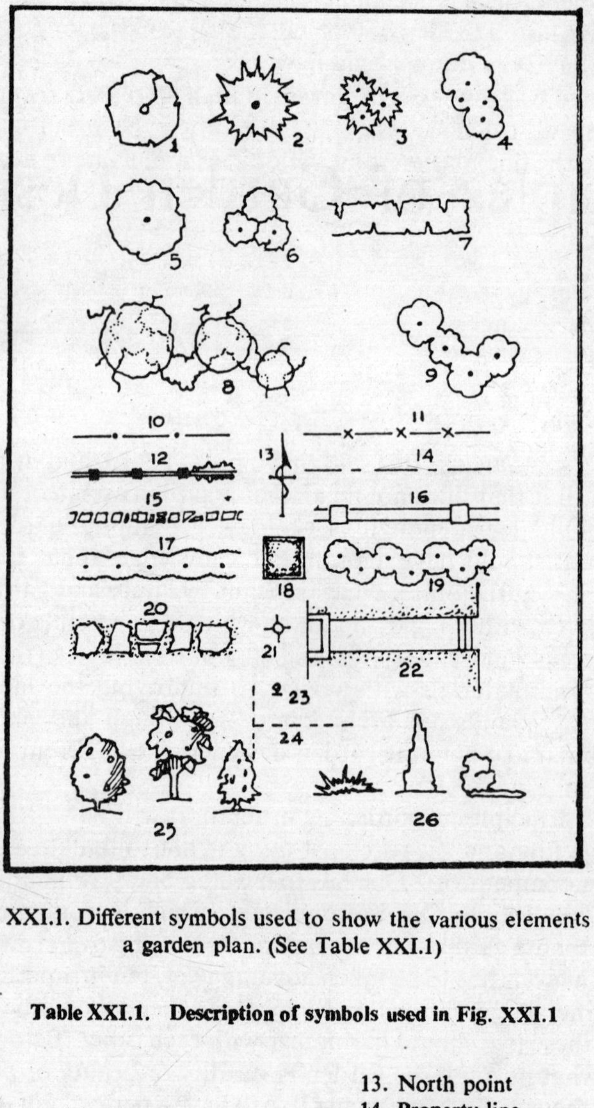

Fig. XXI.1. Different symbols used to show the various elements in a garden plan. (See Table XXI.1)

Table XXI.1. Description of symbols used in Fig. XXI.1

1. Tree
2. Evergreen tree
3. Evergreens
4. Shrub group
5. Tree
6. Shrubs
7. Trimmed hedge
8. Tree with shrub underplantings
9. Shrub group
10. Telephone—electric overhead wires
11. Wire fence
12. Wooden fence (symbol on extreme right indicates a creeper trained on fence)
13. North point
14. Property line
15. Retaining wall
16. Formal wall
17. Untrimmed hedge
18. Ground cover
19. Untrimmed hedge
20. Flagstone walk
21. Pole
22. Steps on metalled path
23. Water
24. Underground pipe (water, gas, sewer)
25. Tree forms
26. Shrub forms

A building or a machinery, once it is complete, takes a final shape. But with living plants it is impossible to visualize how a garden design will look like in the long run when the plants attain full size, even though the design may have been completed. For example, a tree which is normally expected to grow to a height of 10 m, may remain dwarf for some reason or other thus jeopardizing all calculations. One more difference with inanimate objects and living plants is that while the final design for a building or a machinery can be completed on paper this cannot be so with living plants, and as such each design should be adaptable. The man on the spot who will actually implement the design should be given enough scope to change it to adapt to the local needs and personal taste of the owner and/or designer. It is only the formal gardens, not having any trees, that can be drawn on paper and implemented without any appreciable change.

The other terms and principles used in landscape designs are briefly discussed below.

Axis: This is an imaginary line in any garden around which the garden is created striking a balance. In a formal garden, the central line is the axis. At the end of an axis, generally there will be a focal point, although another architectural features such as bird-bath or sundial can also be erected at about the midpoint.

Focal Point: In every garden there is a centre of attraction which is generally an architectural feature focused as a point of interest. A focal point is one of the elements of good landscape design.

Mass Effect: The use of one general form of plant material in large numbers in one place is done to have mass effect. To see that such mass arrangements do not become monotonous, the sizes of masses should be varied.

Unity: Unity in a garden is very important as when this is achieved it will improve the artistic look of the garden. Unity has to be achieved from various angles. First, the unity of style, feeling, and function between the house and the garden has to be achieved. Secondly, the different components of the gardens should merge harmoniously with each other. The aim is to give the visitor an overall impression of the garden rather than blowing up some special features. The last point, which is also very important, is to achieve some harmony between the landscape outside and the garden. A garden laid out in complete defiance of the local conditions may look exotic, but is not a successful garden. As for example, cacti planted in a seashore garden is completely out of place as these are inhabitants of dry localities. To achieve a unity between the house and the garden it is a common practice to train creepers on the front porch which cover the rudeness of the masonry work and also bring the house closer to nature. For the same reasons, foundation plantings are also done. A foundation planting broadly means the planting of bushy plants near the foundation of the house.

Space: The aim of every garden design should be such that the garden should appear larger than its actual size. One way of achieving this is to keep vast open spaces, preferably under lawn and restrict the plantings in the periphery, normally avoiding any planting in the centre. But if any planting has to be done in the centre the choice should be a tree which branches at a higher level on the trunk (or the lower branches are removed), and not a bushy shrub. Such planting will not obstruct the view or make the garden appear smaller than its size. Another way of creating an illusion of distance, which was in vogue in older days but not practised much in recent times, is to make lines converge slightly at a distance. For example, paths in a garden are gradually narrowed or the size of the farthest trees diminish in size. But such tricks will give a reverse impression from the other end. Another suggestion to create the illusion of more space in a large public garden is to alternate large lawns followed by a group of trees. A large open space planted haphazardly all over with trees looks smaller than its size. The technique of creating an illusion of more space is also referred to as 'forced perspective'.

Divisional Lines: In a landscape garden, there should not be any hard and fast divisional lines. But there is the necessity of dividing or rather screening a compost pit or a *mali's* quarter or a vegetable garden from the rest of the garden. In fact areas under lawn, gravel, stone or cement path, and shrubbery border have their natural divisional lines from its immediate neighbour though these are not discreet. This is what is exactly needed. The divisional lines should be artistic with gentle curves and these should also be useful. Above all these lines should harmonize with one another.

Proportion and Scale: Proportion in a garden may be defined as a definite relationship between masses. For example, a rectangle having a ratio of 5 : 8 is considered to be of pleasing proportion. As this ratio comes down the form looks neither a square nor a rectangle, and the design becomes undesirable.

There is no set rule as regards scale or proportion in a garden. But a simple rule is that a design should look pleasant. It is better to have an *ad hoe* design first and then try it out on the actual spot. If the design looks appealing as well as pleasing, it is implemented. When a shrubbery border has to be planted the outer design is marked by arranging a rubber hose or thick wet rope in different designs on the spot and the one which looks best is adopted. Then sticks of different heights, representing the various shrubs, are planted in various positions and by the method of permutation and combination the most proportionate looking arrangement is adopted.

It will be worthwhile to give a few more examples to illustrate the importance of scale in a garden. A narrow step leading from a wide terrace is completely out of scale. The steps in a garden, should not only be broader than those inside the house but should, have deep treads (the stepping) and low risers also. This means the steps are spaced wider, making climbing easier and pleasant. Moreover, a very wide flight of steps dividing two lawn areas, at different levels, make the transition easy and inconspicuous. The common practice of laying out a small rockery at the base of a large tree with small thorny specimens looks not only ugly but is also out of scale and proportion under the large canopy of the tree. A tiny pool in the midst of a large lawn also looks disproportionate.

Texture: The surface character of a garden unit is referred to as texture. The texture of the ground, the leaves of a tree or shrub will all determine the overall effect of the garden. The texture of rugged looking ground can be improved to an appreciable extent by laying meticulously chosen small pebbles from the riverbeds, if establishing a lawn is out of the question.

A *gulmohar* is a fine textured tree when in full leaf, whereas *Spathodea companulata* is a coarse textured tree. The placement of all these various textures with harmony and contrast has to be achieved to get the ultimate desirable effect.

Time and Light: In a garden the time factor is very important. There are three different categories of time in a garden. First comes the daily time, which provides different quantities and qualities of light during the course of the day. As the morning sun is vital for all flowers, the designer has to take this into account while planning. In most parts of India the garden design should be planned in such a way that in the afternoon it is possible to sit in a shaded place from where the best part of the garden should be visible. The second type of time is the seasonal changes in the year. A good planner must roughly take into account the seasonal movement of the sun and where the shade and light are likely to fall during the different parts of the season. As for example, a lawn in Delhi which receives shade during the early parts of the day in the winter will not grow or remain patchy. A good and knowledgeable gardner is he who can visualize such eventualities. The third time, which most people overlook or cannot visualize, is the fact as to what shape and proportion the shrubs and trees will attain in the years to come. Often we can see sickly and lanky shrubs growing near the trunk of a large tree, because of lack of light and possibly nutrients also. The shrubs were planted when the tree was a sapling, without visualizing what would happen to them when the tree attained its full size. The right type of tree should be planted at the proper place so that shade is obtained during the hot days, at the appropriate time. A cleverly planted tree can protect a window from strong afternoon sun or an unplanned tree can ruin the view of the garden from inside a window. The pattern of shade caste by a fine-leaved tree on the ground or lawn can look very artistic. Similarly, straight trunked trees like the Royal Palm (*Oreodoxa regia*) or Eucalyptus when planted in a row along a path will throw oblique bars of shade in the mornings and afternoons and cheer up an otherwise dull walk.

Tone and Colour: For a preliminary knowledge on colours and colour schemes the readers are advised to refer to pages 64 to 65.

A tendency on the part of an amateur gardner is to create a riot of colours by indiscriminately planting flowering annuals of all shades. This practice is not desirable. Moreover, such riot of colours has only temporary effect. In a landscape garden, the permanent backdrop is the green tones of the various trees and shrubs. It is possible to lay out a garden with subtle tone of entirely white or yellow flowers, but at the same time making it charming also. Another important point is that it is better to have masses of a single colour against a mixture of colours. A bed of roses containing only a signle colour of say red, yellow, or pink has a much softer tone and beauty than a bed containing a mixture of colours.

Mobility: In a temperate country, the garden changes colour very sharply and contrastingly from one season to the other thus symbolizing mobility or movement. As for example, many trees in the temperate regions attire themselves with wonderful hues due to the changes in their leaf colour in the autumn. Then suddenly in the winter leaves fall and everything goes to rest bringing an atmosphere of melancholy and dullness all around. Again in the spring the plants spring back to life with the appearance of the new leaves. In most parts of Tropical India, though these contrasting changes cannot be achieved, it is possible to bring in some subtle changes. For example, to create some symbol of movement trees such as Bengal or Indian Almond (*Terminalia catappa*) which changes its leaf colour into striking red twice annually before falling or *Lagerstroemia flos-reginae* which also changes the colour of the leaves to coppery shade in the autumn before shedding, or *Madhuca indica* and *Ficus religiosa*, the new foliage of these appearing as coppery red in the spring, should be planted in some parts of the garden. This, in addition, improves the landscape.

The movement and cluttering of birds also bring life and mobility to the garden, though sometimes some birds may become a menace. Large trees and bird-baths attract birds. For the smaller birds, the safety of shrubberies is needed to protect them from large predator birds. Some plants, bearing berries, such as *Ficus infectoria* and *Syzygium cumini* (Syn. *Eugenia jambolana*), can also be planted in some remote corners though they may not look very ornamental. Flowering trees such as *Bombax malabaricum* (silk cotton) or *Erythrina* also attract birds when in bloom. The seasonal flowers will bring in the motion and movement of colourful butterflies. Fountains or even a lawn sprinkler and streams in a garden also serve the objective of movement. The lily pools should be filled with coloured fish, the movement of which will be an added attraction. The frogs will also come and inhabit these pools uninvited and in the process bring in more movement.

Style: Lastly, one has to decide about the style to be adopted for one's garden. Broadly speaking, every garden lover has to invent his own style of gardening commensurate with his budget, taste, and the nature of the site. But a man can develop his own design only when he studies carefully all the great garden styles of the world and grasps the underlying principles in them. There is no doubt that persons not having enough specialized knowledge will commit mistakes; nevertheless, one should not get deterred by this fact. One word of caution to a novice gardner is that he should not get used to his mistakes, but critically assess every feature and try to correct the mistakes and improve upon the design with the acquiring of new knowledge through experience and learning from others.

Chapter XXII

Styles of Gardening

In the preceding chapter the principles governing landscape gardening have been discussed. There is a misconceived idea among the amateur gardeners that a landscape garden can be laid out only in an undulated land and with this misconceived notion in mind people take great pains to create terraces and mounds in an otherwise flat land. After going through the preceding pages one would be convinced that this need not be so. The goal in a landscape gardening is to improve the landscape with an idea of developing a view or a design.

Besides the term "landscape gardening" the other two familiar terms in gardening are the *Formal* and *Informal* gardens.

Formal Gardens : A formal garden is laid out in a symmetrical or a geometrical pattern. In this garden the design is stiff as everything is done in a straight and narrow way. In such gardens everything is planted in straight lines. Also if there is a plant on the left hand side of a straight road, a similar plant must be planted at the opposite place on the right hand side. The flower beds, borders, and shrubbery are arranged in geometrically designed beds. Trimmed formal hedges, Cypress, *Ashoka* trees, and topiary are typical features of a formal garden.

Informal Gardens: In an informal garden, the whole design looks informal, as the plants and the features are arranged in a natural way without following any hard and fast rules. But here also the work has to proceed according to a set and well-thought-out plan, otherwise the creation will not be artistic and attractive. The idea behind this design is to imitate nature.

Wild Garden : A comparatively recent style of gardening, namely, "Wild Garden" was expounded by William Robinson in the last decade of the nineteenth century. His idea was rather revolutionary and found many admirers to follow this. The concept of wild garden is not only against all formalism but it also breaks the rule of landscape styles. His main idea was to naturalize plants in shrubberies. He also preached that grass should remain unmowed, as in nature, and few bulbous plants should be grown scattered in the grass to imitate a wild scenery. He also suggested that passages should be opened in the woodland, and trees, shrubs, and bulbous plants should be planted among the forest flora to fulfil his idea of a wild garden. His other idea was to allow the creepers to grow over the trees naturally imitating those of the forests.

Before someone starts to venture into designing a garden it will be wise to get an idea about the major gardening styles of the world. This will open up a window to his knowledge on gardening and help him design his own garden by adapting the best from each or any of these. But this does not mean that one should copy any garden style. For example, when a would-be writer studies Shakespeare, Shaw, or Tagore it does not mean that he will translate their ideas in his work of literature. He only studies the styles of writing and forms his own ideas suiting to the situation and time. Similarly, a garden enthusiast has to study the different styles only to gain knowledge to help him form his own ideas suiting the local conditions and limitations such as available space, funds, etc.

Though in India from history and ancient literature we find that gardening was quite in vogue in olden times, but unfortunately there is no garden style called "Indian garden", which can claim a place in the major gardening styles of the world. The famous garden style of India the "Mughal Gardens" are nothing but a replica of the ancient Persian Gardens.

The major garden styles which shall be discussed in the following pages are: English gardens, Mughal gardens, Persian gardens, Italian gardens, French gardens, and Japanese gardens. Out of these, the Mughal, Persian, Italian, and French styles fall in the category of formal gardens, whereas the English and Japanese gardens are classified in the informal style of gardening.

(A) English Gardens

Due to favourable climatic conditions such as high annual rainfall, the natural ground cover in the English countryside is grass. With this in mind the famous British garden architects Repton and Capability Brown advocated the concept that the British gardens should look like the countryside. Their main idea was that the gardens should merge with the countryside without any artificial barriers such as fences, hedges, or the like. But it is only in the eighteenth century that these two gentlemen along with Kent brought the touch of nature in the garden, although the history of gardening in England dates back to fourteenth century. At this time the monks and priests conceived the idea of gardening out of necessity. They started kitchen gardens to supply vegetables to the inmates of the monasteries and grew herb gardens for medicines. But this gradually imbibed the spirit of gardening to the people who realized the goodness of residing in pleasant surroundings. By the middle of the sixteenth century the English gardens saw flower beds, topiary, and terrace gardens. In the middle of the eighteenth century gardens were laid out with more emphasis on architectural features. The main features of gardens during this period were curved paths, informal groups of trees, rivulets or streams, artificial waterfalls, and clipped hedges. The flowering annuals, the main stay of an English garden, came into prominance during the nineteenth century which subdued the architectural features. The main features of an English garden are thus lawn, mixed border especially of herbaceous annuals, as well as herbaceous perennials, shrubbery, and rock gardens. The English climate suits admirably well for the growth of herbaceous annuals. This prompted them to evolve a large number of hybrids of annuals as well as to collect the best flowering plants from all over the world.

The main features of English gardens as we know in India are (a) lawn, (b) herbaceous border, and (c) rockery. Most of the flowering annuals that we see today in the Indian gardens, with the few exceptions of amaranthus, balsam, gomphrena, marigold, etc., were brought here by the British. The details regarding the different features mentioned above have been discussed in the respective chapters.

(B) Italian Gardens

The Italian garden style came into existence at the time of Renaissance. There is a striking similarity between the Persian or the Mughal styles with the Italian style. In all these styles of gardening the similarity was the use of heavy masonry features, though the character of masonry was different in the Italian style. The Italian elites conceived their gardens just as an extension of the lavish palaces, as a glamorous outdoor hall for entertainment and for showing off their wealth as well as status. The most prominent feature associated with Italian gardens was the massive flight of stairs, generally of marble, complete with balustrade to connect the different levels in the garden. Decorative urns, fountains generally in combination with stone sculptures or rather the fountains themselves used to emerge from one part or the other of the statues, are the other equally important features of the Italian gardens. With the fall of the Roman Empire, the art of gardening sharply declined in Italy. The greatest benefit the Italian garden style brought to gardening is that it taught all of Europe that gardening could be a most respectable form of art which demanded thorough and careful study.

(C) French Gardens

In the sixteenth-century France, the court life was shrouded with stiff formality and exactness. Matching with that the French style of garden designs were also very intricate and artificial. Until this time the French gardens were nothing but copies of Italian style. The new style of gardening now known as French style is largely, rather solely, due to the efforts of Le Notre who served in the Royal Garden of Louis XIV from 1643 to 1700. He elevated the art of garden design to a standard which had never since been reached. It was Le Notre who showed to the world the impact in impressiveness of scale, on garden design. His main creations, the gardens at Versailles, have avenues which are memorable for their tremendous length and width. To design a garden at Vaux-le-Vicomte, his first masterpiece, Le Notre had to remove three villages to create his vista. The moral of French garden style of Le Notre seems to teach the lesson "how to think big". The unexampled scale of mass and sweep of design of his style has no parallel. His designs completely defied nature. The style of Le Notre can be termed as an evolution and mastery of the art of formal garden in its perfection. His style dominated the gardens of civilized Europe, for a long time.

(D) Persian Gardens

Persian garden style is one of the oldest. The Persian garden style and the Japanese style both were based on their respective ideas of heaven. Except this similarity the two styles differ widely from one another. The Persian styles were strictly formal and symmetrical. They used for their gardens all crafted materials such as masonry, carved and pierced marble stones, and highly polished stones. The Persian gardens were laid out by cutting terraces on the hill slopes. They also tapped some natural spring to create a straight water-course through the gardens and manipulated the water-course to undertake different movements along its run. If there was no natural source of water this was created artificially by diverting some rivulet or a stream. So the main stay of a Persian garden design was *nahars* (flowing canals) of water – the concept of Persian Paradise, "where cooling water flows". The selection and placing of trees were very judicious. The fruit trees represented the symbol of life while Cypress symbolized death and eternity.

(E) Mughal Gardens

The gardens laid out during the rule of Mughal Emperors in India are known as Mughal gardens. The Mughal gardens are similar to the styles of gardens of Central Asia and Persia. Babar (1494-1531) was the first Mughal ruler to introduce this style in India. All other Mughal rulers and some of the Mughal *Begums* starting from Akbar and followed by Jahangir, Nur Jehan, Shah Jehan, and Aurangzeb all laid their hand on developing one or the other Mughal gardens in India.

The main features of Mughal gardens, which are largely borrowed from the Persian style, are: (a) site and style of the design, (b) walls, (c) gates, (d) terrace, (e) *nahars* or running waters, (f) *baradari*, (g) often a tomb or a mosque, and (h) trees.

Site and Design: Mughals were very choosy about the selection of site and always preferred a site on a hill slope with a perennial rivulet or along the bank of a river. Mughal gardens are generally rectangular or square in shape and different architectural features are the main stay of the design.

Walls and Gates: The Mughals created the gardens not only for pleasure and recreation but also as forts, and residences surrounded by high walls and with-an-imposing wooden gate at the entrance which was studded with bold iron nails and pointed iron spikes. The purpose of the high walls was security from the enemies and shelter against hot winds. The gardens were a place of peace for the Emperors to enjoy with their wives and concubines.

Terrace: The Mughals came from the hilly terrains and so they were fond of terraces in the gardens. For this reason they used to select the location of gardens near hill slopes. Their fascination for terrace was so intense that even in the plains of Punjab they created artificial terraces. According to Islamic faith the Paradise has eight divisions and hence sometimes the gardens have eight terraces corresponding to the eight divisions or occasionally the gardens may be composed of seven terraces also representing the seven planets.

Running Water: The style for having running water by constructing canals and tanks was borrowed from the Persians. The water canals were paved with tiles (or marble) of blue colour to create the illusion of depth. The course of water used to be manoeuvred in various ways taking advantage of each slope, however small it may be, to break up the flow into artificial falls and ripples. The water canals used to have fountains to throw up the water high in the air. In the evenings small lamps used to be illuminated to create beautiful reflexions.

The fascination for water came from the Muslim faith which says that the promised paradise is the place "where cooling waters flow". With this bias for water Mughals selected sites on hill slopes to tap a running rivulet or spring for source of water. In the plains of India, where the summers are hot, Mughals utilized water for its cooling effect.

Baradari: This is nothing but a arbour-like structure, but made of stone and masonry with a *pucca* roof and a raised platform for sitting. These were usually provided with twelve or occasionally more doors on all sides for the Emperors to watch the performance of the dancing girls.

Tomb or Mosque: It was a common practice to have the gardens built around a tomb (e.g., Taj Mahal, Akbar's Tomb at Sikandra). It is often said that the Mughal gardens were at their best when built around a monument.

Trees and flowers: The trees were selected with careful planning and thought, as to Mughals each tree symbolized something, like life, youth, death, etc. Fruit trees were considered symbols of life and youth, while, cypress represented death and eternity. The Mughals had bias for spring flowering trees and flowers. The seasonal flower beds were of geometrical pattern and constructed along the water canals or near the main buildings. The favourite flowers were rose, jasmine, carnation, hollyhock, delphinium, etc.

(F) Japanese Gardens

History: The origin of Japanese style of gardening goes back to the era of Empress Suiko (592-628 A.D.). The recorded history, available today, shows that well-designed gardens existed at that time with artificial hills and ornamental ponds. During the Nara period (646-794 A.D.) the Chinese influences crept into various phases of Japanese life including gardening. In the succeeding period of Heian (794-1185 A.D.), influence of both Korean and Chinese architecture was noticed in Japanese garden styles. The typical features of the gardens of this era consisted of an open space, a pond with an island in the midst of it, a hillock on the south of the pond, and a stream of water. The usual style was to divide the water into two channels, one leading to the pond and the other to form a waterfall from the hill. This style of gardening continued until the middle of Kamakura period (1185-1392 A.D.) when a change was effected by the priests of the Zen sect. The prominent features of these gardens were water, stones, and evergreen plants. The plant materials were selected in such a manner that the garden would not undergo any major changes during the four seasons of the year. During the period of Muromachi (1392-1573 A.D.) both the flat garden (*Hira-niwa*) and the hill gardens (*Tsukiyama-sansui*) developed. The gardens of the Edo period (1603-1867 A.D.) had large lakes and islands surrounded by nicely laid-out paths. These gardens were known as stoll gardens.

What is a Japanese Garden? It has already been mentioned that both the Persian and Japanese garden designs were based on their respective ideas of heaven. One most admirable feature of the Japanese garden is that while other major styles of gardening of the world changed radically or fallen into disfavour, the Japanese continued the same style for centuries but still remained popular. This can be attributed to the special relation of the Japanese gardens to nature. A most important teaching of the Japanese garden is possibly that "unless a garden has an air of peace it's not worth a place visiting. It should be a place where the mind finds rest and relaxation."

Until recently gardens in many countries are meant only flower beds, or large areas under lawn alternating with flowers. Still many people like only the formal type of gardening or the regimented and intricated French or Italian style of gardening. The Japanese gardens which are planned with so much of thought, appear so casual that they do not appeal to many people.

Another strong basis of the Japanese garden is its immutability, i.e., except some seasonal changes in the deciduous trees the Japanese garden hardly goes through any other strong visible changes during different seasons. Even during the winter the Japanese gardens have their own beauty with snow flakes hanging down the stone pieces or the stone lanterns. The immutability is achieved also because rather than a grand mixture of flowers, shrubs, and trees more emphasis is placed on natural elements such as a simple path, a group of rocks, stepping stones, streams, waterfalls, bridges, stone lanterns, and so on. A Japanese garden tries to capture a natural scenery or to imitate a landscape. The three elements most important to achieve these objectives are water, stone, and plants. Low sculptured bushes and trained dwarfed trees look very attractive in a Japanese garden.

Types of Garden: A Japanese garden may either be in the form of a large public park or a small family garden designed for living which is seen usually by members of the family or the family guests. The Japanese gardens are further classified based on positions, shape, and purpose. The important types are : (i) hill garden, (ii) flat garden, (iii) tea garden, (iv) passage garden, and (v) sand gardens.

(A) **Hill Garden:** Out of the various styles, this one is considered to be the ideal by many garden enthusiasts. This style is known in Japanese as *Tsukiyama-niwa* or *Tsukiyama-sansui*, meaning hills and water. Though this garden can be laid out in a space of any size by making use of perspective to maintain a reduced scale, it is obvious that a fairly large area is much suitable as laying out a mountain scenery needs considerable space. The hill garden is made up of one or more hills designed with earth mounds and exposed weathered stones. The other features of this garden are water in the form of a stream or a pond or a waterfalls or all the three with or without islands and also bridges, lantern, stones, and trees. If it is not possible to introduce water in the garden, its presence is effected by a dried bed of stream or a dry shoreline. The important points in the garden are decorated with stones and selected trees. But pine trees may be planted to give the effect of being swept by wind . Untrimmed stepping-stones are placed over the walks.

An island is generally an usual feature in a hill garden. When the island is present it should be decorated with a "Worshipping stone", called *raithai-seki* in Japanese, a "Snow-viewing" lantern and a pine tree.

(B) **Flat Garden:** As the name implies, *Hira-niwa* or flat gardens are laid out in flat ground without hills or ponds. Flat gardens are supposed to represent a mountain valley or a meadowland. These gardens were popular during the era of Muromachi (1392-1573).

A Flat garden is not necessarily as flat as a pan-cake. Since it stimulates a mountain valley, low rounded hills designed with the help of stones or earth mounds or both will look quite appropriate in a flat garden. But some others opine that there should not be any ups and downs in a flat garden.

The usual features to break the monotony of a flat garden are a well, a water-basin made of stone in the shape of an urn, stones lying close to the ground, stepping-stones, and trees. The trees are trained to lie close to the ground. In a flat garden, the principle is to avoid strong vertical lines represented by tall pines.

In olden days the flat gardens were simple in design consisting of a few low-growing trees and flat rocks but in modern times many features of the tea garden, like water-basin, have been incorporated in this type of garden also.

(C) **Tea Garden:** The tea garden is laid out based on certain principles and customs of the Japanese tea ceremony and hence needs a considerable space of at least about 200 square metres, for its designing.

Since the performance of the tea ceremony needs an atmosphere of intimacy it is essential that the garden be enclosed by a fence. But the fence should be rustic in nature, with a gate made of very light material such as bamboo. To protect the tea house from the noise of the outer world, the tea gardens are divided into an outer garden (*soto-roji*) and inner garden (*uchi-roji*).

The outer garden is comparatively a narrow area, with a waiting place where the guests are supposed to wait until the master of the house appears to welcome them. This waiting place has a water-basin for the convenience of the guests who can wash their hands and a stone lantern for illumination, but in present days this serves more as a decoration piece. A stone path, usually of stepping-stones, leads to the inner garden. The inner garden is also separated from the outer garden by a rustic fence and a gate made of light material.

The inner garden contains the tea house. The tea house of the classical time was nothing but a small straw hut with an outside waiting place, a small sideroom for washing the utensils, and the main ceremonial tea house itself having the capacity to accommodate only five persons. The waiting place is utilized by the guests in between two preparations of tea or for the temporary recess between the tea and a meal. The entire path to the tea house is paved with stones or studded with stepping-stones. The most important feature at the entrance of the tea house is a water-basin or a well or both for the visitors to rinse their face before entering for the tea ceremony. To illuminate the water basin and resting place stone lanterns are set in appropriate places. The selection of water-basin and stone lanterns is done with scrupulous care so that both the purposes of utility and beauty and elegance are fulfilled. As far as tree planting is concerned, a contrast is maintained between the outer garden and the inner garden. The outer garden will have simple plantings and stone groupings. The outer garden should be exposed to sun and hence planting of deciduous trees should predominate. The inner garden is a subdued area and hence evergreen trees casting more shadows are used here. The trees used should be of subdued nature and not imposing in character.

The entrance to the tea house is through a low-door so that the guests have to enter in a bending posture, simulating respect and humility.

(D) **Passage Garden:** The passage gardens, the *Roji-niwa*, are those which are laid in narrow passage, as for example a narrow space between two houses or approaches to buildings. As such areas are generally narrow, the garden lay-out should be simple and not overcrowded.

In such gardens there should be hardly any ornaments such as lanterns, basins or other man-made features. The common features of a passage garden are a few key rocks, slabs of stones, and only a couple of types of plant. Bushy shrubs and trees are unsuitable in a passage garden; instead, plants with open form and slender shapes are selected.

The passage gardens generally occur as an accidental necessity in a narrow passage formed by two buildings of a large property. But, occasionally a narrow passage is purposively created to simulate a distant prospect.

(E) **Sand Garden:** It is the simplest style of gardening, though not liked by many as it is totally devoid of plants. The most famous sand garden exists in Kyoto and is known as *Ryoanji* garden. The garden consists of a rectangular area of about 350 square metres adjoining a Zen Budhist temple. The main feature of this style of gardening is to arrange a few vertical and prostrate stones in groups of 2 or 3 and to fill in the gap between the stones with fine white gravel. The gravel is raked in most simple patterns simulating the ripples of flowing water. The raking has to be repeated often to keep the garden in its best shape. This style of garden looks pleasant and effective only when confined to a limited area.

The Features of Japanese Gardens

(a) *Ponds:* The geometric shape does not live up to the Japanese tradition and hence all ponds are of irregular shape. The banks are generally bordered with stone piling work in a regular or an irregular fashion. In olden days the bottom used to be made impervious by puddling clay (for details see water garden). But now-a-days concrete-bottomed ponds are also used. The ponds are generally fed by a stream or a waterfall.

(b) *Streams:* Small streams in Japanese gardens are arranged most naturally with natural stones bordering the banks. Often fanciful stones are arranged in midstream to break the flow of water. The flow of water in the stream may be manipulated by changing the inclination. Sometimes it may be as fast as a mountain rivulet or like a smooth-flowing river of the plains.

(c) *Waterfalls:* The presence of waterfalls in Japanese gardens is another attempt in imitating nature. A waterfall may be made more effective by manipulating it to drop in two or three levels. To make the waterfalls dignified and natural, large-sized stones are arranged around these. To many Japanese, the naked exposer of the cascade of water is not to good taste and hence these are partially screened by planting a group of trees in front of the waterfall. The dense planting of evergreens around the waterfall also symbolizes a mountain scenery where actually a waterfall naturally belongs.

(d) *Fountains:* Often natural fountains are provided near the foot of the hill, on the hillside or in the forest. Often water is conveyed from a hillside by means of bamboo-piping.

(e) *Wells:* In olden days wells in the garden were features of utility. In modern times this feature is more in the nature of an ornament than a utility. The wells may be square, circular, or criss-cross in shape. The frame is generally constructed of stone but occasionally wood is also used. The frame is generally raised above the ground level to a height of 45-60 cm. The mouth of the well is often covered with a mat or a flexible lid knitted out of bamboo-strips. The wells are generally provided with a pulley fitted on a framework. Round or square-shaped buckets are hung on either side of the pulley, suspended from a rope. The well must be made very ornamental.

(f) *Islands:* There are four important garden islands, the first two types representing sea islands. The first of such islands is "Elysian Isle" (*Horai-jima*) which is constructed in the middle of a lake and is never connected by a bridge. The beach is decorated with sand from the sea, shells, and pebbles. No fresh water vegetation is planted over this island. Often this island is given the shape of a tortoise. The second type of island, the "Wind-swept Isle" (*Fukiage-jima*), is also an ocean isle and constructed in a similar way. In addition to sand and shell, the beaches here are decorated with sea rocks.

The other two islands are "Master's Isle" (*Shujin-to*) and the "Guest's Isle" (*Kiakujin-to*) The former is placed in the foreground of the landscape so that it can be easily approached by a bridge from the bank. The latter is located in the background and is accessible by bridges and stepping-stones. These two types of island are generally not used separately. The "Master's Isle" generally has a summer house which is nothing but a thatched arbour. Besides the arbour, some selected trees, stones, stepping-stones, and one or two lanterns are included. Often the islands are in the shape of mounds or hill'ocks. Often the islands are made a peninsula, connected to the mainland by a narrow neck of land instead of a bridge.

According to the strict Japanese style, these are the only islands permitted in a true Japanese garden. But often many people out of fancy introduce a few other types of island, though not according to general tradition, some of which are "Mountain Isle", "Forest Isle", and "Rock Island".

(g) *Bridges:* Bridges are a special feature in a Japanese garden. These are used to reach an island or for crossing a stream. Bridges may be made of stones, polished or unworked, earth, wood, and other materials. The main criterion is that the construction should be pleasing to the eye. The Japanese concept of a bridge is not just a quick crossing of a waterway. The aim is to prolong the crossing time so that the visitor gets enough time to enjoy the scenery around. Even when slabs of stones are used as a bridge, one single span will not connect the destination, but there will be a detour in the midstream and another slab running in a different direction will carry the visitor to the ultimate point.

The earthen bridge is constructed by first putting a wooden framework over which bundles or small logs are laid across and then these are covered with about 15 to 20 cm layer of earth and gravel.

Often to cross a marshy place or a stream a bridge called *Yatsu-hashi* is constructed by placing eight wooden planks in a zig-zag manner supported by arched beams raising from the riverbed. A combination of stepping-stones and bridges is also often used to reach an island or to cross a stream. If there are two or more bridges in the garden, different types should be constructed to bring variety. When the pond or waterway is large enough to permit boating, the bridges should be of arching type to permit the boats to pass under.

(h) *Water-Basins:* The water-basins are fitted near a house meant for the guests to rinse their mouth and wash the hands. But in present days these basins have become nothing but ornamental features. The distance of the water-basin from the house and its size are in proportion to the size of the building. A small house may have a basin 1m tall, whereas in front of a large house the basin becomes as tall as 2 to 2.5 m and thus making useless as a place of washing, but remains there as an ornamental feature. In such cases, if need be, often a smaller basin is provided near the veranda for practical utility. The basins are generally fitted with an ornamental lid.

The water-basin comes in various shapes, the most common ones are in the shape of an urn (see Fig. XXII.1). But square-star shaped, cylindric, stone-bottle shaped, and bowl shaped basins are also quite common. Water basins even resembling the volcanic mountain Fuji are also available. A screen-fence is provided near the water-basin to screen off unwelcome sight. Stones are placed at the base of the water-basin to arrest the splash of water which otherwise may wet the space below the veranda. The sides along these stones are raised either by clay or concrete and a drainage hole is provided which is camouflaged by a number of pebbles. A lantern is provided near by for illumination. Near a tea house the basins are placed low as the persons using these do so in a stooping position as per custom.

(i) *Stone Lanterns:* Stone lantern is an important feature of any Japanese garden. The usual stone used is granite, but sand stone or white marble may also be used. The latter one is specially used for "Snow-Scene" type of lantern.

The usual places of fixing the lanterns are near the base of a hill, on an island, on the banks of a lake, near a water-basin or a well, along a path, on a boat-landing, near waterfalls, and a bridge. The lanterns are used singly but along with a combination of rocks, fences, water-basins, shrubs and trees.

The lanterns are not exactly meant for illumination but as objects of ornaments. Even when a lantern is lit in the night, the light emitted hardly illuminates the place because mostly a group of bushes are grown near it. The object of illumination then becomes a dim mysterious glow. When a lantern is near a river bank, this is lit to have a red reflection on the water surface.

A lantern has six parts, namely, the ornamental top, cap, light chamber, middle stand,

JAPANESE STONE WATER-BASINS

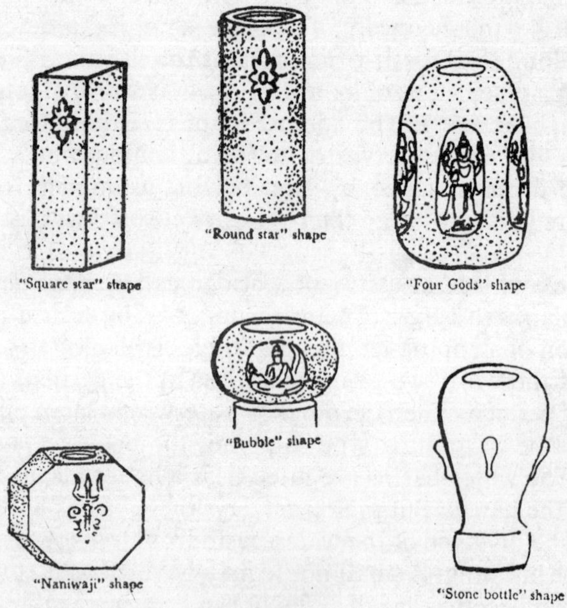

Fig. XXII. 1

post, and base. These may be of various sizes and shapes (*see Fig. XXII.2*). The size should be in proportion to the building or the other garden features around the lantern. One group of stone lanterns are of religious origin, while the other group originated by individuals and these are named after them. The first lantern was erected in Japan in the seventh century by the son of Emperor Suiko.

Lanterns looking of age are valued much and many people make the lanterns look old by artificial means. One such method is to attach moss to the lantern.

(j) *Stones:* Stones are selected according to size, shape, and colour. No stone should be disproportionate to the size of the garden or the features around it, like a lantern or a water basin. The stones are rarely placed in isolation but rather arranged in groups of two to five. Each group will have stones of various heights and shapes. All stones must be arranged with a firm foundation as stones of unstable nature show the weakness of a garden design. The principle governing the arrangement of stones is to make them look natural. For this purpose, low-growing bushes or upright trees are planted near the stone groupings depending upon the size of the nearby stones.

Naturally, the size, shape, and colour of the stones vary according to the purpose and the place of their use. Accordingly, the stones on a hill, on the banks of lakes or streams, and near water cascades vary in their shape, size, and ornamental features.

(k) *Pagodas:* Another favourite feature of Japanese landscape is the stone tower or the pagoda which is a structure consisting of two, three, five, or more separately roofed stages.

(l) *Fences and Gates:* Fences in a Japanese garden are of two types, one is for partition,

DIFFERENT TYPES OF JAPANESE LANTERNS

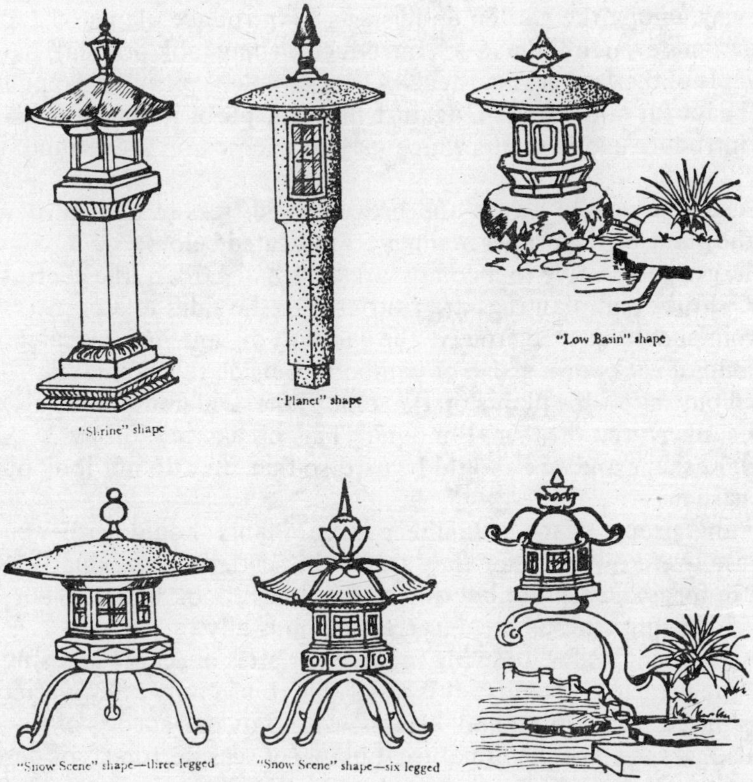

Fig. XXII. 2

while the other is for the purpose of screening which are called "Sleeve Fences". The latter type is so named as its shape resembles that of the long sleeve of a lady's *Kimono*. The partition fences should look light in appearance and hence wood and twigs of bamboo are preferred to stone.

The screen fences are meant for covering something which is not pleasant in sight. This is made of wood or bamboo (whole or split or twigs) woven into patterns thus giving it an artistic look so that it can stand on its own in the garden as a feature of ornamentation.

There are generally two gates, one is the front entrance and second the back entrance. Gates are also made of light materials such as wood or bamboo. Some gates are bare while others are roofed. The roof may be made of bamboo, wood, or simply thatched.

(m) *Vegetation:* The initial step in a Japanese garden is to decide the contours of land and water. Then comes the arrangement of principal rocks which are of primary importance in a Japanese garden as these form the basic structure or the skeleton of the garden. After the stones, the secondary garden framework, i.e., the evergreen plants, are arranged. The trees are of permanent nature and stand as reference points in the garden. Time and again it has been told that the aim of a Japanese garden is to imitate nature by using natural elements, and, hence, there is hardly any bar in using any plant material which serves this purpose. There is a tendency among the garden enthusiasts to introduce plants of Japanese origin in the Japanese gardens even in India. Such introduction may look oriental, but can go contrary to the principle of the Japanese garden. As for example, a pine in a Japanese garden in Delhi may look beautiful but it will go against the principle of imitating nature. But sometimes people do introduce exotic plants which may be excused on the ground of "one time exception".

The Japanese use both needle-leaved and broad-leaved trees in their garden. The broad principles using the plants in the proper manner are indicated below:

(1) To give a larger perspective to a comparatively small garden, the central area is kept clear of trees and shrubs and plantings are restricted to the sides and corners.

(2) An endeavour should be made to give the garden a distant view by catching a natural scenery, say a group of rocks or a grove of bamboos, outside the boundary.

(3) Try to avoid putting all the plants on the same plane. Difference in levels may be created by making a sunken area or a small mound. This breaks monotony.

(4) Plants of right shape and size should be used so that they do not look out of scale to the size of the garden.

(5) While planting, proper spacing should be given. Plants should neither be put in large groups which cause overcrowding, nor they should be scattered individually. Small groups of plants related in form and texture but differing in height look better. Such groups are never put in rigid straight lines, but are arranged naturally.

(6) The Japanese do not prefer to bring in large varieties of plants in a single garden as this goes against the principle of unity. It is advisable to use only a few varieties. In some Japanese gardens the entire planting may be done with only one species of plant.

(7) It will go against the principle of nature if plants of seaside, forest, or desert are combined together.

(8) The Japanese use flowering plants sparingly in their gardens as too many colours go counter to unity. When flowering plants are used, they are planted in between the evergreens.

(9) It is a good practice to use plants which change their leaf colour in the autumn or put forth colourful leaves in the spring. The examples of some such trees and shrubs are: maple, *Cotoneaster*, *Cryptomeria japonica*, *Terminalia catappa*, *Lagerstroemia flos reginae*, and *Ficus religiosa*. Out of different plant materials the trees are planted first which may be either deciduous or evergreen. The deciduous trees are never planted in the front garden except in a tea garden, as the bare branches may not look attractive during the winter. Some typical trees of Japanese gardens are: (a) **Evergreens:** Pines, different species of *Abies*, *Cryptomeria japonica*, *Podocarpus macrophylla*, and *Juniperus chinensis*; (b) **Deciduous:** Maples (*Acer* species), Poplars (*Populus* sp.) Mulbery, (*Morus alba*), and *Salix babylonica* (willow); (c) **Flowering:** The most commonly used plants are different *Prunus* species, besides *Magnolia grandiflora* and others.

Shrubs in a Japanese garden are not introduced just because of the beautiful flowers but mainly on the consideration of shape and foliage. The Japanese never train their trees or shrubs in the topiary form and any such attempt should be curbed. The commonly grown shrubs are: *Aucuba japonica*, Azaleas, *Gardenia florida*, *Nandina domestica*, Camellia, *Lagerstroemia indica*, and *Rhododendrons*.

Bamboos play a special role in the Japanese gardens. The striking patterns of shadow cast by the arching bamboos against paved path, fences, and patios look beautiful.

A paved path in the entrance garden bordered by bamboos simulates a grove.

The traditional Japanese gardens look chiefly green, devoid of any bold colours of flowers. To cultivate the theme of permanence and restfulness trees and shrubs are more appropriate. But, if at all flowers are to be used, the choice should be limited to a few. While planting flowers, remember not to put long beds or borders of flowers in the main garden. In any specific area only one type of flower in a group should be planted, thus avoiding mixed planting. Traditional Japanese will put only one type of flower in one garden, but a non-traditional enthusiast will be well advised to limit his choice only up to three kinds. The flowers the Japanese use more commonly are: chrysanthemums, asters (e.g., *Aster fastigiatus*, *A. glehnii*, *A. microcephalus*), carnation, different lilies, irises, lotuses, peonies, and orchids.

Among the vines, *Clematis*, *Lonicera japonica*, *Ipomoea hederacea* (Syn. *Pharbitis hederacea*), *I. purpurea* (Syn. *Pharbitis purpurea*) *Trachelospermum jasminoides*, and *Wisteria sinensis* are often used.

Chapter XXIII

The Different Features of a Garden

GARDEN WALLS

A garden lover will never like to block the view of his garden by putting a wall along its periphery. A garden as it is enjoyed from inside should also be visible from outside. But, sometimes, from safety or beauty point of view (e.g., to obscure the ugly sight of an open drain), it may be necessary to erect a brick, concrete, or stone wall along the periphery of the garden. A compromise is to have a low brick (or concrete or stone) wall of say 60-90 cm height and to put over it some grills, so that the view is not totally obscured from outside. Alternatively, walls from 1.80 to 5 m may also be erected depending upon the size of the garden.

To break the drabness of a concrete or brick wall, at least in the more prominent places, one may grow creepers such as *Ficus repens* over the wall.

FENCING

Though not exactly a part of the garden in the stricter sense, fences are a utility item essentially needed for marking the boundary and for protection. Sometimes, fences are also put to separate one part of the garden from the other. Several materials such as wood, bamboo, wire, wire nettings, and chain-link fences may be used for this purpose. Fences of wood and bamboo get spoilt quicker than the wire fences. The simplest wire fencing will be to put stout wooden posts at 3 m distance and pass the wires through these posts starting from 22.5 cm intervals at ground level to 30 cm upwards up to a height of 1.8 m. The posts are embedded at least 45-60 cm in the ground and the wires are stretched with the poles with strong stapels or clamps. Wire netting fence is cheaper than chain-link fence but is less durable. Wire netting can also keep away the rabbits, provided the netting is buried at least 22.5 cm in the soil and turned outward at the bottom to stop burrowing. Chain-link fencings of various designs are available and these are fixed in angle iron poles spaced at 3 m distance. Iron grills of various designs can also be used for fencing but the cost, of course, will be prohibitive.

The sight of an expanse of fencing does not suit to the artistic design of a nicely laid-out garden and hence people would like to obscure it from view as quickly as possible by growing quick-growing and elegant creepers such as *Bignonia venusta*, 'Honeysuckle' (*Lonicera periclymenum*), *Vernonia elaegnifolia*, etc.

STEPS

Sometimes it becomes necessary to have steps in a garden, as for example, when a path goes from one level to the other or one has to climb or get down from the terrace garden. Steps can be made of various materials but usually the same material used for the path is used for steps also. The materials used are concrete, stone, wood, or gravel. The gravel can be held in position with a stone-retaining wall. Grass should not be used in the steps as it becomes slippery when watered.

The steps in the garden should be different than those in the building. Here, the treads (i.e., steppings) should be quite broad and the risers low, so that people can have an easier climb. A garden is meant for leisure, pleasure, and comfort and hence steep and narrow steps are unsuitable as they cause discomfort in climbing.

GARDEN DRIVES AND PATHS

A garden must have a carriage drive leading to the house and the garage besides several other paths or walks leading to different parts of the garden. The main criterion of a good garden drive or path is to provide a flat, dry, and pleasant passage for the persons or vehicles and it should harmonize with the other garden features. A path in the garden should be laid out after due consideration is given to its necessity and artistic appeal. Too many paths, especially in a small garden, may spoil the artistic value. A path with graceful and gentle curves will look much better than a straight path.

Materials used for the construction of drives are either gravel, asphalt, or concrete, whereas for garden paths much more attention should be paid for the appearance and as such concrete or asphalt is seldom used. Garden paths are generally made of gravel, paving-stones, crazypaving, bricks, or grass.

(A) Drives

(a) *Gravel drives:* In most of the localities a gravel drive is quite useful for day-to-day work, but if heavy traffic is there the road may not last. A gravel drive should be given adequate foundation to prevent it from sinking. The soil is excavated to a depth of about 30 cm but in heavy clayey soil this should be a little more to ensure proper drainage. The surface should be a little higher at the centre and sloping to the sides. Then the dug up base is soaked and rolled thoroughly. About one-third to half of the excavated depth is filled up with rough stones, well-burnt brick pieces, clinkers, or any other hard material. Water is poured again and the roller applied. The greater part of the remaining excavated depth is filled with coarse gravel mixed with a little earth and rolled. A depth of 5 to 7.5 cm of fine gravel is spread over this, sprinkled with water, and rolled heavily. Too much water should not be used as the gravel will stick to the roller At the time of digging all roots and runners of grass should be removed to keep the drive weed free. Occasionally, a gravel drive will develop a few pot-holes or crack here and there. The damaged area is dug to a depth of about 10 cm with pickaxe and the hole is filled up with stones and gravel. This is then watered and beaten with a heavy beater. The edge of the lawn, if it encroaches onto the road, is cut straight with an edging iron. The drive should be kept free of weeds with the help of some weedicide.

(b) *Asphalt or Concrete Drives:* For asphalt or concrete roads also the foundation is prepared in the same manner and on top of it either asphalt or concrete is laid. A layer of 5 cm concrete will be enough for most roads. For long roads a slope of 1 : 100 may be allowed for drainage. All roads should be a little higher (2.5 cm) than the surrounding ground. The width of the garden roads should be about 3-5 m.

(B) Paths

A garden path should never be less than 60 cm wide but should be preferably between 90 and 120 cm, if sufficient space is available. A wider path allows easy passage for wheel barrows and other garden implements. Sometimes, pathside plants will intrude partly inside the path and if the path is narrow, passage will be obstructed. It will not be easy for a garden lover to chop off a lovely plant; therefore, it will be wiser to have a wider road. Paths should not be generally too high or too low from the adjoining ground except in a marsh or a rock garden.

The principles of construction of garden paths are the same as for drives but since these are narrow and not meant for heavy traffic, the foundation need not be so deep. The excavation will vary from 15 to 22.5 cm, depending upon the nature of the soil. The materials used for garden paths are gravel, bricks, paving stones, and crazy pavings. Although generally concrete is not used for garden paths, it is possible to have concrete paving. This can be made attractive by drawing crazy patterns over the flat surface before the concrete is solidified but not very wet. The artificial crevices should be about 1½ cm deep.

(a) *Gravel Paths:* Gravel paths are made in the same way as gravel drives, except that the excavation and foundation material will be less.

(b) *Brick Paths:* The foundation of a brick path is made in the same way as that of a gravel path but leaving a gap on the top of about 5 to 7 cm for ashes or lime mortar and the depth of the brick to be laid. The bricks chosen should be strong to withstand wear and tear, especially from frost and rain. If the bricks are laid below the ground level no cement will be needed to join them. Bricks are either laid flat or on their edge. Many patterns can be woven such as herring-bone or basket patterns. The bricks in the paths are so arranged that the cross-joints are not continuous. The interspaces between the bricks are tightly packed with sand or if need be with lime mortar.

(c) *Stone Paving:* Paths of stone paving are more popular, especially where stone slabs are easily available, as these are permanent and impart an air of quality. The stones should be between 3 and 5 cm thick and rectangular in shape. The foundation is made in the same manner as for gravel path but need not be very deep. Over this a layer of 5-7 cm of ashes, topped by fine sand, is placed and levelled. To make the path permanent, instead of ash, 5 cm of cement mortar is used. The joint may also be filled with cement mortar to prevent weed growth. Alternatively, these joints are filled in by sand.

(d) *Crazy Paving:* In this type of paving, stones of different sizes and irregular shapes are used. Crazy paving is the most attractive stone path, but sites for such paths should be chosen after giving careful thought. This type of path will not look nice along a modern building, but will be appropriate in a quieter corner of the garden or in the rock garden. Crazy paving is not meant for heavy wear and tear and should not be put to frequent use, although a well-made path will stand moderate usage. Crazy paving must be laid firmly and at one level. Loose stones here and there makes walking unpleasant. This is a disadvantage with crazy paving as the small stones very often dislodge unless cemented. Stones for crazy paving are generally thinner (2-3.5 cm) than those used for stone paths. The foundation is made in the similar manner. Over this hard core about 2.5 cm layer of fine sand or ash is put, levelled properly, and the paving is laid firmly. To make it permanent, the paving should be bedded over 2.5 cm cement mortar. The stones are laid carefully so that no small pieces are needed which get dislodged easily. At the same time there should not be any crevices between stones more than 2.5 cm in width, as walking becomes uncomfortable. The interstices, when cement is not used, are filled with good sandy loam soil so that some rocky plants can be grown here. If this is meant for walking only, the joints are cemented.

(e) *Grass Paths:* It is necessary to have a hard surface for paths which are subject to heavy wear and tear. But if a path is needed for only morning and evening strolling, the most desirable and attractive will be a grass path. Such grass paths are subjected to wear and tear only at short stretches, which are trodden much more than the remaining areas. Paving stones are placed at 60 cm intervals in single row just below the ground level at the places where the maximum usuage will be. This will give a stepping-stone effect. Alternatively, groups of two or three bricks may also be used. The stones are put just below the ground level to facilitate mowing by the machine. But these should not be too deep as the surface will become uneven for mowing.

The best way to make a grass path is from turves. When a garden is already laid, the grass path may be formed out of the existing turf. Grass paths made from turves can stand certain amount of wear and tear from the very beginning. Otherwise, the grass path is to be started from seed of a hardy grass. Under the climatic condition of the Indian plains, *doob* grass will be best suited. For maintenance of grass paths, procedures described under lawn on pages 462 to 468 should be followed.

HEDGE

Shrubs or trees planted at regular intervals to form a continuous screen is called a hedge. The above definition is broad-based, but many other plants such as succulents, cactus, and sub-shrubs also form beautiful and useful hedges.

Purpose: A hedge serves many purposes. A garden hedge can serve the purpose of a compound wall, give shelter from strong gails, ensure privacy, i.e. serve the purpose of a screen, form a background for a floral display such as herbaceous border, as a part of the garden on its own merit, separate one component of a garden from the other (e.g., the vegetable garden from the flower garden), and screen the ugly and unwanted spots such as manure pits, lavatory, servants quarters, etc., in the garden. Though hedge is a useful component in the garden, it should not be overdone, specially in a small garden, by encircling every tree in the garden with a hedge or lining all the paths with hedges. Overcrowding of hedges also robs other plants in the garden with valuable soil nutrients, sunshine, and air. In a very small garden, one should avoid putting any hedge, but instead can put a decent-looking fence on which some light climbers may be trained. A live and impregnable hedge, reinforced with barbed wire, is inexpensive compared to a boundary wall and if maintained properly looks better than a boundary wall. Ornamental hedges are used to demarcate a rose garden or a flower bed or a shrubbery or paths in the garden.

Criterion for selecting a hedge plant: In a garden a hedge is planted with two motives: (a) protective, which means protection against theft, trespass, wind, etc., and (b) for ornamental purposes or screening. For the first category a hedge plant should have the following characteristics - quick growing, hardy, including drought-resistant character, thorny, dense, should respond to frequent pruning and clipping and can be raised quickly by seeds or cuttings. Though attractive foliage and flower is not a criterion in this category, definitely these characters will be an added attraction. Ornamental hedges should have attractive foliage and/or flowers, should be dense in growth habit, and stand regular clipping. The ornamental hedges are generally low in height and they do not obstruct the view completely since the other portion is visible over the hedge.

How to start a hedge: An ideal hedge is that which makes uniform dense growth from base to top. To achieve this objective, each plant in the entire length of the hedge should get more or less uniform supply of water, essential elements, and sunshine. Even if all the parts of the hedge receive uniform supply of the first two factors and does not get equal amount of sunshine, the portion growing in shade will make less and lanky growth.

After the uniformity in the above three factors is ensured, one should next proceed to prepare the land. In a hedge the plants are spaced much closer than they would have been for any other purpose. Moreover, once a hedge is raised it is likely to remain there for about 30 or 40 years. During this period the plants in the hedge have to maintain themselves in that congested position, notwithstanding the frequent clipping and pruning. The preparation of the site is undertaken by digging a trench 30-90 cm wide and 30-75 cm deep, depending upon the type of the hedge. The heavier the hedge the deeper and wider is the digging. Uniformity in digging is to be ensured as variation in digging will result in the fluctuation of growth of the shrub at the cost of symmetry and neatness. Deep digging also encourages the plants to root deeply and prevent sending less number of laterals to compete with the plants growing along the hedge. The digging operation should start several weeks prior to actual planting time. A moderate dose of F.Y.M. is added at the middle layers. Since the intention is to have a bushy growth at the base, over-manuring is to be avoided which results in formation of a few strong vertical growths at the expense of lateral growth. If the manure is spread over the top layer, it may attract white-ants also. Under Indian conditions, the planting of hedge is undertaken at the beginning of the monsoon season in June-July. A hedge is started by sowing seeds or by putting cuttings *in situ* or by planting rooted cuttings, generally in double rows. The spacing will depend upon the nature of growth of the plant in the hedge. Trees such as *Putranjiva roxburghii* and *Polyalthia longifolia* may be planted 75-90 cm apart. The spacing for shrubs will vary from 15 to 45 or 60 cm. If planted in double rows the rows should be spaced from 15 to 45 cm, depending on the nature of growth. The seeds are sown or cuttings are planted 15-60 cm apart in a staggered manner when in double rows, i.e., alternating the plants so that plants are not opposite each other. If sufficient seed is available double the normal seed rate is put and then seedlings thinned out to desired spacing. This is a necessary precaution against failure in germination. Double rows are put when a dense hedge is needed. When rooted cuttings are planted, any unnecessary long tap-roots should be shortened and the evergreens may be tipped to reduce transpiration loss. Immediately after planting or sowing, water is applied copiously. Ample watering space, about 10 cm deep, should be left. A large number of seedlings or rooted cuttings should be kept in reserve and receiving the same treatment for any possible replacement due to casualty.

When the hedge is about 15-25 cm tall, it is topped back to 10 cm with garden shears.

Heading back should not be delayed to achieve quick growth as vertical growth will be at the expense of basal growth. The lateral growth should also be topped in the same way, which in turn produces more laterals and again the procedure is repeated, until a dense growth is obtained. To obtain a cattle-proof hedge, the hedge should be reinforced by three to four rows of barbed wire as the hedge itself may not be sufficient.

Maintenance: Hedges do not like weed growth at the base and these should be uprooted as and when they appear. Deep irrigation is provided once a week, depending on the weather, and all the casualties should be replaced with plants kept in reserve in pots. Keeping reserve plants is essential, otherwise the casualties are to be replaced by seeds or new cuttings, again the uniformity in growth will be lost.

In white-ant-infested localities, the soil should be kept continuously moist as white ants do not like wet conditions. White ants can be controlled by chemicals also (see Chapter XXVI on Diseases and Pests).

Clipping and pruning of hedges are most important to keep the hedges neat and in good shape. They should be done at regular intervals and at no stage the top growth should exceed 15-20 in cm in length.

For details of pruning and clipping consult Chapter IV, page 48.

Manuring: Manuring is one aspect which receives least attention by the gardeners. One may be inclined to believe that manuring of hedges is a sheer waste. But the result of well-manured hedge will speak for itself the benefit of manuring. Once in a year the hedge should receive, before the rains, well-rotten cowdung or horse manure at the rate of 4 kg per running metre. It should be well-incorporated in the soil. Fertilizers can also be added twice a year at the rate of 30 g each of superphosphate, bonemeal, and ammonium sulphate per metre row of hedge.

Selection of suitable plants: There is a wide variety of trees, shrubs, and other plants which can be formed into hedges. Some trees form good tall hedges. A list is provided below for a general guidance.

(A) Palms as Hedge

Some palms form into beautiful ornamental hedges. A few of these are listed below:

Areca lutescens (Syn. *Chrysalidocarpus lutescens*, *Hyophorbe commersoniana*, *H. indica*): A bushy palm with handsome feathery golden coloured leaves.

Ptychosperma macarthurii (Syn. *Actinophloeus macarthurii*) : A dwarf palm sometimes growing up to 4-6 m in height and suckering freely at base, but hardly exceeding 1.5 to 2 m in width.

Rhapis excelsa (Syn. *R. flabelliformis*, *R. aspera*) : Grows to a height of 2-3 m but sends out runners, which should be controlled at regular intervals.

(B) Trees as Hedge

Acacia farnesiana and some other *Acacias* form good thorny hedges in dry localities. This plant should be pruned regularly to keep it thick and dwarf. Seeds are sown 30 cm apart.

Diospyros embryopteris: Forms good hedge if kept low.

Erythrina: Many species of this genus can be trained into a good hedge by pruning and shearing.

Grevillea robusta: An ornamental tree which forms a good hedge when planted closely and trimmed low.

Inga dulcis: A thorny plant which forms a good cattle-proof hedge. It is, however, subject to white-ant attack.

Parkinsonia aculeata: It is a good hedging plant for the dry localities. It has ornamental foliage and nice yellow flowers.

Polyalthia longifolia: It is one of the best among the trees which subjects itself to be trained and pruned into a good hedge, especially for screening and protection against winds.

Pongamia glabra: It is also used as hedge, especially in south India. The seeds are sown *in situ* or seedlings are planted 60 cm apart.

Putranjiva roxburghii: It is another tree with ornamental foliage which is used as a hedge.

Thevetia nerifolia: It forms good hedge especially for screening. It has beautiful flowers. Sometimes this plant is included in shrubs.

(C) Conifers as Hedge

Casuarina equisetifolia: It should be pruned before it forms a woody trunk. It forms a nice hedge and can be kept quite low. It can be also cut to a square top.

Cryptomeria japonica: It forms a good hedge on high hills.

Many species of *Cupressus* are trained as hedge, but these should be limited to a height of 6 m by heading back.

Thuja: Some of the species especially *T. orientalis* can be pruned to shape and thus forming a good hedge.

Trees are generally planted 60-90 cm apart in a hedge.

(D) Shrubs for Hedges

Acalypha: Many species are used as hedges but, species such as *A. tricolor* does not stand pruning well. *A. hamiltoniana*, *A. wilkesiana* var. *musaica* and other small-leaved species can be trimmed regularly to form ornamental hedges. The tall species can be used as protective hedge also. Cuttings are planted 25-30 cm apart.

Aralia: Different species (see page 254) are used as ornamental hedges.

Barleria: Different species with their beautiful flowers subject themselves to be trimmed to beautiful hedges; plants are 60-100 cm high. Rooted cuttings are planted 20-25 cm apart.

Bauhinia acuminata: It can be trimmed to a nice hedge, which looks more beautiful with its white flowers.

Bougainvillea: B. spectabilis, B. glabra, 'Mrs. Fraser', 'Partha', and some other cultivars form good protective as well as ornamental hedges withstanding frequent trimming. Planting is done at 60-90 cm spacing and the plants are trained against barbed wire or wire trellis and the shoots including side shoots are trimmed to form bushy appearance. Plants without support can also be trained, but supported hedge looks neat.

Caesalpinia pulcherrima: It should be pruned tall, harder pruning forces it to ageing faster. The plant flowers freely.

Carissa carandas: A thorny bushy shrub which is used as an effective protective hedge. Seeds are sown 60-75 cm apart in single rows. Cuttings also root easily.

Clerodendron inerme: It is most suitable for ornamental hedge and can also be used as protective hedge. The leaves are ornamental and polished green in colour. It is hardy, drought resistant, stands trimming very well, and does not harbour snakes as the leaves have a repulsive odour. It is rather slow-growing and takes about two years to form into a good hedge. It is planted 15-20 cm apart in double rows spaced at same distance.

Daedalacanthus nervosus: It forms an ornamental hedge which is generally 30-45 cm high. Cuttings are planted 20-30 cm apart in double rows, separated by same spacing.

Dodonaea viscosa (Family, *Sapindaceae*), Origin - Australia, North and South America: It is used as a tall, protective hedge and is quite hardy; hence, used frequently in the dry areas. The shining linear-oblong leaves are ornamental. Seeds or cuttings are planted 30-45 cm apart. It is clipped frequently to form a height of 1.5 m.

Duranta: It forms a good ornamental as well as a protective hedge. The variegated variety looks very ornamental. It is planted 30-40 cm apart and can be kept quite low by trimming.

Eranthemum: Several species (see page 144) are used as ornamental hedge. Cuttings are planted 15 cm apart in double rows.

Galphimia gracilis: A beautiful plant with yellow flowers which forms a good hedge. The plants are planted 30 cm apart.

Hamelia patens: A beautiful shrub which stands trimming and attains a height of 75-90 cm. It is planted 30 cm apart.

Hibiscus: Different types especially *H. rosa-sinensis* are grown as hedge mainly for ornamental purposes. It is frequently pruned to keep a height of 75 to 90 cm.

Justicia gendarussa: It is a quick-growing hedge which is trimmed at 60 to 90 cm. Cuttings are spaced at 30 cm apart.

Lantana: Different species of this genus are used as hedge and these stand trimming and pruning very well. The dwarf species are trimmed at 45-75 cm, while the tall kinds can grow up to 1.5 m. The plants should not be allowed to seed as these spread like weeds.

Lawsonia alba: It is a good hedge for dry localities and is grown from seeds or cuttings spaced at 45-60 cm apart.

Malpighia coccigera and *M. glabra:* Both plants form dense hedge to a height of 60-90 cm. The plants are spaced 20-30 cm apart.

Meyenia (Syn. *Thunbergia erecta*): It forms an ornamental hedge of 45-60 cm high. Rooted cuttings are spaced at 15-30 cm apart.

Murraya exotica: It also forms a good tall attractive hedge when properly trimmed.

Plumbago capensis: It forms a pretty ornamental hedge. Cuttings are put in double rows, 15-20 cm apart.

Punica granatum: It forms a tall hedge, 1.5 to 2 m high. It is drought-resistant.

Poinsettia pulcherrima: It can be used as a beautiful ornamental hedge.

Rosa: Some species such as *R. multiflora* can make a good spiny hedge.

Sesbania aegyptica (Syn. *Aeschynomene sesban*): It grows to a height of 2 m within 2-3 months and is a popular hedge in the drier parts of India. It is raised from seeds and must be replaced after two years.

Strobilanthes anisophyllus: A shade-loving ornamental hedge, which should be trained at 45-60 cm height.

Tecoma stans: It forms an attractive tall (1½ to 3 m) hedge within a span of two years with attractive yellow flowers.

(E) Cactus, succulents, and others as hedge

Agave americana: This and other tall growing agaves are used as barriers to cattle.

Bambusa: Some dwarf species such as dwarf Chinese bamboo (*B. nana*) make untidy hedge. Bamboo hedge may harbour snakes, if kept untidy.

Cereus: Some species of this genus belonging to *Cactaceae* are used as thick spiny hedge.

Euphorbia antiquorum: A spiny succulent shrub which makes a tall hedge up to 3 m.

E. bojeri: With its pretty scarlet flowers this forms a good hedge, 45-60 cm high.

E. splendens: It is similar to above but stems and spines are stouter.

E. tetragona: It is used to form a tall spiny hedge.

E. tirucalli: It is a spineless succulent tree, branches slender, brush-like and cylindrical.

Furcraea selloa: A plant like agave with spines, can be used as cattle-proof hedge.

Opuntia: Many species of *Opuntia*, belonging to *Cactaceae*, form into thorny hedges.

Pandanus: The screw pine is a thorny plant and should not be used as hedge in a small garden. It is used as hedge in large commercial farms growing food crops. The plant bears scented flowers. Because of its thick growth, the plant is a good hiding place for snakes.

Pedilanthus tithymaloides (Syn. *Euphorbia canaliculata*): A spineless plant belonging to *Euphorbiaceae* is commonly used as ornamental hedge. Cuttings are planted 10-15 cm apart in double rows spaced 15cm from each other.

Cactus (*Cereus*, *Opuntia*) when used as hedge outgrows in size and should not be used in home gardens.

The following plants can be used as hedges on elevations of 1000 m and above: *Aucuba japonica*, *Berberis asiatica*, *B. bhutanensis*, *Euonymus* (different species), *Hibiscus syriacus*, *Hydrangea* and *Spiraea* sp.

EDGING

Lining of borders of flower beds, paths, lawn, and shrubbery with brick, concrete, living plants, etc., is known as edging. Edging may be formal, made of stone, bricks, tiles, etc., or informal consisting of living plants. There are certain plants which are very suitable for edging purposes and these are known as edge plants. Edge plants, unlike hedges, should grow very low as the purpose is not for screening but to provide a lining only, for the purpose of decoration or demarcation.

Formal Edgings

All conventional formal English gardens used to have edgings of bricks or tiles diagonally placed along all roads, paths, and flower beds. But in recent years, the trend has changed and with it the fad for edging anything or everything has declined. It is desirable that the practice of edging should not be used too often, but at the same time it should be used with thought wherever necessary. For example, a flower bed adjoining a cinder or a gravel path should be properly edged. The materials used for formal edgings are as follows.

(a) *Bricks:* This is the most commonly used material for edging. The brick selected should be of good quality so that it does not crumble down easily. Brick edgings with worn-out edges look more ugly than beautiful. When the edgings are to be put in a straight line, a rope line is to be fixed in position with pegs. The bricks are buried to half their width and the soils on either side rammed very firmly. If the soil is light in nature these are to be fixed in place with the help of 5 cm cement mortar. When mortar is used only one-third of the brick is buried. Bricks can also be set at 45° angle to give a serrated finish. The top of the bricks can be kept at the same height by placing a straight edge over the top of the bricks and placing a spirit level above this, to adjust alignment.

(b) *Tiles:* Tiles of various shapes and sizes are available which are quite suitable for edging. It is much more difficult to keep the alignment of the tiles than that of bricks. A narrow trench is dug out and the tiles placed in position and earth rammed very firmly. Long and thin wooden boards may be used to ensure fixing the tiles in straight line. It will be better to fix the tiles with cement mortar.

(c) *Stones:* In places as Bangalore and Jaipur where stone slabs are easily available, these can be used for edging. Stone edgings should have nice rounded finish.

(d) *Concrete:* Concrete edgings with rounded tops look quite attractive. Precast concrete edging in lengths of 1 to 1.5 m are available, which pose no problem in fixing. When loose concrete is to be laid *in situ*, a mould 7-8 cm broad is to be placed in the trench, so that at least a thickness of that dimension is obtained. The edge above ground is also between 5-7 cm.

Informal Edgings

(a) *Grass verges:* People who do not like artificial edgings as mentioned above may use living plants as edgings in some places. A grass verge, a strip of grass, is a very attractive edging, especially in front of flower beds. A grass verge should not be less than 60 cm in width as otherwise lawn mower cannot be worked. Whenever a grass verge is used it must be maintained properly.

(b) *Other informal edgings:* An informal edging of stone-crazy path with prostrate plants growing in between serves as a very good edging in front of a border.

Edging of plants is very good, provided these are kept low and maintained by neat and proper trimming. The plants selected should be hardy and should have foliage or flowers of a lasting nature. All casualties are to be replaced regularly. The edging plants may be used either for the foliage or for flowers.

(A) Foliage Plants for Edging

Aerva sanguinolenta; Family: *Amaranthaceae;* Origin: Java. A herb with leaves dark red above and paler beneath; susceptible to insect attack.

Alternanthera; Family: *Amaranthaceae;* Origin: Brazil. The most commonly used herbaceous plant for edging. The species commonly used are *A. amabilis* (Syn. *A, ficoides*), *A. amoenga, A. bettzichiana* (also spelt *bettzickianc*) var. *spathulata,* and *A. versicolor*. The last two are very colourful, shaded red, bronze, and pink. These are low-growing and stand trimming very well. Cuttings are planted 3-5 cm apart in double rows spaced at 5 to 10 cm *in situ*. Bailey (1963) describes the genus as *Telanthera*.

Asystasia coromandeliana; Family: *Acanthaceae;* Origin: India. A straggling herb which forms a good edging for shrubberies. It grows in shade. Flowers are white to pinkish. Should be trimmed to keep within bounds.

Cineraria maritima (Syn. *Senecio cineraria*); Family: *Compositae;* Origin: England, Wales, Ireland. A sub-shrub with a woody base, the whole plant is densely white-woolly. This is a good edge for red *Acalyphas* or similar shrubs in shrubberies.

Coleus; Family: *Labiatae*. Some dwarf coleus (20-45 cm high) can also form edging to ornamental shrubs.

Echeveria; Family: *Crassulaceae*. In the hill stations some species of *Echeveria* are used for edging rockeries.

Eupatorium cannabinum; Family: *Compositae;* Origin: Asia, Europe. A herbaceous erect plant which can be trimmed even to 15 cm. It forms a good green edge.

Iresine; Family: *Amaranthaceae;* Origin: Tropical America. The leaves are colourful, differing in species from rose-carmine, red to deep blood-red. The species are *I. herbstii* var. *brilliantissima* which has rich crimson leaves; *I. lindenii* (Syn. *I. acuminata*) in which the leaves are deep blood-red.

Pilea muscosa (Syn. *P microphylla*); Family: *Urticaceae;* Origin: Tropical America. This species and some other species of *Pilea* are used for edging. *Pileas* are useful for edging rockeries. These are suitable for semi-shade and shade, but can be grown in the open in cool climates.

(B) Flowering Plants Used for Edging

Many of the flowering annuals and some perennials are also used for edging. The most important among these are *Alyssum, Amaryllis, Brachycome, Gerbera, Torenia, Zephyranthes* different types, pansy and perennial verbena. Many cultivars of miniature roses are also used for edging.

ARCHES

A garden may need some arches for training climbers or ramblers. Arches are generally constructed near the gate or over paths in the garden. An arch should be at least 2 to 2.5 m high, so that the branches of creepers hanging down should not interfere with access. The breadth depends upon the path over which it is constructed but should not be less than 1 m. A general rule is to have the same width as the height provided space permits. The cheapest, though not very durable, method is to construct this supported on wooden poles. Over the poles longitudinal rails are placed and cross rails are fixed over

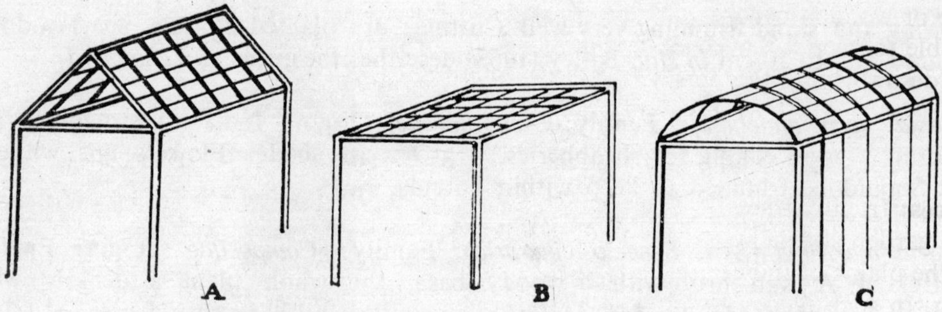

Arches—A. Angular-roofed, B. Flat-roofed, C. Arch-roofed

Fig. XXIII. 1

these at intervals of 45-60 cm. The rails are fixed with the poles by driving in sturdy nails. In Rajasthan and Bangalore, where stones are easily available, the supporting poles can be made of stone pillars. The roof may be made of wood as above or angle iron of suitable section. The angle irons can be fixed with the stones by the help of clamps and the cross rails may be welded or fixed with nuts and bolts. Alternatively, binding wires may be used. The best method is to use angle iron support firmly grouted in the soil by concrete. The roof made of angle irons of different sections may be flat, arch-type, or angular with slopes on either side. (see Fig. XXIII.1). The support can also be made of 7-10 cm diameter galvanized pipes fixed in a similar way. On the top of each pipe a galvanized or steel plate (22.5 X 22.5 cm) with holes should be welded which will facilitate the fixing of the longitudinal angle irons with the help of nuts and bolts or clamps. Galvanized wire nettings may be fixed on the sides of the arches to help the creepers to climb up. The portions of the wooden support going underground is painted with coal tar to prevent damage by termites. The angle iron and G.I. pipes are also coated with rust proof paints. The portions above ground should be painted green.

PERGOLAS

For growing creepers in a row pergolas are ideal structures on which these may be trained. A pergola may be defined as a series of arches joined together. Pergolas are generally constructed over pathways, which add beauty to a garden. Like arches the support can be made of wooden or stone or brick pillars, angle iron and G.I. pipes. The supporting pillars may be connected together by wooden rafters or m.s. rods may be welded to the pillars. The roof may be made of angle irons of different sections with longitudinal and cross rails. Over this base strong galvanized wire-mesh or welded-mesh may be placed for the creepers to spread easily. The width of pergolas is kept generally within 2-2.5 m and the height is also the same but preferably on the higher side. The width may be increased or decreased a little, if necessary. The length will depend upon the area to be covered, the number of creepers available, and of course, the budget. The roof of a pergola may be flat or in the shape of an arch. As in the case of an arch, the pergola is also painted green to merge with the colour of the foliage.

A pergola is a useful resting place during the summer months in a tropical country. The path below remains cool due to the creepers growing above. If the pergola is wide enough concrete or wooden benches may be constructed for sitting. In broad pergolas it may also be possible to keep a few shade-loving plants to protect them against sun, but this may not be desirable all the time.

TERRACES AND TERRACE GARDENING

Terraces: In hilly tracts it is not possible to have a large piece of land in one plane for laying a garden and hence gardens are laid in terraces, where it is a natural phenomenon. But, in the plains of India, the land for gardening may not have any natural undulation for terracing. In such places it will be worth while to have a terracing to break the monotony and bring novelty into the garden. Actually it is a common feature in the English or Japanese gardens to construct a part or a major portion of the garden in terraces. Where natural undulation is there it is not very expensive to create the terraces, but in a plain land it will be little expensive to create one. For a household garden one may remain content with making one terrace only.

In European weather, where sunshine is in great demand, the terrace is best situated in front of the house where the inhabitants can bask in the sun and enjoy the full view of the garden. Under our climatic conditions in the plains also, the same principle may be followed, but the terrace should not be preferably situated on the western side as it will prevent the inhabitants from relaxing in the afternoon because of the hot sun. A terrace may also be constructed in other parts of the garden, say in front of a boundary wall or a dry wall. A terrace should not look out of place but form a natural link between the two and should give a full view of the garden. It is advisable to make the terrace about 45 cm above the general level of the garden supported by a retaining wall or a bank, preferably made of stone blocks. It is not necessary to have a straight outline or to locate the steps in the centre. Ample space should be provided to utilize it as a sitting place. The major portion of the terrace may be paved with stone or brick, but before doing this space should be left for small flower beds or for creepers to climb on the house walls, if there is any plan for these. The soil in these areas should be dug to a depth of 45 cm and bonemeal and F.Y.M. added. A slope of 1 : 60 be allowed towards the house to prevent the run-off water from the house flowing on the terrace. Whenever extra soil has been added to construct the terrace, it should be made firm and allowed to settle before paving is done. If the terrace is sufficiently wide, patches of lawn in between pavings will add to its beauty. In places having a hot summer, it will be advisable to keep the paving to the minimum and leave more space for beds and grass to keep the place cool. Drainage should be ensured, especially in high rainfall areas, by providing drainage holes, which may be connected with the sewerage system.

Terrace Gardening: A terrace is a raised space of ground constructed around a dwelling house or at the corner of a garden or on the sides of a hill. When this terrace is used for some sort of gardening this is known as Terrace Gardening. This type of garden is basically meant for a place of leisure and pleasure. A terrace garden is generally constructed just in front of the house from where a view of the whole garden can be obtained. But this can also be constructed in any other corner of the garden from where a clear view can be obtained.

In western countries and particularly in the U.S.A., hardly there is any house which does not include one or more areas designed for comfortable outdoor living, for leisure or pleasure, through a greater part of the year. And this is the place known as Terrace Gardening.

How the terrace is to be constructed has been described in the proceeding paragraphs. A terrace garden is very often referred to as the outdoor living or drawing room and to make it more cozy and private some sort of walls are to be constructed on the sides or ends. First, a low brick or stone-retaining wall may be built to handle the change of levels while on the top of it about 90-120 cm tall wall of lace patterned concrete blocks may be built which will provide privacy but will not stop breeze or obstruct light. Brick walls can also be raised in the similar pattern to allow enough light and air to pass. Wide-board wooden fences can also be used but this will obstruct light and air. Since a terrace is used as a outdoor livingroom for relaxing, the area should always be nearly dry under the foot. For this particular reason, most people favour a paved terrace. But grass is also suitable for terrace gardens which hardly receives rough usage. To facilitate easy maintenance, for more comfort as a sitting place and to keep the place dry paving is done with local stones, flag-stones, brick, concrete, wood, gravel, and mosaic tiles. Wherever a paving is done first a hard foundation is made over which the paving is laid. If cost is a limiting factor, a gravel paving will be the cheapest but will have certain drawbacks. Gravel pavings subside with the pressure of heavy furniture and children injure themselves while playing. Always locally available materials should be used to reduce the cost. Only well-burnt bricks should be used for durability. Bricks and stones can be laid in simple design or arranged in artistic pattern. A combination of brick and stone paving or a wood and stone paving, etc., can also be tried. Wood pavings are generally available in the form of rectangular blocks or round sections cut from logs. The wooden pavings are suitable for mild climates as in extreme of climates wood disintegrates rapidly. Some rot-resistant woods should be used and further these should be treated with rot-resistant chemicals such as copper naphthenate or pentachlorophenol which are available under many trade names such as cuprinol, naptox, etc., before being fixed in position. The wooden pavings should be kept in position firmly by stone or plank or log treated with preservaties and with edges fixed deeply into the earth.

Planting: Planting in the terrace should be done after much thought and planning. First, it is to be decided if the whole area will be paved or some portions will be left out for lawns, beds, planting of one or two specimen trees and creepers for training over the house walls.

As a general principle, the paved surface itself should have very little plant growth. Wide joints of the pavings allow more growth of grass which is difficult to clip and requires more time in maintenance than narrow unplanted joints. Moreover, such joints are hazards for the ankles of the playing children as well as adults. Too many creeping plants in the joints restrict the movement and reduces the sitting space which is the primary necessity in a terrace garden. Only a few low-growing creeping plants such as *Portulaca*, *Lantana sellowiana* for the tropics and *Thymus serpyllum* and *Veronica repens* for the temperate regions can be recommended for planting in the wider joints. But all such plants should be used with discretion to keep the terrace free from hindrance to comfortable use. Some small specimen dwarf trees or shrubs can be planted amidst grass in the circular, rectangular or square beds left out. But such plants should fit with the architectural design of the house and should also be most ornamental and interesting looking. Some interesting and attractive potted plants in tubs and bowls can also be arranged artistically especially in completely paved terraces. In no case the plants should obstruct vision and screen the architecture of the house from being seen from outside. Hanging baskets can also be displayed in the terrace. Plants can also be displayed in well-designed plant stands. A lily pool may not be out of place in a paved terrace. A sundial or birdbath or a stone sculpture may also be constructed in a suitable corner. Chairs and tables may be arranged in places for resting and comfort. Since the furniture of a terrace garden is not disturbed frequently and remains outdoors, the materials used for its making should be of sturdy nature such as wrought iron.

PAVED GARDEN

A paved garden, if properly laid, can be a very attractive feature of a garden. There are some specific plants, which adapt themselves well to a paved garden. These should be dwarf in nature and stand a considerable amount of wear and tear from shoes of different weights. But a paved garden should be laid in a path which is not used very often. A special paved garden may also be created if a path suitable for this is not available. Ordinarily a paved garden is meant for walking, although not very frequently, and hence the interstices of the paved garden should be planted sparingly.

The foundation of the paved garden should be prepared in the same manner as for an ordinary pathway. Proper drainage should be ensured for the paved garden plants. After the foundation is laid, a good soil, preferably of lighter texture, is laid to a depth of about 10 cm. Before laying the stones over this, 2-3 weeks' time should be given to the soil to settle. The stones selected should not be less than 25 cm in diameter, otherwise the paved garden looks clumsy. For paths, rectangular stones are more suitable, whereas for circles and squares irregular stones of any shape or size can be used. Between two stones gaps of about 4-8 cm are left and the stones are laid in an informal pattern. These spaces are filled with good soil and traces of bonemeal is also added. The boundary of the paved garden is usually supported by some rock work. But if the paved garden is raised or the side stones are not supported by any adjoining feature, these should be cemented as in the case of a crazy path. When the sides are supported by stones, the plants grown to cover these are allowed to grow a little inside the paved garden in an informal manner to give a more natural effect.

Planting should be done with discretion and over-planting should be avoided. The crevices should not be filled completely, rather the plants should be placed irregularly either in small groups or in isolated patches. A dense planting will look unnatural and will have very little appeal. In a sparsely-planted paved garden even the worst clumsy visitor will be able to avoid tramping of a plant in flower.

Some of the plants suitable for a paved garden are listed below :

Achillea rupestris (*Compositae*): A mat-forming, tufted plant with white flower heads.

A. tomentosa (*Compositae*): A compact, mat-forming plant with bright-yellow flower heads.

Alyssum montanum (*Cruciferae*): A dwarf (7-8 cm), prostrate, tufted plant which bears many-flowered racemes with bright-yellow and fragrant flowers.

Dianthus deltoides (*Caryophyllaceae*): Commonly called 'Maiden Pink' is a prostrate dwarf (12-20 cm) herbaceous plant, bearing purple to crimson-red flowers with dark stripes and bright spots.

Euphorbia splendens prostrata: See page 297.

Geranium pylzowianum (*Geraniaceae*): A slender, dwarf (30 cm tall) plant bearing purple flowers.

Gypsophila repens: See page 85.

Hymenatherum tenuifolium: A herbaceous perennial or annual of prostrate habit, the fern-like foliage is finely divided and beautiful. Flowering is round the year.

Lantana sellowiana: See page 154.

Portulaca (Perennial type): See pages. 117-118.

Setcreasea pallida and *S. purpuria:* See page 271.

Verbena erinoides: See page 119.

Vinca rosea: See page 119.

Viola cornuta: See page 109.

Zebrina pendula: See page 273.

Zinnia linearis: See page 120.

A gardener should be able to add many more plants by practical experience.

DRY WALL

The term "Dry Wall" is rather a misnomer, as a wall may be dry but once the plants are planted over it, it no more remains dry but becomes an object of beauty. Whatever may be the name, gardens laid on a wall or walls planted with different plants are termed as "Dry Wall" in the gardening dictionary.

Plants growing in the crevices of the stones and hanging down the face of a dry wall look beautiful and are becoming a common feature of the English gardens. In India, a dry wall is hardly seen in any garden, although this is a very interesting feature of a landscape garden. The dry walls are generally retention walls, with only one surface available for gardening or planting while the other side is supported against the vertical or near-vertical face of two different levels of landscape (see Fig. XXIII. 2). A dry wall, of course, can also be constructed as a column.

Under Indian conditions, specially in towns and cities, there are hardly a few houses which will have a wall suitable for a dry wall. Moreover, the landlord might have never thought of putting up a dry wall while constructing the house and the boundary wall, if any, and as a result there may not be any provision to start a dry wall. But, if space permits, it may be possible to convert a portion of the concrete wall or garage wall with very little additions.

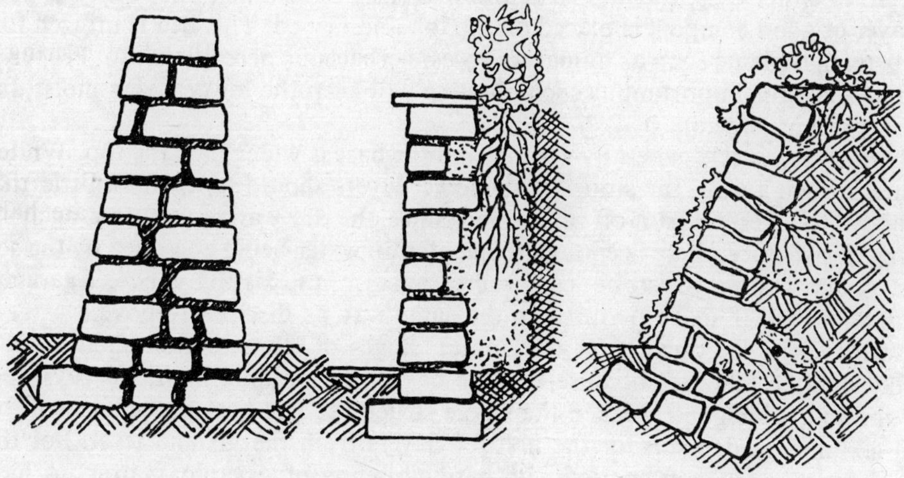

Three methods of constructing a dry wall—A. Laid independently; B. and C. Laid against vertical or near-vertical surface between two different levels of land.

Fig. XXIII. 2 DRY WALL

Construction of the wall in most parts of the plains should start just before the rainy season. The aim is to coincide construction with planting since the rainy season in the plains of India is the best time for planting. Before proceeding with the foundation, the slope of the land supporting the wall is to be prepared. If the total height of the wall does not exceed 75 cm, the wall may be constructed vertically. Where the height is more (generally no wall is more than 2 m in height) the wall should be laid in a slanting position with a slope of 1 : 6. After this is ensured, a firm foundation is laid. Though no mortar is used for the foundation, this should be made firm by ramming the soil. A trench 30-50 cm deep, depending on the type of soil, and about 75-90 cm wide is dug. If necessary the bottom 10-15 cm may be made firm by ramming in old mortar and pebbles and levelled to a more or less flat surface. Then the larger stones at hand are placed flat side upwards over this layer. The stones used for a dry wall should be natural stones, either limestone or sandstone. In the first layer, no gap is left between the stones. The stones in the subsequent layers also should have a flat upper surface. The vertical gaps between two large stones should be about 5-8 cm, while the gap will reduce proportionately with the smaller-sized stones. The horizontal gaps between two stones will be about 2-3 cm. The horizontal gaps are filled in with soil and firmly rammed. Any gap between the back of the wall and the vertical gap is also filled in and rammed. Any air pocket left in any of the gaps will not only weaken the wall structurally but will also be a cause for drying out of the plant roots due to lack of moisture. The laying of stones needs a lot of care. After the first layer is completed, the second layer is started, taking care that the vertical gap in this layer is placed above the centre or as close to the centre as possible of the stone below this. If this is not ensured the wall will not last long. The top of the wall is finished flat, although some people recommend that this should be uneven. In either case, the aim is to allow soaking of as much rain water as possible. A 15 cm layer of good compost is placed on the top and firmed. This bed is utilized for planting the perennial plants such as miniature roses, herbaceous perennials, etc. Proper planting of the upper layer is important as such planting will keep the lower layers moist, firm, and will also prevent erosion.

A columner dry wall is generally vertical and the base is wider than the top. While placing the stones in such a wall, the stones at the lower layers should taper out a little than those placed above to arrest the run-off rain water. Since the dry wall constructed against a slope is slanting in structure, there is enough scope of rain water being absorbed by the soil in the crevices. A dry wall may also be constructed against an existing concrete garage wall or the boundary wall, if space permits, in the same way as discussed above.

The soil used in the crevices at the initial stages of construction should not contain any bulky organic manure, because as time passes the organic matter decays and leaves gaps in the soil, which will weaken the whole structure. The best procedure will be to use medium textured garden soil for the first few layers from the ground level. For the upper layers a richer mixture may be used with bare minimum of organic matter. A handful of bonemeal may be added for every 10-12 kg of soil. Fertilizer mixture can also be applied occasionally, especially for the column wall.

Planting is done at the base of the vertical joint, as, if the plant is planted on the top crevices of the vertical joint, it may soon die for want of moisture. The initial planting should be done with seedlings to have a quick start. For subsequent planting, seeds may be sown directly in the crevices. If moss is available, wet moss is wrapped round the seedlings (or seeds) to prevent drying. Plants in the wall are watered with water cans fitted with a rose. The best method of watering a wall is to shower water from above. The frequency will be at short intervals. While planting it should be borne in mind that the wall should appear to be old. For this, planting of ferns and mosses here and there will be helpful. Planting should be done after proper planning so that the wall looks natural and merges with the general landscape. The plants suitable for dry wall are listed below.

Ageratum, Sweet alyssum, *Bellis perennis* (suitable for cool climate), Brachycome in various colours, *Gazania splendens*, *Gilia* sp., *Gypsophila elegans*, *Hymenatherum tenuifolium*, Mesembryanthemum, *Mirabilis* sp., *Oxalis* sp. Portulaca, Verbena (perennial and annual), Viola, *Zebrina pendula* (Syn. *Tradescantia zebrina*), and *Zinnia linearis*.

Some ferns and other plants are very useful as they give natural effect to the wall. The ferns suitable for dry wall are: ***Adiantum caudatum*, *A. philippinense*, *A. venustum*, *Asplenium compressum*, *Polystichum* sp., *Pteris cretica*, and others, *Lycopodiums* and *Selaginellas*.** As already stated, miniature roses and some other small herbaceous plants can also be grown on the top of the wall.

LAWN

A lawn can be defined as the green carpet for a landscape. It is a basic feature for home ground development and an essential feature for any other type of garden. In a home garden a lawn improves the appearance of the house, enhances its beauty, increases conveniences and usefulness thus adding monetary value to the real estate. The lawn provides a perfect setting for a flower bed, a border, a shrubbery or a specimen tree or a shrub. Besides the material value, a lawn has its spiritual value, too. A lawn is the source of charm and pride and reduces tension of the mind after a day's hard work in the materialistic world.

The view of a lawn should be uninterrupted from the house or from the entrance. If the lawn is spacious, seasonal beds, herbaceous borders, beds of canna, attractive specimen trees or shrubs such as *Araucaria cookii*, *Thuja*, *Callistemon lanceolatus*, *Brya ebenus*, etc., may be planted. Some bulbous plants such as *Zephyranthes* may also be planted in groups in the lawn.

Grass is one of the hardiest perennial herb and it is not very difficult to maintain a lawn, although one has to take the necessary care. Unfortunately, in India the growing of lawn is neglected from its construction stage to its maintenance. A garden lover should do well to remember that 75 per cent of the beauty of a garden depends on a properly maintained lawn.

The Site: It is not always possible to get the best site one would like to choose for the lawn. But a few points should be kept in mind before selecting a site. Grass does not grow under shade, so a site should receive full sunshine. The best situation will be the southern side and the next-best is the south-east and south-west of the building. Most of the grasses also do not grow well under the drip of large trees. Moreover, the dried leaves fall over the grass and make the lawn dirty. Therefore, it is desirable that no big tree should be existing at the site. This does not, of course, mean that a well-placed tree in the lawn is not desirable. Such trees, if selected carefully, will provide the necessary shade in the summer for a sitting place without adversly affecting the lawn. Soil moisture capacity and drainage are also two important points which should be kept in mind. In a poorly-drained soil the grass will perish.

Soil: In India, the common lawn grass, *Cynodon dactylon* (*Doob*), is very hardy and can be grown in any type of soil. But to obtain a most luxuriant lawn, it is desirable to have a fertile loamy soil containing enough humus. The soil should retain enough moisture and at the same time the drainage should also be adequate.

A lawn soil should preferably be a little, but not highly, acidic, the pH range being 5.5 to 6.0. If the pH is very low about half a kilogram of chalk or grounded limestone should be added per square metre area on a sandy soil or a similar quantity of slaked lime should be added to clayey loam soil. In an alkaline soil, gypsum should be added at the same rate.

One reason for a poor quality lawn is the insufficient depth of soil. At least a depth of 25-30 cm of good soil is required for obtaining a good lawn.

Drainage: After selecting the site, the next thing will be to ensure proper drainage. Grasses are shallow-rooted herbs and, therefore, no deep drainage is necessary, but no water should stagnate in the rooting zone. In clayey soils, some kind of drainage must be provided. This may be done by drainage pipes or by putting a layer of broken pieces of bricks and rubbish 90 cm below the surface. Ordinary drainage work can be carried out in conjunction with grading or levelling.

Digging: Rough surface levelling by eye estimation should be done prior to digging. If during rough levelling a lot of shifting and filling of soil is necessitated, the surface soil should first be taken out and kept separately, which should be laid on the top after final levelling.

After rough levelling is completed the digging work will commence. Thorough preparation of the ground is most essential in the success of a lawn. Very often the digging work is done quite hastily. This is a bad practice and contributes to the failure of a lawn. The digging operation should be done by the trenching method, otherwise the gardener may fail to dig to the desired depth. To ensure this a trench of 60 cm deep and 45 cm width is dug at one end of the site along its entire length and the soil is kept outside the lawn area. The rest of the area is dug at the same depth in widths of 45 cm at a time, the earth removed from the freshly dug trench going into the trench dug preceding this. The earth first dug will go into the trench dug at the farthest end. At each stage of digging care should be taken to see that the clods of earth are broken and pulverized thoroughly, otherwise the purpose of digging will be defeated. If clods are left unbroken the ground will take a long time to settle. During the process of digging all stones, old masonry, grass roots, etc., should be removed. Special care should be taken to remove the roots of *motha* grass or nut grass (*Cyperus rotundus*).

In most parts of India, digging is done during the hot months of April and May. After the trenching is completed the soil is left to dry in the scorching sun for a period of 7-15 days to kill the weeds or insects and for sterilizing the soil. The soil should be turned up subsequently 2-3 times at weekly intervals, each time the clods of earth, if any, are broken and roots of weeds removed.

Manuring and Grading: After the digging is over, the soil is to be manured and graded (levelled). If the soil was originally fertile it is better not to apply any organic manure, as organic manure contains weed seeds which may contaminate the lawn with weeds which are difficult to remove. But poor soils will need dressing with organic manure. Night soil manure, F.Y.M., or old stable manure is used for this purpose. The manure is sieved finely and spread over the surface at the rate of 500 kg per 100 square metres of soil. The amount can be reduced depending on soil fertility level. This is then worked up in the soil to a depth of 15-20 cm.

Any freshly dug up soil takes a long time to consolidate. The next step will be to settle the soil thoroughly. In heavy rainfall areas, the work is done by the pouring rains. In other areas, the prepared soil is watered heavily and to check the run-off, bunds should be put up all along the periphery. The accumulation of water in pools will show the area of depression which should be smoothened by shifting soil from one place to another. The flooding should be repeated 2 or 3 times and between each watering the sprouted weeds should be removed. The final levelling is done with the help of levelling pegs, straight-edge, and spirit level. A straight-edge is a straight wooden plank of 3 m length and 10 cm width. Strong split bamboo pegs of 30 cm length and pointed at the bottom are taken and marked by painting at 15 cm from the top. These pegs are driven in the ground up to the 15 cm mark, starting from the centre of the lawn, at parallel lines of 2.9 m and keeping the same distance between the pegs in the line. At a time 4-5 lines of pegs are put. Then the straight-edge is placed on two pegs of the central line, one edge of it resting on one peg and the other on the other peg of the same line. The spirit level is then placed on the centre of the straight edge and the position of the bubble is adjusted in the centre of its run by driving in or lifting the outer peg until the bubble is in the centre. Following the same method from one peg to the other first the level in the central line of pegs is corrected. Then one end of the straight-edge is placed on a corrected peg in the central line and the other end on the corresponding peg in the adjacent line and the level is corrected by lifting or driving the peg in the adjacent line. Proceeding in this way all the pegs are fixed in the correct level. After this, earth is removed round the pegs on which the 15 cm, mark has gone underground to make this visible. Then lines of strings are passed from peg to peg in a criss-cross way through these 15 cm markings and the soil surface is made parallel to the lines in all directions. The operation is time-consuming and needs a lot of attention. The levelling work can be done by drawing slats of wood measuring 45 cm ×7 cm×2.5 cm along the surface. The same procedure is followed until whole area is levelled. The loose soil is then made firm by the use of light rollers. The soil is then lightly irrigated and the levels rechecked when the soil is sufficiently dried up. There may be slight surface settlings after this, which are corrected and the pegs removed. It is always advisable to keep the level of the lawn 5 cm below the levels of paths and drives, the margins along the paths are raised by gradual slope of 15-20 cm, to form a turf edge of 3-4 cm higher. This method will help keep the paths dry when the lawn is flooded with water.

It is not always necessary to have a perfectly levelled lawn. Lawns can be laid in undulated land also and such lawns look very beautiful. But there should not be any depression as then water will collect and kill the *doob* grass. Moreover, the slopes and mounds in a lawn should be gradual and artistic, simulating the nature.

Selection of Grass: The most suitable grass for most parts of India is the *doob* grass (*Cynodon dactylon*) also known as *hariali*, *arukampillu* and *durba* in the vernacular. This also goes by the names Bahama-grass or Couch grass or Bermuda-grass. The grass thrives well under hot, sunny weather. This grass will not grow under shade. In Europe and America many grasses are used for the lawn, some of which may suit for our hill stations. The finer varieties are *Agrostis tenuis*, *A. canina*, *Festuca ovina*, *F. rubra* (Fescues), while the cheaper mixtures consist of perennial rye grass (*Lolium perenne*) and crested dog's tail (*Cynosurus cristatus*). The wood meadow grass (*Poa nemoralis*) grows well under shades of trees. The 'Kentucky bluegrass' is one of the most sought-after lawn grasses in the western countries, which goes by the nomenclature *Poa pratensis*. This is also known as 'June-grass'. The grass is of a very fine texture and gives a soft carpet-like feeling when laid as lawn. The colour is blue-green. This grass may go dormant in the hot summer, and to keep it green and nice heavy watering will be needed. This grass grows best in full sun. In Bangalore this grass is widely used for making lawns. It makes a soft and superb lawn, especially in mild climates and can be tried in other places also. But the grass is difficult to maintain and will need a lot of care to keep it green. Some experts feel that the fine-textured blue-green grass which is often referred to as Kentucky bluegrass or China grass and grown in Bangalore and elsewhere for lawn, in most cases is nothing but a fine strain of *doob*. Any such doubt can be cleared by studying the inflorescence of the lawn grass in dispute.

Method of Planting: If irrigation facilities exist, a lawn can be laid out any time during the year. Under Indian climatic conditions it is better to sow after one or two monsoon showers, while the grass root is planted at the beginning of the monsoon. The different methods for starting a lawn are by (a) seed sowing (b) dibbling, (c) turfing, and (d) turf-plastering.

(a) *From Seed:* If grass-cuttings or roots are not easily available, one should go for the seeds. It is important to secure good quality seeds free from weed seeds. *Doob* grass seed is very light and fine and proper care should be taken during sowing. Prior to sowing, the surface, when relatively dried up, is scratched to a depth of 2.5 cm with the help of a garden rake. The total area should then be divided into equal plots of 200 to 300 square metres to ensure even sowing of seeds. The sowing should be preferably undertaken on a windless day. The seed is divided at the rate of 500 g per 200 square metres and mixed with double the quantity of finely sifted soil and broadcast by hand. After sowing is completed the rake is drawn lightly twice in opposite directions to mix up the seed. The ground should then be rolled with a very light roller. It will be advisable to cover the seeds with a thin layer of finely sifted soil. The plot should be watered at regular intervals with a water can having a fine rose. Watering can also be done with a hose-pipe fitted with a fine rose. Sometimes, ants carry away the seeds and to prevent this the soil should be treated with B.H.C. or heptachlor (for details consult Chapter XXVI). There is a belief that seeds sown within two days of full moon gives best germination and stronger growth, whereas sowing during new moon is the worst. However, there is no experimental evidence to prove this.

The seeds germinate in about 3 to 5 weeks from sowing. When the grass is about 5 cm tall it is clipped with a pair of garden shears. Initially the lawn mower is not used as this will uproot the grass. If the germination is patchy, resowing will be needed to cover such areas.

(b) *Dibbling:* After the land is ready, well-matured unrooted (or rooted) *doob* grass cutting is obtained from a close-cut lawn or nursery or from a lawn-scraping. Grass, growing in shades and having nodes far apart, is not suitable for planting. If grass with short internodes is not available it will be desirable to plant the underground stems. The roots or grass thus obtained are dibbled (planted) in the ground when it is slightly moist at 7-10 cm apart. The soil is kept moist by frequent watering till the grass sprouts. Roots of *doob* grass sprout easily and the cuttings or off-shoots root readily under moist condition and within 5-7 weeks the grass will be ready for first cutting. By this method a lawn will be ready in about four months.

(c) *Turfing:* The quickest method of developing a lawn is by turfing, but the cost is prohibitive. Turf is a piece of earth of about 5 cm thickness with grass thickly grown over it. The pieces may be of small squares or in rolls of small width (30 cm or so). The turves must be free from weed and consist of the required lawn grass. These should be laid closely to each other in a bonded alternate pattern, like bricks in a wall, in the already prepared ground. Any unevenness in thickness can be corrected by underpacking or removing some of the soil before putting in position. Along the joints sandy soil should be filled as packing. Bone-meal is dusted in the prepared ground a few days prior to turfing. The turf thus laid is made firm by a wooden beater made out of heavy block of wood and fitted with a handle. The grass is immediately watered copiously. By this method a lawn will be ready for use in a very short time.

(d) *By turf plastering:* A paste is prepared by mixing garden soil, fresh cowdung, and water. Bits of chopped-up fresh roots and stem or rhizomes of *doob* grass are mixed with this paste and the paste is spread evenly on the surface of the prepared ground after moistening the soil. The paste is then covered by spreading 2 cm of dry soil and watered at regular intervals. This method is not very suitable especially in a dry and variable climate.

Maintenance of Lawn: Having raised a lawn by one of the methods described above, the question of maintenance comes next. If the lawn is not properly maintained, it will become useless within no time. The various aspects of maintenance are discussed below.

(a) *Weeding:* One of the main aspects of maintenance is the control of weeds. Without close attention or care a time will come when weeds will overcome the lawn grass, the soil will become sick, and the grass coarse. Weed is common in both new and old lawns. Therefore, as soon as a lawn is established weeding should start and continue at regular intervals or whenever the weeds come out. The frequency of weeding obviously will be more during the rains than in the colder months. The *motha* (*Cyperus rotundus*) is the most difficult weed to eradicate, because of its deep root system. In a freshly prepared lawn the growth of *motha* is profuse and unless this growth is checked the weed will smother and hamper the growth of *doob*. This should be removed with the roots as deep as possible with a long narrow-bladed (1 – 1.5 cm) *Khurpa*. It is almost impossible to eradicate *motha* completely. But, once the lawn is well established with regular mowing the growth of *motha* is kept in check. A red creeping weed, *dudi* (*Euphorbia thymefolia*), often invades the lawn and is not difficult to eradicate if controlled at the initial stages. All weeds should be removed with the roots and these should never be allowed to seed.

(b) *Liming:* If considerable growth of moss i observed, which is an indication of acidity, powdered chalk or lime should be applied at the rate of 250 gm per square metre area. Even otherwise, application of lime once a year, except in alkaline soils, is beneficial. After liming watering should be done.

(c) *Rolling, mowing, and sweeping:* The object of rolling is to help the grass anchor itself securely and also to keep the surface levelled. Rolling should be avoided when the soil is wet. There is difference of opinion regarding the rolling of *doob* lawns. It is argued that *doob* grass has a growth pattern completely different from the English lawn grass, in that while the latter needs rolling at regular intervals the former does not. Moreover, rolling does not help much in levelling. Instead it leads to caking or hardening of the soil surface. People with the opposite view suggest very light rolling at the time of levelling of the ground and once after the lawn has just established to press down the small stones or pebbles which otherwise will interfere with the mower blades. This view is justified especially for heavy soils. In light sandy soils rolling after each weeding will be helpful to keep the surface levelled. Indiscriminate use of a roller is definitely bad.

Mowing is another important operation. The first thing is to obtain a good machine which will cut evenly at a correct height. The frequency of mowing is determined by the amount of growth and will vary from season to season. But grass should not be allowed to grow more than 5-6 cm in length during any season.

Sweeping the lawn thoroughly after each mowing is essential to clean the cut grasses which might have fallen from the mower box. Sweeping is also done every morning to clean the fallen leaves and other debris. Sweeping may have to be repeated two or three times in a day during the season when the deciduous trees shed their leaves.

(d) *Irrigation: Doob* grass is shallow-rooted and, therefore, frequent light irrigation is better than copious flooding after long intervals. Here again some people prefer flooding at long intervals as this saves labour. The water-hose is left in the soil and shifted occasionally, while the gardener attends to other work in between. Labour as well as water can be saved to a considerable extent if sprinkler irrigation is used. The frequency of irrigation varies with the climate. The winter dew is important and each morning the dew should be brushed into the grass by drawing a hosepipe over the grass, before the dew evaporates. Stagnation of water should not be allowed as it may kill the grass.

(e) *Scraping and raking:* Continuous rolling, treading, and mowing may result in the formation of a hard crust and the lower part of the lawn may get matted and woody. For such lawns, the grass is scraped at the ground level with the help of a *khurpi* in the months of April and May. Scraping is followed by raking to break the crust. Where the condition of the lawn is good, hard and thorough raking is done both ways to loosen the old runners and to aerate the soil. Then the mower blade is lowered and the grass mowed close to the ground.

Top Dressing with Compost and Fertilizer: After scraping or raking a compost consisting of good garden soil, coarse sand, and leaf-mould in the proportion of 1 : 2 : 1 (in sandy soil the proportion of sand should be reduced or eliminated altogether) is spread over the lawn to a depth of 3-5 cm. To cover to such a depth a 100 kg of compost per 100 square metres will be needed. Bonemeal is also applied at the rate of 1 kg per 10 square metres. The same compost is used as top dressing again during September or October. From October to April, ammonium sulphate is applied once every month at the rate of 1 kg per 50 square metres area followed by watering. The authors have found out that the application of fermented town compost in liquid form is very good. This is prepared by fermenting 20 kg of compost in 100 litre of water for a few days. During fermentation ammonium sulphate and superphosphate at the rate of 1 kg and 2 kg are added to this mixture. The concentrated mixture is strained through gunny cloth and diluted to tea colour and added to the lawn with water cans or by siphoning. After the application, the lawn is soaked with water. This can be applied twice a year (October and May-June). Raw cowdung may be fermented and used in the same way.

Frost-injury: In north India, the grass is injured due to frost. This can be avoided to a great extent if the grass is sprayed with water every evening and in the early morning after frost.

Pest and Diseases: See Chapter XXVI.

CARPET BEDDING

The term carpet bedding means covering an area, preferably a bed or a series of beds, with dense low-growing herbaceous plants according to a set design. Carpet beddings look better when arranged in a slope or a slanting position. In such designs, often a figure or some letters are cut out with the help of plants having different growth habits or having different coloured leaves. To have a useful contrast, the usual practice is to grow plants of one colour as background and then run through these plants of other colours in groups, stripes or lines to have the desired effect. Carpet bedding can be designed in innumerable ways depending upon the taste of the designer and the plants available at hand.

In a carpet bed, the plants are planted closely according to a set design. To keep a carpet bedding in right shape needs constant attention. After first planting the plants are trimmed as soon as they are well established. Henceforward, the plants are trimmed now-and-again to keep them within the limits of the design. The plants are generally not pruned during the winter when most of the foliage plants develop good colour under Indian climatic conditions. Whenever there is any gap in the bedding due to mortality, through disease or otherwise, they are filled up by previously grown plants in pots. Once in every 3-4 months the beds should be top dressed with a well-decomposed manure.

The art of carpet bedding was very popular in olden times and used to be a common feature of any formal garden. But, nowadays, this type of fashion is hardly found in any modern garden. However, a well-maintained carpet bedding has its own charm.

The plants which are suitable for carpet bedding are listed below.

Alternanthera (*Amaranthaceae*): There are several species with colour of leaves varying from orange, scarlet, bronze yellow, rose, etc., mostly tinted and mottled with two or more colours. These are compact growing, easily propagated by cuttings and stand trimming very well.

Cineraria maritima (Syn. *Senecio cineraria*, *S. acanthifolius*, *S. maritima*), (*Compositae*): Commonly called 'Dusty Miller' has silvery downy leaves.

Coleus (*Labiatae*): The dwarf species and cultivars are suited for this purpose.

Echeveria secunda (Syn. *Cotyledon secunda*), (*Crassulaceae*): A succulent whose dense rosette of leaves are almost round and grows to a height of 7-15 cm. Suitable for planting at medium and high altitudes.

Irisine in different species: See page 456.

Pilea muscosa (Syn. *P. microphylla*): See page 456.

Portulaca (*Portulacaceae*): A low-growing annual having a trailing stem and brilliant flowers. Raised from seeds. The perennial species are multiplied by cuttings.

Sedum rupestre (*Crassulaceae*): A mat-forming glaucous evergreen creeping plant excellent for this purpose. Suitable for mild climate and semi-shade.

Sempervivum tectorum (*Crassulaceae*): Commonly called the 'Houseleek', forms rosettes varying in height between 5 and 20 cm. Leaves green usually with large purple tip. Suitable for medium to high altitude.

Torenia asiatica (*Scrophulariaceae*): A prostrate perennial herb, with purple flowers.

Vittadinia australis (Syn. *V. V. tricolor, triloba*), (*Compositae*): A compact trailing herb with small leaves and small daisy-like flowers. Suitable for higher altitudes.

FLOWER BEDS

The most essential criterion of a flower bed is to display the flowers in the best way possible. It may be borne in mind that flowers look best when massed in a bed. The flower beds are of special importance under conditions of the plains of India since they can be kept planted throughout the year unlike those in England or other parts of Europe where nothing can be grown in the open between November and March. The flower beds can be planted with winter flowering annuals sown during September/October or summer flowering ones sown during February/March and rainy season flowers sown during April-June. Perennial plants such as canna and michaelmas daisy are also planted in flower beds.

Flower beds should be simple in design as it is easier to maintain them. More elaborate geometrical designs with numerous sharp points and angles found in formal French and Dutch gardens get distorted during digging and are also gradually going out of fashion. The simple designs as circles, rectangles, squares, and ellipses are easier to maintain. Kidney shaped beds in the midst of a lawn also look very attractive. Various other abstract designs can also be made. The easiest way to shape out an abstract design on the ground is to wind a rubber hose or a thick wet rope in the desired shape on the ground and then mark out the outer borders with lime powder, followed by digging.

Flower beds are the most important part of a formal garden and these are marked out most ornamentally in pairs. But in a landscape style garden the number of flower beds are less compared to a formal garden, as the flowers are mostly grown in the borders or in front of the shrubbery, for livening up the landscape design.

The beds should always be placed in full sun except when these are meant for growing shade-loving plants. Rectangular beds are best placed near the building to harmonize with the straight lines of the house. While laying out a rectangular bed on the lawn the longer sides should run parallel to the adjoining straight path or the edge of the lawn. While arranging the beds in a lawn, the grass verges or paths in between beds should not be less than 45-60 cm, otherwise it will be difficult to run a lawn mower. Ideally speaking, flower beds should also be situated away from the competition of roots of larger plants.

Preparing the beds: After the design is marked on the ground the bed is dug. A bed may be meant for growing the most robust or a very weak-growing plant; therefore, it should be prepared thoroughly making it suitable even for the poorest-growing plant. The most important point is proper drainage. In a really heavy soil, pipe drainage may have to be arranged. But it is better to remove 60 to 90 cm of the top soil, depending upon the type of the plant to be grown, in a heavy clayey or stony soil and to replace it with a good loam. For drainage in most other heavy soils 15 cm drainage materials such as broken bricks or cinders or some builders' material is placed at about 60 cm below the surface. Heavy soils are improved by adding sand, while light soils are given a dressing of good garden loam. The bed is double dug to the required depth. A good dressing of well-rotted cowdung or horse manure or leaf-mould is added well before the planting date. About 6-8 kg of manure will be enough for a 16 square metre plot. In addition to manure, about 400 gm of hydrated lime should be added per square metre area except in alkaline soils.

Each bed should receive manure every year before planting or in between planting at different seasons..For perennial plants, manuring is done before the rains. Each bed should be dug up thoroughly at intervals of 3-4 years, and if the soil has turned sour, hydrated lime should be added.

Every year as more doses of soil and manure are added the level of the bed is raised above the surrounding ground which in fact will assist in draings. A 120 cm wide bed should be about 10 cm higher at the centre and slope off to the sides. The gradient will increase or decrease depending on whether a bed is wider or narrower. For beds situated in lawns the edges should be trimmed neatly.

SHRUBBERY

This is an essential feature of any garden, but , unfortunately, the importance of this is not properly understood by the garden lovers in India. Shrubs are of permanent nature and once planted will become a permanent feature unlike the seasonal flowers which are to be replaced every season. Moreover, if properly selected, the shrubs in a shrubbery will provide flowers throughout the year. The beds meant for shrubberies need not be very formal but should appear more natural. Shrubbery can very effectively be used, for hiding one portion of a garden from the other, besides adding beauty. For details regarding planting, maintenance, etc., consult pp. 128–130.

BORDERS

Beds which are more in length than breadth and contain plants of a heterogeneous character are known as borders. There are different kinds of borders and these are named according to the kind of plant material grown in them. There are three different types of borders, namely, herbaceous border, annual mixed border, and mixed border.

Herbaceous Border

This is a characteristic of the English garden, containing hardy perennial herbaceous plants which die down to ground level after flowering, but put up new growth from the roots in the next season. But in actual practice it is quite common to include bulbous plants and annuals also in the herbaceous border. Herbaceous border needs a good depth of soil and a sunny situation. The border may be placed against a wall, a fence, shrubs, a hedge, or form a double border (one behind another) divided by a grass path. A border may be put to mark a division between one part of the garden and another.

It is always desirable to have a living background such as a hedge or a shrub. But care must be taken to see that the roots of shrubs or hedges do not interfere with the border plants. Therefore, shrubs or hedges having limited root-spread should be preferred. If an informal design is desired, the border should have curved margins, provided they are justified under the prevailing circumstances and is not just some thoughtless curves.

(a) *Frontage:* The frontage of a border is perhaps more important than the background, as a border will be viewed from front only. The best frontage is a stretch of neatly kept lawn because this provides a good foreground colour. This must be wide enough and should generally be not less than 2 m across. For a larger border, it may be as wide as 3 m.

A paved brick or stone walk is considered the best because this provides a firm and clean walk throughout the year. The initial cost of laying out will be quite high, but if properly made will last for a long time without any or negligible maintenance. A compromise between the two is, perhaps, the best. The edging, 30-45 cm of width between the border and grass, may be constructed of paved stone or brick. It has some advantage, as some plants in the border overhang the edge and make mowing of grass difficult.

(b) *Height and arrangement:* In a wide border, plants of all heights are accommodated provided the grower can give enough attention on the staking of tallest plants. For a narrow-width or double-faced border, the maximum height should be limited to 3 m.

(c) *Grouping and colour:* In a herbaceous border plants are arranged in bold groups. In an average-sized border each group will consist of 4-5 plants, while in large borders it may vary from 5 to 7 plants. Planting of a large number of plants in one group, particularly those which grow vigorously, should be avoided, otherwise when a group of plants complete flowering early that portion of the border looks patchy. Each group should be planted in drifts of irregular shapes than being put in box or round shaped drifts. Though generally the gradation in a herbaceous border is from dwarf to taller placed towards the back, it can be broken here and there to avoid monotony or too formal a look. For example, a drift of tall but slender plants may well be accommodated in the front or a medium height drifts may be artistically arranged in between the tallest plants.

There are two extreme views regarding colour schemes. One group of gardeners feel that colour schemes should harmonize, i.e. light shades leading to dark shades and dark shades fading to light shades, while the other prevailing opinion is that nature never clashes. Though there is some substance in both the views, one should not be very rigid about this. In olden times, the English gardens used to have exclusive borders of reds, blues, yellows, pinks, and even whites. But today, with the problem of space, very few gardeners can afford to have such a luxury. If, however, adequate space is available, such arrangements should definitely be welcome. In small home gardens, having only one small border, it is not possible to have a full scheme of colour blending. In such cases the advice will be to avoid the extreme contrasting colours. For details of colour contrast, readers may refer to pp. 64-65.

(c) *Planting:* It is always desirable to draw the actual planting plan on a paper according to scale and the scale should not be less than 1 : 15. The drifts should be drawn showing the shape and size with the names of the plants for each drift inserted along with their height, colour of flower, and time and duration of flowering. The plan is then taken to the site to form an idea by imagination how the plan will look. Pegs and canes of same height as the plants to be grown may be placed in position and a rough idea made as to how they will look. After this, the final plan is made and actual planting is undertaken.

In a temperate climate, the planting is done during the autumn or the early spring. Some plants such as *Aster amellus*, if transplanted in the autumn, may not survive the severe winter and hence are planted in the spring only. In the plains of India, planting may be done in September, while in the hills the above pattern is suitable.

Once planted, a herbaceous border will remain in position for about three years after which some replanting is necessary mainly for thinning out and invigoration.

(d) *Choice of plants:* The main aim of a herbaceous border in a temperate country is to have colour in the garden throughout the summer and the autumn, although in practice it is not feasible to have a mass of flowers during all that period. But with careful selection it is possible to have a sufficient number of flowers throughout this period. In a tropical country the aim should perhaps be to exhibit the proper plant materials in a most attractive and artistic way.

Annual Mixed Border

A border consisting of mixed annual flowers are called annual mixed border. There is hardly any difference in principle between a herbaceous border and an annual mixed border, except that the latter is annual in duration. The colour scheme, grouping, etc. will be similar to that of herbaceous border. A border of mixed annuals also is often termed as herbaceous border in the plains of India.

Mixed Border

A mixed border generally includes the following types of plants: (i) herbaceous shrubs (i.e., not woody) with less spreading root system, (ii) sub-shrubs which are of dwarf growth and have soft stems, (iii) herbaceous perennials, (iv) bulbous plants (dahlia, tuberose, zephyranthes, amaryllis, etc.) and (v) annual flowers.

A mixed border is arranged against the background of walls, ornamental trees or a fencing.

Plants for Borders

A list of plants for different borders are given below as a guide. The garden lovers will learn by experience to select more plants suitable for this purpose.

(A) Plants for Herbaceous Border for Plains

Here is a list of plants suitable for herbaceous border for the plains of India. As per actual definition the plants do not die down after the flowering season is over; however, these are quite suitable for the purpose.

Amaryllis (*Hippeastrum*) in different cultivars, *Angelonia* different species, *Asclepias curassavica*, *Asystasia*, different species, *Beloperone guttata*, Canna different cultivars, *Clerodendron balfourii* (Syn. *C. thomsonae*), *Chrysanthemum* (Korean and Single types), *Coreopsis grandiflora*, *Crossandra* different species, *Eranthemum* different species, *Gaillardia perennis Lantana sellowiana*, Michaelmas Daisies, *Plumbago capensis* (including variety *alba*), *Portulaca*, *Rudbeckia* (perennial), *Ruellia* different species, *Russelia juncea* and *R. floribunda*, *Salvia*, different species, *Solidago canadensis*, *Verbena* (perennial), *Vinca* different species and *Zephyranthes* different colours.

(B) Plants for Herbaceous Borders in High Hills

The following plants are suitable for herbaceous borders on the high hills and temperate regions.

(a) Dwarf: *D. allwoodii*, *Dianthus chinensis*, *Campanula carpatica*, *Linum narbonense*, *Sedum spectabile*, *Statice latifolia*, and *Viola cornuta*.

(b) Medium: *A. elegans*, *Achillea eupatorium*, *Anemone lupehensis*, *Aster amellus*, *Campanula persicifolia*, *Chrysanthemum maximum*, *Coreopsis*, *Delphinium belladonna* in cultivars, *Gaillardia*, *Galega*, *Gypsophila elegans*, *G. paniculata*, *Helenium pumilum*, *Lupins*, *Lythrum*, *Oenothera*, *Papaver orientale*, *Phlox decussata*, *Rudbeckia newmanii*, *Scabiosa caucasica* and *Tradescantia*.

(c) Tall: *Anchusa azurea*, *Chrysanthemum uliginosum*, *Delphinium*, *Helianthus annuus*, Hollyhocks, Peony, *Rudbeckia* (tall).

Plants for Mixed Border

A list of shrubs, which do not grow into a tree, sub-shrubs, herbaceous plants, bulbous plants, and annuals arranged according to their height, which are useful in the mixed border, is given below:

(a) Tall (height 1-2 m): *Acalypha* different species with ornamental foliage, *Brunfelsia americana*, *Caesalpinia pulcherrima*, *Cestrum aurantiacum*, *Dahlia* tall cultivars, *Gardenia florida*, *Hibiscus mutabilis* and other tall cultivars, Hollyhock, *Iochroma tubulosa*, *Ixora singaporensis*, *Lagerstroemia indica* in different varieties, *Nerium oleander*, *Poinsettia pulcherrima*, *Stachytarpheta indica*, *Tabernaemontana coronaria*, *Tecoma capensis*, *T. stans*, and *Thunbergia erecta*.

(b) Medium (0.5 to 1 m): African Marigold (tall), *Angelonia grandiflora*, *Antirrhinum majus*, *Beloperone amherstiae*, *B. guttata*, *Browallia eleta*. Canna medium cultivars, *Chrysanthemum* tall cultivars, *Coreopsis grandiflora*, *C. tinctoria*, *Gaillardia*, *Galphimia gracilis*, *Gladiolus*, *Heliotrope*, *Malpighia coccigera*, *Phlox decussata* (Syn. *P. paniculata*), *Plumbago capensis*, and *Salvia* different species.

(c) Dwarf (10 to 50 cm): *Ageratum*, different species, *Alternanthera*, China aster different cultivars, *Brachycome*, Candytuft, *Coreopsis drummondii*, *Dianthus chinensis*, Gerbera, *Lantana sellowiana* Michaelmas Daisy, *Phlox drummondii*, *Portulaca*, *Rudbeckia* (dwarf), *Salvia coccinea*, *Verbena hybrida*, *V. erinoides*, *Zephyranthes*, and *Zinnia linearis*.

Annual Mixed Border

Annuals of different heights and colours are selected for this type of border.

ROCK GARDEN

It is a mistaken idea to think that rock plants grow on nothing but rock. In nature, rocks may get covered with different coloured moss, lichens, etc., and if soil collects in cracks and crevices of rocks some plants grow there naturally. But in a well-planned rock garden the rocks and soil are arranged in such a fashion that conditions are created for the growth of different plants.

A rock garden is artificial in the sense that it is man-made. But a well-prepared and well-managed rock garden should become one of the most attractive and interesting features of the garden. A real rock or alpine garden cannot be created at elevations less than 2,000-2,500 m because most of the true rock or alpine plants do not grow below this altitude.

Aspect

A rock garden should be situated in an open part of the garden. It is often advised that a shaded corner of the garden should be devoted for making the rock garden. With this idea our rock gardens are created round the trunk of a large shady tree with a mound of earth and a few stones placed here and there planting in between a few hardy plants. Such gardens have no beauty, looks artificial, and are out of proportion. The rock garden should get all the sun possible or at least for half the day, preferably in the forenoon. It should be away from large trees and nearby hedges as their roots may interfere with the rock plants. There is no objection if distant trees throw shade for half the day or a hedge is planted a little distance away to protect rock plants from droughts of hot or cold winds. In the tropics it is not a bad idea if a portion of the rock garden remains shaded by a tree for a part of the day especially mid-day or afternoon, as then the rock garden can be planted with plants having various requirements for light. In a shady portion of the garden where lawn grass or any other thing does not grow, one may have no choice but to utilize that land to construct a rockery with shade-loving plants.

If an undulated land is available the rock garden should be situated there, if possible. A miniature rock garden may be all right if it is situated near a wall or a house. But a large rock garden should be situated in a place where this will fit with other garden features. A large rock garden will look out of place in a too formal garden. The site should be well drained and if natural drainage is not there, some artificial means should be provided.

In the hills south-facing slopes or terraces are ideal for sun-loving plants, whereas the shade-living ones can be accommodated on the northern slopes.

Construction of the Rockery

The placement and selection of rock is an important factor. Rocks should be of local origin, porous, and have a weathered look. Granite is unsuitable, whereas a weathered limestone is ideal. Limestones from the quarry can be used as they weather quickly. Stones of uniform size having a diameter around 60 cm should be selected and a few stones should be as large as can be handled without much difficulty.

Rock plants do not grow exclusively on rocks, and, hence, a good soil is needed for every rock garden. In the hills, there is no problem in selecting the soil as the natural soil is the best soil there. The soils obtained from where the paths are laid will make the mounds and ranges. In the plains, a soil full of coarse sand, plenty of leaf-mould and well-rotten farm-yard manure is quite suitable for most of the plants. In addition to leaf-mould, a fine compost suitable for rock garden can be prepared from the cuttings of the turf by the usual method of composting. Only the few centimetres (15-45 cm) of the top layers of the rockery will have this compost. The sub-soil should be well drained and have the property to absorb sufficient moisture. If it is not naturally so, this can be created by adding sand or gravel.

A rock garden is a permanent feature of the garden and it should be laid out after careful planning. If already a depression is there near the vicinity of the rock garden it should be utilized to lay the water garden there along with the rock garden. Even otherwise a water garden goes well with a rock garden, especially if a stream winds its way through the rock garden leading into the pool with a couple of water falls added in the course of its run.

Before placing and setting the rocks the mound should be allowed to settle for some time and firmed by beating, if needed. The placement of stones will be such that it gives protection to the roots, and, hence, placing of stones too perpendicularly will not serve the purpose. The stones are placed in a slightly slanting position gently leaning backwards, not forwards, so that rain water flows towards the plant roots. No rock should hang over a plant as it will deprive the plant below of rain water. The rocks should slope in the same direction with the best faces visible, so that they represent a natural stone strata. Pockets are created in between stones varying in size from 15 cm to 60 cm across. These pockets, where plants are to be grown, should be irregular in shape and constructed in such a manner that the compost inside is not washed out. If there is any chance of the compost inside the pocket becoming sour, rubble or clinder drainage should be provided. While placing two rocks side by side, they should be placed as close as possible or the soil in between them will be washed away. The stones are bedded in the soil starting from the base leading upwards. All stones should be bedded firmly and there should not be any air-pocket or loose soil round the stones. Air-pockets will parch the roots for want of water, whereas loose soils will be washed out. While placing stones the use of concrete should be restricted to the miniumm. The mounds and slopes in the rockery are to be arranged naturally, but artistically, and the peaks should not be very steep. The small ranges and peaks should be interspaced with broad valleys over which occasional spurs from the hillocks should project imitating the nature. A broad path, about 45-60 cm, should wind through the rockery between the gorges and the valleys and make all parts of the rockery accessible. The path may be made of gravel or stepping-stones. A portion of the way leading to the pool may be made into a crazy paving.

The rockery should be planted with grown-up and well-established plants. With a little knowledge about the growth habit of different plants and some artistic taste it is not difficult to establish an attractive rockery. The first thing that is to be kept in mind is that dense planting should be avoided. It will be wise to plant thinly as in course of time the plants will fill up the space as they grow. A pendant plant hanging down a projecting stone or a prostrate plant rambling along a large stone presents a pleasant view.

The best planting season is during the rains in the plains of India. In a climate as that of Bangalore planting is possible any time of the year except very hot weather, but should preferably be done during the rains. In the high hills, planting is done during the spring or the early summer. Potted plants can, of course, be planted any time in the year, but in places having a severe winter this should be completed by the early autumn.

Management of the Rockery

In places having a severe winter delicate plants will need protection. When plants are small a glass sheet resting on the rocks may be placed over the plants. For larger plants four strong galvanized wires may be inserted in the ground and the glass placed over this. If the weather is very cold the sides should also be covered with glass or polythene sheets.

The rock garden should be weeded periodically and all dying or dead leaves and branches are removed. The rockery should be watered by sprinkling water over the rocks as the rains would have watered.

Rock plants need replanting every 4 or 5 years as the soil becomes stale or sour. This is a laborious operation. Instead the plants may be top dressed every year with a compost consisting of good sterilized garden soil, fine sand, and leaf-mould (or peat) in equal proportions. The top few centimetres of soil is scraped out and in its place the new compost is filled in along with a little quantity of bonemeal, about 25-50 g depending upon the plant. Periodically the rock plants should be fed with liquid manure also. But, if the soil in any pocket turns sour, replanting with fresh soil will be a necessity.

Rock Plants

As has been pointed out earlier, a real rock garden can be laid out in the high hills only, since true rock plants can be grown only in such climatic conditions.In the tropical plains of India, most of the ideal rock plants cannot be grown. As such no recommended list of rock plants is available for the tropics. People select the thorny cactus or some other hardy succulents for growing in the rocks. It is wrong idea to crowd the rockery with some cacti. Most of the delicate cacti cannot be grown in the open, especially in high rainfall areas, and the hardy ones which survive do not bring any great beauty to the rock garden. Under tropical conditions, a selection of succulents, shrubs, ferns, and flowering annuals should be made for the rockery. The authors from their practical experience have been able to identify certain plants which are suitable for the rockery.

Cacti and Succulents

Some hardy species of *Opuntia*, *Cereus*, *Mammillaria*, etc.

Among the succulents, many species of *Agave*, *Aloe*, *Euphorbia splendens*, *Furcraea*, *Gasteria*, *Haworthia*, *Kalanchoe*, *Sansevieria*, *Sedum*, *Sempervivum*, *Stapelia*, and *Yucca* can be planted in the rock garden. In a small rock garden large species of *Agave*, *Furcraea*, and *Yucca* should not be planted.

Ferns

Some hardy ferns belonging to the genus *Drynaria*, *Nephrodium*, *Nephrolepis*, and *Polypodium* can be grown in the shady portion of the tropical rock garden.

Shrubs

Many shrubs are suitable for planting in the rock garden, especially for large rock gardens. Shrubs will provide greenery for the rock garden and produce flowers all the year round. The list given below is based on personal experience.

Adeneum obesum, *Asclepias curassavica*, *Asystasia coromandeliana*, *Azalea**, *Brya ebenus* (near water garden or stream), *Calliandra inaequilatera*, *Callistemon lanceolatus* (near water garden), *Cassia alata*, *C. biflora*, *C. laevigata*, *Catesbaea spinosa*, *Clerodendron macrosiphon*, *Cotoneaster** sp., *Crossandra* sp., *Cuphea* sp., *Cytisus** sp., *Daphne** sp., *Duranta plumieri variegata*, *Euonymus** sp., *Jatropha podagrica*, *Juniperus chinensis J. horizontalis**, *Lantana sellowiana*, *Muehlenbeckia platyclados*, *Ruellia* sp., (semi-shade), *Russelia juncea*, *Sanchezia nobilis variegata*, *Thuja* sp., *Vaccinium** sp.

Herbaceous Plants, Bulbs and others

Angelonia grandiflora, *Aster amellus*, *A. novae-angliae*, *A. novi-belgii*, Crocus, *Dracaena sanderiana*, *Fittonia argyroneura* (shade), *Impatiens holstii*, *I. sultanii* (semi-shade), *Oxalis rubra* (semi-shade or shade), *Pedilanthus tithymaloides* var. *variegata*, *Pilea muscosa* and other species (semi-shade), *Portulaca*, 'Rose Miniatures', *Sansevieria cylindrica*, *S. zeylanica*, *Setcreasea*, *Tradescantia zebrina* (Syn. *Zebrina pendula*), *Verbena erinoides*, *Vinca rosea* and other species, *Vittadinia australis**, and *Zephyranthes* sp.

(*These plants are suitable for high elevations.)

Flowering Annuals

Although not considered as true rock plants, many of the dwarf annuals are quite suitable for the rockery. The following may be tried in groups at different pockets to bring colour to the rock garden: *Alyssum*, *Antirrhinum* (dwarf), *Bellis perennis*, *Brachycome iberidifolia*, *Browallia elata*. Candytuft (dwarf), *Dianthus chinensis* (dwarf), *Hymenatherum tenuifolium*, *Mesembryanthemum criniflorum*, *Phlox* (dwarf), *Salvia splendens* (dwarf), *Verbena hybrida*, *Viola tricolor hortensis* (Pansy) and *Zinnia linearis*.

Some Rock Plants for the Hills

Adonis vernalis, *Androsace lanuginosa* and others, *Anemone* sp., *Campanula* species, *Cotyledon oppositifolia*, *Cyclamen* (hardy types), *Draba* species, *Gentiana* species, *Iris* species (dwarf), *Lilium longiflorum*, *L. tigrinum*, etc., *Meconopsis* sp., Narcissus, Penstemon, *Potentilla* sp., *Primula* sp., *Rhododendron* sp., *Selaginella* sp., and *Tulipa* sp.

WATER GARDEN

Water garden is one of the oldest arts of gardening in India. The oldest reference of water garden comes from the Budhist period. It was customary during that period to have lotus pools in temples and *stupas*. Poet Asvaghosa (100 A.D.) wrote that Lord Budha saw lotuses and flowering trees in *Nandanavana*. In our ancient scriptures also lotus occupies a pivotal place. The lotus stem is referred to as the naval cord of Lord Vishnu, whereas the lotus flower is described as the throne of *Pitamaha* Brahma, the Creator.

In recent times, the water gardening has been neglected in most of the gardens in India. This state of affairs is the result of neglect or indifference by our former Mughal and British rulers. The Muslim rulers, of course, developed many water gardens with fountains and other adornments, but they knew almost nothing about water plants such as water-lilies and lotus. The British also neglected the lotus and water-lilies in India, although in many European gardens the water lily finds a place. In contemporary India, water plants can mostly be found in some temple tanks of south and east India and in many village tanks of Bengal. In most of the modern gardens, this part of gardening generally remains neglected.

Informal Pool

The first point to the success of any water garden is to ensure adequate supply of water, as no aquatic plant will survive any length of time in a dried pool. Those who are fortunate enough to have a shallow pond or a natural depression holding sufficient water and not drying up in the summer or overflowing too much in rains, thus maintaining more or less a constant level, can start their water garden conveniently in such a pool. These types of pools are informal and so more natural. Such informal pools are made in a low-lying area of the garden. The shape and size of such pool will depend upon the general landscaping of the area and the slope of the land, but whatever may be the size or design, it should be simple and natural. The curves and bends of such pools must be graceful and artistic and merge with the natural landscape. The larger pools, ponds, and lakes are seldom concreted, as the cost will be prohibitive. The base of such ponds is made impervious to water by the "puddling" method. This can be achieved easily if the sub-soil itself is clayey or by importing clay and laying it to a depth of about 30-45 cm at the bottom. The puddling consists of pummelling the wet clay with feet, or preferably with a ram-rod, continuously till the clay turns to the consistency of a dough. At this stage the clay becomes impermeable to water and henceforth this should never become dry, otherwise the clay cracks and water leaks. After the puddling is complete, the pond is filled with water. The pond or lake proper will be used for planting true aquatics such as lotus, water-lilies, etc., whereas in shallow waters near the banks bog plants are planted. On the banks at intervals, moisture loving plants are planted to bring in more natural effect.

Water-lilies and most of the other water plants do best in still waters. This raises the question whether water-lilies can be grown in running water. There are certain varieties of water-lilies which can be grown in running water, depending upon the velocity of the current. If the flow of water upturns the leaves or injures them, water-lilies will not grow. But a gentle current of water will not much upset the growth of certain varieties.

The Formal Pool

Unlike the informal pool, a formal pool need not be situated in a low lying area, rather it should form a main feature of the garden and should be situated in a prominent place of the garden. A formal pool should never be constructed below a tree, as this robs the pool of its prominence over the tree. Moreover, the shade from the tree interferes with the much-needed sunshine required for the water plants and the falling leaves of the tree foul the water. The roots of the tree may also run through the wall of the pool and cause cracks and crevices.

The shape and size of a formal pool depend upon the size of the garden and, of course, to some extent on the personal liking and disliking. A pool may be formal or informal in shape. In informal pools, the edges are camouflaged by putting plants round the edges, the plants falling over it. Constructing the pool at ground level is the usual practice, but this runs the risk of children getting drowned for which necessary precaution should be taken. The shape may be a simple rectangle or square, round or oval, kidney-shaped or any other ingenious artistic design. A formal pool is made of concrete, without or with a surround for growing the bog plants. The first step in constructing the pool is to mark out the design on the ground (see Fig. XXIII.3) and then digging out the soil of the marked area to a depth of 30-90 cm in proportion to the size of the pool, plus the thickness of the concrete base. When a surround for growing marsh plants is to be constructed this area is to be taken into account while digging, as this will be part of the pool. On completion of digging all loose soil is removed and the base is made firm, as this has to support the tremendous weight of the concrete and water. If the base is not clayey or hard it should be made hard by ramming in rubbish, *morrum*, or cinder and then the concrete base is laid. Before any concrete work is started pipes for inlet of water and draining the pool must be placed in position and an overflow pipe should be provided. The drainage and overflow pipe should be situated at convenient points and the pool should slope towards these outlets. The most ideal solution will be to connect the drainage pipe to the nearest sewer. Alternatively, this pipe may be led to a large pit 120-150 cm deep, filled up to 30 cm from the surface with small pieces of rocks or broken bricks. The drained water may be utilized for irrigation in the garden. But it may not be needed at all to have any elaborate drainage system as a well-constructed pool may need draining only once a year. Draining of water can be done with the help of a fractional horsepower water pump. The overflow pipe will maintain the water level.

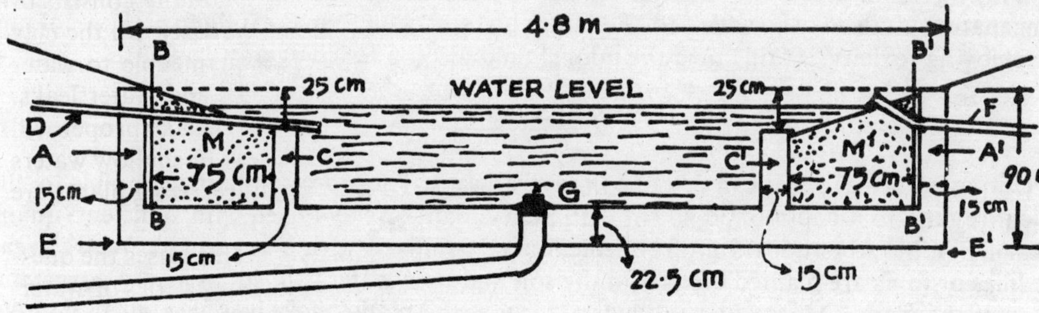

CONSTRUCTION OF A FORMAL LILY POOL

A cross-section drawing

Fig. XXIII.3

B-B[1]-Actual length of Pool, A-A[1]-Outer concrete wall, E-E[1]-The concrete base, C-C[1]-Inner brick or stone wall for surround, M-M[1]-Surround, D-Inter pipe, F-Overflow pipe, G-Drainage pipe.

After arrangement of the pipes are made, a 15 to 22.5 cm thick concrete layer is laid at the bottom of the excavated area. The concrete should be firmed with a shovel or a vibrator to ensure that no air-pockets are left, which will cause leakage of water. The bottom is levelled but not smoothed too much, otherwise the cement lining finish to be provided later will not hold. To avoid any joint the side vertical walls should be constructed, simultaneously with the floor when it is still wet. This can be achieved by dropping in prefabricated wooden moulds and packed firmly. The concrete used should consist of 3 parts aggregates, not larger than 2 cm, 2 parts sand, and 1 part cement. The concrete should be cured by placing wet gunny bags for a few days from the construction date. When the concrete has set, the pool is to be made completely water-tight, by putting a lining of cement mortar consisting of 1 part cement and 3 parts fine sand and a proportionate amount of a water-proofing material. The mortar should be 1.5 to 2.5 cm deep and should preferably put in two coats, one over the other when the first coat is still wet but firm. The coatings are kept wet by covering with wet gunny cloth for a few days to prevent cracking. Special care should be taken to ensure that the corners are completely water-tight. When a surround for growing bog plants is provided, this can be done by providing inner walls made of bricks or stones after the concrete is well set. The inner walls should be constructed at a distance of 45-75 cm from the edge of the pool proper, depending upon the size of the pool and the height should be 20-25 cm lower than the sides. The surround may run all along the periphery of the pool or it may be confined only on two sides. Drainage apertures are provided at the bottom of the walls to allow water to pass at the time of emptying the lily pool. The space between the outer and the inner walls is filled with a good compost consisting of good loam and peat or loam and leaf-mould. The soil is banked up to cover the tops of the outer walls. The edge of the outer walls should slope outwards to prevent surface run-off water entering the pool.

The usual depth of pools is 60 cm, but, for growing lotus, a depth of 1m is needed. In the plains of the tropics many species of water-lilies thrive and flower even in 180-250 cm of water. But in places having a severe winter, such depths will force the lilies to go dormant.

Before filling up the pool finally, it is desirable to fill water and drain it out several times to wash out excess of alkali and other chemicals harmful for plants and fishes. Potassium permanganate crystals may be added to the water just sufficient to make it light pink in colour, for removing toxicity. At this time the pool should also be tested against leakage.

Planting

Planting in formal pools can be done in several ways. The more common method is to fill the entire tank to a depth of 30-40 cm with heavy loam or clay mixed with sufficient quantities of F.Y.M. About 50-75 g of bonemeal may be mixed with every square metre area. The lilies or lotus are planted directly in the soil and then water is filled in just to cover the crown of the plants. More water is filled in stages as the plants make new growth. In no case should the plants be submerged.

The other method is to use large-sized cement boxes or tubs filled with a good compost, that is, heavy loam and F.Y.M. in the proportion of 2 : 1 mixed with liberal quantities of bonemeal. The size of boxes will vary according to the kind of plants. The largest size will be about 75 × 75 × 75 cm for large-sized varieties. The aquatics are then planted in these boxes and the pool is filled in gradually in the same manner as above.

For shallow pools, which are sometimes referred to as reflecting pools, planting by any of the above methods is not possible as these pools have a depth of only 30-45 cm. For such pools while constructing the floor, water-proof concrete boxes are prepared at desired intervals, 45 cm deep from the floor level. The dimensions of such boxes are 45 × 45 cm with 45 cm depth as stated above. These boxes are then filled with the required compost and the water plants planted in them.

In an already filled-up pool the planting procedure is first to plant the water plant in a basket, tub, or pot filled with compost and then lowering the basket in water keeping in view that the crown remains above the water. This can be ensured by keeping the basket on bricks for a week or two and when new growth takes place it is gradually lowered to the floor of the pool. If the pool is quite deep and an aquatic plant needs only shallow water, the basket or the pot can be kept over bricks or an inverted tub to give the desired depth of water. The soil surface in each basket or any other container is covered with 2 cm of washed gravel or pebbles to avoid disturbance of soil while immersing. In an empty pool with a layer of soil, while filling it with water the water should not fall directly in the soil, but the inlet pipe or hose should run into a bucket or basin. The pool is filled in several stages. While filling some plants may get displaced and float up, which are pushed back and kept in position with stones.

While planting, the concrete margins of the pool are concealed by putting compost over them and placing stones at regular intervals. Some moisture-loving plants and bulbs are planted along the margins to create naturalistic effect and to prevent the soil from falling in the pool.

Other Types of Pools

The concrete pools are costlier and difficult to construct. Every person cannot afford to construct such pools. For them several substitutes can be suggested. Large wooden barrels can be cut in halves and buried in soil, keeping the rim a couple of centimetres above the soil level to prevent run-off water coming in. The rims can be concealed with small stones, pebbles, or by growing trailing or low-growing plants along the edges. Barrels which were storing oil or soap are generally not used as it is difficult to free them from the residues. Other materials that can be used are discarded bath-tubs, galvanized water tanks, and wooden tubs. In West Bengal large burnt earthen pots meant for feeding cattle are available which are very useful for making small pools.

Fibreglass Pools: In advanced countries such as the U.S.A ready-made fibreglass pools of various shapes and sizes are available. For installing these pools a hole little larger all round than the size of the pool is dug, all stones are removed, and the bottom is levelled. The pool is placed in position and soil is rammed from all sides. Fibre glass pools are very durable but cost is very high compared to plastic or PVC.

Vacuum-formed Mouldings: There are certain other prefabricated pools available in advanced countries made from such plastics as PVC and polythene. These are inexpensive but often the depths are not sufficient to grow most of the aquatics.

Pool Liners: The simplest method is to line an excavation with a sheet of flexible plastic commonly called pool liner. The commonest material used is 500 gauge polythene in black or blue colour; the former is preferred. The pool is dug in the required shape, but preferably of rectangular shape for polythene liners. The dug-up pool is cleared of all stones and the surface is smoothened with a layer of sand. The polythene sheet is then placed in position with 15-20 cm polythene overlapping the edges of the pool. Two sheets are placed one above the other or one sheet folded double is used as a safety against leakage. The edges are covered with soil or paved with stone or growing dwarf plants over the soil. Though this is the cheapest and the quickest method to make a pool, the major disadvantage is that it may develop leaks at any time and polythene becomes brittle with sunrays above the waterline. Polythene does not stretch, so this must be tucked right down along the excavation. Once punctured, a polythene cannot be repaired and has to be replaced.

The best liner is possibly PVC which is chemically different and superior to polythene as this is not easily damaged by accident. Moreover , PVC is stretchable and can be stretch-fitted. Small accidental leaks can be repaired by putting a patch of PVC with some adhesive such as fevicol. As PVC stretches with the weight of water it fits well along the excavated walls and surface of the pool.

Another good liner is Butyl Rubber which is also stretchable.

Watercourse and Falls

It may sound poetic, but we feel that a placid surface of a pool is dull, and the soft murmur of a running stream or waterfalls and the splash of a fountain create a music bringing a refreshing change in the garden, which no other garden feature could provide. Besides the aesthetic aspect of the above, the running water has several practical applications. The most important is that the running water or splashes from the fountain increase the oxygen content of the pool. The fish in the pool swims playfully against the current. But the strong currents flowing through the pool are disliked by the water-lilies. Therefore, the cardinal point in creating any water course is to circulate the pool water itself. The best way to have a watercourse is to use the dug-up soil of the pool as the base of a rock garden near the pool and run the watercourse through it. The extent of a watercourse will depend upon the ambition of the individual and the area of the overall garden. The watercourse may have one to three drops (falls) depending upon the size of pool and the watercourse. If space permits, three drops, each of 15-20 cm, to a maximum level of 45-60 cm will be quite all right. The watercourse should be zig-zagged to give a natural look. Stones should be placed along the banks and on the beds to give a natural effect. Each drop should fall into a pool to create the splash. The watercourse is made water-proof by concreting. Water-proofing can also be done by using pool liners. The water is pumped on the top of the water-course with the help of a concealed surface pump or a submersible pump. If desired, one may add a fountain, but it is not necessary. Some stone or concrete sculptures may be added in the watercourse.

The pool and the watercourse could be lighted during the night to enhance their grandeur.

Care of Water Garden

Water plants are generally quite hardy and many of them will grow quite well if left undisturbed. Such plants are lifted and replanted only when they appear unhealthy. Some of the water-lilies make too much of leaf growth and deprive the pool of essential sunlight. Such plants are lifted, divided, and replanted.

Often a slimy green growth, popularly known as 'Blanket Weed' covers the surface and the side walls of the pool. This can be controlled by adding 30 g of potassium permanganate to each 100,000 or 125,000 *l* of water (the simplest method of calculating the volume of water in a pool is to multiply the length, breadth and the average depth of the pool in feet and then by multiplying this by six will give the approximate number of gallons contained in the pool.) or 115 g of copper sulphate for the same volume of water. This small amount of chemical is not harmful to the plants or the fish. A second application after a week may be necessary if the first fails.

Fish in Pools

Ornamental fish can be introduced in a pool. But one should keep in mind that the fish are not essential for a pool, No fish should be introduced into the pool at the time of planting as the fish have the habit of nibbling at the oxygenator plants. The fish are introduced after about three weeks when such plants have made sufficient growth and withstand the damage. The coloured gold fish are the best for a pool. The usual rate of fish population initially is 2.5 cm long (excluding tail) fish for every 4.5 to 5 *l* of water. Fish need some hiding place especially in shallow pools from birds such as crow, egret, and crane. This is provided by the densely-growing water plants. Fish, apart from adding beauty to the water garden, eat up the larvae of mosquitoes.

A few water snails should also be introduced in large pools as they act as scavengers eating up the plant and animal residues. But doubt is expressed about the need for deliberately introducing water snails in a pool. Though they consume a good amount of decaying vegetation, they are also capable of feeding on soft growths of certain aquatic plants.

Plants for Water Garden

Plants for water garden can be divided into several groups such as (a) surface flowering aquatics, (b) oxygenators, (c) floaters, and (d) marginals.

(A) Surface Flowering Aquatics

Aponogeton distachyus (*Aponogetonaceae*)

It is a native of South Africa. It is commonly known as 'Cape pond-Weed' or 'Water hawthorn' because of its hawthorn-like scent. The oval leaves are floating and the flowers project just above the foliage. Flowers are snow-white, while the anthers are jet black.

Euryale ferox (*Nymphaeaceae*)

It is a native of East Indies. A large annual aquatic herb, next in size to ***Victoria regia***. Leaves are circular about 60 cm across. It is propagated by seeds.

Linnanthemum indica (Syn. *L. indicum*) (*Gentianaceae*)

It is a native of the Tropics and North Australia. Commonly known as 'Water Snowflake'. Leaves are *Nymphaea*-like and floating. Flowers are white, numerous, short-lived (generally 1 day only), and it flowers most of the season.

Monochoria hastifolia (Syn. *Pontederia dilatata*) (*Pontederiaceae*)

It is a native of S.E. Asia. Flowers are blue in crowded umbels. It is propagated by division.

Nelumbo lutea (*Nymphaeaceae*)

It is a native of southern U.S.A. It is commonly called 'American Lotus', 'Water Chinkapin'. Leaves are 30-60 cm across, raised 60-180 cm out of water. Flowers are yellow and fragrant, 10-25 cm across.

N. nucifera (Syn. *Nelumbium speciosum*)

It is commonly called lotus or *padma* or *kamal*. Flowers are white, tipped rose or pink but with many variants, 30 cm across and overtopping leaves. Flowering is during the summer. Propagation is by cutting of the rhizomatous stems or the runners. It can be grown from seeds also where cuttings fail. While planting cuttings, the cut end is washed in a weak solution of potassium permanganate and air-dried for some hours. Newly propagated plants should not be planted in deep water. Seeds are raised in ordinary seed raising composts in seed pans, just covered with water. When seedlings are about 5-7 cm tall, these are transplanted singly in pots in shallow water and planted in a permanent site when established.

Nymphaea (*Nymphaeaceae*)

These are commonly called water-lilies and have about 40 species and several varieties and cultivars. These are mostly natives of the Northern Hemisphere or the Tropics, while a few belong to South Africa and Australia. There are types which can be grown in very shallow waters (*N. pygmaea helvola*, 'laydekeri', *N. tetragona*, etc.), in medium depth of water (*N. alba*, *N. odorata alba*, etc.), and still others requiring deep water ('Charles de Meurville', 'Colonel Welch', etc.). *Nymphaeas* are propagated by division of clumps, tubers, or by cuttings from the rhizomatous stems of a few species. Seeds germinate easily but the seedlings come to flowering only after about 2 to 3 years.

Orontium aquaticum (*Araceae*)

It is a native of North America and is commonly known as 'Golden Club', deriving the name from the densely-crowded, tiny yellow flowers in the white-stemmed flower spikes. The leaves are velvety, silvered blue-green, which grow either 30 cm above water surface or float, depending on the situation.

Victoria regia (Syn. *V. amazonica*) (*Nymphaeaceae*)

This lily can only be cultivated in large ponds. The leaves are floating, up to 1.5 m in diameter, the margins are upturned and 5-12 cm broad. The leaves have projecting air-filled ribs beneath, which give the leaves great buoyancy to support great weights, and it is said that a grown-up person can seat on a floating leaf. This is an annual and needs 1.2 m of water. The seed is stored in water till sowing. To facilitate germination a portion of seed coat is chipped. The seed is germinated in compost under 5 cm water. The flowers are large and cream coloured.

(B) Oxygenators

The submerged oxygenating plants are vital in any pool, as they maintain hygiene and balance. It is difficult to recommend which of the following plants will grow well in a particular pool. Out of the plants mentioned below *Elodea* has a better chance of survival under most conditions. But it is advisable to grow several types out of which the fittest will survive. Some important types are mentioned below.

Callitriche autumnalis (*Callitrichaceae*)

It is the best oxygenator in the family. This is entirely submerged and grows in running water. It is a suitable plant for sheltering fish and providing food to them. A native of north cool temperate and Arctic zone including Britain.

Ceratophyllum (*Ceratophyllaceae*)

The two species planted are *C. demersum* and *C. submersum*. These slender-branching, fragile, and submerged plants do not need formal planting, instead are just dropped in the pool and grow in wide situations like deep, shaded, or cold pools. This water weed should be removed periodically or they cover the whole tank. Propagation is by small cuttings.

Elodea (Syn. *Anacharis*) *canadensis* (*Hydrocharitaceae*)

It is of vigorous growth but is probably the tidiest and shortest among the good oxygenators. The thick growth provides shelter for the fish.

Ranunculus aquatilis (*Ranunculaceae*)

It is commonly known as water 'Crowfoot' or 'Water butter-cup'. This has fine submerged foliage and in running water the foliage comes in long waving tresses. This bears in May numerous, small, white flowers on the surface.

Vallisneria spiralis (*Hydrocharitaceae*)

It is commonly known as 'Eel-' or 'Tape-grass'. Leaves are long, in deep, and short in shallow waters and about 1 cm broad. The female flowers are borne in coiled-stalks under the water which uncoil to bring the female flowers to the surface, where it is pollinated by a detached floating male flower. It serves as a food for aquatic wild-life, as well as a good oxygenator. Propagated by the rooted runners.

(C) Floating Plants

The speciality of the plants in this group is that no anchorage in soil is needed for survival

Azolla caroliniana (*Salviniaceae*)

It forms a crinkled surface carpet of mossy green and in temperate regions during the autumn it gets beautifully tinted with lilac and red.

Eichhornia speciosa (Syn. *E. crassipes*) (*Pontederiaceae*)

It is commonly called 'Water-Hyacinth'. It is a floating plant which if not kept in control will fill the pond by itself. For academic interest this plant has been mentioned here; but one should avoid growing this dangerous water weed which has been banned in many countries. Blue flowers are borne in erect panicles.

Lemna gibba and *L. minor* (*Lemnaceae*)

These are commonly called the 'Duckweeds' and good as fish food. It should be collected by pond nets occasionally to check rampant growth.

Trapa natans (*Onagraceae*)

It goes by the common names 'Water Chestnut', 'Jesuit's Nut', 'Water Caltrops', and 'Saligot'. The overlapping triangular bronzy-green leaves are attractive. It is an annual regenerated by the production of nut-like seeds. The 4-angled spiny fleshy fruits are edible and known in vernacular as *Pani-Phal*, *Singara* etc.

(D) Marginal Plants

Marginal plants are those which prefer to grow with their roots in shallow water but raise their stems and flowers above the surface. Marginals are mostly hardy and coarser plants and if not kept in check will overgrow the pool plants proper. These plants are not essential for any pool.

Acorus calamus variegatus (*Araceae*)

It is a form of the commonly known 'Sweet Flag' having leaves stripped with green and creamy white. Both *A. calamus* and *A. calamus variegatus* are water-loving herbaceous perennials bearing greenish-white flowers on a 5-7 cm long spadix.

Calla palustris (*Araceae*)

It is called 'Bog Arum' and has shiny round:d leaves projecting 15 cm above water. It blooms arum-like white spathes which if fertilized by water snails develop red fruits.

Cyperus alternifolius (*Cyperaceae*)

The numerous leaves are arranged in an umbrella-like fashion over an angular stem. Leaves are quite attractive and are used for table decoration. In the variety *variegatus* the stems and leaves are mottled white or sometimes completely albino.

Iris (*Iridaceae*)

Several species of iris with their beautiful flowers are suitable for growing in shallow waters.

(a) *I. laevigata:* It is the best among the marginal plants with clear violet-blue flowers.

(b) *I. pseudacorus:* It is commonly called 'Yellow Flag' or 'Water Flag' and is a vigorous-grower having yellow flowers.

(c) *I. kaempferi:* It peforms well only when the soil is just covered with water. Flowers are of various colours, from reddish purple or shades of yellow to white.

Sagittaria sagittifolia (*Alismataceae*, also spelt *Alismaçeae*)

It is the 'Common Arrow-head' with large sagittate leaves. Flowers are white but turns pale blue. A good oxygenator. In var. *florepleno* the flowers are double.

Typha (*Typhaceae*)

A few species of *Typha* are good marginal plants, but in India the most commonly grown species is *T. angustifolia*. The brown cylindrical flower heads are borne in long stalks. This should be kept in check by occasional thinning. The flower heads with long stalks are good for indoor decoration.

MARSH OR BOG GARDEN

Many authors have reservations regarding the name "Bog Garden", as an actual bog is an area where there is stagnant acid water and only plants such as sundew, butterworts, etc., grow. But the bog garden which we find in landscape gardening is actually a marsh where the soil is not sour or acid and a shallow stream or trickle runs through it.

A pond is not an essential feature of any bog garden, but, if there is any adjacent pond, the overflow water can be used to keep the marsh garden moist. The main criterion of a marsh garden is to keep it moist and in a swampy state all throughout. For this reason, the site of the marsh garden should be low-lying where the surface drainage water will collect naturally. A site having a sub-soil of sticky clay is ideally suitable for marsh garden as only a trickle of water will keep this wet. But, if the soil is light in nature, certain amount of digging will be needed before establishing a marsh garden.

The whole area intended to be developed into a marsh garden is dug up to a depth of 60 cm and a thin layer of clay, 10-12 cm deep, is placed at the bottom to form the base of the marsh garden. Then a 12 cm layer of rubble or large stones are placed over this to ensure that the water in the marsh does not become stagnant. The rest of the hollow is filled with a compost of garden loam and leaf-mould in the proportion 1 : 1 and the level made up with the surrounding land. A marsh garden should never exceed a depth of 60 cm although its spread will depend upon the available area and personal choice. A marsh garden should never suffer from drought and be kept sufficiently moist by a trickle of water at the same time avoiding stagnation.

All parts of the marsh garden is made accessible by laying paths of rough stones or bricks, winding round the marsh and putting stepping-stones over them. It must be noted that all marsh plants need water, but in varying degrees. Accordingly, marsh plants can be categorized into three broad groups. In the first group belong plants which love-5-10 cm of standing water and are termed as "swamp dwellers". In the second category are placed the plants which live on or a little above the waterline, with their roots standing in water-logged soil. In the third group are placed those plants which grow above the saturated soil, but near enough to send their roots in search of water. Keeping this in view the paths through the marsh should be laid at different levels so as to divide the marsh into shelving beds of different heights enabling the plants of the above three groups to find the right type of growing conditions.

A marsh garden can accommodate any moisture-loving plant, those which are found commonly growing on the banks of streams or ponds. Some of the plants which can stand in 12-15 cm of water are also included in a marsh garden. In some pockets of the marsh even plants growing in shallow water can be accommodated. While planting, the main point one should bear in mind is not to overcrowd the bog. It is better to plant one type of plant in a group of 3 to 6 allowing sufficient space in between before another group of different colour, size, and height is planted. This variation in planting will break the monotony.

The actual plants selected for a marsh will, however, depend upon the size of the garden and the undulation of the marsh. If the marsh is in one plane it will not be possible to grow the plants of all the groups mentioned earlier. In a small area it will be advisable to create a greater variety by growing the dwarf marsh plants. But in a vast area the vigorous growing ones should be used more freely, especially to create background effects. While planting one should also keep in mind the natural surroundings and should visualize what shape the plant will take in 2-3 years.

As already stated, a pool is not essential for a marsh garden, but one of the effective ways of creating a marsh is to incorporate the pool design into this. This can be done by having a broad surround around the pool or if a pool already exists without a surround, tack the marsh garden next so that it appears as a part of it. The plants suitable for bog garden are listed in the following table.

Bog or Marsh Garden Plants

Code: S-roots in soil submerged by a few centimetres of water.
W-roots in moist soil, but crown at or not much above water level.
M-crown above saturated soil, but roots within the reach of sufficient moisture.

Name and common name	Zone	Height (cm)	Characteristics	Season of flowering
1	2	3	4	5
Acorus calamus and *A. calamus* var. *variegatus* Sweet flag	S	60	The leaves 60-90 cm long 1-2 cm wide, margins wavy. The second species having leaves broadly stripped green and cream.	Summer
Alocasia Different Spp.	M	Varies upon variety	Beautiful broad leaves of Sagittate-cordate shape.	June-August
Arundinaria (Bamboo) Dwarf species	WM	30 cm to 2.5 m	Leaves bamboo-like.	All the year round
Arundo donax and *A. donax* var. *variegata* Giant Reed	M	Up to 3.5 m	Leaves 30-60 cm long, 3-5 cm broad. The variegated species has white margins.	Autumn
Asplenium Fern	MW	10-60	Fronds are of evergreen nature.	—
Calla palustris Bog Arum	SW	15	Leaves shiny, flowering spathe white, Arum-like	June-July
Caltha palustris var. *plena* Double Marsh Marigold	WM	25	The showiest *Caltha* with bright yellow double flowers.	March-May
Cypripedium spectabile (Syn. *C. reginae*) An orchid	M	45-90	Flowers white or flushed rose.	June
Eriophorum sp. Cotton Grass	W	45-60	Plumes silky white.	Summer

1	2	3	4	5
Filipendula palmata Syn. *Spiraea digitata*	M	60-90	Flowers pale pink turning white.	Summer
F. ulmaria var. *aurea* (*variegata*), Syn. *Spiraea ulmaria* var. *aurea* Meadow Sweet, Queen of the Meadows	M	45-60	Leaves gold-splashed, flowers cream.	Summer
Hemerocallis in different species and cultivars Day-Lily	M	30-60	Leaves narrow, grass-like. Flowers funnel-shaped, in many shades.	Summer to Autumn
Hosta (Syn. *Funkia*) sp. Plaintain-Lily	M	60	Bold foliage plants for group planting. *H. glauca* have blue-green leaves *H. undulata variegata* have dark green leaves splashed with pale-green and cream.	Summer
Iris kaempferi described in water garden	WM	60		June-July
I. laevigata described in water garden	S	60		June-July
Lysimachia nummularia Creeping Jenny or Charlie Moneywort	WM	Prostrate	Stem spreading studded with yellow flowers.	June-August
Lysimachia punctata Loose-strife, Spires of yellow	WM	60	Less-branched perennial herb, flowers yellow. Needs controlling.	June-July
Osmunda regalis Royal Fern	WM	90-180	A good plant for making clumps at borders of water.	
Pandanas odoratissimus, Syn. *P. elegantissimus*, *P. utilis*, Screwpine (*Keora, Ketaki, Keya*)	SW	2-15 m	Bears scented flowers. To be pruned to keep dwarf.	Summer

(*Contd.*)

1	2	3	4	5
Primula Candelabra types	M	30-60	*P. aurantiaca*, flowers orange-red, *P. japonica*, crimson. Suitable for mild and flowers-temperate climates.	May-July
Primula florindae Giant Cowslip	WM	90	One of the best primulas for a wet place. Suitable for mild and temperate climate. Flowers sulphur-yellow.	July-August

Besides the above plants, the following trees also love moist soil and can be grown in the background; *Acacia farnesiana*, *Barringtonia acutangula* (Syn. *B. samoensis*, *B. insigne*, *B. acutangula rubra*), and *Salix*.

SUNKEN GARDEN

As the name implies, this garden is laid below the ground level. One may question the utility or the purpose of laying a garden below the ground level. The answer is simple: It brings variety in the garden. In a long stretch of land it may be useful to reserve a portion for sunken garden. If a garden is laid in one plane, all the garden will be visible to the visitor at a glance or two from any one point. But if there is a sunken garden or a mound the curiosity of the visitor is aroused and he moves to the place to find out what surprise is in store for him. Moreover, an existing ditch or a dried-out pond may be utilized for laying out a sunken garden.

A sunken garden can be laid out in an already existing depression of land or a portion of the garden may be specially dug to lay out such a garden. The cost of laying will be much cheaper in the former than in the latter case. Since a sunken ground is likely to collect water, adequate drainage arrangement should be made. The simplest method will be to have one or two drainage outlets and connect them to the sewerage system. It is not advisable to have a sunken garden in a very heavy soil because of the drainage problem. If the garden is likely to be flooded rapidly, special arrangements may have to be made to drain out the excess of water accumulated. But in most cases this may not be necessary, especially in drier tracts. Some drainage materials such as stones, rubbles, etc., may be placed below the surface of the sunken garden at a depth of about 1 m from the surface. The thickness of these materials may be about 60 cm. Steps should be taken to prevent the surface run-off water to fall in the sunken garden.

There is no hard and fast rule as to what is to be planted in a sunken garden. This will depend mainly on the area available and the choice of the grower. If the area is small only some annuals and herbaceous perennials are grown in formal beds. But in a large sunken garden an elaborate formal or informal garden can be laid with paths, lawns, flower beds, shrubbery, trees, rose garden, and greenhouse. The ideal depth of a sunken garden is about

120 cm. In case sufficient depth is available one or all the walls of the sunken garden can be developed into dry wall. This will help support the raised sides of the sunken garden in addition to beauty. In dry localities, fountains and streams may be provided to increase the humidity. It is always advisable to have an attractive central feature such a sundial or bird-bath or stone figure. This helps tie the composition of the garden together. While planning the garden one must take into account the portions receiving or not receiving the morning sun and the portions that are under full sun or in complete shade. A sunken garden is likely to be more humid, especially in high rainfall areas, than the portions above the ground. Hence, it is likely that moisture-loving plants such as ferns, Ixora, *Gardenia*, etc., will do better in such a garden. The approach to a sunken garden is generally by a staircase, preferably made of stones. The surface of the sunken garden can be decorated with crazy paving, gravel, and steps in addition to neatly arranged beds and borders.

It is also possible to have a sunken rock garden. When the paths in a rock garden are dug deep to create the mounds and hillocks, it takes the appearance of a sunken rock garden.

GARDENING IN THE SHADE

The home gardens, a portion of which may be shaded, is a blessing in disguise if one can take full advantage of it. But many people consider shade in the garden as a nuisance as no plants are likely to flower properly under it. This is a hasty conclusion, as some people might have grown few sun-loving plants such as sunflower, marigold, or zinnia, etc., which naturally might not have done well under shade. But with a thorough knowledge about the habits of different plants, it is possible to lay out a beautiful shade garden with shade-loving plants in such shaded places.

Before starting a shade garden one must study the nature of the shade. After a thorough study conducted, throughout the course of a day and during various seasons, a comprehensive picture will arise about the nature of the shade. The conclusion may be that the garden has three types of shade. One in the category of dense shadow in north or north-west corner under a high wall or below low-spreading trees such as *Ficus benghalensis*, *Brownea ariza*, beach, etc., where the ground is semi-dark. There are other places where though there is shade, but the sun penetrates to a considerable extent. Finally, there are places which are in half shade like on the east or west sides of the buildings.

Problems Under Shade

The first thing to be investitaged from a shaded portion is the nature of the soil. An indication will be available from the health of the natural undergrowth. If these are lanky or weak the plants may either be suffering from lack of water or plant food or both. While gardening under shade of trees, the tree roots will pose a problem. They compete with the plants grown under it for water and food. Therefore, the soil under the trees is to be dug deeply (25-30 cm) to cut the surface feeding roots. This operation should be repeated every year by digging deep along the borders of the beds. The beds are to be fertilized more liberally as some of these fertilizers will be utilised by the tree itself. The soil pH should also be tested.

One general notion is that shade gardens need less watering. This is true as far as a shade from an inert object such as a wall is concerned. But if the shade is from a living tree it may be necessary to water the shade garden more frequently, as a major part of this water will be utilized by the tree itself. For this rationing of water and plant food, shade garden plants under trees are spaced much wider than in the open.

What to Grow?

The selection of the plants for the shade garden should be done after much thought. If the garden is in a shaded terrace, only few shade-loving plants and possibly some flowering shrubs should be grown. In a milder climate, hydrangeas can be tried. Azaleas may be a possibility, but this and camellias will do better in a much colder climate. In tropics, there are not many flowering shrubs suitable for growing in pots in shade. But some shrubs with beautiful foliage can be grown in semi-shade. These are *Sanchezia nobilis variegata*, *Aphelandra cristata* (Syn. *A. tetragona*) which bears orange-scarlet flowers also, *Strobilanthes* species, etc. Some shade-loving flowering annuals can also be grown (see page 64). A few herbaceous perennials will do well in semi-shade. These include : *Impatiens sultanii*, *Coleus* (should preferably be grown as annuals), geranium, etc. Few non-flowering type creepers such as *Hedera helix baltica*, *Ficus repens*, *Pothos aureus* (possibly *Scindapsus aureus*), commonly called 'Moneyplant' do very well in semi-shade. The beautiful yellow markings on foliage in the moneyplant may, however, disappear under deep shade.

In the ground a flower border may be planned with shade-loving annuals or bulbs (e.g., football lily, crocus). Many ornamental ferns can also be grown in shade. A shaded place in between two buildings can be developed into a wild garden with a winding path. Ferns, bulbs (e.g., *Caladiums*), *Impatiens sultanii*, etc., could be included in the wild garden.

ROOF GARDENING

There is a misconception in India between roof gardening and terrace gardening. In many articles the gardening on the roof is often termed as terrace gardening which is not strictly correct according to British practice. Terrace gardening is completely different (see Terraces and Terrace gardening discussed earlier in this Chapter) from roof gardening and in this book by roof gardening we shall only refer to gardening on the roof.

Why Roof Gardening?

In modern times, homes with a compound and lawn specially in cities and towns, are becoming rare and skyscrapers are replacing such homes. As a result, the private home gardens are vanishing and the only places left for gardening are the roofs of houses and the balcony. A spacious and well-planned roof garden can be a place of joy and recreation. In bigger cities of India, many of the large hotels and public buildings are developing this type of gardening. Even in a small city as Lucknow a hotel has a roof garden with lawn, though not well-planned. Many private houses in Delhi, Calcutta, Bombay, and Baroda have roof gardens, including a lawn. The roof garden of Dr. B.P. Pal in Delhi, dominated by a collection of roses, besides other plants such as bougainvillea, pansy, etc., is a charm by its own right. There are many other private homes in our cities having charming roof gardens with rare collections of cacti, orchids, dahlia, chrysanthemum, and other plants.

Limitations

The difficulties encountered in a roof garden is completely different from that of gardens at the ground level. Since the garden is at a high level from the ground, the cost of maintenance may go up as everything has to be carried up. But the more fundamental difference is in the depth of the soil. At ground level the depth may be unlimited with a source of groundwater, whereas on the roof the depth of the soil is shallow, not exceeding 90 cm, but generally varying between 20 and 60 cm. Not only the depth is shallow in roof garden, but the drainage is also good and hence water has to be replenished constantly by frequent watering. Large trees and shrubs are generally not grown on the roof as the growth of the tap root is limited by the roof below. Moreover, because of shallow depth, good drainage and frequent watering, the plant food in the soil leaches off more rapidly and is to be made good by more frequent replacement.

Planning

Before planning a roof garden certain points are to be checked. Flower beds can be made directly on the roof. This can be achieved by placing the soil between the outer wall or the parapet and a wall built on the inner side with the help of moisture-proof wood shuttering, stones, or bricks. The width of such boxes will vary according to the available space, the nature of the plants to be grown, and other factors. Such beds directly constructed on the roof surface are not preferred always, as they are of permanent nature and may also damage the roof. Alternatively, the best method will be to arrange a series of containers holding soil, such as boxes, pots troughs, tubs, etc. These are placed over wedges or bricks to leave a clear gap between them and the roof for proper drainage and passage of air.

The first thing to check before starting a roof garden is whether the roof surface can bear the weight of the soil. Soil, especially wet soil, is much heavier than one can usually conceive. Secondly, the roof should be made water-proof to prevent seepage of water into the room below; and, finally, it should have adequate drainage so that rain and irrigation water drains off quickly.

The concept of roof gardening itself is artificial; therefore, while planning and designing this point should be kept in mind. As for example, it is not possible to create natural scenery with hills and valleys or a natural forest scenery. As for any other gardening a roof garden should also have a dominant focal point, like lawn or water garden or rock garden or a prominent climber (e.g., Bougainvillea, *Clerodendron splendens*, *Bignonia venusta*, climbing rose). The vision of a visitor in the roof garden tends to travel outwards in the horizon; therefore, the endeavour on the part of the planner will be to hold his view by a barrier of masses of bright colour and tall-growing plants at the outer edges of the roof.

On a roof the space available for gardening is limited; therefore, it will be a good idea to encourage vertical growth. This is the reason why a roof garden lover should use more climbers and trailers in his garden, some of which can climb over a supporting pillar while others can grow rampantly over the boundary wall and spill over the top and hang down the side of the building camouflaging the brick and concrete. *Ficus repens* is a good root climber which can cover a wall. Other attractive creepers include *Cobaea scandens* (annual) railway creeper, *Vernonia elaegnifolia*, passion flower, *Thunbergia alata* (annual), etc.

Hanging baskets should play a major role in a roof garden. Various types of basket either containing flowering or foliage plants can be displayed with great advantage at suitable places (for details see Chapter XXV). One can also display with advantage the collection of *bonsai* and few alpine plants. One or a few vertical gardens can also be displayed in a planned manner in the roof garden.

It is possible to grow a wide range of plant materials in a roof garden excluding, of course, the large shrubs and big trees. But even these can be grown till they are young. Depending upon the sun and the shade, the climate, the size of the roof, etc., the following plants are recommended for growing in the roof garden. This is only a suggestive list and not an exhaustive one.

Flowering Annuals: Antirrhinum, stocks, dwarf sweet peas, pansy, dahlia, chrysanthemum, marigold, sweet alyssum, phlox, pinks (Dianthus) and verbena.

Herbaceous Perennials: Pelargoniums, michaelmas daisy, canna, *Mirabilis jalapa*, portulaca, *Solidago canadensis*, *Vinca rosea*, and perennial verbena.

Shrubs: Many of the dwarf and medium shrubs can be grown.

Trees: One or two dwarf trees such as *Plumeria* sp. *Callistemon lanceolatus*-and-*Gliricidia maculata* can be grown as specimen plants. Some large to medium trees such as *Araucaria cookii*, *Mimusops elengi*, *Brassaia actinophylla*, etc., can also be grown till they are young.

Creepers: The creepers that can be grown are mentioned in the text.

Bulbs: A variety of bulbous plants of annual or perennial nature can be grown.

Water plants: Water-lilies and other water plants can be grown in the lily pool, if any, or in cement tubs.

In a railway guest house in Bangalore a lawn has been laid out on a roof by making the roof water-proof in an ingenious way. A polythene sheet was laid out first over the roof before putting the soil. The joints in the polythene may be sealed by heat or a glue. But before trying this method, one must ensure that the roof can take the load. A roof garden with lawn and other features were laid in similar manner, under the guidance of the authors in a factory near Bangalore.

GREENHOUSE, CONSERVATORY, AND LATH HOUSE

There is actually very little difference between a greenhouse and a conservatory structurally but these differ as to the use to which they are put. Originally and etymologically the term greenhouse meant a house in which plants are kept there to grow. But sometimes a greenhouse may be used for a different purpose also. For example, in severe winter many plants are kept inside for winter protection only and are not likely to grow. The Americans use the term greenhouse to mean any type of glass building inside which plants are grown, except the structures such as cold frames and hotbeds. In England, the term is defined as a glass structure in which plants not requiring a high temperature are cultivated or displayed. In a greenhouse, plants are exclusively grown in pots and tubs. According to the British terminology the term has wide application as it may include houses in which plants are grown from seed to its maturity, or may mean a house exclusively used for displaying plants which were grown up to the flowering stage in another house.

A *conservatory* differs from a greenhouse in which permanent plants are planted in central beds and the side beds are used for continuous floral display. A conservatory is situated very near the dwelling house and should actually be included in the same architectural plan. This may be connected with the house by a corridor or a pergola. In India a conservatory is also known as *Fernery* as in olden days ferns occupied predominant position in such houses.

A *forcing-house* is a structure where plants are grown out of their normal season of growth by artificial means.

It will be seen from the above discussion that there is no one terminology which will include all the glass plant houses. Some people may use the term glasshouse to include these, but there are other glasshouses also in which plants are not grown. Hence, the best terminology is greenhouse meaning a structure used for growing living or green plants.

Under Indian climatic conditions, the term greenhouse includes a glasshouse for plants and any other structure including Lath house for growing shade loving or greenhouse plants. In the tropical plains of India hardly any glasshouse is used for growing plants except for researches on plant viruses. Even if glass structures are made say for growing cactus and for breeding purposes it is generally only the roof which is made of glass and the sides are covered with fine wire-mesh. Otherwise, a full glasshouse has to be airconditioned to keep it cool during the summer months.

In most parts of India a greenhouse and conservatory or fernery, as it is called popularly, is meant for providing shade and cool temperature for the plants growing within it. A greenhouse provides a cool and pleasant retreat for the dwellers and friends in the hot summer. In high hills, a glass greenhouse is necessary to grow the more tropical plants to protect them from severe cold weather.

Location

Under cool temperate conditions the criterion is to admit plenty of light and air inside the house and hence the construction is made north to south so that the broader sides are exposed more to the morning and the evening sun and the structures get less of the midday sun in the summer. In temperate countries, it is often advised to position the greenhouse from east to west also on the ground that the house gets the slanting rays of sun in the winter when the plants need it more. In Indian plains the idea is to have plenty of light but little of direct sunlight. For this purpose it will be best to run the greenhouse from north to south so that the longer sides face east and west to avoid the scorching midday sun. Even in such cases it will be only the mild morning sunlight which will be preferred and the afternoon sun is excluded by some lattice work of bamboo or wood or growing creepers on the western side. It is not advisable to situate the house in the drips of large trees as this may interfere with plant growth. However, it is not a bad idea to grow small trees and shrubs round the greenhouse especially on the western and the northern sides provided these do not interfere too much with the life inside the house. Such plants will cut off the hot summer wind and the cool winter breeze detrimental to plants in a tropical greenhouse.

A location good for laying out a garden is usually best suited for a greenhouse also. It should be within the easy reach to attend to the daily necessities. Where a greenhouse or a conservatory is meant for private use, i.e., the plants are kept in display for the pleasure of the family and visitors, it should be situated as near the house as possible. For commercial greenhouses other considerations such as cost of land, easy access to market, plenty of water supply, and cheap fuel, especially in places with a severe winter, should be taken into account.

A greenhouse should be located on a slightly raised ground and should preferably not be situated along the boundary as large trees of the neighbour growing along the boundary may interfere with the greenhouse life.

The size of a greenhouse will depend upon the space available, the financial capacity, and the taste of individual growers.

Types of Greenhouses

In temperate countries the greenhouses and conservatories are made of glass on frames of brick stone, wood, steel, or aluminium or frames combined of wood and either of these materials. Where a combination of wood and iron or aluminium or brick is used, the main structures bearing the weight is made of either of these materials and the frames for fitting the glass is made of wood, as in colder countries, the greenhouses are artificially heated and wood serves as a non-conductor. The construction from foundation to plinth level may be either of brick or of stone and above this the iron or aluminium or a combination frame-work is constructed on which glass is put. Aluminium is rust-proof requiring no painting and is light in weight. Steel and wood need painting to prevent decay. All brick work should be 22.5 cm thick. An alternative to glass is fibre-glass, which is though expensive initially does not incur recurring expenses. But fibre-glass will not admit as much light as a glass-pane. Of course, glazed glasses are also available for those who want less light through the roof. Transparent plastic sheets of different grades are also available but these develop scratches easily. The quality of glass-panes used is usually called "double-thick". Wire-mesh reinforced glasses are also available for the roofing for greater durability. As an additional protection against hail-storms a wire net of 1 cm mesh should be fixed10-15 cm above the roof fitted on projections over the roof. The glass-panes are fitted to wooden frames with the help of best quality putty to make the joints water-proof. In metal frames these are fixed with the help of chips or by nuts and bolts, putting washers round the holes made in the glass to prevent leakage.

Several designs can be followed for the construction of glass greenhouses. These are the span-roofed house with path in the centre, span-roofed house with staging in the centre, hip-roofed or three-quarter span house, and the lean-to house (see Fig. XXIII.4). The lean-to house needs a wall to form the tall side. For growing plants on the ground such as carnation or chrysanthemum the glasses are put up to the ground level, whereas in houses where plants are grown on staging, the glasses are fixed on low brick walls constructed a little above (about 60 cm) the ground level. The paths in the greenhouse should be of cement or stone to keep it dry.

The staging for displaying pots may be flat but preferably in gallery style. These may be made of wood or cement. Cement is easy to construct and durable. For raising cuttings cement trays 120 cm wide may be constructed with sufficient depth (15-30 cm) to hold the soil.

The width and the length of a house depend upon the factors mentioned earlier. But the height should not be less than 150 cm at the eaves and not less than 210 cm at the ridge (i.e., centre of the roof). The doors in a green house should be broad enough to admit the wheel barrow and open outside.

Ventilation and Heating

Provision for ventilation should be made at the time of construction for admitting the required quantity of air whenever needed. The ventilators should open from inside and should not cause the inside temperature to drop too much in the cold season.

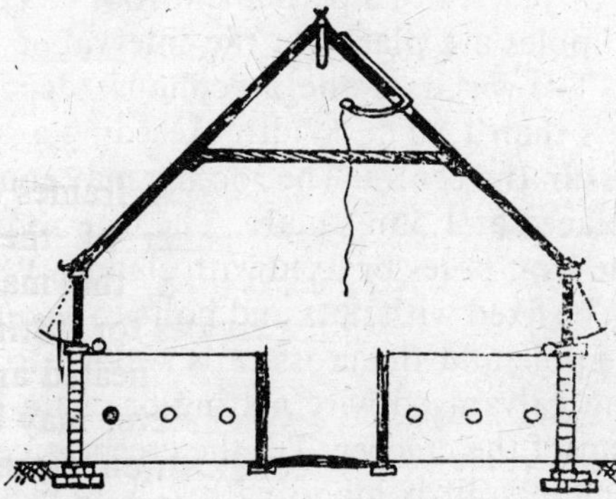

Section of Span-roofed Greenhouse with staging on sides

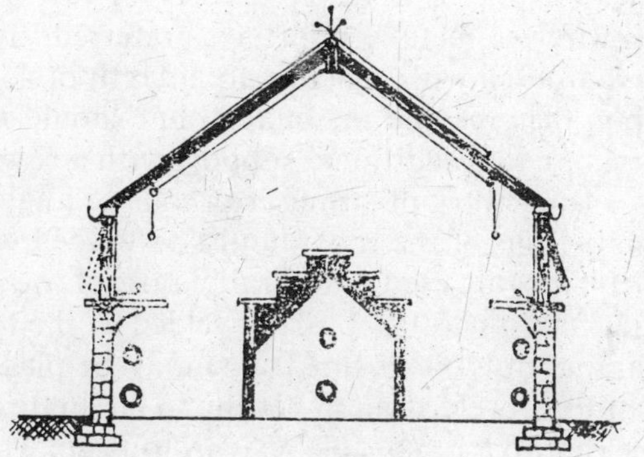

Section of Span-roofed Greenhouse with central staging

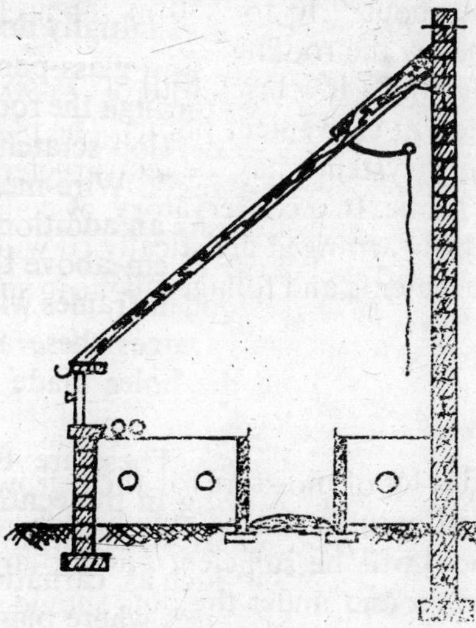

Section of Lean-to Greenhouse

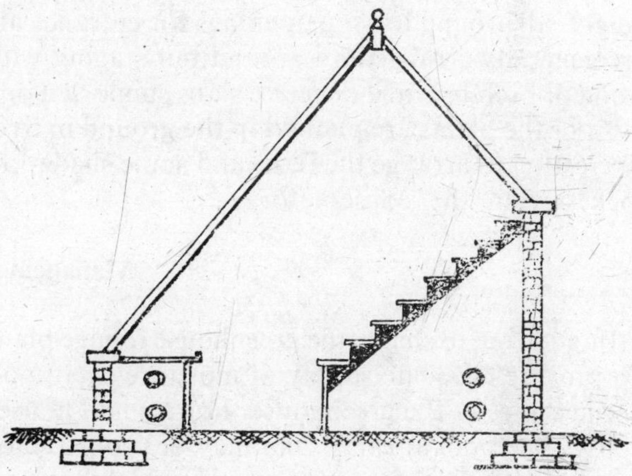

Section of Hip-roofed Greenhouse

Fig. XXIII.4

Heating is also an important factor in cold countries. This can be achieved by circulating hot air and steam or by electricity. The hot water or the steam and hot air is run through pipes kept at convenient places.

Greenhouses for Tropical Countries

In the above paragraphs the construction and requirements for glass greenhouses have been discussed. But for tropical countries, as already mentioned, greenhouses are needed for keeping the shade-loving plants in cool shaded places. For this, structures other than glass are needed. Greenhouses for tropical countries are made of stone, wood, brick, angle iron or 7.5 cm diameter G.I. or steel pipe pillars in double parallel rows over which a framework of roof is made of wood or split-bamboo overlaid with thick wire-mesh to facilitate training creepers over this. If the roof is made of lattice work of any of the above materials no creeper may be trained over it, as a lattice work permits only diffused light. Lattice work of split bamboo and plaited matting of coconut or date palm leaves supported by a strong frame are cheap but needs to be replaced every year or every few years. For a permanent roof creepers grown over metal support are preferred. The metal poles are placed at the interval of 3 m and grouted with concrete to a depth of 45-60 cm. The width of the greenhouse depends upon the area and the budget, but should not be less than 1.80 m. Width exceeding a span of 6 m needs additional support with a row of pillars in the centre. The roofing may consist of a framework of strong iron rods or angle irons placed at 1.5 m square. The iron rods or angle irons of the roof should be welded with angle iron poles or fixed with clamps. When G.I. pipes are used, the parallel rails of the roof may be fixed with nuts and bolts to a square (22.5 cm square) G.I. plate welded at the top of the pole and the cross rails welded to the parallel rails. Over this frame may be placed strong galvanized wire netting or chain-link netting or welded-mesh netting to facilitate the spread of the creeper. Till the creepers cover the roof to a reasonable extent, to begin with matting of split-bamboo with 2-3 cm meshes or plaited matting made of coconut or toddy palm or date palm leaf may be placed over the roof for shading. The height of the greenhouse should be about 2.10 to 2.70 m for a clear passage keeping enough margin for the creepers to sag below the roofing.

The greenhouse or conservatory may be enclosed by putting a low brick wall of 75-90 cm height, all around leaving openings for entrance and outlets at convenient points. The plants are generally arranged on ground but staging with gallery system may be constructed of wood or prefabricated concrete slabs supported on metal frame. In a conservatory, of course, most of the plants are planted in the ground in irregular beds arranged artistically. It will be interesting to arrange the ferns and some shade-loving succulents and foliage plants in small rockeries in the conservatory.

Management

In addition to shade, the greenhouse foliage plants need a lot of moisture for their growth. To provide adequate supply of moisture small pools fitted with fountains may be provided in the centre of the greenhouse. For a small house one pool will be sufficient but for larger houses one pool at every 20m may be constructed. The ground under the pots should be covered with such material as cinder which holds moisture.

There cannot be any general rule for watering as different categories of plants have different water requirement. But one general rule to be followed is that light as well as too heavy watering is bad. Again potted plants will need more frequent watering than plants grown in the ground.

In a conservatory, when the plants are grown in ground, a general compost consisting of 1 part each of cinder and sand and 2 parts leaf-mould are spread over the top 15-20 cm layer. For plants requiring a richer compost, one part of well-rotten cowdung or horse manure may be added. Proper drainage should be ensured. When planting in groups particular care is taken not to plant all the deciduous plants in one group, otherwise that patch will look barren during the winter. Flowering plants such as geranium, African violet, begonia (fibrous rooted), *Impatiens*, etc., should be kept ready in pots and sunk in the ground wherever it looks barren.

Good-looking hanging baskets containing orchids, asparagus, *Peperomias*, etc., should be fixed on poles or suspended from the roof at prominent places to beautify the conservatory. Ornamental vases and tubs containing plants may also be placed at some places for beautification.

In a greenhouse, the plants tend to grow towards the direction of light and in the process bend in that direction. To keep them straight the pots should be rotated regularly and the foliage should be sprayed once in a while to wash the dust that accumulates.

Greenhouse plants benefit from the application of liquid manure prepared by fermenting oil cake, or fresh cowdung, or made from dissolving ammonium sulphate in water (30 g in 5 L), once every fortnight.

In a single temperate greenhouse it is not possible to grow all sorts of plants. Only a few plants requiring the same temperature and humidity can be accommoded, whereas in a tropical greenhouse it is possible to grow a wide range of foliage and flowering plants.

In a climate as that of Bangalore, the creeper *Bignonia gracilis* with its bright green foliage and bright yellow flowers is very suitable for growing over the roof. But it should not be allowed to make a very dense growth. Other creepers suitable are the railway creeper (*Ipomoea cairica*, Syn. *I. palmata*), *Vernonia elaegnifolia*, etc. *Antigonon leptopus* is also capable of spreading quickly but this looks dirty with the dried leaves and twigs. Many people, especially around Bangalore, prefer to grow woodrose, a species of *Ipomoea*, over the roof as it is quick-growing. But this creeper is unwieldy in growth and too heavy and should preferably not be grown. Inside the conservatory some partial shade-loving climbers such as *Trachelospermum jasminoides* may be allowed to climb over some wooden pillars.

Greenhouse Plants

Most of the foliage plants described on pages 251-273 and the indoor and house plants included in Chapter XXV are suitable for growing in a tropical greenhouse. In addition, palms, ferns, and orchids can also be grown in the conservatory and greenhouse. Shade-loving herbaceous flowering plants such as geranium, African violets, *Impatiens*, fibrous-rooted begonias, etc., can also be grown with great advantage.

GARDEN ADORNMENTS

In a garden several adornments and necessities should be provided to make it look more ornamental and useful. Garden seats, boxes, tubs, urns, ornamental vases, fountains, bird baths, sundials, floral clocks, Japanese lanterns, statues, ornamental stone pillars or pieces, pergolas, and arches are some of the examples of such garden ornaments.

Garden Seats

Garden seats are a necessity in any garden. But what type of seats have to be provided in a garden should be decided judiciously. The seats should be comfortable, durable, artistic-looking, and should not look out of place. Wooden and fabric seats are comfortable to seat compared to those built in stone or iron. When wooden seats are selected they should be made of good timber like teak to withstand the vagaries of weather especially moisture as the furniture has to remain outdoors. The wood is treated with preservative and painted with moisture-proof chemicals. The chair or bench should have an appropriate design. As for example, a chrome seated chair will look completely out of place in the midst of a woodland scenery. Similarly, a rustic furniture in front of a modern sophisticated house may be completely out of place. Iron or stone or concrete seats get easily heated in the summer and become cool in the winter thus making them uncomfortable to seat. But for durability and to prevent from vandalism and theft concrete or stone seats are preferable in public parks. Though iron furniture is also durable it is more likely to be stolen because of the value of metal unless firmly fixed.

Ornamental Tubs, Urns, and Vases

These add beauty to a garden. Plants displayed in ornamental tubs or urns at suitable places look beautiful. The tub or the vase can be made of timber or preferably of brick, concrete or, carved out of stone. These can be built permanently in position or may be kept just like that and moved whenever needed. When fixed permanently, these should generally be positioned over an ornamental pillar keeping in view the proportion between the two. If there is a paved path an ornamental vase or tub can be placed at the end of it or one each on both sides. Similarly, these can be placed near the gate or near the staircase of the main entrance. Ornamental urns preferably of metal with carvings outside look beautiful in the terrace, near the staircase, or even inside the house. Suitable ornamental plants should be grown in such containers. Window boxes (for details, Chapter XXV) are also useful garden ornaments.

Bird Baths

These are nothing but large bowl-shaped containers, generally made of concrete, fixed over a pillar or column which is about one metre tall (see Fig. XXXIII.5). Water is stored in the bowl for the birds to come and drink or bath in it. Bird baths should be constructed at a quiet corner of the garden.

Sundials

A sundial has many roles to play in a garden. It can be used as a focal point in a garden, can form a centre-piece of a formal flower bed, can be placed in the centre or at the end of the lawn and at the junction or termination of a path. A sundial is also a good feature in a sunken garden. The sundial should be positioned in a place where the shadow from a tree or building does not fall for a long duration, otherwise the feature becomes meaningless. There should be free access to the base of the column, so that one can closely observe the dial (see Fig. XXIII.6).

The column of the sundial should be fixed firmly in the ground and is generally made of bricks, tiles, stones, or a combination of these without a mortar coating to look more artistic. The top is generally square where the sundial with the compass is fixed. This is also made of stone or concrete. In western countries, metal sundial plates with compass are available for sale. Even dials for specific districts are available there. In India one has to get it made with the digits. The digits can be carved on the cement or the stone itself. Before fixing the dial, the movement of the sun should be observed and then the dial is screwed to the base. The height of the column should not exceed 60-90 cm or the dial will not be visible easily. Live sundials made of living plants complete with the figures, compass etc., in the form of topiary can be seen in some western countries.

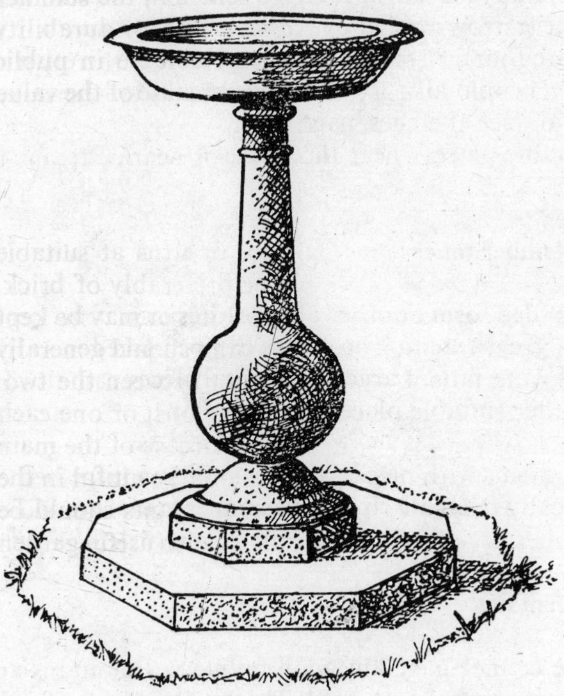

A BIRD BATH
Fig. XXIII.5

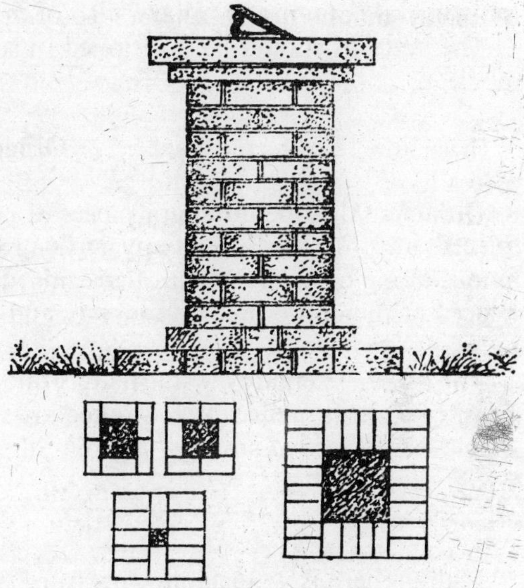

A SUNDIAL
The drawings show the designs of the flat-top without the dial.
Fig. XXIII.6

Floral Clocks

These are nothing but huge clocks, generally operated by electricity, having huge hands for showing the seconds, minutes, and hours. The machinery of the clock is concealed in an underground chamber with only the hands showing above the ground against a dial of carpet bedding plants or flower beds. A carpet bedding of green plants (e.g., *Pilea muscosa*, *Sedum*) with figures made of red plants (e.g., *Alternanthera*, *Iresine* sp.) or *vice-versa* looks quite ornamental. Similar effects can be obtained by various coloured flowering annuals. The problem with annuals is that these are to be replaced seasonally, whereas carpet bedding plants are of permanent nature. Instead of living plants the dial can be decorated with various coloured (natural or artificially coloured) pebbles also.

Japanese Lanterns

These are highly sought-after garden ornaments. The lanterns should preferably be carved in stone and should be low and decorative as the Japanese would prefer. The column may be cylindrical or square faced with ornamental carvings. Similarly, the fire box should also be carved in an ornamental manner. The roof may be broad-roofed, commonly called "Snow View" or *Yukimi*, so called because it gives a charming effect to the landscape when it is snow-bound. Obviously these types of lanterns are made of white stone or white marble and are suitable for areas having a snowfall. There is no harm even if these are preferred in a tropical country especially near a pool. The narrow-roofed or *Kasuga* lanterns have a six-facet fire box. Hundreds of these lanterns are found in the Kasuga temple near the town of Nara in Japan and hence the name. Many people also fix decorative metal lanterns over ornamental columns or pillars, also of metal. (see sketches, page 443).

The lanterns should be positioned in suitable places – near the house or near a stream or pool.

Ornamental Stones

Ornamental stone pillars or pieces of rounded form or other abstract designs, if properly placed, improve the look of any garden to a great extent. The stones can be placed near lily pools, along or in the midst of streams and waterfalls. Large ornamental pieces can also be placed at the corner of the doorway and other suitable places.

The Japanese create most artistic stone gardens with pieces of stones. First a large bed of crushed stone is made which is raked artistically to create ripples to simulate sea, over which single large piece of stone or a group of stones are placed artistically at intervals.

Fountains

Unless there is water under pressure one cannot have any good fountain. Fountains are made to work by circulating the same water contained in a pool. There are various designs of fountains. There may be a straight upright water jet or a number of finer water jets converging in the form of an umbrella. Water jets can also be designed in the form of rainbows. The jets and pipes are made of anti-corrosive material. To make the fountains more colourful during the nights, coloured lights are provided under water, with water-proof fittings. With an automatic switch it is possible to change the colours at regular intervals which further adds to the beauty of the fountains.

Statues

Statues of animate or inanimate objects are also used in the garden to improve the looks. The statues can be carved out of stone or made of bronze. The concrete statues can be placed in the midst of a running stream or a pool, at the inter-section of two roads, or at the end of a road or near the doorway of the house. The statues should be of good taste and artistic value.

Pergolas and arches also serve as good ornaments in a garden. For details see pages 457-459.

Plant Stands

The plant stands are also very useful structures for decorating the gardens. Plant stands are generally made of mild steel rods moulded in different fashions with various-sized rings attached to hold the pots growing the plants. These can be displayed in the terrace garden, at the entrance of the house, in the roof garden or in any other advantageous position in the garden where these will add to the beauty of the garden. Some plant stands are specially made for fixing in the walls of the house where indoor plants can be displayed. The plant stands should be painted pleasantly either in white or in green to make them more presentable. If the plant stands are to be placed indoors it is advisable to put G.I. sheet or plastic buckets, just one size bigger than the pots to be displayed, and attached to the rings, to arrest the seepage water from the pots.

Chapter XXIV

Landscaping Places of Public Importance

LANDSCAPING HIGHWAYS

The landscaping of the national and State highways with trees is an important aspect of beautifying our countryside. Landscaping of highways does not mean only planting of trees; it is only one part of it. Landscaping of a highway also includes all other measures which help enhance the beauty and fits it into the natural landscape of the area. Besides its engineering perfection, a highway must look aesthetic, and should not disturb the ecological aspect of the area too much. But, unfortunately in India, this aspect has been either neglected or done in a clumsy fashion. Planting of trees on highways is necessary not only for the purpose of beautification but also for utility and necessity. The main purpose of roadside trees is to provide shade during the summer. For this purpose, evergreen trees with spreading crowns should be selected For wider roads, double rows can be planted, with the outer rows having shade trees and the inner rows with flowering trees.

The planting of roadside trees started during the time of Emperor Asoka (268-231 B.C.). The Mughals also planted roadside trees. But there was no planning in planting the trees. Trees were planted in a haphazard manner, with a mixture of trees such as *pipal*, banyan, etc., with other kinds such as *neem*, mango, tamarind, etc., which are not good combinations and do not enhance the beauty of the landscape. Even in the present day the roadside trees are being planted in an unplanned manner by the highly paid staff, who are neither trained in the subject nor have great love for plants, with, of course, a few exceptions. The maintenance is generally left to a *mali* who lacks imagination. If there is any casualty, the *mali* replaces it with whatever plant he can get hold of easily, as a result we get a highway planted with a mixture of plants.

The roadside trees on the national highways should not only provide shade but preferably also have some economic value. With this in view many of the highways have been planted with trees such as tamarind, mango, *Eugenia operculata* (Syn. *Syzygium operculatum*), etc. The roadside fruit trees will be subjected to damage by urchins and anti-social elements and in the process, may have less life expectancy. *Eugenia operculata* has one nuisance value as when its fruits ripe they drop on the road and make it dirty. The stain of this fruit is also difficult to remove from the clothes.

Neem (*Azadirachta indica*; Syn. *Melia azadirachta*), *Mahua* (*Madhuca indica;* Syn. *Bassia latifolia*), *Dalbergia sissoo* (*Sisham*), and *Shorea robusta* (*Sal*) are roadside trees of economic value.

The highway trees should never be planted in mixed avenues, but only one species should be planted for a long distance of the road. For example, if *neem* trees are planted in pure avenue for a-long stretch, an oil extracting industry can be started on the roadside. Similarly, *Sal* yields valuable timber and its seeds yield an edible oil which is used as an ingredient in the manufacture of chocolates and other purposes. If a single species is planted in a pure avenue for miles together this looks more beautiful and gives a wavy appearance to the skyline. In such a case the management and gap-filling also become easier.

The trees should be planted 12 m apart in the row and at least 5-6 m away from the edge of the roads, so that they get enough space for spreading and do not interfere with the traffic. If a road is as wide as 30 m or more, double rows of trees should be planted, rows being spaced 10-12 m apart. The inner row may be of a flowering tree. If a roadside tree is intended for timber, replacements should be planted well ahead, (4-6 years) of the cutting time.

The selection of trees for a particular locality is done giving due considerations to sub-soil water, soil, climate including rainfall, locality, etc. As for example, if *Mahua* is planted along a highway near the tribal belts of Madhya Pradesh or Bihar, the flowers can be marketed easily as these are in great demand among the tribals. Trees with shallow root system such as *Millingtonia hortensis* and brittle wood as in the case of *Eugenia jambolana*, *Albizzia lebbek*, *Cassia siamea*, and Eucalyptus should never be planted on highways, as during storms they get uprooted or branches are broken and casualties may result on the unaware road-users. *Neem* and tamarind can grow very well in dry localities. *Samanea saman* (Syn. *Pithecolobium saman*) and *Dalbergia sissoo* grow better in places having a rainfall of 100 cm or above. Often the banyan (*Ficus benghalensis*) is planted on highways, which is not appropriate as its growth is unwieldy. However, banyans can be planted singly and a little away from the road at some distances, for its cool shade. Similarly, *Ficus religiosa* is also not a very good roadside tree. On the other hand, *Albizzia procera* is a good roadside tree. *Polyalthia longifolia*, though not a tree of great economic value, is a very good shade tree and thus suitable for roadside planting. Thorny trees should not be planted along roadside as the falling thorns may damage the tyres of vehicles. Though not planted on our highways, shrubbery borders with suitable shrubs can be maintained along the highways. This will improve the scenery.

Though so far we have, in general, suggested avenue plantation it is also possible to landscape the highways by other methods. One such method is "mixed plantation" which consists of planting different varieties of trees in a mixed avenue against the planting of a single species. This has got a couple of advantages. The first is that the different plants flower and fruit at various seasons thus enhancing the aesthetic view for a longer time of the year. The other advantage is that this planting avoids monotony.

The other method is "group planting" where a group of 3-4 trees or more are planted at specified intervals instead of planting in avenue.

In rural areas the formal planting of avenues may look quite unnatural which act against the principle of landscaping. For such areas, it is suggested to landscape by the "informal planting" method, which consists of planting informally without following any regular pattern.

Some other methods of improving the roadside landscape may be suggested here. While constructing a new road it becomes necessary to excavate earth from the roadside for the road embankment, as a result the roadside develops some scars in the form of irregular ditches and burrow areas, which not only look ugly but also serve as breeding grounds for mosquitoes. It is suggested that such burrows and pits should be reshaped and formed into roadside lakes, if needed by outlining the borders with brick or cement lining. The banks can be landscaped with a few beautiful trees. Alternatively, several burrows can be connected and reshaped to form into a stream so that no water stagnates. If neither of these are possible, such ugly sights should be screened off by proper planting.

If already some feature such as a group of trees or a temple or historic ruin exists on the roadside, this should be incorporated in the overall landscaping of the highway.

LANDSCAPING RAILWAY STATIONS AND RAILWAY LINES

One important aspect of bio-aesthetic planning is to landscape public places on a priority basis compared to private places. For this specific reason, the railway station and the railway lines should attract attention at the very outset for landscape gardening. The railway station is the first touching point for a visitor to the town. A well-kept and well-planted railway platform gives a visitor or a passerby the first impression about the town. A barren or ghastly-looking platform will create bad impression about the place to passengers in a running train. For most of the people living in small towns the station platform is a place for evening recreation or a place for morning walk. It will be a social obligation on the part of our railway authorities to keep the platforms planted with beautiful trees. The second important consideration why a platform needs to be planted with shade or flowering trees is that these trees offer the passengers a resting place under the hot sun, while waiting for the train. Concrete seats or benches could be constructed around the trunk or under the canopy for the benefit of the commuters.

Railway Stations

Besides flowering and foliage trees, the railway authorities can also improve and beautify the platforms with tubs and troughs planted with palms and other attractive plants such as bougainvilleas. Even hanging baskets can be displayed near the booking office or on the pillars of resting sheds and in similar other places.

One more aspect of why the railway platforms should be beautified with trees is that many more people pass through a railway station compared to those who visit a public park. Therefore, in addition to beautifying the area it will be possible to bring home the importance of bio-aesthetic planning to a large proportion of the population by landscaping a railway platform.

The tourist traffic in India is on the increase and with the introduction of super-fast luxury trains many of the tourists are now travelling by train. If the railway stations are landscaped with avenues of such flowering beauties as *Cassia fistula* (*amaltas*), *Cassia nodosa*, *Peltophorum ferrugineum*, *gulmohar*, etc. the tourist will carry back pleasant memories from India.

Railways should have their own nurseries at suitable places for raising the plants needed for such landscaping work. The transportation of these trees to the different railway stations will be an easy task for the railways. To inculcate the habit of growing trees and to impart a love for trees and flowers the station-masters at the time of their training should be given a short course on landscaping and bio-aesthetic planning.

Railway Lines

It will be a little hard task to landscape the vast stretches of railway lines compared to landscaping the platforms, especially in the drier tracts of the country. Fortunately, in India we have got beautiful flowering trees such as *Butea monosperma* (*palas*), *Cassia fistula*, *Erythropsis colorata*, etc., which can withstand considerable drought conditions. *Palas* or *dhak* can be seen growing naturally in wide stretches in the arid climate of Rajasthan.

The landscaping of railway lines poses some problems, the basic being watering. This can be partly overcome by planting drought-resistant trees during the rains. Another way to solve this difficulty is to grow the plants in the nursery for 3-4 years in large tubs and then planting them at site, by tossing out the tree with a ball of earth, without disturbing the soil round the roots. The larger plants will have more chances of survival than a small sapling. However, in landscaping railway lines more emphasis can be placed on planting trees of economic importance, although ornamental plants are not excluded either. Other points which should be considered are that plants should be deep-rooted and non-spreading. A shallow-rooted plant will be easily uprooted by a storm and may cause obstruction on the track, whereas a spreading plant has to be trimmed now and again, to keep the track free. The trees are to be planted at a specified distance from the tracks as per railway regulations. These plants, besides breaking the monotony and drabness of the landscape, will offer shade to the gang workers.

The innumerable railway level-crossings are the places, which can be landscaped with much ease than the railway lines, as the gate-man will be there to look after the plants. The gate-men who live there in the houses provided for them can be asked to plant a row of a few flowering trees on either side of the lines on both sides of the gate or in a small group-near the gate. The beauty of such trees will not only be enjoyed by the railway passengers but also by the people who pass by the road or wait for the train to pass.

The following trees are recommended for planting on railway platforms and lines.

(A) Ornamental

Albizzia procera, *Bauhinia variegata*, *Cassia fistula*, *C. javanica*, *C. nodosa*, *Delonix regia*, *Gliricidia maculata*, *Jacaranda mimosifolia*, *Lagerstroemia* (different species), *Michelia champaca*, *Peltophorum ferrugineum*, *Polyalthia longifolia* and *Saraca indica*.

(B) Economic

Ailanthus excelsa, *Anacardium occidentale*, *Averrhoa carambola*, *Dalbergia sissoo*, *Mangifera indica*, *Melia azedarach* (Persian Lilac), *Shorea robusta*, *Swietenia mahagoni*, *Tamarindus indica*, *Tectona grandis*, *Terminalia arjuna* and *T. catappa*.

Bus Terminus and Airport

Now-a-days, with the increase in road traffic many State governments are constructing large bus terminals and bus stations, but except in a few States such as Haryana these stations consist of nothing but concrete and tar. The bus terminals should be beautified on the lines of railways stations. Airports should also be beautified by planting ornamental trees, lawns, flower beds, and displaying plants in tubs and troughs. Fortunately, many of the airports in India are quite nicely landscaped, but there is some scope to improve them.

LANDSCAPING ALONG BANKS OF RIVERS AND CANALS

The banks of the rivers and canals passing through towns and villages are well-suited for landscaping. The plants themselves will be objects of beauty and their reflection on the water surface is an additonal attraction. The moisture-loving trees will do better along the water-front compared to others. In India, the rivers Jamuna, Ganga, Kaveri etc. are regarded as sacred and on the banks of these rivers are situated many old Hindu temples and *ghats*. The banks along these places should be planted with flowering trees, especially the sacred trees such as *Kadam* (*Anthocephalus cadamba*) and *Asoka* (*Saraca indica*) which are associated with Lord Krishna and Sita.

The bank of the river Hoogly on the Calcutta side has been beautifully landscaped by the Calcutta Port Commissioners with shrubs and trees and has become a place for recreation for the citizens of the metropolis who can breath fresh air away from the congested city. Recently, some good landscaping work has been done by the State Horticulture Department near the river Gomti at Lucknow. Besides beautification, planting of trees along banks of rivers and canals help soil conservation also.

The dam sites which generally look drab should be converted into a place of beauty by landscaping with such beautiful trees as *Cassia nodosa*, *C. javanica*, *Gulmohar*, etc. A garden or park can also be planned here, if space permits. The Brindavan Gardens in Mysore constructed below-a-dam site has become a place of great tourist attraction. Since irrigation is not a problem plants along canals and rivers once established will not need much care.

River banks near places of historic importance, such as Taj on the bank of Jamuna, should also be landscaped with beautiful trees. The following flowering trees are recommended for planting along the banks of rivers and canals.

Amherstia nobilis*, *Anthocephalus cadamba*, *Barringtonia acutangula*, *Bauhinia variegata*, *Browneas*, *Cassia fistula*, *C. javanica*, *C. marginata*, *C. nodosa*, *Delonix regia*, *Largerstroemia flos-reginae*, *L. thorelli*, *Melia azedarach (Persian Lilac), ***Peltophorum ferrugineum*, *Samanea saman***, and ***Saraca indica***.

LANDSCAPING CITY, TOWN, AND COUNTRYSIDE

In India most of the old cities and towns are very much unplanned without any consideration given to landscaping with trees and plants and as a result they look nothing but like brick and concrete jungles. The great architect Corbusier says that "unless the conditions of nature are re-established in man's life, he cannot be healthy in body and spirit". The vast majority of our population in the congested cities are cramped inside the four walls of their dwellings and are constantly injected with the fumes of petrol and other polluting agents. The city dwellers grossly lack in sun, space, and verdure and are thus deprived of the essential joys of life.

Our towns and cities can be made more livable, healthy, and beautiful by resorting to bio-aesthetic planning. This can be achieved by planting roadside trees and establishing parks planted with green plants. In the old congested cities this can be achieved by planning garden suburbs. Garden subrubs are nothing but satellite townships developed among parks and avenues. The cities can further be beautified by accommodating the slum dwellers in multi-storey buildings and utilizing the surplus land thus obtained for developing parks. The planning of new towns poses no problem as this can be planned in advance even before construction starts.

In a tropical country like India, the planting of roadside avenues are not only important for beautification, but also from utility point of view. Trees provide cool shade which is so much needed in the long hot summer months. Trees also help bring down air pollution, act as windbreak, and cut down noise. Parks are places of relaxation, entertainment, and fresh air in an otherwise congested city.

India possibly has the largest number of flowering trees, both indigenous and exotic which can be used for landscaping town roads. Except a few cities such as New Delhi, Chandigarh, Bangalore, etc, no other city has utilized these vast resources to beautify the roads. Unfortunately, in Bangalore also many of the small roads in the new housing colonies have not been properly planted with trees though there was scope to do so. In Calcutta, many existing old avenues out of the few are being cut down for expanding the roads for traffic.

In selecting trees for cities and towns the main criteria should be beauty and shade. The broad roads in the cities should be planted with double rows of trees, the outer rows should consist of shade trees, so as to provide shade to the footpath users and the inner row consisting of flowering trees adding beauty. M.S. Randhawa suggested long back that wide roads in the cities and towns should have two lanes on each side, one for fast-moving vehicles and the other for slow-moving vehicles flanked by footpaths. Each lane should be separated by a patch of land planted with grass and shrubs, while the outer rows are planted with double rows of flowering and shade trees. Unfortunately, this advice from an eminent horticulturist is not always followed while constructing new industrial townships even in the public sector. Planting the roads in a town or a city is a difficult proposition and this should be done in coordination with the electricity, telephone, and sewerage departments. The tree should not interfere with the telephone or electric lines or the underground sewers. For this reason, tall trees (*Eucalyptus*, *Araucaria*, *Millingtonia hortensis*) and trees with spreading habit (banyan, *Cedrus deodara*, *Ficus benjamina*, etc.) are not suitable for city or town roads. The tropical flowering trees are generally deciduous in habit and flower in leafless condition, which may be at a time when shade is needed to the maximum. The solution is, wherever feasible, to have an outer row of small shade trees such as *Ficus infectoria*, *Putranjiva roxburghii*, Silver oak, *Mimusops elengi*, etc., and inner row of flowering trees. The trees in both the rows, or when there is only a single row, should be planted in pure avenues, i.e., with one kind of tree only. The spacing should be 10 m between the trees and in the case of double rows the distance between the rows will vary between 3 and 10 m, depending upon the trees planted and the situation. The trees in the opposite rows are planted in a staggering (alternating) fashion. For dwarf trees and non-spreading trees (e.g., *Tecoma argentea*, *Cochlospermum gossypium*, etc.) the spacing can be a little less. Though it has been suggested to plant flowering trees in pure avenue for beauty, it is possible to combine two different species flowering in the same season and whose flower colours also harmonize. For example, the scarlet *gulmohar* combines well with the yellow *amaltas*. Similarly, in Bangalore the scarlet *gulmohar* matches well with the blue jacaranda but flowering season may not coincide always. Again

C. marginata and *C. nodosa* can be combined together. Another excellent combination for Bangalore will be jacaranda and *Tecoma argentea.*

In many cities a piece of land is left out at the intersection of roads as a safeguard against accidents. These plots should be planted with grass and shrubs. One more novel way of beautifying the city roads is to train creepers on iron structures at regular intervals on footpaths. This has been done with bougainvilleas in cities such as New Delhi and Baroda. It is often seen that citizens in the congested cities wait on the hot footpath under the blazing sun for buses. It is suggested that instead of erecting sheds of concrete or iron pillars and G.I. sheets, arbours with creepers may be planned at bus stops which will not only present more aesthetic look but will also be cool compared to a concrete structure.

Trees for Countryside

The planting of trees in the countryside is done on a completely different criterion. Here more emphasis should be given on economic consideration and utility. The trees planted should provide the villager either with fuel, timber, fruit, or fodder.

Babul is a common tree in the village, which is a cheap source of fuel and is used for the manufacture of wheels of bullock-carts. The bark of this tree is used for tanning leather. *Desi* (seedling) mango and jackfruit trees also yield timber for making doors and windows and other household articles such as *charpais*, benches, etc. *Dalbergia sissoo* is also a good source of timber and fuel. *Sal* (*Shorea robusta*) tree is good for both fuel and timber.

Some minor fruit trees should be planted along the country road in open spaces either in a formal way or in an informal fashion. This will not only provide shade but will yield valuable fruits for the villagers. Seedling mangoes of different cultivars such as Langra, Dasehri, and Chausa can be planted. Jackfruit is an important fruit tree for arid regions. The Toddy or Palmyra palm (*Borassus flabellifer*) is an important tree, which yields a minor edible fruit but the leaves are used extensively for thatching roofs of dwellings. Both date and palmyra palms look beautiful when planted in rows along country roads or on the borders of cultivated fields or when planted in informal groups. *Mahua* and large-fruited *Jamuns* (*Syzygium cumini*) can also be planted.

The foliage of the following plants yield valuable fodder for the cattle and goats, *babul*, *Kachnar* (*Bauhinia*), *neem*, *toot* (*Morus alba*) and *ber* (*Zizyphus* species). *Ber* also produces edible fruits.

However, ornamental trees should also be planted near the *Panchayet*, schools, churches, mosques, temples, and gurdwaras and along the field boundaries and near the village well or pond.

City Parks

In city there may be parks of several sizes from very large to medium size and also squares or small gardens are generally found at street intersections. The small gardens or squares are planted with a view to relieving the eyes of the people passing by them or for a short resting period for those who care to use them. Therefore, these may be planted with a patch of grass, few flower beds, one or two shade or flowering trees or a group of shrubs and trees. The medium to large parks are meant for a place of recreation and these are considered as lungs of the cities. These should be a place of beauty as well as utility.

Most of the cities in India have developed unplanned and hence there is hardly any space for a large park. But to give the citizens of such cities a breathing place garden or large parks may be laid out in the nearest vicinity of the countryside easily approachable by car or bus or even by bicycle. But in the present-day conditions in a developing country like India, it may not be possible to take up such ventures in the immediate future. Even in advanced countries such as the U.S.A. only a few very large cities have parks like this which are referred to as large rural parks. The Prospect Park at Brooklyn is such an example. The main features of this park is given below which will give an insight as to how such parks are constructed. The Prospect Park has (1) a long meadow, which is a park-like open space; (2) an undulated and rising hilly section with woodland scenery; (3) a lake and its surroundings; (4) a number of sceneries and objects of beauty. The Budha Jayanti Park in New Delhi somewhat falls into this category. The vast park called "Rabindra Kanan" in south Calcutta with large grassy open spaces, a huge lake with islands, stadium, lily pool garden with a miniature zoo, flowering trees, shrubs, and a hanging bridge may come into this category. These rural parks are really very large varying from 200 to 1,000 hectares. Such parks have several self-contained sectors each having its special features and one need not have to see the whole park unless specially interested.

The small city park may be an area anything between 5 and 100 hectares or little more. A large rural park gives a degree of seclusion from the city but the small city park, as it is situated within the city, has no such characteristic although the features may be the same as that of a large rural park. In the small parks, the scenery created will not look so natural as those of a large rural park because of the limitation of space. A small park should have enough strolling space for the citizens. Good flowering and shade trees should be planted in groups or singly in some corners or other suitable places for creating beauty as well as a place for resting. Garden benches should be constructed at regular intervals especially under the shade of the trees. Few interesting and rare shrubs should also be included. Besides these, some garden adornments such as statues and fountains can also be planned in appropriate parts of such parks.

The third category of city parks may be called as pleasure grounds which have large reserve areas for playing games, and often this is the main feature of these parks. A restricted swimming pool is also often a feature of a pleasure ground. If it is meant for the children, features such as swings, see-saw, sliding chute, merry-go-round, etc., should form part of the park. Due to hard usage it is almost impossible to maintain grass area within this park. Some trees and shrubs are planted aesthetically to keep the place pleasing to the eyes.

Ornamental Trees recommended for Town Roads

Ornamental Shade Trees: *Alstonia scholaris*, *Anthocephalus cadamba*, *Averrhoa carambola*, *Ficus infectoria*, *Melia azadirachta* (Syn. *Azadirachta indica*), *M. azedarach*, *Polyalthia longifolia*, *Putranjiva roxburghii*, *Sterculia alata*, *Swietenia mahagoni*, *Tamarindus indica*.

Flowering Trees: *Amherstia nobilis*, *Bauhinia purpurea*, *B. variegata*, *Brownea ariza*, *Cassia fistula*, *C. javanica*, *C. marginata*, *C. nodosa*, *Colvillea racemosa*, *Erythropsis colorata*, *Gliricidia maculata*, *Jacaranda mimosaefolia* (*mimosifolia*), *Lagerstroemia flos-reginae*, ***Peltophorum** ferrugineum*, *Poinciana* (*Delonix*) *regia*, *Saraca indica*, and *Spathodea campanulata*.

LANDSCAPING PUBLIC BUILDINGS

There are several buildings in towns and cities which are frequently visited by a large number of people. The government and private offices, courts, auditoria, cinemas, hotel, and travellers' bungalows fall under this group. A large cement and concrete building by itself is not very attractive unless the harshness of this structure is mellowed by appropriate landscaping. In large cities, due to paucity of space, hardly any compound is left around these buildings for gardening. Therefore, in most of such buildings one has to remain content by arranging some potted plants near the entrance and a few indoor plants may be arranged to decorate the corridors. But where space is available a lawn should be laid with a few flower beds and possibly some shrub border may be added. The entrance and exit roads may be lined with flowering trees. The kind and the size of the trees will depend upon the size of the building. Dwarf trees in front of a sky-scraper will be a total misfit. Majestic-looking non-spreading large trees will be the right choice for such buildings.

Some flowering climbers may be trained over the portico. One or few bougainvilleas or some other flowering creepers can also be trained over the front walls.

Many of these public buildings like courtrooms get over-crowded and a lot of people spill over to the compound. If the compound is properly planted with a few shade and flowering trees and a nice lawn, the people will find a resting place and their waiting will be less tiring.

Dak bungalows and travellers' bungalows generally have sufficient compound area. A miniature garden with lawn, flower beds, shrubbery, lily pool, fountain, and a few specimen trees should be planted in such gardens. Hotel buildings should also have gardens of the same style if space is available. Fountains fitted with coloured lights can also be added. Sufficient parking space should also be provided.

Since space is a problem in large cities, the scope of gardening around the compound is limited. To overcome this, the public buildings, especially hotels and banks, can have roof and indoor gardens. As in educational institution gardens, the aim of planting trees around public buildings should also be to cut down noise and dust and to provide shade.

LANDSCAPING OF EDUCATIONAL INSTITUTES

The compounds around our educational institutes such as schools and colleges are generally grossly neglected. It is a pity that the importance of landscaping in schools and colleges is not properly understood by the authorities. As a result, the campuses present a bare and neglected appearance. A planned and properly landscaped school building brings a world of difference in appearance and beauty than an unplanned one. Moreover, a good garden in the campus inculcates aesthetic sense to our younger generation.

It is imperative that our administrators and citizens should take landscaping of campuses seriously and while planning any architectural design for educational buildings it should incorporate in it a landscape design also. Instead of taking up gardening of the campus in bits, it is necessary to have a master plan of the whole area and to implement it in phases, according to the resources available. The main aim of the master plan will be to create a barrier against noise, storm, dust, and to provide shade. It may also be necessary to screen some ugly places with the help of plants. Even in unplanned school buildings it is possible to improve the landscape by careful planting. But, unfortunately, in most of the congested cities, due to paucity of space, there is hardly any scope for planning a garden or planting a few trees.

The general recommendation is to plant large trees in the school compound. In the periphery of the school campus, along the rear and wings, a thick belt of large shady trees should be planted to bring down noise and cut down dust and storms. This plantation will also help keep down severe heat and cold. The front should be planted with medium-sized flowering trees for beauty. The trees should not completely obstruct the view of the building from outside. For enhancing the scenic beauty it is also suggested to plant a row of flowering trees, with different blooming seasons, in front of the large trees along the periphery. It is difficult to give any general recommendation regarding the types of such trees as this will vary according to the architectural design, situation, and climate. The object is to provide beauty and comfort depending on convenience. A thickly planted belt of *Eucalyptus* (notwithstanding the objections in certain quarters against the genus especially for the reason that these plants draw enormous quantities of groundwater and release this in the atmosphere), for peripherial planting is considered ideal. Silver oak (*Grevillea robusta*), *Polyalthia longifolia*, *Putranjiva roxburghii*, seedling mango, and rain tree (*Samanea saman*) are also very suitable for this purpose. *Cassia fistula*, *C. nodosa*, *gulmohar*, *Tecoma argentea*, *Erythrina indica*, *Lagerstroemia flos-reginae*, and *Bauhinia variegata* are suitable for planting in the front and in the front row of the border planting.

The roads and paths are to be formally planted with medium to tall flowering plants. Before planting provision should be made for overhead wiring and sewarage so that these do not interfere with the avenue planting. Where the electric wires limit the choice of avenue trees, small flowering trees such as *Cochlospermum gossypium*, *Callistemon lanceolatus*, *Bauhinia variegata*, and *Tecoma argentea* can be planted. The trees should be planted in pure avenues. Again, selection of trees will be different for a wide road and a narrow path, the former being planted with large while the latter with small trees.

Shrubs play an important part in the school landscaping. Border planting of shrubbery on large grounds or at the back of the school campus serves useful purpose of filling the gaps between the trees and lawns. Moreover, small paths can be planted with shrubs arranged in formal beds or rows, thus improving the beauty of the path. Shrub borders round parks or playgrounds is very effective and can replace hedges as the maintenance is minimum. A hedge needs regular trimming. Shrubs, with their flowers, are more colourful than hedges. A school garden should be planned with plant materials which need minimum of maintenance. The foundation planting can also be undertaken with evergreen dwarf and trailing shrubs. Because of the problems of maintenance, flower beds, especially the seasonal ones, should be disfavoured.

A lawn looks nice in an educational institution, but it is very difficult to maintain. The playground can be planted with lawn, if this can be maintained or should be left bare.

A bougainvillea creeper trained over the wall of the building can change the whole look. Similarly, a *Bignonia venusta* supported against a wall also looks beautiful. Creepers climbing with their rootlets such as *Ficus repens*, *Campsis radicans* (Syn. *Tecoma radicans*) can also be trained over some stone or brick wall or pillar. A Principal, having imagination and resources, can add many more things besides these. For example, a bird bath or two provided amidst the trees will attract many birds which will be educative to the students. It is desirable that the maintenance of the campus should be left to the students as this will be quite educative also. The authorities should run competitions between institutional gardens and award trophies for the best kept garden. Similarly, students should also be rewarded for their good garden work. If the students take the charge of maintenance themselves they will identify themselves with the creation and the gardens will be spared from destruction by the students. Different trees, shrubs, roses, flowering annuals, etc., may also be properly labelled so that students could learn their names. This knowledge will be subsequently helpful for the students to lay out their home gardens and help their parents in gardening when on vacation.

Besides an ornamental or a landscape garden, universities and colleges can also maintain a botanical garden or a student garden, where the plants are arranged in groups, family-wise, so that such gardens become educative.

LANDSCAPING FACTORIES

In modern times, a factory should not become a place of only machinery, dust, pollution, and noise, but should also be provided with nicely laid-out parks and gardens. This is not only needed from the point of beautification, but also to fight pollution and dust. The factories may be broadly categorized into two groups. The first group comprises comparatively neat factories such as a plywood factory or a fruit processing plant which emit less dust and other polluting materials. The second group consists of factories such as cement, steel, fertilizer, etc., which emit a lot of dust, smoke, and harmful chemicals. If gardens are to be laid out in the factory premises itself one has to be choosy in selecting the plant materials depending upon the type of the factory. In factories belonging to the second category, comparatively hardy plants are to be selected. For the factories of the first group a wider range of plant materials can be used. The primary aim in a factory garden will be to plant trees to arrest the drifting dust and smoke and to cut down noise. Tall and hardy trees such as *Casuarina equisetifolia*, *Eucalyptus*, *Polyalthia longifolia*, and silver oak should be planted all around or in the direction of the winds to stop the spread of dust and smoke. It is generally accepted that several spaced rows of plants planted in a staggering manner bring down the noise from the factory to the surroundings outside this barrier. Another important aim is to provide ample shade and coolness so that the workers get a respite under the coolness of trees from the hostile hot interior of the factory. Moreover, in a well-planted factory, the trees bring down the temperature in the factory premises to a considerable extent. In a large factory a buffer zone may be created by afforestation between the factory and its residential colony. Afforestation can be done with hardy ornamentals such as *Acacia auriculiformis*, *Casuarina equisetifolia*, *Dalbergia sissoo*, and some other shade trees. Besides planting of trees, a factory area can also be beautified with rockeries, statues, water pools or lakes, fountains, etc. Bougainvillea should be used freely to beautify a factory area. Hardy cultivars suitable for the locality should be selected for this purpose. These may be planted in rows along all the roads of the factory spaced at 3-5 m and trained as half standards. Planting in groups in vacant places is also very effective and useful. The colourful flowers of the bougainvilleas will bring life to the otherwise drab landscape of a factory area. Canna is another hardy and beautiful plant suitable for planting in beds or groups in a factory area. Hardy shrubs also play an important role in beautifying factories when planted in shrubberies. Lawns laid in vacant lands not only add to beauty but also cut down dust.

The other places where gardens can be laid in the factory area are canteen, rest-shed, hospital, administration building, etc. The gardens in the industrial colony should be on similar lines as recommended for a town. For the bungalows and residential houses, recommendation given for small home gardens should be followed, keeping in mind that some delicate plants may not do well because of the pollution from the factory.

LANDSCAPING PLACES OF HISTORIC IMPORTANCE

India has many relics of the past which are very famous, some being of world-fame, as places of tourist attraction. In India, the maintenance of such relics is the job of the Archaeological Department. It is heartening to note that this department is doing a fine job, on the whole, in beautifying and landscaping such places. The deer park at Akbar's Tomb in Sikandra, Agra, or the lovely landscape with vast stretches of lawns around the *Golgumbaj* in a dry place as Bijapur in Karnataka, and many other such examples are a pointer to the good work done by the Archaeological Department. The beautiful landscaping, around the historic monument "Victoria Memorial" in Calcutta run by a trust, laid out with vast stretches of lawns, large ponds, canna beds, flowering trees, etc, has become a place of great scenic beauty which attracts thousands of visitors everyday. The objective of landscaping such historic monuments is only a secondary one, keeping in view that the planting should not overshadow the monument. The best way to do this is possibly laying vast stretches of lawns around the monuments and planting some shade and flowering trees for beauty and places of rest. The trees should be selected depending upon the climate of the region. As for example, the approach roads to the historic Amber Fort near Jaipur which is a dry area can be planted with drought-withstanding flowering trees such as *palas* or *dhak*, *amaltas*, *Erythrina indica*, etc.

LANDSCAPING PLACES OF WORSHIP

The places of worship such as temples, gurdwaras, mosques, churches, etc., offer a good opportunity for landscaping. Here again, the scope is very limited in large cities because of space problem. But in the villages and towns the places of worship offer good scope for bioaesthetic planning. In a temple the offering is made with flowers and hence it is important to have a garden with the right type of flowering plants. The association of *Plumeria acutifolia* is so common with Budhist temples that it is commonly called as "Temple tree" or "Pagoda tree". Shrubs such as Jasmines, Crossandra, Barleria etc., are quite useful in a Hindu temple. Seasonal flowers such as marigold and small-flowering chrysanthemums yellow-flowering type) can also be planted around Hindu temples. Generally, there will be some water tank in the vicinity of a Hindu temple. This should be planted with water-lilies and lotus and on the banks moisture-loving trees may be planted. Some sweetly-scented flowering trees such as *Mimusops elengi* and *Michelia champaca* should be planted around the temple to create an atmosphere of serenity and sanctity. The flowers are useful for worship also. There should be enough open space in the front for the devotees to assemble there under the shade of trees. A tree of *Aegle marmelos* (*bael*) may also be planted in a temple as the foliage of this tree is sacred to Hindus and used for offering. Similarly, *Saraca indica*, the Asoka, is another sacred tree which should find a place in a Hindu temple.

Churches also offer good scope for planting of ornamental trees. Foundation planting with shrubbery is quite useful. Potted plants can also be arranged near the portico. There is scope for displaying hanging baskets especially for those Gothic-type buildings.

In front of Mosques, Mughal-type gardens can be laid out with stress on formal water pools fitted with fountains.

LANDSCAPING A SMALL HOME GROUND

There are many people who think that landscape gardening relates to only gardening in large public parks or palaces of the rich. Landscaping as it is done for larger estates or public parks can also be implemented in a tasteful and artistic way for a small home ground, though on a smaller scale. The term "small" is a misleading one so far as it relates to gardens. The simplest definition of "small", as suggested by some authors quite appropriately, is an area which can be effectively managed and maintained physically as well as financially by the owner and his family with occasional hired labour for such hard work as digging, mowing, and shearing of hedges. Here, ways will be suggested for landscaping only small residential houses. For larger estates, a combination of landscaping effects suggested for parks and home landscaping may be followed.

There are some basic guidelines for a home landscape. But personal preference plays a considerable role in developing a home garden. The home including its surroundings should be an outward expression of the inner personality and individuality of the owner. Often a common mistake is made by many to copy a successful competitor of a garden competition or a neighbour. This may not suit your own home for various reasons. For example, location aspect of your own garden may be quite different compared to the one which you want to copy. It is advisable to think a lot before even a single digging work starts. It is a matter of great regret that in our country sometimes we spend a fortune in decorating the interior of the house to make it attractive but ignore the outside compound.

Basic Principles

Before actually drawing the master-plan the following points should be kept in mind in home landscaping.

(a) *Background:* The background in a garden, whether a wall, tall trees or a hedge should be neutral in nature; that is to say this should not become a distracting feature over the main features of the garden.

(b) *Contrast:* The design should be such that it should break monotony. To achieve this a variation in form, texture, or colour has to be brought in.

(c) *Balance or Proportion:* A balance has to be maintained between different components (masses, forms, colour, etc.) of a garden. In a formal garden, this is achieved by balancing the quantity or by objects, whereas in an informal garden this can be achieved by planting a small mass of colour in front of a large neutral mass. Overcrowding of plants or other garden features should be avoided.

(d) *Open Centre:* The central area of the garden should be left out of any items of major interest. The best way to achieve this is to have a lawn, which also gives an effect of largeness to the property. A specimen shrub in the centre of the lawn is unsuitable as this counters the principle of a spaciousness, but a tree branching at higher levels from the ground could be planted.

(e) *Repetition:* The repetition or duplicating some features of a garden helps achieve rhythm, balance, and unity. In a formal garden, generally the same feature is repeated. But for an informal design this need not be so. Here one may repeat the colour tone without disturbing the texture, form, or quantity. If there is a circular path this can be repeated by having two or three consecutive circular shaped beds of annual flowers, hedges, and shrubbery border. Thus, though the shape is repeated, the variation in texture, colour, and form ensure that the design does not look monotonous.

(f) *Rhythm:* A landscape designer should have an artistic sense to understand how to bring in rhythm in the design. Arranging the different elements haphazardly, without harmony, does not enhance beauty. Harmonious lines, often artistically curved, bring in rhythm to the landscape. A group of shrubs in front of a rockery breaks the rhythm. Repetition of certain elements, such as form, enhances the rhythm.

(g) *Variety:* To break the monotony in a garden, variety is essential. This is achieved by contrast of colour, form, and texture. Planting of different seasonal flowers all in red colour, does not necessarily mean variety. But red salvias contrasted against bright green hedge or shrubbery is a good contrast and may look very attractive.

Besides these, a design should be simple, easy to maintain, and provide comfort for inmates.

Making a Plan

Before any actual garden work is undertaken a master plan has to be prepared according to a scale (1 : 15 or 1 : 20) in which all the features such as house wall, drive-way, paths, flower beds, shrubbery, etc., are plotted. Any possible shaded area due to large trees or the building itself has to be marked on the plan. A plan prepared on a printed graph paper is of great help. The plan thus prepared should be studied again and again keeping in view what shape a plant will take in the long run. It is frequently observed that people attracted by the graceful form of a young *Araucaria cookii*, plant this in the centre of a lawn or near the house without possibly knowing the gigantic form and height it will attain after some years. Perhaps the owner of the house will cut this tree when overgrown or it may be retained to the detriment of other plants growing below it. Either way, this is not a good planning. Perhaps, one way of satisfying the urge of a garden lover to grow such beautiful trees in a small compound, is to grow them in large concrete tubs and bury the tub growing the tree in the appropriate place, thus giving the impression that the plant has actually been grown on the ground. When this attains a considerable height, say 3-6 m, the tree along with the pot should be lifted and given to someone who can afford to use such a grown-up tree. But it is better not to include such controversial items.

If the garden area is sufficiently large, this can be divided into three areas.

(1) *Approach or Public Area:* This is the area from the street side extending to the entrance of the house. The area may be small or quite large depending upon where the building is situated. The aim is to harmonize or blend the surroundings with the house proper. The approach area should not be overcrowded with large trees. It is better to have doorway or "foundation" plantings with low growing shrubs and evergreens. Floribunda and miniature roses are also suitable for foundation planting provided sufficient sun, at least during the morning hours, is available. It is important to note that planting in front of the house should neither obscure it nor cut off light and air nor block the windows thus obstructing view of the garden from indoors.

Big trees, if space permits, can go in the backyard but should not be overcrowded in the front. But a few low-growing trees can be accommodated at the appropriate places as next to entrance, if space is available or somewhere in the front lawn. An open spacious lawn with some annuals (cut-and-come again zinnias, salvias, and petunias) or herbaceous perennials (chrysanthemum, canna, and *Impatiens* in shade) can be planned in addition to the foundation plantings.

(2) *Work or Service Area:* Wherever feasible this and the living area should be situated at the back of the house as these need seclusion or privacy. This area includes the kitchen garden, compost bin, nursery, tool shed, and garage. Some people like to include the children's swings and the slide in this portion as the children can be kept under surveillance from the kitchen. This should be separated from view by planting a thick hedge or a row of bushy shrubs, as this is considered as the shabbiest part in any garden.

(3) *Private Garden Area or Living Area:* In the western countries, this is generally termed as the outdoor living area, where people sit out in the winter to enjoy the sun or rest in the summer under a arbour or shade of tree. This area should be easily approachable and visible from the living (drawing-room) or dining-room, screened from unsightly objects and for privacy. In the western countries people prefer a terrace and this is the place where it should come. There should be some shaded sitting spot such as a tree or arbour with garden benches. A wide stretch of lawn with shrub border or few annual beds or a rose garden can also be included in this section. A tennis court or a play area has to be included here, if there is enough room.

But before actual planning one has to first decide what one wants for one's house. A choice has to be made from the following. Whether the garden is needed (a) as an outdoor living-room with a long stretch of lawn and terrace, (b) as a fenced-in playground, (c) as a show-piece with collection of exotic and rare plants, or (d) a yielder of vegetables and fruits or cut flowers for the house, is to be determined first. Some may like to add to the list a large tree for shade or trees to attract birds. It is to be considered first what should be the major theme of the garden. If somebody is fascinated with flowers, the borders are to be planned wide to fulfil his desire. People fond of vegetables and fruit may like to reserve the major portion of the area for this purpose with possibly a little area left around the house for a pleasure garden. But, if the garden is desired as a place for outdoor living, a vast expanse of lawn with minimum of beds and borders has to be planned. Some novices may like to combine the good qualities of all these themes and incorporate in his garden. This is bound to create a mess of everything and the ultimate result will be a garden good for nothing.

Many people advise not to include any pool or formal rock garden or the kind in a home garden. But there is no harm if a formal or informal lily pool can fit in with the overall design, with or without a fountain or a rock garden. A statue or sun dial can also be wellfitted in some spacious compounds.

Some Points to Ponder

In designing a house some more thinking is necessary. To keep down maintenance cost and time, an untrimmed hedge should be preferred over a trimmed one, open lawns and shrubs need less attention than annual flower beds. If the beds and borders in a lawn are edged with stone or brick no hand-clipping of grass will be required. A pool needs to be cleaned occasionally and one should ponder twice before including this in the plan. The faucets in the garden should be fixed at appropriate places so that the hoses are not be dragged long distances. The above suggestions are for reducing the labour cost which is specially relevant in industrially advanced countries where labour is costly. Fortunately in India, labour is not so costly and one can include one or two features needing help of manual labour.

To create privacy, trees, hedges, shrubs, fences, or creepers trained on wire-mesh structure supported by angle iron or G.I. pipe pillars can be grown. Trees are used when height is needed, otherwise hedges and other types of screens should be preferred.

Lighting is needed in the night specially for terrace area and paths. The same electricity points can be utilized for running an electric lawn mower.

How to Proceed

When everything has been decided, it is time to tentatively select the plants needed. The different features are then drawn on the paper with a pencil so that this can be erased if alterations are to be made. After thorough study and several additions and omissions a plan is finalised. For an experienced man, this would not be much of a problem. But a novice must visit the neighbours and see some of the local parks to know what can be grown. The first thing is to select the materials for the basic framework such as background, screens, trees needed for shade, the doorway, and the corner of the house. To this the features needed for effects and beauty as for example plants for foundation planting, flower beds, specimen shrubs or trees, are added.

After everything is finalised on paper these are put into practice on the ground with the help of split-bamboo stakes and rubber hose. The trees are represented by bamboo stakes, while the beds and borders can be plotted by bending a rubber hose in the desired pattern. Paths, hedge, or screen area can also be marked with stakes. When everything is plotted the design is studied again and last-minute changes are effected if required. After this, digging and planting work are started according to plan. Before implementing the plans, some compounds may need a little dressing-up like cleaning, levelling, and tidying-up.

Salvaging an Old Garden

If a property has been purchased which already had some garden, it is to be studied whether the old garden can be re-made. This is a complicated job which has its advantages and disadvantages. It is often difficult to adjust some existing features into the new plan. The aim is to incorporate in the new design every interesting existing feature and remove others not needed. A bird bath, water garden, and terraces should be retained and improved upon. The remaking process has to be completed slowly after watching every feature carefully and evaluating their utility in the new design.

Some Problems and How to Solve Them

Often it so happens that a double-storey house has an one-storey garage attached at one side of the building thus disturbing the whole balance. The solution in correcting this imbalance lies in planting tall trees with rounded canopy along the garage end. It is also important to select the proper corner plants near the house to soften and broaden the view of the house. A medium tree with low-branching habit and with a rounded or little oval-shaped top is planted near the corner with some low-growing shrubs planted around it. Trees such as silver oak, *Amherstia nobilis*, *Cassia nodosa*, *Dillenia indica*, *gulmohar*, *Magnolia grandiflora*, and *Saraca indica*, can be used for this purpose. If the tree selected is deciduous in nature the evergreen shrubs below should form the contrast. For a two-storey house, a high-branching rounded canopy tree such as *Anthocephalus cadamba*, *Erythropsis colorata*, *Michelia champaca*, and *Polyalthia longifolia* should be planted farthest from the corner of the house and in between a fairly large second tree or shrub is planted, besides some other low-growing shrubs.

A doorway near the house needs special attention as this is the place which receives maximum attention from a visitor. Depending upon the approach a doorway can be planted informally, formally, or in a semi-informal pattern. It can be arranged with a garden-type design consisting of an ornamental shade or flowering tree perennial and annual flowers, climbing roses, some bulbs such as *Zephyranthes*, *Amaryllis*, and daffodils (for temperate regions). Where it is not possible to plant the annual and perennial flowers in ground these can be put in tubs and arranged artistically. An ornamental light post, an urn, or an artistically shaped boulder will be an object of interest when placed near the shade tree or in an appropriate corner. A bed of roses can also be a spot of beauty provided it receives the morning sun. Symmetrical plants with pyramidal form such as *Thuja*, *Juniperus chinensis*, and *Cupressus macrocarpa* are preferred by many near the doorway for a formal treatment.

If a plot is rectangular in shape, where the length is far greater than the breadth, the best way to develop such plot will be to divide the whole area into a series of garden compartments enclosed by hedges with continuous walk of gravel paving, or grass connecting each segment with the other. But some people may prefer to have long stretches of open lawn, when this arrangement will not do. In such cases the plot should at least be subdivided into two segments as too much of a rectangular land with narrow width is not pleasing to the eye.

Many people inherit irregular-shaped plots specially those who are allotted the corner plots. It is far more difficult to plan a garden for such plots compared to a rectangular or a square plot. Inept handling may ruin such plots but a man with imagination and artistic sense can develop a garden which will be far more interesting than a garden in a regular-shaped plot. Such plots should receive informal treatments. An irregular-shaped corner may be tackled by constructing an informal lily pool. Similarly beds and borders should be shaped irregularly according to the contour of the plot. It may be far more convenient to develop a rock garden in a shapeless corner than to have a lawn or flower bed. It may be wise not to have formal paths of any kind; instead, the various areas may be reached by stepping-stones placed in artistically-curved fashion over the grass. These are only a few suggestions. Much depends on the imagination and tact of the man doing the job on the spot as the situation demands.

Plans for Very Small Compounds

What we have so far discussed suits the plots which are relatively large. For very small plots which cannot be divided into different segments such as public area, living area, and work area, one has to depend upon one's own imagination to landscape such plots. But it should be remembered that the majority of the flowering plants and Calcutta *doob* do not flourish well in a shaded place. For such plots situated under shade it is wise to put shade-loving foliage plants and flowering plants preferring semi-shade such as *Impatiens sultanii*, geranium, day lily and footfall lily. Otherwise, a lawn planted with a few specimen shrubs or roses or one or two small beds of flowering annuals will be more than sufficient for small compounds situated in the open. In all probability it will not be possible to have any large tree in such compounds.

What we have discussed so far are some possibilities and nothing is sacrosanct. One can use one's own imagination to alter one plan or other. Actually landscape design has a wide flexibility and the same plot can be landscaped in two or more different ways. Moreover, opinion varies between one landscape designer to another and, hence, the controversy whether a design is perfect will nerver end. But the basic theories must be followed and mistakes such as overcrowding, monotony, and placing of plants in wrong situations (e.g., a sun-loving plant placed under the shade of a tree) should be avoided.

Once the design is decided, the different features such as paths, walls, pools, lawn are constructed as per the procedures suggested in this book. The basic necessities such as irrigation and drainage should also be taken care of.

Trees Suitable for Small Gardens

While selecting trees for the home garden the following questions must be answered. First of all, why the tree is needed? Is it for a background or corner planting to frame the house; whether this is needed for shade for sitting or for the terrace and if so, whether grass will grow under shade? Once the questions are answered, the right type of tree has to be selected. Enough room has to be left for the tree to grow. As for example, a 25 × 50 m plot has room only for a large shade tree and two to three small flowering trees. Shallow-rooted trees such as *Millingtonia hortensis* should not be planted as they are surface feeders and may be uprooted by storms.

Some trees suitable for the garden has already been mentioned in the course of discussion, but some more may be added. *Bauhinias* in different species are quite suitable. Bottle brush is suitable for many situations. *Tecoma argentea* is a wonderful flowering tree for home gardens around Bangalore. The following trees are also suitable for planting in the home grounds: *Mimusops elengi*, *Gliricidia maculata*, *Cochlospermum gossypium*, *Cassia fistula*, and *Cassia spectabilis*.

There are some beautiful trees for the temperate regions some of which (marked with*) can also be tried in the plains of India. Cherries, especially *Prunus sargentii*, is an outstanding flowering tree. Some ornamental peaches (*P. persica** 'Clara Meyer') look beautiful when in flowering. Some of the plums bear beautiful flowers out of which *Prunus cerasifera nigra* is possibly best for a home garden. Many of the maples are very ornamental in form and deserve planting in medium compounds. The weeping willow (*Salix babylonica*) and *S. purpurea* var. *pendula* are also very ornamental.

Some shrubs may be grown as specimens in the lawn. A few suggested shrubs are *Ixora singaporensis*, *Brya ebenus*, *Sophora tomentosa*, *Mussaenda philippica*, *Azalea*, *Cotoneaster horizontalis*, and *Rhododendrons* in different species. The last three are suitable for temperate climates. For shrubbery border a list of shrubs may be made from the chapter on ornamental and flowering shrubs, depending upon situation.

LANDSCAPING A COUNTRY HOME

A villager in India may not need a sophisticated garden as has been discussed in the foregoing pages. However, a countryman will need as much privacy as a city dweller. A village home has to be planned with more utility items. A villager would like to grow more vegetables and fruit for his family consumption and consequently more area should be earmarked for this purpose. But a shade tree or two and some area reserved for children's playground is definitely needed. Some utility flowering trees such as *Michelia champaca*, *Plumeria acutifolia* and shrubs such as *Hibiscus rosa-sinensis*, *Tabernaemontana coronaria*, *Barleria*, *Crossandra*, and jasmines yielding flowers for worshipping and hair decoration should be included for planting.

The common flowering annuals such as marigold, zinnia, balsam, and sunflower can be grown for beauty and cut flowers for various purposes.

LANDSCAPING CEMETERIES AND BURNING GHATS

The landscaping of cemeteries and burning *ghats* may sound paradoxical under Indian conditions. This is a country where a large number of people do not even get the minimum amount of food to keep them alive during their life-time, so the idea of having a garden around their tombs after death may sound ridiculous at the first thought. But there is a need for this also, not only from the point of peace, tranquillity, and beauty, but also to fight pollution and provide shade. Whether the body is buried or burnt in an ugly surrounding or a beautiful landscape matters nothing to the departed person (unless, of course, we believe in life after death), but it certainly has some sobering effect on the sorrow-laden hearts of the near and dear ones who attend the funeral. While bidding final farewell one would definitely like the last resting place of one's dearest in a quiet, peaceful surrounding. For this reason also, a cemetery needs to be landscaped with trees, shrubs, and grass. If properly landscaped with vegetation a cemetery can be used as a park or breathing place, when not in use, especially near congested cities. In western countries, the idea of making the cemeteries park-like places with minimum of stone structures such as tomb-stones, head-stones, etc., and more of trees, shrubs, and flowers are catching up. Rules have been framed to keep the burial grounds as natural as possible, planted with trees.

The cemetery is sub-divided into several sections and each section is made accessible by providing driveways at a distance of 50-100 m depending upon the size of the place. In small places of 3-4 hectares no driveway is needed and the walks in such cemetery may be left in grass only. Most of the area in the cemetery should be left under grass with few trees grouped here and there for immediate effect. More trees, especially the evergreen types, are planted along the boundaries. Some flowering trees are, of course, selected to bring a bright and cheerful effect. The buildings should be modest in appearance and be planted with shrubbery and vines.

Under Indian conditions also it is possible to keep the cemeteries in a park-like fashion without much elaborate arrangements. The basis will be the same as stated above with emphasis on modesty and sobriety. In India many cemeteries have been planted with trees mainly for the purpose of shade and to create some formal effect. Very often there will be some shade-trees such as seedling mango, banyan, *neem*, etc., and the roads are lined with *Polyalthia longifolia* var. *pendula*. *Plumeria acutifolia* may also be planted in burial grounds, which flowers freely under most Indian conditions.

The burning *ghats* in India present a picture of horror and sorrow, which can definitely be improved upon by proper landscaping. It is not possible to have a formal garden with grass, flower beds, etc. in a conventional burning *ghat*, for obvious reasons, but the picture can be changed by planting in groups, some shade and flowering trees. In most cases these *ghats* are located near water tanks or ponds or rivers. The ponds can be planted with water-lilies and lotus while the banks can be beautified by planting shade and flowering trees. In modern electric crematoriums it is possible to layout a formal or an informal garden around it with important features such as trees shrubbery, flower beds, lawns, etc.

The "Towers of Silence" of the Parsi community afford good scope for gardening. Since these towers have long stretches of ground all around them, a nice garden with lawn, shrubbery, and flowering trees along the periphery, can be laid out. For instance, the Tower of Silence near Bangalore has a long avenue of jacarandas which, when in bloom, present a wonderful scenery.

Chapter XXV

Indoor Gardening

HOUSE PLANTS AND INDOOR GARDENING

The House Plants, or Indoor Plants, have become a necessity in the homes of the affluent West, but even in some affluent Indian homes these types of plant are also now finding a prominent place. It is less costlier to decorate the interior of a room with live plants compared to flowers, which are becoming costlier day by day and besides they are to be replaced frequently. On the other hand, with little care, a well-chosen house plant will continue to decorate a room for a period of time. The initial investment may be comparatively high but it proves economical in the long run.

Though the fashion of growing house plants became universally popular during the past three or four decades, definitely it is not a new art. In the ancient civilizations of Egypt, India, and Rome it was not unusual to bring pot-grown or tub-grown plants inside a room for the purpose of decoration. In Europe, particularly in Britain, during the seventeenth, eighteenth, and nineteenth centuries it was a common practice for the well-to-do people to grow exotic house plants for interior decoration. But it is only in this century the practice of growing house plants has spread to millions of homes in Europe and America. The pioneer among the house plants is *Aspidistra* which was popular in all walks of life.

Selection of House Plants

The plants which are generally grown in the houses are of two kinds. In the first category are included what we commonly call "the flowering plants" such as African violets, azaleas, geraniums, etc., which are spectacular in appearance by virtue of their colourful flowers. But once the flowering is over, these plants have very little use inside a room and hence their usefulness as house plants is only for a limited period. To the other category belong plants which provide permanent display with their graceful foliage and sometimes with their architectural or unusual form (e.g. *bonsai*).

One point has to be taken into account while choosing the house plants. Even though a room may appear to be well lit to our eyes, the available light may not be enough for proper growth and development of the plant. Hence, the majority of the house plants should have the capacity to tolerate shade of varying intensity. One more quality important to any house plant is that it should remain evergreen to retain its permanent decorative character, with possible exceptions of *bonsai* which are quite attractive even without the foliage because of the attractive outline of the tree. Though green leaves can also be very attractive, especially if the shape is unusual or interesting (e.g., *Monstera deliciosa*), but leaves with some colour other than green are considered to be more attractive. In some plants the leaves are naturally-coloured as in *Gynura aurantiaca*, *Caladiums*, Rex begonias, etc., while in others coloured forms of the natural green-leaved types are available as in *Peperomia magnoliaefolia variegata*, *Ficus radicans variegata* and others. Another quality a house plant is expected to have is compactness of its growth habit as space becomes a limiting factor in any house in a congested city.

Now it is possible to describe in a nutshell the qualities expected of an ideal house plant. A house plant should be compact in growth habit, evergreen in nature, and should stand some amount of shade around its growing environments. In addition, the leaves should be attractive by virtue of their shape or colour. Though the emphasis should be on permanance of attraction, handsome flowers produced by house plants should be regarded as a valuable trait. But the combination of good foliage and flowers is unfortunately very rare.

Before procuring a house plant one has to consider many points. The first consideration is that under what conditions a plant has to grow, i.e., whether there is sufficient light or the humidity is adequate or the temperature is favourable. Secondly, due thought has to be given as for what purpose the plant is needed. For example, if it is for decoration of a small table, the plant should be compact and bushy in nature. Another important point is the experience of the grower in handling a house plant. A beginner must try his hand with hardy plants only.

Again, a busy man who cannot spend much time in the care of the plants must also choose plants which are easy to grow. Lastly, the plants should be purchased from a reputed nursery. A house plant grown in the humid and warm atmosphere of a greenhouse should be hardened off before selling to a customer. It goes without saying that a grower should select a plant with firm and healthy-looking foliage and which is also free from disease and insect pests.

Besides the ornamental foliage and flowering plants, cacti and succulents, palms, ferns, *bonsai*, and some bulbous plants can also be grown inside a house. A miniature waterlily *Nymphaea* × *pygmaea helvola* can also be grown indoors in a bowl of at least 30 cm diameter.

Cultivation

After a house plant is procured it should be handled with care for the first one or two weeks. When a plant is brought indoors directly from the favourable conditions of a greenhouse, it shows signs of yellowing of leaves or even defoliation as a result of reduced light, less humidity, and unfavourable temperature conditions.

The plant passes through a period of readjustment after which it gets acclimatized in its new home. A useful hint in house plant culture is that, whenever possible they should be taken outdoors once in a while, during favourable weather conditions. This will invigorate the plants. But this should not be practised too frequently also as again the problem of readjustment will arise. After each such treatment the plants should be first hardened off before introducing again indoors. The hardening may be done by exposing the plants to the indoor conditions gradually, alternating with the favourable conditions of the greenhouse. Initially the plants may be introduced indoors only for a couple of hours, gradually increasing the time. The plants may finally be shifted to indoors in about a fortnight or after a month. Even after this period there may be a setback initially, which most plants will withstand after the readjustment period is over.

The key to success in growing house plants depends on various factors which are discussed below.

Light: Light is an important factor in the cultivation of house plants. The demand for light, however, varies from plant to plant. While *Hedera helix*, a climbing house plant thrives well under a relatively dark corner, plants such as *Sansevieria trifasciata* require a good amount of light. Though as a general rule no house plant should be exposed to direct sunlight, there are a few exceptions such as *Sansevieria trifasciata*, most of the cacti and a few trees grown as house plants such as *Grevillea robusta* which do not mind exposure to direct sun provided it is not very scorching. Modern houses with broad glass windows admit sufficient light inside the rooms. But there may be some corners which need decoration with live plants but the place is not sufficiently lit for this purpose. Such dim corners need supplementary artificial illumination. The best artificial source is fluorescent lighting. The advent of fluorescent lighting has opened up a wide frontier for house plant culture which has proved to be an ideal substitute for natural day light. But the ideal illumination is a combination of a balance of red and blue rays. This can be achieved by illuminating an area with one 120 cm long and 40 watt 'daylight' fluorescent tube (for blue waves) and one similar sized 'warm white' (for red waves) tube. In the U.S.A. different electrical companies market these indoor lights under different trade names as 'Plant Lights', 'Vita-Lites', 'Agro-Lites' and 'Gro-Lux'. It is difficult to stipulate the optimum light conditions for indoor plants as this depends on a number of factors including the varietal difference. The plants should not be exposed for more than 16 hours per day under two 40 watt 120 cm fluorescent tubes, placed not more than 90 cm from the plants. It is also difficult to specify at what distance the plants should be kept from the tubes. The plants exposed to fluorescent lighting themselves react to the amount of light received and send out signals. A plant exposed too much will turn yellow when this has to be removed further away from the source of light or the duration of exposure has to be reduced. A plant receiving insufficient amount of light will grow lanky, when the plant should be moved closer to the source or the illumination time has to be increased. At the initial stages when a rapid growth is desired the plants are placed as close as 10 cm from the tubes, gradually increasing the distance as they grow. The tubes lose power as they become older and it is advisable to remove them after 5,000 hours of burning, when the ends show dark rings, for the benefit of the plants.

A general recommendation is that 15-20 watts of fluorescent light is needed for each 30 square centimetres of plant area. A 60 cm long tube produces 20 watts of light, while a 120 cm tube emits 40 watts of light. A 20 watt tube will provide sufficient light for a 75 cm by 25 cm wide area.

Temperature: The temperature requirement of house plants is much more variable than that of light, because of the diverse climatic conditions of origin of the different house plants. Some plants belong to the tropics while others hail from subtropics and a few from the temperate zones. The most important point about temperature is that house plants, whether from the tropics or from temperate zones, do not like any great fluctuations in the course of a day. The ideal range of temperature should be around 15°-21°C (59°-70°F) during day time and the night temperature never falling below 70°F (20°C). But most of the hardy house plants can withstand much higher temperatures than this provided the humidity is good. It is in the temperate countries with severe winter, that the role of temperature is much more important than the tropical countires. But fortunately with the advent of central heating the rooms in western houses provide the near-ideal conditions for the culture of house plants. In houses with central heating, plants may dry out, so it is advisable to resort to misting of the foliage with water during day time, if necessary.

Humidity: Many of the house plants, with few exceptions such as *Sansevieria*, *Aspidistra*, desert cacti, and a few other succulents, in their natural habitat receive high amount of humidity. In a house it is not possible to maintain high humid conditions because that will be rather uncomfortable for the inmates. As a result plants requiring high humid conditions cannot be grown properly indoors. But it is possible to improvise methods by which humidity can be increased to some extent in the vicinity of the plants. One such method is to put the potted plants in another empty pot one size larger than itself and to pack the surrounding empty space with peat or sphagnum moss which is kept moist by constant watering. The other method is to stand the pots in a shallow waterproof metal tray containing small-sized gravels or pebbles. Water is then poured until it is just short of the top of the pebbles at which level this should be maintained constantly. Pots should just stand on the top of the pebbles so that even a little water does not enter the pots. The evaporating water will maintain the humidity. An affluent plant lover can grow the more moisture-loving species such as *Alocasias*, *Anthuriums*, orchids, etc., in a "*Wardian Case*" with waterproof base, which is nothing but a miniature glass greenhouse with the inside temperature controlled thermostatically. The plants may be set inside this over a tray of water, as suggested above, for humidity. The less affluent people can encase them in a frame which is covered from all sides with transparent polythene with few ventilator holes. The humid atmosphere is created in the same fashion. In hotter days, the plants can be kept moist by spraying the foliage with water.

Watering: The water requirement of house plants varies from one kind to another. An *Alocasia* needs more water than a cacti or some succulents. Plants kept in a comparatively cooler atmosphere need less water than when kept in a warmer house. A plant growing in a porous clay pot will need more water compared to one growing in a plastic or glazed container. Similarly, a smaller pot will need more frequent watering. Whether or not a pot needs water will be indicated by the dry surface of the compost. Other method is to strike the pot with a wooden hammer gently and if the resulting sound is of a ringing tone the plant needs watering, while a dull sound indicates that the compost is moist. A plant needs more water during its growing period and the amont is reduced appreciably during the winter which is a resting period for most plants. It is always better to water in good amounts at reasonable intervals, which may vary from 3 to 10 days depending on the climate and species, rather than small daily doses. It is a safe practice to under-water rather than over-watering. Rain water, if available, should be preferred to tap-water which contains lime.

Fresh Air: The air in the room may become polluted due to fumes, high concentration of carbon dioxide, etc., which is detrimental for most of the house plants. The windows can be opened occasionally to let in fresh air or the plants can be put outdoors in mild days.

Other Cultural Fractices: Cleanliness is one important factor. The large-leaved specimens can be sponged regularly with clean water to improve the appearance. The smaller leaved plants can be kept clean of dust by regular spraying.

The plants should be fed with liquid manure during their growing season, and never during the rest period. To make the fertilization effective it should be added when the compost is moist. Some plants which are not easy to repot can be sustained for some time by top dressing with fresh compost replacing a portion of the old compost from the top. It may be necessary to repot most of the plants every alternate year. For details of repotting refer page 335.

An useful hint to a house plant enthusiast is to grow house plants together in a group to increase humidity around them and also to keep around them a few small glass tumblers containing water. An occasional shower bath is welcome as this cleans the foliage thoroughly. In temperate countries, only luke-warm (not hot) water should be used for this purpose and also for irrigation.

Sometimes house plants may get sick and show distress signals. The following chart will be helpful in detecting the causes of the ailment and growers may take suitable remedial measures.

Symptoms (1)	Probable causes and remedies (2)
1. Stunted growth	Inadequate feeding or excessive watering, but in the winter it may be due to resting period of the plant. Probable remedies for the first cause is to feed with liquid manure and to stop watering for a few days or a reporting may have to be done. But in the winter this symptom may not be unusual if that be the resting period of the plant.
2. Wilting	Lack of moisture in the compost or over-watering or bad drainage. Shade-loving plants kept in too sunny position. Rotting of roots. Remedial measures are to regulate watering, check drainage, and ensuring correct light exposure.
3. Dropping of flowers, buds and leaves or yellowing	Bud formation takes place in good light, and if the plant is not receiving good light, buds will fall. Flowers, buds, and foliage may also drop as a result of fluctuations in light and temperature, lack of humidity, excess or under-watering, gas fumes, and drought. Red spider mites may also cause the falling of leaves.
4. Leaves turning yellow but not dropping	Plants preferring acid soil may show such symptoms if there is excess of lime.

(1)	(2)
5. Leaves drooping	If the soil is too wet at this time this must be due to rotting of roots as a result of excess watering. The affected plants should be removed to a warm place immediately and watering stopped for a couple of days and resumed thereafter only moderately.
6. A variegated species becoming green	Insufficient illumination.
7. Tips of foliage turning brown or brown spots	Fungal disease, over-heating, drought conditions, too much fertilizers, and scorching due to water drops emaining too long on foliage.

The above chart is also useful for ornamental foliage plants growing in greenhouses.

Pest and Diseases: See Chapter XXVI.

Modes of Indoor Gardening

The art of growing house plants or indoor plants inside a house is known as indoor gardening. The simplest method is to grow individual plants in pots and keeping them in appropriate places in the house. In western countries some affluent people may even have an elaborate indoor roof gardening. The other popular methods of growing indoor plants are in hanging baskets, China baskets, troughs made of concrete, wood, cane, etc., bowls and dishes, bottle gardening, miniature gardens, window gardens, and vertical gardens, all of them have been discussed in the subsequent pages. In addition to these, gardening in trays and trollies are also part of indoor gardening.

Trollies fixed with flexible wheels meant for serving tea and snacks are quite suitable for growing indoor plants. One advantage of growing plants on trollies is that it can be moved from one corner to the other or even from one room to the other without much difficulty. To grow plants in a trolly it has to be improvised a little first by making the base water-proof and raising all the four sides to conceal the pots kept in it. To maintain humidity a shallow tray containing small pebbles and filled with water can be kept at the base of the trolly over which the plants are kept. In no case should water enter the pots. Moist sphagnum moss may be used to conceal the pots which also preserves humidity. The trolly should be placed below a fluorescent tube to get the best results

Trays of different make and size can be used for growing indoor plants. For details readers may see pages 542 to 544. Growing of house plants in China baskets made of cane or even plastic gives an artistic and elegant touch to the decor of the room. The basket is first lined with polythene sheets to make it water-proof and then the pots are placed inside. Troughs made of cane standing on its own legs, similar to that of cradles for children, are very artistic and pleasing for growing plants indoors. Here also the trough is first lined with polythene to arrest seapage from the pots.

When specimen plants grown in clay pots are used for indoor decoration these may be plunged in a larger sized ceramic, glazed pot, or in any other good-looking metallic container to improve the look.

Putting the plants in various-shaped plant stands also improves the look of display. Plant stands may be made to accommodate only one plant or it may be branching to hold several pots together. These are generally made of mild steel rods or plates with a heavy base and having a ring to hold the pot. The branching types generally have several protruding hands from the main support at the end of which there will be rings to hold the pots. Plants kept in plant stands should be watered just enough so that there is no drip, or the pots are watered outside and the excess of water is allowed to drain off after which the pots are put back in the stand. Even after these precautions it is most likely that the pots may drip. So it is advisable to fix buckets made of G.I. sheets or plastic at each ring and to place the pots inside the bucket to arrest the drips. The seepage water collected in the buckets should be drained off occasionally.

There are some other methods of displaying house plants. A live screen can be created in a window by growing light indoor creepers such as different *Hedera helix*, *Scindapsus aureus*, the "golden pothos" (money plant) and others. The dining space in a drawing or living-room can be separated by growing a screen of creepers in between or placing a vertical garden. Plants grown in bowls or metal hanging baskets can be fixed on the walls by using brackets which will bring a relief to an otherwise empty expanse of a wall. However, one should be cautioned not to overdo the practice of decorating with house plants.

Some House Plants

For the sake of convenience, the house plants are classified into several groups. The different groups are mentioned below and some important house plants belonging to each group are listed below. For details regarding the characteristic features of the plants and their cultivation the respective chapters in this book may be consulted.

I. Climbing and Trailing Foliage Plants

The climbing and trailing plants will need support for growing. For light creepers or trailers split-bamboo cane support driven deep into the compost will be enough. The creeper or trailer should be tied to the cane taking care that the knot round the stem is loose enough for the future growth of the plant. A trellis can be made by driving into the soil three or more canes at slightly inclined angle across the centre of the pot. Then crossbars made of split-bamboo cane are tied across these verticle canes with thin wires at intervals of 10 cm. For making a screen, thin chicken wire-mesh may be used supported on wooden frame. Moss sticks, also called totem poles, are made by tying sphagnum moss all round and along the entire length of a thick cane and supporting against it generally creepers such as *Monstera deliciosa*, *Philodendron scandens*, money plant, and also *Hedera helix* having arial roots.

The following are examples of few house plants belonging to this group.

(a) Climber: *Ficus pumila*, *Ficus radicans variegata* is a beautiful variegated creeper suitable for hanging baskets. The creamy-white variegation starts from the margins, *Asparagus plumosus*, *A. sprengeri*, *Hedera helix*, *Philodendron elegans*, *P. laciniatum*, *P. melanochrysum*, *Scindapsus aureus* ('Pothos"), *S. aureus* 'Marble Queen", and *S. aureus* "Tricolour", and *Syngonium podophyllum* (*Nephthytis liberica*).

(b) Trailers: The important plants in this group are: *Chlorophytum comosum variegatum*, *Fittonia verschaffeltii*, *Tradescantia fluminensis*, *T. fluminensis variegata*, *Zebrina pendula*, and *Z. purpusii* (*Tradescantia purpurea*).

II. Bushy and Upright Foliage Plants

Generally, the plants of this group possess beautiful foliage and are suitable for display both as individual specimen and in arrangements. Some important plants belonging to this group are mentioned below. Though *Araucaria excelsa* and *Brassaia actinophylla* are not strictly foliage plants, but are used so until they grow up.

Aglaonema commutatum, *Aralia elegantissima* (Syn. *Dizygotheca elegantissima*), *Araucaria excelsa*, *Aspidistra lurida*, Begonia 'Rex', *Brassaia actinophylla*, *Caladium* in different colours, *Calathea insignis*, *C. makoyana*, *Codiaeum* (crotons) in different colours, *Cordyline terminalis*, *Cryptanthus bivittatus*, *C. zonatus*, *Cyperus alternifolius*, *C. alternifolius variegatus*, *Dieffenbachia exotica*, *D. picta*, *Dracaena godseffiana*, *D. sanderiana*, *Fatsia japonica*, *Ficus benjamina*, *F. elastica decora*, *Fittonia verschaffeltii*, *F. verschaffeltii argyroneura*, *Grevillea robusta*, *Gynura×sarmentosa*, *Maranta leuconeura*, *Monstera deliciosa*, *Neoregelia carolinae*, *Nidularium fulgens*, *Pandanus veitchii*, *Peperomia caperata* and its cultivars, *P. hederaefolia*, *P. magnoliaefolia*, *Philodendron bipinnatifidum*, *P.* 'Burgundy', *P. dubium*, *P. selloum*, *P. wendlandii*, *Pilea cadierei*, *P. muscosa* (*microphylla*), *Sansevieria trifasciata* and *Setcreasea purpurea*.

III. Flowering House Plants

The flowering house plants generally have attractive foliage which remains on the plant all the year round and in addition they bear attractive flowers.

(a) Climbing and training types: *Begonia glaucophylla*, *B. glabra*, *Passiflora caerulea*, *Trachelospermum jaminoides*.

(b) Bushy and upright types: *Aphelandra squarrosa*, *Begonia maculata*, *B. manicata*, *Beloperone guttata*, *Billbergia nutans*, *B.×windii×Impatiens petersiana*, and *Saintpaulia ionantha* (African Violet).

IV. Flowering Pot Plants

The difference between a flowering house plant and pot plant is that while the former has decorative leaves even when flowering is over, the latter generally shed the leaves after flowering thus losing its attractiveness as house plant. Any pot plant which prefers partial shade for their growth can be grown indoors. A few examples are *Azalea indica*, tuberous rooted begonias, *Calceolaria*, *Chrysanthemums*, cineraria (*Senecio cruentus*), Coleus, *Pelargoniums* (geranium) and others.

V. Bulbs

Most of the bulbs can be grown indoors only once, after which they are to be planted out. The bulbs are generally grown in special bulb bowls and special bulb fibre. The method of growing has been discussed on pages 230-31. The hyacinths and crocuses can be grown in special bulb glasses having an annular shelf in the upper deck where the bulb is placed. The lower deck is filled with water in which a few pieces of charcoal are dropped. At first the water level will be such that it just touches the bulb. It is then removed to a dark place for about 7-8 weeks by which time the bulb sprouts and roots. When the sprout is at least about 2-4 cm long it is removed to semi-shade for about a week and from this time onwards water should never touch the base of the bulb, otherwise it will rot. Then the bulb is removed into a lighted room. Clear rain water or distilled water is used for bowl culture, which may be changed once a fortnight with the help of a-siphon without disturbing the roots. For water culture and fibre culture both, liquid manure should be applied once in 10 to 14 days after the flowering stalk appears. The bowls are turned around occasionally so that all the plants receive equal light.

After flowering the annual bulbs exhaust themselves and the bulbs are then removed for outdoor culture only.

(a) *Annual bulbs:* Most of the annual bulbs recommended for indoor culture are suitable only for temperate regions. In the plains, where the weather is cool enough, one may try some of these but the success will not be satisfactory. The bulbs recommended are Daffodils (*Narcissus*), *Crocus*, Hyacinths, and Tulips.

(b) *Permanent bulbs:* Only *Amaryllis* (*Hippeastrum*) grows well indoors year after year. These can be grown in the tropics also.

VI. Ferns and Palms

Many ferns and palms are suitable for indoor culture. For cultivation refer to the respective chapters. (XV and XVI).

(a) *Suitable Ferns: Adiantum*, *Asplenium nidus*, *Nephrolepsis exaltata*, *N. exaltata bostoniensis*, *Platycerium*, and *Pteris cretica*, *Selaginella* in different species, (though not ferns) can also be grown.

(b) *Suitable Palms: Chamaerops humilis*, *Cycas revoluta* (not palm but similar looking), *Howea* (also spelt *Howeia*) *belmoreana*, *Phoenix roebelenii* and *Rhapis excelsa.*

VII. Rose for Indoors

Even roses can be grown indoors provided proper attention is given. One should select the suitable cultivars by a trial and error method. But with proper fluorescent lighting it may be possible to grow a large number of cultivars. The cultivars 'All Gold', 'Lilac Charm', and 'Red Wonder' do well.

VIII. Cacti and Succulents

Most of the cacti need sufficient light or they will become lanky. The epiphytic types need less light and so do well under indoor conditions. For culture, consult chapter on Cacti and Succulents. A selected list is given below.

Aeonium haworthii, *Agave americana marginata*, *Aloe variegata* and some other *Aloes* can also be grown; *Astrophytum myriostigma*, *Cephalocereus senilis*, *Chamaecereus silvestrii*, *Cotyledon undulata*, *Crassula arborescens*, *Euphorbia splendens*, *Kalanchoe blossfeldiana*, *Opuntia microdasys*, and *Sedum* (various species).

Some epiphytic cacti such as *Aporocactus flagelliformis*, *Epiphyllum* sp., *Rhipsalis*, and *Zygocactus* also do well as house plants.

IX. Bonsai or Miniature Plants

The *Bonsai* are very good for indoor decoration. Even when they are devoid of leaves, their architectural shapes are attractive enough for decorative display.

GARDENING IN TUBS OR URNS

Gardening in tubs or urns gives immense pleasure to the garden lover. These are meant for display in the terrace, roof garden, backyard, veranda, and on the doorsteps. Tubs and urns are portable and hence can be used for temporary decoration indoors or elsewhere. Wooden tubs are best suited for hot sunny position. Rot resistance woods such as teak, oak, poplar, etc., are best suited for this purpose. In addition wooden tubs are treated with rot resistant chemicals such as cuprinol. Besides wood there are limitless other materials which can be used as tubs or urns. The common materials available in India are the coal-tar drums cut into half, old packing cases, earthenware pitchers, etc. In western countries, empty beer barrel tubs are frequently used for this purpose. Tubs and urns of different shapes and designs are made to order from fine grade concrete. In western countries, fibre glass urns are also available. Wooden tubs are made to order with ar istic designs. All tubs should stand on their leg or should be placed over bricks for easy drainage Each tub or urn should have one or more drainage holes. Wooden or cement tubs are painted to match with the colour of the house.

The main criterion in selecting a tub is to ensure that it holds sufficient amount of soil and have provision for proper drainage; though a container without drainage hole is also useful if sufficient drainage material such as gravel is put at the bottom and over-watering is avoided. At the same time, the tub should be quite aesthetic in look.

Culture

The tubs or urns are initially filled with a good compost as suggested for window gardening. The tub plants need frequent manuring and it is advisable to feed them with liquid manure every fortnight or so depending upon the type of plant. Liquid manure may be prepared by fermenting 2-4 kg of fresh cowdung or oil-cake (mustard, neem, or castor) in five litres of water. This fermented liquid should be diluted to tea colour and to each 5 *l* of solution 30-35 g of ammonium sulphate and a handful of superphosphate should be added. About 1-2 *l* of this liquid manure should be added to each pot depending on the size and type of the plant grown. One shall be able to standardize the quantity required for each kind with experience. Besides this, each tub has to be top dressed with fresh compost every six months or once a year. This is done by scrapping out 7-15 cm of the used up top soil and replacing this with new compost. All plants in tubs or urns need to be replanted every two or three years to replenish plant food. In India, under tropical climate, repotting should be done during the rainy season. But in a milder climate as that of Bangalore and Puna this can be extended till December or January. In temperate regions, repotting is done during the spring. Repotting should be avoided in the summer. At the time of planting a stick may be inserted deeply in the centre of the tub and during watering this may be pulled out and the channel used for watering. After watering, the stick should be replaced. Frequency of watering depends upon the weather and type of plants. As for example, Bougainvilleas and Jasmines need dry conditions at the time of flowering.

Types of Plant

For tubs and urns large type plants are preferred. These may be large annuals such as African marigold, chrysanthemum, sunflower, hollyhock, etc., or herbaceous perennials and bulbous plants such as canna, bird of paradise, datura, *Vinca rosea*, *Impatiens sultanii*, *Fuchsia* (for temperate climates) geranium, etc. Many of the shrubs, such as hibiscus, oleander, lantana, camellia, and azalea (the last two for temperate climates) can be grown in tubs. Some ferns (e.g., Birds' nest) and most of the ornamental palms are also suitable for growing in tubs. Some other plants such as *Fatsia japonica*, *Aralia elegantissima* (Syn. *Dizygotheca elegantissima*), *Dracaena* sp., *Yucca*, etc., are also suitable for this purpose. Bougainvilleas also do very well in tubs and urns in most parts of India. Some beautiful trees such as *Araucaria cookii* and *Brassaia actinophylla* can also be grown in tubs when they are young.

The most advantage of plants in tubs and urns is that these are movable and can be shifted from one place to another for display. Suppose a plant in the tub has completed flowering, this can be replaced by another tub-grown plant from the nursery which is in flowering.

In houses having window garden, tubs and urns can also be matched with flowers of similar or complimentary colours.

Topiary

Urns and tubs offer a good opportunity for exhibiting topiary and espaliering work. For training an espalier, trellis or a frame made of galvanized wire may be fixed in the tub itself. The following plants offer good opportunity for such works: (a) *Clerodendron inerme*, (b) *Duranta plumieri*, (c) Bougainvillea, (d) Camellias (for temperate climate), (e) Juniper, and (f) *Thuja* different species. The topiary works should be displayed against a wall or a screen in the terrace.

GARDENING IN HANGING BASKETS

Gardening in hanging baskets has become very popular after the Second World War in Europe, especially England, but again went into oblivion. In recent times, this has once more become a popular hobby with the garden lovers.

Hanging baskets with trailing or cascading plants are suited for indoors as well as outdoors. These can be hanged at the entrance of the house to welcome the visitor. Hanging baskets can also be placed in the hall or the drawing room besides a well-lit window, or in the bathroom above a fluorescent light. In western countries, hanging baskets even decorate the kitchen walls. A plain wall in a sunny room or a passage can be also artistically decorated with hanging baskets.

Out of doors, hanging baskets can be suspended from trees, electric poles, fences, etc. Blank walls along a city path can be decorated with hanging baskets by arranging them in pattern, provided this does not interfere with foot traffic. In a conservatory also shade-loving trailing plants are suspended from the poles and roof, grown in hanging baskets.

Types of Container

The container should be attractive, light, and easy to hang and preferably with a drainage hole (see Fig. XXV.1). Wire baskets with three hooks, for hanging is a good container as they provide good aeration. In hot climates such baskets dry out rapidly and, therefore, is not recommended for outdoor. Wooden pattern baskets are better than wire baskets

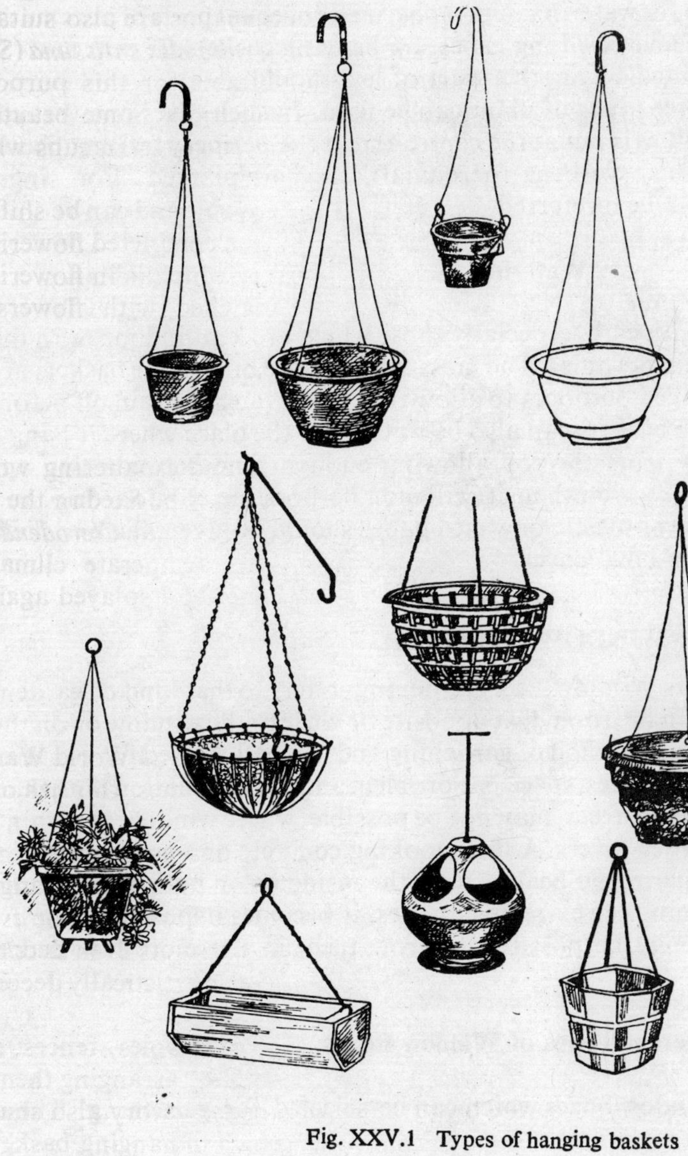

Fig. XXV.1 Types of hanging baskets

in dry places and are better looking. Earthen pots of various sizes and shapes are available for use as hanging baskets. Hanging baskets can also be made out of dried gourd and coconut shells. Empty tin cans, though not attractive in appearence, serve the purpose of hanging basket. For indoor purposes baskets of brass, copper and glazed pottery may be used.

Many people often put potted plants inside the hanging baskets and camouflage this by packing sphagnum moss around the pot. Even ordinary earthenware or ceramic pots can be hanged in brackets on the wall. A wire or wooden pattern basket is to be lined properly to prevent the soil being washed out. First a inner lining of a gunny cloth is put over which about 2-3 cm thick sphagnum or sheet moss is spread. Over this the soil is filled up to the rim leaving a gap of 2-3 cm for watering. A compost made of equal parts of peat moss, leaf-mould, garden loam, sand, and cowdung is best for hanging baskets. If peat moss is not available, it can be substituted by another part of leaf-mould.

For a larger basket two or three types of plant can be used. In such cases one vertical plant such as *Dracaena* or geranium is put at the centre and at the peripery trailing plants such as *Zebrina pendula*, petunias, verbena (perennial), etc., are planted. For single planting, trailing plants should be preferred.

Watering

Hanging baskets are watered with care, especially those which are kept indoor or in the case of brass pots having no drainage holes. The moss-packed wire or wooden baskets are dipped in a tub of water and hanged outdoors to allow the excess water to drain off before putting them back indoors. Such baskets can also be watered at the place where it hangs, provided water is poured slowly from the top allowing enough time between two application for the water to soak. But even then it will not be possible to stop drip and hence a protective outer covering of plastic or waxed paper should be given. Such baskets can also be placed over a plastic bowl cover.

WINDOW GARDEN

Window garden, also known as Window-Box Gardening, refers to that kind of gardening where plants are grown within the room just opposite or close to the window or on the window-sill outside. This is a novel method of gardening and is ideally suited for congested cities and flat-dwellers. In large cities, space is a problem and the conventional method of gardening such as planting avenue trees, may not be possible, where window gardening opens an opportunity for the garden lovers. A drab looking concrete or a metal road can be converted into a place of charm and beauty if all the residents on both sides of the street take up to window gardening. In extreme climates, it becomes impossible to grow certain plants outdoors, but it may be possible to grow them in the more favourable situation in the window.

Different Types of Window Boxes

There are different types of window boxes which can be selected for gardening. These are briefly described below.

(1) *Fibre-glass Box:* In India, this type of box is not easily available but in countries

such as England or the United States of America it is widely used. This is light, decorative, and durable, but most expensive too.

(2) *Pottery Box:* Terracotta boxes of different size and design are available in some parts of India These are generally smaller in size.

(3) *Iron Boxes:* Iron boxes can be made by order. These are heavy but durable.

(4) *Cast Cement and Asbestos Boxes:* These are long lasting. Engraving with all sorts of figures is possible to decorate such boxes. While constructing new buildings cast cement boxes on the window-sill can become a part of the building.

(5) *Plastic Boxes:* Plastic boxes of different sizes can be used. These are light in weight and cheap but not very durable.

(6) *Wooden Boxes:* Wooden boxes of various sizes are most suitable for window gardening. Teak and oak timbers are suitable for this purpose. The timber of *Arjun* (*Terminalia arjuna*) tree is also moisture resistant and may be used for making window boxes.

The boxes can be filled with compost directly and plants grown in it, or pot plants grown earlier can be kept inside the boxes. People not having the knack or time to grow plants can purchase grown-up plants from nurseries with buds ready to open for display in the window boxes.

The size of boxes should be comparatively small as these may be hanging outside the houses in most cases. Even small boxes containing soil, especially wet soil become heavy. A manageable size is 20 cm deep, 25 cm wide, and the length varying according to situation but never longer than 180 cm.

Drainage: All these types of boxes should have at least one drainage hole at the bottom, but it may be necessary to provide two or more holes for larger boxes. For wooden boxes the drainage hole is made with a red-hot poker, as charred wood is more resistant to water and rot. Water will drip from the boxes and to catch this a metal or tile tray should be placed below the boxes keeping a gap of 1-2 cm from the box.

Protecting the inner side; Constant contact with water will cause rotting in wooden boxes and to prevent this the inside is charred by burning. Some of the present-day wood preservatives such as cuprinol is harmless to plants, but creosote oil is not. For metal and cement boxes, moisture and corrosion-proof coal tar or moisture proof paint may be used.

Linings for window boxes: There is a difference of opinion regarding lining of window boxes. It is widely believed that in places having a hot summer window boxes need double walls to protect the plant roots from heating and drying out. But others feel that there is no need for this. In most parts of India, the summer being hot, it is felt that double walls will be helpful. The lining may be of sheet zinc or lead and wood for wooden boxes. To have a double wall one has to make two boxes, the inner one bottom-less and little smaller than the outer one, to fit inside, leaving a gap of 1.5-2 cm all round. This space between the two boxes can be packed with vermiculite or sphagnum moss, which are kept moist. This will help keep the soil inside the box cool.

Security and Precaution

As has been stated earlier, boxes containing wet soil is pretty heavy and needs to be secured firmly in the window-sill for one's own safety and that of the passer-by. This can be achieved by fastening the boxes with the window frame by a hook-and-eye arrangement, or the boxes can rest over metal or wooden brackets attached to the frame of the window. Aluminium nails or screws should be used to prevent rusting.

In most of the modern houses the windows open outwards and there may not be any sill at all. In such homes window boxes hang outside, below the ledge of the window, resting on brackets. In addition to brackets such boxes can be secured further by tying them round with galvanized wires and fixing the ends of the wires with the window-frame. This type of box has two advantages; first, this does not obstruct light to the room and, secondly, garden operations are easier. The most objectionable point is that drip is almost impossible to control from such a type of box. The boxes should rest on wooden wedges to keep a gap between the sill and the box for drainage. The boxes are also fitted about 2 cm away from the frame or wall to prevent rotting and attack by termites.

Decoration

To keep the boxes showy and attractive, these should be painted every two years. This also helps protect them from rot or rusting. A decorative tile panel may be hung in the front with the help of hooks. But whatever kind of decoration that may be, it should not be very gaudy as this distracts attention from the plants.

Type of Compost

The success or failure in window gardening depends to a large extent on the compost used. In the window garden, the density of planting is more and the nutrients are drained off more quickly. For these reasons, a rich compost must be used. A single box generally contains different kinds of plant requiring separate types of compost. But it is not possible to satisfy them all in a window garden. So the object will be to have a general compost which meets the requirements of most of the plants. In most of the advanced countries in the West, readymade window box composts are available. But in India, the cit ydwellers may not find a readymade compost available. The following compost suiting most of the plants, may be prepared at home: 3 parts good fibrous loam, 1 part well-decayed coarse leaf-mould, ½ part well-rotten cow manure, and a handful of bonemeal. An alternative is to grow the plants in individual pots each having the right type of compost and then place these pots in the box. The pots may be camouflaged with moss.

In cities having the problem of severe air-pollution, the compost is likely to be replaced every year. But in cleaner cities and towns the replacement may be carried out every third year provided the plants are well fed.

Management of Window Gardens

The first important thing is the aspect of the window garden, i.e., whether it is in shade or sun. Depending upon the aspect either shade-or sun-loving plants are grown. In places having hot summer, windows on the eastern side will be preferred for window garden. Watering is done according to the type of the plants grown and necessity, depending on weather. Both over and under-watering are undesirable. The first causes water-logging and the second leads to wilting. Syringing during the hot summer is helpful. Drainage is an important point and this should be looked into properly. Sufficient drainage materials such as crocks, small stones, or rubbles should be placed round the drainage holes. The drip tray placed below the box is meant for collecting the superfluous water and as soon as the water has run through this should be emptied.

Manuring the plants every three or six months is helpful. Liquid manure made by fermenting fresh cowdung or oil-cake and then applied after diluting is more desirable than fertilizers. Fertilizers, if applied, should be given in liquid form. Liquid fertilizer doses recommended for chrysanthemum may be generally followed.

Other operations required are pruning, trimming, pinching, disbudding, and staking. In places having the danger of frost, the window box may be placed temporarily on the floor of the room unless there is fire inside.

Window Rock Garden

A window garden specialist can build a miniature rock garden in the window box ready with peaks, a pool, winding paths, etc. As in rock garden, the rocks here also should be embedded in the soil. The principle for building the garden is the same as that for a rock garden. Using too many rocks in such a small space should be avoided. Plants such as *Sedum*, pinks (*Dianthus chinensis*), *Haworthia* (dwarf species), sweet alyssum, *Saxifraga* (temperate) portulaca, and pansy can be planted in such gardens.

Suitable Plants

Plants for window boxes are generally dwarf in size, otherwise they will obscure the view and the container will appear top-heavy. Trailing plants are generally preferred to upright plants, but there is no restriction so far as the plants do not prevent light getting in. Small climbers such as Morning Glory, *Thunbergia alata*, and *Clitoria ternatea* can also be grown.

The following plants are recommended for growing in window boxes depending upon the climate:

Antirrhinum (dwarf), *Begonia* (tuberous and fibrous rooted), Cacti (*Opuntia microdasys*, *O. tunicata*, *Cereus peruvianus* var. *monstrosus minor*, *Mammillaria elongata*) candytuft, carnation, celosia, China aster, chrysanthemum (annual), *Crocus*, *Echeverias*, ferns (some dwarf types), *Gaillardia*, *Impatiens*, Kochia marigold (French), michaelmas daisy, pansy, *Pelargoniums* (ivy-leaved and zonal types) petunias, phlox, pinks (dianthus), portulaca salvia, *Sedum* sp., statice, verbena (including perennial), and *Zebrina pendula*.

It should be kept in mind that plants varying widely in their water requirements cannot be grown in the same box. As for examples, a cactus cannot be grown along with a flowering annual such as phlox which will need more frequent watering than the former. This problem can be overcome to some extent by growing the plants in individual pots and placing them in the box. Similarly, it will be difficult to grow a sun-loving plant such as carnation along with some shade-loving ferns.

MINIATURE GARDEN

In sprawling cities, struggling for space, a flat dweller hardly gets any place for the conventional type of gardening. For him only a few alternatives such as hanging baskets, bottle gardening, and miniature gardening are left for landscaping. Miniature gardening is a little complicated in the sense that many garden features such as paths, rocks, hills, pool (arranged with the help of mirrors), well heads, flowers, shrubs, trees, etc. are to be designed in the miniature form. In practice, a miniature garden is nothing but a rock garden done on a miniature scale or what is sometimes referred to as "show-case size." There is hardly any difference between a window box rock garden and a miniature garden. The only difference is that the former is placed permanently in the window-sill, whereas the latter can be placed anywhere inside the house or in the portico.

This type of garden is best laid in troughs of concrete, stone or wood. In advanced countries, even troughs of plastics and fibre-glass are available. The true English tradition demands that miniature gardens should preferably be laid in old wooden troughs. But under most Indian conditions, the best container will be of concrete. For this purpose, a box 25-30 cm deep, 60 cm wide and 90-100 cm long with drainage holes, will be ideal.

The trough is first filled to a depth of 7-8 cm with pebbles or gravels for drainage. Over this a compost consisting of equal parts of good garden loam, sand and leaf-mould or peat is filled leaving a gap of 1-2 cm from the rim for watering. The miniature garden has to be placed in full sun. This should be protected from adverse weather conditions such as hot winds and torrential rains.

The surface of the miniature garden is laid to create small valleys, peaks, paths and pools (arranged by the reflection of mirror). The plants are planted in between. The surface is scattered with coloured pebbles and stones which not only give a natural effect but also prevent the soil from being washed off.

Most of the plants suitable for the rock garden can also be used for miniature garden. Many people like to put their miniature plants (*bonsai*) in this garden. For treatment of the plants consult recommendations given in Trough and Dish Gardening.

GARDENING IN TROUGHS, DISHES, BOWLS, AND TRAYS

This type of landscaping is again most suited for the city flat dwellers, but, of course, there is no harm if others also take interest in this. Gardening in troughs and dishes is generally done with indoor plants only. The use of single plants to decorate a room is very attractive unless, of course, too many of them are used at a time. But a display, in the truest sense, is not possible with single individual plants in pots. For this purpose, large or small plants or a combination of both types are arranged in dishes or troughs. Though the principle is the same, trough garden differs a little from the dish and bowl gardening. A trough is larger in size which may contain more number of plants and can be displayed in the window-sill, on the floor or at the entrance of the house under a covered veranda. When placed in a window-sill, venetian blinds should be used to exclude hot sun and frost. Bowls and dishes are meant for decorating dining-table, television set, rodio set, tea table, etc. A trough generally has drainage holes while bowls and dishes have generally. none. Hence watering has to be done more carefully for dish and bowl gardens. Dish and bowls can be kept on any shining surfaces without the fear of spoiling them from seepage water.

Troughs can be of many kinds as has been stated in miniature gardening. The difference between miniature and trough gardens is that in the former many miniature garden features are created, whereas in a trough garden indoor plants are just displayed in an artistic manner.

In selecting containers for a bowl and dish gardens a visit to the antique shop will be very helpful to get the right type of containers. Fruit bowls, conical baskets, wooden dishes or bread trays, pottery, china and low plastic bowls, and pedastal vases used for flower arrangements are very suitable for starting a dish or bowl garden.

When a large trough is used, the bottom 2-3 cm is spread with gravel for drainage. Over this a compost, consisting of equal parts of good garden loam, peat, leaf-mould, and soil, is filled up to the rim leaving a gap of 2-3 cm for watering. For bowl and dishes, the potting mixture, consists of an initial thin layer of gravel and few charcoal chips, followed by a compost similar as above.

The selection of subjects for troughs and dishes are to be done with extreme care. Since the growing conditions either in a trough or in a dish is the same, plants liking identical growing conditions are to be planted in each container. For example, a cactus needs sunny and dry growing conditions and, therefore, if this is grouped in the same bowl with a shade and moisture-loving plant such as *Fittonia verschaffeltii* or *Peperomia argyreia* the results will be disastrous.

The second point to be remembered is that the plants are to be arranged in an aesthetic manner. To achieve this plants of various heights should be used, with the tallest one being used as the leader. Each plant in the container should be equally vigorous in growth so that one does not over-run the other. The shy growers should be grouped in a separate container. One should not forget to include one or a few trailing plants such as *Hedera helix*, *Syngonium podophyllum*, *Zebrina pendula*, or *Pilea muscosa* which will hang down the sides of the bowl.

The compost in the dish or the bowl should preferably slope from the centre towards the periphery. One may like to create a forest scene, a lonely tree with bushy undergrowth or an Island scenery with sand and some tropical plants. On the surface of sand, gravel or pebbles may be arranged. Many people arrange ceramic tiles, stones, or stone figures while others keep the landscape untouched. One advantage of mulching with stones is that the top soil is not washed off while watering. The plants are put in the pattern decided in advance. Water is added just enough to moisten the compost, over-watering will create drainage problems.

An alternative to planting direct on the container is to place a thick layer of peat moss at the bottom of the container and over this individual pot-grown plants are placed. These are then packed round with additional damp moss up to the rim of the pots. The rims can also be camouflaged with stone, moss or driftwood. Only large containers can hold potted plants even though the pot sizes may be pretty small. One advantage in growing potted plants in containers is that the individual plants receive compost of their liking, but the environmental conditions such as moisture or sun remains the same. Good-looking posts can be arranged in dishes even without being camouflaged by moss as is generally done in arranging indoor plants.

Even though only indoor plants are grown in dish gardens, they need light which sometimes may be in the diffused form. Sometimes, after planting, the dish garden should be fertilized with weak liquid manure every fortnight or third week in small quantities with a view not to starving the plants and at the same time not to allowing fast growth. As the space is limited it will be necessary to prune the plants occasionally to keep them in check. Dish gardens are a costly affair as plants in the bowls and dishes spent themselves very quickly and replanting may have to be done yearly or even every six months. But the idea of dish gardening is so fascinating that it is worth taking the trouble.

One more novel idea in Dish or Trough Garden is to insert glass or metallic florists' tubes into the compost for display of cut flowers which will enhance the beauty of the garden. A florist's tube is nothing but a narrow hollow tube of glass or metal which holds water to enable the cut stem of the bloom to draw water and thus keep the bloom fresh.

The following plants are suitable for Dish Gardening.

Group A: *Cacti* and *succulents; Agave* (dwarf types), *Aloes*, *Cacti* (*Opuntia*, *Echinocactus*, *Chamaecereus*, etc.), *Crassula argentea*, *Echeveria coccinea*, *E. haageana*, *Gasterias*, *Haworthias*, *Sedum sieboldii*, *S. multiceps*, and *Sempervivum tectorum*.

Group B: For large size arrangement, the following combinations are suggested.

Combination I: Aralia elegantissima (Syn. *Dizygotheca elegantissima*), *Dracaena sanderiana*, *Hedera helix* cv. 'Scutifolia', *Peperomia obtusifolia* var. *variegata*, and *Syngonium podophyllum*.

Combination II: Aphelandra squarrosa, *Dracaena sanderiana*, *Ficus benjamina*, *Fittonia verschaffeltii*, *Peperomia argyreia*, *Pilea muscosa*, and *Zebrina pendula*.

Group C: For small size arrangement *Chlorophytum comosum* 'Vittatum', *Begonia* rex, *Peperomia rotundifolia*, and African violets are recommended.

The principle of gardening in trays is also the same. Any tray which is used for serving will be suitable for tray garden, provided the margins are raised at least about 3-4 cm to accommodate the compost. Alternatively, plants grown in decent-looking pots may be displayed in trays.

VERTICAL GARDEN

This new concept of gardening has developed in Switzerland and is not normally found in Indian gardens. This consists of a wooden frame of thickness varying from 15 cm to 30 cm depending upon convenience. The height and the breadth of the frame varies depending upon the available space. The broader faces are enclosed with wire netting whereas the two vertical sides (i.e., the thickness of the frame) and the bottom are covered with wooden flats. The top is kept open. Sphagnum moss or coarse peat is pressed within the wire frame which serves as the growing medium. It is possible to grow many dwarf flowering and foliage plants in such frames provided the medium is supplied with nutrient and watered regularly (see Fig. XXV.2).

This type of gardening has some advantages. In cities people living in flats have very little space for the conventional type of gardening, but can easily afford to put up a vertical garden. A vertical garden can be shifted from place to place and even used as an ornamental partition in the drawing-room. Since the aeration and the drainage of the medium are perfect, the shallow-rooted plants needing very little anchorage will grow well. The vertical garden shou'd be planted with either sun-loving dwarf and trailing flowering annuals such as Alyssum, pansy, nasturtium, etc., or shade-loving foliage or flowering plants such as Semperflorens begonias, rex or ornamental-leaved begonias, African violets, *Fittonia*, *Peperomia*, *Oxalis*, *Zebrina pendula*, etc., on the broader sides, which are expected to grow horizontally with the tendency to go vertical. On the vertical top also some plants may be planted. A single vertical garden should not have a mixed planting, i.e., a combination of shade-loving and sun-loving plants. Plants are planted on both faces of the frame. The vertical garden is provided with legs on the sides to enable it to stand on its

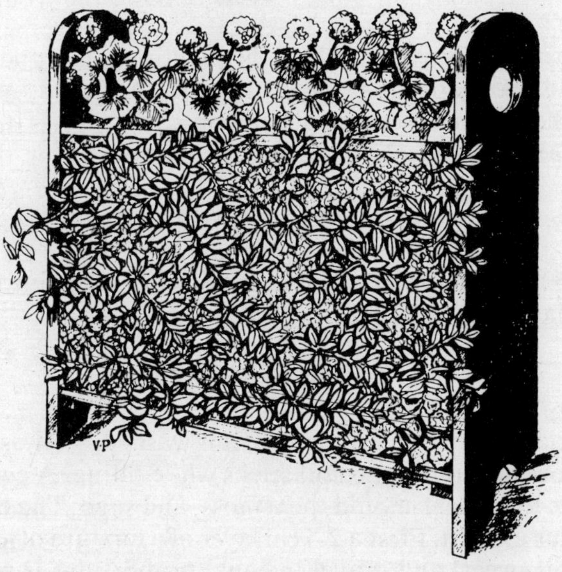

Fig. XXXV. 2. A vertical garden

own. A rot-resistant wood such as teak should be used for the frame and in addition should also be treated with a rot-resistant chemical such as cuprinol.

JAR, BOTTLE, AND TERRARIUM GARDENING

The principles involving these three types of gardening are more or less the same; the only difference is in the shape and style of the container. A jar made of glass having tall sides with or without lid is good for bottle gardening as this conserves moisture. The jar has a wider mouth than a bottle. A bottle should be wide at the bottom with taller sides and a narrow but wide enough (2-3 cm diameter opening) mouth for enabling the plants to be inserted. A terrarium is a glass bowl of 25-30 cm diameter having an air-tight lid. One advantage of bottle is that it can be placed either ver.ically or horizontally on a stand. When placed horizontally it offers more space for landscaping.

Bottle gardens may be called as miniature greenhouse in which liliputian landscapes are planned. Any size of bottle, a jar, or a jug is suitable for gardening provided these are made of clear glass. Bottle gardening is a novel way of growing certain moisture-loving indoor plants, which may be otherwise impossible to grow in a dry house. To a layman, it may become a matter of conjecture, how such large-sized plants could have gone inside through such a narrow mouth. It is as much a curiosity as a "ship in a jug". These types of gardening offer a good display on a table, a stand, or a well-lit window. Besides jars, bottles, and terrariums, any other clear glass container which is large enough to hold plants can be used for this purpose. Even a wide-mouthed wine-glass can grow tiny plants. Fish aquariums can also serve this purpose though this cannot be considered as bottle gardening, rather it should be included in trough or box gardening.

Tools

Certain tools are to be improvised before the planting job is undertaken. These are

(a) *Trowel:* An old tea spoon from the kitchen is shortened by breaking the handle to half and then attached to a 60 cm long bamboo cane.

(b) *Fork-cum-rake:* A narrow old table fork is improvised for the purpose in the same way.

(c) *A Pair of Tongs:* Some metal tongs are cut down and fitted to two 60 mm long bamboo canes.

(d) *Rammer:* A 60 cm bamboo cane is fixed with adhesive in the central hole of cotton reel or any similar material.

Planting Procedures

After the appropriate container is selected, it is washed spotlessly clean inside and outside and dried thoroughly so that no soil sticks while filling. A compost is made of equal parts of good garden loam, leaf-mould, peat moss, and sand. The compost should be sterilized to prevent fungus growth. First a 2.5 cm layer of a mixture of crushed charcoal (which helps to keep the soil sweet) and gravel in equal proportions is placed for drainage. The compost is made slightly damp and poured inside with the hlep of a long-tubed funnel. After the soil is poured it is moulded into the desired landscape with the help of the fork and rammed with the rammer. A sloping compost offers increased display. A 25 *l* bottle needs 5-7 cm drainage material and 10-12 cm of compost. After the soil is ready, holes are made at the desired positions with the help of the trowel. Planting is done from outer periphery.

The rooted cuttings or potted plants are uprooted and the roots are cleaned of the compost, if necessary, by washing. Some excessive growth such as a large leaf may be trimmed off. The inserting of plants and planting need a lot of patience. It is possible even to insert fairly large-sized specimens of *Pilea*, *Dracaena*, and *Sansevieria* which have soft stems. A plant is dropped in carefully, leaf by leaf. Specimens with delicate foliage such as *Fittonia* and rex begonia are wrapped spirally in a tissue paper at the time of insertion and the paper is removed afterwards. After a plant is dropped manoeuvre this with the help of the tongs into the intended hole. The roots are covered with the trowel while the plant is held in position with the tongs. Then the soil around the roots are rammed firmly with the rammer. To bring back a delicate leaf in position, sometimes even a blow through the drinking straw may be helpful. The larger plants at the centre are planted last. It is advisable to drop and fix the plants one by one. To add colour and variety to a bottle garden, coloured pebbles may be arranged in the form of a path, or a shell may be positioned to represent a pool, or even small stone figures of animals may be displayed.

After planting is over every bits of soil, that might be sticking to the sides of the bottles and on the leaves are removed. Since the planting is done on a wet compost no watering is needed, at the time of planting, but to give a good start to the newly planted plants, the leaves should be moistened with a mist of clear water. After this the mouth is closed tightly with a stopper and the bottle is not watered for the next two months. Bottle gardens need very little watering as whatever evaporates, condenses and goes back to the compost. Sometimes, it may not be required to water as long as for a year. An open-mouthed jar needs more frequent watering than a narrow mouthed bottle with stopper. Watering is done drop by drop with the help of a long tube so that there is no over-watering and the soil does not splash on the glass. Misting may be done at intervals of two months. At this time liquid manure may be added. This may be prepared by dissolving 30 g of ammonium sulphate, 30 g potassium sulphate, and 60 g of disodium hydrogen phosphate in 10 *l* of water. About two table spoonfuls of this solution may be sufficient for an average-sized bottle as overfeeding will result in cramped growth.

Bottle garden plants need a reasonable amount of light but direct sunlight will overheat the plants. One good idea is to convert a bottle garden after planting into a table lamp, to ensure proper lighting. But in such a case, the plants should remain accessible for future care. The lamp is to be fitted in the neck of the bottle, with the help of a clamp.

Often it may be necessary to remove a dead leaf or a overgrowth from a plant in the bottle garden. This can be achieved by attaching a broken razor blade to a stick.

Mealy bugs and aphids my sometimes infect bottle plants. A stick with some cotton at one end soaked in alcohol is inserted in the bottle and brushed against the insects.

Chapter XXVI

Diseases and Pests

A garden lover would like to have healthy plants by following proper cultivation practice and keeping the surroundings clean, rather than try to cure them after the appearance of diseases and pests. Regular prophylactic measures such as fungicidal and insecticidal sprays are beneficial and should be followed.

DISEASES

One of the first essential steps in ensuring plant health is to start with good soil. A scientist considers the soil as something which is living rather than dead. A large number of bacteria are present in a good fertile soil which are beneficial to plant growth, whereas a few others may be harmful. Beneficial bacteria can grow and multiply in a soil which is properly drained, well aerated, and contains enough decayed organic matter. Harmful bacteria also flourishes in the soil containing humus, but only when the soil is not properly aerated and is over-saturated with moisture. Similarly, fungi also prefer humid conditions for their growth. Therefore, the first essential step in ensuring the healthy growth of plants is to obtain well-aerated and well-drained soil with abundant organic matter. For this purpose, it is always desirable to plough the soil thoroughly.

In acidic soils lime is useful for maintaining soil health. Most microbial population prefer either an alkaline, a neutral or a slightly acidic soil, but will perish in a highly acidic soil. Though lime prevents some diseases such as club rot or pests such as slugs, it may encourage some other disease such as scab disease of potato. Therefore, while advocating application of lime one should not overdo it. If needed, lime should be applied at the rate of 200-225 g per square metre once in three years. It should be remembered that some plants such as rhododendrons thrive quite well in acidic soils and application of lime may not be necessary.

Water-logging in the soil not only helps proliferation of the growth of harmful micro-organisms but also causes some physiological disorders. With most of herbaceous ornamentals and shrubs the immediate effect of water-logging will be yellowing of the foliage and ultimate shedding of leaves. If proper drainage is not provided the affected plants may soon perish. In extreme cases of water-logging even large trees are killed. Often a person not familiar with symptoms of diseases and other physiological disorders, may consider physiological disorders such as this, as diseases.

Excessive heat, drought, or severe cold often produce symptoms which may be mistaken as disease. Sometimes, the symptoms of drought may appear to be very similar to that of water-logging, i.e., yellowing of foliage and sudden collapse of the whole plant. The scorching rays of the sun may cause the edges of leaves to curl or the edges may turn brown. Sometimes, brown spots may also develop all over the surface of the foliage. Generally, young tender leaves suffer more. Foliage of certain plants such as *Magnolia grandiflora*, which prefer mild climate for their growth, gets scorched in hot summer when grown in the northern plains. Similarly, severe cold and frost also cause damage to the foliage of many plants. The foliage turns blue, yellow, or even brown. In Delhi, during severe cold, the petals of many rose cultivars develop yellowish brown spots. An overnight frost even kills a plant like dahlia or salvia and the whole plant may become black by early morning. Many shrubs or trees also show various symptoms of frost injury and the names of such plants have been included in the respective chapters. These symptoms also should not be confused with disease symptoms.

Another physiological disorder which produces disease-like symptoms in plants is caused due to deficiency of one or more plant nutrients. The details about the various symptoms produced as a result of deficiency of plant nutrients have been discussed in the chapter on commercial floriculture. Such physiological disorders are sometimes known as deficiency diseases. With experience one may be able to distinguish between nutrient deficiency symptoms and real disease.

Symptoms of Diseases

The diseases are caused by bacteria, fungi, viruses, and microplasma like organisms. General symptoms caused by the first three are mentioned below.

Fungi: Fungal diseases are very well recognized by the prominent and visible external symptoms they produce. Rust diseases in many plants are easily recognized by the characteristic rusty spots and pustules on either surface of leaves. These symptoms can be easily distinguished from other causes. Magnifying glass is useful in observing the spore-bearing structures protruding outside the leaf surface. Mildew disease fungi produce powdery coating of spores and mycelia on the surface of leaves, stems, and fruit. Black spots produced by fungi can be distinguished from other spots upon examination by lens which will show the presence of fungi.

Bacteria: The diagnosis of diseases produced by bacteria is not so easy as these do not produce any external growth and usually remain confined within the plant tissue. The general symptom is often the rotting of mass of tissues or sudden wilting and collapse of the plant.

The sure way of detecting the presence of a particular fungus or bacterium is to isolate in pure culture in the laboratory from the affected part.

Viruses: The symptoms produced by viruses are rather different. Bos (1963) of the Netherlands has described virus symptoms in detail. His book *Symptoms of Virus Diseases in Plants* may be consulted. Local virologists may also be consulted. Viruses which exist in the sap of plants are very minute. Viruses can be detected only by an electron microscope. Some of the symptoms produced by viruses resemble physiological disorders. Common symptoms are curling and distortion of leaves, mottling, yellowing, and dry spotting. Grafting of the diseased tissue on a virus-free plant is one of the techniques to detect viruses. Symptoms are transmitted to virus-free plants if these were due to viruses.

PESTS

Any creature belonging to the animal kingdom and causing damage to crop plants or ornamentals is considered as a "pest". Rabbits gnaw the tender bark of many ornamentals at ground level and eat away tops of tender ornamental plants. Similarly, sparrows are great enemies of the tender seedlings of sweet peas. Therefore, these are considered as pests in the garden. Snails and slugs also cause damage to plants and are considered as important pests in a garden. The majority of the pests, however, belong to the group "Insects".

Damages caused by pests are varied. If any above-ground part of a plant shows some irregular holes or cuts, they could be due to caterpillar attack. Other insects such as the leaf-cutting bee or chaffer bettle of rose also eat foliage and cause cuts and holes.

Another group of pests such as aphids and thrips do not cause any conspicuous damage visible to the naked eye, but they make minute punctures in foliage and tender stems and suck the sap. As a result of their activity, tender shoots get deformed and leaves curl up. Another pest, the red spider mite, which is actually a mite and not a spider, also feeds on sap. While the aphids are easily visible to the naked eye, the thrips and the mites are of the size of a pinhead or like a minute dot. For proper identification of these creatures the help of a lens is necessary.

There are other groups of pests which are soil dwellers and include caterpillars, grubs and others. If any seedling or plant breaks at ground level, it may be either due to damping-off or collar rot at ground level by fungi or as a result of insect damage. If the toppling is due to insect attack, the tissues at both the cut ends are clean in nature, whereas fungal-damaged tissues are rotten and soft.

Borers bore through the stem and branches. One such example is the digger wasp of roses which makes tunnels through the cut ends of the pruned stem. But a great majoriy of the borers belong to the groups of some beetles and weevils. With experience a gardener may be able to recognise and differentiate between the various damages.

GENERAL CONTROL MEASURES

(a) *Diseases:* It is more economical to keep plants healthy than trying to cure plants after attack of diseases. There are several methods to keep plants healthy. Some diseases develop in damp wet soil and hence it is advisable to maintain proper drainage and regulate irrigation to check such diseases. To prevent damping-off of seedlings, seed treatment with appropriate fungicides and thin sowing is recommended. Sterilization of nursery soil by heat or chemicals (e.g., formalin 2 per cent) is effective in controlling many diseases of seedlings. Some soil-borne diseases can also be checked to a great extent by soaking or drenching the soil with fungicides such as blitox, brassicol, captan, bavistin, etc. Various fungicides are also sprayed and dusted on the plants to prevent the onset of diseases or check their growth. One important point to remember is that no fungicide should be used arbittrarily without properly identifying the casual organism. As for example, if diseases caused by *Phytophthopa* are treated wrongly by benlate or bavistin and other related fungicides because of wrong diagnosis, the disease may increase as these fungicides are not effective against *Phycomycetes*.

Crop rotation is another method by which some diseases can be controlled. Several fungi are specific for their hosts. For example, a particular fungi can attack only gladiolus and not carnation. Therefore, if a land growing gladiolus for one or two years is either kept fallow or some other crop is grown for some years the disease/diseases can be eliminated as pathogonic fungi cannot survive in the absence of desired hosts. This method also has a limitation as spores of some fungi (e.g., *Stromatinia gladioli*) are reported to have survived in the soil as long as 10 years or more, even without a host. Clean cultivation is another way of checking the severity of many fungal diseases. In Chrysanthemum, many of the foliar and floral diseases can be kept in check if all the infected plant parts are collected and destroyed.

The best method for disease control is to develop varieties which are resistant to various diseases. Cultivars resistant to severe wilt disease of China Aster have been evolved. Similarly, there are types of Zinnia which are resistant to the leaf curl virus. In antirrhinum, cultivars resistant to rust have been developed through breeding.

The virus and bacterial diseases are generally considered to be incurable as they cannot be easily cured with available chemicals. Viruses spread from diseased to healthy plants by insect vectors or through the garden tools (e.g., budding or pruning knives). It is advisable that while working with diseased plants the hands or tools should be washed with disinfectant solutions such as Lysol or weak carbolic acid solutions. It is also suggested that since while taking cuttings from virus infected chrysanthemum or dahlia with knife or razor blade, virus may spread through these contaminated implements from the diseased to new healthy plants; therefore, it is better to snap the cuttings from the mother plants with a gentle downward pull of the hand. This method will avoid spread of the disease. The trouble with virus is that many cultivars are symptomless carriers. In such cases the cultivars will carry the virus but will not show any apparent symptoms. There are some dahlia cultivars which are symptomless carriers of viruses, and they are potential danger to the healthy plants. It is possible to check the spread of the disease by controlling the insect vector. For example, spotted wilt of dahlia is spread by thrips and the recommended spray schedule for thrips can check the spread of virus.

Some other methods are also used in controlling viruses on a limited scale. These methods are heat treatment, use of chemotherapeutant chemicals, and cross-protection.

The use of heat, either in the form of hot air or hot water, has been used successfully in inactivating viruses in living plants and curing virus-infected plants. There are several instances of curing viruses in this way. Kunkel (1936) was successful in eliminating aster yellows virus from diseased plants of *Nicotiana rustica* and *Vinca rosea* by placing the plants in a hot room (38°C) for a fortnight.

The possibility of using chemotherapeutant chemicals for the control of virus disease was demonstrated by Stoddard (1942), who was said to have inactivated the X-virus of peach by soaking the infected buds in solutions of quinhydrone, urea, and.sodium thiosulphate. The other chemicals used are 2-thiouracil against multiplication of tobacco mosaic virus and 8-azaguanine against several viruses.

The inoculation of a host plant with a mild strain of virus has been found to protect it against the attack of a virulent strain of the same virus. This method is also referred to as cross-protection.

(b) *Pests:* There are several ways of tackling pest attack. One general rule is to follow clean cultural practices. This includes weed-free fields and keeping the surroundings clean by removing plant debris so that the pests do not get any hiding place. It is sometimes observed that overcrowded plants or plants growing in shade harbour some insects, the infestation of which could have been avoided if the plants were in full sun. The infestation of scales in cycus plants is more when these are in shade or partial shade compared to plants grown in the open. A pest like mite does not like wet conditions and if plants are sprayed with jet of water the menace of mite can be reduced to a great extent.

The method of controlling pests by using one living being against the other is known as "Biological Control". The growing of marigolds around fields affected with nematodes is said to be effective in reducing nematode population. There are other methods of biological control. These methods consist of using one living being in destroying the other which is a pest. At the Indian Institute of Horticultural Research, Bangalore, scientists have used effectively one predatory beetle (*Crytolaemus montronzuri*) in destroying mealy bugs in citrus and coffee. Similarly, bacteria and viruses are also used as parasites on pests.

Some insects may multiply under specific optimum weather conditions. It may be possible to avoid growing a susceptible crop under such climatic conditions. Such practice may avoid the attack of pests.

The above-mentioned methods may not be enough. Therefore, often the use of insecticides and pesticides may be necessary. It is possible to handpick the insects in the case of ornamental plants on a small scale and destroy them. It may be easy to destroy insects mechanically without resorting to chemical spraying or dusting when a few plants are grown as house plants. However, on a large scale in the field, in the greenhouse or elsewhere preventive and curative applications of pesticides are required. While applying pesticides one should be acquainted with the mode of action of different pests. Certain group of insects which are called "biting pests" such as caterpillars cause damage by biting plant parts. Such insects are killed by spraying stomach poisons which enter their body as they feed upon pesticide sprayed over plant parts. Sucking insects such as greenflies and thrips feed on plant sap by making minute holes. Such pests can be killed by "systemic" insecticides which enter the system of the plant when either sprayed on the foliage or applied to the soil. The contact insecticides kill insects when these come in contact with such insecticides. Pests such as slugs, snails, and rabbits are killed by using poisoned baits. Several chemicals do not actually kill the pests but drive them out when used as fumigants. These groups of chemicals include para-dichlorbenzene and naphthalene. Some pesticides act both as contact and stomach poison. Contact insecticides are in one way bad as they kill harmful as well as beneficial insects.

CONCLUSIONS

From the above discussions it is evident that the methods of control of diseases can be divided into several categories, as follows:

(a) *Eradication:* This consists of eradicating the disease by means of (i) crop rotation, (ii) elimination of alternate hosts, (iii) removal of infected plant parts, (iv) eradication of weeds and wild hosts, and (v) removal of virus-infected plants (rogueing). The last method consists of removal and destruction of affected plants especially in case of viruses.

(b) *Soil Sterilization:* Soil sterilization is done to eliminate soil-born diseases. This can be achieved by either of the following methods: (i) exposing the soil by repeated digging and harrowing to summer heat, (ii) burning straw or any other combustible material on the surface of the ground, thus raising the temperature of the soil, (iii) heating the soil electrically, (iv) heating of small amounts of soils in pans over fire, (v) heating soil with steam, conveyed through underground perforated pipes, and (vi) application of chemicals such as formalin at low concentrations, which are harmful to fungi and other organism.

(c) *Seed Treatment:* Many of the seed-born diseases can be eliminated or checked to a great extent by sowing treated seeds, either by heat or chemicals.

(d) *Fungicides:* Application of fungicides and antibiotics (e.g., penicillin, streptomycin) to control diseases are well known. Fungicides are applied as prophylactic spray to prevent disease attack or as curative sprays to eradicate the disease already established.

(e) *Resistant Varieties:* The ideal method to control the disease is to obtain disease-resistant varieties through breeding.

(f) *Quarantine:* Quarantine or legislative measures is another method, which has not been mentioned so far, in checking the spread of disease. This includes compulsory inspection, quarantine and embargo of plant materials or seeds brought from abroad, to check the spread of disease, especially those which are not prevalent in India. If a plant material brought from abroad is infected with pathogens which are not prevalent in India, the material is destroyed by the quarantine authorities.

PRECAUTIONS

Certain precautions should be taken in handling and spraying insecticides and fungicides as these are highly toxic in nature. Insecticides such as Chlordane, Dimethoate, Endosulfan, and fungicides such as Benlate can be absorbed directly through the skin in harmful quantities. While spraying such chemicals one should use protective clothing, gloves, and respiratory devices. In a windy day spraying or dusting should be done along the direction of the wind and the person should stand away so that the drifts of spray or dust do not fall over the body. Some insecticides are toxic to certain plants. Malathion may cause leaf injury to ferns, petunias, and lantana. Nuvacron may be toxic to tender foliage of roses which show signs of scorching. Spomer and Langhans (1965) reported injuries caused to tulip and hyacinth bulbs in storage as a result of application of *para*-dichlorbenzene, a repellant, instead of naphthalene. The fungicide Captan, when applied to germinating petunia seedling was reported to have killed 95 per cent of the seedlings (Bartos, 1964). Carnation seedlings are also killed due to toxicity of copper-based fungicides, especially when the seedlings have mechanical injury at the time of application (Shiegeta).

Other precautions include washing of hands with soap and water. It is advisable to keep the recommended antidotes handy for use in case of accidental swallowing. All insecticides and fungicides should be kept under lock and key and away from the reach of children. All the used-up cans and containers should be destroyed or buried, and should never be re-used for any other purpose.

The names of various diseases and pests attacking the ornamental plants and their possible control measures are given in Tables I and II. In the case of the insecticides Nauvacron and Rogcr, the concentrations mentioned in the table are in terms of the active ingredients. But the dilution of the commercial preparation of Nuvacron will be 1 ml per litre, and in Rogor it will be 1-2 ml per litre.

Table I

Diseases of ornamental plants and control measures

Name of disease (1)	Control measures* (2)
Alternaria dianthi Blight or leaf spot of carnation	Spray with Dithane Z-78 (0.2 per cent) Dithane M-45 (0.2 per cent), Blitox (0.2-0.3 per cent), and Captan (0.2 per cent).
Aphanomyces euteiches Root-rot of sweet pea and *Viola odorata*	Caused due to damp soil condition. Remedial measures consist of well-drained soil and long crop rotation. Not common in India.
Bacterium fascians Leafy gall of *Pelargoniums*	All the affected plants should be destroyed as the disease may affect Carnation, Chrysanthemum, dahlia, gladiolus, petunia, sweet pea, etc.
Bacterial Wilt	Long crop rotation with plants not susceptible; breeding of resistant varieties.
Black leg of *Pelargonium*	Caused by a species of *Pythium* due to damp soil conditions. Follow proper drainage, drench soil with Blitox (0.3 per cent).
Blight	See *Alternaria dianthi*.
Blight (general)	
(a) *Pseudomonas syringae* Bacterial Blight of Lilac	Prune young diseased and dead stems, and spray with Bordeaux mixture (5:4:50) to prevent secondary fungal attack.
(b) *Botrytis* (Fungal blight)	Dust with Bordeaux mixture (5 : 4 : 50) spray with Dithane Z-78 (0.2 per cent).
(c) *Cercospora* (Fungal blight)	Spray with Benlate (0.1 per cent), Bavistin (0.1 per cent) and Bordeaux mixture (5 : 4 : 50).
(d) *Phytophthora* (Fungal blight)	Soil sterilization, spray with Dithane Z-78 (0.2 per cent) and Bordeaux (5 : 4 : 50).
Botrytis Bulb rot of lilies	Destroy affected plants, soak bulbs for 30 minutes in 0.1 per cent Benlate solution before planting and storing and use disease-free planting material.

*Spray only one of the suggested fungicides, unless otherwise rccommended.

(1)	(2)
Botrytis cinerea Petal blight	Fortnightly spray with Dithane Z-78 (0.2 per cent).
Botrytis gladiolorum	A leaf disease of gladiolus which also affects stem, flowers and corms. Affected plants should be uprooted and burnt. Benlate spray (0.1 per cent) at fortnightly interval for prevention; soaking corms for 30 minutes with Benlate (0.1 per cent) or Captan (0.2 per cent) solution before planting and storing.
Crown rot (*Thielaviopsis basicola*)	The affected parts should be destroyed. Dust with a good copper-lime formulation. (Bordeaux)
Curvularia prasadi Leaf disease of jasmine	Spray with Difolatan (0.3 per cent) and Dithane Z-78 (0.2 per cent).
Damping-off (*Pythium and Phytophthora*), Collar and Foot rot.	A group of diseases primarily of seedlings in the nursery. Control measures include sterilization of nursery soil, seed treatment with Captan (2 g/kg), Ceresan (2 g/kg) and Thiram (3 g/kg), drenching soil with Blitox (0.2-0.3 per cent) and Dithane M-45 (0.2 per cent) or Cheshunt compound* (28.350 g dissolved in 9 *l* of water).
Die-back of pines	Spray with Bordeaux mixture (5 : 5 : 50), Dithane Z-78 or M-45 (0.2 per cent) or Bavistin (0.1 per cent). Prune and burn affected twigs.
Die-back of roses (*Diplodia rosarum* often associated by a *Colletotrichum*)	Apply a paste composed of 4 parts copper carbonate, 4 parts red lead, and 5 parts linseed oil with 0.1 per cent BHC and DDT at pruned end of branch.
Diplocarpon rosae Black leaf spot of rose	Spray with Captan (0.1 to 0.2 per cent), Benlate (0.05 per cent), Bavistin (0.05 per cent), at 7-14 days intervals. Collect diseased leaves and burn.
Downy mildew	Spray with Dithane M-45 (0.2 per cent), Difolatan (0.3 per cent), Blitox (0.2 per cent), Fycol 8E (1.0 per cent) and Ridomil (0.1 per cent).
Dropsy	See "Oedema".
Dry rot by bacteria in succulents	Clean cultivation. Avoid over-watering.

(*Contd.*)

*For preparation of cheshunt compound see page 561.

(1)	(2)
Entyloma dahliae Leaf smut of Dahlia	Use disease-free planting material. Fortnightly application of Blitox (0.2 per cent) or Bordeaux mixture (5 : 4 : 50).
Erysiphe cichoracearum Powdery mildew of chrysanthemum—	Spray with Calixin (0.05 per cent) and Karathane (0.2 per cent).
E. polygoni Powdery mildew of sweet pea	Spray with Sulfex (0.3 per cent) and Benlate (0.1 per cent.)
Fomes igniarius Bracket Fungi of Salix	Affected portions should be removed and burnt and the wound treated with a disinfectant such as Lysol or dilute carbolic acid. Severely affected trees cannot be saved and should be destroyed.
Foot rot (*Fusarium*)	Seed treatment with Bavistin (2-3 g/kg) soil drench with Benlate (0.1 per cent) and Blitox (0.2 per cent). Prevalent in heavy damp soil which should be avoided.
Foot rot (*Phytophthora cryptogea* and others)	Soil sterilization, seed dressing with Ceresan (2 g/kg) and Captan (2 g/kg), drench soil with Blitox (0.2-0.3 per cent), Difolatan (0.2-0.3 per cent) or Dithane M-45 (0.2 per cent).
Fusarium dianthi Wilt of carnation	Seed treatment with Bavistin (1-2 g/kg), drench soil with Blitox (0.2-0.3 per cent) or Benlate (0.1 per cent).
F. oxysporum Stem rot and wilt of Chrysanthemum	Soil fumigation with Methyl bromide and Methyl bromide-chloropicrin mixtures. Use disease-free planting material and avoid planting in infected soil. The disease proliferates in warm wet weather.
F. oxysporum f. sp. *gladioli*	Same as in chrysanthemum. The fumigants are toxic and great care should be taken during fumigation.
F. roseum Rot of carnation	Same as in *F. dianthi*. Cuttings free of wilt pathogens should be used.
Gymnosporangium Rust of junipers	Not common in India. Affected parts are removed and burnt. See *Puccinia* and *Rust*.

(1)	(2)
Leaf spot	See *Septoria*.
"Oedema"	Also known as "Dropsy". A physiological disorder of *Pelargoniums* caused as a result of excess watering. Remedial measure consists in judicial watering.
Oidium Powdery mildew of dahlia	Sulphur dust, or spray with wettable sulphur (0.3 per cent), Karathane (0.2 per cent) or Benlate (0.1 per cent).
Oidium hortensiae Powdery mildew of hydrangea	As above.
Peronospora viciae	See downy mildew.
Pestalozia hartiggi Crown Gall, Stem Canker of *Cupressus and Pines*	Remove and burn affected parts. Spray the affected parts with Blitox (0.3 per cent) after removing the canker.
Phragmidium mucronatum Rust of rose	Remove and burn all the diseased portions. Spray with Dithane Z-78 (0.2 per cent).
Phytophthora (General)	Soil sterilization, seed dressing with Ceresan (2 g/kg), Soil drenching with Blitox (0.2 per cent), Dithane M-45 (0.2 per cent), or Cheshunt compound. Avoid thick sowing and over-watering.
Phytophthora cactorum Damping-off in cactus	Dissolve 28.350 g of Cheshunt compound in 9 *l* of water and pour this over the affected plants. For details see pp. 560-61.
Phytophthora parasitica Crown and Root rot of gerbera	Excessive moisture in soil favours the disease. The affected plants are uprooted and burnt. Drench soil with Blitox (0.2-0.3 per cent), Dithane M-45 (0.2 per cent) or Cheshunt compound.
Polyporus sulphureus Bracket Fungi of Salix	Badly affected plants cannot be saved and should be destroyed.
Powdery Mildew (General)	Spray with Benlate (0.1 per cent), Calixin (0.05 per cent), Karathane L.C. (0.05 per cent), Sulfex (0.2-0.3 per cent) or with any other wettable sulphur.

(*Contd.*)

(1)	(2)
Pseudomonas caryophylli Bacterial wilt of carnation	Crop rotation, use of disease-free propagation material and destruction of affected plants.
Puccinia chrysanthemi Rust of chrysanthemum	Spray Dithane Z-78 (0.2 per cent), Dithane M-45 (0.2 per cent) Vitavax DMOC (0.2 per cent) or 0.2 per cent Plantavax (Oxycarboxin).
Pythium (General)	The fungus thrives best in damp and wet soil conditions and hence proper drainage should be arranged and excess watering avoided. Seed treatment with Ceresan dry (2 g/kg), Damosan (2 g/kg) or Captan (2 g/kg) is advised. Drench soil with Blitox (0.2 per cent) or Dithane M-45 (0.2 per cent). But copper fungicides (e.g. Blitox) may prove toxic to some seedlings such as carnation.
Rhizoctonia solani Rot or wilt of carnation	Seed treatment with Ceresan dry (2 g/kg), Captan (2 g/kg), Thiram (2 g/kg), Benlate (2 g/kg) or Bavistin (2 g/kg). Use disease-free cuttings. Drench soil with Brassicol (0.2 per cent), before sowing.
Rhizopus necans Dormant bulb rot of lily	Lily bulbs are often attacked in transit or storage by this fungus as a result of damp or poorly ventilated conditions. Bulbs may be soaked in 0.1 per cent Benlate for 30 minutes after harvest and before planting.
Root rot (Black), of Sweet pea and Primula (*Thielaviopsis basicola*)	The disease results due to heavy damp soil, which should be rectified. Seeds should be sown in sterilized soil. The affected plants are removed and burnt. Other measures include crop rotation and use of resistant varieties.
Root rot (General)	All nursery soil should be sterilized with steam or formalin. In porous, well-drained and well-aerated soil occurrence of root rot is less. Affected plants should be uprooted and burnt and the vacated soil disinfected with 2 per cent formalin, but formalin should not come in contact with healthy plants.

(1)	(2)
Rust (General)	Many ornamentals, especially annuals, are infected with rust. The symptoms are easily recognizable by the appearance of rust-coloured spots, principally on the leaves. The badly infected leaves are removed and burnt. Spray with Dithane Z-78 (0.2 per cent), Dikar (0.3 per cent), Difolatan (0.3 per cent), Dithane M-22 (0.2 per cent), or wettable sulphur such as Sulfex or Thiovit (0.2 per cent) and dusting with sulphur is recommended. Prophylactic spray with Blitox (0.2 per cent) is recommended.
Sclerotium rolfsi Stem rot of dahlia	Treat tubers, in Benlate solution (0.1 per cent) for 30 minutes after harvest and before planting. Treated tubers are dried before storing. Drench soil with Brassicol (0.2 per cent).
Sclerotinia sclerotiorum Stem rot of dahlia and marigold	Seed treatment with Bavistin (2 g/kg), Benlate (2 g/kg), or Ceresan (2 g/kg). Drench soil with Brassicol (0.2 per cent).
Septoria dianthi Leaf spot of carnation	Affected plants should be uprooted and burnt. Spray with Difolatan (0.3 per cent), Bavistin (0.1 per cent), Benlate (0.1 per cent), or Blitox (0.2 per cent).
S. hydrangeae Leaf spot of hydrangea	Spray with Dithane Z-78 (0.2 per cent), Difolatan (0.3 per cent), or Bordeaux mixture (5: 4: 50).
Sphaerotheca pannosa Powdery mildew of rose	Spray with wettable sulphur (0.2 per cent) and repeat at weekly intervals if needed. There are resistant cultivars. Sulphur may cause injury if atmospheric temperature is above 32°C. Karathane (0.05 per cent) spray is also effective.
Smut	See *Entyloma* and *Ustilago*.
Stem Rot (Collar Rot)	This may be caused by *Phytophthora* (foot rot), *Rhizoctonia*, *Sclerotinia*, and damping-off fungi. For control measures see respective headings.
Thielaviopsis basicola	See Root rot, Black and Crown Rot.
Uromyces lilii Rust of lilies	Spray with Dikar (0.3 per cent) at 7-15 days' intervals, depending upon the severity of infection.

(Contd.)

(1)	(2)
Ustilago violacea Smut of *Dianthus*	Seed treatment with Thiram (3 g/kg) and Ceresan (2 g/kg). Affected plants should be uprooted and burnt. Crop rotation with plants not infected with smut is recommended.
Wilt (a) Bacterial	Crop rotation.
(b) Fungal	Generally caused by *Fusarium* and *Verticillium*. See respective headings.
(c) Viral	Virus also cause wilt of foliage or plants in dahlia, petunia, nasturtium, cineraria etc. See virus.
Verticillium Wilt of phlox	Soil should be sterilized with steam or formalin (2 per cent). Infected plants should be uprooted and burnt. Soil drainage has to be improved. Wherever possible soil should be changed or crop rotation should be followed. Good cultivation should always be given to improve soil aeration and drainage.
Virus	Virus diseases are incurable and hence the infected plants should be destroyed. It is now possible to destroy some viruses by heat treatment without injuring the plant. But this is a very specialized job and can be handled by professionals only. The research findings of Stoddard has opened the possibility of the use of chemotherapeutant chemicals for control of virus. For example, 2-thiouracil has been found to be effective in checking multiplication of tobacco mosaic virus and 8-azaguanine against many viruses including tobacco mosaic virus. The viruses live in the sap of plants and these are spread from diseased to healthy plants by such sucking pest as aphids, thrips, capsid bugs, and mites. Control of these pests will lessen the liklihood of virus infection.

SOME IMPORTANT DISEASES OF ORNAMENTAL PLANTS

Cacti and Succulents

Cacti: Most of the cacti can be spared from the attack of serious diseases provided over-watering is avoided and plants are sheltered from rain. Sometimes, a damping-off disease is caused to young seedlings by the fungus *Phytophthora cactorum*. This happens mostly as a result of over-crowding and over-watering. The disease can be prevented by (a) sterilizing the compost, (b) clean cultivation, (c) checking over-watering, and (d) avoiding overcrowding of the seedlings. Drenching of cacti compost with cheshunt compound and pouring this over affected plants go a long way in controlling this disease. The cheshunt compound is prepared by thoroughly mixing 2 parts of finely-powdered copper sulphate and 11 parts of finely-powdered ammonium carbonate, and storing this in a tightly stoppered vessel (not of zinc or tin). The mixture should not be used for the first 24 hours. One ounce (28.350 g) of this mixture is dissolved in 2 gallons (about 9 *l*) water and poured over the affected plants. It is easier to dissolve the mixture first in a little hot water and then the volume is made up with cold water. The solution does not injure plants but controls damping off effectively.

Rhizoctonia and *Fusarium* may cause rot in some isolated cacti. Unless detected at early stages, the affected plants may perish. Prevention of the diseases is much easier than control. Among the preventive measures, clean cultivation and controlled watering are both important. Drenching the soil and spraying the plants with Blitox (0.2 per cent) and Brassicol (0.2 per cent) alternatively once a fortnight is recommended. If only a portion of a large specimen is affected, the diseased portion may be cut off with a sharp knife and the exposed tissue treated with 0.2 per cent Blitox solution. If it is not possible to save a specimen from the disease, a portion of a healthy tip, if available, should be grafted over a suitable stock to perpetuate the variety.

Succulents: Succulents are not so delicate plants as far as disease attack is concerned, provided sanitation in cultivation is followed and over-watering is avoided. But occasionally a few succulents, especially some members of the mesembrianthemum group, are attacked by a dry rot caused by certain bacteria. The affected plants start shrivelling all of a sudden and dry up or collapse. No effective remedial measure is known. It may be desirable to cut off the affected portion or destroy the plant altogether to save other healthy plants.

Carnation

Carnation is affected by a number of fungal, bacterial, and virus diseases. The virus diseases can be prevented by using virus-free planting material and controlling the insect vector. Virus-affected plants should be uprooted and destroyed.

The bacterial and *Fusarium* wilts caused by *Pseudomonas caryophylli* and *Fusarium dianthi* respectively, can play havoc with carnation. Both the diseases are vascular in nature. The symptoms produced by both are remarkably similar. The early symptom is wilting of a stray branch on one side of the plant. The foliage in the affected branch becomes greyish-green, then turns yellow, and ultimately becomes straw-coloured when the branch and foliage are dead. Under hot climate, the disease spreads rapidly and may kill plants extensively. One way of controlling the disease is to sterilize the soil by fumigation or chemicals and the use of disease free propagation stock. Affected plants should be destroyed.

Fusarium rot caused by *Fusarium roseum* is also very common. Warm and humid weather are favourable for disease development. Therefore, under these conditions affected plants collapse and die. The preventive measures consists of soil fumigation, controlled watering, and proper soil drainage.

Rhizoctonia solani and *Sclerotium rolfsii* both cause rotting of roots and stem. The affected plants die. *Phytophthora* causes severe damage to carnation seedlings under moderate climate prevailing in Bangalore.

Tiny purple spots on leaves are caused by *Alternaria dianthi* which is commonly known as blight or leaf spots. In severe cases the disease spreads to stems and flowers.

The application of copper-based fungicides causes toxicity to young carnation seedlings at the time of transplanting. Severe toxicity resulting in massive mortality has been observed by the authors at the research farm of the Indian Institute of Horticultural Research, Hessaraghatta, Bangalore.

China Aster

Damping-off is a serious disease in China Aster. It is advisable to sow the seeds treated with Ceresan dry in sterilized soil. Seeds should be sown thinly and over-watering should be avoided. Fusarium wilt is another serious disease of aster. Drenching the soil with Blitox (0.2 per cent) and Benlate (0.1 per cent) is helpful in checking the disease. Fortunately, cultivars resistant to wilt disease have now been developed for cultivation.

Chrysanthemum

Chrysanthemum is affected by a serious soil-borne disease caused by *Fusarium oxysporum*. The foliage of the affected plants turns yellow, the internal vascular tissues of the stem and roots turn brown and the plant collapse. The disease is favoured by warm and wet weather. The best method is to avoid planting in infested soil. It is important to use disease-free planting material only. A soil fumigation with a mixture of methyl bromide and methyl bromide chloropicrin is very effective in not only controlling fungi but also nematodes. The fumigated soil should be covered with plastic sheets for 24-48 hours. Crop rotation is another method to check this disease.

The other diseases are *Botrytis* petal blight caused by *Botrytis cinerea*, which is favoured by cool and moist conditions. The fungus produces a grey mould on the stems, foliage, and flowers.

Chrysanthemum rust, caused by *Puccinia chrysanthemi*, manifests in the form of small pustules or blisters on the stems and the underside of leaves. Under-surfaces of the leaves get completely covered with rusty-brown spores when the pustules burst.

Erysiphe cichoracearum occasionally affects chrysanthemum causing the disease commonly known as powdery mildew. This can be recognized by the white powdery fungal growth present on stems and leaves. The affected plants do not flower properly.

Virus diseases may also occasionally invade chrysanthemum field. The symptoms generally appear on the foliage which may curl or show mosaics.

The severity of leaf and floral diseases can be minimized by removing all plant residues from the cultivated area and destroying them. If plants are not watered in the late afternoon and night, these diseases can be reduced to a considerable extent.

Cupressus

The common root-rot disease of conifers also affects species of *Cupressus*. To prevent the occurrence of this disease the plants should be grown in well-drained soil and irrigation should be given in a controlled manner. Young plants sometimes suffer from crown gall and stem canker caused by *Pestalozia hartigii*, which causes swollen canker in the stems.

In general, badly affected trees are uprooted and destroyed, although in some cases the canker may be removed and the plants sprayed with bordeaux mixture (5: 4: 50) or blitox.

Dahlia

Dahlia is attacked by a number of diseases, some of which are quite serious. Possibly virus is by far the most serious disease. There are three types of virus disease in Dahlia. The "spotted wilt", also known as "tomato spotted wilt", appears in the form of ringed or spotted areas on the leaves, which are lighter in colour than the leaf proper. The second type of virus disease is known as "dahlia mosaic" where the symptoms are yellow-green banding of the vein and in extreme cases these coalesce and form into mosaic. This disease is transmitted by the greenfly. This disease is also known as stunt, as the affected plants develop bushy growth and remain stunted. The third type of virus disease is "cucumber mosaic" which is less severe than the above two and is transmitted by black fly and certain aphids. The spotted wilt is generally transmitted by thrips. Controlling the insect vectors will check the spread of virus diseases. Sometimes, the dahlias are symptomless carriers of virus diseases. Different cultivars infected by virus grow almost normally and flower quite satisfactorily. But some other cultivars fail to grow properly and flower poorly. It is advised to grow only those cultivars which in spite of attack by the virus grow satisfactorily and flower almost normally. The susceptible ones should be uprooted and burnt. It is not possible to cure plants affected with stunt virus. The virus persists in the tubers and hence it is imperative to uproot and burn the affected plants. Only disease-free materials should be used for propagation to keep the dahlias free of virus.

Powdery mildew may also become quite serious in dahlias which is caused by a species of the fungus *Oidium*. The leaves and young stems at first develop circular spots which gradually increase in size and ultimately the whole leaf gets covered with a white powdery growth. Generally, warm humid conditions favour the growth of this disease. The plants should be dusted fortnightly with sulphur dust or sprayed with wettable sulphur. In hot weather, fungicides containing sulphur may damage the flowers or young foliage.

Stem rot, caused by the fungi *Sclerotinia sclerotiorum* and S. *rolfsii*, is a serious disease which causes rot in the stem and as a result the plants suddenly collapse. Both are soil-borne diseases. To avoid the occurrence of this disease it is advised to change the dahlia growing area. A wide spacing and good drainage will help keep the disease in check. Affected plants should be uprooted and burnt. For chemical control consult Table I given earlier.

Ferns

The first step to ensure healthy growth in a fernery is to sterilize all planting materials such as pots, compost, and crocks thoroughly. If this is done properly, the majority of the ferns will remain free from diseases. Over-watering may cause rotting of rhizome as a result of fungal growth. To prevent this the compost used should be very porous in nature and the pots or beds should have enough drainage materials.

Flowering Annuals

(1) *Damping-off:* It is not a single disease but a group of closely allied diseases all of which affect seedlings and sometimes even grown-up plants at ground level, and the ailment

is hence, also known as foot rot or collar rot. The fungi causing this are *Pythium* sp. and *Phytophthora cryptogea*. Affected seedlings suddenly collapse in large numbers in broad patches. The fungi causing the disease either remain in the soil or seed or carried by irrigation water. The conditions favourable for the spread of the disease is infested soil, thick sowing resulting in over-crowding of seedlings, excessive moisture in the soil, bad soil aeration, and insufficient light. The above conditions are to be avoided to prevent the occurrence of the disease. One simple way of reducing the chances of damping-off disease is to sprinkle sharp sand particles or charcoal powder on the surface of seed pans or seed beds. The preventive measures consist of using fresh soil or steam sterilized soil. Soil can also be sterilized with formalin (2 per cent). The soil is well soaked with formalin and covered with polythene or tarpaulin for 48 hours. Then the formalin is allowed to evaporate for about 7-10 days and sowing is done when there is no smell of formalin left in the soil. The old seed pans should also be cleaned with soda water or steam sterilized. Seed dressing should be done with chemicals such as Ceresan dry. If in spite of these precautions seedlings are affected with damping-off, the affected seedlings should be removed with little surrounding soil and the rest of the seedlings watered with Cheshunt compound or Blitox. Copper fungicide could be toxic to certain seedlings.

(2) *Leafspot and Blight:* These are common diseases of, chrysanthemum and some other annuals, caused by some fungi. The symptoms are necrotic spots on the leaves. Fortnightly spray of Blitox or any other copper fungicide is helpful. The carnations are attacked with leaf blight caused by two fungi, *Septoria dianthi* and *Alternaria dianthi*, resulting in the drying up of the leaves. Destruction of affected plants and spraying with Blitox is recommended. Leaf Smut of dahlia is caused by the fungus, *Entyloma dahliae*, as a result of which the leaves die. Smut is also reported in *Dianthus* sp. caused by *Ustilago violacea*. The control measures consist of breeding resistant varieties, seed treatment with Ceresan, crop rotation, and burning of affected plants.

(3) *Stem Rot:* This is caused by a fungus *Sclerotinia sclerotiorum* and affects antirrhinum, carnation and poppy. Soil sterilization and burning of the affected twigs or plants are recommended. Deep ploughing of the soil and crop rotation will also help check the disease. Soaking of soil with Brassicol also helps to check the disease.

(4) *Root Rot:* It is a soil-borne disease caused by various soil fungi and the annuals affected are antirrhinum, nasturtium, and nicotiana. Application of Cheshunt compound (see damping-off) and Blitox (0.2 per cent) is useful.

(5) *Rust:* It is caused by a group of fungi. Rusty or reddish brown symptom appears on the leaves and branches. All reddish brown symptoms should not be confused with the symptoms of rust. Antirrhinum, celosia, gomphrena, hollyhock, lineria, and marigold are attacked by this disease. Use of disease-resistant cultivars, as in antirrhinum, is the best way to overcome the problem of rust. However, the disease can be controlled by spraying with Dithane Z-78 or Dithane M-22 or dusting with sulphur (300 mesh) at 17 kg/ha. Thorough spraying with wettable sulphur, Thiovit or Sulfex or Sultaf, is also effective in controlling rust. Prophylactic spray with Bordeaux mixture (5 : 4 : 50) or Blitox is also helpful.

(6) *Wilt and Foot Rot:* See Table I given earlier.

(7) *Virus diseases:* Virus diseases attack a number of annuals. Zinnias are killed by the mosaic virus. Viruses are also found in China aster, lobelia, marigold, petunia, and schizanthus. The symptoms are crinkling, yellowing, and mosaic of leaves. The affected plants should be destroyed.

Some diseases of flowering annuals are described below.

Celosia: Rust and rot are quite serious in celosia. The symptoms of rust are very conspicuous by the presence of rusty spots on the surfaces of leaves and branches. Affected plants remain stunted and do not flower properly. The rot may be caused by various fungi. The difference between a stem or a root rot can be distinguished by the nature of damage. In stem rot the rotten tissues can be noticed above ground, whereas in the case of root rot the plants wilt without any apparent damage to the tissues above ground.

Cineraria: Powdery mildew and rust are two common diseases in cineraria. Powdery mildew appears as greyish white spots and in severe cases whitish powdery growth appears. Affected tissues die and turn brown. Leaves also get crinkled as a result of the disease attack. In the case of rust attack, the rusty growth of fungal bodies in the shape of spots becomes evident on the surface of leaves, principally on the under-surface.

Impatients balsamina: Balsam is affected by powdery mildew throughout the year. The fungus attacks all the above ground parts such as foliage, buds and branches. The presence of the fungus can be detected by the greyish-white spots which coalesce and form into greyish patches covered with a powdery mass. The affected plants remain crippled and do not flower properly.

Petunia: The most serious disease is foot rot, caused by *Phytophthora cryptogea*, which causes the blackening of the lower part of stems and roots. Affected plants are uprooted and burnt. Seedlings should be raised in sterilized soil to void damage by this disease.

Phlox: A wilt caused by *Verticillium* sp. damages the roots and the affected plants collapse. These should be uprooted and burnt.

Powdery mildew, caused by *Erysiphe cichoracearum*, affects the foliage and tender branches, first appearing as greyish-white spots and then spread rapidly when the affected portions show powdery growth. In severe cases the leaves, tender shoots, and buds get crippled and die.

Sweet Pea: Sweet pea is attacked by a number of root diseases such as Foot Rot (*Fusarium* sp.) Black Root Rot (*Thielaviopsis basicola*), Root Rot (*Aphanomyces euteiches*), and Damping-off due to *Pythium* and *Phytophthora*. All these diseases become prevalent due to damp heavy soil conditions. In the case of foot rot the plants collapse at ground level. In balck root rot affected plants, the roots turn black.

The other serious diseases are powdery and downy mildews caused by *Erysiphe polygoni* and *Peronospora viciae* respectively. These diseases affect the foliage, tender branches, and buds. In severe attack affected portions die. In powdery mildew the shiny powdery coating is prominent over the affected parts, whereas in downy mildew faintly powdered white mould is evident.

Zinnia: Zinnias are badly-affected with spotted wilt, a virus disease, as a result of which the leaves get curled. The disease does not manifest in the hills but is very common in the northern plains and heavy rainfall areas. Early or late sowing in such localities may reduce the incidence of this disease. Spotted wilt is spread by thrips, which should be kept under control by fortnightly spray of Nuvacron. Plants affected by this disease cannot recover and hence it is advisable to uproot and destroy them. Persian Carpet is resistant to this disease. The other diseases of Zinnia are powdery mildew and wilt

Gerbera

The most serious diseases of gerbera are crown and root rots caused by *Phytophthora parasitica* which result in the blackening of the base (crown) of the plant. The affected plants wilt and die since there is no control measure. Excessive moisture in the soil favours the disease. Preventive measures against both the diseases are proper drainage, avoiding over-watering, and excessive mulching. Affected plants should be uprooted and burnt.

Gladiolus

A few fungal diseases can cause real havoc in a gladiolus plantation. *Botrytis gladiolorum* is by far the most troublesome leaf disease in gladiolus. This also affects the corms, stem and flowers when the humidity is high. Regular spraying with benomyl (benlate) can check this disease. Once affected the plants cannot be saved, and hence diseased plants should be uprooted with corms and destroyed by burning.

The major problem in gladiolus cultivation is caused by two soil-borne fungus diseases, viz. *Stromatinia gladioli* and *Fusarium* yellow (*Fusarium oxysporum* f. sp. *gladioli*). It is reported that *Fusarium* can survive in the soil for 3-4 years while *Stromatinia* remains for 10-12 years in the soil even if gladiolus has not been grown in that plot for that many years. Partial success in controlling the diseases can be achieved by corm treatment with Benlate or Captan. Fumigation of soil with methyl bromide and methyl bromide-chloropicrin mixtures are used in Israel and California (U.S.A.) to control these diseases. Special fumigation machinery for large areas is available. These chemicals are highly toxic and should be used with great precaution.

In advanced countries, disease-free propagation stocks are available. The commercial growers harvest the flowers along with corms and destroy the corms so that no debris is left. Then clean stock is purchased again and planted. It is also advised that cultivators should remove all plant debris and destroy it by burning so that no source of the disease is left. Cormels can be destroyed by herbicide sprays over a period of time.

Often crop rotation is suggested as a method for controlling these diseases but it is not a very satisfactory method. No commercial grower can afford to use the land for gladiolus production only once in 4 or 10 years. Even then, crop rotation is quite effective under certain circumstances. The optimum temperatures for growth of *Fusarium* and *Stromatinia* are 28°C and 15-16°C, respectively and if a grower can afford to avoid growing seasons having these temperature ranges it will be possible to control the diseases to a great extent. It is suggested to grow gladiolus alternatively in the winter and the summer wherever feasible (e.g., Bangalore) to overcome the problem of one or the other disease on a short term basis.

There are few virus diseases which cause chlorotic mottling of the leaves. In cases of severe infection the plants remain stunted. Virus affected plants are uprooted along with corms and cormels and these are destroyed.

Greenhouse Diseases

Possibly the most serious disease in a greenhouse is damping-off of seedlings. Always sterilized soil should be used to raise seedlings and sowing should be done thinly. Older plants may also be affected by foot rot caused by similar type of fungi as in case of damping off. Proper drainage and controlled watering will go a long way to check foot rot. Mildew is another disease which may appear on a limited scale on the foliage of some plants.

Virus diseases can be controlled by maintaining cleanliness and destroying the affected plants. Preventive fortnightly spray with Rogor or Nuvacron may help to check the vectors.

Herbaceous Perennials

Pelargoniums: There are a couple of fungal diseases in *Pelargonium* caused as a result of excess of watering. The disease Blackleg, caused by a *Pythium* species as a consequence of excess of watering, results in blackening and softening of the bases of stems. Excess of watering also encourages growth of grey mould fungus (*Botrytis cinerea*) which causes a lot of damage to the cuttings and even foliage of grown-up plants. The best remedial measure is to control over-watering which is the root cause of both the diseases. Use of sterilized potting mixture will also help to some extent in controlling the disease. A leafy gall caused by *Bacterium fascians* produces the symptom of a mass of proliferated shoots at the base.

Leaf curl virus also sometimes attack the plants. It is advised to use disease-free propagating material.

A physiological disorder known as "Dropsy" or "Oedema" manifests itself by producing pale pimple-like or wart-like outgrowths on the leaves or even stems. This takes place as a result of roots drawing more water than the plants can utilise or leaves can transpire, and as a result the excess water is stored in certain cells which swell. Judicious regulation of water supply is recommended to control this disorder.

Primulas: A root and crown rot caused by the fungus *Thielaviopsis basicola* is quite common in *Primula*. The affected roots and portion of stem turn black.

Downy mildew (*Peronospora* sp.) causes much serious damage. This affects the foliage and buds producing spots and faintly powdered appearance especially on undersurface of foliage.

The tomato spotted-wilt virus affects *P. malacoides* causing stunting, yellowing, and loss in flower production. In *P. sinensis* the disease manifests itself in a more virulent form. A mosaic virus also affects *P. obconica* and *P. sinensis*. Use of virus-free propagation materials is the only practical method of eliminating the virus disease. Insecticidal spray to control the insect vectors will help contain the spread of the disease.

Viola odorata: The common diseases of *V. odorata* are leaf spots caused by a *Cercospora* sp. and *Septoria violae*. The leaf spot becomes troublesome only when the plants are over-crowded and there is lack of free air movement or high humidity conditions. The collar rot is caused by *Phytophthora* and *Rhizoctonia* species as a result of which the affected plants collapse at ground level. The root rot is caused by the fungus *Aphanomyces euteiches* as a result of which affected plants wilt. Both these diseases are favoured by excessive moisture in the soil.

Hippeastrum

A virus disease causes yellowing and withering of foligae. The symptoms at first appear as pale yellow or whitish spots which enlarge gradually and coalesce, ultimately covering the whole leaf blade. The only remedy is to obtain virus-free planting material and to destroy the affected parts or better to destroy the whole plant if the manifestation of the disease is serious.

House Plants

Most common diseases of the house plants are rots and powdery mildew. Rot is caused by several fungi as a result of excessive watering. It is, therefore, advisable to water moderately to minimise the diseases. The compost should be porous and the pots should have enough drainage materials in the form of broken crocks and pieces of stones or gravel. Powdery midlew fungi affect mainly the foliage and remain covered with massive spore masses which spoil the appearance of the indoor plants. Flower buds are also affected, which do not open properly. Damp and warm environment favours the disease. For control, refer to Table I (p. 557).

Hydrangea

The common diseases of *Hyderangea* are leaf spot and powdery mildew caused by *Septoria hydrangeae* and *Oidium hortensiae*, respectively. In the former case, brown spots appear on the leaves which may drop when severely infected. Symptoms of powdery mildew appear on the foliage first in the form of minute greyish-white spots which increase in size, coalesce, and finally powdery mass becomes visible. Severely affected leaves are crinkled, curled, and stunted.

Jasmines

Sometimes *Curvularia* may become a serious problem, especially in places such as Rajasthan. The disease is caused by *Curvularia prasadi* which results in complete defoliation. In waterlogged soil *Sclerotium* may attack roots, and cause serious damage.

Juniperus

Different species of *Gymnosporangium* cause rust in *Juniperus*. This fungus is not found in India. It attacks the branches. Affected branches should be removed and burnt.

Lillies

A number of diseases attack various kinds of lillies. The most common amongst these is *Botrytis* which affects all the species. This can be checked by using disease-free stocks, clean cultivation, and crop rotation. The bulbs may be soaked in Benlate (0.1 per cent solution) before storage and at the time of planting. Rusts caused by *Uromyces lilii*, though not very common, produce coloured spots on leaves, stems, and buds. Spraying with Dikar is recommended to control the disease. Dormant bulbs are often attacked in transit by *Rhizopus necans* which cause rotting. This happens mainly due to damp and poorly ventilated conditions which should be avoided. Soaking the bulbs in Benlate for 30 minutes before transporting is recommended. Bulbs should not be packed in damp and poorly ventilated containers.

Fusarium basal rot often appears which is soil-borne. Sterilization of soil either with formalin (2 per cent solution) or steam should be done before planting.

Many species are also attacked by viruses which cause spots, mottles, and streakes on leaves and bending the tip of stem. Virus-free planting material should be planted to avoid loss. Lilies grown from seeds are free from virus unless these are infected at a later stage. Affected plants should be uprooted and burnt. Systemic insecticides such as Nuvacron or Rogor are recommended to prevent insect vectors.

Marigold

Marigolds are hardy plants and are rarely attacked by diseases. In poorly-drained soil, foot rot caused by *Phytophthora* may occur. Sometimes stem rot caused by *Sclerotinia sclerotiorum* also affects marigolds. The diseased plants are uprooted and burnt. Prophylactic spray or drenching with Blitox is helpful in checking foot rot, while *Sclerotinia* is controlled by soil drench with Brassicol. Flooding the soil will cause rotting of fruiting structures of fungi.

Pinus

Among the pine diseases, die-back and rust are common. Rust affected twigs get twisted and killed. In die-back the infection takes place through buds. The best remedial measure is to cut back the infected shoots.

Rose

Though rose is considered to be a hardy plant, it is still attacked by a number of diseases, die-back being the most serious one. Other diseases are rust, black spot, powdery mildew, and virus diseases.

Die-back: This can cause a serious problem in the rose plantation unless necessary precautions are taken. The disease kills the plant from top downwards. The disease starts mainly from the pruned ends. While pruning roses, the cut should be clean and transverse without splitting the stem. For this purpose a sharp and nice secateur is needed. Bad cut may cause more chances of this disease, although a clean cut end may also get infected. Initially the disease spreads only a few centimetres from the pruned end, in the form of black tissues on the bark, either partially on one side or engulfing the stem all round. A partially diseased stem if cut open to half will show the spread of the disease deep inside the tissues, showing brown or black zones. Under favourable conditions, the whole branch may blacken and the disease may spread downwards to the main stem and ultimately to the roots and kill the whole plant. This disease is more prevalent in older plants than in younger ones. Die-back infection starting from the withered bloom has also been reported, but death from such infection of an older plant is not likely as the infection remains confined to within a few centimetres (see Fig. XXVI.1).

The disease is caused by a fungus *Diplodia rosarum*. Often a species of *Colletotrichum* also remains associated. The fruiting bodies of the fungus may hibernate on pruned dead

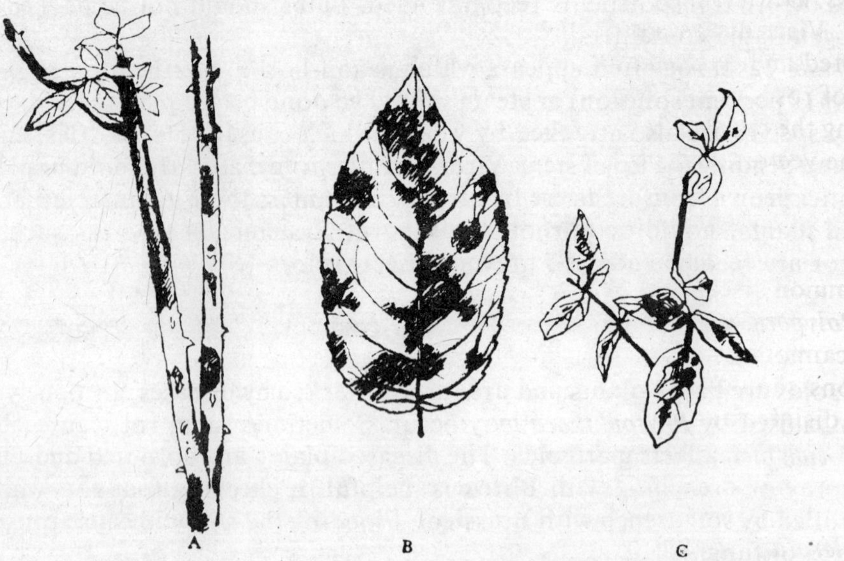

Fig. XXVI. 1. A. Die-back affected twig. B. Black spot of rose. C. Powdery mildew of rose.

twigs and flower stalks and hence all pruned wood should be cleared off and burnt. In general, week plants are affected more by die-back than healthy ones. Therefore, it is important to keep the plants healthy. To prevent the spread of disease, the sharp edge of the secateur should be dipped in 70 per cent alcohol or formaldehyde for disinfection after each cut. All pruned ends should be painted with the paste suggested for digger wasp of roses (p. 585). There is hardly any cultivar which is not affected by die-back. It is better to avoid growing one which is more susceptible in a particular area.

Diplocarpon rosae (Black spot): This disease is widespread in nature and is commonly known as Black spot. The disease is distinguishable from other spots by its very dark colour, which occurs on both surfaces of the leaves and the margins of the spots are frilled. In some cultivars defoliation starts immediately after infection, while in others even severely damaged leaves still remain attached. It is reported that the fungus emits ethylene gas which is responsible for defoliation.

Sphaerotheca pannosa (Powdery Mildew): The disease is quite widespread in India except in Delhi where the roses are not much affected. The disease affects all the aerial parts. The tender leaves get twisted and some raised areas may be noticed on the affected leaves. Powdery coating of spores also develop over the affected parts and in severe cases the affected tissues die causing necrotic spots similar to black spots. The young plant parts also remain coated with powdery spore masses of the fungus. The affected buds do not open properly. Wettable sulphur is effective in controlling powdery mildew.

Phragmidium mucronatum (*Rust*): Rust of roses is more common in the northern hills than anywhere else. Roses in other places also get rust but not frequently. The pustules of rusts which are small yellowish (rusty) to blakish swellings appear on leaves, petioles; and stems. The disease is not serious in the plains. For control, refer to Table I.

Virus: Occasionally roses get affected by virus. Mosaic disease is one of the virus diseases. Virus diseases are difficult to control. Therefore, virus affected plants should be uprooted and burnt. It was demonstrated by F.O. Holmes in the U.S.A. that mosaic virus of roses can be cured by heat treatment. This method is practicable and useful in treating the budding eyes to get a virus-free propagation stock. Other methods are to control the vectors with pesticide sprays and to use virus-free propagation material.

Salix

Common diseases of *Salix* are crown gall and bracket fungi caused by *Fomes igniarius* and *Polyporus sulphureus*. The latter two fungi generally attack the old trees. Affected trees cannot be saved and hence should be destroyed. If the attack is not severe, affected portions should be cut out thoroughly and burnt and the wound should be dressed with some disinfectant such as dilute carbolic acid.

Tuberose

Sclerotium species of fungus attacks the leaves and flower stalks at the ground level and causes rot which results in defoliation and toppling of spikes. Soaking of soil around the plants with 0.2 to 0.3 per cent Brassicol at the rate of 500 to 1000 ml per square metre can give effective control. The soil should be irrigated lightly prior to fungicide application for proper penetration of fungicide.

Table II

Pests of ornamental plants and control measures

Name of pest (1)	Control measures (2)
Ant	The anthills in lawn look bad. Ants also make nests over trees and shrubs. Ants extract honey from aphids and drag these insects to new areas thus spreading the infestation. Spread BHC dust on ant nests and visible population. The rate is 5 per cent to 10 per cent BHC powder at 20-25 kg per hectare. Heptachlor or Aldrin dust as recommended for white ants could also be used. For other control measures see lawn later in this chapter (p. 577).
Aphid	Aphids are small lice-like insects either winged or wingless, varying n colour from green to bluish-grey or black. Spray 0.03 per cent Rogor (dimethoate) to kill the insects. Malathion (maldison) also controls the insects but this is toxic to ferns and petunia.

(1)	(2)
Beetle	As recommended for weevil.
Borer	See stem borer.
Bug	Spray of Nuvacron (0.04 per cent of active incredient), at fortnightly intervals is recommended. Rogor (Dimethoate) 0.03 per cent (active ingredient) spray is also effective.
Caterpillar (e.g. *Hendecasis duplifascials*, bud worm, *Elasmopalpus jasminophagus*, bud and shoot worm and *Contarinia maculipennis* maggot of a gall fly called blossom midge and many others.	There are various types of caterpillars in the garden, but most of these are killed by spraying Sevin (Carbaryl) 50 per cent wettable powder at 0.15 to 0.2 per cent. Spray with Endosulfan (0.05-0.07 per cent) is also effective.
Chafer-bettle	The larvae can be driven out from soil by the application of paradichlorbenzene, or killed by Thimet at 4-10 kg/ha or 1 to 1.5 g/sqm. The adult beetles are controlled by Nuvacron spray at 0.05 per cent (active ingredient).
Cricket	Apply 10 per cent BHC dust for control.
Cutworm	As in grub.
Earthworm	The wormcasts formed by earthworms in lawns look ugly. Earthworm is normally considered to be beneficial, but sometimes they become a nuisance. Control measures consist of application of Chlordane 75 per cent, e.c. applied at 30 ml/20 sqm, or granules (5 percent at 10-20 kg/ha. This treatment can control 90 per cent of the worms, four weeks after application. For further details, consult earthworm later in this chapter.
Grasshopper	Grasshoppers are destroyed by BHC (10 per cent) or Sevin (10 per cent) dust at 10-20 kg/ha. Birdbaths or other arrangements for attracting birds should be made which may eat the hoppers.
Greenbug	Weekly or fortnightly spray of Rogor (dimethoate) at 0.03 per cent (active ingredient) may kill the bugs.

(1)	(2)
Grub	Grubs of some beetles stay in the soil and damage roots. These can be controlled by the application of Thimet at 4-10 kg/ha. Regular hoeing and forking can bring out grubs which can be hand picked and killed or birds may eat them.
Jassid	Jassids are also known as leafhoppers, and hence the control measures suggested below for leaf hoppers should be followed.
Leaf hopper	Spraying with Rogor (dimethoate) at 0.03 per cent (active ingredient) at 10-15 days intervals is recommended to control the hoppers. Malathion (maldison) spray is also effective. Sevin (Carbaryl) spray as recommended for caterpillars is also effective.
Leaf miner	This name applies to the larvae of several kinds of insects. They attack interior tissues of leaves and cause snaky channels or blister such as excavations. Carnations and Chrysanthemums are frequently attacked. The miners can be controlled by spraying Endosulfan (0.05-0.07 per cent), Fenitrothion (0.05-0.01 per cent) or Rogor (dimethoate) at 0.03 per cent (active ingredient). Badly affected foliage should be picked up and burnt.
Leafroller	Generally, the larvae of sawfly which are grey or green in colour attack leaves which roll up. The larvae live within the leaf roll. They form webs on the lower surface. Some other larvae or caterpillar may also cause similar damage. Affected leaves should be picked up and burnt. Spray with Malathion (maldison) is effective.
Mealy bug	These are small bug-like creatures which remain covered with white or greyish waxy material. They suck juice and weaken the plants. Some ornamentals such as croton and coleus may be killed if attacked at young stage. Spraying with Dichlorvas (Nuvan) 0.1 per cent plus 1 per cent Agricultural spray oil at fortnightly intervals is recommended. Some bettles which feed on mealy bugs are effective in controlling this pest.

(*Contd.*)

(1)	(2)
	Agricultural spray oil may be injurious to some delicate ornamentals. A new chemical should be tried first on a limited scale to see any possible adverse effect. In a small garden the insects could be hand picked or brushed off and burnt.
Mite (often called Spider mite or Red spider mite).	These are very minute in size, barely visible to naked eye and not a true spider. Fortnightly spray with Dicofol (Kelthane) at 0.025-0.04 per cent or Nuvacron at 0.04 per cent or Rogor at 0.03-0.06 per cent (a.i.) are recommended.
Moth, Pine shoot	Young caterpillars bore into the base of pine needless (foliage) and developing shoots. They cause distortion and kill them. Control measures recommended for caterpillars should be followed. Spray with Rogor (0.03 per cent) is also effective. Affected shoots and needless should be removed and burnt.
Nematodes	These are worm-like minute pests which attack the roots and often induce nodules in roots. For control, see Thrips.
Rabbit and Rats	Traps and poison baits as suggested for slugs should be used. For the first 5-7 days the baits should be without poison and then poison is added.
Red Spider Mite	See Mite.
Sawflies	The larvae feed on needles (foliage) and young shoots. Spray Sevin (Carbaryl) 50 per cent Wettable powder at 0.15 to 0.2 per cent to control the larvae.
Scale	Young insects remain firmly attached to bark and sometimes to foliage as in the case of cycas. Mature insects get covered with a hard shell. They suck sap and thus weaken the plant or cause its death. Once the infestation is severe, these are difficult to control. Spraying with Nuvacron (0.05 per cent) plus 1 per cent Agricultural Spray Oil at 10 to 15 days' intervals is recommended. Metacid spray (0.1 to 0.15 per cent), is also effective. Application of Temik or Thimet in the soil is effective. See thrips for details. In case of severe infestation both Nuvacron spray and application of Thimet is advised.

(1)	(2)
Slugs	These are crawling slimy creatures ranging in colour from grey to black and measuring from 1.5 to 10 cm when crawling. They devour small seedlings *in toto* and make holes in leaves of large seedlings. They move after dusk and can be collected with the help of a flash light and destroyed. They may be trapped by heaping vegetable refuse in the ground during evening. The heaps are turned over each morning to collect the slugs sheltering beneath and destroyed. But the most effective method will be to use poisonous baits. A poison bait is prepared by using 250g wheat bran, 50 g jaggery (*gur*), and 6-8 g Metaldehyde. The contents are made into mash by adding sufficient water. The mash is made into small pellets and kept in places likely to be visited by the slugs. The baits should be used with great precaution to safeguard household pets and children. Other poisons used for baits are Zinc phosphide (5-10 per cent), Barium carbonate (16-20 per cent) and warfarin (0.025 per cent). All poisons should be used with great care. Empty containers should be destroyed and unused poison should be kept under lock and key and away from the reach of children.
Snail	Same as above.
Stem Borer	Since the borers make tunnels or holes and live inside, the control measures should be thorough and effective. A thorough soaking of stem and bark with Dichlorvas (Nuvan) 0.1 per cent is quite effective in controlling the borer. Endosulfan (0.05-0.07 per cent) spray is also effective. Sometimes the borer may live deep inside the tunnel and surface sprayed insecticide may not reach there. In such cases it is advisable to put a few drops of kerosene or petrol in the hole and plug the opening with putty, wax, or mud. Also application of drops of Carbon disulphide or Carbon tetrachloride and plugging the hole may be more effective. Borers can also be killed by probing the hole or tunnel with a firm but flexible wire or pulled out with the help of a hooked wire. The presence of the borers in trees and shrubs

(*Contd.*)

(1)	(2)
	can be detected by the presence of saw-dust near the trunk. The badly damaged twigs should be cut back to healthy green tissue and the wound treated with weak carbolic acid solution followed by Blitox paste and sealed with cement mortar or mud or putty.
Thrips	Minute insects of a pin-head size which are yellow to black in colour. They suck sap from leaves, stems and flowers and cause damage. Silver and brown streaks can also be noticed on foliage and stem. The control measures consist of foliar spray with Nuvacron (0.04 per cent of active ingredient). It is also possible to control thrips by soil application of Aldicarb (Temik) at 550-850 g of active ingredient per 100 sqm area or 8-10 kg per hectare. Thimet at 4-10 kg per hectare is also effective. Soil application of Furadan (Carbofuran), 3 times the quantity of Thimet is also effective. Since the last-mentioned three insecticides/nematicides are quite costly, it is better to use Nuvacron as long as it is effective. If already either of these insecticide is added to the soil, it may not be necessary to apply additional insecticide to control thrips. Soil application of either of the insecticides will be economical only when these are used simultaneously to control nematode infestation.
Weevil	The adults are controlled by 5-10 per cent Sevin (Carbaryl) dust, while, the grubs are controlled by soil application of Aldrin (0.3 per cent) or Chlordane (0.1 per cent). The weevils are also killed by Nuvacron 0.05 per cent (active ingredient) spray. If the weevils bore tunnels in the branches of trees and shrubs, control measure suggested for stem borers should be followed.
White Ants (Termites)	Drenching with Aldrin (0.3 per cent) or Chlordane (0.1 per cent) or 5 per cent dust is recommended.
White Fly	Wedge-shaped, white winged flying insect and fly off in a cloud when disturbed. Nymphs are pale-green in colour. Both suck sap from leaves which turn pale-yellow, becomes sticky with honey-dew and blackened by sooty mould. Nuvacron spray (0.05 per cent of active ingredient) once a week for 2 to 3 weeks is effective to control the fly.

SOME IMPORTANT PESTS OF ORNAMENTAL PLANTS

A number of pests attack different ornamental plants and their control measures have been standardized (see Table II). Some important pests of ornamental plants are described below:

Ant

Plants affected by ants are cacti and succulents, China aster, greenhouse plants, and lawn.

Ants may cause various problems in many ornamental plants. It is generally the red ants which are more prevalent in the gardens, although other types of ants are found occasionally and may also be harmful. The ants burrow tunnels near the root zones and cause excessive aeration near the root. The roots, therefore, dry up. Moreover, fibrous root development is also hampered as a result of activity of the ants near the root zone. Ants carry with them live aphids and mealy bugs for the sweet secretions excreted by these insects which ants relish. These transported insects may infest plants and cause damage. If the compost remains too dry it favours ant growth. The affected compost in pots should be replaced. At the time of planting in ground or pots 5 to 10 per cent BHC should be added.

The anthills in lawn look ugly. Moreover, their activity near the root zones is not desirable from the point of plant health. Ants do not like wet conditions and, therefore, the lawn should not be allowed to remain dry for too long.

Aphid

Plants affected by aphids are cacti, carnation, China aster, chrysanthemum, dahlia, flowering annuals (antirrhinum, cineraria, larkspur, lupins, petunia, stock, sweet pea and many others), gerbera, greenhouse plants, lilies, rose, shrubs (many), trees [e.g. *Cupressus* affected by *Cinara* (*Cupressobium*) *cupressi* and *Salix*] and *Viola odorata*.

Aphids are sucking insects which are blackish or brownish or green in colour and suck sap from tender shoots, leaves and flower buds as a result of which shoots or foliage get twisted and distorted. The affected plants become weak and the buds may not open or open into deformed flowers. The insect-injured plants may lose their ornamental look.

The aphids excrete honey-dews on the foliage causing sooty-mould to grow over the foliage. As a result the affected plants lose vitality on two counts, firstly, due to loss of sap to the insect, and, secondly, the soot impairs normal leaf activity. The soot-covered plants look ugly also. Sometimes severely affected twigs may die.

Beetle

Plants affected by beetles are China aster, flowering annuals (gomphrena, marigold, morning glory, petunia) rose, and *Salix*.

The adult beetles cause damage to a number of ornamentals. In chrysanthemum, the beetles may feed on leaves or cut young plants at ground level. In many flowering annuals also, similar damage is caused. In rose, the beetles generally feed on foliage, the details of which have been described later in this chapter. The damages done by beetles to *Pinus* and *Salix* have been dealt separately later in this chapter. In some plants the beetles (e.g., blister beetle in rose) bore through opening flower buds and also feed on open flowers.

Bug

Plants affected by bugs are herbacious perennials such as *Phlox decusata* and *Viola odorata*.

A type of green bug occasionally attacks the above plants. But many other ornamental plants may also be infected. This is a sucking type of insect, which feeds on plant sap. As a consequence of their sucking activity plants become weak.

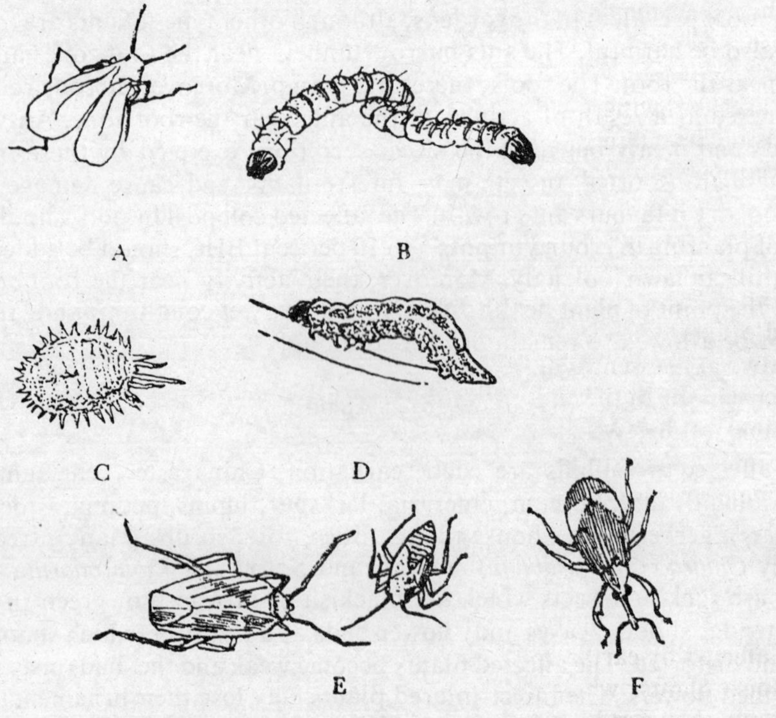

Fig. XXVI.2. Some Insects: A. White fly; B. Cutworms; C. Mealy bug; D. Winter moth caterpillar, E. Capsid bug cleft, right-nymph; F. Weevil.

Caterpillar

Plants affected by caterpillars are canna, carnation, China aster, ferns, flowering annuals (antirrhinum, celosia, hollyhock) herbaceous perennials (*Pelargonium*), jasmines, shrubs (*Nyctanthes arbor-tristis*), and trees (e.g., *Cassia fistula*).

Caterpillars of many insects may affect several ornamentals. It has been found at the Indian Institute of Horticultural Research, Hessaraghatta, that some caterpillars bore holes in the growing flower buds of carnation. Such buds either do not open or opened flowers are badly deformed. Preventive sprays should be given to stop any caterpillar attack (Table II).

Some green caterpillars are occasionally noticed on ferns in greenhouse. They feed on foliage. Caterpillars may also feed on the foliage of some flowering annuals. Pelargoniums are also attacked by some leaf-eating caterpillars.

Nycatanthes arbor-tristis frequently harbour large colonies of hairy-caterpillars covering completely the stem and the branches. Unless these are controlled in time they will spread all over the garden and inside the dwelling house. The damages done by other types of caterpillars have been described separately under different plants.

Cricket

Plants affected by crickets are generally the greenhouse plants.

Though crickets may affect many plants they prefer more the sheltered atmosphere of the greenhouse. They devour seeds and damage seedlings and also foliage and stems of grown-up plants. The presence of the cricket is easily detected by the joyous sustained shrill chirp. Crickets are burrowing creatures of the garden and field. But there is a house-dwelling species also. They shun sunlight and live in cracks and crevices. They resemble somewhat the grasshopper.

Cutworm

Plants affected by cutworms are carnation, flowering annuals such as gomphrena, marigold, and others.

The cutworm is essentially an insect of the nursery, especially that of flowering annuals. The insect remains in the soil and attacks the seedlings at ground level as a result of which the seedlings topple. Whether a seedling has toppled at ground level because of cutworm attack or infection due to fungi (collar rot or damping off) will be evident from the nature of tissue damage. In case of an insect damage, both ends of the cut will be clean and healthy, whereas fungal-damaged tissues will be soft and rotten.

Earthworm

Plants affected by earthworm are generally the house plants and lawn grass. In case of potted house plants, wherever possible, the earthworms should be picked up by hand. Alternatively, they can be forced out of the soil by the application of lime water. The lime water is prepared by dissolving a tablespoonful of hydrated lime into a quart (1.136 *l*) of water and allowing it to stand for 24 hours, after which only the top clear portion is used for applying. Too much of lime water should not be used.

In lawns the earthworms form with their excrement "worm-casts" which look ugly. Though earthworms are helpful to a limited extent, as with their burrowings they improve aeration and drainage of the soil, steps should be taken to reduce their population in lawns. Both poisonous (copper sulphate, colloidal lead arsenate and, corrosive sublimiate, i.e., mercuric chloride) and non-poisonous (lime water and potassium permanganate) chemicals are available for control. Both lead arsenate and corrosive sublimate are deadly poisonous and are used at dilutions of 1 : 1000 or 1 : 2000, i.e., 1 g and 500 mg in 1 *l* of water, respectively. Another effective treatment is to dissolve 15 to 28 g of potassium permanganate crystals in 4.5 *l* of water and apply this to about 1 square metre of lawn area, with a fine-rose watercan. The worms come out on the surface, which may be swept up and destroyed.

For further information consult Table II.

Grub

Plants affected by grubs are: China aster, chrysanthemum, flowering annuals (Celosia, gomphrena, marigold, morning glory, petunia), herbaceous perennials (*Pelargonium*), and trees.

The grubs of chaffer beetle and some other beetles, which are soil dwelling in habit feed on the roots or attack the stem at ground levels. In flowering annuals, the seedlings are cut at ground level either by the grub or the adult beetle. As a result of root damage by the grubs the plants do not grow properly and remain stunted, or may be killed occasionally.

Leaf Miner

Plants affected by leaf miner are carnation, chrysanthemum, and greenhouse plants. There are several insects the larvae of which make snaky channels and blister-like excavations on leaf surfaces. The affected leaves look ugly.

Mealy Bug

Plants affected by mealy bugs are bottle garden plants, cacti and succulents, ferns, greenhouse plants (*Solanum*, *Streptosolen*, *Trachelospermum*, and croton), herbaceous perennials (e.g., *Phlox decusata*), hippeastrum, house plants (coleus), and shrubs (barleria, crossandra, hibiscus).

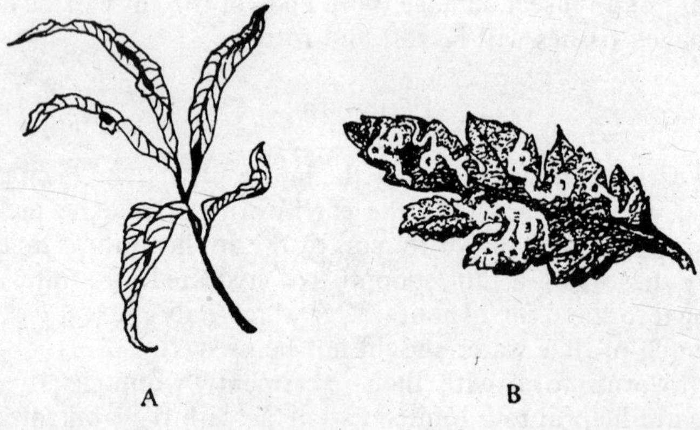

Fig. XXVI. 3. A. Rose leaf-rolling sawfly (damaged leaf); B. Chrysanthemum leaf showing damage by leafminer.

Mealy bugs are small fleshy insects which remain covered in a white, woolly and waxy secretion. Though this is a dangerous pest, occasionally it is also beneficial to mankind. About 40 years ago the cactus *Opuntia* formed impenetrable barricade in the fields of south India. Then the mealy bug, *Dactylopius tomentosus*, naturally bred on this weed in large numbers and the weed was almost eradicated.

In some cacti and succulents (e.g., *Stapelia*) mealy bugs, attacking both aerial parts and roots, may cause a lot of damage. In greenhouse and indoor plants mealy bugs may create a lot of problems. Generally, plants of older age are affected by mealy bugs. The insects suck the sap as a result of which the plants weaken. If not checked in time the affected plants may die.

In *Hippeastrum* the mealy bugs infest the foliage as well as the bulbs. Tender plants of *Phlox decusata* may be killed by severe mealy bug infestation. Many shrubs are also affected by mealy bug and the nature of damage is same as in the case of house plants or greenhouse plants. The mealy bugs secrete honey dews which stick to the foliage and as a result of which sooty mould covers the leaf surface impairing the normal activity of the leaves. The sooty mould infestation in any plant will either indicate the presence of aphids or mealy bugs.

Nematode

Plants affected by nematodes are cacti and succulents, chrysanthemum, dahlia, ferns, rose and tuberose.

Nematodes are microscopic worms which live inside plant roots and sometimes in the tissue of leaves. These generally affect the roots of the plants, although several instances of foliar infestation in ornamentals have also been reported. Roots infected with nematodes remain stunted in growth. The affected roots in most of the cases get swollen in the form of knots and hence is commonly called root-knot nematode. Among the succulents the member of the family *Crassulaceae* and *Echeveria* in particular are affected more.

Red Spider Mite (Spider Mite, Mite)

Plants affected by this mite are cacti and succulents, carnation, chrysanthemum, ferns, gerbera, greenhouse plants, herbaceous perennials (*Viola odorata*, etc.), house plants, jasmines, rose and shrubs (e.g., *Hibiscus rosa-sinensis*).

The mite is a minute dot-sized pest, almost invisible to the naked eyes, resembling the spider but not a true spider. The affected parts in cacti and succulents appear to be bleached and develop spots. *Euphorbias* are more susceptible. This is a sucking type of pest and hence leaves or tender shoots affected by mites develop spots and get distorted.

If established once, the red spider mite may become a serious problem in the greenhouse. This common pest of the greenhouse should be kept in check. The mite does not like wet conditions and if the plants are syringed with a jet of water this pest generally disappears.

Scale

Plants affected by scales are cacti and succulents, cycus, ferns, flowering annuals (few types), greenhouse plants (crotons, cycus, and some others), herbaceous perennials (*Phlox decusta*), house plants, rose, and shrubs such as *Crossandra* and *Daedalacanthus nervosus*.

Red scales are dangerous insects of the garden. These remain covered inside a hard coating which looks like a waxy scale and appears as pox marks on the affected parts. These insects are sucking in nature and suck plant sap as a result of which plants become weak. If not controlled in time, the affected leaves, the branches, or the whole plant may die. It has been observed that weak or overcrowded plants or plants growing in shade or semi-shade are attacked more than others. As for example, leaves of cycus plants get infested more with scales when these are growing in shade or semi-shade compared to plants growing in sun. Since the insects remain covered inside a hard scale it is difficult to control by ordinary insecticides. The pest should be controlled early before the infestation is severe. In case of isolated attacks on indoor plants and others the insects may be brushed off with a hard brush, collected in a newspaper, and destroyed. Among the cacti, *Opuntia*, *Oreocereus* and *Trichocereus* are more affected than others. In the greenhouse plant *Cycus*, both surfaces of the leaves, are infested. The tender flowering annuals if attacked by scales die very quickly and hence special attention should be taken to eradicate this pest. Sometimes the ferns are affected by scales, quite similar to the San Jose Scale. Some shrubs like *Hibiscus rosa-sinensis* are affected with a different type of scale which is plump in nature (green or rusty) and look like beatles.

Slug and Snail

Plants affected by these pests are greenhouse plants, ferns, and flowering annuals especially young seedlings in the nursery.

Slugs and snails are essentially the pests of greenhouse, fernery, and nursery. They cut the young plants at ground level and devour seedlings. Any refuse should not be allowed to accumulate in the corners of the greenhouse where these pests can find a hiding place. Though the snails act as scavengers in the garden, especially in the water garden, their presence is not always welcome, particularly in large numbers.

Sparrow

Sparrows are a real menance to the sweet pea seedlings and to a lesser extent to sweet alyssum also. Sometimes it becomes almost impossible to save the sweet pea seedlings unless special measures are taken. To prevent damage from birds, black threads or pieces of cloth may be stretched across the row of plants. Often the seeds are also eaten away by birds or rats after sowing. This can be prevented by soaking the seeds in solution of red lead and linseed oil before sowing.

Spider

The plants affected by spider are some shrubs such as *Hibiscus rosa-sinensis* and some trees.

Spiders infest many plants, and cause curling and rolling of the leaves which become very small, the plants get stunted and may stop flowering. Moreover, the webs formed by the spider make the plants look ugly. The control measures suggested for mites should be followed.

Thrip

The plants affected by thrip are cacti and succulents, carnation, chrysanthemum, dahlia, flowering annuals such as cineraria, phlox and nasturtium, gerbera, gladiolus, hippeastrum, lilies, rose, and tuberose.

Thrips are minute insects of the size of pin-head and barely visible to the naked eye. These are sucking insects which affect a large number of plants. Both the nymph and the adult can affect plant parts. As a result of their sucking activity the plants become weak and the affected parts develop minute spots. The tender twigs, foliage, flower buds, and spikes get distorted and twisted. The thrips as a result of their sucking activity on the upper surface of the foliage of dahlia cause pinprick-like symptoms which are of lighter colour than the leaf colour. Similar symptoms are produced on other plants also. In dahlia and some other plants the thrips transmit virus diseases.

In gladiolus both nymph and adult thrips cause damage to leaves, flowers, and corms and the symptoms produced are silvering and russetting of the feeded zones. Thrips affect the corms also, especially during storage. The lily thrips (*Liothrips vaneecki*) cause considerable damage to the bulbs. In tuberose, the spikes get distorted and flowers fail to open.

Weevil

The plants affected are the flowering annuals such as lupins, primulas and a few others, rose, shrubs, (hibiscus and others), and trees.

In *Hibiscus* the weevil bore through the tender shoots, especially during the rainy season. Some other shrubs may also be affected. Several trees also get weevil attack which has been dealt later in this chapter.

White Ant

The plants generally affected by white ant are China aster, chrysanthemum, rose, shrubs and trees.

White ants cause damage to a number of ornamental plants. They damage roots and attack the trunk thus causing injury. In severe cases of infection plants may succumb to the injury. White ants do not like wet conditions and wherever possible the soil should be kept wet to keep them away.

PESTS OF SOME SPECIFIC PLANTS

Bottle Garden

Occasionally even bottle garden plants are subjected to pest attack. It may cause surprise to an amateur how pests can enter the completely sealed arena of the bottle garden. The eggs of pests enter the bottle through unsterilized compost or with the plants themselves and hatch afterwards. The common pests are aphids and mealy bugs. The mealy bugs, if limited in number, can be controlled by spotting (touching) them with a cotton ball saturated with rectified spirit and attached at the end of a long stick. Aphids may be controlled by a mist spray of Rogor. Possibly a better method would be to inject chloroform vapours into the bottle.

Jasmine

Jasmines are comparatively hardy plants and normally insect attack is not a major problem. However, Sometimes caterpillars such as *Hendecasis duplifascials* (bud worm), *Elasmo-*

palpus jasminophagus (bud and shoot worm), and *Contarinia maculipennis* (maggot of a gall fly blossom midge) damage buds by boring a hole at the base of the bud. Therefore, affected buds produce deformed and damaged flowers. In severe cases a portion or the whole bud may be eaten up by the worm. For control consult Table II, under caterpillar. *Eriophide*, a mite, also attacks jasmine plants and may become a nuisance if not controlled in time.

Lilies

Mice and rabbit are natural enemies of lilies. While the former eats the bulbs the latter eats the above ground portions. Snails and slugs also cause damage. Poisonous baits should be scattered near the plants for destroying these pests.

Considerable damage is caused to bulbs by lily thrips (*Liothrips vaneeckei*). Aphids cripple the young growth and distort the general look of the plants. Control measures as suggested for these insect pests in Table II should be followed.

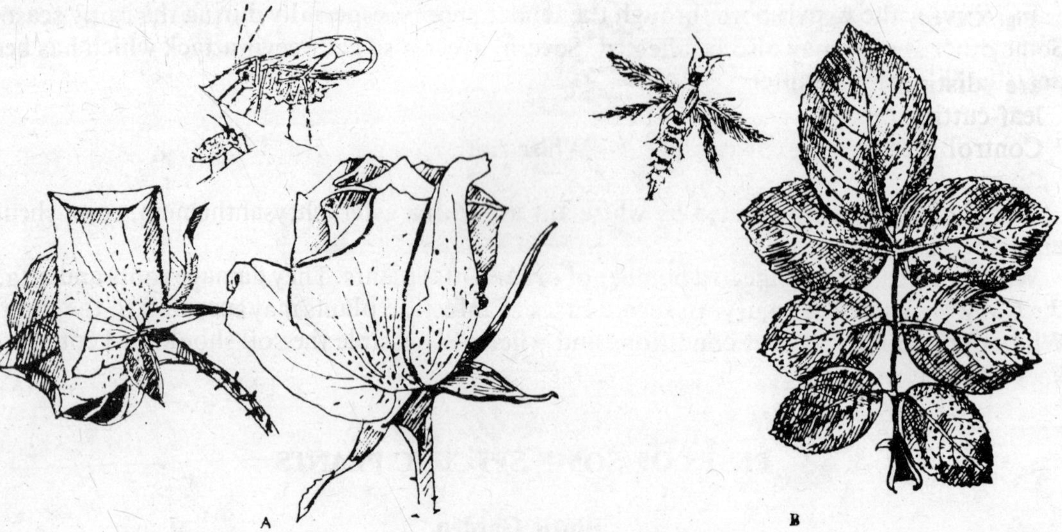

A B

Fig. XXVI. 4. A. Aphids (winged adult and nymph) and aphid affected rose flowers. B. Thrip and injured leaves of rose. (Both the insects highly magnified).

Rose

The rose has a number of pests and some of these may create a real problem. The important pests are white ants, chafer beetle, red scale, digger wasp, aphids, thrip, mealy bug, and nematode.

White ants by their activity can kill newly planted as well as established plants. At the time of land preparation and also once in 3 months or when white ant attack is expected, 5 per cent BHC or aldrin or chlordane dust (alternatively drenching with aldrin 0.3 per cent or chlordane 0.1 per cent) should be applied to the rose beds.

The chafer beetle feeds on leaves and these show as irregular cuts and punctures which

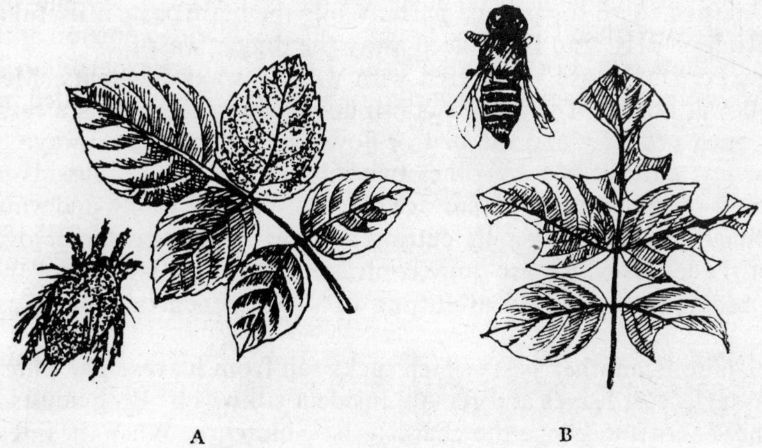

Fig. XXVI.5. A. Red spider mite (magnified) of rose affected leaves, B. Leaf cutting bee and characteristic, neat, circular or oval cuts in leaves of rose.

are distinctly different from the neat circular or oval cuts near the margins done by leaf-cutting bee. The grub of this beetle remain in the soil and cause damage to the roots. Control measures have been suggested for both in Table II.

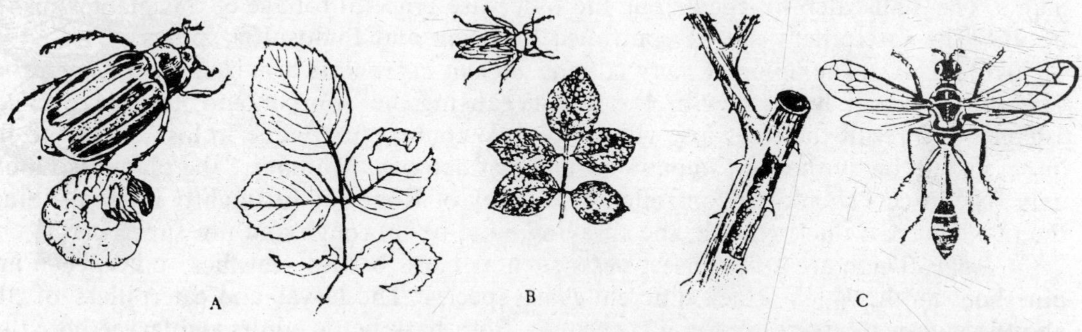

Fig. XXVI.6. A. chafer beetle adult (top) and grub, with rose leaves showing irregular cuts by the beetle (actual size 11mm); B. Jassid (magnified) and affected leaves; C. Nest of digger wasp at cut end of stem and the wasp at right (siz) 1.5 cm)

The red scales attack the tender as well as older branches which look like waxy scales and appear as pox marks on the attacked branches. Scales are really troublesome insects which may kill the affected branches or the whole plant if not controlled in time.

The damage caused by thrips and aphids are similar in nature. The first causes distortion of tender branches, leaves and blooms. The aphids, commonly known as plant lice, are minute globular, light green or deep green to blackish-green insects which thrive under semi-shade and suck the tender parts of the plants including flower buds or even petals of open flowers.

Digger wasps usually tunnel through the cut end of pruned branches. The injury caused by this insect generally leads to the die back disease of roses. A paste consisting of a mixture of DDT and BHC in proportion of 1 : 1 (0.1 per cent each) is mixed with the fungicidal paint consisting of 4 parts of red-lead, 4 parts of copper carbonate, and 5 parts of linseed-oil, is pasted to all the pruned parts. While the main paste is useful against die back disease, addition of BHC and DDT keep away the digger wasp.

Mealy bugs are small insects covered with whitish waxy filaments. Mealy bugs attack flowers and flower stalks. The insects continuously suck the sap as a result of which the buds do not open properly and the bud or flower may just wither away.

Jassids and leaf cutting bee are other two minor insect pests. Jassids suck sap from leaves as a result of which the upper surface of leaves become yellow and white. The leaf cutting bees damage the rose leaves by cutting regular, neat, circular or oval patches from the border of the leaves which are quite contrast to the irregular cuts of the chaffer beetle. Except this minor damage, the leaf cutting bees are beneficial insects as they are good pollinators.

Red spider-mite is another pest which sucks sap from leaves. The mites live in groups on the under-surface of leaves and remain inside a silky web. Both adults and young ones suck from the leaves and cause the characteristic blotches. When the infestation is acute, the foliage dries up.

Nematodes (Eelworms) attack the roots. Affected plants remain stunded and look sickly. Application of nematicides in the soil control the nematodes.

Trees

(a) *Cassia fistula:* Occasionally some caterpillars feed on the foliage and defoliate the plants. The half-eaten foliage among the otherwise graceful foliage of this plant ruins the beauty. The caterpillars can be controlled by Sevin and Endosulfan sprays.

(b) *Cupressus:* The aphid, *Cinara* (*Cupressobium*) *cupressi* is a nuisance on *Cuppressus*. The insects excrete honey dew on the foliage causing the sooty-mould to grow over the foliage. As a result the trees lose vitality on two counts, firstly due to loss of sap to the insect and secondly the soot impairs normal leaf activity. Moreover, the plant also looks ugly. The insects should be controlled by a spray of Rogor at fortnightly intervals. Since the plants have a thick growth, the spraying must be thorough and not superficial.

(c) *Pinus:* There are a few insect pests such as bark bettles, sawflies, pine-weevil and pineshoot moth, which attack different *Pinus* species. The larvae and caterpillars of the above-mentioned insects are the real enemies. Both bark bettle adults and larvae bore tunnel in bark and cause damage. The beetles generally invade the weak plants. The most damage is done to the young shoots. The larvae of sawfly are found in colonies on pines which defoliate the branches and eat up bark. The presence of the larvae will be evident from the green excreta found on the ground. The females of the pine-shoot moth lay eggs singly or in groups over pine bud needles (foliage) and shoots. On hatching, the caterpillars start feeding and a number of small shoots take the shape of a "Witches Broom" type (branches without foliage) injury. Plants grown on bad soil are mainly attacked. The adult weevils feed on young stem and cut irregular holes up to the cambial region. The soil dwelling grubs make longitudinal tunnels in the roots and stems. The presence of borers (whether beetle or weevil) can be detected by the deposition of sawdust on the ground around the trunk. All affected branches should be cut to healthy tissue and the larvae or adult killed. The wound should be treated with Lysol or weak carbolic acid solution and pasted with a copper fungicide like Blitox or Bordeaux mixture.

(d) *Rhododendron:* A rhododendron bug (*Stephanitis rhododendri*) and a white fly (*Dialeurodes chittendeni*) occasionally attack the foliage of some species, but neither insect is serious in India and hence there is no cause for alarm.

(e) *Salix:* The common pests are willow aphids, willow leaf beetles, willow moths (the caterpillars), scale insects, and willow weevils or borer. The aphides by their sucking activity cause considerable damage to the foliage, shoot and wood which get deformed and may die ultimately. Both adults and grubs of willow leaf beetle feed on leaves, the grubs feeding mainly on the lower surface, while the terminal shoots are specially attacked by the beetles.

The caterpillars of a number of moths attack willows, some feeding on leaves, other spiinning the terminal shoots, and still others burrowing in rods and stumps. Both the spinning and burrowing types are difficult to control since they are not easy to reach because of the fortification around them. However, measures suggested in Table II are helpful in controlling the caterpillars either living in tunnels (borer) or surface dwelling. It will be a wise thing to prune all the spun tender branches and burn them along with the caterpillars. The grubs of willow weevil (borer) burrow in stems and branches and through these wounds fungi and bacteria can invade the plants. The measures suggested for controlling grubs, borers and weevils in Table II should be followed.

The scales remain firmly attached to bark and suck sap which ultimately weaken and kill branches. These should be controlled early before insect population becomes unmanageable.

(f) *Others:* Besides the above mentioned trees many other trees such as *Casuarina* and *gulmohar* may also be infested by borers. The presence of borer is evident from the sawdust collected on the ground around the trunk.

Tuberose

The tuberose is attacked by some pests such as slugs and grasshoppers, which feed on the leaves, and thrips which damage and cause distortion of the spikes.

Sometimes nematodes may become a serious problem, especially in places such as Bangalore where the climate is moderate. The roots of affected plants rot and the plants remain stunted. Temik is an effective nematicide which is applied at the rate of 8-10 kg/ha. Individual plants may be treated with only a pinch of this nematicide. Furadane is also very effective but the quantity required is three times more than Temik.

Water Garden

Some water plants may be attacked by various larvae and caterpillars. Avoid insecticides to control these insects when ornamental fishes are there in the pool. However, these can be hand-picked and destroyed.

Sometimes leach may invade the lily pool. An ingenious method of controlling the leech is suggested here. A tin-can (Bournvita or Baby food or Cigarette tin) with a lid is used. Few small holes (2 mm diameter) are made on the sides near the base. Then the lid is fitted tightly after a large piece of raw liver is placed inside the container. Afterwards the can is lowered in the pool and left there overnight. Unfed leaches will enter the can through the small holes and feed on the liver and increase in size. The fattened leaches are unable to come out through the small holes. The can is taken out and the leaches are killed. To immerse the can, a large stone may be placed inside.

Chapter XXVII

Some Famous Gardens of India

The government, municipalities, city corporations, and some other agencies maintain some sort of gardens and parks in many parts of India. These gardens generally fall into two categories. To the first category belong gardens which are meant for research as well as recreation, while in the second category come gardens catering to the needs of the ordinary citizens. The gardens of the first category are generally referred to as botanical gardens, where all kinds of botanical species are maintained for the benefit of the students of botany and general public. These plantations are designed in such a way that they serve the purpose of a public park also. The second category of gardens can be called as pleasure gardens, where the common man comes, enjoys the grass, plants, and flowers, and gets relaxed from the otherwise drab city life. Unfortunately, the number of parks and gardens in Indian cities and towns are not sufficient and whatever exist are sometimes not looked after well.

Some of the important Indian parks and gardens and their important features are discussed below.

LALBAGH, BANGALORE (KARNATAKA)

Lalbagh is the Karnataka State botanical garden situated at Bangalore. Bangalore with an altitude of about 900 m has got a mild climate and is, therefore, suitable for growing a wide range of plants. The initial layout of the garden was started in 1760 by Hyder Ali. Presently, the garden is the seat of the Directorate of Horticulture of the State.

Possibly the most attractive features of the garden is a large glasshouse where the annual flower shows are held. The glasshouse is also used for holding important meetings, conferences, and for receiving important dignitaries. The next important attraction is possibly the tall majestic-looking *Araucarias*, especially *Araucaria excelsa* (Syn. *A. cookii*) which steal the eyes of the visitors. The avenue of the large *Ficus benjamina* near the glasshouse is definitely a special feature of the garden. Plants such as *Tecoma argentea*, *Tabebuia avalandii*, and *Tabebuia spectabilis* are prize collections of the garden. The garden has good collections of different *Cassias*. The collections of bougainvillea and hibiscus are also very rich. The other attractive features of the garden are collections of foliage plants, pergolas, arbours, statues, fountains, water garden, aquarium, rose garden, a beautiful natural rock formation and others. The garden with its well-laid-out features, paths, open spaces, shade and flowering trees attracts a large number of visitors everyday.

Besides ornamental trees, the garden has a fruit section. The total area of the garden is above 50 hectares.

BRINDAVAN GARDENS, MYSORE

This garden is one of the most important tourist spots in India. The garden is famous mainly for its illuminated running waters and innumerable fountains decorated by coloured lightings. In the evening when all the fountains and running water start working and are illuminated with changing colour of lights, the whole place looks like a paradise. The fountain and lightings, apart from the garden, have open spaces under lawn, illuminated flower beds and other ornamental plantings. The river Cauvery below the giant dam divides the garden into two parts and visitors enjoy a boat ride in the river.

The garden is a public park and not meant for research.

GOVERNMENT BOTANIC GARDENS, OOTACAMUND

This garden is situated at an altitude of 2,175-2,280 m above sea level in the Nilgiris Hills. It actually started functioning in 1848. The garden covers an area of 20 hectares in ascending terraces. The garden has six major sections: (i) the lower gardens consisting of the main entrance, the lower lawns, and the new gardens; (ii) the Bandstand; (iii) the Conservatory, the Bog Garden, and the Tennis Court Gardens; (iv) the terrace with fountains and picnic spots; (v) the new Bandstand and its surroundings; and (vi) the nursery area.

The garden is the pioneer in introducing potato, cabbage, cauliflower, carrot, beetroot, etc., and many fruits in the Nilgiris. It is also a pioneer in introducing Cinchona and different species of *Eucalyptus* in this region. Many essential oil-yielding plants were introduced by this garden, out of which scented geranium is the most important.

SIM'S PARK, COONOOR (TAMIL NADU)

The park was established in 1874 by J.D. Sim after whom it was named. This park in the Nilgiri Hills is the centre of attraction for the people of Coonoor and Wellington. The park is spread over an area of 15 hectares at an altitude of about 1,740 m above the sea level.

Some of the important features of the garden are the terraces with winding paths, lawns, beds of annuals, rockeries, a pond with two islands, arches, and pergolas training creepers. The park has a good collection of roses, trees of ornamental as well as economic value, shrubs, and creepers.

BOTANIC GARDEN, COIMBATORE (TAMIL NADU)

The botanic garden situated on the campus of Tamil Nadu Agricultural University, against the background of hills, presents a unique look. The layout in the garden is a harmony between the formal and the informal designs. In the background of this garden is a bluish-green hill range which has been incorporated in the designing of different garden features.

The garden is comparatively young compared to other botanical gardens. It was established in 1908 by the Department of Agriculture, Madras. The main function of the garden is to raise and maintain trees, shrubs, herbaceous plants, etc., which are useful from economic as well as ornamental and bioaesthetic point of view. The garden meets the needs of public, scientists, and government department in the supply of plants.

The garden has many features out of which possibly the sunken garden is most interesting. The topiary work in the garden is most interesting which attracts the attention of any visitor. Besides these, the children's corner, greenhouses, rockery, *Hibiscus* varietal collection, large stretches of lawns, and rosary are also important features of the garden. The botanic garden has large collections of various types of plant materials.

THE BYRANT PARK, KODAIKANAL (TAMIL NADU)

The park has a total area of 10 hectares. The park was actually laid out in 1909, but it suffered due to the lack of trained personnel and funds until 1961 when it was taken over by the Agricultural Department of Tamil Nadu government and development work started in right earnest. The park has terrace gardens, lawns, children's park, a sunken garden, besides a good collection of roses, chrysanthemums, trees, and shrubs. The park is a centre for supplying ornamental plants.

THE INDIAN BOTANIC GARDEN, SIBPUR, CALCUTTA (WEST BENGAL)

Actually speaking, the garden is not situated in Calcutta proper, but is in the twin city of Howrah, on the opposite side of the river Hoogly. The garden was established in 1787 on a 150 hectare land. It ranks among the great botanical gardens of the world and is the seat of the Botanical Survey of India. The garden was established on the suggestion of Robert Kyd, an army man. The garden had a unique privilege of having famous scientists as its superintendents such as Dr. William Roxburg, Sir J. Hooker, Dr. N. Wallich, Dr. Hugh Falconer, Sir George King, Sir William Wright Smith, Sir David Prain and Dr. C.C. Calder. The first Indian to occupy this post was Dr. K. Biswas.

The feature which attracts most visitors first is the giant 200-year-old banian tree (*Ficus benghalensis*). The diameter of the crown is more than 330 m and the tree had 666 aerial roots in 1938 which are still growing in number. The main trunk measuring about 15 m in girth in 1885 decayed long back but the tree is flourishing as before. The large collection of palms in the Palmetum with a pond in the foreground is another important feature of the garden. The garden has 26 lakes occupying about one-ninth of the total area and inter-connected by underground pipes.

Another important feature is the giant lily, *Victoria regia* which has giant disc-like leaves raised at the margins. It is reported that the leaves floating on the surface of water can withstand the weight of a baby.

The garden has 15,000 trees and shrubs in the open, representing 2,500 species. In addition, the palm houses, orchid houses, and ferneries house several thousand herbaceous plants. The Royal Palm Avenue near the river gate is another beautiful feature of the garden. The garden houses the best herbarium in the country and the library has more than 25,000 volumes.

The activities of the garden are distribution of seeds and seedlings, conducting botanical explorations, collection, preservations and identifications of plants, besides research on botany and horticulture. The garden has to its credit the introduction of economic plants such as coffee, India rubber, tobacco, cardamom, cocoa, flax, hemp, etc.

THE AGRI-HORTICULTURAL SOCIETY GARDENS, CALCUTTA (WEST BENGAL)

It is possibly the oldest Horticultural Society of India. The society first started its garden in 1825, but the garden in the present site was established in 1872.

The Society's gardens are well laid out and a metalled path goes round the major parts of the garden. Near the main entrance at Alipore Road, there is a collection of Bougainvilleas and behind this a large pond houses good collections of water-lilies. Besides, the garden has feature gardens complete with rockeries and lily ponds, a large independent rock garden, sunken garden, and a long beautiful pergola with a rich collection of creepers and climbers, a huge open lawn, children's corner, auditorium, a research laboratory with air-conditioned rooms and mist chamber, and rose garden. The garden has a large nursery for propagation of plants and every year a large number of trees, shrubs, creepers, seedlings, and seeds of annuals are sold from here.

The society has possibly the largest collection of trees, shrubs, and creepers including some hybrids developed in this garden by the famous horticulturist, S. Percy Lancaster. Among the hybrid trees mention may be made of a white flowering *Cassia fistula*, *C. lancasteri*, and *Lagerstroemia lancasteri*. There are 38 species and cultivars of *Plumeria* many of which are hybrids developed at this garden. The garden has three greenhouses having rich collections of shade-loving plants and house plants.

The society also conducts a diploma course in Horticulture, which is recognized by the Calcutta University. The Birla laboratory with a well-equipped laboratory and a mist chamber is the centre of advanced research on plant propagation and other aspects of physiology of plant growth.

The society conducts several flower shows and garden competitions out of which the annual flower show held in February is a big event in the life of Calcutta city. The society publishes *Horticultural Bulletin* containing useful articles on different aspects of gardening. Several books on gardening have also been published from here.

With its rich collection of plants and the nicely laid-out gardens, this garden is positively the most important centre for ornamental horticulture in the country.

LLYOD BOTANIC GARDEN, DARJEELING (WEST BENGAL)

Established in 1878 and situated at an altitude of about 2,100 m in the midst of the Himalayas, it is one of the most picturesque botanic gardens of India. The garden was laid out under the expert supervision of the famous scientist, Sir George King, the then superintendent of the Royal Botanic Garden (now Indian Botanic Garden), Calcutta, on a piece of land donated by William Llyod, an old resident of the town. The garden has a total area of 24 hectares laid out in beautiful terraces and provided with metalled approach roads. The garden has the following sections:

(1) Upper indigenous section consisting of dominant species of the eastern as well as of the western Himalayas and Burma.
(2) A central or middle section containing coniferous plants.
(3) A miscellaneous section predominantly occupied by species from the eastern Himalayas and other parts of India and abroad.
(4) A rock garden, naturally displaying flora of high altitudes and of the alpine region of this locality.
(5) The lower exotic section houses plants mainly from the temperate regions of the world.
(6) Orchidarium: There are two large greenhouses housing about 2,500 orchids of different species.
(7) Students' garden consists of the Sikkim Himalayan flora, arranged family-wise.
(8) Cacti and other succulents are housed in a greenhouse.
(9) In the bulbous section there are collections of various species of lilies, amaryllis and tuberous begonias.
(10) There is a rosary containing many cultivars of roses.
(11) The acquatic garden is housed in two pools consisting of species of *Nymphaeaceae* and other submerged acquatics.
(12) The Herbarium and seed section contains 30,000 herbarium sheets arranged according to Bentham and Hookers system of classification. The seed section has collection of fresh seeds of the Sikkim Himalayan plants, which are supplied on request to educational and research institutions in India and abroad.

The garden has about 1,800 botanical species representing regions such as Burma, Malaysia, Central Asia, Japan, North and South America, Europe, and Africa. The garden has a rare collection of exotic conifers representing the genera *Araucaria*, *Callitris*, *Cephalotaxus*, *Cunninghamia*, *Cupressus*, *Picea*, *Retinispora* (also spelt *Retinospora*) *Thujopsis*, etc. Possibly the most interesting plant in the garden is the Dawn Redwood, *Metasequoia glyptostroboides*, popularly known as "Living Fossil Tree". The genus was believed to have been extinct and only fossil records were available. In 1941, this species was found to be growing in wild in China and thus this is the only species of plant having both fossil record and living plants. *Ginkgo biloba*, another plant of the garden, belongs to prehistoric age while its contemporaries have long become extinct. The garden also has collection of exotic maples which change the colour of the leaves during autumn. There is a *Piecea* species in the garden which can be called as a living greenhouse, as a host of orchids, ferns, *Hedychiums* and some other epiphytes can be seen growing all over the plant.

The garden distributes plant and seeds during *Vana-mahotsava*. Every year thousands of tourists from home and abroad visit this garden.

NATIONAL BOTANICAL RESEARCH INSTITUTE, LUCKNOW (UTTAR PRADESH)

The National Botanical Research Institute popularly known as Sikander Bagh, was laid out by Nawab Saadat Ali Khan (1789-1814) which was further improved upon by Nawab Wajid Ali Shah, the latter naming it after his wife Sikander Mahal Begum. The present area of the garden is 27 hectares.

The garden has a library, a large herbarium and has well-equipped laboratories on different branches of botany, chemistry, and a tissue culture laboratory. The main objective of this garden is to extend knowledge for establishing scientifically planned gardens, to carry on research on horticultural (including ornamental) plants, essential oil-bearing and other economic and medicinal plants. The garden has done good work in plant introduction, improvement of ornamentals through breeding and developed improved techniques of ornamentals for rose-oil extraction from *Rosa damascena*, besides other useful researches on economic plants. The technique developed by this garden in preserving flowers and ornamental foliage in dry form for dry flower arrangements and preparing greeting cards has earned it wide acclaim.

The gardens with fine lawns, rose gardens, conservatory, cactus house, and lily pools are very well maintained and attracts streams of visitors from all over the country.

HORTICULTURAL RESEARCH INSTITUTE, SAHARANPUR (U.P.)

The garden was possibly laid out well before 1750. The garden was taken over by the East India Company in 1817. The well-known botanists such as Falconer, Jameson and Duthie worked in this garden, one time or the other.

The garden mainly conducts research on sub-tropical fruits. The garden has a good collection of fruit trees. The collection of ornamental plants such as trees, shrubs, cacti and succulents are also worth mentioning.

RASHTRAPATI BHAVAN GARDEN, NEW DELHI

Rashtrapati Bhavan is the official residence of the President of India and one of the largest buildings of its kind in the world. It was completed in 1929. The architecture of the palace is a mixture of Indian and western style.

The gardens inside this palace was laid on the pattern of Mughal gardens with conventional arrangement of squares, terraces, water channels, etc. The main garden area is roughly 134 square metres and is bounded from all sides by a paved red stone path. Two canals each of 5.40 m width run from north to south and two similar canals intersect these to form a 60 m square island in the centre. This island is the venue for the most of the receptions held at *Rashtrapati Bhavan*. There is a sunken or circular garden which is a beautiful spot especially during the winter when innumerable seasonal flowers bloom. The garden is famous for quantity and quality of seasonal flowers. There are good collections of Bougainvilleas, different bulbous plants as well as flowering trees which ensure enough of colour throughout the year. There is a large collection of roses also. Other important features are greenhouses with collection of orchids, cacti, succulents and ferns. The pergolas are laden with fine creepers. There are a large number of trees, especially cypress (*Cupressus*) which are clipped in the topiary style. The garden remains open for about a month for the general visitors during the winter months when the seasonal flowers are in full bloom. This garden is popularly referred to as Mughal Gardens.

BUDDHA JAYANTI PARK, NEW DELHI

To commemorate the 2500th birth anniversary of Lord Buddha a park was laid out in the rugged and rockey "Ridge" of New Delhi which turned out to be one of the best laid-out landscape gardens in the country. The garden is of about 30 hectares in size. The most praiseworthy feature of this garden is that most of the existing natural flora such as *Butea monosperma*, *Acacias*, etc., were not destroyed and the natural hillocks and rock formations were incorporated in the garden without trying to level or distort them. The plantation was done keeping large open vistas and usable areas. The barren rocks and woodland areas have been planted with bulbs such as *Narcissus*, *Oxalis*, *Zephyranthes* and hardy annuals such as *Coreopsis*, *Cosmos*, and *Tithonia* which seed themselves and grow year after year giving a natural effect. There is a Japanese style of garden in one portion. There is a floral clock also. This garden is a favourable spot of picnickers and other people and children. The garden has a good collection of trees and shrubs.

JAPANESE STYLE GARDENS

Dr. M.S. Randhawa, when he was the Vice-President of the Indian Council of Agricultural Research, once decided to lay out a chain of Japanese style gardens and with this in view he visited the famous Japanese landscape architect Professor K. Mori in 1958. Prof. Mori stayed in India for four months and selected some sites for this purpose and prepared the designs. The first Japanese style garden was laid out in the Roshanara Park, Delhi. The garden was laid out complete with rivulets, water falls, pond with stepping-stones, wooden bridge and the like which are common features of a Japanese garden. Another garden with rivulet, stones, bridge, etc., was laid out in the official residence of the Prime Minister's house at Safdarjung Road. Similarly, a small Japanese style garden was laid out in the residence of the Lt.-Governor of Delhi. The Japanese garden at "Kamala Retreat", the private residence of Sri Padampat Singhania at Kanpur, is also worth visiting. There is a Japanese style garden in the residential complex of Poet Tagore at Santiniketan, which was developed by his son, a U.S.A.-trained horticulturist himself. This garden complex is very rich with wide collections of ornamental trees and shrubs. Fruit trees like mango, guava, sapota etc. has been trained as creepers.

BOTANICAL GARDEN, FOREST RESEARCH INSTITUTE, DEHRADUN (U.P.)

The botanical garden occupies about 10 hectare in this 500-hectare Forest Research Institute's estate. The botanic garden was started in 1934 at an altitude of 663 m. The garden is well laid out and is a place of attraction for the visitors.

The primary function of the garden is to introduce new plants from all over the globe. The garden played an important part in the introduction of forest trees, ornamental and other economic plants. The collection of indigenous flora is also quite rich. The important features of attraction are a greenhouse containing large collection of terrestrial and epiphytic orchids, ferns and other shade-loving plants, a greenhouse containing cacti and succulents; and the plant introduction nursery including a net-house.

In addition to the botanic garden, there is an arboretum of 50 hectares containing mostly forest plants yielding timber and other economic products.

The garden has a herbarium containing 300,000 sheets. The collection of water plants is also quite good. It has a plant and seed exchange programme.

MUGHAL GARDENS OF KASHMIR

Emperor Akbar was the first Mughal ruler to visit Kashmir. Jehangir along with his famous Queen Nur Jahan visited Kashmir several times. Shah Jahan also went there thrice. The credit of developing the Mughal gardens in Kashmir goes to these three rulers. All these gardens have a series of descending terraces, following the tradition of Mughal style, to facilitate the flow of water which is another main stay of the Mughal gardens.

The gardens on the bank of the Dal Lake, Shalimar, Nishat Bagh, and Chasma-e-Shahi are well preserved and mostly frequented by visitors. Some other popular gardens are at Achabal, Verinag, and Bijbehara.

A most spectacular feature of these gardens are the planting of the majestic *chenar* trees in groups.

Shalimar Garden: This garden was initiated by Jehangir in 1619 and extended in 1630 by Zafar Khan, the then Governor of Kashmir, under the instructions of Emperor Shah Jahan.

The garden is connected with the Dal Lake by a 1.6 km canal which is about 10.8 m wide. On both sides of the canal there are broad green paths lined by majestic *chenar* trees. The garden consists of three terraces, the first having a *baradari*, the *Diwan-e-Am*, and the second contains the *Diwan-e-Khas*. But, unfortunately, both these buildings do not exist today, but only their stone bases are left surrounded by fountains. Along the centre of the garden there are a series of water reservoirs inter-connected by a wide canal. The canals and the reservoirs are paved with polished limestone. The source of running water is a stream which flows through these reservoirs and canals and sometimes through beautiful chutes of various designs.

The third terrace containing a magnificent black stone pavilion was meant for ladies. The pavilion is surrounded by a reservoir containing 140 large fountains. Maharaja Hari Singh of Kashmir provided electricity to the garden.

Nishat Bagh: The literary meaning of *Nishat* is "pleasure". This pleasure garden was built by a Persian, Asaf Jah, who happened to be the brother of Nur Jahan. The garden has twelve terraces each representing the sign of the Zodiac. The two wooden doorways and the *baradari* in the third terrace was added by Maharaja Ranbir Singh. The water runs all along the garden through a limestone lined wide canal which has a number of fountains. In the second terrace, there is a large reservoir containing 25 large fountains. There are two main pavilions, one at the lower end and the other at the upper end of the garden. The lower one is a double-storey structure of wood and plaster, in the middle of the floor of which there is a reservoir with numerous fountains. The other pavilion is situated at the end of the upper terrace from where one can have a thorough view of the entire garden. The beautifully designed flower beds and the avenue of *Chenars* attract immediate attention of the visitor. Morning is the best time for visiting this garden.

Chasma-e-Shahi: The garden situated about 8 km away from Srinagar is the smallest among the three gardens of the valley. The main attraction of this garden is a spring spewing clean, ice-cold water having appetizing properties. The garden was constructed by Ali Mardan Khan, the Governor of Kashmir, under the instructions of Emperor Shah Jahan. The garden consists of three terraces, a central aqueduct, water tanks, waterfalls or chute, and fountains. The two *baradaris* of the garden are of recent origin, although the plinth was provided by the Mughal designer. The garden provides a panoramic view of the Dal Lake.

Verinag: The centre of attraction of this garden is the Verinag spring, the water of which is exceedingly pure and highly transparent. The reservoir around it was commenced by Jehangir and completed by Shah Jahan. The garden here has a charming beauty. Emperor Jehangir who expired on his way back from Kashmir wished that he be burried there. But his followers buried him near Lahore. Thus his last wish to have eternal rest in a beautiful surrounding remained unfulfilled.

Achabal: This garden on the outskirts of the town Anantnag was built by Nur Jahan in 1620. The spring at Achabal is the largest in Kashmir, which flows through this garden traversed by three aqueducts. Along the central aqueduct there are two large tanks and the upper one of these has a wooden pavilion constructed on a wooden base. There are six vertical waterfalls in the garden, three on the upper part and three in the lower. The banks of the tanks and the canals are lined with stone and these abound in fish. There is a ruin of *hammam* (Turkish bath) and some other buildings. The garden has a good collection of different fruit trees besides the large majestic *chenars*.

There are two objects of engineering ingenuity—one is a wooden conduit pipe and the other a network of earthen pipes carrying water to the Royal bathrooms. Both objects originated during the Mughal time.

Bijbehara: This garden constructed by Dara Shikoh, the eldest son of Shah Jahan, is situated about 48 km from Srinagar. There are magnificent *chenars* in this garden; the girth of one of the tree measures about 16.2 m.

The garden was divided in two parts which were connected by a stone bridge, only the ruins of which remain today. The design of the garden is typical of the Mughal style.

THE MUGHAL GARDEN, PINJORE (HARYANA)

This garden situated 5 km below Kalka on the Ambala-Simla road, was laid out by Fidai Khan, the foster-brother of Aurangazeb during his reign in the seventeenth century. The original name of the place Panchapura or Panjpur has association with the five *Pandavas* of our epic *Mahabharata*.

It is one of the best preserved gardens of north India and is famous for its beauty. The garden is uniquely laid out in an area of 25 hectares and is divided into six terraces. During its heyday a dignitary paying a visit to the garden saw about 40 *maunds* (each *maund* is equal to about 37 kg) of red roses being sent to the perfumery (*gulab khana*) of Fidai Khan in a single day.

As customary with all Mughal gardens, this garden is also encircled by an embattled wall. The main gate is at the highest terrace while the remaining five appear in a descending way. Because of this, in spite of the formal layout the full garden is not visible to the visitor at one glance.

As in other Mughal gardens there is a central water channel. Water falls from one terrace to the other and into the tanks. The tanks and the water channels have numerous fountains. On other side of the central channel the paths are paved. Along these paths there are lawns, flower beds, trimmed hedges, rows of bottle palms and many other ornamental shrubs and trees. There is a good collection of fruit trees, especially of mango, litchi, and sapota. There are three magnificent buildings—the *Shish Mahal,* the *Rang Mahal*, and the *Jal Mahal*. The credit for the fine collection of ornamental plants goes to the erstwhile rulers of Patiala, especially the former Maharaja Yadavindra Singh. The garden is situated at the foothills of the Himalayas at an altitude of about 600 m.

CHANDIGARH ROSE GARDEN

Rose is a very important ornamental flower for its beauty, elegance, grace, as well as fragrance. Though all the modern roses do not possess fragrance, still the name of rose reminds one of fragrance.

Though rose is so beautiful a flower and grows so well in many parts of India, especially in northern India, there is hardly any rose garden worth its name. The rose garden at Chandigarh fills up this void.

Among the biggest rose gardens of the world the top position goes to Jackson and Perkins of Newark near New York. The garden has an area of 10 hectares and contains more than 36,000 roses of all types including hybrid tea, floribunda, polyanthas, miniatures, climbers, and standards.

The rose garden at Parque Bagatelle in Paris is also a famous garden. Parque Del Oeste, Madrid in Spain, has a rose garden having about 30,000 roses. The rose gardens in Parc Qe La Grange, Geneve, the Queen Mary rose garden in Regent Park, London, and the Golden Walk, Cartwell, England, are also very well known.

The Chandigarh rose garden is situated in the centre of the city on a 15-hectare plot. It is designed to contain about 60,000 roses when completed. It is also contemplated to collect about 5,000 outstanding cultivars of roses. The garden is situated in a valley and a natural stream runs through it. The land on the banks of the stream is undulating and has natural curves. To preserve this natural landscape, the garden has been laid in the most informal and natural manner. The area adjoining the stream has been planted with a large number of scented cultivars, which fill the whole area with exquisite fragrance. A section will be developed as a moonlight garden planted with only pure-white scented roses.

The target of collecting 5,000 named cultivars for the rose museum, however, has not been fulfilled yet, although the garden started in December 1966. At present about 1,500 named roses are there in the museum. Similarly, the target of planting 60,000 roses also remains unfulfilled. Although the ambitious plans and targets of the garden have not yet been fulfilled, still this garden is a wonderful place to visit.

MANDOR GARDEN, JODHPUR (RAJASTHAN)

This garden is situated about 5 km away from the heart of Jodhpur. It is one of the most beautifully laid-out landscape gardens in our country but unfortunately many people do not know about it. The credit of laying the garden goes to Raja Abhai Singh (1724-1749 A.D.). It is highly creditable that such a beautiful garden has been laid out in the desert region of Rajasthan. The garden is fairly large in size and terraces have been constructed on the slopes of the hillocks to extend the garden up to the peaks. The large spaces provided under lawn, the flower beds, trees, and shrubs add beauty to the garden. There are fountains fitted with coloured lights which are run on a payment of Rs. 20 per hour. The tea stall situated at a higher terrace adds beauty to the garden.

It is a pity that this beautiful garden is not getting the publicity and the share of visitors that it so richly deserves.

SAYAJI PARK, BARODA (GUJARAT)

The Sayaji Park in Baroda is named after its founder the former Maharaja Sayajirao III who established it in 1879 over an area of 40 hectares. The garden is a beautiful belt of greenery in the otherwise dry climatic region of Baroda.

The garden has about 8,000 various ornamental trees with rare plants such as *Hyphaene thebaica*, the branching palm, forming an avenue in one corner. The most attractive part of the garden is the bandstand spread over an area of little over one hectare. This formal portion of the garden has the arbour-like bandstand, paved walks, lush green lawns, four fountains, some illuminated with changing multicolour lights, and piped-music.

The garden has a toy train, a giant wheel, childrens' park, and a traffic training centre for children.

The Baroda Museum opened in 1894 is situated in this park. Recently, a planetarium has been added to the park. The park has a zoo also with rare animals such as white tiger.

Chapter XXVIII

Flower Shows and Garden Competitions

In recent times, the holding of flower shows has spread widely and now these are not restricted only to the capital cities of a few States, but are being held in all State capitals as well as in many district and sub-divisional towns. It is because of the good work done by some Agri-Horticultural Societies in India that the growing of flowers by amateurs, exhibiting these in flower shows, and taking part in garden competitions have become so popular in many parts of India. The pioneer among them is the Agri-Horticultural Society of India, Calcutta, which was established in 1820. The society had maintained records from which we know that the first show was arranged in 1827, though it was predominantly a vegetable show.

FLOWER SHOWS

The primary objective of flower show is to create interest among the general public to grow quality flowers and maintain beautiful gardens around their house compounds. The display of quality exhibits inculcates the spirit of healthy competition among the participants. Flower shows give an opportunity to the people to know the wide range of plants that can be grown in the locality. The other advantage is that a visitor gets a chance to see all the best materials at a time in one place. A flower show should be a place for discussing the various garden problems and to find out ways for solving each other's difficulties. Each participants should share his knowledge and experience with others. But, unfortunately, due to professional jealousy people sometimes do not want to divulge their secret to success, lest others copy and score over them. It is also an occasion to demonstrate how a perfect exhibit can be grown. Besides all these, the aesthetic value of a show cannot be overlooked.

The success of a show depends much upon the co-operation between the exhibitor and the members of the Flower Show Committee. The judging should be above criticism though one hundred per cent success cannot be achieved on this last score, as controversies are often bound to crop up even with the best judgement. Some preliminary spadework has to be done to make a show successful especially if it is a maiden one. Firstly, the interested garden lovers of the locality should get together, elect a president, honorary secretary and form some committees to run the show. For example, a general or a ground committee should prepare the general arrangements such as selection of site, arranging tents, *shamianas*, chairs, etc. An editorial committee should prepare the schedules well in advance, prepare the rules and regulations, printing admission tickets, undertake the necessary propaganda and send invitations to V.I.Ps. The finance committee should look after the financial aspect. The other committees may be the plant committee whose duty will be to look after the different plant sections; the floral committee will look after the cut flowers and floral arrangements, and so on.

Besides the duties mentioned above, the different committees should also divide among themselves the following works so that the exhibitor is not put to hardship. Arrangement should be made for watch and ward to check unauthorized visitors and theft. There should be provision for enough water, for drinking as well as for irrigation purposes. Flower vases, bamboo stakes, show passes, benches, and tables for exhibits should be made available to the participants. Arrangements should be made to open refreshment stalls. Nurserymen, seedsmen, companies selling agricultural implements, chemicals, etc., should be allowed to open their stalls on rental basis. Different educative charts on horticulture should be displayed and demonstration on some horticultural operations such as budding, grafting, pruning, etc., should also be arranged. A film show on flowers and gardens, preferably in colour, could be arranged in the evening accompanied by a supporting talk. Many flower show committees also make arrangements for bringing in exotic exhibits from abroad.

Tips to Exhibitors: Most of the amateur exhibitors do not know the proper procedures and the schedule of the flower show and get dejected when their entries are refused. Instead of learning through the trial and error method, the prospective participants will benefit themselves if they go through the following check points.

The first and foremost thing is to get a schedule of the flower show well in advance and to convince oneself of the requirements. For example, roses are grouped into several classes for show purposes, such as Hybrid Tea, Floribunda, etc. and one has to exhibit the right type of rose in the appropriate class. Similarly, it may be specified that the annuals are to be displayed in groups of 6, 12, or 18 pots in as many different varieties.

After going through the schedule it has to be decided in which groups the entries are to be made. Once the decision is made the plants are raised accordingly and seeds of flowering annuals procured. It is to be remembered that only best quality seeds win prizes and so the seeds are to be purchased from reputable nurseries or seedsmen at least a fortnight ahead of the sowing date. The sowing is to be staggered at intervals of 4-7 days so as to avoid disappointment as a result of casualty when planted in one lot. One more reason for staggering the sowing dates is to ensure that at least one group of plants is in perfect condition during the show time, as it may so happen that one group from a particular sowing date may fail to open the flowers on the scheduled date for climatic or other reasons. The plants for exhibition needs extra feeding, but not overfeeding, from the date flower buds start to appear, with liquid manure. To obtain large flowers all auxilliary buds in flowers such as carnation, marigold, dahlia, etc., should be disbudded as soon as they appear leaving only the apical bud to bloom. The seasonal flowers or any other plant should be grown in the appropriate-sized pot and not in an under- or over-sized pot. To make the plants bushy pinching should start at an early date for flowers such as brachycome, carnation, marigold, zinnia etc., and the operation repeated frequently, especially for brachycome, candytuft, petunia, *Schi-*

zanthus, French marigold, etc. Only one annual should be grown in a pot. In some shows it is often seen that 3-4 seedlings are grown in one pot to obtain a compact effect. This is absolutely not needed if the proper size of pot is selected and proper cultural procedure is followed. Some annuals such as zinnia produce flower buds at a very early stage of growth which should be removed constantly until the plants attain sufficient vegetative growth. Some extra pots are to be raised than required as per schedule as few plants may die or fail to bloom on the appropriate date. If annuals are first grown in ground and then transplanted in pots, the operation should be undertaken at least one month before the date of show to enable the plants to spread their roots and overcome the transplanting shock. One should not try to hoodwink the judges by lifting ground-grown plants in pots a few days ahead of the show. Before displaying, the pots are to be cleaned properly or lightly painted with terracota red (*geru*). No plants, either annuals or foliage, should be displayed in a crowded fashion. The judges will like to see the plants individually and will also see the foliage. The foliage and other perennial plants such as bougainvillea and hibiscus are also grown in a similar fashion. They should be fed with liquid manure prepared from oil cake once in a fortnight or 20 days. The leaves should be washed regularly by syringing.

Precautions: A few extra plants are to be taken to the show than required as plants may get damaged while in transit. The branches should be staked properly during transportation to prevent damage. Some sort of staking is permitted in shows for certain category of plants but this should not be obtrusive. During transit delicate flowers may be covered with tissue paper as a precaution against being bruised. Often plants get stolen from shows and one should do well to keep a vigil on his exhibits. Identification marks should be put on the pots, but this should not show prominently.

Staging the exhibits: The staging of exhibits is some sort of an art and the value of artistic display cannot be minimized in winning prizes. A stage rehearsal of grouping of plants and flowers may be undertaken at home to get the best combination and placement. In the tension ridden show ground people commit blunders in arranging the exhibits. In the shows at Bangalore, the exhibits are arranged in a gallery fashion and one should arrive well ahead of show time as this is a time-consuming process. All plants are displayed distinctly to make the task easy for the judges. The colour arrangement should be artistic and harmonize with each other and the background. For example, crimson or scarlet colour shows well against a white background or yellow and white suits well against a black background.

The judgement is made on the rarity of the specimen, the bloom, the cultivation, and the impact or effectiveness of the group. A plant if not well grown should not be staged, even if it is a rare plant. Dead leaves and dried flowers should be removed carefully without leaving a scar. After the plants are arranged, the leaves should be washed with clean water. Artificial polishing of leaves is forbidden. The exhibits are to be correctly labelled with indelible ink.

Cut blooms: The blooms for exhibiting in the cut flower section are raised with extra feeding with liquid manure and disbudding. It is, sometimes, advisable to cover rose blooms in polythene or paperbags a few days before the show date to prevent injury from sun, frost, or rain. A pertinent question is: When should the blooms be cut for the show? In temperate countries, a rose bloom is cut in the previous afternoon. But under Indian conditions it is advisable to cut the bloom in the early morning of the day of the show unless these are to be carried long distances, in which event this has to be cut earlier. Often exhibitors cut promising buds which are just opening a few days ahead of the show and store them in a refrigerator at 1.5 to 3°C. The stem is kept dipped in water in some receptacle. To prevent freez-drying, buds may be wrapped in polythene. The refrigerator should not contain any food stuff during this period. A little portion of the stem should be recut everyday under water to reopen the water absorbing channels. The flowers should be taken to the show with stems dipped in water and blooms wrapped with tissue paper. It is advisible to take the blooms in hard cardboard boxes wrapped in wet cloth or moss if these are to be taken to long distances. It is often necessary to dress up a little, a flower like rose. A small lower petal or an insect-damaged petal may have to be removed. This can be done by gently pulling the petal downwards and sideways without leaving a trace. Specks of dust or odd insects should be removed with a camel hair brush. Some extra flowers should always be taken as a few may get damaged or wither. In most of the shows no artificial aid is allowed to prop up the blooms. Otherwise also it is best to show flowers with their own foliage. It is also to be found out from the schedule the length of the stem permitted for roses and other blooms. A rose bloom should at least be half open and not more than three-quarters open, except if specified otherwise, in the specimen bloom classes. To keep the blooms upright in the vase it may be padded up with leaves and cut stems below the vase level. All blooms are to be labelled in indelible ink.

Judging: Only competent persons having sound knowledge on the subject should be appointed judges. The integrity of judges should be above suspicion. In a small show a group of three judges will be more than enough to judge the exhibits in all the groups. But in shows of larger size the exhibits are classified in several groups such as plants in pots (other than flowering annuals), flowering annuals, cut flowers (excluding floral decoration), and table decoration and floral arrangement, and a group of two to three judges are assigned for each group.

The duties of a judge are manifold. He should penalize a crowded display which robs the individuality of a plant or tries to obscure a defective plant from viewing. Plants showing signs of wilting which are lifted from ground and potted should be disqualified outright. Plants grown in over- or under-sized pots should be penalized. Plants not grown in the locality should not be entertained. An artificial colouring of foliage or bloom should not be permitted. Biennials or perennials impatiens, carnation, etc. which flower naturally in one year may be allowed in the annual section, but dahlias and carnations raised from cuttings should not be allowed as annuals. Specimens not labelled or incorrectly labelled should be rejected. A group should contain only the specified number of plants and varieties mentioned in the schedule. Only one plant per pot should be allowed.

Score cards: There is a set pattern for judging different categories of plants and a specimen blooms and marks are awarded accordingly.

(A) *Foliage plants:* Foliage plants are judged on the basis of variety of the specimen, culture, appearance or health of the plant. Points may be awarded for display also. The marks may be distributed as follows:

Culture	4
Quality (including rarity)	3
Appearance of foliage	2
Display	1

(B) *Annuals:* Annuals lifted from ground just before show should be disallowed. Generally, only one plant per pot should be allowed (in some shows more than one plant is allowed). The marks are awarded as follows:

Quality and Quantity of bloom	4
General appearance	3
Rarity of the specimen	2
Display and colour scheme	1

(C) *Cut flowers:* Marks should be deducted if cut flowers are supported artificially unless specified otherwise. The substance, i.e. quality of petals in relation to symmetry, colour, length of stem, vigour, size, decorative effect and form of the flower are important points to be considered. The usual score-card is:

Size and form	5
Colour and rarity	3
Arrangement	2

The Rose Society of India has divided the marks for judging specimen rose blooms in vase or bowl, as follows:

Colour	20
Form	20
Size	20
Substance (Texture, firmness and thickness of petals)	20
Foliage	10
Stem	10

Foliage and stem carry no marks when blooms are exhibited in boxes. But a weak neck should lose points on this count.

Floral display: The quality of a floral display is judged mainly by its aesthetic and artistic value. Hence, the superiority of a particular exhibit over the other depends largely on the personal judgement of the person judging rather than on specific rules and points. For this reason the judgement on this section is questioned more often than in the other sections. The marks are distributed as follows:

General set-up	5
Colour scheme	3
Novel ideas	2

Bouquet: There are mainly two types of bouquet, namely, posy and shower, and each should be placed in separate groups while judging. The usual marks allotted on different counts are:

Size and artistic value	5
Arrangement	3
Novelty	2

Button-holes, Sprays, Garlands, etc: In judging these items it should be noted if the contrasting shades have been used judiciously and whether the flowers have been cramped against one another thus spoiling the effect. The quality of a button-hole or a garland depends largely on the right type of backing used and this should be checked. The freshness of the flower should also form a criterion for judgement. The usual marks allotted are as follows:

Tindiness of the arrangement	5
Colour or appeal and freshness	3
Novelty	2

In the above score cards wherever the total marks are 10, this should be multiplied, in multiples of ten (e.g., 20, 50, etc.) for the convenience of allotting marks. Accordingly, the marks for different items under one group will also increase. For example, if the mark in any one item is only 1 as it is in the item "display" of the first score card above (item A), it will be difficult to allot marks for a bad display compared to a good one. But if the total marks are increased to 50 in this case, then the marks for display become 5, and thus make it convenient to allot marks for different qualities of display.

GARDEN COMPETITION

Along with the flower show, many horticultural societies run garden competitions, generally a week ahead of the actual show. All amateurs are invited well in advance to enter themselves for the competition and are advised to register their names three to four months in advance of the judging date. The gardens should be placed in different categories such as private residential gardens, factory gardens, institutional or government gardens etc., and judged separately. The aim of such competition is to make the people garden-minded and correct the deficiencies by constructive criticism.

Generally, the amateur tries to pack his garden with more colour by planting masses of annual flowers. But a garden without the proper landscaping effect is no garden at all. A garden should have an "all the year round garden" character so that the judges visiting it anytime of the year find the place worth visiting. In his earnestness to copy others, an amateur may crowd his garden with many features, which may look out of place. For instance, a sundial in the lath-house where sun never enters or a pergola leading to nowhere, or a Japanese lantern in a formal garden, etc., may look odd and spoil the naturalness of the garden. Similarly, a large shade tree such as the rain tree or a flowering tree such as *Peltophorum ferrugineum* will be completely out of place in a small home garden.

Some features in a garden are definitely needed, but this should be created in such a manner as to enhance the beauty of the garden, and not added just as a personal pride of the owner. The lawn should be kept trimmed and fed with liquid manure to make it lush green. The borders of the beds should be cut neatly, and the garden kept tidy by removing dead wood, leaves, grasses, and other rubbish. The judges will visit all the corners of the garden, and any artificial camouflaging such as putting potted plants in an empty flower bed, will be detected and penalized. It is rather advisable to keep empty beds neat and clean. Any specialities in a garden like rare collections of palm, cacti and succulents, and *bonsai* will receive special attention from the judges. The following score card could be used to judge the small and large gardens:

	Small Garden	*Large Garden*
(I) *Overall effect of the garden*		
1. General landscape and planning (including lawn)	75	100
2. Shrubbery	30	30
3. Proper use of climbers	20	20
4. An all-the-year-round effect for the garden (i.e. less annuals)	30	30
5. Features such as bird bath, statute, green-house, arbour, etc.	30	30
6. Specialities of a garden, e.g. collection of *bonsai*, palms, cacti and succulents	40	40
(II) *Lawns and grasses in other places*		
Places such as tennis court, round the annual garden, crazy paving, etc. (Points to be considered are quality and colour of grass, mowed or not, weed-free or not)	75	100
(III) *Flowering annuals*		
1. Uncommon cultivars and varieties	10	10
2. Design of the beds, selection of site and its suitability in the place chosen (e.g., sunny annuals sown under a tree is out of place), harmony in colour selection	50	50
3. Quality and quantity of bloom	40	40
(IV) *Maintenance and attention*	50	50
	450	500

Chapter XXIX

Floral Ornaments and Flower Arrangements

The human nature is such that it does not remain content by only growing the flowers in the garden but is tempted to use the flowers for interior decoration, hair decoration, making garlands, bouquets and for worship. The staunch nature lover may advocate that flowers look best on the plants only. While this is true to a great extent, the practice of using flowers for decoration is an age-old tradition, which is followed today and may remain so far ages to come. Hence, it is considered necessary to include a chapter on floral decoration in this book.

GARLANDS AND OTHER FLORAL ORNAMENTS

Possibly floral garland is one of the oldest method of using flowers for decoration. In our epics *Mahabharata* and *Ramayana* we often read that the princess proceed to the court of the king for her *Sayamvar Sabha* with a floral garland in her hand to choose a husband of her liking from among the kings assembled there for this occasion. Even today the tradition of putting a floral garland round the neck of men and women on special functions such as marriage, welcome ceremony, etc., is very much prevalent.

Sweet scented flowers are more preferred than others for making a garland. The most sought-after flowers are jasmine and tuberose, though other flowers such as marigold, chrysanthemum, *Tabernaemontana*, and rose are also used for garland-making. Often a flower of lotus is used as the locket of the garland or in between the garland of other flowers. In West Bengal devotees offer garlands of red Hibicus (*H. rosa-sinensis*) to Goddess Kali and the same is worn by a *Tantrik*, i.e., a sage, who worships this particular Goddess.

Garlands are made of only one type of flower or a combination of different flowers. The flowers are held together with the help of a cotton, nylon or a silk thread, inserted in it with the help of a needle. For heavy garlands fine wire strings are used. To make the garlands more gorgeous, in between the flowers threads of *jari* are also inserted. Decorative foliage of some ornamental plants such as ferns are also used in the garland to improve its look. Garlands often have a pendant or locket which is also made of flowers and is generally given a heart shape.

Besides garlands, the other important floral ornaments are bangles and floral crowns. In certain Indian dances the dancer, especially the female dancer, also wears floral bangles along with garlands. The bangles may be made of either jasmine or tuberose among the fragrant flowers, and *Tabernaemontana* and marigold among the non-fragrant flowers. In West Bengal, a floral crown is also made to be used in ceremonies by the central character, especially during *Annaprasana*, a ceremony when the newborn takes its first cereal food, generally cooked rice, at the age of 6-9 months-wearing a floral crown. A floral crown may also be worn by a female character such as a queen or a princess in a drama. The flower most used for this purpose is tuberose. At first a crown-shaped structure of cardboard is made over which the flowers are sewn or glued. Fine wire strings may be used to make the structure and the flowers may be sewn inside the wires. In the countryside, generally, the frame is made of a large leathery leaf, such as jackfruit or *Kadamba*, on which the flowers are sewn with the help of cotton thread and needle. Ear-rings and *Bajubandhs*, made of flowers, are also used in ceremonial dances and other functions. It is quite common to see photographs or paintings of the epical heroes and heroines such as Rama and Sita wearing floral ornaments such as garlands, bangles, and *bajubandhs*.

Hair Decoration: The temptation of women to decorate their hair with flowers is an age-old custom. The classic of poet Kalidasa, the *Meghaduta*, written during the rule of Chandragupta II (A.D. 500), mentions that the women of Alakapuri in monsoon decorate their hair with *Kadamba* flowers, wear *Sirisha* flowers (*Albizzia lebbek*), in the ears and carry pink lotuses in their hands. Though the present-day elite women may not put *Sirisha* flowers in the ears, their tradition of hairdo with flowers still continues. But among the tribal women it can be found even today that they decorate their ears with flowers. Here again, the fragrant flowers get preference over others in hair decoration. But this is not the case always, as in south India flowers such as *Crossandra* and *Barleria* are widely used for hair style (*gajra*). The practice may be to put one or more flowers singly in the plaits of hair or to make minigarland-like short chain of flowers (*gajra*) and dangle this from the hairdo or to wrap around the hairdo. The chain may be made out of a single fragrant species of flowers such a jasmine, tuberose, or *Michelia champaka* or a mixture of flowers and decorative foliage. In south India, flowers of *Barleria* and *Crossandra* are widely used for making the chains either singly or in combination with jasmines. Roses are used generally singly in decorating the hair.

During season many flowering annuals or perennials such as chrysanthemum, especially the small-flowered ones, *Gomphrena* (globe amaranth), carnation, snapdragon, sweet pea, dianthus (pink), etc., are used for hair decoration. Though it is not very common in our country, in other countries, especially in Hawaii and in the islands of Tahiti, flowers of Cattleya orchids are commonly worn in the hair.

Veni, a special kind of flower arrangement, is widely used in south India to decorate the long plait of hair (*veni*) during some ceremonies such as marriage or at the time of *Bharat Natyam* dance recital. This flower arrangement consists of about 90 cm (depending upon the length of the plait of hair) long and 7-10 cm broad *veni*-like structure which may be made of a frame of a comparatively hard cardboard or some leathery and tough leaves over which different flowers are arranged decoratively covering completely the background frame. Several decorative materials such as *jari* thread may be used to make the *veni* more gorgeous. This is attached to a hair style in which the tress of hairs are made into a plait by interlacing. This hair style is known variously in the vernacular as *veni*, *choti*, and *binuni*. Often the ladies take photographs wearing this floral *veni*. To make the beauty of the face as well as the *veni* to be captured in the photograph, the lady seats with her back facing a dressing table mirror or any other large mirror, so that the direct view of her face and the mirror image of the floral *veni* come in the focus of the camera.

Rangoli: Rangoli, the drawing of pattern with the help of dry colour or a coloured paste on the floor, is a common practice especially among the Hindus in India. This custom of *rangoli* has a religious overtone and is done during *puja* in front of temples or in some social functions such as marriage. *Rangoli* with flowers is also a common practice in India, especially in the south. Generally, the petals of different flowers are taken out and are arranged in various patterns. Intact flowers of small-flowered chrysanthemums and other flowers can also be used for this purpose. Though the *rangoli* of flowers may look quite attractive for a short duration, the authors feel that this is not a very deserving way of using flowers for decoration, as the flowers or flowering petals devoid of any water supply wither away in no time. Similarly, putting flowers for hair decoration does not find favour with many flower lovers, as the flowers wither away within a short time. The best way of putting flowers for decoration is to use them in flower arrangements in vases containing water, where they may last for a period in time.

Floral Bouquets: It is a usual practice to present a nice floral bouquet on the occasion of birthdays, marriages, or to welcome important dignitaries. Bouquets may be of many types and shapes. The usual shapes are the flat and the round types, but arranging bouquets in decorative baskets is also considered as of high taste and beauty.

For making a flat bouquet a hard poster paper in white or any other pleasing colours matching with the flowers is used having length of 45-75 cm. It is given a conical shape by folding the base or cutting the paper in that shape. Over this poster paper aluminium foil (white or coloured) may be laid to make the bouquet colourful. The spikes are laid flat over this paper, with the end having the flowers placed on the broader end, and the bare stems converging on the narrower side. The stems may be held in position with the help of cellotape stuck in 2-3 places against the paper backing. Sometimes, ornamental foliage such as *Thuja* and *Polyalthia longifolia* are also spread over the paper before arranging the flower stems. Flowers such as gladious can be arranged in one plane since the numerous florets can cover the whole spread of the paper. Flowers such as roses should be arranged in layers on different planes so that a lot of blank space is not left uncovered. *Jari* threads can also be spread over the flowers. Once the flowers are arranged they are covered with a cellophane paper which is fixed with the backing poster paper at several places on the top and side with the help of staples, as this arrangement will prevent the flowers from withering quickly as well as enhance beauty and prevent damage. The cut end of the flower stems at the narrow end of the bouquet should preferably be wrapped with wet moss and the moss camouflaged with the help of some aluminium foil, which will also prevent leakage of water. The stems converging at the base of the bouquet are tied tightly with twine/thread to prevent displacement. The twine is covered by wrapping around a nice silken ribbon which is patterned into a floral motif. A card may be attached near the ribbon announcing the name of the well-wisher along with the message of goodwill.

It is somewhat easier to prepare a round-shaped bouquet, but this is not as appealing as the flat one, although it has a beauty of its own. The flowers are arranged in whorl, which takes the shape of a cone, the stem-end becoming tapering while the flower-end comes in a round whorl. Once the flowers are arranged, on the outer whorl some ornamental foliage is backed up to make the bouquet firm as well as decorative. The foliage commonly used for this purpose are fronds of ferns (the long tapering types), *Thuja*, *Polyalthia longifolia*, *Nandina domestica*, and crotons. The base of the bouquet is tied firmly with gunny twine which is then camouflaged with a silk ribbon. The ribbon should be tied artistically. Wet moss may also be used as in the previous type.

Bouquets in the form of flower and foliage arrangements in a basket are also used for presentation or decoration. Here the person preparing the bouquet is much more free to use his ideas. A nice basket is selected in which the foliage and flowers are arranged. Other inert materials such as coloured festoon papers (tissue papers), ribbons, *jari*, drift wood, etc., may also be used for this type of bouquet. Even two or three large chrysanthemum flowers with long stem along with a few of its foliage will look most beautiful in a well-designed and nicely-patterned cane basket. When only one type of flower such as rose is used, it is suggested to cut the stems of the individual flowers at different lengths so that the flowers in the basket are at various levels adding artistic touch to the bouquet. Various types of ornamental leaves can also be added to the bouquet. A useful suggestion to keep the flowers fresh for longer periods is to put a small wet ball of cotton at the cut end of the stem and wrap the ball with a small piece of polythene, the open ends of which are secured with the flower stem by the help of a rubber band to prevent the moisture escaping out of the cotton. The whole arrangement may also be encased with a cellophane paper, if possible.

Buttonholes: This is another fascinating item in floral decoration. These are generally worn by males in their coat collars near the chest, where a provision is made by criss-crossing two threads to hold the buttonhole. But there is no harm if the ladies decorate their hairdo with a buttonhole. Of course, the ladies can also wear this in their coats as the males.

The most suitable flowers for a buttonhole are possibly roses and orchids. But other flowers such as carnation, camellia, and daisy may also be used. This is prepared by first taking a small portion (about 5-10 cm long and 4-7 cm broad) of a leaf of a *Thuja* or *Asparagus plumosus*, the most commonly used materials for this purpose, and then fixing the desired flower or flower bud in position with the help of threads. Often a small coconut broomstick is attached at the base of the buttonhole to facilitate insertion in the appropriate place. Here also *jari* threads may be used for brightening up the look. To camouflage the broomstick aluminium-foil may be wrapped around.

FLOWER ARRANGEMENTS

In a book like this it is impossible to discuss the various types of floral arrangement, the rules, the regulations, and the customs associated with these, in detail. Some guidelines and useful hints are given here for a workable knowledge. Arranging flowers is a kind of art and a successful flower arranger is a born artist. But one can become reasonably conversant about this art with proper training, a lot of patience and perseverance.

What is meant by a design, whether it is a floral design or designing a city? The answer is that any design is meant to bring orderly form. The Japanese method of flower arrangement, popularly known as *Ikebana*, is possibly the most popular method of floral arrangement now in vogue. One of the major differences between the western style of arrangement and the Japanese is that while the former lays emphasis on mass arrangement, the latter uses only a few blooms and branches. Of course, the more basic difference is that the Japanese style of arrangement has a spiritual and religious background, while the western style is rather a form of art only.

Japanese Flower Arrangement

The first school of *Ikebana*, called the *Ikenobo*, meaning "The Arrangement of Flowers of the hermitage along a pond", was started by the Buddhist Monk Semmu around A.D. 621. Previous to his time the flower offerings were in the form of standing arrangements, which was tall in nature and was called *Tatebana*. The Japanese flower arrangement has its root in the custom of offering flowers to the image of the Buddha. As a result of this, the *Ikebana* has some rigid rules. But later some grand masters of the art rebelled against the rigid rules of *Ikebana* and allowed the students to use their creative imagination, without deviating much from the basic rules. One such new style is the *Sogetsu* style. In the present-day Japan, there are numerous *Ikebana* schools following their own styles and designs. But all such schools follow three basic lines of the traditional *Ikebana* school, namely, "Heaven", "Man", and "Earth".

A few hints about the *Ikebana* styles of one such school and the methods they follow are discussed here.

The basic style in this school is the *Moribana*, meaning "Piled Flowers" in which the arrangements are made in shallow containers. The flowers and branches are supported with the help of pin-holders (called *Kensan* in Japanese). This arrangement looks very natural and is often referred to as the natural style of *Ikebana*. Another style called *Nageire*, meaning literally "thrown in", needs tall upright vases for making the arrangements. Here also the flowers are arranged as naturally as possible. But in this arrangement the flowers must have sufficient stem length taking into consideration that, a considerable part will remain hidden in the tall container. In this style the flower stems are not held by pin-holders in the vase but are supported with the help of "cross-bar fixture", "single-bar fixture", and "split-bar fixture" (see below for details).

There is another method called *Jiyu-bana* (Free flowers) which can be arranged in both *Moribana* and *Nageire* styles. While in the latter two styles generally only the materials available in the season are used, in *Jiyu-bana* one can use combinations of wood, metal or, any other material. In this style, though the prescribed rules about angles and measurements are ignored to a great extent, the basic themes of "Heaven", "Man", and "Earth" are never overlooked. Though one is free to use one's own hand in this arrangement, the total effect should create an out-of-the-way and impressive arrangement. In *Zen'ei-ka* (Abstract style) the arrangement does not simulate nature. In this arrangement plant materials are used in combination with any other inert material that a person can conceive of. Abstractness does not mean that one is free to do whatever he likes. The arrangement should be pleasing and in agreement with the surroundings of the room. In abstract arrangement, generally straight materials of uneven heights are used. There is still another style, the *Zen'ei-bana* (*Avant-garde*), which is a piece of art where a beautiful sculpture is created using wood, stone, rocks or metal, may be depicting a seaside or any other natural scenery.

Then there is the combination of fruit and flower arrangement which the Japanese call *Morimono*. Though the Japanese are very fastidious about rules in their floral arrangements, for this particular type they hardly follow the rules for lines and angles as long as the arrangement is more or less triangular in shape and having height, width, and depth. Besides these Japanese styles, there can be an all foliage arrangement, using various beautiful foliage types in combination.

Materials Required

The first thing is to have a suitable container. The container must be good by itself, as an ugly looking container can spoil the whole show. It is not necessary to have only Japanese type vases, as any similar vases available in the locality will be suitable for *Nageire* arrangement. Any good looking flat saucer-type container which holds water is good enough for *Moribana* arrangement. Natural materials such as a hollow stone or a sea shell can also look beautiful as a container.

A pair of sharp-edged secateurs is needed to trim the flower stems into desired shape. Alternatively, a sharp pair of scissors may be used, but this may not work with hard stems. For *Moribana* arrangement sharp, somewhat heavy, metal pin-holders are essential. A fine saw should be kept ready to cut hard stems and another heavier saw or a butcher's knife is needed to cut hard branches. A sharp knife and a fine sprayer, as the barber's sprayer, are also needed. While the former is used for scraping barks or making a slanting cut or preparing a split-bar, the latter is used to spray fine mist of water over the finished flower arrangement, to give a dew-washed effect.

No *Ikebana* school will use pin-holders for stem support in *Nageire* arrangement. Different types of bar are used for the support of the flowering stem in the vase. There. are three types of fixuture, namely, *split-bar* fixture, *single-bar* fixture, and *cross-bar* fixture, the mechanism of each will be easily understood from the illustrations in Fig XXIX.1. The bar fixtures are especially useful when the stem to be used is short in length. In such a case the stem need not have to rest at the bottom of the vase, instead it can rest inside and across the wall of the vase. In a split-bar fixture both the supporting bar and the branch are given wedge shaped split and fitted with each other. For making the joint more firm, these two may be tied with a fine wire. For additional support the cut end of the branch can rest against the vase wall. Additional branches may get support from this main branch. In cross-bar fixture two bars are crossed and fitted a little below the mouth of the vase. Such a bar will easily get stuck halfway in a wide-mouthed narrow-based vase. Once these bars are fixed, the branches need not rest at the bottom but can take support from the bars and against the vase wall. Sometimes, it becomes necessary to bend a thick branch in one or more places for the sake of an artistic display. Normally, it is not possible to bend a thick branch inside a tall vase and hence the wedging method is employed. At the desired point of the branch a deep incision (about two-thirds of the thickness) is made with a knife and the branch is bent in the required shape. The curve can be made permament by inserting a small triangular wedge cut out from a similar branch in this incision, (Fig. XXIX.2.). The bark on the otherside of the branch should not be injured as that will impair flow of water.

In *Moribana* arrangement the pin-holder can be made firm inside the container by placing some florist clay at the base on which the pin-holder is fixed. Water is added

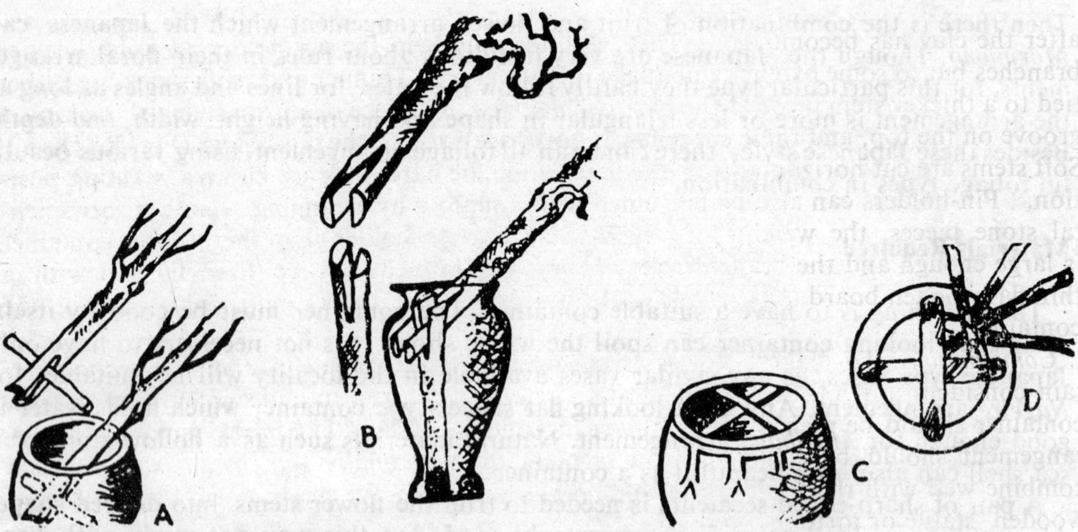

Fig. XXIX.1 Different Kinds of Fixture for Nageire

A. Single-bar-fixture; B. Split-bar; C. Cross-bar; D. Cross-bar fixture showing the procedure of flower stem insertion.

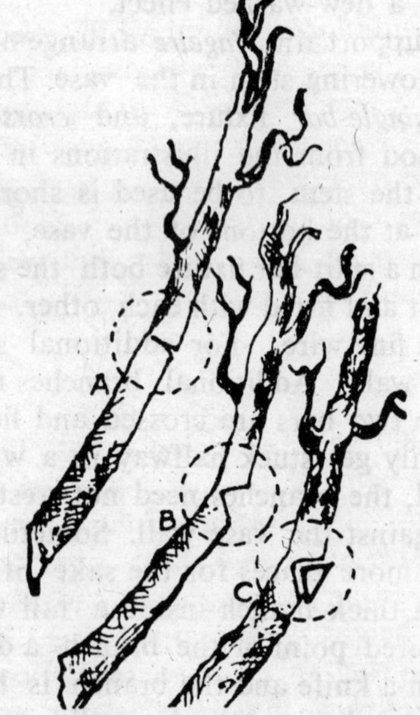

Fig. XXIX.2 Method of Wedging thick Branches for Bending.

A. Deep incision; B. Stem is bent in required shape; C. A small triangular wedge inserted in curve.

Fig. XXIX.3 Supporting Thin Stems

A. A thin or soft stem is tied at base to a thicker stem or B. Inserted to a thicker stem by making groove.

after the clay has become dry. This will not only help the pin-holder to hold heavier branches but to some extent camouflage it from view. Thinner and soft stems are either tied to a thicker stem piece at the base or inserted in a piece of sturdy stem by making a groove on the top, and these supports, in turn, are fixed to the pin-holder (Fig. XXIX. 3.). Soft stems are cut horizontally at the base, while the hard ones are cut in a slanting position. Pin-holders can also be prevented from toppling by arranging around it ornamental stone pieces, the weights of which will keep the holder in place. If the container is large enough and the branch is very heavy, the branch can be fixed by nail with a thin flat wooden board (12 cm long × 4 cm wide × 1 cm thick) and then placed in the container.

Combinations: To make a flower arrangement artistic, beautiful and composite, certain considerations other than the flowers should be taken into account. The shallow container should be placed over wooden or bamboo bases. The background of the arrangement should be complementary or matching and finally the arrangement should combine well with the other ornamental objects used in the arrangement like a curved wooden statue or tortoise or frog or a small bird, drift-wood, stone pieces, slim bamboo or wooden pieces, either coloured or left with natural colour. An experienced *Ikebana* artist can add many other objects of art to his or her arrangements. The aspects of placement is another point which is very important for any arrangement. Placing a beautiful arrangement in the right place of the room itself is an important consideration. The background colour should be complementary to the colour of the flowers or it should be a pleasing contrast colour.

As has already been stated that *Ikebana* arrangement should look natural while selecting plant materials use things which are appropriate for the season as are available in the gardens or parks during that period. In the winter, winter flowers in combination with bare branches look most appropriate. Likewise in the spring, branches with new foliage look most natural. The Japanese also use wall-hanging in combination with the arrangements. These wall-hangings should depict the theme of the appropriate season. A summer wall-hanging is out of place with a spring arrangement and so on. But wall-hangings with images of the Buddha or his life can go with any arrangement.

General Rules and Basic Styles

It has been told time and again that *Ikebana* arrangement should look natural. The whole arrangement should present a look as if it is a flowering bush or tree planted in the container itself. In any style of *Ikebana* arrangement there are three main lines, viz., "Heaven", "Man", and "Earth". According to *Sogetsu* school these lines are named *Shin* (Heaven), *Soe* (Man), and *Hikae* (Earth). In any arrangement the *Shin* should always be the tallest stem, followed in terms of height by *Soe* and *Hikae*. In addition, there are the *Jushi* stems or the fillers attached to each of the three main lines these associate. In the Moribana arrangement additional *Jushi* on dwarf stems are often used to conceal the pin holder. But this should not be used as a rule but rather resorted to if need be and the thing goes with the arrangement. The *Jushi* stems are placed close to the main line and when two *Jushi* are used for one main line they should be arranged in the form of a triangle with the main line. The angles at which these three main lines are placed is illustrated in Fig. XXIX.4. The lengths of the different main lines depend upon the size of the vase or the container used. In a large vase, the length of the *Shin* is equal to twice the combined measurement of the diameter and height of vase, i.e. 2 × (diameter + height), while *Soe* should be three-fourths of *Shin*, and *Hikae* one half of *Soe*. In a standard-sized vase and a small-sized vase the length of *Shin* will be 1 ½ × (diameter + height) of the vase and diameter + height of the vase, respectively. In both the above-mentioned vase sizes the other main lines (*Soe* and *Hikae*) should be only three-fourths of its predecessors.

The *Ikebana* arrangement generally uses only a few flowers in odd numbers except when two flowers are used. The flowers and the branches should be selected after careful thought. Faded flowers, torn or damaged leaves, and injured branches are never included in an arrangement. It is always advantageous to use half-opened buds as then the arrangement will last longer. The pin-holders are generally not placed in the centre of the container in the *Moribana* arrangement unless there is no other way out. The probable positions of the pin-holders are illustrated in Fig. XXIX.5 for containers of different shapes. It must be noted that a flower arrangement is viewed from the front only. The correct angles of the three main lines in an arrangement is shown in Fig. XXIX.4. The main lines may either be on the left or the right hand side of the vertical zero degree. The measurement of angles are taken only from the tip of the branch and from no other position; otherwise a curved branch if measured from any other position will give a wrong measurement. In the *Moribana* arrangement all the materials used must be fixed firmly in the pin-holder. In *Nageire*, the branches are very often allowed to rest against the wall of the vase by using fixtures.

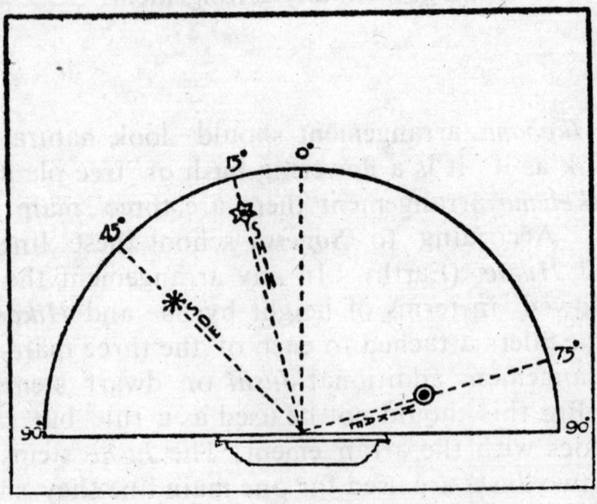

Fig. XXIX.4. Angles of shin, soe and hikae.

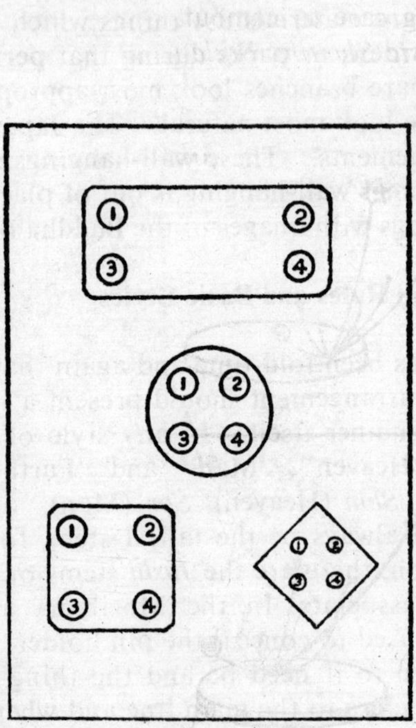

Fig. XXIX.5. The position of the pin-holder in various shaped containers.

In *Sogetsu* school both for the *Moribana* and the *Nageire* arrangements there are only two basic styles, namely, the "Basic upright" and the "Basic slanting". There are several variations from these basic styles which include the flat, floating, hanging, and cascade arrangements.

Basic Upright Style in Moribana: It will be easier to understand the basic four styles from Fig. XXIX.6. For a better understanding, a short description of the arrangement is given below. The pin-holder is placed at the frontal left corner of the container. The *Shin* is fixed in the pin-holder at the rear end at an angle of 10° to 15° on the left of the vertical 0° and bend forward by about 10°. *Soe* is then arranged on the left of *Shin* at 45° from vertical and caused to lean forward by 20°. The *Hikae* finds its position on the right side of the vertical at the angle of 75° with a forward inclination of 30°. The *Jushi* are now fixed in position as per rule described earlier.

Basic Slanting Style in Moribana: In this arrangement the pin-holder is positioned at the rear of the container towards your right end. Here, all the angles and inclinations remain the same as above, except that the angle and inclination of *Shin* is interchanged with that of *Soe* as above and *vice versa*. *Hikae* remains in the same position as above. The *Jushi* are arranged as per rules. As the tallest main line *Shin* is placed at a relatively slanting angle of 45°, the total effect also becomes slanting.

Basic Upright Style in Nageire: If a cross-bar fixture is used for this arrangement, the opening of the vase will be divided into four quadrants and only one such quadrant is used at a time for arranging the flowers. In this style of arrangement all the angles and inclinations for the three main lines are exactly similar to that of basic upright style of *Moribana*. The *Jushi* are arranged in the same quadrant close to the respective lines taking care to camouflage the fixture with their help.

Basic Slanting Style in Nageire: Here the tallest main line *Shin* is fixed firmly in a

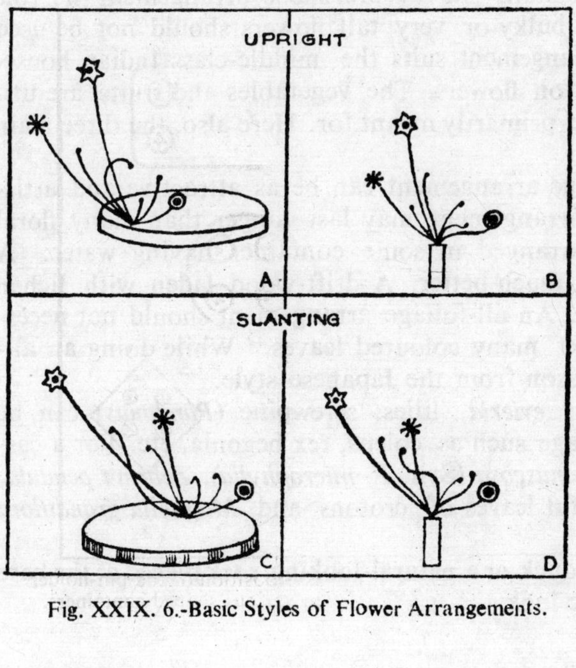

A. Moribana : basis upright;
B. Nageire : upright;
C. Moribana : slanting style;
D. Nageire : slanting style

Fig. XXIX. 6.-Basic Styles of Flower Arrangements.

split-bar fixture and the cut end of *Shin* should rest against the opposite inside wall of the vase for additional support. The *Soe* may either have its own split-bar fixture or draw support from *Shin*. *Hikae* being the smallest in length does not need its own fixture, but can be held in position with the support of the other stems. Here also, the angles and inclinations of the main lines remain the same, as in the basic slanting style of *Moribana*.

There are eight variations which differ slightly from these basic styles. The variations are mainly in the changing of positions and inter-changing of angles of the main lines. In one variation the main line *Soe* is not represented. While in most of the arrangement the three main lines are arranged in one corner of the container fixed in the same pin-holder, in one variation the *Shin* and *Soe* are arranged in one pin-holder in one corner but the *Hikae* is arranged from another pin-holder placed in the diagonally opposite corner of the container. Another variation is the floating style. One of these is done with water flowers in a flat container, with normal angles for the three main lines. The second type is arranged without a vase with the help of fresh flowers, leafy twigs, and branches arranged directly on the table. In still another arrangement a combination of *Moribana* and *Nageire* is arranged side by side in two different containers (flat type and vase) keeping in view the artistic value.

Morimono: The arrangement in combination with fruit and or vegetables and flowers are termed by the Japanese as *Morimono*. These are arranged on a wooden base, called *Dai* in Japan, or a tray, or in a basket, or in a shallow bowl. Sometimes, a small shallow bowl is used for the flowers only, while the fruits and vegetables are arranged in such a way as to camouflage this bowl.

Fruits such as grape, orange, grapefruit, pineapple, apple, etc., can be selected for this arrangement. Vegetables such as red tomatoes, capsicums, brinjals, and carrots are quite suitable for this purpose. Too many of fruits or vegetables should not be used as that may look like a fruit or a vegetable stall. In the *Morimono* arrangement the role of the flower is only secondary. Large, bulky or very tall flowers should not be used for this arrangement. This type of arrangement suits the middle-class Indian housewives who cannot afford to spend much on flowers. The vegetables and fruits are utilized ultimately for the purpose they are primarily meant for. Here also, the three main lines of *Shin*, *Soe*, and *Hikae* are used.

All-Foliage Arrangement: An all-foliage arrangement can be as attractive and artistic as a floral arrangement. A foliage arrangement may last longer than many floral arrangements. The foliage should be arranged in some container having water. A natural-looking container of stone, looks much better. A drift-wood laden with lichen will make an arrangement very natural. An all-foliage arrangement should not necessarily be a green one, as in nature we get many coloured leaves. While doing an all-foliage arrangement, one may draw a lesson from the Japanese style.

Tall leaves such as palms, *Cycas*, *Sansevieria*, lilies, screwpine (*Pandanus*) can be combined with dwarf oval colourful foliage such as coleus, rex begonia, etc. For a cascade arrangements foliage such as *Pilea muscosa* (Syn. *P. microphylla*), *Zebrina pendula*, and *Hedera helix* are very useful. Colourful leaves of crotons and *Magnolia grandiflora* are also very useful.

The figure of a squirrel or a bird or a duck or a natural-looking rock placed at the base of an arrangement will bring naturalistic look.

Appendix to Commercial Floriculture

In addition to the information already provided in the main chapter on "Commercial Floriculture", it was felt necessary to provide further knowledge on growing and marketing of cut flowers and pot plants, including export of cut flowers. Due to the new awareness generated amongst the researchers and commercial growers during the last 5 years or so, as a result of market survey by several delegations deputed by Government of India to various European flower markets, some new experiences have been gained concerning commercial floriculture.

It is estimated that the annual consumption of flowers in the world is worth $13 billion (U.S. dollars). In the European flower market the demand of cut flowers is met by import to supplement the shortfall in internal production. West Germany (F.R.G.) consumes and imports the largest quantity of cut flowers. The annual domestic production (Swarup, 1984) is worth about DM 1,000 million, which is just enough to meet about 42 per cent of the internal market. It imports the other 58% of the demand mainly from the Netherlands (Holland), Italy, France, South Africa and Israel. The import of cut flowers in F.R.G. and other countries in Europe is maximum during winter months (November to April), when the price of the produce is also about 50 per cent higher compared to other seasons. The production during winter in Europe has to be confined to the limited climate-controlled glass-house space, which becomes very expensive due to high cost of energy and labour. During this part of the year many developing countries of the tropical region (e.g., Brazil, Colombia, Israel, Ivory Coast, Kenya, Thailand, etc.) send their products to Europe. The total imports of ornamental products (Swarup, 1984) to Europe are to the extent of about $2.32 billion.* The split-up of the figure indicates that the share of cut flower is $1.04 billion, cut foliage $142.89 million, live plants $810.74 million, and bulbs etc. $324.7 million. However, a major portion of this demand is met by imports from within Europe (91.7 per cent), and the Netherlands accounts for 64.3% of all floricultural products except cut foliage. Italy is a major supplier of cut foliage.

According to one source (Anon. 1981) the following quantities of different cut flowers and live plants were imported in Western Europe during 1980:

1. Roses	19,080 tons
2. Carnations	23,638 tons

*Currency is U.S.A. dollar.

3. Gladioli 4,266 tons
4. Other cut flowers (other than roses & carnation) 109,766 tons
5. Import of live plants (excluding nursery plants) 156,464 tons

According to Swarup (1984), India exported ornamentals worth Rs.2.4 million in 1976-77, which rose to Rs.8.4 million in 1980-81. A declining trend was noticed in 1981-82, when the value of export was only Rs.6 million. It is estimated that out of the total export, foliage plants account for 48 per cent, cut flowers 38 per cent and other items (e.g., bulbs, seeds, etc.) worth 14 per cent. The Indian export of live plants were mainly to some Gulf countries. A major part of the cut flower export was directed towards a few countries in Europe (Italy, F.R.G. & Holland), whereas U.S.A. also imported a part of this product. India has not yet been able to enter the sophisticated European market for export of foliage plants (except a few small consignments), because only very high quality plants are accepted there for sale. There are several constraints for exporting cut flowers or foliage plants. The flowers which are in demand in the importing countries are carnation, rose, gladiolus, chrysanthemum, etc. The flowers like anthuriums and orchids are also popular.

The cut flowers produced in India are generally of inferior quality and do not meet the strict international standards. In European markets, flowers grown under glass-house conditions are preferred for flowers like rose, carnation, chrysanthemum, etc. Perhaps only gladiolus spikes grown under field conditions will be accepted for sale. Flowers grown under plastic-houses could also be exported, but these normally fetch much less prices compared to flowers grown under climate-controlled greenhouses. The cost of building such greenhouses is very high (1 acre of such greenhouse may cost around Rs.45-50 lakhs). The return from such houses is also very good. Moreover, India does not have the expertise or technology of construction of computerised temperature-controlled greenhouses. In the Netherlands highly sophisticated computer-controlled greenhouses with desired temperature, light and humidity conditions are used for cut flower production. If India has to enter successfully into the cut flower market of Europe, the Government of India should allow some enterprising entrepreneurs to import the climate-controlled greenhouses without charging any duty, to give a fillip to the cut flower export trade. The inferior quality flowers are rejected at the auction in Europe and these are then destroyed. Although he air-freight for cut flowers and live plants were reduced recently from Delhi/Bombay to some points in Europe, it is felt that the rate should be further reduced and the concession should be allowed for more cities in Europe as an incentive for exporting live plants.

The live plants for export to most European countries are to be grown in sterilised peat, as these countries would not permit soil-grown plants for import. Since peat is not available in India, liberal import of this substrate should be allowed. It is reported that Soviet Union will allow import of live plants grown only in 'Perlite', a synthetic growing medium. Peat and Perlite are very light in weight and hence air freight will be economised.

The packaging technology has to be developed. Unless the cut flowers are packed properly it will not reach the destination in an unblemished condition. Any damaged material will not be accepted for sale. The cut flowers and live plants are to be moved to the international airports in air-conditioned vans. Near these airports cold storage facilities are to be built up for storing the produce meant for export, till these are loaded in the aircraft. Research in commercial floriculture has to be updated to support the export trade. A constant market study has to be undertaken to keep abreast about the changes taking place in the European cut flower/plant market and changing preferences of consumers. Because of a lack of study of the European market, until recently we were not aware that the preference for commercial cultivars of roses, carnation, gladiolus, etc., had undergone a sea change, than what we used to know. For example, in India we were under the impression that the export quality rose cultivars are 'Queen Elizabeth', 'Sea Pearl', 'Super Star', 'Christian Dior', etc. Actually, none of these cultivars (varieties) are grown extensively as cut flowers in Europe at present.

It is not possible to go into great detail about the cut flower/pot plant production and pre- and post-harvest physiology of cut flowers in a general book on floriculture like this. Hence, besides the broad information given above concerning commercial floriculture, some useful hints regarding a few major cut flower producing crops are also discussed below.

Rose

The rose cultivars presently in demand in the cut flower market of Europe are: 'Sonia' (Hybrid Tea), 'Ilona' (H.T.), 'Golden Times' (H.T.), 'Sandara' (H.T.), 'Pasedena' (H.T.), 'Belinda' (Floribunda) and 'Mercedes' (F). The Indian cultivars like 'Raktagandha', 'Arjun' and 'Sindoor' are being evaluated as cut flower for export market.

Rose cut flower must be cut at tight bud stage with at least 60 cm stem length for export. The buds must not open before reaching the auction centre in Europe, as even slightly opened buds are rejected in the auction.

For commercial cultivation 3 rows of plants should be planted 30 cm apart in a bed, and then 75 cm should be allowed between 2 beds. This method of planting will accommodate 61,750 roses per hactare (Gill, 1984).

Fertilizer doses suggested are nitrogen 20-60 g/sqm area, P_2O_5 (phosphorus) 10-20 g/sqm area and K_2O (Potash) 20 g/sqm area.

Gladiolus

Based on information collected recently on the varieties (cultivars) in demand in foreign markets, seven cultivars of gladiolus are being recommended for commercial cultivation for export. These are: 'Vink's Glory', 'Hunting Song', 'Carmine', 'Spick and Span', 'Happy End', 'Charles de Gaule' and 'Peter Pears'.

Gladiolus corms should be treated with 0.1% Benlate or Bavistin for about 15 minutes and then air-dried before planting. A procedure for treating cormels to get disease-free material has been suggested by Gill (1984). The cormels are treated with hot water at 48°C for 30 minutes and dried under cool air in thin layer for one week. These are then stored in cold storage after treatment for 20 minutes in solution containing 0.5 per cent Bavistin +0.4 per cent Blitox + 0.4 per cent Zineb.

The spacing suggested for commercial cultivation is 30 cm between rows and 20 cm between corms. A high density planting of 60 corms/m² has been suggested by Gill (1984). In this method in a row 10 corms should be planted 10 cm apart and the rows spaced about

15 cm apart. Gladiolus spikes should be cut when 1-2 lowermost florets show colour but are still in tight-bud stage (Kaicker, 1984). It has been advised that spikes without leaves should always be shipped and stored in vertical position to prevent geotropic bending.

Fertilizer doses suggested are: nitrogen 40-60 g/sqm area, phosphorus (P_2O_5) 10-30 g/sqm area and potash (K_2O) 20 g/sqm area.

Carnation

According to Mukhopadhyay (1984), the commercial cultivars in demand for cut flower in export market are 'Scania Red', 'Arthur Sim', 'William Sim' and 'Shocking Pink' in the standard type, and 'Orange Elf', 'Peachy', 'Goldilocks' and 'Exquisite' in spray type.

It has been observed that Sim type carnations are doing very well under Punjab and Delhi conditions (Gill, 1984). Rooted cuttings (terminal cuttings) give better results and these should be planted in sterilised beds to avoid incidence of fusarium wilt and stem rot. It is suggested to plant carnation on raised beds. The spacing suggested for commercial planting for standard and spray carnations are 15×15 m (row$\times$line) and 20×15 cm, respectively.

The fertilizer doses suggested are: nitrogen 20-40 g/sqm, P_2O_5 (phosphorus) 10-20 g/sqm and K_2O (potash) 10 g/sqm.

Carnation flowers should be cut immediately after unfolding of petals.

Pre- and Post-Harvest Physiology

The vase life of a cut flower depends on several pre-harvest or post-harvest factors. It has been observed that carnation and chrysanthemum flowers produced under low light intensity (pre-harvest condition) last for less number of days than those produced under high light intensity. This difficulty could be overcome by harvesting at bud stage and then treating the buds with bud-opening solutions, which include metabolic sugar (Halevy and Mayak, 1979). This may be due to the fact that pre-harvest light intensities affect inherent carbohydrate levels. Growing plants at high temperatures will reduce carbohydrate levels and adversely affect flower longevity. The quality of cut flowers, like pigmentation and colouration of rose, is affected by pre-harvest light and temperature conditions. It has been observed that in rose, even short term shading (3 days) of flower buds, or exposing the buds to high temperatures for about 7 hours at the sensitive stage (about 4-7 days before harvest), can to a great extent reduce the anthocyanin content and increase 'blueing' of 'Baccara' roses (Halevy and Mayak, 1979). The form of nitrogen applied before harvest appears to influence the vase life. It has been reported (Carter *et al.*, 1973) that chrysanthemum flowers receiving nitrate forms of nitrogen lasted longer than those receiving ammonium or urea.

Several post-harvest treatments help in promoting keeping quality. Four main uses of water and chemical solutions are common for promoting longevity of flowers: conditioning (or hardening), pulsing, bud-opening and holding.

The main purpose of *Conditioning* or *Hardening* is to restore the turgidity of flowers by saturating them with water, after these suffer due to water stress at different stages of handling. Previously, conditioning was done with warm water at room temperature, which treatment was later not found to be of much advantage (Systema, 1975). It is advised to condition flowers at cooler temperature throughout (Halevy and Mayak, 1981). Roses are conditioned at 0-2.2°C for 6-12 hours before shipment (Kaicker, 1984). For orchids hardening in water for a day at room temperature, and then packaging dry is good. Further, conditioning should be done with deionized water containing some germicide, but without sugar (Lancaster, 1975).

Pulsing or *Loading* is a preshipment treatment employed by growers or shippers, the beneficial outcome of which normally lasts for the total shelf-life of a flower (Halevy, 1976). The pulsing formulations are specific for different flowers, and this may even vary for different cultivars of a flower. Sucrose is the main constituent of a pulsing solution, the concentration of which may be much higher than normally used in different preservative solutions. The optimum concentration of sucrose for pulsing ranges from 20 per cent or more for gladiolus and gerbera, 2 to 5 per cent for roses and chrysanthemum and 10 per cent for carnation. It should be borne in mind that besides the duration of treatment, the temperature and lighting conditions at the time of pulsing also plays a vital role. The specific conditions needed for a number of flowers have also been standardised (Halevy and Mayak, 1981). The time needed for pulsing is generally between 12-24 hours, under a light intensity of ca. 1000 lux and a temperature range of 20-27°C. Varying the relative humidity between 35 to 100 per cent did not have much effect on the efficacy of carnation pulsing (Neumannova and Paulin, 1974). For cut roses if the pulsing temperature is high the buds will open much faster than desired. Halevy *et al.* (1978) recommended that for pulsing roses, these should be done first at 20°C for 3-4 hours followed by 12-16 hours in the cooler. It has been observed that pulsing is of immense benefit in prolonging cut flower life, promoting opening and improving the size and colour of petals in roses, gladiolus, miniature and standard carnations and chrysanthemums. However, the procedure of pulsing should be followed strictly specified for each flower or there may not be any benefit or rather damage may take place.

Bud-opening is a method in which flowers are harvested at a stage than what is considered as 'cutting stage', and then opening them off the plant. In this method the environmental conditions and chemicals used are identical to those used in pulsing. But since the procedure for bud-opening is much longer (a number of days), so the concentration of sugar solutions used should be much lower than those used in pulsing. High sugar concentrations may cause damage to flowers. However, the stage of harvesting for bud-opening is critical for each type of flower or sometimes even for cultivars. The techniques for bud-opening had been standardised amongst others for roses, chrysanthemums, carnations and gladiolus also (Halevy and Mayak, 1981).

Many commercial preparations of *Holding* (*Vase*) *Solutions* are available for prolonging the vase life. Most of these contain low concentrations (0.5 to 2 per cent) of sugar, besides some other chemicals like 8-Hydroxyquinoline citrate (HQC) or 8-HQ sulphate (HQS).

The Low Pressure Storage method (Hypobaric method) has been discussed to some extent in the main chapter. Though this method was first described by Burg and Burg (1966) quite some time back, but till the year 1978 the method was not much used commercially for cut flower storage (Lougheed *et al.*, 1978). The recent trend about this method is not much known in our country.

For further details on cut flower preservation the works of Halevy and Mayak (1979, 1981) may be consulted.

ADAMOVICH, J. 1978. C
AGNIHOTRI, B.N. and
Bose, S. and Bh
but and Bal, J.
AGNIHOTRI, 1960.
ALEXANDER, M.
ARMEY, J.R. 1972.
STARRETT (Ed.)
ANDERSON, R.E.
ANONYMOUS. 1960. F
ANONYMOUS. 1967
London.
ANONYMOUS. 1975.
of Agric
ANONYMOUS. 1976.
Ministry of Agric
ANONYMOUS. 1981.
from India 1970
ASAHIRA, T. et al.
ASEN, S. and G.H.
BAILEY, L.H.
BAILEY, L.H. 1963.
New York.
BALFOUR, A. 1959.
BAKER, J.E. 1972. Flo
BARTOSE, G. 1964. The
BASU, R.N., T.K. Bose, B.N.

Bibliography

ADAMOVIC, I. 1977. 'Crossing for Fragrance in Garden Gladiolus'. *Gladiolus Annual*: 38-39.

AGNIHOTRI, B.N. and J.C. SRIVASTAVA. 1964. Contribution to our Knowledge on the Effect of B-indol Butyric Acid on the Germination, Flowering and Fruiting of *Impatiens balsamina* Linn. *Indian Agric.* **8** (2): 183-84.

AILINCAI, N. 1960. The Influence of Some Fertilizers on Flowering of *Polianthes tuberosa*. *Lucr. Sti. Inst. Agron. Lasi.* 355-60.

ALEXANDER, M.P. 1977. *Personal Communication on Rose Hybridization.*

ALMEY, J.R. 1972. The Mechanics of Hybridizing. The World of the Gladiolus. In STEFFEK, F. (Ed.). North American Gladiolus Council Inc., Edgewood Press, Maryland.

ANDERSON, E.B. *The Small Rock Garden*. Pan Books Ltd., London.

ANONYMOUS. 1969. *Handbook of Agriculture*. Indian Council of Agricultural Research, New Delhi.

ANONYMOUS. 1969. *The Marshall Cavendish Encyclopaedia of Gardening*. Paul Hamlyn, London.

ANONYMOUS. 1972. Landscape for Living. *The Year Book of Agriculture*. U.S. Department of Agriculture.

ANONYMOUS. 1976. *Report of the National Commission on Agriculture*. Govt. of India, Ministry of Agriculture and Irrigation.

ANONYMOUS. 1981. Western Europe – Market Possibilities of Floriculture Production from India *PROFODCIL Bulletin*. **XVI** (1 & 2): 3-49.

ASAHIRA, T.; H. IMANISHI and Y. TSUKAMOTO. 1968. Studies on the Cormel Formation in Gladiolus. *Mem. College Agric.* Kyoto Univ. **93**: 21-34.

ASEN, S. and C.L. HAMNER. 1953. Effect of Growth Regulating Compounds on Development of Basal Shoots of Green House Roses. *Bot. Gaz.* **115**: 86-89.

BAILEY, L.H. 1947. *The Nursery Manual*. The Macmillan Company, New York.

BAILEY, L.H. 1951. *Manual of Cultivated Plants*. The Macmillan Company, New York.

BAILY, L.H. 1963. *The Standard Cyclopaedia of Horticulture*. The Macmillan Company, New York.

BALFOUR, A.P. 1959. *Annual and Biennial Flowers*. Penguin Books, England.

BAKER, J.G. 1877. *Flora of Mauritius and the Seychelles*. Reeve and Co., London.

BARTOSE, G. 1964. The Use of Disinfection Materials at Sowing. *Zierplanzenbau.* **4**: 941.

BASU, R.N.; T.K. BOSE; B.N. ROY and A. MUKHOPADHYAY. 1969. Auxin Synergists in

Rooting of Cuttings. *Physiologia Plantarum.* **22**: 649-52.

BEACH, R.G. and A.C. LEOPOLD. 1933. The Use of Maleic Hydrazide to break Apical Dominance of Chrysanthemum. *Proc. Amer. Soc. Hort. Sci.* **61**: 543-47.

BENJAMIN, LOVELL. 1973. *The Love of Indoor Plants.* Octopus Books Limited, London.

BHADRI, B.; B. SING and B.L. DESAI. 1962. *Water Plants.* Indian Council of Agricultural Research, New Delhi.

BHATTACHARJEE, S.K.; K. DASGUPTA and T.K. BOSE. 1971. B-Nine—An Effective Growth Retardant on Dahlia, Ineffectiveness of CCC. *Phyton.* **28** (1): 61-65.

BHUJBAL, B.G. and V.R. ZURANGE. 1973. Effect of Pinching and TIBA on Growth and Yield of Marigold. *Ind. J. Orna. Hort.* **4** (1): 25-27.

BIRLA, L.N. *Planning A Landscape Garden.* The Royal Agri-Horticultural Society of India, Calcutta.

BONNER, J. and R.S. BONDURSKI. 1952. Studies of the Physiology, Pharmacology and Biochemistry of Auxins. *A. Rev. Pl. Physiol.* **3**: 59-86.

BOS, L. 1963. *Symptoms of Virus Diseases in Plants.* Centre for Agricultural Publications and Documentation, Wageningen, p. 132.

BOSE, S. and A. Hati. 1969. Breeding and the Use of Gibberellic Acid to increase Seed Production in Pansy. *Bulletin Botanical Survey India.* **11** (3-4): 445-47.

BOSE, T.K. 1972. Dwarfing of Ornamental Plants by using Chemicals. *Ind. J. Orn. Hort.* **3**: 30-37.

BOSE, T.K. 1972. Floriculture in India in the Last Two Decades. *Indian Farming.* **22** (8): 13-18.

BOSE, T.K. and B. HORE. 1967. Effect of Growth Retardants on Some Varieties of Bougainvillea. *Sci. Cult.* **33** (18): 379-80.

BOSE, T.K.; B.N. ROY and R.N. BASU. 1972, Synergism between Auxins and Phenolic Compounds in the Rooting of Cuttings. *Indian Agric.* **16** (2): 171-76.

BOSE, T.K. and D. MUKHERJEE. 1972. *Gardening in India.* Oxford and IBH Publishing Co., Calcutta, India.

BOSE, T.K.; D. MUKHERJEE and B.K. HORE. 1968. Cycocel An Effective Growth Retardant on Malvaceous Plants. *Science and Culture.* **34** (7): 306-7.

BOSE, T.K. and D.P. MANDAL. 1973. Rooting of Cutting under Mist. *Indian J. Agric. Sci.* **43** (3): 229-33.

BOSE, T.K. and D.P. MUKHERJEE. 1968. Improvement in the Seed Formation in Cineraria by using Growth Substances. *Bull. Bot. Soc. Beng.* **22** (1): 41-44.

BRITTON, N.L. and J.N. ROSE. 1963. *The Cactaceae.* Dover Publications Inc., New York.

BUCH, P.O. 1972. The Species. The World of the Gladiolus. In STEFFEK, F. (Ed.). North American Gladiolus Council Inc., Edgewood Press, Maryland.

BURG, S.P. 1973. Hypobaric Storage of Cut Flowers. *Hort. Sci.* **8** (3): 202-5.

BURG, S. and E.A. BURG. 1966. Fruit Storage at Subatmospheric Pressure. *Science.* **153**: 314-15.

CAJLAHJAN, M.H. 1970. Flowering Hormones in Plants. *Bot. Z.* **55**: 913-30.

CARTER, G.D.; D.C. KIPPLINGER; H. TAYAMA and G. STABY. 1973. Nitrogen Source affects Keeping Quality of Standard Chrysanthemum. *Ohio Flower Asso. Bull.* **529**: 5.

CATHEY, H.M. and H.E. HEGGESTAD. 1973. SADH Reduces Ozone Pollution of Petunias. *Hort. Sci.* **8** (1): 7.

CHITTENDEN, FRED J.; and PATRICK M. SYNGE. 1956. *The Royal Horticultural Society*

Dictionary of Gardening. Oxford, Great Britain.

CHYLAH, A. 1972. Bud and Root Neoformation on Leaf Fragments of *Begonia rex* Putz: Effect of Different Growth and Trophic Substances and Environmental Conditions. *Biol. Plant*. **14**: 204-12.

COATS, PETER. 1962. *Roses*. Weidenfeld and Nicolson, London.

CONDER, JOSIAH. 1964. *Landscape Gardening in Japan*. 1964. Dover Publications Inc., New York.

COOKE, T. 1905. *The Flora of Presidency of Bombay* – Vol. **2**. Taylor and Francis, London.

COWEN, D.V. 1965. *Flowering Trees and Shrubs in India*. Thacker and Co. Ltd., Bombay.

DAKERS, J.S. 1970. *The Modern Greenhouse*. Cassel & Company Limited, London.

DAWSON, OLIVER. 1964. *Planning and Planting the Small Garden*. W.H. & L. Collingridge Ltd., London, 1964.

DENISEN, ERRIN L. 1965. *Principles of Horticulture*. The Macmillan Company, New York.

DIONNE, L.A. 1974. Hybrids from *Gladiolus* and *Acidanthera*. *Plants and Gardens*. **30** (1): 39.

DUTHIE, J.F. 1911. *Flora of the Upper Gangetic Plain and Adjacent Sewalik and Sub-Himalayan Tracts*. Superintendent, Govt. Printing, Calcutta, India.

EDLAND, H. 1973. *The Pocket Encyclopaedia of Roses in Colour*. Blandford Press Ltd., London.

EDMOND, J.B.; T.L. SENN and F.S. ANDREWS. 1957. *Fundamentals of Horticulture*. McGraw-Hill Book Company, New York.

EINERT, A.E. 1971. Reduction in last internode Elongation of Cut Tulips by Growth Retardants. *Hort. Science*. **6** (5): 459.

EINERT, A.E. 1975. Effects of Ancymidol on Vase Behavior of Cut Tulips. *Acta Hortic*. **41**: 97.

FIELD, XENIA. 1965. *Window Box Gardening*. Pan Books Ltd., London.

FISHER, PAUL. 1960. *Variegated Foliage Plants*. Blandford Press, London.

FOSTER, GORDON F. 1964. *The Gardener's Fern Book*. D. Van Nostrand Company Inc., Princeton, New Jersey.

FUTEHALLY, LAEEQ. 1978. *Gardens*. National Book Trust, India, New Delhi.

GAMBORG, G.L.; T. MURASHIGE; T.A. THORPE and I.K. VASIL. 1976. Plant Tissue Culture Media *in vitro*. **12**: 473-78.

GANGULEE, H.C.; K.S. DAS and C.T. DATTA. 1961. *College Botany*. Books and Allied Pvt. Ltd., Calcutta.

GILL, A.P.S. 1984. Modern Trends in Production of Important Cut-Flower Crops. *Indian Hort*. **29**(2): 37-42.

GINZBURG, C. and M. ZIV. 1973. Hormonal Regulation of Cormel Formation in Gladiolus Stolons *in vitro*. *Ann. Bot*. **37**: 219-24.

GOH, C.J. and H.C. SEETOH, 1973. Apical Control of Flowering in an Orchid Hybrid *Aranda* "Deborah". *Ann. Bot*. **37**: 113-19.

GOPALASWAMIENGAR, K.S. 1970. *Complete Gardening in India*. G. Kasturi Rangan, Bangalore, India.

GOSLING, STANLEY. 1971. *The Pocket Encyclopaedia of Chrysanthemums in Colour*. Blandford Press Ltd., London.

GRAF, A.B. 1973. *Exotica* 3. Roehrs Company Inc., E. Rutherford, N.J. 07073, U.S.A.

GREEN, G. GILBERT. *Cacti for Everyone*. Faber and Faber, London.

GRIESBACH, R.A. 1972. Breeding: The World of the Gladiolus. In STEFFEK F. (Ed.). North American Gladiolus Council Inc., Edgewood Press, Maryland.

GRINDAL, E.W. 1963. *Everyday Gardening in India*. D.B. Taraporevala Sons & Co. Pvt. Ltd., Bombay.

GROUNDS, ROGER. 1974. *Ferns*. Pelham Books Ltd., London.

GUPTA, RAJENDRA and V.S. SHARMA. 1972. Jasmine Cultivation. *Indian Hort*. **16** (4): 23-27.

HACKETT, W.P. and R.M. SACHS. 1967. Chemical Control of Flowering in Bougainvillea 'San Diego Red'. *Proc. Amer. Soc. Hort. Sci*. **90**: 361-64.

HADFIELD, MILES. 1962. *Gardens*. Weidenfeld and Nicolson, London.

HALEVY, ABRAHAM H. 1976. Treatments to improve Water Balance of Cut Flowers. *Acta Hort*. **64**: 223: 30.

HALEVY, A.H.; T.G. BYRNE; A.M. KOFRANEK; D.S. FARNHAM; J.F. THOMPSON and R.E. HARDENBURG. 1978. Evaluation of Post-Harvest Handling Methods for transcontinental Truck Shipments of Cut Carnations, Chrysanthemums, and Roses. *J. Amer. Soc. Hort. Sci*. **103**: 151-55.

HALEVY, ABRAHAM H. and SHIMON MAYAK. 1979. Senescence and Post-Harvest Physiology of Cut Flowers, Part I. *Hort. Reviews*. **1**: 204-36.

HALEVY, ABRAHAM H. and SHIMON MAYAK. 1981. Senescence and Post-Harvest Physiology of Cut Flowers, Part II. *Hort. Reviews*. **3**: 59-143.

HARRIS, CYRIL C. 1973. *House Plants and Indoor Gardening*. 1973. Octopus Books Ltd., London.

HAWKES, ALEX D. 1961. *Orchids: Their Botany and Culture*. Peter Owen Ltd., London.

HAWKES, ALEX D. 1965. *Encyclopaedia of Cultivated Orchids*. Faber and Faber Ltd., London.

HELLYER, A.G.L. 1964. *The Amateur Gardener*. W.H. & L. Collingridge Ltd., London.

HELLYER, A.G.L. 1966. *Garden Pests and Diseases*. W.H. & L. Collingridge Ltd., London.

HERITAGE, BILL. 1973. *The Lotus Book of Water Gardening*. The Hamlyn Publishing Group Ltd., London, New York, Sydney, Toronto.

HOGG, N.J. and H.P. ORR. 1969. Effect of Gibberellic Acid on Germination of Seeds from Selected Native Trees and Shrubs. *Propagator*. **15** (2): 8-9.

HOLDGATE, D.P. 1977. Propagation of Ornamentals by Tissue Culture. In REINERT, J. and Y.P.S. BAJAJ (Eds.), *Applied and Fundamental Aspects of Plant Cell, Tissue and Organ Culture*. Springer-Verlag, Berlin.

HUNT, PETER. 1972. *Greenhouse and Indoor Plants*. Ward Lock Limited, London.

JACOBSEN, H. (Translation by VERA HIGGINS). 1959. *Succulent Plants*. Ernest Benn Ltd., London.

JANA, B.K.; S. ROY and T.K. BOSE. 1974. Studies on the Nutrition of Ornamental Plants – III. Effect of Nutrition on Growth and Flowering of Dahlia and Tuberose. *Indian J. Hort*. **31** (2): 182-85.

JINDAL, S.L. 1968. *Ornamental Bulbous Plants*. Indian Council of Agricultural Research, New Delhi.

KACHROO, P. 1968. *Handbook of Rock Gardening on the Hills*. Indian Council of Agricultural Research, New Delhi.

KACHROO, P. 1975. *Introduction to Gardening*. Indian Council of Agricultural Research, New Delhi.

KAICKER, U.S. 1984. Pre- and Post-Harvest Handling and Marketing of Cut Flowers. *Indian Hort*. **29** (2): 47-52.

KAINS, M.G. and L.M. MCQUESTEN. 1947. *Propagation of Plants*. Orange Judd Publishing Co. Inc., New York.

KHER, M.A. 1969. Photoperiodic Response in Chrysanthemum. *Plant Science*. **1**: 200-6.

KHER, M.A. 1975. *Chrysanthemum*. National Botanic Gardens, Lucknow, India.

KHOSHOO, T.N. 1974. Prospects of Ornamental Horticulture in India. *Indian J. Hort*. **31**(4): 295-300.

KUNKEL, L.O. 1936. Heat Treatment for the Cure of Yellows and Other Diseases of Peach. *Phytopathology*. **26**: 809-30.

LANCASTER, D.M. 1975. Post-Harvest Handling of Cut Carnations from harvest through Consumer Use. *Colorado Flower Growers' Assoc. Bull*. **295**: 1-5.

LEEBURN, M.E. 1963. *Garden Lilies*. W.H. and L. Collingridge Ltd., London.

LEOPOLD, A.C. 1967. *Auxins and Plant Growth*. University of California Press, Barkley and Los Angeles.

LEONARD, H. 1974. *Roses*. Hyonlyn.

LETOUZE, R. and G. BEAUCHESNE. 1969. Action d'eclairements monochromatiques sur la rhizogenese de tissue Topinambour. *C.R. Acad. Sci*. Paris. **269**: 1528-31.

LOUGHEED, E.C.; D.P. MURR and L. BERARD. 1978. Low Pressure Storage for Horticultural Crops. *Hort. Sci*. **13**: 21-27.

MAROUSKY, FRANCIS J. 1973. Recent Advances in Opening Bud-Cut Chrysanthemum Flowers. *Hort. Sci*. **8**(3): 199-202.

MAROUSKY, F.J. and S.S. WOLTZ. 1971. Effect of Fluoride and A Floral Preservative on Quality of Cut Gladiolus. *Proc. Fla. Sta. Hort. Soc*. **84**: 375-80.

MARTH, P.C. 1942. Effects of Growth Regulators on Development of Roses during Common Storage. *Bot. Gaz*. **104**: 26-49.

MCCANN, CHARLES. 1966. 100 *Beautiful Trees of India*. D.B. Taraporevala Sons & Co. Pvt. Ltd., Bombay.

MCDOWELL, JACK. 1972. *Sunset Ideas for Japanese Gardens*. Lane Books, Menio Park, California, U.S.A.

MCQUOWN, F.R. 1963. *Plant Breeding for Gardeners*. W.H. & L. Collingridge Ltd., London.

MASSINGHAM, BETTY. 1972. *Flower Arrangement in Colour*. Paul Hamlyn, England.

MENNIGER, EDWIN A. 1970. *Flowering Vines of the World*. Heartside Press Inc., New York.

MILITU, A.; E. VIERASU and A.F. ILIESCU. 1970. The Influence of Chemical Fertilizers on Flower Quality and Bulb Production in Tuberose Growing. *Lucrari Stintifica, Institutul. Agronomic "N. Balacesen", B*. **13**: 115-20.

MITTAL, S.P. 1967. Effect of Gibberellin on Growth and Flowering of Dahlia. *Madras Agric. J*. **54**(3): 103-7.

MORTON, J.W. 250 *Beautiful Flowers and How To Grow Them*. W. Foulsham and Co. Ltd., London.

MUKHERJEE, H. and A.K. GANGULEE. 1964. *Plant Groups*. New Central Book Agency, Calcutta.

MUKHOPADHYAY, AMITABHA. 1984. Carnation Culture in India. *Indian Hort*. **29**(2): 43-46.

MUKHOPADHYAY, A. and G.J. BANKAR. 1978. Effect of N, P and K on Flower and Bulb Production and Keeping Quality of Flowers and Bulbs of Tuberose (*Polianthes tuberosa*

Linn). Twentieth International Hort. Congress, Sydney, 15, 23, Abst. No. 1913.

MUKHOPADHYAY, A.; G.J. BANKAR and K.S. SHUKLA. 1980. A Thornless Rootstock for Roses. *Indian Horticulture*. **25**: 1-2.

MUKHOPADHYAY, A.; P. DAS and T.K. BOSE. 1975. Effect of Spacing, Pinching, Disbudding and Treatment with GA_3 on Growth and Flowering in Carnation. *Plant. Science*. Symp. Volume: 132.

MUNDKUR, B.B. 1967. (Revised by S.B. CHATTOPADHYAY.) *Fungi and Plant Disease*. Macmillan and Co. Ltd., Calcutta.

MURASHIGE, T. 1974. Plant Propagation through Tissue Cultures. *Ann. Rev. Plant Physiol.* **25**: 135-66.

MURASHIGE, T. 1977. Clonal Crops through Tissue Culture. In W. BARZ, E. REHNHARD, M.H. ZENK (Eds.), *Plant Tissue Culture and Its Bio-Technological Application*. Springer-Verlag, Berlin, pp. 392-403.

MURASHIGE, T. and R. NAKANO. 1967. Chromosome Complement as a Determinant of the Morphogenic Potential of Tobacco Cells. *An. J. Botany*. **54**: 963-70.

MURASHIGE, T.; M. SERPA and J. JONES. 1974. Clonal Multiplication of Gerbera through Tissue Culture. *Hort. Sci*. **9**: 175-80.

MUTHUSWAMY, S. and V.N. MADHAVA RAO. 1972. Tamil Nadu's Fragrant New "Pari Mullai". *Indian Hort*. **17** (3): 28-29.

NANDA, K.K.; H.N. KRISHNAMOORTHY; K.L. TOKY and K. LATA. 1969. Effects of Gibberellins A_3, A_{4+7}, A_{13} and of (—) Kaurene on Flowering and Extension Growth of *Impatiens balsamina* under different photoperiods. *Planta*. **86**: 134-41.

NANDA, K.K.; M. KUMAR; S. SAWHNEY and N. SAWHNEY. 1973. Effect of Cycloheximide on GA_3 Response and on Photoperiodic Reactions of *Impatiens balsamina* L. *Ann. Bot*. **37**: 107-11.

NEUMANNOVA, B. and A. PAULIN. 1974. Conditions de temperature et d'humidite relative a appliquer dans les chambros de floraison d'ocillets. *Ann. Tech. Agr*. **28**: 467-80

PAL, B.P. 1960. *Beautiful Climbers of India*. Indian Council of Agricultural Research, New Delhi.

PAL, B.P. 1972. *The Rose in India*. Indian Council of Agricultural Research, New Delhi, India.

PAL, B.P. 1973. *All About Roses*. Vikash Publishing House, Delhi, India.

PAL, B.P. and V. SWARUP. 1974. *Bougainvilleas*. Indian Council of Agricultural Research, New Delhi, India.

PEARSON, C.E. 1968. *Everyday Gardening*. Ward Lock Limited, London, Sydney.

PEARSON, C.E. 1970. *Ward Lock's Complete Gardening*. Ward Lock Limited, London & Sydney.

PIERIK, R.L.M. 1975. Vegetative Propagation of Horticultural Crops *in vitro* with special attention to Shrubs and Trees. *Acta Horticulturae*. **54**: 71-82.

POWELL, L. 1974. Growth Regulators can extend Plant Shelf-Life. *Commercial Grower*. **4081**: 465.

PRESTON, G.H. 1970. *All About Gardening*. Ward Lock Limited, London.

RAM, K.; S.L. ABBAS and B.P. SACHAN. 1970. Influence of Gibberellic Acid on Growth and Flowering of Antirrhinum (*Antirrhinum majus*). *Lal-Baugh*. **15**(4): 9-12.

RANDHAWA, G.S. 1973. *Ornamental Horticulture in India*. Today & Tomorrows Printers

& Publishers, New Delhi.

RANDHAWA, G.S.; K.L. CHADHA and DALJIT SINGH. *The Famous Gardens of India.* Malhotra Publishing House, India.

RANDHAWA, M.S. 1961. *Beautifying Cities of India.* Indian Council of Agricultural Research, New Delhi, India.

RANDHAWA, M.S. 1965. *Flowering Trees.* National Book Trust of India, New Delhi.

RANDHAWA, M.S. 1976. *Gardens Through the Ages.* The Macmillan Co. of India Ltd.

RAO, A.S. 1979. *Orchids of India.* National Book Trust of India, New Delhi.

RAVINDRAN, P. and B. NARISIMHAN. 1970. Regulation of Flower Size and Petal Retention in Rose, Variety "Edward". *Andhra Agric. J.* **17**(6): 181-89.

READ, P.E. and V.C. HOYSLER. 1969. Stimulation and Retardation of Adventitious Root Formation by Application of B-Nine and Cycocel. *J. Amer. Soc. Hort. Sci.* **94**: 314-16.

REISTER, DOROTHY W. 1971. *Design for Flower Arrangers.* Van Nostrand Rainhold Company, New York.

ROGERS MARLIN, N. 1973. A Historical and Critical Review of Post-Harvest Physiology Research on Cut Flowers. *Hort. Sci.* **8**(3): 189-94.

SAGAWA, Y. 1976. Potential of *in vitro* Culture Techniques for Improvement of Floricultural Crops. *Acta Horticulturae.* **63**: 61-66.

SANTAPAU, H. 1966. *Common Trees.* National Book Trust of India, New Delhi.

SANTAPAU, H. and A.N. HENRY. 1973. *A Dictionary of the Flowering Plants in India.* Council of Scientific and Industrial Research, New Delhi, India.

SCOTT, S.H. 1958. *The Observers Book of Cacti and Other Succulents.* Fredrik Warne & Co. Ltd., London.

SEARLE, S.A. and B.J. MACHIN. 1968. *Chrysanthemums the Year Round.* Blandford Press, London.

SEIBERT, M.P.; P.J. WETHERBEE and D.D. JOB. 1975. The Effects of Light, Intensity and Spectral Quality on Growth and Shoot Initiation in Tobacco Callus. *Plant Physiol.* **56**: 130-39.

SEVERN GILLIAN, E. 1967. *Miniature Trees in the Japanese Style.* Faber and Faber Ltd., London.

SHANMUGAN *et al.* 1973. Effect of Growth Substances on *Chrysanthemum indicum* Linn. *Madras Agric. J.* **60** (1): 1-5.

SHEPHERD, R. 1977. ABC of hybridization. *Rose World.* **3**: 14-27.

SHIGETA, DANIEL T. *Carnation Culture in Hawaii.* University of Hawaii Cooperative Extension Service Circular 395.

SKOOG, F. and C.O. MILLER. 1957. Chemical Regulation of Growth and Organ Formation in Plant Tissues cultured *in vitro. Symp. Soc. Biol.* **11**: 118-30.

SNYDER, RACHEL. 1964. *The Complete Book for Gardens.* D. Van Nostrand Company Inc., New Jersey.

SPOMER, L.A. and R.W. LANGHANS. 1965. Paradichlorobenzene (Mothball) injury to Bulbs. *Bull. N.Y. St. Flower Grs.* **233**: 3-4.

STEWARD, F.C.; M.O. MAPES and P.V. AMMIRATO. 1969. Growth and Morphogenesis in Tissue and Free Cell Cultures. In STEWARD, F.C. (Ed.), *Plant Physiology: A Treatise.* Academic Press, London-New York, pp. 329-76.

STODDARD, E.M. 1942. Inactivating *in vivo* the Virus of X-disease of Peach by Chemotherapy. *Phytopathology.* **32**: 17.

STUTTE, CHARLES A. 1977. *Plant Growth Regulators*. American Chemical Society, Washington D.C.

SUDELL, RICHARD. 1966. *Practical Gardening in Pictures*. Odhams Books Ltd., London.

SWARUP, VISHNU. 1967. *Garden Flowers*. National Book Trust, New Delhi.

SWARUP, VISHNU. 1984. Export of Ornamentals: Problems and Prospects. *Indian Hort*. **29**(2): 53-56.

SWARUP, V.; U.S. KAICKER; R.S. MALIK and A.P. SINGH. 1973 (a). Rose Breeding in India. *Indian J. Genetics*. **33**: 43-51.

SWARUP, V.; R.S. MALIK and M.P. SINGH. 1973(b). Some New Indian Roses. *Indian Hort*. **19**: 15-19.

SWARUP, V.; R.S. MALIK and M.P. SINGH, 1972. New Indian Roses Registered. *Delhi Garden Magazine*. Flower show Number: 17-22.

SYSTEMA, W. 1975. Conditions for measuring Vase Life of Cut Flowers. *Acta Hort*. **41**: 217-25.

TAMBERG, T.G. 1967. Polyploidy in the genus *Gladiolus*. *Genetika*. **3** (**6**): 123-28.

TATSUI, MATSUNOSUKE. 1968. *Japanese Gardens*. Japan Travel Bureau Inc., Tokyo.

THOMPSON, P.A. 1969. Comparative Effects of Gibberellins A_3 and A_4 on the Germination of Seeds of Several Different Species. *Hort. Res*. **9**: 130-38.

THOMPSON, P.M. 1969. Germination of Species of *Labiatae* in response to Gibberellins. *Physiol. Plant*. **22**: 575-86.

THROWER, PERCY. 1965. *Roses*. W.H. and L. Collingridge Ltd., London.

TRIVEDI, DEVIKA. 1975. *Ikebana*. Vikash Publishing House, Delhi.

UNDERWOOD, C.J. 1971. *The Time Life Encyclopaedia of Gardening "Bulbs"*. Time Life Books, New York.

VALMAYOR, H.K.; F.I.(Jr.) HILARIO and O.K. BANTISTA. 1963. The Effects of Maleic Hydrazide on the Growth and Flowering of Bougainvillea 'Snow White'. *Philipp. Agric*. **47**: 23-31.

VELLUSWAMY, P.; S. MUTHUSWAMY and T. THANGARAJ. 1973. Origin and Distribution of Jasmines. *Indian J. Orn. Hort*. **4**(2 & 3): 45-55.

WENT, F.W. and K.V. THIMANN. 1937. *Phytohormones*. The Macmillan Co., New York.

WILFRET, G.J. 1974. Gladiolus Breeding. *Plants and Gardens*. **30**(1): 35-39.

WRIGHT, J. 1971. A New Fragrant Gladioli Hybrid *X Gladanthera* "Lucky Star". *Amer. Hort*. **50**(2): 79-83.

ZEEUW DE D. and A.C. LEOPOLD. 1956. The Promotion of Floral Initiation by *Auxin*. *Am. J. Bot*. **43**: 47-50.

Index

The index covers botanical nomenclature of plants. Pages in bold-face indicate that the plant concerned has been dealt in detail in that page.

*Presently called *Asteraceae*.

D

E

*Recently changed to *Fabiaceae*.

U

V

W

X

Y

Z

ERRATA

Some errors crept in due to oversight. An errata is given below. While preparing errata more attention was given to errors in botanical names. If any further error is noticed the same may be brought to the notice of the second author.

Page	Line from top	Line from bottom	Incorrect	Correct
8	18		*J. grandiflorum*	*Jasminum grandiflorum*
16	18		to be	to the
19		2	*Phlox decusata*	*Phlox decussata*
63		14	*Centauria*	*Centaurea*
64	17		*Centauria*	*Centaurea*
64		15	*Lupin*	Lupin
74	4		dwarf annual,	dwarf annual
75	1		*Centauria*	*Centaurea*
89		16	at a distance	at a spacing
109	1		*Viola tricolor hortensis* should appear after *Viola cornuta* in the same page as per alphabetical sequence.	
116	19		*Geraniacea*	*Geraniaceae*
122		10-11	*P. wittamanniana*	*P. wittmanniana*
127	2		*nerifolia*	*neriifolia*
127		20-21	*Malvaviscus conzatii*	*Malvaviscus conzattii*
127		19-20	*Mussaeanda*	*Mussaenda*
133	8		*A. violaceae*	*A. violacea*
133		16	*Ericacea*	*Ericaceae*
142	5		60 m	60 cm
146	6		Var. *fortuniana*	var. *fortuniana*
146	8		Var. *veitchiana*	var. *veitchiana*
146		15	*Acanthaceze*	*Acanthaceae*
156	1		*Malvaviscus conzatii*	*Malvaviscus conzattii*
166		8	*T. lumosa*	*T. plumosa*
167		9	*nerifolia*	*nereifolia*
172	9		*'cypheri', 'sanderiana',* and *'variegata'.*	*cypheri, sanderiana,* and *variegata*.
182	1		Var.	var.
189	15		*P. grandiflora*	*P. grandifolia*
189	16		*P. grandiflora*	*P. grandifolia*
197		3	*Pinaceae*	*Pinaceae,*
205	13		*Anonaceae*	*Annonaceae*
228	17		(Miscanthus)	(*Miscanthus*)

(Contd.)

Page	Line from top	Line from bottom	Incorrect	Correct
228		4	own depth	own diameter
233		20	fibrous-rooted,	fibrous-rooted
237		17	*C × powellii*	*C. × powellii*
242		1	*L. flavescens*	*I. flavescens*
243	9-10		*I. kamaonensis*	*I. kumaonensis*
252		10	Var. *grandis*	var. *grandis*
254	5		For cultur	For culture
254		15	Var. *victoriae*	var. *victoriae*
256		18	*Maranta arundinaceae*	*Maranta arundinacea*
258	21		*C. antiquorum illustiris*	*C. antiquorum illustris*
264		10	*Melstomaceae*	*Melastomaceae*
264		5	*H. hookeriana*	*M. hookeriana*
272		10	304	305
278		4	*Aprorocactus*	*Aporocactus*
280	18 (right column)		*mihan ovichii*	*mihanovichii*
286	12		*M. bicolour*	*M. bicolor*
287	1		*Cereu*	*Cereus*
287	6		*clavariodes*	*clavarioides*
301	9		*Echeveria aducna*	*Echeveria adunca*
301	11		*Crossulaceae*	*Crassulaceae*
301		20	*Pedilanthus carniatus*	*Pedilanthus carinatus*
301		16	*Geraniacae*	*Geraniaceae*
301		9	*'Gouty Geranium'*	'Gouty Geranium'
301		9	*'Knotted Geranium'*	'Knotted Geranium'
302		16	*S. trifasciata 'Hahnii'*	*S. trifasciata* 'Hahnii'
310	9		*Howea belmoriana*	*Howea belmoreana*
314	16		*Phoenicophorum*	*Phoenicophorium*
315		3	*Cycus cercinalis*	*Cycas circinalis*
324	17		*Gymnogramma tartarica*	*Gymnogramma tartarea*
324		11	*Nephrolepsis*	*Nephrolepis*
324		5	*N. davallioidies*	*N. davallioides*
325		3	*Schellolepsis*	*Schellolepis*
325		4	*Schellolepsis*	*Schellolepis*
329		15	*B. nana 'Suochiku'*	*B. nana* 'Suochiku'
331		3	*Phyllostachys kumasaca*	to be deleted.
332	6		*'St. Augustine Grass'*	'St. Augustine Grass'
337	16		*Criterion*	*Criteria*
357	Page No. on top		457	357
385	8		*J.pubescens* var.*rubescense*	*J. pubescens* var. *rubescens*
385		23	*G. grandiflorum*	*J. grandiflorum*
400	20		Ldl. and *Paxt.*	Ldl. and Paxt
400	21		Ldl. and *Paxt.*	Ldl. and Paxt

Page	Line from top	Line from bottom	Incorrect	Correct
401	19-20		(*Aerides Vanda*)	(*Aerides* × *Vanda*)
401	21		*Rhyncostylis*	*Rhynchostylis*
425	2		(Family:	Family:
426		5	Allincai	Ailincai
429	15		features	feature
430		10	*companulata*	*campanulata*
436	18		with-an-imposing	with an imposing
440		7	hill'ocks	hillocks
444		16	*Lagerstroemia flos reginae*	*Lagerstroemia flos-reginae*
452	15		*Thevetia nerifolia*	*Thevetia nereifolia*
454	12		(Syn. *Thunbergia erecta*)	(Syn. *Thunbergia*) *erecta*
454	18		*Poinsettia pulcherrima* will appear before *Punica granatum* as per alphabetical order.	
456		2	*A. amoenga*	*A. amoena*
456		2	*bettzickiane*	*bettzickiana*
471	4		*Irisine*	*Iresine*
471	4		456	457
471	5		456	457
471	14		(Syn. *V. V. tricolor, triloba*)	(Syn. *V. tricolor, V. triloba*)
475	6-7		*Gaillardia perennis*	*Gaillardia perennis,*
475	9		*Salvia,*	*Salvia*
475	14		*D. allwoodii*	*Dianthus allwoodii*
475	14		*Dianthus chinensis*	*D. chinensis*
475	16		*A. elegans*	*Anemone elegans* (This name should appear just after *Achillea eupatorium*.)
475	18		*Lupins*	Lupins
475		6	*Heliotrope*	Heliotrope
475		2	*Lantana sellowiana*	*Lantana sellowiana,*
477		21	clinder	cinder
478		8	The word '*Opuntia*' to go after '*Cereus*'.	
479	5		*Adeneum obesum*	*Adenium obesum*
481	Fig. XXIII.3		D-Inter pipe	D-Inlet pipe
487		3	*Pani-Phai*	*Pani-Phal*
488		19	*I. pseudacorus*	The name to go below *I. kaempferi*.
496	14		*Plumeria* sp.	*Plumeria* sp.,
496	14		*Callistemon lanceolatus-*	*Callistemon lanceolatus*
496	14		*-Gliricidia*	*Gliricidia*
503	Illustrations		The drawings show the...	The drawings below show the...
510		13	*Largerstroemia*	*Lagerstroemia*
512	1		*C. marginata*	*Cassia marginata*
517		11	yellow-flowering type)	(yellow-flowering type)

(Contd.)

Page	Line from top	Line from bottom	Incorrect	Correct
519	20		such as house wall	such as house, wall
520	18		sitting spot	sitting spots
523		4	*Cotoneaster*	*Cotoneaster**
523		3	*Rhododendrons*	*Rhododendrons**
533		15	*B.* × *windii* ×	*B. windii,*
544	7		*Cacti*	Cacti
557	10 (left column)		*and Pines*	and Pines
565	14		*Impatients*	*Impatiens*
578	3		*Phlox decusata*	*Phlox decussata*
578	Fig. XXVI.2.E		Capsid bug cleft	Capsid bug left
580	17		*Phlox decusata*	*Phlox decussata*
581	7		*Phlox decusata*	*Phlox decussata*
585	Fig. XXVI.6		(siz) 1.5 cm.)	(size 1.5 cm)
604	7		Tindiness	Tidiness
615	Fig. XXIX.6		A. Moribana: basis upright;	A. Moribana: basic upright;